Frontiers in Parasitology

(Volume 1)

Molecular and Cellular Biology of Pathogenic Trypanosomatids

Edited by:

Marcelo Santos da Silva
Instituto Butantan São Paulo
Brazil

&

Maria Isabel N. Cano
Universidade Estadual Paulista
"Júlio de Mesquita Filho" (UNESP)
Botucatu, São Paulo
Brazil

Frontiers in Parasitology

Volume # 1

Molecular and Cellular Biology of Pathogenic Trypanosomatids

Editors: Marcelo Santos da Silva and Maria Isabel N. Cano

eISSN (Online): 2542-422X

ISSN (Print): 2542-4211

eISBN (Online): 978-1-68108-405-3

ISBN (Print): 978-1-68108-406-0

BENTHAM SCIENCE PUBLISHERS LTD.

End User License Agreement (for non-institutional, personal use)

This is an agreement between you and Bentham Science Publishers Ltd. Please read this License Agreement carefully before using the ebook/echapter/ejournal (**"Work"**). Your use of the Work constitutes your agreement to the terms and conditions set forth in this License Agreement. If you do not agree to these terms and conditions then you should not use the Work.

Bentham Science Publishers agrees to grant you a non-exclusive, non-transferable limited license to use the Work subject to and in accordance with the following terms and conditions. This License Agreement is for non-library, personal use only. For a library / institutional / multi user license in respect of the Work, please contact: permission@benthamscience.org.

Usage Rules:

1. All rights reserved: The Work is the subject of copyright and Bentham Science Publishers either owns the Work (and the copyright in it) or is licensed to distribute the Work. You shall not copy, reproduce, modify, remove, delete, augment, add to, publish, transmit, sell, resell, create derivative works from, or in any way exploit the Work or make the Work available for others to do any of the same, in any form or by any means, in whole or in part, in each case without the prior written permission of Bentham Science Publishers, unless stated otherwise in this License Agreement.
2. You may download a copy of the Work on one occasion to one personal computer (including tablet, laptop, desktop, or other such devices). You may make one back-up copy of the Work to avoid losing it. The following DRM (Digital Rights Management) policy may also be applicable to the Work at Bentham Science Publishers' election, acting in its sole discretion:

- 25 'copy' commands can be executed every 7 days in respect of the Work. The text selected for copying cannot extend to more than a single page. Each time a text 'copy' command is executed, irrespective of whether the text selection is made from within one page or from separate pages, it will be considered as a separate / individual 'copy' command.
- 25 pages only from the Work can be printed every 7 days.

3. The unauthorised use or distribution of copyrighted or other proprietary content is illegal and could subject you to liability for substantial money damages. You will be liable for any damage resulting from your misuse of the Work or any violation of this License Agreement, including any infringement by you of copyrights or proprietary rights.

Disclaimer:

Bentham Science Publishers does not guarantee that the information in the Work is error-free, or warrant that it will meet your requirements or that access to the Work will be uninterrupted or error-free. The Work is provided "as is" without warranty of any kind, either express or implied or statutory, including, without limitation, implied warranties of merchantability and fitness for a particular purpose. The entire risk as to the results and performance of the Work is assumed by you. No responsibility is assumed by Bentham Science Publishers, its staff, editors and/or authors for any injury and/or damage to persons or property as a matter of products liability, negligence or otherwise, or from any use or operation of any methods, products instruction, advertisements or ideas contained in the Work.

Limitation of Liability:

In no event will Bentham Science Publishers, its staff, editors and/or authors, be liable for any damages, including, without limitation, special, incidental and/or consequential damages and/or damages for lost data and/or profits arising out of (whether directly or indirectly) the use or inability to use the Work. The entire liability of Bentham Science Publishers shall be limited to the amount actually paid by you for the Work.

General:

1. Any dispute or claim arising out of or in connection with this License Agreement or the Work (including non-contractual disputes or claims) will be governed by and construed in accordance with the laws of the U.A.E. as applied in the Emirate of Dubai. Each party agrees that the courts of the Emirate of Dubai shall have exclusive jurisdiction to settle any dispute or claim arising out of or in connection with this License Agreement or the Work (including non-contractual disputes or claims).
2. Your rights under this License Agreement will automatically terminate without notice and without the need for a court order if at any point you breach any terms of this License Agreement. In no event will any delay or failure by Bentham Science Publishers in enforcing your compliance with this License Agreement constitute a waiver of any of its rights.
3. You acknowledge that you have read this License Agreement, and agree to be bound by its terms and conditions. To the extent that any other terms and conditions presented on any website of Bentham Science Publishers conflict with, or are inconsistent with, the terms and conditions set out in this License Agreement, you acknowledge that the terms and conditions set out in this License Agreement shall prevail.

Bentham Science Publishers Ltd.
Executive Suite Y - 2
PO Box 7917, Saif Zone
Sharjah, U.A.E.
Email: subscriptions@benthamscience.org

BENTHAM
SCIENCE

CONTENTS

FOREWORD

Since 1970, due to important financing programs, Brazilian scientists have been able to practice Biology while studying trypanosomatids, mainly *Trypanosoma cruzi* and *Leishmania* spp. The Annual Meeting in Caxambu since 1974, focusing on *Trypanosoma cruzi* at the beginning, and later extending the interest to other parasites, particularly *Leishmania,* were the catalysts of a scientific boom in Brazilian biology. Many young and enthusiastic students have been trained by the pioneers in the morphology, physiology, biochemistry, immunology and molecular biology of trypanosomatids, including vectorial transmission and clinical aspects of the diseases. The Caxambu meetings also brought to Brazil renowned scientists of the field who established important and long-lasting connections with Brazilians. After 2005, the field has gained momentum after the publication in Science of the complete genome of three species of the order Kinetoplastida, namely *Leishmania major*, *Trypanosoma cruzi* and *Trypanosoma brucei*. This e-book is a consequence of the Brazilian scientific activities in the field in the last 45 years. Maria Isabel Cano is my scientific grand-daughter and Marcelo Santos da Silva got his Ph.D. under her supervision.

The first chapter deals with the cellular organization of trypanosomatids during the life cycle. This group diverged early in evolution and due to this fact their members conserved certain characteristics not found in other eukaryotes as organelles like glycosomes, reservosomes and acidocalcisomes, among other cell compartments. Differences between species and between the distinct life cycle forms of each organism are comprehensively discussed.

Chapter 2 unveils the fascinating area of the trypanosomatid genomes. Information has been made available by the Tritryps project that provided major insights into the genome structure and organization of these parasites. While *T. brucei* presents subtelomeric expansion of genes related to antigenic variation, *T. cruzi* and *Leishmania* genomes contain species-specific genes related to cellular invasion and survival inside the mammalian host cells. As control of gene expression operates mainly at a post-transcriptional level in trypanosomatids, gene copy number variation is probably an efficient mechanism to enhance gene expression and increase sequence variability. Among the Tritryps, *T. cruzi* presents the most striking expansion of species-specific multigene families, which could be related to the ability of the parasite to infect any nucleated cell of a broad range of mammals. Chromosomal copy number variation is also well tolerated by these parasites, allowing the expansion of a whole set of genes simultaneously. The functional implications of these chromosomal expansions to the parasite biology are still to be determined.

Chapter 3 clarifies the function of chromosome ends and telomeres in trypanosomatid biology. These structures, in addition to the preservation of chromosomal integrity, play a major role in survival. The telomeric repeat (5'-TTAGGG-3')$_n$ is conserved among trypanosomatid species, but adjacent subtelomeric sequences vary between species and chromosomes within the same cell. For example, size and gene content of *T. cruzi* subtelomeres differ in each chromosome due to differences in the abundance and organization of these genes, whereas in *Leishmania* spp. subtelomeres show a more conserved organization.

Chapter 4 is a pleasant travel on nuclear and kinetoplast DNA replication in trypanosomatids. These organisms exhibit both conserved and unique non-conserved features in the DNA replication machinery. Curiously, the trypanosomatid pre-replication complex differs from other eukaryotes, having features similar to those of Archaea. The completion of DNA replication, at trypanosomatid telomeres, apparently is similar to other eukaryotes, although the processing of the leading and lagging telomeres required to generate the 3' overhangs, which serve as telomerase substrate, remains unknown. DNA replication in trypanosomatids initiates almost simultaneously in the nucleus and the kinetoplast, suggesting that regulation of DNA synthesis in the two DNA-containing organelles may be coordinated. The kinetoplast DNA consists of mini- and maxi-circles, which are replicated by many proteins with, as yet, unknown mechanisms of action. The complex DNA replication mechanisms, independently acting in both kinetoplast and nucleus, are reviewed.

Chapter 5 discusses mechanisms by which trypanosomatids maintain genome integrity and preserve faithful DNA replication despite multiple environmental aggressions. A growing body of evidence on how trypanosomatids recognize and repair damages is reviewed.

Chapter 6 explores one of the most fascinating features that make trypanosomatids unique organisms in the biological world. Despite being eukaryotic organisms they constitutively synthesize polycistronic mRNAs from separate gene clusters. Control of gene expression is not carried out at the DNA level but relies upon post-transcriptional mechanisms. This chapter aptly discusses the common post-transcriptional pathways for most genes, although many regulatory strategies within species of the group differ from each other. It is argued that these complex and diversified regulatory machineries allow rapid responses of these organisms to drastic environmental changes during their life cycle.

Chapters 7 and 8 describe virulence factors and the immune evasion in *Leishmania* spp and *Trypanosoma cruzi*, respectively. These parasites co-evolved with their hosts – mammalian and insects – for several millions of years and developed specialized strategies to evade the immune system by overcoming both innate and adaptive immune responses. The role of different species of molecules as virulence factors is discussed.

Diagnosis, chemotherapy, and potential recombinant vaccines are discussed in the ensuing chapters 9, 10 and 11, respectively. The diseases caused by kinetoplastids, called neglected diseases since they are predominant in poorer tropical countries having scarce resources, are responsible for thousands of deaths per year. No vaccines are available for these diseases. Presently, drug therapies are not very effective because the few available drugs are toxic, and treatment is costly. Comprehensively, methods of vector control are insufficient, despite the Southern Cone initiative in the nineties by which transmission of *Trypanosoma cruzi* in the domicile and outdoors by *Triatoma infestans,* at least in some countries, has been controlled. Chapter 9, stresses the need for reliable and specific diagnostic tests for epidemiological surveys, surveillance for vectorial transmission, blood screening, screening of pregnant women and their newborns, and in individual patients. The chapter summarizes the most commonly used molecular tools described to date to detect *T. cruzi* infection and to identify and genotype *Leishmania* spp. Chapter 10 discusses the present status of the chemotherapy area, emphasizing the screening assays that led to a few drugs reaching the stage of a clinical trial after a selection from millions of molecules tested. A milestone has been set for the year 2020, by the London Declaration: control of Chagas disease and leishmaniasis, and elimination of Human African Trypanosomiasis. Advances to achieve these goals are presented in this chapter. Finally, Chapter 11 deals with the research efforts on the development of recombinant vaccines against trypanosomatids, most of which are at the stage of preclinical experimentation. It is hoped that these efforts can be translated into efficient human vaccines.

This book is being edited in electronic form. In addition to the PDF edition accessible online, it will be aired in open access electronic media. Any reader can also have access to a printed version and also to individual chapters offered by the authors themselves. Thus, it is hoped that biologists, graduate students and post-doctoral researchers benefit with this up-to-date and competent review of the present literature on trypanosomatids.

Prof. Walter Colli
University of São Paulo
Brazil

PREFACE

Among the pathogenic trypanosomatids are the etiological agents of leishmaniasis, African trypanosomiasis and Chaga's disease, protozoa parasites belonging respectively, to the genus *Leishmania* and *Trypanosoma*. The diseases are high prevalent in tropical areas of the globe and according to WHO cause thousands of new cases and deaths every year. It is worth noting though that international migration made Chaga's disease and leishmaniasis becoming an issue also in many developing countries in Europe, Canada, USA, Australia, and Japan. Although the number of new cases and deaths from Chaga's disease had decreased in the last decade, the millions of chronically infected persons who are at risk for developing cardiovascular and/or digestive pathology make Chaga's disease one of the leading causes of cardiovascular morbidity and premature death in Latin America. Human African trypanosomiasis (HAT), also known as sleeping sickness, is endemic in sub-Saharan countries and is caused by *T. brucei gambiense* and *T. b. rhodesiense*. The number of HAT cases has been decreasing in the last years due to very active control efforts although according to a recent WHO report, country or regional averages may be misleading since the burden of the disease falls very heavily on some areas. HAT patients require lots of care and disease diagnosis and treatment can be costly and time-consuming. However, in many Africa countries diagnostics and HAT drugs are provided free of charge. Leishmaniasis in its turn is still considered a dangerous menace, principally to the poor, with about 12 million people currently infected. Depending on the species leishmaniasis can be expressed in different clinical forms, with the cutaneous form being the most common, causing self-limiting skin ulcer or a highly disfiguring scar, to the disfiguring and mutilating mucocutaneous form, and the visceral form is the most severe and fatal if not treated. All these diseases are still challenges to overcome since the absence of effective vaccines, and the toxicity of current anti-parasite drugs, in addition to the emergence of drug-resistant parasite strains and HIV co-infections are non-transposable barriers to disease control. Thus, many research initiatives have been direct to understand parasite biology and its interactions with different hosts, as well as the mechanisms of disease pathogenesis, of drug resistance and genome organization and maintenance, with the aim of the development of more efficient diagnostic tools and non-toxic and effective drugs and vaccines.

In this eBook, experts review and explore current knowledge about the molecular and cellular biology of trypanosomatids, highlighting the most important and actual discoveries in each research field. Topics covered include cell organization during development; genome organization and maintenance; control of gene expression; nuclear and kinetoplast DNA replication; mechanisms of DNA damage repair; virulence factors and immune evasion; new methods for molecular diagnosis; new therapeutic tools and recombinant vaccine biology.

We believe that the eBook content will be of keen interest to undergraduates, graduate students and principally to the Parasitology community and researchers working in related fields.

We would like to thank the contributing authors of this book for their time, expertise, and for making this eBook novel, educational, and informative and Professor Walter Colli for written the Forward.

Finally, we would like to express our gratitude to Mr. Shehzad Naqvi, the Senior Manager Publications from Bentham Science Publishers for his continuous help.

Dr. Marcelo Santos da Silva
Instituto Butantan
São Paulo
Brazil

Dr. Maria Isabel N. Cano
Universidade Estadual Paulista "Júlio de Mesquita Filho" (UNESP)
Botucatu, São Paulo
Brazil

List of Contributors

Alba Marina Gimenez	Universidade Federal de São Paulo, São Paulo, Brazil
Alejandro G. Schijman	Instituto de Investigaciones en Ingeniería Genética y Biología Molecular, Ciudad de Buenos Aires, Argentina
Bessy Gutiérrez	University of Antofagasta, Antofagasta, Chile
Bruna M. Valente	Universidade Federal de Minas Gerais, Minas Gerais, Brazil
Camila B. Storti	Universidade Estadual Paulista "Júlio de Mesquita Filho" (UNESP), Botucatu, São Paulo, Brazil
Carlos R. Machado	Universidade Federal de Minas Gerais, Minas Gerais, Brazil
Christiane Araújo	Instituto Butantan, São Paulo, Brazil
Cristiane Regina Antonio	Universidade Federal de São Paulo, São Paulo, Brazil
Daniella Castanheira Bartholomeu	Universidade Federal de Minas Gerais, Minas Gerais, Brazil
Edna Gicela Ortiz Morea	Universidade Estadual Paulista "Júlio de Mesquita Filho" (UNESP), Botucatu, São Paulo, Brazil
Gonzalo Cabrera	Universidad de Chile, Santiago, Chile
Hugo Oswaldo Valdivia	Universidade Federal de Minas Gerais, Minas Gerais, Brazil
Isabela C. Mendes	Universidade Federal de Minas Gerais, Minas Gerais, Brazil
Jair L. Siqueira-Neto	University of California San Diego, La Jolla, USA
João Luís Reis-Cunha	Universidade Federal de Minas Gerais, Minas Gerais, Brazil
Jorge Araya	University of Antofagasta, Antofagasta, Chile
Jorge González	University of Antofagasta, Antofagasta, Chile
José Franco da Silveira	Universidade Federal de São Paulo, São Paulo, Brazil
José L. Veja	University of Antofagasta, Antofagasta, Chile
Jose M. Requena	Universidad Autónoma de Madrid, Madrid, Spain
Juan M. Burgos	Universidad Nacional de San Martín, Provincia de Buenos Aires, Argentina
Julia Pinheiro C. da Cunha	Instituto Butantan, São Paulo, Brazil
Leonardo da Silva Augusto	Universidade Federal de São Paulo, São Paulo, Brazil
Loyze P. Oliveira de Lima	Instituto Butantan, São Paulo, Brazil

Manuel Soto	Universidad Autónoma de Madrid, Madrid, Spain
Marcelo Santos da Silva	Instituto Butantan, São Paulo, Brazil
Maria Alejandra Viviescas	Universidade Estadual Paulista "Júlio de Mesquita Filho" (UNESP), Botucatu, São Paulo, Brazil
Maria Carolina Elias	Instituto Butantan, São Paulo, Brazil
Maria Isabel Nogueira Cano	Universidade Estadual Paulista "Júlio de Mesquita Filho" (UNESP), Botucatu, São Paulo, Brazil
Mariana de Camargo Lopes	Instituto Butantan, São Paulo, Brazil
Marjorie Mendes Marini	Universidade Federal de São Paulo, São Paulo, Brazil
Maurício Martins Rodrigues	Universidade Federal de São Paulo, São Paulo, Brazil
Miguel Angel Chiurillo	Universidade de Campinas, Campinas, Brazil
Nilmar Silvio Moretti	Universidade Federal de São Paulo, São Paulo, Brazil
Paula L. Marcet	Center for Diseases, Control and Prevention (CDC), Atlanta, United States of America
Priscila Martins Andrade Denapoli	Universidade Federal de São Paulo, Brazil
Raphael Souza Pavani	Instituto Butantan, São Paulo, Brazil
Renata Torres de Souza	Universidade Federal de São Paulo, São Paulo, Brazil
Richard McCulloch	University of Glasgow, Glasgow, UK
Santuza M. R. Teixeira	Universidade Federal de Minas Gerais, Minas Gerais, Brazil
Sergio Schenkman	Universidade Federal de São Paulo, São Paulo, Brazil
Simone Guedes Calderano	Instituto Butantan, São Paulo, Brazil
Teresa Cristina Leandro	Teresa Cristina Leandro de Jesus Instituto Butantan, São Paulo, Brazil
Viviane G. Silva	Universidade Federal de Minas Gerais, Minas Gerais, Brazil

Frontiers in Parasitology

CHAPTER 1

The Cellular Organization of Trypanosomatids During Life Cycle

Simone Guedes Calderano[1], Nilmar Silvio Moretti[2], Christiane Araujo[3,4], Marcelo S. da Silva[3,4], Teresa Cristina Leandro de Jesus[3,4], Loyze P. Oliveira de Lima[3,4], Mariana de Camargo Lopes[3,4], Leonardo da Silva Augusto[2], Julia Pinheiro Chagas da Cunha[3,4], Maria Carolina Elias[3,4,*] and **Sergio Schenkman[2,*]**

[1] *Laboratório de Parasitologia, Instituto Butantan, São Paulo, Brazil*

[2] *Departamento de Microbiologia, Imunologia e Parasitologia, Universidade Federal de São Paulo, São Paulo, Brazil*

[3] *Laboratório Especial de Ciclo Celular, Instituto Butantan, São Paulo, Brazil*

[4] *Center of Toxins, Immune Response and Cell Signaling – CeTICS, Instituto Butantan, São Paulo, Brazil*

Abstract: *Trypanosoma cruzi* , *Trypanosoma brucei* and *Leishmania* spp. are etiological agents of the following neglected diseases: African sleeping sickness (*T. brucei*), Chagas' disease (*T. cruzi*) and leishmaniasis (*Leishmania* spp.). These parasites are eukaryotic cells that diverged early in evolution and therefore harbor modified organelles, such as glycosomes, and present subcellular compartments with unusual characteristics. This chapter aims to overview the most striking features of the structures and functions of these organelles, which ensure the existence of these parasites, and to discuss the differences between species and between the distinct life cycle forms of each organism.

Keywords: Acidocalcisomes, Cellular compartmentalization, Endoplasmic reti-

* **Corresponding authors Sergio Schenkman:** Universidade Federal de São Paulo, Brazil; E-mail: sschenkman@unifesp.br; **Maria Carolina Elias:** Instituto Butantan, São Paulo, Brazil; E-mail: carolina.eliassabbaga@butantan.gov.br

Marcelo Santos da Silva & Maria Isabel N. Cano (Eds.)

culum, Flagellum, Glycosomes, Golgi, Kinetoplast, Nucleolus, Nucleus, Reservosomes.

INTRODUCTION

Trypanosomatids are unicellular flagellated eukaryotes that belong to the Kinetoplastida class, the members of which are characterized by the presence of a kinetoplast, which is a structure that contains the mitochondrial genetic material. The Kinetoplastida class includes the Trypanosomatidae family, which comprises human pathogens, such as *Trypanosoma cruzi* , *Trypanosoma brucei* and *Leishmania* spp. These are the etiological agents that cause Chagas' disease, African sleeping sickness and leishmaniasis, respectively. It is currently estimated that 6.5 million people are infected with *T. cruzi,* that there are 1.3 million new cases of leishmaniasis and approximately 6,000 cases of sleeping sickness every year (www.who.int).

These protozoans have life cycles stages that possess different morphologies and cellular structures when living in mammalian and insect hosts. The *T. cruzi* epimastigote is a form that is defined by the lateral exit of the flagellum from the cell body; it is also the non-infective life cycle stage of the parasite. It proliferates *via* binary fission in the gut of *Triatoma infestans* insects, which are more commonly known as kissing bugs. In the insect hindgut, epimastigotes transform into the metacyclic-trypomastigotes in which the flagellum is inserted along almost the entire length of the protozoan. When the insect vector ingests blood from a mammalian host, the trypomastigotes are then eliminated with the feces. Released parasites can penetrate the mammalian host through contact with different mucosal tissues. They enter cells by forming a parasitophorous vacuole, which is then disrupted and the parasites transform into amastigotes, which are spherically shaped and have a very short flagellum. Amastigotes proliferate inside the cell cytosol and then transform into non-replicative trypomastigotes, which are released *via* cell lysis to reach the bloodstream. The life cycle is completed by the ingestion of the blood by insect vectors. The trypomastigotes are able to transform into epimastigotes that replicate inside the insect gut [1].

Leishmania, in contrast, alternates between a promastigote and an amastigote form. Promastigotes are protozoan with a flagellum attached to one extremity that develops in the digestive tract of sand flies. Promastigotes differentiate into metacyclic forms that are eliminated *via* regurgitation when the insect feeds. The parasites are then phagocytized by macrophages before transforming into amastigotes that divide inside vacuoles. After several rounds of division, the macrophages are disrupted, and the new amastigotes are released to infect adjacent macrophages. Insects ingest the infected macrophages when they feed on mammalian blood. Amastigotes transform into promastigotes in the insect midgut, continuing the life cycle [2].

Unlike what occurs in *T. cruzi* and *Leishmania*, the life cycle of *T. brucei* is entirely extracellular. An infected tsetse fly (*Glossina* spp.) bites a mammalian host; inoculating metacyclic-trypomastigote forms into the circulatory system. The injected metacyclic-trypomastigotes transform into bloodstream trypomastigotes, which then proliferate in the hemolymphatic system as slender trypomastigotes. *T. brucei* species survive the immune defenses of the host by continuously changing their coat, which is formed by a single variant surface glycoprotein (VSG) through a process known as antigenic variation. When a high density of parasites is achieved in the blood, some of these parasites transform in a non-proliferative, stumpy form. This form is able to differentiate into procyclic trypomastigotes when ingested by new tsetse flies. In the fly midgut, the procyclic form proliferates and then migrates to colonize the salivary glands of the insect, where they transform into epimastigotes that can proliferate by binary fission. After some rounds of duplication, the epimastigotes become metacyclic trypomastigotes, which are then injected into a new mammalian host during a tsetse fly's bite [1].

These organisms contain organelles that are common to all eukaryotes, but they also harbor unique organelles, such as the kinetoplast, reservosomes, glycosomes and flagellum-related structures, all of which present peculiar features as a consequence of their earlier evolutionary origins and the requirements of adaption (Fig. **1**).

This chapter aims to present recent developments that have increased our understanding of how these organelles ensure the survival of the organism in different hosts during parasite life cycles. We have also compared the features that are unique to each trypanosome.

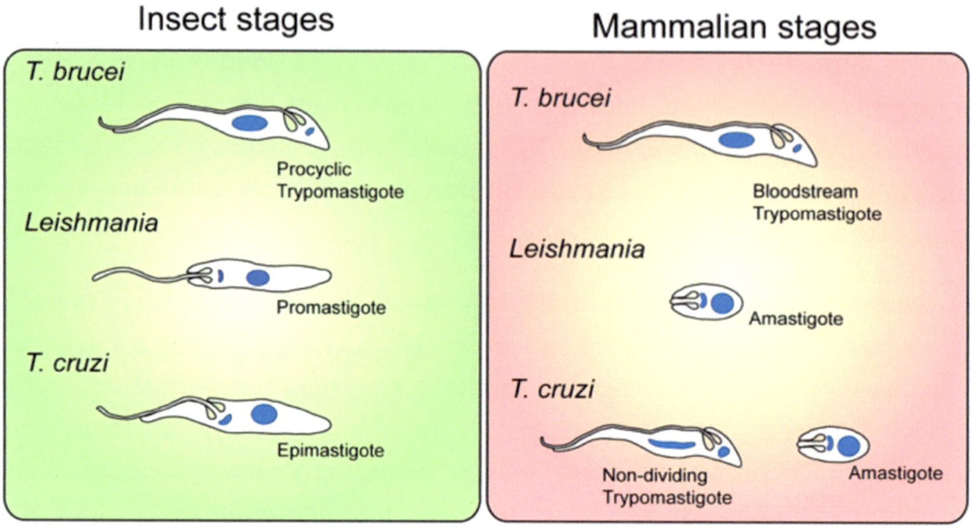

Fig. (1). Schematic representation of the Trypanosomatid forms found in insects and mammals. The figure on the left illustrates the main replicative forms, which in *T. brucei* are procyclic-trypomastigotes, in *Leishmania* are promastigotes, and in *T. cruzi* are epimastigotes. The right panel shows the corresponding forms that are found in the mammalian hosts. The nucleus is larger than the kinetoplast, and both are shown in blue.

NUCLEUS

The nucleus is the largest organelle within the cell, and it stores the genetic material. The first nucleated cell, referred to as the first eukaryotic common ancestor, emerged approximately 2.5 billion years ago and is the ancestor from which all eukaryotic cells, including the Kinetoplastida protozoan parasites, diverged [3, 4].

The nuclear environment is delimitated by the nuclear envelope, which is a double bilayer membrane. The inner membrane faces the nuclear lumen, while the outer membrane faces the cytoplasm. The nuclear envelope is mechanically supported by a nuclear lamina underneath the inner membrane and by another filament network that surrounds the outer membrane. Nucleus-cytoplasm communication occurs through nuclear pore complexes (NPCs).

The nucleus of trypanosomes undergoes several morphological changes during the life cycle. For example, while the replicative forms of *T. cruzi* (epimastigotes and amastigotes) have a rounded nucleus with a clear nucleolus and with heterochromatin at the nuclear periphery, the non-replicative and infective bloodstream trypomastigote and metacyclic trypomastigote forms have an elongated nucleus without nucleolus. Trypomastigotes also lose peripheral heterochromatin [5]. It remains unknown what mechanisms drive these structural changes, but they might be related to modifications in the NUP-1 protein, a protein that is associated with the inner nuclear membrane [6, 7].

Nuclear Organization

Many processes occur in the nuclear space at sites that are non-randomly distributed, meaning that all nuclear events are organized in the space during cell cycle progression. In higher eukaryotes, the formation of heterochromatin and euchromatin are closely linked to the transcriptional status of each DNA region. In addition, the transcriptional status dictates the timing of DNA replication. In other words, while highly transcribed genes are replicated at the beginning of S phase, silent regions and sequences with low transcriptional rates are replicated at the end of S phase. In addition, the location of early and late replication regions varies during S phase, during which discrete replication sites are found throughout the nuclear space at the beginning of S phase, and a more peripheral replication pattern is observed during the final stages of S phase [8].

Trypanosomes display a peculiar nuclear organization. In replicating *T. cruzi* epimastigotes, RNA polymerase II transcription sites are dispersed throughout the nucleus, with the specific transcriptional site of the spliced leader (SL) genes concentrated in a region adjacent to the nucleolus [9]. SL is transcribed at a high rate because its mRNA serves as the precursor for the first 39 nucleotides that are added to all pre-RNAs. These are transcribed as polycistronic units, and they form mature mRNAs *via* a trans-splicing reaction. The site of SL transcription in the nucleus is different between *T. cruzi,* and *T. brucei* . While in *T. cruzi* , only one SL transcription site has been observed in the nucleus [10], in *T. brucei* , two major sites are observed, probably as a consequence of the ploidy of the SL gene repeats in this parasite [11, 12]. This is not observed in *Leishmania* spp., which, like *T. cruzi* , has only one SL transcription site. Details about gene transcription in trypanosomatids are described in Chapter 6.

The replication machinery is located in the trypanosome nuclear periphery, revealed by the presence of the sliding clamp of the DNA polymerase machinery and by the incorporation of thymidine analogs that were detected using fluorescence techniques [13, 14]. Moreover, the distribution of chromosomes varies according to the cell cycle stage. At the onset of S phase, *T. cruzi* chromosomes are dispersed; they then progressively migrate to the nuclear periphery, where replication takes place, and remain there until mitosis is completed. The chromosomes then disperse again throughout the nucleus [14]. It is interesting to note that the same replication organization pattern is not observed in *T. brucei* and *L. donovani*. In both parasites, the sliding clamp of the replication machinery is found at several sites distributed throughout the nucleus, indicating a distinct type of nuclear organization [15, 16]. These different localizations of transcription and replication machinery could explain how DNA replication and transcription are maintained separately in the nuclear space, and they may imply that trypanosomes display a rudimentary level of nuclear organization compared to that observed in more complex eukaryotes [17, 18].

Another difference in nuclear organization is the expression of the VSG and procyclin genes, which encode the coat protein of procyclic form in *T. brucei* . Active VSG genes are located in the telomeric regions of *T. brucei* chromosomes, and only one gene is expressed at a time. Transcription of this single VSG gene is catalyzed by RNA polymerase I [19] in a locus present in a nuclear region called the expression site body [20]. Procyclin genes, are also transcribed by RNA polymerase I and its transcription occurs at the nucleolus periphery. In contrast, the silent loci for the VSG and procyclin genes remain in the nuclear periphery within the heterochromatin [21, 22]. Differences in telomere distribution have also been observed in *L. major,* in which telomeres were found to be dispersed in promastigotes but concentrated in the center of the nucleus in amastigotes [23]. These observations collectively indicate the presence of variable nuclear structures and chromosomal localizations, which reflect adaptations in each parasitic stage to the conditions found in the host during their life cycle.

Nuclear Pore Complex

Nuclear pore complexes (NPCs) allow nucleus-cytoplasm communication through the nuclear envelope. In eukaryotes, approximately thirty proteins called nucleoproteins (Nups) participate in the formation of NPCs. The NPC structure consists of a central transport channel, a core scaffold, a cytoplasmic ring, a nuclear ring and eight filaments attached to each ring [24]. The nucleoproteins from the core scaffold in most eukaryotes are arranged in an octagonal framework that surrounds a central transport channel, while the phenylalanine-glycine (FG) repeats from these nucleoproteins occupy the center of the transport channel and are involved in transport and interactions with cargo-carrying transport factors [25]. The nuclear basket proteins associate with the chromosomes influence gene expression by regulating the transport of mRNA and gene regulatory factors. They bind to specific transcription factors and transcriptionally active genes, promoting the assembly and/or maintenance of gene loops, in which the promoter and terminator regions are juxtaposed, accelerating the reactivation of gene transcription [26].

The NPC octagonal symmetric framework structure is highly conserved among eukaryotes [27, 28] and differs mainly in its size. In yeast, it is approximately 60 MDa, while in humans, it is 90-120 MDa. However, the proteins forming the complex are not highly conserved among eukaryotes. In *T. brucei* and *Leishmania major* , 22 Nups were identified through proteomic analyses [29, 30]. From these, 20 and 16 were localized in the nuclear envelope in *T. brucei* and *L. major,* respectively. Similar genes were found in *T. cruzi.* However, the precise location and function of each protein in the NPC have not been established. TbNup92 and TbNup110 from *T. brucei* and LmMlp2 from *L. major* are analogs to the *Saccharomyces cerevisiae* nuclear basket proteins Mlp1 and Mlp2 that were found to be involved in chromosome segregation during the closed mitosis observed in these organisms [24, 29 - 31]. It is worth mentioning that TbNup92 is larger than its yeast analog because it contains an extra domain that is typically found in DNA repair and cell cycle checkpoint proteins. These data indicate another possible function of NPCs in trypanosomes.

Nuclear Lamin

Lamins (lamin A, B and C) are intermediate filament proteins that form a meshwork between the nuclear envelope and the nuclear matrix. Lamins associate with peripheral heterochromatin and have diverse functions in Metazoans. They provide support for NPCs and control gene expression during development [32 - 36]. In non-Metazoan organisms, no lamin orthologues have been identified. For example, lamins are absent in *S. cerevisiae* [37, 38]. However, *T. cruzi* and *T. brucei* contain a coiled coil protein called NUP-1 that resembles lamin and is located in the inner nuclear envelope [6, 7]. This protein lacks similarity with metazoan lamins but has a conserved structure. It has a conserved pattern of coiled-coil domains, suggesting that it is capable of self-assembly, which is characteristic of lamin proteins in metazoans. In *T. brucei* and *T. cruzi,* NUP-1 forms a net-like structure at the inner surface of the nuclear envelope and interacts with some chromosomal regions, similar to what has been observed in Metazoan lamins. In *T. brucei* , NUP-1 also plays a role in telomeric silencing, in the regulation of VSG gene expression and chromatin organization. *T. cruzi* NUP-1 interacts with some chromosome regions that encode clusters of genes for surface proteins. Therefore, it appears that these interactions are important not only for chromatin organization but also for the control of gene expression in trypanosomes [6, 7, 39].

Nucleolus

The nucleolus is a subcompartment of the nucleus wherein processes including rRNA transcription, pre-rRNA processing and ribosome assembly occur. In addition, the nucleolus is involved in protein sequestration in response to stress, cell cycle control and RNA processing [40 - 42]. The nucleolus has a tripartite architecture: a fibrillar center and dense fibrillar and granular components. In trypanosomes, a single nucleolus is found in the nucleus, and it has a bipartite fibrillar center and granular component structure [43]. Furthermore, in trypanosomes, the nucleolus is smaller than in most eukaryotes, having a diameter of approximately 1 μm and some peculiar characteristics. In *T. cruzi,* it is not present in trypomastigote forms, which are a non-dividing stage, while it is present in the *T. cruzi* replicative epimastigote and amastigote forms. In

Leishmania and *T. brucei* , in contrast, the nucleolus is present throughout the major stages respectively promastigotes and amastigotes, bloodstream and procyclic trypomastigotes. Proteins and mRNA are sequestered in the nucleolus under stress conditions, such as the induction of severe heat-shock in *L. major* and *T. cruzi* [44], but this does not occur in *T. brucei* [45 - 47].

Chromatin

Chromatin is composed of tightly associated nuclear DNA and proteins. The fundamental unit of chromatin is the nucleosome, in which two copies of each core histone (H4, H3, H2A and H2B) form an octamer of proteins that wraps around every ~154 base pairs (bp) of DNA. This regular array of nucleosomes, which forms 10-nm fibers, has been observed under electron microscopy to form a "beads-on-a-string" structure. A fifth histone called histone H1 (also known as linker histone) associates with the DNA located between two nucleosomes and compacts chromatin into 30-nm fibers known as solenoids. The majority of eukaryotes, however, further compact their chromatin (approximately 15,000 x), as has been observed in fully compacted chromosomes during mitosis. Thus, multiple levels of folding and compaction have been detected in eukaryotes and is used to fit the genomic information into a tiny nucleus (for a recent review, see [48].

Trypanosome chromatin is organized into nucleosome filaments, which form the same 10-nm fibers that have been observed in eukaryotes. However, neither 30-nm fibers nor condensed chromosomes during mitosis have been observed. A possible explanation for this is the lack of a typical N-terminus and the globular domain of histone H1 and/or the absence of a phosphorylation site at H3S10, all of which are associated with condensed chromosomes during mitosis [49]. Trypanosome histones diverged from fungi and other metazoan histones, and a disproportionate amount of this divergence has been observed in their N-terminal domains [50]. The migration patterns of histones in polyacrylamide gels made at acid pH, in urea and in non-ionic detergent, which differentiate histones by their charges and modifications, have also revealed differences between parasites such as *T. cruzi* and *T. brucei* [49].

Histone H1 is the most divergent histone. It comprises only the C-terminal domain of the eukaryotic histone H1. In *T. cruzi* , histone H1 is constitutively synthesized throughout the cell cycle but at an increased rate during S-phase, while the other histones are synthesized concomitant with DNA replication [51]. In contrast, the expression of histone H1 is related to the progression of the cell cycle and cellular differentiation in *L. major* [52], and histone H1 is not essential to *T. brucei* growth *in vitro*. *T. brucei* lacking H1 live longer than wild-type *T. brucei* in infected animals, probably because H1 depletion resulted in significant changes in the expression of RNA-Pol I genes (VSG and ESAGs). There is evidence showing that trypanosome histone H1 may also play a role in DNA repair because H1-depleted clones were clearly more resistant than the wild-type clones to drugs that induce DNA damage [53, 54].

In addition to the canonical histones (H2A, H2B H3 and H4), variant forms of these histones are present in trypanosomes. These variants are present in reduced amounts but may have unique functions that affect gene expression, centromere function and DNA damage/repair [55]. They include the histone variants H2A, H2B and H3. *T. cruzi* and *T. brucei* also encode H4 'orphans' that have 85% and 96% identity to canonical H4, respectively [56]. Despite this finding, trypanosomes do not contain homologs for H3.3, a variant that replaces histone H3 at highly transcribed sites, or CenH3, a variant that is found at the centromeres of eukaryotes. *T. brucei* H3v shares some similarity with CenH3, but it localizes to telomeres and is not required for viability or chromosome segregation [57].

T. cruzi and other trypanosomes do contain H2AZ and a new variant, H2B (H2ABv), which is associated with transcriptional activation, gene silencing and the avoidance of heterochromatin formation in euchromatin regions [55]. Studies performed on *T. brucei* have shown that H2Bv, which has been identified in few organisms and only under specific situations [58], dimerizes with H2AZ and is essential for cell viability. Interestingly, these dimers are absent from highly transcribed sites [59].

Histone Post-Translational Modifications (PTM) in Trypanosomes

Histones can, in general, be chemically modified both at the tail, which extends from the nucleosome, and inside the octamer. Modifications, such as phosphorylation, acetylation, methylation (mono, di or tri), and sumoylation, have been identified in histones and have, more importantly, been associated with gene expression, DNA repair and replication, chromosome condensation and other important regulatory processes in cells. Although trypanosome histones are highly divergent from the histones identified in other eukaryotes, their sequences, including their tails, are conserved among trypanosomatids, and specific (as well as unusual) PTMs have been identified.

T. cruzi histone H1 is phosphorylated in a cell cycle-dependent manner at a typical cyclin-dependent kinase site (^{12}SpPKK), most probably by TzCRK3 (a *T. cruzi* CDK-like protein), as has been observed in higher eukaryotes, mainly from S to mitosis [60]. Phosphorylated histone H1 is homogenously distributed in foci in *T. cruzi* epimastigote nuclei, which appear in the G2 phase of the cell cycle, increase until mitosis and disappear during cytokinesis. Interestingly, the majority of non-phosphorylated histone H1 is located in the central regions of the nucleus, but when cells progress to G2 phase, it becomes phosphorylated and begins to diffuse in the nuclear space [61, 62].

PTMs at histone H4 of *T. cruzi* were identified using mass spectrometry (MS). Lysines (K) 4, 10, 14 and 57 are acetylated (ac); K18 is mono-methylated (me1), and arginine (R) at position 53 is dimethylated (me2) [63]. In epimastigotes, the majority of histone H4 is acetylated at K4, while less than 5% of H4ac proteins are acetylated at K10 and K14, suggesting that these modifications play a regulatory role [64]. H4K4ac decreases following DNA damage and in non-proliferative forms. In contrast, H4K10ac and H4K14ac increase at these times. Histone acetylation is well known for its role in transcription. However, no colocalization was observed between acetylations of histone H4 and the major RNA Pol II labeling site in *T. cruzi* , which corresponds to SL transcription. Nevertheless, H4K10ac and H4K14ac were found at eu- and heterochromatin boundaries, which are involved in events related to the active metabolism of DNA. No MS analyses have been performed on the remaining histones of *T. cruzi* , although radiolabeling experiments in parasite cultures have indicated that histone H2A is mainly acetylated, while H3 and H2B are mainly methylated [62, 63].

A complete scenario for histone PTMs has been described for *T. brucei* . Electrophoresis analyses indicated that H3 and H2A are highly modified compared to H4 and H2B [65 - 67]. For example, the methylation states of H3K76 in *T. brucei* depend on two methyltransferases called DOT (Disruptor Of Telomeric silencing, DOT1A and DOT1B), which is unlike to other eukaryotes that have a single DOT1 enzyme. The methylation of H3K76 appears to be analogous to the methylation of H3K79 observed in other eukaryotes, which has been found to be associated with the prevention of heterochromatin formation [68]. H3K76me2 is regulated by the cell cycle and occurs in *T. brucei* during mitosis, while K76me3 is present throughout the cell cycle. Furthermore, DOT1A is essential for viability, while DOT1B is not. However, DOT1B is necessary for differentiation during the transition to procyclic forms and for VSG switching and silencing [69]. In contrast to other organisms, the *T. brucei* histone H2A and H2B N-terminal domains show few modifications, while the C-terminal domain of H2A is highly acetylated. To date, no function has been attributed to these PTMs.

One of the earliest markers of DNA damage is the phosphorylation of the H2AX histone. In mammals, serine 139 is phosphorylated in the histone variant H2AX, while the replication-dependent H2A histone is phosphorylated in yeast. In *T. brucei* , it has been shown that the majority of gamma H2A becomes phosphorylated at threonine 130 in response to DNA damage [70]. This phosphorylation could play a role in signaling the presence of damaged DNA. The region surrounding the phosphorylated residue is highly conserved in trypanosomes, suggesting that they might also be modified in other species.

Chromatin Regulation at Specific Genomic Regions

Trypanosomatid genes are distributed in large co-directional clusters that are expressed as polycistronic transcripts. These clusters, which may be located in different DNA strands or in the regions between clusters, have been proposed as transcription initiation sites [71, 72]. A TATA-binding protein and a Small Nucleolar-activating protein complex (SNP50) that is involved in initiating the transcription of small nuclear RNA were found to be associated with tRNA, snRNA, and SL RNA gene promoter regions [73]. Whether and how these factors are involved in the initiation of transcription in trypanosomatids is unknown.

Furthermore, non-canonical DNA regulatory sequences have been identified in trypanosomes, suggesting that their chromatin structure and the presence of histone modifications may play a prominent role in the initiation of transcription [74]. Chromatin immunoprecipitation (ChIP) assays performed in *T. cruzi* showed enrichment in H3K4me3, acH3 and acH4 at divergent DNA strand regions [75]. A similar scenario was observed in a genome-wide ChIP-chip analysis of *L. major* promastigotes that showed that the majority of the acetylated H3-enriched regions are at divergent strand-switch regions [73]. The probable transcription start sites of *T. brucei* are enriched with H4K10ac and H3K4me3. Interestingly, H4K10ac represents only 10% of the total H4. Moreover, H2AZ, H2BV, and the bromodomain factor BDF3 are enriched up to 300-fold in the same regions. In contrast, the histone variants H3V and H4V are enriched at probable transcription termination sites. No acetylated histones H3 and H4 were detected in the promoters of highly expressed genes, such as 18S rRNA and SL RNA, at which the depletion of nucleosomes was detected. A similar observation was made in *L. tarentolae*, in which the SL promoter lacks nucleosomes [10]. Supporting its potential presence at the heterochromatin, the highly repetitive satellite DNA was found to be depleted of H4ac [75]. These results suggest that chromatin structure might regulate the transcription of polycistronic genes by RNA Pol II, as indicated by the presence of special PTMs and variants at its initiation and termination sites [76, 77].

DNA Modifications

Trypanosome DNA is uniquely modified by the addition of glucopyranosyl residues to nucleotides, resulting in the formation of β-*d*-glucopyranosy--oxymethyluracil. This modification was called 'base J', and it was discovered in *T. brucei* [78, 79] because silenced telomeric VSG sequences were found to be resistant to some nucleases. J-base predominates in the telomeric repeats but is also present within sequences flanking the polycistronic units. The exact role of this modification remains unknown. Decreased levels of base J increase transcription by RNA Polymerase II, resulting in genome-wide increases in global gene expression [80, 81]. Similarly, 99% of DNA base J is found at telomeric sites in *Leishmania*, and the remainder is located at RNA Pol II termination sites. Knockouts of JBP2 (J Binding Protein 2) in *Leishmania*, one of the enzymes responsible for catalyzing the formation of base J, reduced the level of internal J and generated a massive read-through of transcriptional termination sites. Therefore, inducing the reduction of J in *L. major* resulted in genome-wide defects, likely as a consequence of the generation of antisense RNAs by avoiding transcriptional termination at the end of polycistronic gene clusters. In *T. brucei* , the loss of J base also affected the termination of gene expression at specific sites within polycistronic gene clusters, but the termination is only affected by depletion of the histone H3V [82].

CYTOSKELETON

The cytoskeleton of trypanosomes is quite simple compared to that of free-living flagellates. A subpellicular corset of microtubules cross-linked to each other underneath the plasma membrane maintains the peculiar cell shape of different trypanosome stages and forms. This cytoskeleton forms a dynamic matrix that also includes the mitotic spindle and encloses some specific regions of the trypanosomatid cell, such as the flagellum attachment zone, the paraflagellar rod, the flagellum, and basal bodies [83 - 85]. Trypanosome microtubules have an intrinsic polarity and are formed mainly by α and β-tubulin heterodimers, which undergoes several post-translational modifications [83, 84]. These organisms also contain structures similar to microfilaments, which are composed mainly of actin and are distributed throughout the cytoplasm as rounded and punctuated structures [86]. Actin is present along the flagellum but is more concentrated at the base and the tip [87]. In *Leishmania* species, actin is also associated with the nucleus, kinetoplast, plasma membrane, and flagellum, and it is found associated with subpellicular microtubules [88]. Knockdown of actin genes in *T. brucei* affected endocytosis, indicating a role for microfilaments in trypanosome membrane trafficking [89].

Trypanosomes display mitotic spindles, which are composed mainly of microtubules. Their spindles include longer 'pole-to-pole' microtubules and a classical kinetochore, which is a protein structure that is associated with chromosomes, to which the fibers of the mitotic spindle attach during cell division [83]. Unlike other eukaryotes, the microtubule organizing centers associated with spindle poles are often anchored in the nucleoplasm rather than to the internal face of the persistent nuclear envelope [83]. During mitosis, the nucleolar material persists and is stretched along spindle microtubules, as can be visualized using indirect immunofluorescence with an antibody against tubulin [83, 90].

Trypanosomes contain at least five different centrins, which are calcium-binding proteins that are associated with basal bodies. Centrins are involved in organelle duplication and segregation. Depletion of centrins 2 and 4 in *T. brucei* caused abnormal cell division, which resulted in enlarged cells that contained duplicated basal bodies and multiple nuclei and displayed new flagella that were detached along the length of the cell body [91]. *T. brucei* centrin 3 was detected in the flagellum, and its depletion affected motility by disrupting its interactions with flagellar dynein [92]. Furthermore, centrins participate in the formation of a unique structure called a bilobe, which is associated with the flagellum and is probably involved in the formation of the flagellar pocket [93].

It is possible that the mechanisms by which the different trypanosomes have adapted to obtain different nutrients have shaped the evolution of these various structures. For example, *T. cruzi,* unlike other trypanosomes, has retained an additional feeding apparatus called a cytostome that is located, outside the flagellar pocket [94]. The structural features of this element and the molecular basis of its retention should be further investigated in *T. cruzi.*

FLAGELLUM

Phylogenetic evidence indicates that it is likely that the last common ancestor of eukaryotes had a flagellum that provided motility and sensory information [95, 96]. This flagellum may have been set as a tail-like structure that emerged from the cell. Although motility and sensory perception functions have already been established for trypanosomatids flagella, recent functional studies have shown that they also have other roles, for example, in cell signaling, cellular morphogenesis, the direction of cytokinesis during cell division, and immune evasion [97 - 101]. Therefore, the trypanosome flagellum is considered an essential and multi-functional organelle that is typical of unicellular organisms [98, 101, 102].

Trypanosomes present a single flagellum that normally emerges from a specialized invaginated region of the plasma membrane that is called the flagellar pocket [103, 104]. Nine outer ring microtubule doublets and a central pair of microtubules form the inner core of the flagellum, and these are collectively called the axoneme [98, 103]. The axoneme is assembled on the basal body, which is itself attached to the kinetoplast [103, 105]. The basal body is an organelle that corresponds to the centriole of higher organisms and that acts as a microtubule organizing center that forms the flagellum axoneme. Each cell contains two basal bodies, but only one gives rise to the flagellum [103, 104] (for details regarding the flagellar structure and its composition, see the figures in the review [98]). In trypanosomes, the flagellum contains a large paracrystalline filament structure known as the paraflagellar rod (PFR) [106]. The PFR is a stable lattice-like structure that is formed mainly by two major proteins (PFR1 and PFR2) and several minor proteins [104, 106, 107]. Some studies have suggested that the PFR provides a scaffold for the assembly of the regulators of axoneme motility [98, 108, 109]. The flagellum plasma membrane is covered by specialized domains and has distinct biophysical properties compared to the plasma membrane of the rest of the cell body [110 - 112].

The flagella of organisms within the genus *Trypanosoma* are the most well-studied and well-characterized among the trypanosomatids parasites [113 - 116]. *T. cruzi* epimastigote and trypomastigote flagella are long but become reduced in size in amastigotes, which are adapted to survive inside host cells. The paraflagellar rod is absent in amastigote flagella [83, 105]. The small amastigote flagellum may be involved in sensory perception and cellular organization, which are critical for intracellular survival during the intracellular parasite stages within host cells [100, 104, 105].

Unlike *T. cruzi* , all *T. brucei* stages possess a long flagellum [117]. By using genetic modifications, mainly in *T. brucei* , flagella have been shown to be a key structure in the coordination of trypanosome morphogenesis [101, 118], host-parasite interactions [101, 119], mitochondrial DNA segregation [120], virulence [101] and cell division [101, 118, 121]. Furthermore, knockout and knockdown of some paraflagellar rod components showed that the paraflagellar rod is a key element in flagellar motility [108, 109, 122], which is necessary for parasite survival in the host [102]. Proteins located at the base of the flagellar axoneme are also involved in functions related to virulence and cytokinesis [123]. Therefore, the trypanosome flagellum can be considered a critical interface between the parasite and its hosts [111, 124].

MITOCHONDRION

Mitochondria were discovered as a collection of granules that form threads inside the cell. Mitochondria house the machinery that produces energy and synthesizes specific metabolites and iron-sulfur clusters. These functions make it one of the most important organelles in eukaryotic cells. This importance is reflected in their wide distribution, their number and their sizes in different organisms [125]. Multicellular organisms typically contain a vast number of mitochondria, reaching approximately 2,000 mitochondria per cell in the liver. Single-celled organisms, such as Apicomplexa and Kinetoplastida, possess a single ramified mitochondrion that has peculiar ultrastructural aspects that are different from those of multicellular organisms, especially in the density of its matrix and the number and shape of its cristae [114].

Paulin confirmed the unitary nature of the mitochondrion in trypanosomatids in a series of papers based on 3D reconstructions [126, 127]. The models generated by Paulin have a fidelity and a resolution that remains impressive, even today. In general, the mitochondrion of the trypanosomatids has a dense and sparse matrix that is distributed in branches under the subpellicular microtubules; but this structure can vary depending on the species and strain of the trypanosome. Moreover, changes in environmental and nutritional conditions can affect the size of the mitochondrion, which, depending on the situation can fill up to 12% of the volume of the parasite [128].

The mitochondrial structure of *T. cruzi* varies during its life cycle and assumes different shapes depending on its form. In epimastigotes, the mitochondrion has the shape of an inverted triangle. The base of the triangle encloses the nucleoid of the kinetoplast [126]. Two closely appressed finger-like projections extend from this base into the lancet-like apex of the cell. The mitochondrion of the trypomastigote form of *T. cruzi* is unbranched, appearing as a sausage-shaped structure with a smooth outer surface. The kinetoplast nucleoid, which contains the mitochondrial DNA, is found at the posterior end of the mitochondrion within a bulbous swelling. A short, finger-like projection of the chondriome pervades the truncated posterior pole of the cell, with the major portion coursing distally through about two-thirds of the cell body, anterior to the kinetoplast [129]. Amastigote mitochondrion is morphologically dissimilar to the structure found in the trypomastigotes (sausage-shaped) or epimastigotes (reticulated), instead presenting with a circular or horseshoe-like shape [127]. These variations in the mitochondria may reflect differences in the physiological state of the cell or the morphogenetic activity of the mitochondria, which are associated with cell division and differentiation. Although the mitochondrial structural features that characterize the different *T. cruzi* forms have been described, an extensive analysis of the molecular and metabolic events that cause such changes in mitochondrial structures during the differentiation of the parasite is still required.

Organization of Mitochondrial DNA

Mitochondria possess their own genetic system – a vestigial genome that originates from an endosymbiotic proteobacterial ancestor. In most eukaryotes, mitochondrial genes are involved in five basic processes: respiration and/or oxidative phosphorylation, translation, transcription, RNA maturation and protein import.

The mitochondrial DNA of trypanosomatids is enclosed in an unusual structure known as the kinetoplast [130]. It is located within the mitochondrial matrix, perpendicular to the axis of the flagellum. Kinetoplast DNA (kDNA) comprises approximately 30% of the total cellular DNA and differs from nuclear DNA in its buoyant density, base ratio, and degree of thermal denaturation. The kDNA of trypanosomatids is composed of circular molecules that are topologically relaxed and interlocked to form a single network. Two types of DNA rings, minicircles and maxicircles, are present in the kinetoplast. Each kinetoplast has several thousand minicircles, which range in size from approximately 0.5 to 2.5 kb (depending on the species), and a few dozen maxicircles, which usually vary between 20 and 40 kb [131 - 134]. The maxicircles encode the traditional mitochondrial genes including ribosomal RNAs and proteins that are components of the respiratory complexes. Of the 18 open reading frames in the maxicircle, 12 require post-transcriptional editing by uridine insertion and deletion to generate translatable mRNAs. Editing is specified by minicircle-encoded guide RNAs and is catalyzed by 20S editosomes, which perform the required endonuclease, exonuclease, terminal uridylyl transferase, and ligase activities [135].

Differential RNA editing appears to aid the adaptation of the variable metabolic processes during the different life stages of *T. brucei* . For example, mRNAs encoding the NADH dehydrogenase complex I (ND7 [136], ND8 [137], and ND9 [138]) are mainly edited in the bloodstream form of the parasite, while those encoding cytochrome b (CYb; complex III) [139] and cytochrome oxidase II (COII; complex IV) [140] are predominantly edited in procyclic parasites. Some mRNAs, such as the one encoding ATPase subunit 6 (A6), show no stage-specific editing [141]. *T. brucei* 20S editosomes contain at least 20 proteins [142] and were isolated *via* TAP-tagging, which allowed the purification and the characterization of their different components [143 - 145]. The editing- and life cycle stage-specific regulation of mitochondrial functions is much less characterized in *T. cruzi* . Annotated *T. cruzi* genomes have been found to contain orthologues for all characterized editosome proteins, but the mechanisms by which they function and whether their activity is important to the various parasite stages remain unknown.

Biochemical studies indicate the existence of proteins that are involved in replication and DNA repair and that are associated with the kinetoplast [130, 132]. The following proteins have been detected in association with kDNA: type II topoisomerase, DNA polymerase B, minicircle origin-binding protein, kDNA condensing proteins, kinetoplast-associated protein, a nicking enzyme, primase, mitochondrial heat shock proteins, and a set of kinetoplast-binding proteins that can bind both minicircles and maxicircles. A few examples of such genes that have been found in *T. cruzi* are described in [146 - 149].

The morphology of the kinetoplast varies depending on the developmental stage of the parasite. In the *T. cruzi* epimastigote and amastigote forms, the kinetoplast forms a bar-like structure. In contrast, the kinetoplast has a rounded-like shape in the non-dividing metacyclic trypomastigotes and cell-derived trypomastigotes [114]. Studies performed using fluorescence techniques have shown that a) kDNA replication takes place nearly concurrently with the nuclear S phase; b) before replication, the minicircles that are covalently closed are released vectorially from the network face near the flagellum; c) replication initiates in the region localized between the flagellar face of the disk and the mitochondrial membrane and at sites where the universal minicircle sequence-binding protein is located; and d) the replicating minicircles then move to two antipodal sites that flank the network [150, 151]. More details about kDNA organization and replication are found in Chapter 4.

Protein Synthesis in the Mitochondrion

As in all other organisms, only a small number of proteins are synthesized within the mitochondria of Trypanosomes. The majority of the 1000 mitochondrial proteins are encoded in the nucleus, synthesized by the cytoplasmic ribosomes, and imported into the mitochondrion [152]. Proteins synthesized in the mitochondrion include subunits of cytochrome oxidase (COX) and NADH dehydrogenase, cytochrome b (Cytb), subunit 6 of the mitochondrial ATPase, a ribosomal protein and a few other proteins [153]. No tRNA is synthesized in the mitochondria of trypanosomes, and this is compensated by the import of approximately 10% of the cytosolic tRNAs. In *T. brucei* mitochondria, a nuclear-encoded elongator, tRNAMet, is used for both elongation and initiation. Once imported into the organelle, part of tRNAMet is formylated by a methionyl-tRNAMet formyltransferase, and it must then interact with mitochondrial initiation factor 2 (mrIF2) before it can be assembled in the ribosome [154]. Similar to what has been observed in most eukaryotes, the stop codon UGA has been reassigned to the tryptophan codon CCA in the mitochondria of trypanosomatids. This reassignment occurs *via* a specific RNA-editing event that converts the CCA anticodon of the cytosolic imported tRNATrp into UCA, which allows the decoding of both UGG and UGA triplets [155 - 157].

The mitochondrial ribosomes of trypanosomes are also different from other eukaryotes. The large and the small subunits are very short, composed of only 1,150 nucleotides and 610 nucleotides, respectively [158, 159]. The mitochondrial ribosomes of *T. brucei* are constituted by 133 proteins [160], which is an extremely high number compared to the 77 subunits that are present in the mammalian mitochondrial ribosomes [161], suggesting that some of these proteins do not belong to the stable ribosome complex and could therefore be participating in accessory functions. Approximately 70% of the mitochondrial ribosome subunits are unique to trypanosomatids [160].

ENDOPLASMIC RETICULUM

The endoplasmic reticulum (ER) is an endomembrane compartment that extends throughout the cell. As in higher eukaryotes, the ER of trypanosomatids is the organelle that is responsible for protein folding, co and post-translational modifications and transport, the synthesis of lipids, calcium storage, and maintenance of cellular homeostasis [162]. The ER is distributed throughout the cell body in different trypanosomes.

An important function of the ER in trypanosomatids is the incorporation of glycosylphosphatidylinositol (GPI) anchors, a PTM modification that is present in most of their surface proteins. Examples include the family of mucin-like glycoproteins and members of the large family of trans-sialidase and 85 kDa glycoproteins in *T. cruzi* [163, 164], VSG and procyclin in *T. brucei* [165], and the major surface proteins of several *Leishmania* species [166]. These proteins are fundamental to parasite virulence and host immune system evasion, and the addition of GPI to these proteins is essential to ensuring their exit from the ER and their incorporation into the plasma membranes [167].

In most eukaryotes, proteins synthesized in the ER are properly folded by different molecular chaperones and folding-facilitating enzymes. A major ER luminal chaperone family of binding proteins (BiPs) [168] is also present in trypanosomes [169]. Moreover, the binding of BiPs to nascent polypeptides contributes to efficient and unidirectional transport inside the ER because they interact with the Sec61 translocon in the ER membrane [170]. In higher eukaryotes, secreted proteins usually undergo N-glycosylation, an important step in proper control of protein folding. This process is assisted by calnexin and/or calreticulin lectin chaperones [171]. *T. cruzi* and *T. brucei* encode only calreticulin. Furthermore, they lack the $Glc_3Man_9GlncNAc_2$ that is added to nascent chains, and they therefore use $Glc_1Man_9GlcNAc_2$ instead [172]. In *Leishmania* spp., more differences have been observed. For example these parasites synthesize a shorter version of the polyisoprenoid lipid carrier dolichol than mammals, and they lack homologs to several ER luminal glycosyl-transferases [173].

If improper folding occurs, in the majority of eukaryotes organisms, there are three ER-associated degradation pathways, luminal, membrane and cytosolic [174], which target the protein for proteasomal degradation. These events are part of the unfolded protein response (UPR). In eukaryotic cells, endoplasmic reticulum stress results from a number of factors, such as oxidative stress, nutrient insufficiency, hypoxia, changes in pH and temperature, calcium level disturbances, and failed glycosylation, that can result in the accumulation of misfolded or unfolded proteins, leading to the induction of the UPR [175]. Markers of proteins that are not properly folded include the exposition of hydrophobic regions, unpaired cysteine residues, changes in the glycan code and the increase in BiP activity [172].

Three important transmembrane ER-resident proteins are involved in the UPR in mammals: a) the endonuclease inositol-requirement 1 (IRE1), b) the activating transcription factor 6 (ATF6), and c) PKR-like ER kinase (PERK). The activation of IRE1 by *trans*-autophosphorylation promotes the degradation of ER-associated mRNAs [162]. ATF6 activates the transcription of genes that are involved in ER stress, such as genes that encode chaperones, and it also up-regulates proteins related to ER-associated degradation (ERAD), leading to the proteasomal degradation of misfolded proteins [172]. The third protein, PERK, is a protein kinase that when activated, phosphorylates the eukaryotic initiation factor 2 (eIF2), leading to the inhibition of mRNA translation and thereby reducing the level of newly synthesized proteins and decreasing the ER luminal protein burden [173]. Prolonged ER stress can lead to cell death by apoptosis in a process that involves IRE1, ATF6 and PERK mediating the production of common cell death mediator C/EBP homologous protein (CHOP) [176].

Trypanosomatids lack typical UPR response machinery because they have non-conventional gene expression regulation mechanisms and less transcriptional regulation of individual genes [173]. They do not possess homologs of IRE1, ATF6 and CHOP. Nevertheless, these parasites can respond to improper protein folding because they express BiP, suggesting a minimal capacity for quality control [177]. The trypanosome ER-folding machinery is capable of upregulating alternative mechanisms that can increase folding efficiency in the absence of lectin-N-glycan interaction [178], a typical mechanism that facilitates ER glycoprotein folding.

A homolog of PERK that phosphorylates eukaryotic initiation factor 2 (eiF2) is present in trypanosomes and has been characterized in *T. brucei* , *T. cruzi,* and *L. infantum* [179 - 181]. However, this protein, named eIF2K2, is located in the ER only in *Leishmania* [182], whereas it is localized in the flagellar pocket in *T. brucei* and the endosomal compartment of *T. cruzi.* It promotes eiF2 phosphorylation and the arrest of protein synthesis, similar to its functions in most eukaryotes. Protein synthesis arrest is required for epimastigotes to differentiate into metacyclic trypomastigotes [183], which probably signals changes in gene expression through eIF2 phosphorylation. The signals that induce kinase activation and eIF2 phosphorylation have been recently found to involve the absence of heme in *T. cruzi,* which links the control of oxidative species to protein synthesis in this parasite [179]. Phosphorylation of eIF2 also occurs *via* other kinases, one of which is homologous to the general control non-derepressible 2 (GCN2) and is named eIF2K1 and a third kinase (eIF2K3) that responds to extensive heat shock and phosphorylates TATA-binding proteins that are involved in the control of SL RNA transcription in the nucleus [162, 184]. Therefore, despite the lack of a typical UPR response system, the different trypanosomatids have developed the capacity to respond to stress conditions, which ensures their adaptation to new environments and the success of infection.

GLYCOSOMES

All members of Kinetoplastida have organelles that are surrounded by a single bilayer membrane that have peroxisome characteristics [185]. These are designated glycosomes or microbodies. They are approximately 0.2 - 0.3 µm diameter [186] and contain, in addition to typical peroxisome components, enzymes that are involved in glucose and sometimes glycerol metabolism, which usually occur in the cytosol in other organisms [185, 187]. The number and composition of glycosomes is variable among species and across different lifecycle stages. *T. cruzi* contain approximately 50 glycosomes that are dispersed throughout the cytosol [188], while the bloodstream form of *T. brucei* has approximately 65 glycosomes. These numbers are smaller than those in *L. major,* which contain 20 glycosomes in the promastigote stage but only 10 in the amastigote stage. Moreover, the amount of glycosomes in *L. major* is controlled during the life cycle by processes related to autophagy and that involve glycosome-containing autophagosomes, which are translocated to lysosomes for degradation [189]. Procyclic *T. brucei* undergo changes in glycosome composition depending on environmental conditions, such as glucose levels [190, 191]. Approximately 90% of the glycosome content is composed of glycolytic enzymes in bloodstream *T. brucei* , whereas in the procyclic, this number drops to 40% [188], reflecting the important metabolic changes that occur in this parasite.

The glycosomes of *T. cruzi* epimastigotes show a peripheral 75 kDa membrane protein that is abundant in exponentially growing epimastigotes but absent in stationary phase parasites. Other molecules characterized in *T. cruzi* glycosomes include a pair of integral membrane proteins that are approximately 85-100 kDa in size and that are present only in the stationary phase and two proteins that are 36 and 41 kDa in size. The lipid composition of these organisms is very similar to that in the membranes of whole cells, except for the increased presence of endogenous sterols (ergosterol, 24-ethyl-5,7,22-cholesta-trien-3b-ol) and their precursors [192].

Glycolysis occurs mainly inside the glycosomes and is an essential event in these organisms. The inhibition of glycolysis in *T. brucei* bloodstream form results in the loss of motility and the lysis of the cell [193]. The bloodstream form of *T. brucei* guarantee the production of ATP by obtaining blood glucose through an efficient glycolysis, whereas other trypanosomatids produce energy in the mitochondria *via* oxidative phosphorylation [194 - 196]. In *T. brucei* , for instance, only three glycolytic enzymes are found outside of the glycosome, and they are involved in the conversion of 3-phosphoglycerate (3-PGA) [197]. Moreover, other enzymes involved in several cellular processes are located in these specialized peroxisomes. Examples include components of the gluconeogenic and pentose phosphate pathways, components of pyrimidine biosynthesis and purine salvage, and, similar to the peroxisomes of other eukaryotes, enzymes involved in fatty-acid oxidation and peroxide metabolism and ether-lipids [198, 199]. Furthermore, the biosynthesis of and the ability to synthesize vitamin C is also a function of the *T. brucei* and *T. cruzi* glycosomes [200].

Proteins that constitute the intracellular matrix of glycosomes are generated in the cytoplasm and then imported to the glycosome through a sequence motif called peroxisomal-targeting signal (PTS). Some transporters, which are designed like peroxins (PEX), are responsible for carrying the cargo to the organelle. Approximately 32 PEXs have been identified in animals ranging from yeasts to mammals . In *T. brucei* and *L. donovani,* two proteins that are essential for the initial events have been identified and named peroxisomal targeting proteins PEX5 and PEX14 [201 - 204]. Three other carriers that are homologous to peroxins have been named PEX6, PEX10 and PEX12. These proteins are involved in the final steps of the import process [199]. In addition, isoforms of peroxins, including PEX13.1 and PEX13.2, have been associated with glycosome biogenesis [205]. In the absence of some of these proteins, reduced parasite growth, changes in the size of glycosomes and, in some cases, motility changes and cell death have been observed [206 - 209].

Some glycosomal enzymes are regulated by phosphorylation, which allows the control of metabolic flux because changes in nutrition are required throughout the parasite life cycle [210, 211]. Phosphatases PIP39 and PTP1 are involved in this regulation. The absence of PIP39 in the glycosome impairs parasite differentiation, which indicates the importance of compartmentalization to the activity of these enzymes [212]. In addition, *T. brucei* proteomic data identified more than 98 proteins in procyclic that could be located in the glycosome. This discrimination was possible through the use of advanced techniques, such as epitope-tagged glycosomes and stable isotope labeling in cell cultures [213]. Similarly, 789 proteins were identified in *L. donovani* glycosomes. Many of them appeared to be involved in metabolic events, such as lipid, carbohydrate and nucleic acid metabolism [214]. The unique compartmentalization inside this organelle is crucial to maintaining the integrity of metabolism-involved enzymes and therefore the infectivity of trypanosomatids [215]. For this reason, a focus has been placed on the design and synthesis of compounds that can inhibit glycosome and control trypanosome growth and infection [216, 217].

MEMBRANE TRAFFICKING AND THE GOLGI COMPLEX

Trypanosomes have a well-defined Golgi apparatus that is involved in the production of the secretory organelles that release soluble and membrane components (cargos) to the flagellar pocket and the lysosomal apparatus [218, 219]. In *T. cruzi* epimastigotes, a single Golgi complex is located in the perinuclear space of the cell, close to the flagellar pocket [220]. Enriched Golgi fractions isolated from this parasite were shown to contain enzymes that are involved in the formation of *O*-linked glycoproteins. The presence of a unique polypeptide-α-N-acetyl-glucosaminyltransferase and β-galactosyl transferase, which are involved in the biosynthesis of mucins, is of particular importance because they represent the major surface proteins that are present in all *T. cruzi* stages [163]. Two genes encode this enzyme [221]. In addition, several enzymes required for the addition and processing of N-glycans and GPI-anchors have been characterized in several Kinetoplastida, and striking differences have been observed between species [222, 223].

As in most eukaryotes, trypanosomes mediate vesicle transport *via* several cytosolic (*e.g.*, *N*-ethylmaleimide-sensitive factor and SNP] and receptor (*e.g.*, soluble NSF Attachment Protein and SNARE) proteins. These include COPI (ER to Golgi) and COPII (Golgi to ER), adaptor proteins (Aps), clathrin, and small GTPases, named Rabs [224]. Several Rabs have been identified and characterized in *T. brucei* , where they act to promote the formation and fusion of intracellular vesicles [225, 226]. A few of them have been identified in *T. cruzi.* One of these is TcRab7, a typical component of late endosomes in *T. brucei* [227] that has also been found to be associated with the Golgi apparatus in *T. cruzi* epimastigotes [228]. This unusual localization might reflect the specific features of *T. cruzi,* which contains soluble cargo that originates from endocytosis and that accumulates in organelles called reservosomes, which are found in proliferative epimastigote forms [229]. Reservosomes also accumulate recently synthesized forms of the major lysosomal cysteine protease cruzipain, and the inhibition of protease maturation affects the structure of the Golgi and membrane trafficking [230]. In non-proliferating trypomastigotes, membrane traffic is redirected towards the secretory pathway [165, 225, 226]. Trypomastigotes release soluble cruzipain and incorporate larger amounts of GPI-anchored proteins into the parasite surface, such as members of the mucin-like glycoprotein and trans-sialidase families. This transport also involves other Rabs, including a Rab11 homolog that is located in the contractile vacuole of the parasite [231]. Interestingly, *T. cruzi* trypomastigotes preferentially release vesicles containing major surface proteins, while dividing *T. brucei* trypomastigotes perform endocytosis and membrane recycling to obtain some nutrients and release bound and damaging antibodies [232].

Clathrin, a major coat protein that is involved in eukaryotic cell endocytosis, has been described in trypanosomes. In trypanosomes, typical clathrin coats have been visualized and found to be associated with membrane trafficking and endocytosis in vesicles that are connected to the flagellar pocket of different trypanosomes, including *T. cruzi* [233], *T. brucei* [234] and *L. major* [235]. Significant differences in clathrin localization are observed, depending on the parasite stage and thc species, and these differences have been found to be associated with differences in their endocytic requirements [236]. Clathrin vesicle coat formation depends on assembly polypeptides (AP) that anchor clathrin to the membrane. Four types of AP complexes have been found in trypanosomes, a smaller number than in higher eukaryotes. The absence of AP2 in *T. brucei* is probably related to the unique endocytic recycling of VSG observed in this parasite [237].

Therefore, the different adaptations observed in membrane trafficking are certainly vital to the survival of the parasite under different environmental conditions. It remains to be determined how membrane trafficking is differentially regulated and what mechanisms control sorting to the plasma membrane or intracellular organelles.

LYSOSOMES AND ACIDIC ORGANELLES

Lysosomes are digestive organelles that represent the final destination of the endocytic pathway. They contain acidic hydrolases and unique proteins that are involved in the organelle acidification. The lysosome is well characterized in *T. brucei* by the presence of the protein p67, which contains dileucine motifs in its

C-terminus and faces the cytosol, as found in the lysosomal-associated membrane protein (LAMP) of mammalian cells [238]. p67 colocalizes with *T. brucei* cysteine protease, also called trypanopain [239], which corresponds to the cruzipain in *T. cruzi* that has been found in multiple vesicles in epimastigotes [179]. The characterization of lysosomes in other trypanosomes is, however, less clear. A p67 lysosomal protein homolog is present in *T. cruzi* in not fully acidic organelles known as reservosomes. Reservosomes are most like endosomal structures that accumulate proteins in *T. cruzi* epimastigotes and that probably represent immature lysosomes [240]. Trypanosomes also present acidic organelles known as acidocalcisomes. We will describe these structures separately because they have specific roles.

Reservosomes

Reservosome morphology varies according to environmental conditions and the strain of *T. cruzi* [241]. Unlike lysosomes, the reservosome lumen presents a pH of 6.0, which is maintained by a P-type H+ adenosine triphosphatase [242]. It contains several hydrolases, including cruzipain [243]. Therefore, it seems to have a lysosomal function, and some authors have referred to them as pre-lysosomes. Reservosomes can also be considered late endosomes, but it was shown that TcRab7, a common marker of late endosomes, is absent in this organelle [244]. Hundreds of proteins have been detected in purified reservosomes using mass spectrometry. Among these are the expected hydrolytic enzymes, proton pumps and transporters [244]. The Rab11 GTPase, which is involved in vesicle recognition and fusion, has also been found in reservosomes [245]. Reservosomes contain proteins that are associated with the secretory pathway, indicating that this organelle combines storage and degradation functions. Reservosomes accumulate molecules by ingestion *via* fluid-phase endocytosis, which are substrates for homeostasis and parasite differentiation. Reservosomes also store heme [246] and contain several lipid inclusions that are rich in sterols [247]. For these reasons, they can appear in transmission electron microscopy as large electron-dense structures or empty vesicles that contain intramembranous structures, depending on the environmental conditions. Importantly, reservosomes are degraded during differentiation from the epimastigote form to the metacyclic trypomastigote form [248]. Inversely, when a trypomastigote differentiates into a replicative epimastigote form, reservosome appear to originate from the Golgi apparatus [241].

Acidocalcisomes

Acidocalcisomes are also lysosome-related organelles that accumulate calcium and phosphorus compounds in the form of pyrophosphate (PPi) and poly-phosphate (polyP), which are complexed with other cations. These organelles were initially described in trypanosomatids [249, 250] and later found to be conserved in organisms ranging from bacteria to humans cells [251].

Acidocalcisomes appear as an empty vacuole with small electron-dense punctae under conventional electron microscopy [252, 253]. In trypanosomatids, they are easily detected by staining the cells with dyes such as acridine orange [250] and cycloprodigiosin [254] or by the presence of polyP, which can be observed under a fluorescence microscope using 4',6-diamidino-2-phenylindole (DAPI), a common DNA stain [255, 256]. The organelles are generally spherical and have a mean diameter of 0.2 µm in *T. cruzi* and *T. brucei* [252, 257] and 0.6 µm in some *Leishmania* species [258, 259]. They are randomly distributed throughout the cells, some of them localized in proximity of the contractile vacuole [251]. *T. cruzi* amastigotes contain more acidocalcisomes than epimastigotes and trypomastigotes [252], and in *L. major*, amastigotes and stationary promastigotes present more acidocalcisomes than replicative promastigotes [260]. In *T. brucei* , acidocalcisomes are more abundant in the procyclic form than in the bloodstream form [257].

Trypanosome acidocalcisomes contain high concentrations of polyP, a polymer of inorganic phosphate (Pi) that ranges in length from a few to several hundred residues [261]. PolyP is essential for the responses of trypanosomatids to different stress conditions. For example, they are associated with the ability to respond to osmotic or nutritional stresses and their function can be implicated in the parasite virulence [262, 263]. In fact, because trypanosomes alternate between different hosts they are exposed to extreme changes in osmolarity [251]. For example, hydrolysis, or the synthesis of polyP occurs under conditions of hyposmotic or hyperosmotic stress in *T. cruzi* epimastigotes [264] and conditions involving changes in ionic concentrations in *L. major* promastigotes [265, 266].

The osmotic regulation in trypanosomes requires aquaporin, a water channel protein that is present in acidocalcisomes membranes and the contractile vacuole complex (CVC) [267]. CVCs are present in both *Leishmania* and *T. cruzi* [266, 268]. It is proposed that in *T. cruzi* epimastigotes that are exposed to hyposmotic stress, the acidocalcisomes swell and fuses with the CVC allowing aquaporin translocation to extrude the accumulated water [269]. The fusion process is controlled by the small GTPases of the RAB family, particularly TcRab32, a protein that also modulates the trafficking of proteins during acidocalcisome biogenesis [268].

Trypanosomes express four enzymes that are related to the target of rapamycin (TOR) kinases. One of these, TOR3 or TOR-like 1 kinase, has been found to be also involved in the osmotic regulation control in *T. brucei* [270] and *L. major* [271]. RNAi against TOR3 in *T. brucei* for example, led to growth arrest, the formation of larger acidocalcisomes, the accumulation of polyP and PPi, and increased sensitivity to hyperosmotic conditions [270]. *L. major* TOR3 knockout mutants had smaller acidocalcisomes that contained less polyP, and the knockout parasites were less responsive to hyposmotic stress [271]. These data suggest that TOR3 may play different roles in acidocalcisome functions in each parasite [266].

In addition, acidocalcisomes also function as a cation storage organelle by presenting the elements O, Na, Mg, P, K, Ca and Zn. Iron was also found in these organelles in *T. cruzi* trypomastigotes and in *L. amazonensis* promastigotes that were cultivated in complex media [251]. Free amino acids are also present in acidocalcisomes, where they predominantly (90%) contained arginine and lysine [272].

The acidocalcisome matrix contains a polyphosphate kinase [264], an exopolyphosphatase [273, 274], an inorganic pyrophosphatases (VSP1) [262, 263, 275] and a metacaspase [276]. Trypanosome acidocalcisomes membranes contain several pumps. Among these, there are two proton pumps, which are important for the acidification of acidocalcisome. One is the multi-subunit vacuolar type H^+-ATPase (V-H^+-ATPase), which is sensitive to bafilomycin A [249, 250]. The other is the vacuolar proton pyrophophatase (V-H^+-PPase or VP1), which uses PPi instead of ATP as an energy source [277, 278]. These enzymes are required for growth [279], and they allow trypanosomes to recover after external pH changes [251, 280].

Acidocalcisomes are also involved in calcium signaling, which is necessary for host cell invasion by trypanosomes [281]. A calcium pump (Ca^{2+}-ATPase) of the PMCA-type that is sensitive to vanadate and is involved in calcium uptake was characterized in the acidocalcisome membranes of both *T. cruzi* [250] and *T. brucei* [249]. In *T. brucei* , RNAi of the Ca^{2+}-ATPase decreased the levels of calcium in acidocalcisomes, affecting cell growth [282]. Recently, an inositol 1,4,5-trisphosphate receptor (IP_3R), the primary target for the initiation of intracellular Ca^{2+} signaling in most eukaryotic cells was identified in the acidocalcisomes of *T. brucei* [283]. The ablation of the *T. brucei* IP_3R using RNAi resulted in a significant reduction in IP_3-mediated Ca^{2+} release in trypanosomes, defects in growth and reduced virulence in mice [283]. While IP_3R is localized to the ER in eukaryotes, *T. cruzi* IP_3R was also detected in acidocalcisomes [279, 284]. Knockouts of $TcIP_3R$ resulted in the deficient growth of epimastigotes and amastigotes, deficient metacyclogenesis, and deficient host cell invasion by *T. cruzi* trypomastigotes and was associated with reduced Ca^{2+}release upon their attachment to host cells and defects in virulence *in vivo* [281, 284].

Fig. (**2**) illustrates the major organelles approached in this chapter present in the replicative and non-replicative forms of *T. cruzi* .

CONCLUDING REMARKS

From the nucleus to the flagellum, and including intracellular traffic and energy generator organelles, these organisms present unique features. Many of these have been described in this chapter, and these include: a) the presence of a bipartite nucleolus, b) the condensation of chromatin only into 10-nm fibers; c) the presence of unusual chromatin regulatory mechanisms for gene expression and base J; d) the presence of the flagellum and the flagellar pocket coordinated with the cell cycle and morphogenetic adaptations in different species; e) unconventional mechanisms for the secretion of GPI-anchored proteins and the biosynthesis of glycoproteins; f) the presence of unique organelles called glycosomes, which are involved in glucose metabolism; g) the presence of exclusive and ramified mitochondria, which contain the genetic material in kinetoplast; and h) unique mechanisms for the control of membrane trafficking that contribute to parasite adaptation.

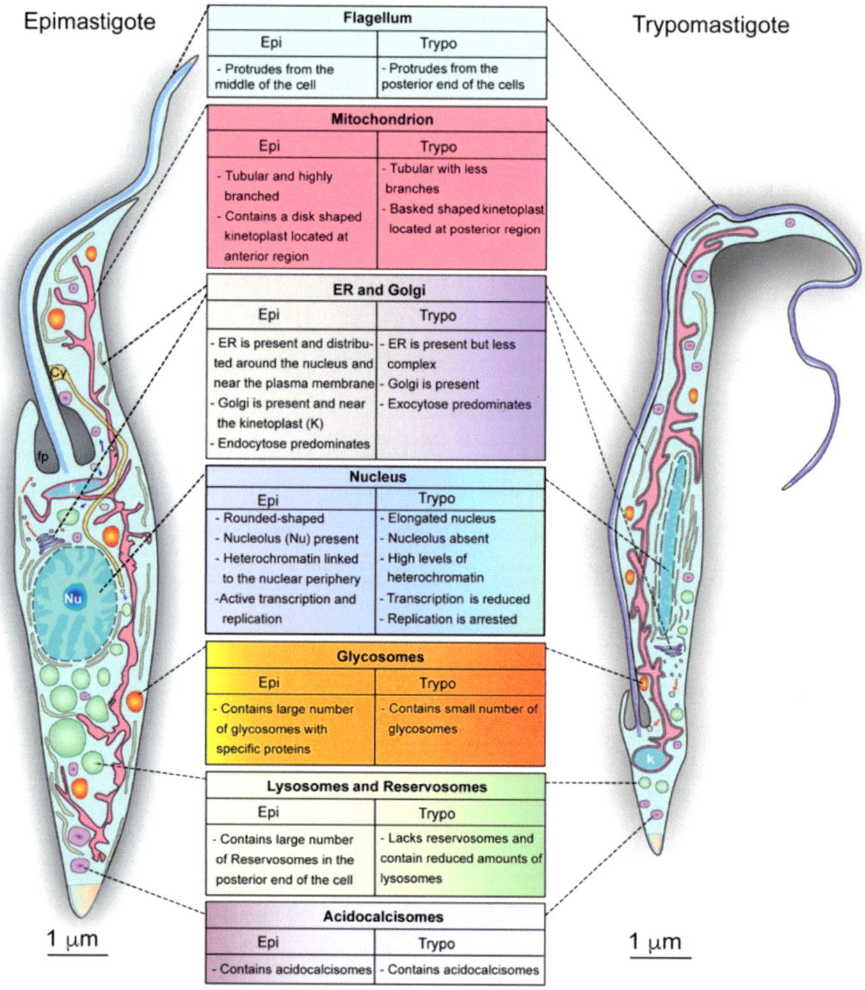

Fig. (2). Schematic representation of the major organelle in the replicative (epimastigote) and infective and non-replicative (trypomastigote) forms of *T. cruzi.* The central boxes indicate the different characteristics of each of these forms. The nucleus is shown in blue, and the heterochromatin is highlighted in dark blue. The mitochondrion is shown in pink. The glycosomes are shown in orange/yellow, the acidocalcisomes are shown in magenta, the Golgi apparatus is shown in purple, the ER is shown in light brown, and the lysosomes and reservosomes are shown in green. The figure also indicates the location of the nucleolus (Nu), the kinetoplast (k), the flagellar pocket (fp) and the cytostome (Cy). The blue lines in the flagellum correspond to the microtubule arrays, the red arrows indicate exocytic traffic, and the black arrows indicate endocytic traffic.

These unusual features provide valuable insights into how each species evolved and how they differentiate during their life cycles to increase their fitness. In the case of parasitic species, the different adaptions that are observed reflect the diverse strategies used by these species during survival and transmission. Therefore, increasing our understanding of these mechanisms provides us with new perspectives from which we can better address each disease.

CONFLICT OF INTEREST

The authors confirm that they have no conflict of interest to declare for this publication.

ACKNOWLEDGEMENTS

This was supported by different grants from the Fundação de Amparo à Pesquisa do Estado de São Paulo (FAPESP) and the Conselho Nacional de Desenvolvimento Científico e Tecnológico (CNPq), from Brazil.

ABBREVIATIONS

3-PGA	3-phosphoglycerate
AP	Assembly polypeptides
ATF6	Activating transcription factor 6
ATP	Adenosine triphosphate
BDF	Bromodomain factor
BiPs	Binding proteins
Ca2+-ATPase (PMCA)	Plasma membrane calcium ATPase
ChIP	Chromatin Imunopreciptation
CHOP	C/EBP homologous protein
COP	Coated vesicle membrane protein
COX	Cytochrome oxidase
CVC	Contractile vacuole complex
Cytb	Cytochrome b
DAPI	4',6-diamidino-2-phenylindole

DOT	Disruptor of telomeric silencing
eIF2	Eukaryotic initiation factor 2
ER	Endoplasmic Reticulum
ERAD	Endoplasmic reticulum associated degradation
ESAG	Expression site-associated gene
FG	Phenylalanine-glycine
GCN2	General control non-derepressible 2
GPI	Glycosylphosphatidylinositol
IP3R	Inositol 1,4,5-trisphosphate receptor
IRE1	Endonuclease inositol-requirement 1
JBP	J Binding Protein
kDNA	Kinetoplast DNA
LAMP	Lysosomal-associated membrane protein
mrIF2	Mitochondrial initiation factor 2
NPC	Nuclear Pore Complex
NSF	N-ethylmaleimide sensitive factor
Nup	Nucleoprotein
PERK	PKR-like ER kinase
PEX	Peroxins
PFR	Paraflagellar rod
PFR1	Paraflagellar rod protein 1
PFR2	Paraflagellar rod protein 2
Pi	Inorganic phosphate
PolyP	Polyphosphate
PPi	Pyrophosphate
PTM	Post-translational modifications
PTP1	Protein-tyrosine-phosphatase
PTS	Peroxisomal-targeting signal
RNAi	RNA interference
SL	Spliced Leader
SNARE	Soluble N-ethylmaleimide-sensitive factor attachment protein receptor
SNP	Small Nucleolar-activating protein complex
TOR	Target of rapamycin
TzCRK	T. cruzi CDK-like
UPR	Unfolded protein response

V-H+-ATPase	Vacuolar type H+-ATPase
V-H+-PPase or **VP1**	Vacuolar proton pyrophosphatase
VSG	Variant surface glycoprotein
VSP1	Type 1 soluble pyrophosphatase

REFERENCES

[1] Calderano SG, de Melo Godoy PD, da Cunha JP, Elias MC. Trypanosome prereplication machinery: a potential new target for an old problem. 2011. [http://dx.doi.org/10.4061/2011/518258]

[2] Kaye P, Scott P. Leishmaniasis: complexity at the host-pathogen interface. Nat Rev Microbiol 2011; 9(8): 604-15. [http://dx.doi.org/10.1038/nrmicro2608] [PMID: 21747391]

[3] Dolan MF, Melnitsky H, Margulis L, Kolnicki R. Motility proteins and the origin of the nucleus. Anat Rec 2002; 268(3): 290-301. [http://dx.doi.org/10.1002/ar.10161] [PMID: 12382325]

[4] Koumandou VL, Wickstead B, Ginger ML, van der Giezen M, Dacks JB, Field MC. Molecular paleontology and complexity in the last eukaryotic common ancestor. Crit Rev Biochem Mol Biol 2013; 48(4): 373-96. [http://dx.doi.org/10.3109/10409238.2013.821444] [PMID: 23895660]

[5] Elias MC, Marques-Porto R, Freymüller E, Schenkman S. Transcription rate modulation through the *Trypanosoma cruzi* life cycle occurs in parallel with changes in nuclear organisation. Mol Biochem Parasitol 2001; 112(1): 79-90. [http://dx.doi.org/10.1016/S0166-6851(00)00349-2] [PMID: 11166389]

[6] DuBois KN, Alsford S, Holden JM, *et al.* NUP-1 Is a large coiled-coil nucleoskeletal protein in trypanosomes with lamin-like functions. PLoS Biol 2012; 10(3): e1001287. [http://dx.doi.org/10.1371/journal.pbio.1001287] [PMID: 22479148]

[7] Picchi GF, Ferreira AM, Souza FS, *et al. Trypanosoma cruzi* : identification of DNA targets of the nuclear periphery coiled-coil protein TcNUP-1. Exp Parasitol 2011; 127(1): 147-52. [http://dx.doi.org/10.1016/j.exppara.2010.07.009] [PMID: 20647012]

[8] Hiratani I, Takebayashi S, Lu J, Gilbert DM. Replication timing and transcriptional control: beyond cause and effect part II. Curr Opin Genet Dev 2009; 19(2): 142-9. [http://dx.doi.org/10.1016/j.gde.2009.02.002] [PMID: 19345088]

[9] Dossin FdeM, Schenkman S. Actively transcribing RNA polymerase II concentrates on spliced leader genes in the nucleus of *Trypanosoma cruzi*. Eukaryot Cell 2005; 4(5): 960-70. [http://dx.doi.org/10.1128/EC.4.5.960-970.2005] [PMID: 15879530]

[10] Hitchcock RA, Thomas S, Campbell DA, Sturm NR. The promoter and transcribed regions of the *Leishmania tarentolae* spliced leader RNA gene array are devoid of nucleosomes. BMC Microbiol 2007; 7: 44. [http://dx.doi.org/10.1186/1471-2180-7-44] [PMID: 17517143]

[11] Uzureau P, Daniels JP, Walgraffe D, *et al.* Identification and characterization of two trypanosome TFIIS proteins exhibiting particular domain architectures and differential nuclear localizations. Mol Microbiol 2008; 69(5): 1121-36.
[PMID: 18627464]

[12] Chaves I, Rudenko G, Dirks-Mulder A, Cross M, Borst P. Control of variant surface glycoprotein gene-expression sites in *Trypanosoma brucei.* EMBO J 1999; 18(17): 4846-55.
[http://dx.doi.org/10.1093/emboj/18.17.4846] [PMID: 10469662]

[13] Calderano SG, de Melo Godoy PD, Motta MC, Mortara RA, Schenkman S, Elias MC. *Trypanosoma cruzi* DNA replication includes the sequential recruitment of pre-replication and replication machineries close to nuclear periphery. Nucleus 2011; 2(2): 136-45.
[http://dx.doi.org/10.4161/nucl.2.2.15134] [PMID: 21738836]

[14] Elias MC, Faria M, Mortara RA, *et al.* Chromosome localization changes in the *Trypanosoma cruzi* nucleus. Eukaryot Cell 2002; 1(6): 944-53.
[http://dx.doi.org/10.1128/EC.1.6.944-953.2002] [PMID: 12477795]

[15] Kaufmann D, Gassen A, Maiser A, Leonhardt H, Janzen CJ. Regulation and spatial organization of PCNA in *Trypanosoma brucei.* Biochem Biophys Res Commun 2012; 419(4): 698-702.
[http://dx.doi.org/10.1016/j.bbrc.2012.02.082] [PMID: 22387477]

[16] Kumar D, Minocha N, Rajanala K, Saha S. The distribution pattern of proliferating cell nuclear antigen in the nuclei of *Leishmania donovani.* Microbiology 2009; 155(Pt 11): 3748-57.
[http://dx.doi.org/10.1099/mic.0.033217-0] [PMID: 19729406]

[17] Ninio J. Connections between translation, transcription and replication error-rates. Biochimie 1991; 73(12): 1517-23.
[http://dx.doi.org/10.1016/0300-9084(91)90186-5] [PMID: 1805967]

[18] Liu B, Alberts BM. Head-on collision between a DNA replication apparatus and RNA polymerase transcription complex. Science 1995; 267(5201): 1131-7.
[http://dx.doi.org/10.1126/science.7855590] [PMID: 7855590]

[19] Landeira D, Navarro M. Nuclear repositioning of the VSG promoter during developmental silencing in *Trypanosoma brucei.* J Cell Biol 2007; 176(2): 133-9.
[http://dx.doi.org/10.1083/jcb.200607174] [PMID: 17210949]

[20] Navarro M, Gull K. A pol I transcriptional body associated with VSG mono-allelic expression in *Trypanosoma brucei.* Nature 2001; 414(6865): 759-63.
[http://dx.doi.org/10.1038/414759a] [PMID: 11742402]

[21] Daniels JP, Gull K, Wickstead B. Cell biology of the trypanosome genome. Microbiol Mol Biol Rev 2010; 74(4): 552-69.
[http://dx.doi.org/10.1128/MMBR.00024-10] [PMID: 21119017]

[22] Navarro M, Peñate X, Landeira D. Nuclear architecture underlying gene expression in *Trypanosoma brucei.* Trends Microbiol 2007; 15(6): 263-70.
[http://dx.doi.org/10.1016/j.tim.2007.04.004] [PMID: 17481901]

[23] Dossin FdeM, Dufour A, Dusch E, *et al.* Automated nuclear analysis of *Leishmania major* telomeric clusters reveals changes in their organization during the parasites life cycle. PLoS One 2008; 3(6):

e2313.
[http://dx.doi.org/10.1371/journal.pone.0002313] [PMID: 18545650]

[24] Holden JM, Koreny L, Kelly S, *et al.* Touching from a distance. Nucleus 2014; 5(4): 304-10. [http://dx.doi.org/10.4161/nucl.32228] [PMID: 25482119]

[25] Terry LJ, Wente SR. Flexible gates: dynamic topologies and functions for FG nucleoporins in nucleocytoplasmic transport. Eukaryot Cell 2009; 8(12): 1814-27. [http://dx.doi.org/10.1128/EC.00225-09] [PMID: 19801417]

[26] Raices M, DAngelo MA. Nuclear pore complex composition: a new regulator of tissue-specific and developmental functions. Nat Rev Mol Cell Biol 2012; 13(11): 687-99. [http://dx.doi.org/10.1038/nrm3461] [PMID: 23090414]

[27] Yang Q, Rout MP, Akey CW. Three-dimensional architecture of the isolated yeast nuclear pore complex: functional and evolutionary implications. Mol Cell 1998; 1(2): 223-34. [http://dx.doi.org/10.1016/S1097-2765(00)80023-4] [PMID: 9659919]

[28] Akey CW, Goldfarb DS. Protein import through the nuclear pore complex is a multistep process. J Cell Biol 1989; 109(3): 971-82. [http://dx.doi.org/10.1083/jcb.109.3.971] [PMID: 2475512]

[29] Morelle C, Sterkers Y, Crobu L, *et al.* The nucleoporin Mlp2 is involved in chromosomal distribution during mitosis in trypanosomatids. Nucleic Acids Res 2015; 43(8): 4013-27. [http://dx.doi.org/10.1093/nar/gkv056] [PMID: 25690889]

[30] DeGrasse JA, DuBois KN, Devos D, *et al.* Evidence for a shared nuclear pore complex architecture that is conserved from the last common eukaryotic ancestor. Mol Cell Proteomics 2009; 8(9): 2119-30. [http://dx.doi.org/10.1074/mcp.M900038-MCP200] [PMID: 19525551]

[31] Field MC, Horn D, Alsford S, Koreny L, Rout MP. Telomeres, tethers and trypanosomes. Nucleus 2012; 3(6): 478-86. [http://dx.doi.org/10.4161/nucl.22167] [PMID: 22992703]

[32] Mekhail K, Moazed D. The nuclear envelope in genome organization, expression and stability. Nat Rev Mol Cell Biol 2010; 11(5): 317-28. [http://dx.doi.org/10.1038/nrm2894] [PMID: 20414256]

[33] Crisp M, Burke B. The nuclear envelope as an integrator of nuclear and cytoplasmic architecture. FEBS Lett 2008; 582(14): 2023-32. [http://dx.doi.org/10.1016/j.febslet.2008.05.001] [PMID: 18474238]

[34] Goldman RD, Gruenbaum Y, Moir RD, Shumaker DK, Spann TP. Nuclear lamins: building blocks of nuclear architecture. Genes Dev 2002; 16(5): 533-47. [http://dx.doi.org/10.1101/gad.960502] [PMID: 11877373]

[35] Lenz-Böhme B, Wismar J, Fuchs S, *et al.* Insertional mutation of the drosophila nuclear lamin Dm0 gene results in defective nuclear envelopes, clustering of nuclear pore complexes, and accumulation of annulate lamellae. J Cell Biol 1997; 137(5): 1001-16. [http://dx.doi.org/10.1083/jcb.137.5.1001] [PMID: 9166402]

[36] Liu J, Rolef Ben-Shahar T, Riemer D, *et al.* Essential roles for *Caenorhabditis elegans* lamin gene in nuclear organization, cell cycle progression, and spatial organization of nuclear pore complexes. Mol

Biol Cell 2000; 11(11): 3937-47.
[http://dx.doi.org/10.1091/mbc.11.11.3937] [PMID: 11071918]

[37] Cohen M, Lee KK, Wilson KL, Gruenbaum Y. Transcriptional repression, apoptosis, human disease and the functional evolution of the nuclear lamina. Trends Biochem Sci 2001; 26(1): 41-7.
[http://dx.doi.org/10.1016/S0968-0004(00)01727-8] [PMID: 11165516]

[38] Melcer S, Gruenbaum Y, Krohne G. Invertebrate lamins. Exp Cell Res 2007; 313(10): 2157-66.
[http://dx.doi.org/10.1016/j.yexcr.2007.03.004] [PMID: 17451683]

[39] Rout MP, Field MC. Isolation and characterization of subnuclear compartments from *Trypanosoma brucei* . Identification of a major repetitive nuclear lamina component. J Biol Chem 2001; 276(41): 38261-71.
[PMID: 11477078]

[40] Audas TE, Jacob MD, Lee S. Immobilization of proteins in the nucleolus by ribosomal intergenic spacer noncoding RNA. Mol Cell 2012; 45(2): 147-57.
[http://dx.doi.org/10.1016/j.molcel.2011.12.012] [PMID: 22284675]

[41] Tsai RY, Pederson T. Connecting the nucleolus to the cell cycle and human disease. FASEB J 2014; 28(8): 3290-6.
[http://dx.doi.org/10.1096/fj.14-254680] [PMID: 24790035]

[42] Pederson T. The nucleolus. Cold Spring Harb Perspect Biol 2011; 3(3): a000638.
[http://dx.doi.org/10.1101/cshperspect.a000638] [PMID: 21106648]

[43] Thiry M, Lamaye F, Lafontaine DL. The nucleolus: when 2 became 3. Nucleus 2011; 2(4): 289-93.
[http://dx.doi.org/10.4161/nucl.2.4.16806] [PMID: 21941117]

[44] Názer E, Verdún RE, Sánchez DO. Severe heat shock induces nucleolar accumulation of mRNAs in *Trypanosoma cruzi* . PLoS One 2012; 7(8): e43715.
[http://dx.doi.org/10.1371/journal.pone.0043715] [PMID: 22952745]

[45] Gluenz E, Taylor MC, Kelly JM. The *Trypanosoma cruzi* metacyclic-specific protein Met-III associates with the nucleolus and contains independent amino and carboxyl terminal targeting elements. Int J Parasitol 2007; 37(6): 617-25.
[http://dx.doi.org/10.1016/j.ijpara.2006.11.016] [PMID: 17239886]

[46] Názer E, Sánchez DO. Nucleolar accumulation of RNA binding proteins induced by Actinomycin D is functional in *Trypanosoma cruzi* and *Leishmania mexicana* but not in *T. brucei*. PLoS One 2011; 6(8): e24184.
[http://dx.doi.org/10.1371/journal.pone.0024184] [PMID: 21904613]

[47] Názer E, Verdún RE, Sánchez DO. Nucleolar localization of RNA binding proteins induced by actinomycin D and heat shock in *Trypanosoma cruzi* . PLoS One 2011; 6(5): e19920.
[http://dx.doi.org/10.1371/journal.pone.0019920] [PMID: 21629693]

[48] Li G, Zhu P. Structure and organization of chromatin fiber in the nucleus. FEBS Lett 2015; 589(20 Pt A): 2893-904.
[http://dx.doi.org/10.1016/j.febslet.2015.04.023] [PMID: 25913782]

[49] Hecker H, Betschart B, Bender K, Burri M, Schlimme W. The chromatin of trypanosomes. Int J Parasitol 1994; 24(6): 809-19.

[http://dx.doi.org/10.1016/0020-7519(94)90007-8] [PMID: 7982743]

[50] Thatcher TH, Gorovsky MA. Phylogenetic analysis of the core histones H2A, H2B, H3, and H4. Nucleic Acids Res 1994; 22(2): 174-9.
[http://dx.doi.org/10.1093/nar/22.2.174] [PMID: 8121801]

[51] Sabaj V, Díaz J, Toro GC, Galanti N. Histone synthesis in *Trypanosoma cruzi*. Exp Cell Res 1997; 236(2): 446-52.
[http://dx.doi.org/10.1006/excr.1997.3749] [PMID: 9367629]

[52] Smirlis D, Bisti SN, Xingi E, Konidou G, Thiakaki M, Soteriadou KP. Leishmania histone H1 overexpression delays parasite cell-cycle progression, parasite differentiation and reduces Leishmania infectivity *in vivo*. Mol Microbiol 2006; 60(6): 1457-73.
[http://dx.doi.org/10.1111/j.1365-2958.2006.05205.x] [PMID: 16796681]

[53] Povelones ML, Gluenz E, Dembek M, Gull K, Rudenko G. Histone H1 plays a role in heterochromatin formation and VSG expression site silencing in *Trypanosoma brucei*. PLoS Pathog 2012; 8(11): e1003010.
[http://dx.doi.org/10.1371/journal.ppat.1003010] [PMID: 23133390]

[54] Pena AC, Pimentel MR, Manso H, *et al. Trypanosoma brucei* histone H1 inhibits RNA polymerase I transcription and is important for parasite fitness *in vivo*. Mol Microbiol 2014; 93(4): 645-63.
[http://dx.doi.org/10.1111/mmi.12677] [PMID: 24946224]

[55] Santoro SW, Dulac C. Histone variants and cellular plasticity. Trends Genet 2015; 31(9): 516-27.
[http://dx.doi.org/10.1016/j.tig.2015.07.005] [PMID: 26299477]

[56] Alsford S, Horn D. Trypanosomatid histones. Mol Microbiol 2004; 53(2): 365-72.
[http://dx.doi.org/10.1111/j.1365-2958.2004.04151.x] [PMID: 15228519]

[57] Lowell JE, Cross GA. A variant histone H3 is enriched at telomeres in *Trypanosoma brucei*. J Cell Sci 2004; 117(Pt 24): 5937-47.
[http://dx.doi.org/10.1242/jcs.01515] [PMID: 15522895]

[58] Malik HS, Henikoff S. Phylogenomics of the nucleosome. Nat Struct Biol 2003; 10(11): 882-91.
[http://dx.doi.org/10.1038/nsb996] [PMID: 14583738]

[59] Lowell JE, Kaiser F, Janzen CJ, Cross GA. Histone H2AZ dimerizes with a novel variant H2B and is enriched at repetitive DNA in *Trypanosoma brucei*. J Cell Sci 2005; 118(Pt 24): 5721-30.
[http://dx.doi.org/10.1242/jcs.02688] [PMID: 16303849]

[60] Hohmann PG. Phosphorylation of H1 histones in normal and transformed mouse cells. Cancer Biochem Biophys 1983; 7(1): 45-51.
[PMID: 6667453]

[61] Gutiyama LM, da Cunha JP, Schenkman S. Histone H1 of *Trypanosoma cruzi* is concentrated in the nucleolus region and disperses upon phosphorylation during progression to mitosis. Eukaryot Cell 2008; 7(4): 560-8.
[http://dx.doi.org/10.1128/EC.00460-07] [PMID: 18281601]

[62] da Cunha JP, Nakayasu ES, Elias MC, *et al. Trypanosoma cruzi* histone H1 is phosphorylated in a typical cyclin dependent kinase site accordingly to the cell cycle. Mol Biochem Parasitol 2005; 140(1): 75-86.

[http://dx.doi.org/10.1016/j.molbiopara.2004.12.007] [PMID: 15694489]

[63] da Cunha JP, Nakayasu ES, de Almeida IC, Schenkman S. Post-translational modifications of *Trypanosoma cruzi* histone H4. Mol Biochem Parasitol 2006; 150(2): 268-77. [http://dx.doi.org/10.1016/j.molbiopara.2006.08.012] [PMID: 17010453]

[64] Nardelli SC, da Cunha JP, Motta MC, Schenkman S. Distinct acetylation of *Trypanosoma cruzi* histone H4 during cell cycle, parasite differentiation, and after DNA damage. Chromosoma 2009; 118(4): 487-99. [http://dx.doi.org/10.1007/s00412-009-0213-9] [PMID: 19396454]

[65] Janzen CJ, Fernandez JP, Deng H, Diaz R, Hake SB, Cross GA. Unusual histone modifications in *Trypanosoma brucei*. FEBS Lett 2006; 580(9): 2306-10. [http://dx.doi.org/10.1016/j.febslet.2006.03.044] [PMID: 16580668]

[66] Mandava V, Fernandez JP, Deng H, Janzen CJ, Hake SB, Cross GA. Histone modifications in *Trypanosoma brucei*. Mol Biochem Parasitol 2007; 156(1): 41-50. [http://dx.doi.org/10.1016/j.molbiopara.2007.07.005] [PMID: 17714803]

[67] Figueiredo LM, Cross GA, Janzen CJ. Epigenetic regulation in African trypanosomes: a new kid on the block. Nat Rev Microbiol 2009; 7(7): 504-13. [http://dx.doi.org/10.1038/nrmicro2149] [PMID: 19528957]

[68] Nguyen AT, Zhang Y. The diverse functions of Dot1 and H3K79 methylation. Genes Dev 2011; 25(13): 1345-58. [http://dx.doi.org/10.1101/gad.2057811] [PMID: 21724828]

[69] Figueiredo LM, Janzen CJ, Cross GA. A histone methyltransferase modulates antigenic variation in African trypanosomes. PLoS Biol 2008; 6(7): e161. [http://dx.doi.org/10.1371/journal.pbio.0060161] [PMID: 18597556]

[70] Glover L, Horn D. Trypanosomal histone γH2A and the DNA damage response. Mol Biochem Parasitol 2012; 183(1): 78-83. [http://dx.doi.org/10.1016/j.molbiopara.2012.01.008] [PMID: 22353557]

[71] El-Sayed NM, Myler PJ, Blandin G, *et al.* Comparative genomics of trypanosomatid parasitic protozoa. Science 2005; 309(5733): 404-9. [http://dx.doi.org/10.1126/science.1112181] [PMID: 16020724]

[72] Martínez-Calvillo S, Yan S, Nguyen D, Fox M, Stuart K, Myler PJ. Transcription of *Leishmania major* Friedlin chromosome 1 initiates in both directions within a single region. Mol Cell 2003; 11(5): 1291-9. [http://dx.doi.org/10.1016/S1097-2765(03)00143-6] [PMID: 12769852]

[73] Thomas S, Green A, Sturm NR, Campbell DA, Myler PJ. Histone acetylations mark origins of polycistronic transcription in *Leishmania major*. BMC Genomics 2009; 10: 152. [http://dx.doi.org/10.1186/1471-2164-10-152] [PMID: 19356248]

[74] Vermeulen M, Timmers HT. Grasping trimethylation of histone H3 at lysine 4. Epigenomics 2010; 2(3): 395-406. [http://dx.doi.org/10.2217/epi.10.11] [PMID: 22121900]

[75] Respuela P, Ferella M, Rada-Iglesias A, Aslund L. Histone acetylation and methylation at sites

initiating divergent polycistronic transcription in *Trypanosoma cruzi*. J Biol Chem 2008; 283(23): 15884-92.
[http://dx.doi.org/10.1074/jbc.M802081200] [PMID: 18400752]

[76] Wright JR, Siegel TN, Cross GA. Histone H3 trimethylated at lysine 4 is enriched at probable transcription start sites in *Trypanosoma brucei*. Mol Biochem Parasitol 2010; 172(2): 141-4.
[http://dx.doi.org/10.1016/j.molbiopara.2010.03.013] [PMID: 20347883]

[77] Siegel TN, Hekstra DR, Kemp LE, *et al.* Four histone variants mark the boundaries of polycistronic transcription units in *Trypanosoma brucei*. Genes Dev 2009; 23(9): 1063-76.
[http://dx.doi.org/10.1101/gad.1790409] [PMID: 19369410]

[78] Gommers-Ampt J, Lutgerink J, Borst P. A novel DNA nucleotide in *Trypanosoma brucei* only present in the mammalian phase of the life-cycle. Nucleic Acids Res 1991; 19(8): 1745-51.
[http://dx.doi.org/10.1093/nar/19.8.1745] [PMID: 1674368]

[79] Gommers-Ampt JH, Van Leeuwen F, de Beer AL, *et al.* beta-D-glucosyl-hydroxymethyluracil: a novel modified base present in the DNA of the parasitic protozoan *T. brucei*. Cell 1993; 75(6): 1129-36.
[http://dx.doi.org/10.1016/0092-8674(93)90322-H] [PMID: 8261512]

[80] Ekanayake DK, Minning T, Weatherly B, *et al.* Epigenetic regulation of transcription and virulence in *Trypanosoma cruzi* by O-linked thymine glucosylation of DNA. Mol Cell Biol 2011; 31(8): 1690-700.
[http://dx.doi.org/10.1128/MCB.01277-10] [PMID: 21321080]

[81] Ekanayake D, Sabatini R. Epigenetic regulation of polymerase II transcription initiation in *Trypanosoma cruzi*: modulation of nucleosome abundance, histone modification, and polymerase occupancy by O-linked thymine DNA glucosylation. Eukaryot Cell 2011; 10(11): 1465-72.
[http://dx.doi.org/10.1128/EC.05185-11] [PMID: 21926332]

[82] Reynolds D, Cliffe L, Förstner KU, Hon CC, Siegel TN, Sabatini R. Regulation of transcription termination by glucosylated hydroxymethyluracil, base J, in *Leishmania major* and *Trypanosoma brucei*. Nucleic Acids Res 2014; 42(15): 9717-29.
[http://dx.doi.org/10.1093/nar/gku714] [PMID: 25104019]

[83] Gull K. The cytoskeleton of trypanosomatid parasites. Annu Rev Microbiol 1999; 53: 629-55.
[http://dx.doi.org/10.1146/annurev.micro.53.1.629] [PMID: 10547703]

[84] Rodrigues JC, Godinho JL, de Souza W. Biology of human pathogenic trypanosomatids: epidemiology, lifecycle and ultrastructure. Subcell Biochem 2014; 74: 1-42.
[http://dx.doi.org/10.1007/978-94-007-7305-9_1] [PMID: 24264239]

[85] Vedrenne C, Giroud C, Robinson DR, *et al.* Two related subpellicular cytoskeleton-associated proteins in *Trypanosoma brucei* stabilize microtubules. Mol Biol Cell 2002; 13(3): 1058-70.
[PMID: 11907282]

[86] De Melo LD, SantAnna C, Reis SA, *et al.* Evolutionary conservation of actin-binding proteins in *Trypanosoma cruzi* and unusual subcellular localization of the actin homologue. Parasitology 2008; 135(8): 955-65.
[PMID: 18477418]

[87] Cevallos AM, Segura-Kato YX, Merchant-Larios H, *et al.* *Trypanosoma cruzi*: multiple actin iso

variants are observed along different developmental stages. Exp Parasitol 2011; 127(1): 249-59. [http://dx.doi.org/10.1016/j.exppara.2010.08.003] [PMID: 20705070]

[88] Sahasrabuddhe AA, Bajpai VK, Gupta CM. A novel form of actin in *Leishmania*: molecular characterisation, subcellular localisation and association with subpellicular microtubules. Mol Biochem Parasitol 2004; 134(1): 105-14. [http://dx.doi.org/10.1016/j.molbiopara.2003.11.008] [PMID: 14747148]

[89] García-Salcedo JA, Pérez-Morga D, Gijón P, Dilbeck V, Pays E, Nolan DP. A differential role for actin during the life cycle of *Trypanosoma brucei* . EMBO J 2004; 23(4): 780-9. [http://dx.doi.org/10.1038/sj.emboj.7600094] [PMID: 14963487]

[90] Sasse R, Gull K. Tubulin post-translational modifications and the construction of microtubular organelles in *Trypanosoma brucei.* J Cell Sci 1988; 90(Pt 4): 577-89. [PMID: 3075618]

[91] Selvapandiyan A, Kumar P, Salisbury JL, Wang CC, Nakhasi HL. Role of centrins 2 and 3 in organelle segregation and cytokinesis in *Trypanosoma brucei.* PLoS One 2012; 7(9): e45288. [http://dx.doi.org/10.1371/journal.pone.0045288] [PMID: 23028908]

[92] Wei Y, Hu H, Lun Z-R, Li Z. Centrin3 in trypanosomes maintains the stability of a flagellar inner-arm dynein for cell motility. Nat Commun 2014; 5: 4060. [http://dx.doi.org/10.1038/ncomms5060] [PMID: 24892844]

[93] Esson HJ, Morriswood B, Yavuz S, Vidilaseris K, Dong G, Warren G. Morphology of the trypanosome bilobe, a novel cytoskeletal structure. Eukaryot Cell 2012; 11(6): 761-72. [http://dx.doi.org/10.1128/EC.05287-11] [PMID: 22327007]

[94] Okuda K, Esteva M, Segura EL, Bijovsy AT. The cytostome of *Trypanosoma cruzi* epimastigotes is associated with the flagellar complex. Exp Parasitol 1999; 92(4): 223-31. [http://dx.doi.org/10.1006/expr.1999.4419] [PMID: 10425150]

[95] Richards TA, Cavalier-Smith T. Myosin domain evolution and the primary divergence of eukaryotes. Nature 2005; 436(7054): 1113-8. [http://dx.doi.org/10.1038/nature03949] [PMID: 16121172]

[96] Silverman MA, Leroux MR. Intraflagellar transport and the generation of dynamic, structurally and functionally diverse cilia. Trends Cell Biol 2009; 19(7): 306-16. [http://dx.doi.org/10.1016/j.tcb.2009.04.002] [PMID: 19560357]

[97] Ginger ML, Portman N, McKean PG. Swimming with protists: perception, motility and flagellum assembly. Nat Rev Microbiol 2008; 6(11): 838-50. [http://dx.doi.org/10.1038/nrmicro2009] [PMID: 18923411]

[98] Ralston KS, Kabututu ZP, Melehani JH, Oberholzer M, Hill KL. The *Trypanosoma brucei* flagellum: moving parasites in new directions. Annu Rev Microbiol 2009; 63: 335-62. [http://dx.doi.org/10.1146/annurev.micro.091208.073353] [PMID: 19575562]

[99] Rotureau B, Morales MA, Bastin P, Späth GF. The flagellum-mitogen-activated protein kinase connection in Trypanosomatids: a key sensory role in parasite signalling and development? Cell Microbiol 2009; 11(5): 710-8. [http://dx.doi.org/10.1111/j.1462-5822.2009.01295.x] [PMID: 19207727]

[100] Gluenz E, Ginger ML, McKean PG. Flagellum assembly and function during the *Leishmania* life cycle. Curr Opin Microbiol 2010; 13(4): 473-9.
[http://dx.doi.org/10.1016/j.mib.2010.05.008] [PMID: 20541962]

[101] Langousis G, Hill KL. Motility and more: the flagellum of *Trypanosoma brucei*. Nat Rev Microbiol 2014; 12(7): 505-18.
[http://dx.doi.org/10.1038/nrmicro3274] [PMID: 24931043]

[102] Broadhead R, Dawe HR, Farr H, *et al.* Flagellar motility is required for the viability of the bloodstream trypanosome. Nature 2006; 440(7081): 224-7.
[http://dx.doi.org/10.1038/nature04541] [PMID: 16525475]

[103] Field MC, Carrington M. The trypanosome flagellar pocket. Nat Rev Microbiol 2009; 7(11): 775-86.
[http://dx.doi.org/10.1038/nrmicro2221] [PMID: 19806154]

[104] Gadelha AP, Cunha-e-Silva NL, de Souza W. Assembly of the *Leishmania amazonensis* flagellum during cell differentiation. J Struct Biol 2013; 184(2): 280-92.
[http://dx.doi.org/10.1016/j.jsb.2013.09.006] [PMID: 24041804]

[105] Gluenz E, Höög JL, Smith AE, Dawe HR, Shaw MK, Gull K. Beyond 9+0: noncanonical axoneme structures characterize sensory cilia from protists to humans. FASEB J 2010; 24(9): 3117-21.
[http://dx.doi.org/10.1096/fj.09-151381] [PMID: 20371625]

[106] Portman N, Gull K. The paraflagellar rod of kinetoplastid parasites: from structure to components and function. Int J Parasitol 2010; 40(2): 135-48.
[http://dx.doi.org/10.1016/j.ijpara.2009.10.005] [PMID: 19879876]

[107] Moreira-Leite FF, Sherwin T, Kohl L, Gull K. A trypanosome structure involved in transmitting cytoplasmic information during cell division. Science 2001; 294(5542): 610-2.
[http://dx.doi.org/10.1126/science.1063775] [PMID: 11641501]

[108] Bastin P, Sherwin T, Gull K. Paraflagellar rod is vital for trypanosome motility. Nature 1998; 391(6667): 548.
[http://dx.doi.org/10.1038/35300] [PMID: 9468133]

[109] Rocha GM, Teixeira DE, Miranda K, Weissmüller G, Bisch PM, de Souza W. Structural changes of the paraflagellar rod during flagellar beating in *Trypanosoma cruzi*. PLoS One 2010; 5(6): e11407.
[http://dx.doi.org/10.1371/journal.pone.0011407] [PMID: 20613980]

[110] Tetley L, Coombs GH, Vickerman K. The surface membrane of *Leishmania mexicana* mexicana: comparison of amastigote and promastigote using freeze-fracture cytochemistry. Z Parasitenkd 1986; 72(3): 281-92.
[http://dx.doi.org/10.1007/BF00928737] [PMID: 3716538]

[111] Fridberg A, Buchanan KT, Engman DM. Flagellar membrane trafficking in kinetoplastids. Parasitol Res 2007; 100(2): 205-12.
[http://dx.doi.org/10.1007/s00436-006-0329-2] [PMID: 17058110]

[112] Tyler KM, Fridberg A, Toriello KM, *et al.* Flagellar membrane localization *via* association with lipid rafts. J Cell Sci 2009; 122(Pt 6): 859-66.
[http://dx.doi.org/10.1242/jcs.037721] [PMID: 19240119]

[113] Porto-Carreiro I, Attias M, Miranda K, De Souza W, Cunha-e-Silva N. *Trypanosoma cruzi* epimastigote endocytic pathway: cargo enters the cytostome and passes through an early endosomal network before storage in reservosomes. Eur J Cell Biol 2000; 79(11): 858-69. [http://dx.doi.org/10.1078/0171-9335-00112] [PMID: 11139150]

[114] De Souza W. Basic cell biology of *Trypanosoma cruzi*. Curr Pharm Des 2002; 8(4): 269-85. [http://dx.doi.org/10.2174/1381612023396276] [PMID: 11860366]

[115] Gull K. Host-parasite interactions and trypanosome morphogenesis: a flagellar pocketful of goodies. Curr Opin Microbiol 2003; 6(4): 365-70. [http://dx.doi.org/10.1016/S1369-5274(03)00092-4] [PMID: 12941406]

[116] Rocha GM, Seabra SH, de Miranda KR, Cunha-e-Silva N, de Carvalho TM, de Souza W. Attachment of flagellum to the cell body is important to the kinetics of transferrin uptake by *Trypanosoma cruzi*. Parasitol Int 2010; 59(4): 629-33. [http://dx.doi.org/10.1016/j.parint.2010.07.005] [PMID: 20670692]

[117] Lee SH, Stephens JL, Englund PT. A fatty-acid synthesis mechanism specialized for parasitism. Nat Rev Microbiol 2007; 5(4): 287-97. [http://dx.doi.org/10.1038/nrmicro1617] [PMID: 17363967]

[118] Kohl L, Robinson D, Bastin P. Novel roles for the flagellum in cell morphogenesis and cytokinesis of trypanosomes. EMBO J 2003; 22(20): 5336-46. [http://dx.doi.org/10.1093/emboj/cdg518] [PMID: 14532107]

[119] Tetley L, Vickerman K. Differentiation in *Trypanosoma brucei*: host-parasite cell junctions and their persistence during acquisition of the variable antigen coat. J Cell Sci 1985; 74: 1-19. [PMID: 4030903]

[120] Robinson DR, Gull K. Basal body movements as a mechanism for mitochondrial genome segregation in the trypanosome cell cycle. Nature 1991; 352(6337): 731-3. [http://dx.doi.org/10.1038/352731a0] [PMID: 1876188]

[121] Moreira-Leite FF, de Souza W, da Cunha-e-Silva NL. Purification of the paraflagellar rod of the trypanosomatid Herpetomonas megaseliae and identification of some of its minor components. Mol Biochem Parasitol 1999; 104(1): 131-40. [http://dx.doi.org/10.1016/S0166-6851(99)00143-7] [PMID: 10589987]

[122] Santrich C, Moore L, Sherwin T, *et al.* A motility function for the paraflagellar rod of *Leishmania* parasites revealed by PFR-2 gene knockouts. Mol Biochem Parasitol 1997; 90(1): 95-109. [http://dx.doi.org/10.1016/S0166-6851(97)00149-7] [PMID: 9497035]

[123] Tran KD, Vieira DP, Sanchez MA, Valli J, Gluenz E, Landfear SM. Kharon1 null mutants of *Leishmania mexicana*are avirulent in mice and exhibit a cytokinesis defect within macrophages. PLoS One 2015; 10(8): e0134432. [http://dx.doi.org/10.1371/journal.pone.0134432] [PMID: 26266938]

[124] Webster P, Russell DG. The flagellar pocket of trypanosomatids. Parasitol Today (Regul Ed) 1993; 9(6): 201-6. [http://dx.doi.org/10.1016/0169-4758(93)90008-4] [PMID: 15463754]

[125] Kakkar P, Singh BK. Mitochondria: a hub of redox activities and cellular distress control. Mol Cell

Biochem 2007; 305(1-2): 235-53.
[http://dx.doi.org/10.1007/s11010-007-9520-8] [PMID: 17562131]

[126] Paulin JJ. The chondriome of selected trypanosomatids. A three-dimensional study based on serial thick sections and high voltage electron microscopy. J Cell Biol 1975; 66(2): 404-13.
[http://dx.doi.org/10.1083/jcb.66.2.404] [PMID: 1095599]

[127] Newberry LB, Paulin JJ. Reconstruction of the chondriome of the amastigote form of *Trypanosoma cruzi*. J Parasitol 1989; 75(4): 649-52.
[http://dx.doi.org/10.2307/3282922] [PMID: 2668492]

[128] Fidalgo LM, Gille L. Mitochondria and trypanosomatids: targets and drugs. Pharm Res 2011; 28(11): 2758-70.
[http://dx.doi.org/10.1007/s11095-011-0586-3] [PMID: 21935742]

[129] Paulin JJ. Conformation of a single mitochondrion in the trypomastigote stage of *Trypanosoma cruzi*. J Parasitol 1983; 69(1): 242-4.
[http://dx.doi.org/10.2307/3281310] [PMID: 6338196]

[130] Lukes J, Guilbride DL, Votýpka J, Zíková A, Benne R, Englund PT. Kinetoplast DNA network: evolution of an improbable structure. Eukaryot Cell 2002; 1(4): 495-502.
[http://dx.doi.org/10.1128/EC.1.4.495-502.2002] [PMID: 12455998]

[131] Riou G, Delain E. Abnormal circular DNA molecules induced by ethidium bromide in the kinetoplast of *Trypanosoma cruzi*. Proc Natl Acad Sci USA 1969; 64(2): 618-25.
[http://dx.doi.org/10.1073/pnas.64.2.618] [PMID: 5261038]

[132] Liu B, Liu Y, Motyka SA, Agbo EE, Englund PT. Fellowship of the rings: the replication of kinetoplast DNA. Trends Parasitol 2005; 21(8): 363-9.
[http://dx.doi.org/10.1016/j.pt.2005.06.008] [PMID: 15967722]

[133] Liu Y, Englund PT. The rotational dynamics of kinetoplast DNA replication. Mol Microbiol 2007; 64(3): 676-90.
[http://dx.doi.org/10.1111/j.1365-2958.2007.05686.x] [PMID: 17462016]

[134] Shapiro TA, Englund PT. The structure and replication of kinetoplast DNA. Annu Rev Microbiol 1995; 49: 117-43.
[http://dx.doi.org/10.1146/annurev.mi.49.100195.001001] [PMID: 8561456]

[135] Stuart K, Panigrahi AK, Schnaufer A, Drozdz M, Clayton C, Salavati R. Composition of the editing complex of *Trypanosoma brucei*. Philos Trans R Soc Lond B Biol Sci 2002; 357(1417): 71-9.
[http://dx.doi.org/10.1098/rstb.2001.0994] [PMID: 11839184]

[136] Koslowsky DJ, Bhat GJ, Perrollaz AL, Feagin JE, Stuart K. The MURF3 gene of *T. brucei* contains multiple domains of extensive editing and is homologous to a subunit of NADH dehydrogenase. Cell 1990; 62(5): 901-11.
[http://dx.doi.org/10.1016/0092-8674(90)90265-G] [PMID: 2393904]

[137] Souza AE, Myler PJ, Stuart K. Maxicircle CR1 transcripts of *Trypanosoma brucei* are edited and developmentally regulated and encode a putative iron-sulfur protein homologous to an NADH dehydrogenase subunit. Mol Cell Biol 1992; 12(5): 2100-7.
[http://dx.doi.org/10.1128/MCB.12.5.2100] [PMID: 1373807]

[138] Souza AE, Shu HH, Read LK, Myler PJ, Stuart KD. Extensive editing of CR2 maxicircle transcripts of *Trypanosoma brucei* predicts a protein with homology to a subunit of NADH dehydrogenase. Mol Cell Biol 1993; 13(11): 6832-40.
[http://dx.doi.org/10.1128/MCB.13.11.6832] [PMID: 8413276]

[139] Feagin JE, Jasmer DP, Stuart K. Developmentally regulated addition of nucleotides within apocytochrome b transcripts in *Trypanosoma brucei.* Cell 1987; 49(3): 337-45.
[http://dx.doi.org/10.1016/0092-8674(87)90286-8] [PMID: 3568129]

[140] Benne R, Van den Burg J, Brakenhoff JP, Sloof P, Van Boom JH, Tromp MC. Major transcript of the frameshifted coxII gene from trypanosome mitochondria contains four nucleotides that are not encoded in the DNA. Cell 1986; 46(6): 819-26.
[http://dx.doi.org/10.1016/0092-8674(86)90063-2] [PMID: 3019552]

[141] Bhat GJ, Koslowsky DJ, Feagin JE, Smiley BL, Stuart K. An extensively edited mitochondrial transcript in kinetoplastids encodes a protein homologous to ATPase subunit 6. Cell 1990; 61(5): 885-94.
[http://dx.doi.org/10.1016/0092-8674(90)90199-O] [PMID: 2140530]

[142] Panigrahi AK, Schnaufer A, Ernst NL, *et al.* Identification of novel components of *Trypanosoma brucei* editosomes. RNA 2003; 9(4): 484-92.
[http://dx.doi.org/10.1261/rna.2194603] [PMID: 12649499]

[143] Panigrahi AK, Ernst NL, Domingo GJ, Fleck M, Salavati R, Stuart KD. Compositionally and functionally distinct editosomes in *Trypanosoma brucei.* RNA 2006; 12(6): 1038-49.
[http://dx.doi.org/10.1261/rna.45506] [PMID: 16611942]

[144] Lerch M, Carnes J, Acestor N, Guo X, Schnaufer A, Stuart K. Editosome accessory factors KREPB9 and KREPB10 in *Trypanosoma brucei.* Eukaryot Cell 2012; 11(7): 832-43.
[http://dx.doi.org/10.1128/EC.00046-12] [PMID: 22562468]

[145] Carnes J, Soares CZ, Wickham C, Stuart K. Endonuclease associations with three distinct editosomes in *Trypanosoma brucei.* J Biol Chem 2011; 286(22): 19320-30.
[http://dx.doi.org/10.1074/jbc.M111.228965] [PMID: 21474442]

[146] Duhagon MA, Pastro L, Sotelo-Silveira JR, *et al.* The *Trypanosoma cruzi* nucleic acid binding protein Tc38 presents changes in the intramitochondrial distribution during the cell cycle. BMC Microbiol 2009; 9: 34.
[http://dx.doi.org/10.1186/1471-2180-9-34] [PMID: 19210781]

[147] Cavalcanti DP, Shimada MK, Probst CM, *et al.* Expression and subcellular localization of kinetoplast-associated proteins in the different developmental stages of *Trypanosoma cruzi* . BMC Microbiol 2009; 9: 120.
[http://dx.doi.org/10.1186/1471-2180-9-120] [PMID: 19497120]

[148] Rajão MA, Passos-Silva DG, DaRocha WD, *et al.* DNA polymerase kappa from *Trypanosoma cruzi* localizes to the mitochondria, bypasses 8-oxoguanine lesions and performs DNA synthesis in a recombination intermediate. Mol Microbiol 2009; 71(1): 185-97.
[http://dx.doi.org/10.1111/j.1365-2958.2008.06521.x] [PMID: 19007414]

[149] Lopes DdeO, Schamber-Reis BL, Regis-da-Silva CG, *et al.* Biochemical studies with DNA

polymerase beta and DNA polymerase beta-PAK of *Trypanosoma cruzi* suggest the involvement of these proteins in mitochondrial DNA maintenance. DNA Repair (Amst) 2008; 7(11): 1882-92. [http://dx.doi.org/10.1016/j.dnarep.2008.07.018] [PMID: 18761429]

[150] Klingbeil MM, Drew ME, Liu Y, *et al.* Unlocking the secrets of trypanosome kinetoplast DNA network replication. Protist 2001; 152(4): 255-62. [http://dx.doi.org/10.1078/1434-4610-00066] [PMID: 11822657]

[151] Morris JC, Drew ME, Klingbeil MM, *et al.* Replication of kinetoplast DNA: an update for the new millennium. Int J Parasitol 2001; 31(5-6): 453-8. [http://dx.doi.org/10.1016/S0020-7519(01)00156-4] [PMID: 11334929]

[152] Niemann M, Schneider A, Cristodero M. Mitochondrial translation in trypanosomatids: a novel target for chemotherapy? Trends Parasitol 2011; 27(10): 429-33. [http://dx.doi.org/10.1016/j.pt.2011.03.011] [PMID: 21531629]

[153] Feagin JE. Mitochondrial genome diversity in parasites. Int J Parasitol 2000; 30(4): 371-90. [http://dx.doi.org/10.1016/S0020-7519(99)00190-3] [PMID: 10731561]

[154] Tan TH, Bochud-Allemann N, Horn EK, Schneider A. Eukaryotic-type elongator tRNAMet of *Trypanosoma brucei* becomes formylated after import into mitochondria. Proc Natl Acad Sci USA 2002; 99(3): 1152-7. [http://dx.doi.org/10.1073/pnas.022522299] [PMID: 11792845]

[155] Alfonzo JD, Söll D. Mitochondrial tRNA import the challenge to understand has just begun. Biol Chem 2009; 390(8): 717-22. [http://dx.doi.org/10.1515/BC.2009.101] [PMID: 19558325]

[156] Charrière F, Tan TH, Schneider A. Mitochondrial initiation factor 2 of *Trypanosoma brucei* binds imported formylated elongator-type tRNA(Met). J Biol Chem 2005; 280(16): 15659-65. [http://dx.doi.org/10.1074/jbc.M411581200] [PMID: 15731104]

[157] Crain PF, Alfonzo JD, Rozenski J, Kapushoc ST, McCloskey JA, Simpson L. Modification of the universally unmodified uridine-33 in a mitochondria-imported edited tRNA and the role of the anticodon arm structure on editing efficiency. RNA 2002; 8(6): 752-61. [http://dx.doi.org/10.1017/S1355838202022045] [PMID: 12088148]

[158] Eperon IC, Janssen JW, Hoeijmakers JH, Borst P. The major transcripts of the kinetoplast DNA of *Trypanosoma brucei* are very small ribosomal RNAs. Nucleic Acids Res 1983; 11(1): 105-25. [http://dx.doi.org/10.1093/nar/11.1.105] [PMID: 6306559]

[159] de la Cruz VF, Lake JA, Simpson AM, Simpson L. A minimal ribosomal RNA: sequence and secondary structure of the 9S kinetoplast ribosomal RNA from *Leishmania tarentolae*. Proc Natl Acad Sci USA 1985; 82(5): 1401-5. [http://dx.doi.org/10.1073/pnas.82.5.1401] [PMID: 3856267]

[160] Zíková A, Panigrahi AK, Dalley RA, *et al. Trypanosoma brucei* mitochondrial ribosomes: affinity purification and component identification by mass spectrometry. Mol Cell Proteomics 2008; 7(7): 1286-96. [http://dx.doi.org/10.1074/mcp.M700490-MCP200] [PMID: 18364347]

[161] Sharma MR, Koc EC, Datta PP, Booth TM, Spremulli LL, Agrawal RK. Structure of the mammalian

mitochondrial ribosome reveals an expanded functional role for its component proteins. Cell 2003; 115(1): 97-108.
[http://dx.doi.org/10.1016/S0092-8674(03)00762-1] [PMID: 14532006]

[162] Goldshmidt H, Matas D, Kabi A, Carmi S, Hope R, Michaeli S. Persistent ER stress induces the spliced leader RNA silencing pathway (SLS), leading to programmed cell death in *Trypanosoma brucei*. PLoS Pathog 2010; 6(1): e1000731.
[http://dx.doi.org/10.1371/journal.ppat.1000731] [PMID: 20107599]

[163] Buscaglia CA, Campo VA, Frasch AC, Di Noia JM. *Trypanosoma cruzi* surface mucins: host-dependent coat diversity. Nat Rev Microbiol 2006; 4(3): 229-36.
[http://dx.doi.org/10.1038/nrmicro1351] [PMID: 16489349]

[164] Freitas LM, dos Santos SL, Rodrigues-Luiz GF, *et al.* Genomic analyses, gene expression and antigenic profile of the trans-sialidase superfamily of *Trypanosoma cruzi* reveal an undetected level of complexity. PLoS One 2011; 6(10): e25914.
[http://dx.doi.org/10.1371/journal.pone.0025914] [PMID: 22039427]

[165] Manna PT, Boehm C, Leung KF, Natesan SK, Field MC. Life and times: synthesis, trafficking, and evolution of VSG. Trends Parasitol 2014; 30(5): 251-8.
[http://dx.doi.org/10.1016/j.pt.2014.03.004] [PMID: 24731931]

[166] Olivier M, Atayde VD, Isnard A, Hassani K, Shio MT. *Leishmania* virulence factors: focus on the metalloprotease GP63. 2012.
[http://dx.doi.org/10.1016/j.micinf.2012.05.014]

[167] McDowell MA, Ransom DM, Bangs JD. Glycosylphosphatidylinositol-dependent secretory transport in *Trypanosoma brucei* . Biochem J 1998; 335(Pt 3): 681-9.
[http://dx.doi.org/10.1042/bj3350681] [PMID: 9794811]

[168] Mayer MP, Bukau B. Hsp70 chaperone systems: diversity of cellular functions and mechanism of action. Biol Chem 1998; 379(3): 261-8.
[PMID: 9563820]

[169] Bangs JD, Uyetake L, Brickman MJ, Balber AE, Boothroyd JC. Molecular cloning and cellular localization of a BiP homologue in *Trypanosoma brucei* . Divergent ER retention signals in a lower eukaryote. J Cell Sci 1993; 105(Pt 4): 1101-13.
[PMID: 8227199]

[170] Zimmermann R, Eyrisch S, Ahmad M, Helms V. Protein translocation across the ER membrane. Biochim Biophys Acta 2011; 1808(3): 912-24.
[http://dx.doi.org/10.1016/j.bbamem.2010.06.015] [PMID: 20599535]

[171] Aebi M. N-linked protein glycosylation in the ER. Biochim Biophys Acta 2013; 1833(11): 2430-7.
[http://dx.doi.org/10.1016/j.bbamcr.2013.04.001] [PMID: 23583305]

[172] Michaeli S. Spliced leader RNA silencing (SLS) - a programmed cell death pathway in *Trypanosoma brucei* that is induced upon ER stress. Parasit Vectors 2012; 5: 107.
[http://dx.doi.org/10.1186/1756-3305-5-107] [PMID: 22650251]

[173] Dolai S, Adak S. Endoplasmic reticulum stress responses in *Leishmania*. Mol Biochem Parasitol 2014; 197(1-2): 1-8.

[http://dx.doi.org/10.1016/j.molbiopara.2014.09.002] [PMID: 25224909]

[174] Field MC, Sergeenko T, Wang YN, Böhm S, Carrington M. Chaperone requirements for biosynthesis of the trypanosome variant surface glycoprotein. PLoS One 2010; 5(1): e8468. [http://dx.doi.org/10.1371/journal.pone.0008468] [PMID: 20052285]

[175] Schröder M, Kaufman RJ. ER stress and the unfolded protein response. Mutat Res 2005; 569(1-2): 29-63. [http://dx.doi.org/10.1016/j.mrfmmm.2004.06.056] [PMID: 15603751]

[176] Li Y, Guo Y, Tang J, Jiang J, Chen Z. New insights into the roles of CHOP-induced apoptosis in ER stress. Acta Biochim Biophys Sin (Shanghai) 2015; 47(2): 146-7. [http://dx.doi.org/10.1093/abbs/gmu128] [PMID: 25634437]

[177] Gosline SJ, Nascimento M, McCall LI, *et al.* Intracellular eukaryotic parasites have a distinct unfolded protein response. PLoS One 2011; 6(4): e19118. [http://dx.doi.org/10.1371/journal.pone.0019118] [PMID: 21559456]

[178] Conte I, Labriola C, Cazzulo JJ, Docampo R, Parodi AJ. The interplay between folding-facilitating mechanisms in *Trypanosoma cruzi* endoplasmic reticulum. Mol Biol Cell 2003; 14(9): 3529-40. [http://dx.doi.org/10.1091/mbc.E03-04-0228] [PMID: 12972544]

[179] da Silva Augusto L, Moretti NS, Ramos TC, *et al.* A membrane-bound eIF2 alpha kinase located in endosomes is regulated by heme and controls differentiation and ROS levels in *Trypanosoma cruzi* . PLoS Pathog 2015; 11(2): e1004618. [http://dx.doi.org/10.1371/journal.ppat.1004618] [PMID: 25658109]

[180] Moraes MC, Jesus TC, Hashimoto NN, *et al.* Novel membrane-bound eIF2alpha kinase in the flagellar pocket of *Trypanosoma brucei* . Eukaryot Cell 2007; 6(11): 1979-91. [http://dx.doi.org/10.1128/EC.00249-07] [PMID: 17873083]

[181] Chow C, Cloutier S, Dumas C, Chou MN, Papadopoulou B. Promastigote to amastigote differentiation of *Leishmania* is markedly delayed in the absence of PERK eIF2alpha kinase-dependent eIF2alpha phosphorylation. Cell Microbiol 2011; 13(7): 1059-77. [http://dx.doi.org/10.1111/j.1462-5822.2011.01602.x] [PMID: 21624030]

[182] Chow C, Cloutier S, Dumas C, Chou MN, Papadopoulou B. Promastigote to amastigote differentiation of *Leishmania* is markedly delayed in the absence of PERK eIF2alpha kinase-dependent eIF2alpha phosphorylation. Cell Microbiol 2011; 13(7): 1059-77. [http://dx.doi.org/10.1111/j.1462-5822.2011.01602.x] [PMID: 21624030]

[183] Tonelli RR. Augusto LdS, Castilho BA, Schenkman S. Protein synthesis attenuation by phosphorylation of eIF2alpha Is required for the differentiation of *Trypanosoma cruzi* into infective forms. PLoS One 2011; 6(6): e27094. [PMID: 22069492]

[184] Hope R, Ben-Mayor E, Friedman N, *et al.* Phosphorylation of the TATA-binding protein activates the spliced leader silencing pathway in *Trypanosoma brucei* . Sci Signal 2014; 7(341): ra85. [http://dx.doi.org/10.1126/scisignal.2005234] [PMID: 25185157]

[185] Opperdoes FR, Michels PA. The glycosomes of the Kinetoplastida. Biochimie 1993; 75(3-4): 231-4. [http://dx.doi.org/10.1016/0300-9084(93)90081-3] [PMID: 8507685]

[186] Opperdoes FR, Baudhuin P, Coppens I, *et al.* Purification, morphometric analysis, and characterization of the glycosomes (microbodies) of the protozoan hemoflagellate *Trypanosoma brucei*. J Cell Biol 1984; 98(4): 1178-84.
[http://dx.doi.org/10.1083/jcb.98.4.1178] [PMID: 6715405]

[187] Hannaert V, Michels PA. Structure, function, and biogenesis of glycosomes in kinetoplastida. J Bioenerg Biomembr 1994; 26(2): 205-12.
[http://dx.doi.org/10.1007/BF00763069] [PMID: 8056787]

[188] Szöör B, Haanstra JR, Gualdrón-López M, Michels PA. Evolution, dynamics and specialized functions of glycosomes in metabolism and development of trypanosomatids. Curr Opin Microbiol 2014; 22: 79-87.
[http://dx.doi.org/10.1016/j.mib.2014.09.006] [PMID: 25460800]

[189] Cull B, Prado Godinho JL, Fernandes Rodrigues JC, *et al.* Glycosome turnover in *Leishmania major* is mediated by autophagy. Autophagy 2014; 10(12): 2143-57.
[http://dx.doi.org/10.4161/auto.36438] [PMID: 25484087]

[190] Bauer S, Morris JC, Morris MT. Environmentally regulated glycosome protein composition in the African trypanosome. Eukaryot Cell 2013; 12(8): 1072-9.
[http://dx.doi.org/10.1128/EC.00086-13] [PMID: 23709182]

[191] Bauer S, Conlon M, Morris M. Using fluorescent proteins to monitor glycosome dynamics in the African trypanosome. J Vis Exp 2014; (90): e51647.
[PMID: 25177828]

[192] Quiñones W, Urbina JA, Dubourdieu M, Luis Concepción J. The glycosome membrane of *Trypanosoma cruzi* epimastigotes: protein and lipid composition. Exp Parasitol 2004; 106(3-4): 135-49.
[http://dx.doi.org/10.1016/j.exppara.2004.03.006] [PMID: 15172221]

[193] Aronov AM, Verlinde CL, Hol WG, Gelb MH. Selective tight binding inhibitors of trypanosomal glyceraldehyde-3-phosphate dehydrogenase *via* structure-based drug design. J Med Chem 1998; 41(24): 4790-9.
[http://dx.doi.org/10.1021/jm9802620] [PMID: 9822549]

[194] Mottram JC, Coombs GH. *Leishmania mexicana*: enzyme activities of amastigotes and promastigotes and their inhibition by antimonials and arsenicals. Exp Parasitol 1985; 59(2): 151-60.
[http://dx.doi.org/10.1016/0014-4894(85)90067-0] [PMID: 2982638]

[195] Hart DT, Vickerman K, Coombs GH. Respiration of *Leishmania mexicana* amastigotes and promastigotes. Mol Biochem Parasitol 1981; 4(1-2): 39-51.
[http://dx.doi.org/10.1016/0166-6851(81)90027-X] [PMID: 7322186]

[196] Burchmore RJ, Rodriguez-Contreras D, McBride K, *et al.* Genetic characterization of glucose transporter function in *Leishmania mexicana*. Proc Natl Acad Sci USA 2003; 100(7): 3901-6.
[http://dx.doi.org/10.1073/pnas.0630165100] [PMID: 12651954]

[197] Opperdoes FR, Borst P. Localization of nine glycolytic enzymes in a microbody-like organelle in *Trypanosoma brucei*: the glycosome. FEBS Lett 1977; 80(2): 360-4.
[http://dx.doi.org/10.1016/0014-5793(77)80476-6] [PMID: 142663]

[198] Hannaert V, Bringaud F, Opperdoes FR, Michels PA. Evolution of energy metabolism and its compartmentation in Kinetoplastida. Kinetoplastid Biol Dis 2003; 2(1): 11. [http://dx.doi.org/10.1186/1475-9292-2-11] [PMID: 14613499]

[199] Krazy H, Michels PA. Identification and characterization of three peroxinsPEX6, PEX10 and PEX12involved in glycosome biogenesis in *Trypanosoma brucei*. Biochim Biophys Acta 2006; 1763(1): 6-17. [http://dx.doi.org/10.1016/j.bbamcr.2005.11.002] [PMID: 16388862]

[200] Wilkinson SR, Prathalingam SR, Taylor MC, Horn D, Kelly JM. Vitamin C biosynthesis in trypanosomes: a role for the glycosome. Proc Natl Acad Sci USA 2005; 102(33): 11645-50. [http://dx.doi.org/10.1073/pnas.0504251102] [PMID: 16087875]

[201] de Walque S, Kiel JA, Veenhuis M, Opperdoes FR, Michels PA. Cloning and analysis of the PTS-1 receptor in *Trypanosoma brucei*. Mol Biochem Parasitol 1999; 104(1): 106-19. [http://dx.doi.org/10.1016/S0166-6851(99)00144-9] [PMID: 10589985]

[202] Jardim A, Liu W, Zheleznova E, Ullman B. Peroxisomal targeting signal-1 receptor protein PEX5 from *Leishmania donovani*. Molecular, biochemical, and immunocytochemical characterization. J Biol Chem 2000; 275(18): 13637-44. [http://dx.doi.org/10.1074/jbc.275.18.13637] [PMID: 10788481]

[203] Jardim A, Rager N, Liu W, Ullman B. Peroxisomal targeting protein 14 (PEX14) from *Leishmania donovani*. Molecular, biochemical, and immunocytochemical characterization. Mol Biochem Parasitol 2002; 124(1-2): 51-62. [http://dx.doi.org/10.1016/S0166-6851(02)00164-0] [PMID: 12387850]

[204] Choe J, Moyersoen J, Roach C, *et al.* Analysis of the sequence motifs responsible for the interactions of peroxins 14 and 5, which are involved in glycosome biogenesis in *Trypanosoma brucei*. Biochemistry 2003; 42(37): 10915-22. [http://dx.doi.org/10.1021/bi034248n] [PMID: 12974625]

[205] Brennand A, Rigden DJ, Michels PA. Trypanosomes contain two highly different isoforms of peroxin PEX13 involved in glycosome biogenesis. FEBS Lett 2012; 586(13): 1765-71. [http://dx.doi.org/10.1016/j.febslet.2012.05.019] [PMID: 22641036]

[206] Banerjee SK, Kessler PS, Saveria T, Parsons M. Identification of trypanosomatid PEX19: functional characterization reveals impact on cell growth and glycosome size and number. Mol Biochem Parasitol 2005; 142(1): 47-55. [http://dx.doi.org/10.1016/j.molbiopara.2005.03.008] [PMID: 15907560]

[207] Furuya T, Kessler P, Jardim A, Schnaufer A, Crudder C, Parsons M. Glucose is toxic to glycosome-deficient trypanosomes. Proc Natl Acad Sci USA 2002; 99(22): 14177-82. [http://dx.doi.org/10.1073/pnas.222454899] [PMID: 12386344]

[208] Guerra-Giraldez C, Quijada L, Clayton CE. Compartmentation of enzymes in a microbody, the glycosome, is essential in *Trypanosoma brucei*. J Cell Sci 2002; 115(Pt 13): 2651-8. [PMID: 12077356]

[209] Kalel VC, Schliebs W, Erdmann R. Identification and functional characterization of *Trypanosoma brucei* peroxin 16. Biochim Biophys Acta 2015; 1853 (10 Pt A): 2326-37.

[http://dx.doi.org/10.1016/j.bbamcr.2015.05.024] [PMID: 26025675]

[210] Urbaniak MD, Guther ML, Ferguson MA. Comparative SILAC proteomic analysis of *Trypanosoma brucei* bloodstream and procyclic lifecycle stages. PLoS One 2012; 7(5): e36619. [http://dx.doi.org/10.1371/journal.pone.0036619] [PMID: 22574199]

[211] Urbaniak MD, Martin DM, Ferguson MA. Global quantitative SILAC phosphoproteomics reveals differential phosphorylation is widespread between the procyclic and bloodstream form lifecycle stages of *Trypanosoma brucei*. J Proteome Res 2013; 12(5): 2233-44. [http://dx.doi.org/10.1021/pr400086y] [PMID: 23485197]

[212] Szöor B, Ruberto I, Burchmore R, Matthews KR. A novel phosphatase cascade regulates differentiation in *Trypanosoma brucei via* a glycosomal signaling pathway. Genes Dev 2010; 24(12): 1306-16. [http://dx.doi.org/10.1101/gad.570310] [PMID: 20551176]

[213] Güther ML, Urbaniak MD, Tavendale A, Prescott A, Ferguson MA. High-confidence glycosome proteome for procyclic form *Trypanosoma brucei* by epitope-tag organelle enrichment and SILAC proteomics. J Proteome Res 2014; 13(6): 2796-806. [http://dx.doi.org/10.1021/pr401209w] [PMID: 24792668]

[214] Jamdhade MD, Pawar H, Chavan S, *et al.* Comprehensive proteomics analysis of glycosomes from *Leishmania donovani*. OMICS 2015; 19(3): 157-70. [http://dx.doi.org/10.1089/omi.2014.0163] [PMID: 25748437]

[215] da Silva MF, Zampieri RA, Muxel SM, Beverley SM, Floeter-Winter LM. *Leishmania amazonensis* arginase compartmentalization in the glycosome is important for parasite infectivity. PLoS One 2012; 7(3): e34022. [http://dx.doi.org/10.1371/journal.pone.0034022] [PMID: 22479507]

[216] Barros-Alvarez X, Gualdrón-López M, Acosta H, *et al.* Glycosomal targets for anti-trypanosomatid drug discovery. Curr Med Chem 2014; 21(15): 1679-706. [http://dx.doi.org/10.2174/09298673113209990139] [PMID: 23834165]

[217] Melisa Gualdrón-López PA. Wilfredo Quiñones, Ana J. Weinheim, Germany: Cáceres, Luisana Avilán, Juan-Luis Concepción. Function of glycosomes in the metabolism of trypanosomatid parasites and the promise of glycosomal proteins as drug targets 2013.

[218] De Souza W. Basic cell biology of *Trypanosoma cruzi* . Curr Pharm Des 2002; 8(4): 269-85. [http://dx.doi.org/10.2174/1381612023396276] [PMID: 11860366]

[219] He CY, Ho HH, Malsam J, *et al.* Golgi duplication in *Trypanosoma brucei*. J Cell Biol 2004; 165(3): 313-21. [http://dx.doi.org/10.1083/jcb.200311076] [PMID: 15138289]

[220] Figueiredo RC, Soares MJ. The Golgi complex of *Trypanosoma cruzi* epimastigote forms. J Submicrosc Cytol Pathol 1995; 27(2): 209-15. [PMID: 7757947]

[221] Koeller CM, van der Wel H, Feasley CL, *et al.* Golgi UDP-GlcNAc:polypeptide O-α-N-Acetl-d-glucosaminyltransferase 2 (TcOGNT2) regulates trypomastigote production and function in *Trypanosoma cruzi*. Eukaryot Cell 2014; 13(10): 1312-27.

[http://dx.doi.org/10.1128/EC.00165-14] [PMID: 25084865]

[222] Hong Y, Kinoshita T. Trypanosome glycosylphosphatidylinositol biosynthesis. Korean J Parasitol 2009; 47(3): 197-204.
[http://dx.doi.org/10.3347/kjp.2009.47.3.197] [PMID: 19724691]

[223] McConville MJ, Mullin KA, Ilgoutz SC, Teasdale RD. Secretory pathway of trypanosomatid parasites. Microbiol Mol Biol Rev 2002; 66(1): 122-54.
[http://dx.doi.org/10.1128/MMBR.66.1.122-154.2002] [PMID: 11875130]

[224] Silverman JS, Bangs JD. Form and function in the trypanosomal secretory pathway. Curr Opin Microbiol 2012; 15(4): 463-8.
[http://dx.doi.org/10.1016/j.mib.2012.03.002] [PMID: 22445359]

[225] Ackers JP, Dhir V, Field MC. A bioinformatic analysis of the RAB genes of *Trypanosoma brucei*. Mol Biochem Parasitol 2005; 141(1): 89-97.
[http://dx.doi.org/10.1016/j.molbiopara.2005.01.017] [PMID: 15811530]

[226] Overath P, Engstler M. Endocytosis, membrane recycling and sorting of GPI-anchored proteins: *Trypanosoma brucei* as a model system. Mol Microbiol 2004; 53(3): 735-44.
[http://dx.doi.org/10.1111/j.1365-2958.2004.04224.x] [PMID: 15255888]

[227] Silverman JS, Schwartz KJ, Hajduk SL, Bangs JD. Late endosomal Rab7 regulates lysosomal trafficking of endocytic but not biosynthetic cargo in *Trypanosoma brucei*. Mol Microbiol 2011; 82(3): 664-78.
[http://dx.doi.org/10.1111/j.1365-2958.2011.07842.x] [PMID: 21923766]

[228] Araripe JR, Cunha e Silva NL, Leal ST, de Souza W, Rondinelli E. *Trypanosoma cruzi* : TcRAB7 protein is localized at the Golgi apparatus in epimastigotes. Biochem Biophys Res Commun 2004; 321(2): 397-402.
[http://dx.doi.org/10.1016/j.bbrc.2004.06.159] [PMID: 15358190]

[229] SantAnna C, de Souza W, Cunha-e-Silva N. Biogenesis of the reservosomes of *Trypanosoma cruzi*. Microsc Microanal 2004; 10(5): 637-46.
[http://dx.doi.org/10.1017/S1431927604040863] [PMID: 15525436]

[230] Engel JC, Doyle PS, Palmer J, Hsieh I, Bainton DF, McKerrow JH. Cysteine protease inhibitors alter Golgi complex ultrastructure and function in *Trypanosoma cruzi*. J Cell Sci 1998; 111(Pt 5): 597-606.
[PMID: 9454733]

[231] Niyogi S, Mucci J, Campetella O, Docampo R. Rab11 regulates trafficking of trans-sialidase to the plasma membrane through the contractile vacuole complex of *Trypanosoma cruzi*. PLoS Pathog 2014; 10(6): e1004224.
[http://dx.doi.org/10.1371/journal.ppat.1004224] [PMID: 24968013]

[232] Torrecilhas AC, Schumacher RI, Alves MJ, Colli W. Vesicles as carriers of virulence factors in parasitic protozoan diseases. Microbes Infect 2012; 14(15): 1465-74.
[http://dx.doi.org/10.1016/j.micinf.2012.07.008] [PMID: 22892602]

[233] Corrêa JR, Atella GC, Menna-Barreto RS, Soares MJ. Clathrin in *Trypanosoma cruzi*: *in silico* gene identification, isolation, and localization of protein expression sites. J Eukaryot Microbiol 2007; 54(3): 297-302.

[http://dx.doi.org/10.1111/j.1550-7408.2007.00258.x] [PMID: 17552985]

[234] Hung CH, Qiao X, Lee PT, Lee MG. Clathrin-dependent targeting of receptors to the flagellar pocket of procyclic-form *Trypanosoma brucei*. Eukaryot Cell 2004; 3(4): 1004-14. [http://dx.doi.org/10.1128/EC.3.4.1004-1014.2004] [PMID: 15302833]

[235] Denny PW, Morgan GW, Field MC, Smith DF. *Leishmania major*: clathrin and adaptin complexes of an intra-cellular parasite. Exp Parasitol 2005; 109(1): 33-7. [http://dx.doi.org/10.1016/j.exppara.2004.10.007] [PMID: 15639137]

[236] Natesan SK, Peacock L, Matthews K, Gibson W, Field MC. Activation of endocytosis as an adaptation to the mammalian host by trypanosomes. Eukaryot Cell 2007; 6(11): 2029-37. [http://dx.doi.org/10.1128/EC.00213-07] [PMID: 17905918]

[237] Manna PT, Kelly S, Field MC. Adaptin evolution in kinetoplastids and emergence of the variant surface glycoprotein coat in African trypanosomatids. Mol Phylogenet Evol 2013; 67(1): 123-8. [http://dx.doi.org/10.1016/j.ympev.2013.01.002] [PMID: 23337175]

[238] Kelley RJ, Alexander DL, Cowan C, Balber AE, Bangs JD. Molecular cloning of p67, a lysosomal membrane glycoprotein from *Trypanosoma brucei*. Mol Biochem Parasitol 1999; 98(1): 17-28. [http://dx.doi.org/10.1016/S0166-6851(98)00155-8] [PMID: 10029306]

[239] Peck RF, Shiflett AM, Schwartz KJ, McCann A, Hajduk SL, Bangs JD. The LAMP-like protein p67 plays an essential role in the lysosome of African trypanosomes. Mol Microbiol 2008; 68(4): 933-46. [http://dx.doi.org/10.1111/j.1365-2958.2008.06195.x] [PMID: 18430083]

[240] de Souza W, de Carvalho TU, Benchimol M, Chiari E. *Trypanosoma cruzi*: ultrastructural, cytochemical and freeze-fracture studies of protein uptake. Exp Parasitol 1978; 45(1): 101-15. [http://dx.doi.org/10.1016/0014-4894(78)90050-4] [PMID: 352715]

[241] Cunha-e-Silva N, SantAnna C, Pereira MG, Porto-Carreiro I, Jeovanio AL, de Souza W. Reservosomes: multipurpose organelles? Parasitol Res 2006; 99(4): 325-7. [http://dx.doi.org/10.1007/s00436-006-0190-3] [PMID: 16794853]

[242] Vieira M, Rohloff P, Luo S, Cunha-e-Silva NL, de Souza W, Docampo R. Role for a P-type H+-ATPase in the acidification of the endocytic pathway of *Trypanosoma cruzi*. Biochem J 2005; 392(Pt 3): 467-74. [http://dx.doi.org/10.1042/BJ20051319] [PMID: 16149915]

[243] Souto-Padrón T, Campetella OE, Cazzulo JJ, de Souza W. Cysteine proteinase in *Trypanosoma cruzi*: immunocytochemical localization and involvement in parasite-host cell interaction. J Cell Sci 1990; 96(Pt 3): 485-90. [PMID: 2229199]

[244] SantAnna C, Nakayasu ES, Pereira MG, *et al.* Subcellular proteomics of *Trypanosoma cruzi* reservosomes. Proteomics 2009; 9(7): 1782-94. [http://dx.doi.org/10.1002/pmic.200800730] [PMID: 19288526]

[245] Mauricio de Mendonca SM, Nepomuceno da Silva JL, Cunha e-Silva N, de Souza W, Gazos Lopes U. Characterization of a Rab11 homologue in *Trypanosoma cruzi*. Gene 2000; 243(1-2): 179-85. [http://dx.doi.org/10.1016/S0378-1119(99)00480-1] [PMID: 10675626]

[246] Cupello MP, Souza CF, Menna-Barreto RF, *et al.* Trypanosomatid essential metabolic pathway: new

approaches about heme fate in *Trypanosoma cruzi* . Biochem Biophys Res Commun 2014; 449(2): 216-21.
[http://dx.doi.org/10.1016/j.bbrc.2014.05.004] [PMID: 24824181]

[247] Pereira MG, Nakayasu ES, SantAnna C, *et al. Trypanosoma cruzi* epimastigotes are able to store and mobilize high amounts of cholesterol in reservosome lipid inclusions. PLoS One 2011; 6(7): e22359.
[http://dx.doi.org/10.1371/journal.pone.0022359] [PMID: 21818313]

[248] Sant'anna C, Pereira MG, Porto-Carreiro I, Jeovanio AL, de Souza W. Reservosomes: multipurpose organelles? Parasitological Research 2006; 99(4): 325-7.

[249] Vercesi AE, Moreno SN, Docampo R. Ca2+/H+ exchange in acidic vacuoles of *Trypanosoma brucei*. Biochem J 1994; 304(Pt 1): 227-33.
[http://dx.doi.org/10.1042/bj3040227] [PMID: 7998937]

[250] Docampo R, Scott DA, Vercesi AE, Moreno SN. Intracellular Ca2+ storage in acidocalcisomes of *Trypanosoma cruzi.* Biochem J 1995; 310(Pt 3): 1005-12.
[http://dx.doi.org/10.1042/bj3101005] [PMID: 7575396]

[251] Docampo R, de Souza W, Miranda K, Rohloff P, Moreno SN. Acidocalcisomes - conserved from bacteria to man. Nat Rev Microbiol 2005; 3(3): 251-61.
[http://dx.doi.org/10.1038/nrmicro1097] [PMID: 15738951]

[252] Miranda K, Benchimol M, Docampo R, de Souza W. The fine structure of acidocalcisomes in *Trypanosoma cruzi.* Parasitol Res 2000; 86(5): 373-84.
[http://dx.doi.org/10.1007/s004360050682] [PMID: 10836511]

[253] Scott DA, Docampo R, Dvorak JA, Shi S, Leapman RD. *In situ* compositional analysis of acidocalcisomes in *Trypanosoma cruzi.* J Biol Chem 1997; 272(44): 28020-9.
[http://dx.doi.org/10.1074/jbc.272.44.28020] [PMID: 9346954]

[254] Scott DA, Docampo R. Characterization of isolated acidocalcisomes of *Trypanosoma cruzi.* J Biol Chem 2000; 275(31): 24215-21.
[http://dx.doi.org/10.1074/jbc.M002454200] [PMID: 10816577]

[255] Ruiz FA, Marchesini N, Seufferheld M, Govindjee , Docampo R. The polyphosphate bodies of *Chlamydomonas reinhardtii* possess a proton-pumping pyrophosphatase and are similar to acidocalcisomes. J Biol Chem 2001; 276(49): 46196-203.
[http://dx.doi.org/10.1074/jbc.M105268200] [PMID: 11579086]

[256] Docampo R, Moreno SN. Acidocalcisomes. Cell Calcium 2011; 50(2): 113-9.
[http://dx.doi.org/10.1016/j.ceca.2011.05.012] [PMID: 21752464]

[257] Rodrigues CO, Scott DA, Docampo R. Characterization of a vacuolar pyrophosphatase in *Trypanosoma brucei* and its localization to acidocalcisomes. Mol Cell Biol 1999; 19(11): 7712-23.
[http://dx.doi.org/10.1128/MCB.19.11.7712] [PMID: 10523660]

[258] Rodrigues CO, Scott DA, Docampo R. Presence of a vacuolar H+-pyrophosphatase in promastigotes of *Leishmania donovani* and its localization to a different compartment from the vacuolar H+-ATPase. Biochem J 1999; 340(Pt 3): 759-66.
[http://dx.doi.org/10.1042/bj3400759] [PMID: 10359662]

[259] Moreno SN, Docampo R. The role of acidocalcisomes in parasitic protists. J Eukaryot Microbiol 2009;

56(3): 208-13.
[http://dx.doi.org/10.1111/j.1550-7408.2009.00404.x] [PMID: 19527347]

[260] Zhang K, Hsu FF, Scott DA, Docampo R, Turk J, Beverley SM. Leishmania salvage and remodelling of host sphingolipids in amastigote survival and acidocalcisome biogenesis. Mol Microbiol 2005; 55(5): 1566-78.
[http://dx.doi.org/10.1111/j.1365-2958.2005.04493.x] [PMID: 15720561]

[261] Kornberg A, Rao NN, Ault-Riché D. Inorganic polyphosphate: a molecule of many functions. Annu Rev Biochem 1999; 68: 89-125.
[http://dx.doi.org/10.1146/annurev.biochem.68.1.89] [PMID: 10872445]

[262] Lemercier G, Espiau B, Ruiz FA, *et al.* A pyrophosphatase regulating polyphosphate metabolism in acidocalcisomes is essential for *Trypanosoma brucei* virulence in mice. J Biol Chem 2004; 279(5): 3420-5.
[http://dx.doi.org/10.1074/jbc.M309974200] [PMID: 14615483]

[263] Espiau B, Lemercier G, Ambit A, *et al.* A soluble pyrophosphatase, a key enzyme for polyphosphate metabolism in *Leishmania*. J Biol Chem 2006; 281(3): 1516-23.
[http://dx.doi.org/10.1074/jbc.M506947200] [PMID: 16291745]

[264] Ruiz FA, Rodrigues CO, Docampo R. Rapid changes in polyphosphate content within acidocalcisomes in response to cell growth, differentiation, and environmental stress in *Trypanosoma cruzi*. J Biol Chem 2001; 276(28): 26114-21.
[http://dx.doi.org/10.1074/jbc.M102402200] [PMID: 11371561]

[265] LeFurgey A, Ingram P, Blum JJ. Elemental composition of polyphosphate-containing vacuoles and cytoplasm of *Leishmania major*. Mol Biochem Parasitol 1990; 40(1): 77-86.
[http://dx.doi.org/10.1016/0166-6851(90)90081-V] [PMID: 2348832]

[266] Docampo R, Jimenez V, Lander N, Li ZH, Niyogi S. New insights into roles of acidocalcisomes and contractile vacuole complex in osmoregulation in protists. Int Rev Cell Mol Biol 2013; 305: 69-113.
[http://dx.doi.org/10.1016/B978-0-12-407695-2.00002-0] [PMID: 23890380]

[267] Montalvetti A, Rohloff P, Docampo R. A functional aquaporin co-localizes with the vacuolar proton pyrophosphatase to acidocalcisomes and the contractile vacuole complex of *Trypanosoma cruzi*. J Biol Chem 2004; 279(37): 38673-82.
[http://dx.doi.org/10.1074/jbc.M406304200] [PMID: 15252016]

[268] Niyogi S, Jimenez V, Girard-Dias W, de Souza W, Miranda K, Docampo R. Rab32 is essential for maintaining functional acidocalcisomes, and for growth and infectivity of *Trypanosoma cruzi*. J Cell Sci 2015; 128(12): 2363-73.
[http://dx.doi.org/10.1242/jcs.169466] [PMID: 25964650]

[269] Rohloff P, Montalvetti A, Docampo R. Acidocalcisomes and the contractile vacuole complex are involved in osmoregulation in *Trypanosoma cruzi*. J Biol Chem 2004; 279(50): 52270-81.
[http://dx.doi.org/10.1074/jbc.M410372200] [PMID: 15466463]

[270] de Jesus TC, Tonelli RR, Nardelli SC, *et al.* Target of rapamycin (TOR)-like 1 kinase is involved in the control of polyphosphate levels and acidocalcisome maintenance in *Trypanosoma brucei*. J Biol Chem 2010; 285(31): 24131-40.
[http://dx.doi.org/10.1074/jbc.M110.120212] [PMID: 20495004]

[271] Madeira da Silva L, Beverley SM. Expansion of the target of rapamycin (TOR) kinase family and function in *Leishmania* shows that TOR3 is required for acidocalcisome biogenesis and animal infectivity. Proc Natl Acad Sci USA 2010; 107(26): 11965-70. [http://dx.doi.org/10.1073/pnas.1004599107] [PMID: 20551225]

[272] Rohloff P, Rodrigues CO, Docampo R. Regulatory volume decrease in *Trypanosoma cruzi* involves amino acid efflux and changes in intracellular calcium. Mol Biochem Parasitol 2003; 126(2): 219-30. [http://dx.doi.org/10.1016/S0166-6851(02)00277-3] [PMID: 12615321]

[273] Fang J, Ruiz FA, Docampo M, *et al.* Overexpression of a Zn2+-sensitive soluble exopolyphosphatase from *Trypanosoma cruzi* depletes polyphosphate and affects osmoregulation. J Biol Chem 2007; 282(44): 32501-10. [http://dx.doi.org/10.1074/jbc.M704841200] [PMID: 17827150]

[274] Rodrigues CO, Ruiz FA, Vieira M, Hill JE, Docampo R. An acidocalcisomal exopolyphosphatase from *Leishmania major* with high affinity for short chain polyphosphate. J Biol Chem 2002; 277(52): 50899-906. [http://dx.doi.org/10.1074/jbc.M208940200] [PMID: 12393865]

[275] Galizzi M, Bustamante JM, Fang J, *et al.* Evidence for the role of vacuolar soluble pyrophosphatase and inorganic polyphosphate in *Trypanosoma cruzi* persistence. Mol Microbiol 2013; 90(4): 699-715. [http://dx.doi.org/10.1111/mmi.12392] [PMID: 24033456]

[276] Lee N, Gannavaram S, Selvapandiyan A, Debrabant A. Characterization of metacaspases with trypsin-like activity and their putative role in programmed cell death in the protozoan parasite *Leishmania.* Eukaryot Cell 2007; 6(10): 1745-57. [http://dx.doi.org/10.1128/EC.00123-07] [PMID: 17715367]

[277] Scott DA, de Souza W, Benchimol M, *et al.* Presence of a plant-like proton-pumping pyrophosphatase in acidocalcisomes of *Trypanosoma cruzi*. J Biol Chem 1998; 273(34): 22151-8. [http://dx.doi.org/10.1074/jbc.273.34.22151] [PMID: 9705361]

[278] Hill JE, Scott DA, Luo S, Docampo R. Cloning and functional expression of a gene encoding a vacuolar-type proton-translocating pyrophosphatase from *Trypanosoma cruzi*. Biochem J 2000; 351(Pt 1): 281-8. [http://dx.doi.org/10.1042/bj3510281] [PMID: 10998372]

[279] Huang G, Ulrich PN, Storey M, *et al.* Proteomic analysis of the acidocalcisome, an organelle conserved from bacteria to human cells. PLoS Pathog 2014; 10(12): e1004555. [http://dx.doi.org/10.1371/journal.ppat.1004555] [PMID: 25503798]

[280] Lemercier G, Dutoya S, Luo S, *et al.* A vacuolar-type H+-pyrophosphatase governs maintenance of functional acidocalcisomes and growth of the insect and mammalian forms of *Trypanosoma brucei*. J Biol Chem 2002; 277(40): 37369-76. [http://dx.doi.org/10.1074/jbc.M204744200] [PMID: 12121996]

[281] Docampo R, Huang G. Calcium signaling in trypanosomatid parasites. Cell Calcium 2015; 57(3): 194-202. [http://dx.doi.org/10.1016/j.ceca.2014.10.015] [PMID: 25468729]

[282] Luo S, Rohloff P, Cox J, Uyemura SA, Docampo R. *Trypanosoma brucei* plasma membrane-type

Ca(2+)-ATPase 1 (TbPMC1) and 2 (TbPMC2) genes encode functional Ca(2+)-ATPases localized to the acidocalcisomes and plasma membrane, and essential for Ca(2+) homeostasis and growth. J Biol Chem 2004; 279(14): 14427-39.
[http://dx.doi.org/10.1074/jbc.M309978200] [PMID: 14724285]

[283] Huang G, Bartlett PJ, Thomas AP, Moreno SN, Docampo R. Acidocalcisomes of *Trypanosoma brucei* have an inositol 1,4,5-trisphosphate receptor that is required for growth and infectivity. Proc Natl Acad Sci USA 2013; 110(5): 1887-92.
[http://dx.doi.org/10.1073/pnas.1216955110] [PMID: 23319604]

[284] Hashimoto M, Enomoto M, Morales J, *et al.* Inositol 1,4,5-trisphosphate receptor regulates replication, differentiation, infectivity and virulence of the parasitic protist *Trypanosoma cruzi*. Mol Microbiol 2013; 87(6): 1133-50.
[http://dx.doi.org/10.1111/mmi.12155] [PMID: 23320762]

CHAPTER 2

Trypanosomatid Genome Organization and Ploidy

João Luís Reis-Cunha, Hugo Oswaldo Valdivia and **Daniella Castanheira Bartholomeu***

Universidade Federal de Minas Gerais, Instituto de Ciências Biológicas, Departamento de Parasitologia, Belo Horizonte, Minas Gerais, Brazil

Abstract: In 2005, the draft genome sequences of the parasites *Trypanosoma cruzi*, *Trypanosoma brucei* and *Leishmania major*, also known as the Tritryps, were published providing major insights into their genome structure and organization. Even though these parasites diverged around 200 to 500 million years ago, their core genomes are highly syntenic and conserved. These conserved regions are interspersed by retroelements, structural RNAs and species-specific genes related to host-parasite interactions. While *T. brucei* presents a subtelomeric expansion of genes related to antigenic variation, *T. cruzi* and *Leishmania* genomes contain species-specific genes related to cellular invasion and survival inside the mammalian host cells. Duplication events have also shaped the genome architecture of these parasites. As control of gene expression operates mainly at a post-transcriptional level in trypanosomatids, gene copy number variation is probably an efficient mechanism to enhance gene expression and increase sequence variability. These parasites also explore gene conversion mechanisms to generate variants and increase their surface complexity. Among the Tritryps, *T. cruzi* presents the most striking expansion of species-specific multigene families, which could be related to the parasite's ability to infect any nucleated cell of a broad range of mammals. Chromosomal copy number variation is also well tolerated by these parasites, allowing the expansion of the whole set of genes simultaneously. The functional implications of these chromosomal expansions to the parasite biology are still to be determined.

Keywords: Aneuploidy, Chromosomal variation, Chromosomes, Comparative genomics, Copy-number-variation, Genome organization, Genome structure, Multigene families, Polycistron, Synteny.

* **Corresponding author Daniella Castanheira Bartholomeu:** Universidade Federal de Minas Gerais, Instituto de Ciências Biológicas, Departamento de Parasitologia, Belo Horizonte, Minas Gerais, Brazil; E-mail: daniella@icb.ufmg.br

Marcelo Santos da Silva & Maria Isabel N. Cano (Eds.)

INTRODUCTION

Trypanosomatids are a group of kinetoplastid parasitic protozoa that belong to the Trypanosomatidae family, which includes several species that cause disease in humans.

Trypanosoma and *Leishmania* genera stand out due to their importance in public health being the etiological agents of African trypanosomiasis (*Trypanosoma brucei*), Chagas disease (*Trypanosoma cruzi*) and Leishmaniasis (various species of *Leishmania*). These parasites have a digenetic life cycle alternating between invertebrate and vertebrate hosts and present different developmental stages adapted to survive in each of their hosts. Another important characteristic of these organisms is the presence of a network of circular mitochondrial DNA, also known as kinetoplast that is unique in terms of structure, replication and functionality and gives the order Kinetoplastida its name. In the following sections of this chapter, we will provide an overview of the epidemiology, lifecycle, taxonomy and genomic organization of *Leishmania* and *Trypanosoma*.

EPIDEMIOLOGICAL CONTEXT OF THE TRYPANOSOMATIDS

Leishmania: The Causative Agent of Leishmaniasis

Leishmaniasis is a group of complex tropical diseases caused by protozoan parasites of the genus *Leishmania*. These parasites are transmitted by the bite of infected phlebotomine sand flies from the *Lutzomyia* genus *in* the New World (Central and South-America) and *Phlebotomus* in the Old World (East-Africa, the Mediterranean region, and the Indian subcontinent).

Epidemiological studies have shown that leishmaniasis is currently spread in 98 countries on five continents, with 12 million people infected, 350 million people at risk and more than 1.5 million new cases documented each year. The incidence of this disease is increasing due to its progressive adaptation to urban environments, an increase in its dispersion due to human migration, climate change and co-infection with other diseases, especially in poor regions and developing countries (Fig. **1**) [1, 2].

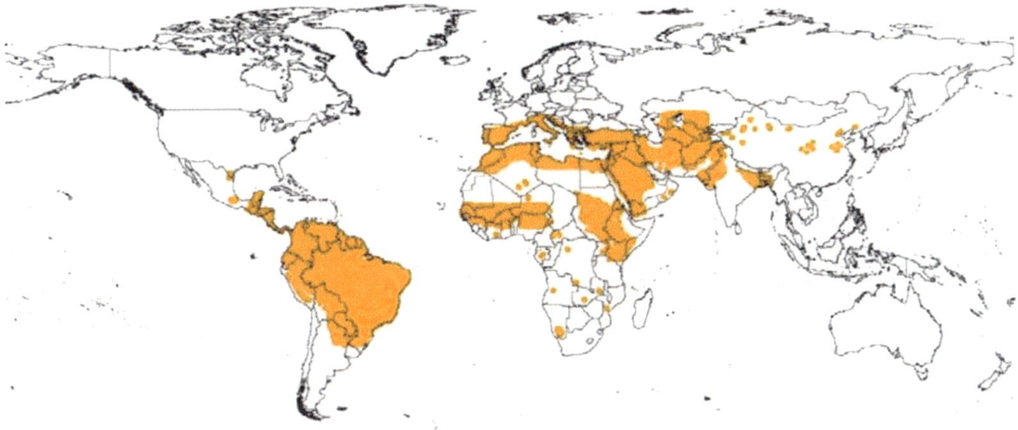

Fig. (1). Geographic distribution of leishmaniasis. Source: World Health Organization, October 2010.

There are at least 20 *Leishmania* species that can cause disease in humans, and are divided into two subgenera: *Leishmania* and *Viannia*. These species are responsible for a wide spectrum of clinical manifestations that are grouped into tegumentary leishmaniasis (TL), and visceral leishmaniasis (VL) (Table **1**) [3].

Table 1. Clinical manifestations of leishmaniasis and associated species.

Clinical Manifestation	Characteristics	Species
Cutaneous leishmaniasis	Papules, nodular lesions with keratotic plates that can ulcerate	*L. (Viannia) braziliensis* *L. (V.) peruviana* *L. (Leishmania) major* *L. (L.) mexicana*
Mucocutaneous leishmaniasis	Lesions in mucosal tissues at the nares, mouth and throat	*L. (V.) braziliensis* *L. (V.) guyanensis* *L. aethiopica*
Diffuse cutaneous leishmaniasis	Disseminated non-ulcerative skin lesions in the body	*L. (L.) amazonensis* *L. (L.) mexicana*
Visceral leishmaniasis	Fever, anemia, weight loss and swelling of visceral organs	*L. (L.) infantum* *L. (L.) donovani*
Post-kala-azar dermal leishmaniasis	Macular or nodular rash after treatment of visceral leishmaniasis	*L. (L.) infantum* *L. (L.) donovani*

TL encompasses three distinct types of diseases that are cutaneous leishmaniasis (CL), mucosal leishmaniasis (ML) and diffuse cutaneous leishmaniasis (DCL), which are widespread in the Old and New World and can lead to disfigurement and social stigma [4] (Fig. **2**). CL usually begins with a papule that appears near the area of the sand fly bite after an incubation period of 2 to 3 months. Occasionally papules heal spontaneously or develop into a nodule, progressing into an ulcer with keratotic plates. ML tends to appear years after the first cutaneous lesion and is characterized by a progressive destruction of the pharyngeal cavity and nares, leading to nasal septum perforation, breathing complications and disfigurement. This type of leishmaniasis is associated with infection by *L. (V.) braziliensis* and *L. (V.) guyanensis* [5, 6]. DCL is an uncommon form of TL that has been associated with infection caused by *L. (L.) amazonensis* and *L. (L.) aethiopica* in the New and Old World, respectively [7, 8]. DCL is the result of an inefficient immune response against the parasite leading to the appearance of non-ulcerative lesions on the body that could be unresponsive to treatment [8, 9].

VL is considered as the most severe form of the disease and is characterized by the preference of parasites that invade visceral organs such as the liver, spleen and bone marrow. Clinical symptomatology of VL infected patients can vary from sub-clinical, oligosymptomatic to fully established infection that initiates with general weakness, fever and weight loss and can progress into hepatomegaly and splenomegaly [3]. If it remains untreated, VL can produce cachexia, multisystem failure and death [3, 10]. Additionally, unsuccessful treatment of VL patients is associated with a chronic form named post-kala-azar dermal leishmaniasis [11] that is characterized by a macular or nodular rash and has been associated with an exacerbated immune response or related to co-infection with human immunodeficiency virus (HIV) [2, 12]. The two etiological agents of VL are *L. (L.) donovani* and *L. (L.) infantum* that are distributed in several regions of the Americas, Europe, Asia and Africa [1, 3]. Around 90% of VL cases are concentrated in six countries: India, Bangladesh, Sudan, South Sudan, Brazil and Ethiopia, while 75% of CL cases are reported in Afghanistan, Algeria, Colombia, Brazil, Iran, Syria, Ethiopia, North Sudan, Costa Rica and Peru [1, 2].

Pentavalent antimonials such as sodium stibogluconate and meglumine antimoniate are the first line of treatment for leishmaniasis [13 - 16]. However, due to their high toxicity, several side effects can occur including vomiting, myalgia, arthralgia, hepatitis and cardiotoxicity. Other drugs including amphotericin B, pentamidine and miltefosine are currently used in response to emerging resistance to first line drugs [17 - 20].

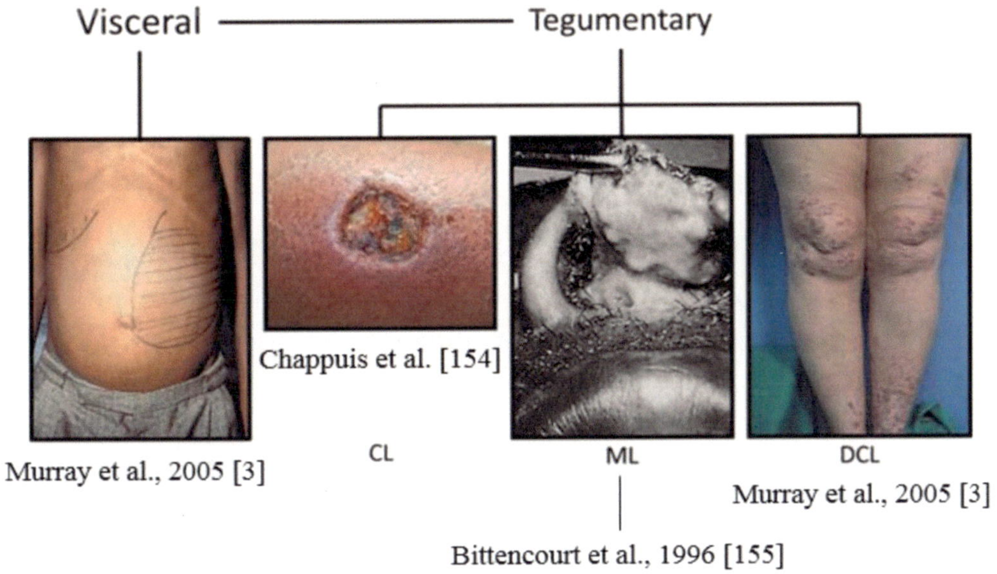

Fig. (2). Clinical symptomatology of the leishmaniasis. CL: cutaneous leishmaniasis; ML: mucosal leishmaniasis; DCL: diffuse cutaneous leishmaniasis.

T. cruzi: The Causative Agent of Chagas Disease

Trypanosoma cruzi is the etiological agent of American trypanosomiasis, also known as Chaga's disease. American trypanosomiasis is endemic in 21 countries in South and North America posing a risk to more than 90 million people and resulting in an infected population of 12 million people (Fig. **3**) [21, 22].

The parasite is mainly transmitted through the feces or urine of infected triatomines, which are released during the insect's blood meal, when it encounters skin lesions or mucosae in the host. However, other non-vectorial mechanisms such as blood transfusion, congenital and oral transmission have been described [23].

Currently, there is an increased number of imported cases in non-endemic countries due to the migration of Chagasic patients further spreading the disease and constituting a novel important health problem in these countries [24]. The disease is characterized by an acute phase that is usually unnoticed or can be manifested with an edema known as "Romana's sign" at the periocular region, or a local inflammation at the infection site, named "chagoma" [23]. Chagas disease acute phase resolves within 4 to 6 weeks after infection and leads to a long asymptomatic chronic phase that can last the entire life of the patient [21, 23]. Around 30% of infected patients can develop cardiac, digestive or mixed chronic manifestations that can lead to death by cardiac failure or obstruction in the esophagus or colon [21, 23].

Currently, the available drugs for Chagas disease, nitroimidazole and benznidazole [24], are only effective during the acute phase and therefore, there is no efficient treatment for the chronic phase of the disease.

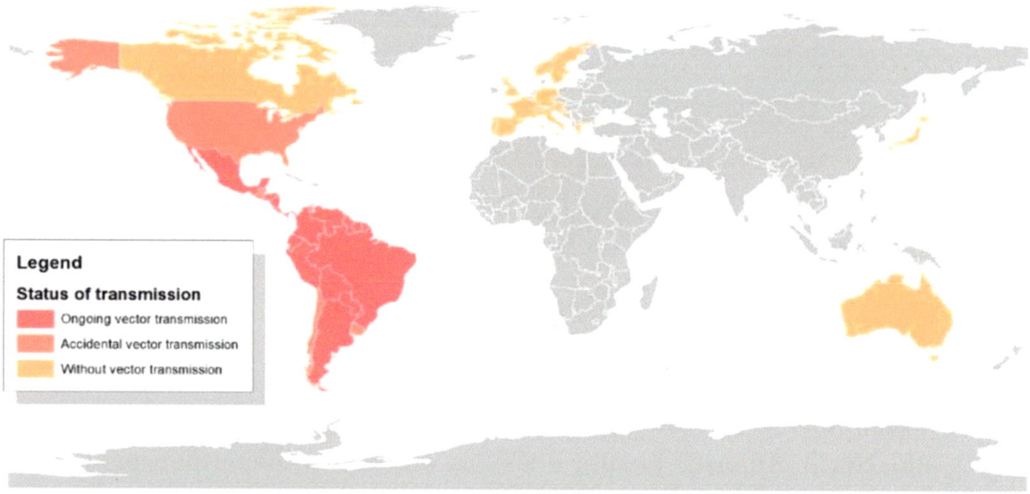

Fig. (3). Geographic distribution of *T. cruzi*.

T. brucei: The Causative Agent of Sleeping Sickness

Trypanosoma brucei is the etiological agent of African trypanosomiasis or sleeping sickness that is transmitted to the mammal host through the bite of an infected tsetse fly of the genus *Glossina* (Diptera).

Three distinct subspecies of *T. brucei* (*T. b. brucei*, *T. b. gambiense* and *T. b.*

rhodesiense) have been implicated in the disease and have different distribution and importance for the occurrence of African trypanosomiasis [25].

T. b. gambiense is responsible for more than 90% of human cases of African trypanosomiasis in 24 countries in central and west Africa (Fig. **4**) [25, 26]. *T. b. rhodesiense* is responsible for acute disease in east and southern Africa and *T. b. brucei* is usually associated with animal African trypanosomiasis although sporadic human infections have been reported [27].

Clinical manifestations of sleeping sickness are very similar for both subspecies (*T. b. gambiense* and *T. b. rhodesiense*) with the difference in disease evolution. *T. b. rhodesiense* infection is usually characterized by an acute disease and rapid appearance of clinical manifestations that can kill the host within a period of 6 months [28] if untreated, whereas *T. b. gambiense* is characterized by a slow and chronic course of the disease that can last 3 years [29].

The clinical progression of African trypanosomiasis can be divided into two phases: haemolymphatic and meningoencephalitic [25]. During the haemo-lymphatic phase, parasites are mostly restricted to the circulatory and lymphatic systems. Very unspecific symptoms such as headaches, intermittent fever, swelling of lymph nodes and joint pain can occur at this stage [25], which can easily be misdiagnosed by health surveillance systems.

The meningoencephalitic or neurological phase starts when the parasite reaches the host central nervous system. The most important symptoms during this phase are the disruption of the sleep cycle (hence the name "sleeping sickness"), paralysis, muscle weakness and mental deterioration that can inevitably conduct to death if it remains untreated [30].

A number of drugs are available for African trypanosomiasis that can be employed depending on the stage and infecting species: pentamidine, eflornithine, nifurtimox, melarsoprol, and suramin [31].

LIFE CYCLE

T. brucei, *T. cruzi* and *L. major* are digenetic protozoan parasites involving an insect vector and a mammalian host including humans, sylvatic and domestic animals. These Tritryps have adapted to the distinct environments inside the invertebrate and vertebrate hosts by presenting different developmental stages (Fig. **5**), which will be further described in the following sections.

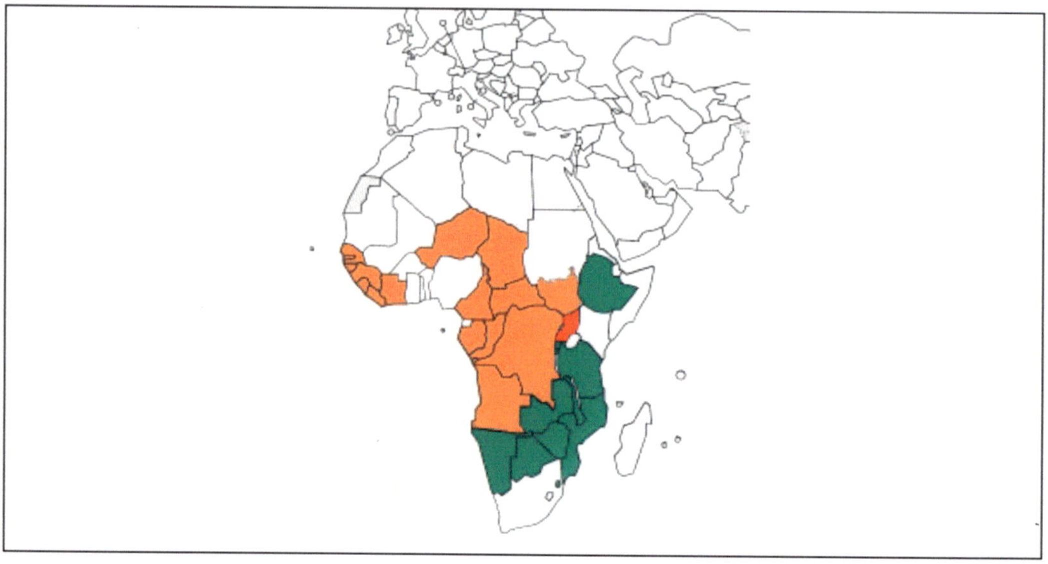

Fig. (4). Geographic distribution of *T. brucei* based on reported cases of 2014. Green: *T. brucei rhodesiense*; orange: *T. brucei gambiense*; red: areas of co-occurrence. Source: WHO, National Human African trypanosomiasis control programs. Annual country reports, 2014.

The Life Cycle of *Leishmania* spp.

Leishmaniasis is transmitted to the mammalian host by the bite of an infected female phlebotomine sand fly belonging to the genus *Lutzomyia* in the New World and *Phlebotomus* in the Old World. While there is a high diversity of sand flies (500 different species in South America) about 30 species have been identified as putative vectors of leishmaniasis [32, 33].

The lifecycle begins when a non-infected female sand fly ingests *Leishmania* amastigotes in infected blood or tissue from a reservoir host. Amastigotes undergo a rapid differentiation process into a motile flagellated form called promastigote inside the sand fly intestinal tract [10]. There are up to five different stages in the development of promastigotes: procyclic promastigotes, nectomonads, hapto-monads, paramastigotes and metacyclic promastigotes [32].

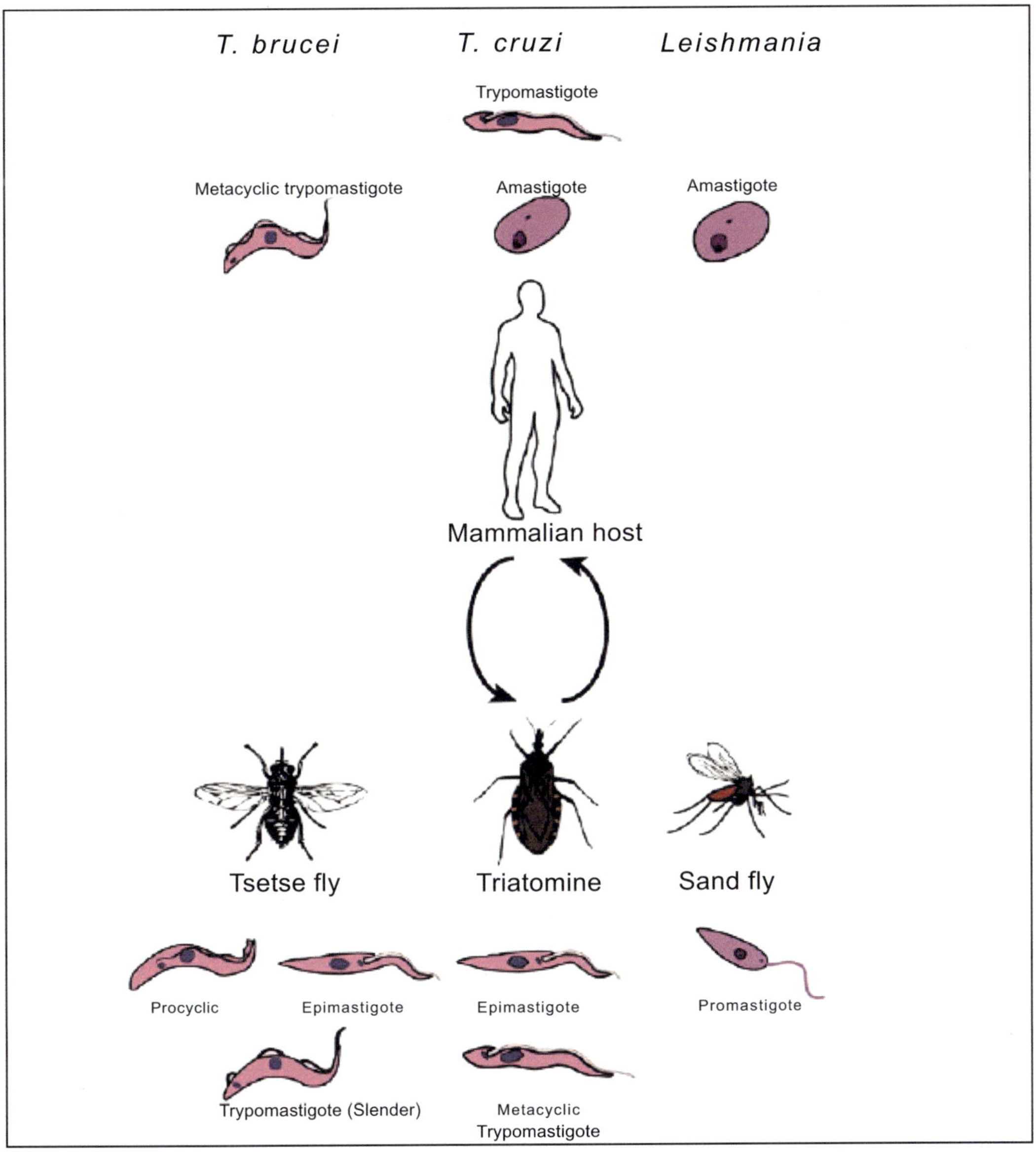

Fig. (5). Summary of the life cycle and developmental stages found in the insect vector and mammalian hosts for *Trypanosoma brucei*, *Trypanosoma cruzi* and *Leishmania* spp.

During the migration to the sand fly proboscis, promastigotes differentiate into metacyclic promastigotes, the infective form of the vertebrate host. The cycle in the vertebrate host starts when an infected female sandfly takes a blood meal from a non-infected mammalian host. During this process it will regurgitate from the proboscis the metacyclic promastigotes that rapidly trigger phagocytosis by the resident host macrophages.

Inside the macrophages, promastigotes remain in parasitophorous vacuoles where they transform into a non-motile form named amastigote that is adapted to survive and replicate inside its acidic pH conditions [10]. Amastigotes undergo binary fission multiplication generating a large number of parasites that eventually burst the cell and invade other macrophages. The cycle starts again when a non-infected female sand fly bites an infected host and ingests amastigotes or infected macrophages.

The Life Cycle of *Trypanosoma cruzi*

The life cycle of *T. cruzi* begins when a Triatominae of the genera *Triatoma*, *Rhodnius*, or *Panstrongylus* takes a blood meal from an infected host [34]. *T. cruzi* trypomastigotes that are in the peripheral blood are ingested and differentiate into epimastigotes that reproduce through binary fission inside the insect midgut. As epimastigotes migrate to the vectors rectal region, they undergo metacyclogenesis, differentiating into the infectious metacyclic trypomastigote [35].

Metacyclic trypomastigotes are excreted with the vectors feces or urine during a blood meal, infecting the mammalian host if they encounter a mucosal or skin lesion. Unlike *Leishmania*, trypomastigotes are capable of invading different types of host cells including macrophages, epithelial cells, fibroblast, muscle, adipocyte, and nervous cells [36].

Like *Leishmania*, *T. cruzi* trypomastigotes form a parasitophorous vacuole inside the host cell. However, trypomastigotes escape from the vacuole to the cytoplasm, where they differentiate into amastigotes, which reproduce through binary fission. Amastigotes subsequently transform into trypomastigotes and burst the host cell, releasing parasites that can infect neighbor cells or enter the host bloodstream where a non-infected triatomine can ingest them.

The Life Cycle of *Trypanosoma brucei*

The infection starts when an infected fly from the *Glossina* genus injects metacyclic trypomastigotes during its blood meal on a non-infected host [37]. Trypomastigotes invade the lymphatic system and pass into the bloodstream where they transform into slender replicative trypomastigotes that migrate through the body and multiply by binary fission. It is important to notice that unlike *T. cruzi* and *Leishmania* sp., *T. brucei* is exclusively extracellular. The lifecycle concludes when a non-infected tsetse fly ingests trypomastigotes during blood meal on an infected host. Inside the tsetse digestive tract trypomastigotes differentiate into procyclic trypomastigotes that multiply by binary fission and subsequently transform into epimastigotes. Epimastigotes migrate to the salivary glands where they multiply and transform into the infective non-proliferative metacyclic trypomastigotes [37].

TAXONOMY OF THE TRITRYPS

Leishmania genus is composed of 35 different species, of which at least 20 are pathogenic to humans. At first, *Leishmania* was classified based on ecobiological criteria like vectors, geographic distribution and clinical manifestations [38, 39]. Based on this primary classification and the recent use of molecular markers, *Leishmania* was divided into three distinct subgenera: *Leishmania, Viannia* and *Sauroleishmania*. *Leishmania* subgenus comprises both Old and New World species that are associated to VL or CL and develop in the midgut and foregut of the sand fly vector [38]. *Viannia* subgenus comprises exclusively New World species that are causative agents of CL and are characterized by their development in the sand fly hindgut [38]. *Sauroleishmania* subgenus is composed of species that infect lizards and are incapable of replicating within macrophages [40]. Although there have been significant advances in the taxonomy of *Leishmania,* the status of several *Leishmania* species is still on debate.

Trypanosoma genus is composed of two distinct clades, which group the American and African trypanosomes. Phylogenetic analyses have suggested an ancient split of trypanosomes into a branch of Salivaria and Stercoraria trypanosomes. Among salivaria trypanosomes are those that are transmitted directly by the bite of infected tsetse flies like *T. brucei* [41]. Stercoraria trypanosomes are those transmitted mainly through the feces of infected triatomines like *T. cruzi*. *T. cruzi* taxa are currently divided into six discrete typing units (DTUs), named TcI – TcVI, due to its high genotype and phenotype variability [42 - 44]. Besides the usual clonal replication [45], there is strong evidence of natural hybridization and genetic exchange between strains, resulting in the formation of TcV and TcIV hybrid DTUs [46 - 49]. Although the occurrence of hybridization is well recognized, there are still discrepancies regarding the evolutionary origin of hybrid DTUs, leading to the hypothesis of different hybridization events during *T. cruzi* evolution [46, 50, 51]. More details about natural hybridization and genetic exchange between Tritryps strains are described in Chapter 5.

GENOME ORGANIZATION

The development of high-throughput sequencing technologies has allowed the study of the genome organization of several trypanosomatids. These genomic endeavors have resulted in an increased understanding of the complexity of the trypanosomatids and the discovery of unique features in the genomes of these organisms.

The genome size of trypanosomatids is highly variable among different species and even within the same species. The haploid genome content of different *Leishmania* species is in the range of 31-33Mb, organized into 34-36 chromosomes [52, 53]. *T. cruzi* and *T. brucei* present an estimated haploid genome of 41-55Mb [54, 55] and 26Mb [56], organized into ~41 and 11 chromosomes, respectively.

A remarkable feature of the trypanosomatid genomes is their organization into large directional gene clusters (DGCs) that comprise a few to thousands of non-overlapping genes present in the same DNA strand and orientation, which are separated by strand switch regions [53, 57] (Fig. **6**). These DGCs are transcribed polycistronically in a similar way as prokaryote operons. Polycistronic pre-mRNAs are rapidly processed in the nucleus to generate mature monocistronic transcripts by the means of two coupled reactions, *trans*-splicing and polyadenylation. *Trans*-splicing lead to the addition of a species-specific 39-nucleotide miniexon, also known as spliced leader, to the 5' end of every trypanosomatid mRNA. Different from prokaryote operons, genes belonging to the same transcriptional unit are functionally unrelated and have distinct levels of expression, highlighting the fact that control of gene expression in trypanosomatids operates mainly at the post-transcriptional level. Polycistronic RNA processing, mRNA stability and storage, translation and post-translation controls are important mechanisms to achieve different levels of expression among genes belonging to the same polycistronic pre-mRNA [58]. Details about control of gene expression in Tritryps are described in Chapter 6.

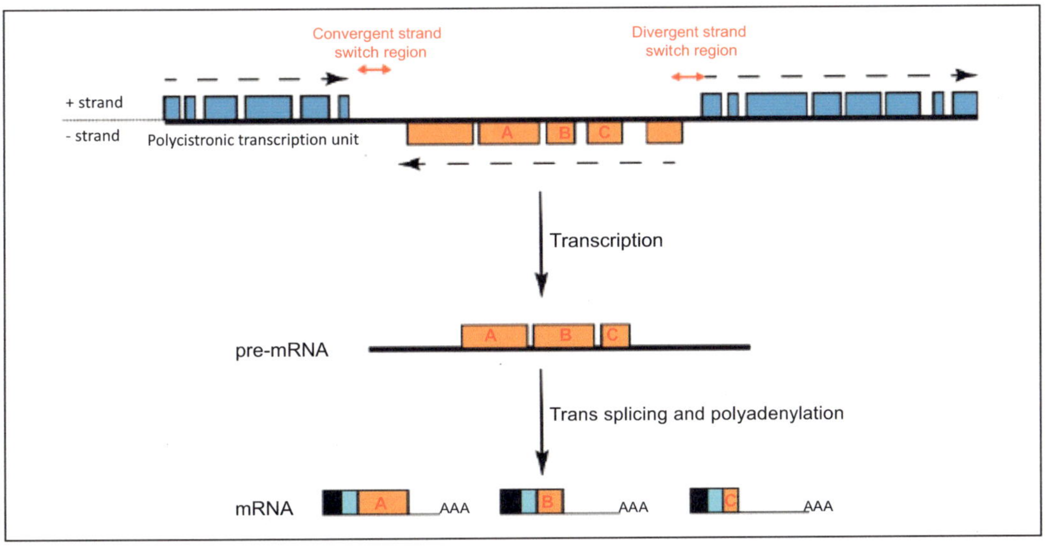

Fig. (6). Polycistronic transcription and mRNA processing in trypanosomatids. Most genes are organized into large bidirectional gene clusters that are separated by strand-switch regions. Polycistronic RNAs are processed to generate monocistronic RNA by *trans*-splicing and polyadenylation reactions. Mature RNA consists of the spliced leader sequence (black boxes), 5'UTR (light blue boxes) and coding sequence (orange boxes), 3'UTR (black line) and the poly (A) tail.

Trypanosomatid Comparative Genomics

Since 2005, a large number of trypanosomatid genomes have been sequenced, which provided important datasets to comparative genome analysis that had shaded light into our understanding of the parasitism evolution among the Kinetoplastida [52 - 55, 59 - 64]. These analyses have shown that, even though they diverged around 200 to 500 million years ago, the genomes of trypanosomatid species are highly syntenic, meaning that the order of ortholog genes is highly conserved among their genomes. This is especially remarkable in genomic regions that encode housekeeping genes [59]. Comparative analyses of the genomes of the Tritryps, *T. cruzi* clone CL Brener [54], *T. brucei* [56], and *Leishmania major* [53], have shown that they share a conserved core of ~6,200 genes, 94% of which are localized in syntenic regions [59]. This corresponds to the large proportion of the Tritryps genomes, since the haploid genome of *T. cruzi* CL Brener, *L. major* and *T. brucei* have 12,000, 8,300 and 9,000 genes, respectively [59]. Alignments of amino acid sequences of a large subset of these 6,200 genes revealed an average of 57% identity between *T. cruzi* and *T. brucei*, and 44% of identity between *L. major* and both *Trypanosoma* species, which reflects the phylogeny rather than the intra or extracellular parasitic lifestyle [59, 65]. Thus, it appears that there is a strong selective pressure to keep the synteny of the conserved genes despite their sequence variation. This could be a result of the low incidence of sexual recombination in these organisms [66], or caused by the need to maintain the order/orientation of the genes in the polycistronic units.

The syntenic regions are interspersed by species-specific genes, structural RNAs and retroelements, which vary greatly in size and content among species and even within species [54, 55, 59, 67, 68]. While in *T. brucei* these clusters are enriched in subtelomeric regions, they are evenly distributed in internal chromosomal and sub-telomeric regions in *T. cruzi* CL Brener and *L. major* [59]. These species-specific genes are mostly composed of multigene families that encode surface proteins, which are related to host-pathogen interactions and to the lifestyle adopted by each parasite, reflecting their adaptation to the vertebrate host [69].

Recently, the sequencing of a larger set of trypanosomatid genomes [55, 70], including the mammalian non-pathogenic *Trypanosoma rangeli* [61] and the bat-parasite *Trypanosoma marinkellei* [67] further confirmed the conservation of the core proteome among the *Trypanosoma* genus. Comparative analysis showed that *T. rangeli* shares 5,178 of its 7,613 coding sequences with the Tritryps [61], and that the whole set of *T. marinkellei* housekeeping genes is present in Sylvio X10 *T. cruzi* genome, with the exception of an acetyltransferase that was acquired by *T. marinkellei* after speciation, possibly by horizontal gene transfer [71].

The genome sequencing of a larger number of *Leishmania* species [52, 63, 64] also confirmed the broad core conservation within the genus. Interestingly, representatives of subgenus *Viannia*, such as *L. braziliensis*, *L. guyanensis* and *L. panamensis*, kept the RNA interference (RNAi) machinery [52, 72]. This machinery is also found in other trypanosomatids as *T. brucei* and *T. congolense*, however is absent in *T. cruzi*, *L. major* and *L. donovani* [72 - 74]. *T. rangeli* has the majority of the RNAi pathway as pseudogenes, which suggest a recent loss of this function in this parasite [61]. As RNAi is a mechanism to control gene expression, transposon duplication and anti-virus defense in a sequence specific manner, the presence/absence of this machinery could have a substantial impact on the differential evolution of these parasite genomes [72].

Comparative genomics among trypanosomatids that have a distinct mode of life may also shed light on other different aspects of the biology of these parasites, such as genes involved in host cell invasion and intracellular survival. Differently from housekeeping genes, the analysis of genes present exclusively in intracellular parasites have shown that, from 3,340 orthologs clusters identified by OrthoMCL, only 37 (~1%) are shared among the CL Brener and Sylvio X10 *T. cruzi strains*, *T. marinkellei* and six species of *Leishmania*, further supporting the diversification of these sequences after speciation [69]. The expansion of multigene families is more prominent in *T. cruzi* [54, 59], and accounts for its larger genome size, 55 Mb, when compared with *L. major*, 33 Mb, or *T. brucei* 26 Mb [59]. This relative expansion in *T. cruzi* could be a consequence of the large range of mammalian hosts that this parasite can infect, and also to its unique ability, among the trypanosomatids, to actively infect any nucleated host cell [75 - 77]. This vast number of host cells and reservoirs exposes *T. cruzi* to more diverse intracellular and extracellular environments, subjecting its proteins to a broader range of selective pressures when compared to *T. brucei* and *Leishmania* [78]. In fact, Flores-Lopez and Machado have shown that a significantly larger fraction of *T. cruzi* protein-coding genes shows evidence of positive selection for diversification than in *Leishmania* spp [78]. This evolutionary pressure also favors the fixation of gene duplications events, creating paralog genes and resulting in 50% of the *T. cruzi* CL Brener genome being composed of repetitive sequences [54]. As mentioned above, the control of gene expression in trypanosomatids is mainly based on post-transcriptional mechanisms, where the majority of the genome is constitutively transcribed [68]. Therefore, gene duplication seems to be an efficient method to increase the mRNA levels of key genes [79, 80]. Copy number variation (CNV) is more frequent in *T. cruzi*, but also present in *T. brucei* and *Leishmania*. CNV is a well-documented mechanism to increase gene expression and variability, allowing the generation of novel genes without loss of function [59, 69, 81]. More recently, Chromosomal Copy Number Variation (CCNV), were identified in several species of the *Leishmania* genus [63, 64] as well as in *T. cruzi* strains from different DTUs [82, 83]. CCNV is another level of gene expansion mechanism, allowing the diversification and rapid adaptation from the parasite to different environments and pressures driven by the host [64, 83]. Variations in chromosome numbers were already associated with increased fitness in stress conditions and drug resistance in *Saccharomyces cerevisiae* and *Candida albicans* [84, 85], and could also be explored by the trypanosomatids to allow natural selection of favorable phenotypes, resulting in an enhanced adaptability to the parasite [82, 86]. Aneuploidy appears to be well tolerated in trypanosomatids, due to their predominantly asexual reproduction. The occurrence of CCNV in other related parasites such as *T. brucei*, *T. rangeli* and *T. marinkellei* still needs evaluation.

Leishmania Genomic Features

Sequencing projects in *Leishmania* have led to the availability of several genomes including *L. (L.) major*, *L. (L.) infantum*, *L. (V.) braziliensis* [52], *L. (L.) amazonensis* [87], *L. (V.) panamensis* [88] and other draft assemblies that are available to the scientific community [89].

Comparative analyses in these genomes have shown that there are important inter-specific differences in *Leishmania* in terms of chromosomal organization. The genomes of *L. (L.) donovani, L. (L.) major and L. (L.) infantum* are composed of 36 chromosomes [90] whereas the genomes of New World *L. (V.) braziliensis* and *L. (L.) mexicana* consist of 35 and 34 chromosomes, respectively. The differences in chromosome numbers between these species are thought to be caused by fusion events between chromosomes 20 and 34 in *L. (V.) braziliensis* [52] and chromosomes 8 and 29 and 20 and 36 in *L. (L.) mexicana* [91].

Gene copy number studies have suggested that gene duplication and the presence of expanded tandem gene arrays could be one form to increase gene expression at the transcriptional level [63]. Moreover, analyses of chromosomal content revealed a high variability at inter and intra-specific level that is referred to as mosaic aneuploidy [92]. Although generally diploid, *Leishmania* species present several chromosomal duplications, which patterns vary greatly among species and even within the same species. While two isolates of *L. major* (Friedlin and LV39) have a similar pattern of chromosome duplication with only one extranumerary chromosome, isolates from *L. mexicana* (U1103 and M379) as well as from *L. donovani* (BPK206/0 and LV9) have several intra-species differences in chromosomal expansions [63]. Interestingly, *L. braziliensis* M2904 present an overall triploid genome, been the only trypanosomatid identified to date with this ploidy [63]. However, the implications of the triploidy for the biology of the parasite are still unknown. Fluorescence *in situ* hybridization (FISH) in *Leishmania* during mitosis have shown a high rate of asymmetrical chromosomal segregations, generating aneuploidies in the population, suggesting that the generation of CCNV is a common event in the parasite biology [93, 94]. An unequal chromosomal segregation during mitosis could generate cells with one or three copies of a given chromosome, generating highly divergent phenotypes and could contribute to a rapid adaptation to a new environment.

As previously described, the clinical manifestations of the leishmaniasis are mainly associated with the infecting *Leishmania* species and the host immune response. Some virulent parasite factors are derived from large gene families that are distributed in the *Leishmania* genomes and can vary between species, stage within the host and mediate the interaction with the sand fly and mammal hosts [95]. Amastin is one of the major gene family in *Leishmania.* It encodes transmembrane glycoproteins that are located on the cell surface of the parasite [96]. Although a clear function of amastins has not been yet identified, it has been suggested that these highly abundant surface protein could act as a signal transducer or transporter molecules allowing parasite sense and survival. Moreover, it has been previously shown that amastins are expanded in all *Leishmania* species compared to *Trypanosoma*, suggesting a functional adaptation [96].

A second important gene family is comprised of GP63 zinc-metallopeptidases that present wide substrate specificity and stage-specific expression in *Leishmania* promastigotes. Among the diverse roles of this gene family are midgut attachment, protection against trypsin and quimiotrypsin mediated digestion inside the sand fly midgut [97]. GP63 has also been incriminated in inactivation of complement mediated lysis and with subsequent macrophage interaction by direct cleavage of C3b into iC3b that can act as an opsonin, but lacks the ability to promote the membrane attack complex assembly [98]. Moreover, there is consistent evidence that members of GP63 family have a potent immuno-regulatory role acting on host transcription factors involved in protective immune responses like myristoylated alanine-rich kinase and JAK kinases. Both factors play important roles in cell shape, membrane trafficking and production of antimicrobial agents [99, 100].

Another important expanded gene family in *Leishmania* is composed of Cysteine Proteases (CPs). This family participates in different parasite processes like differentiation, host immune evasion and during the initial steps of infection through interaction with proteoglycans at the host cell surface [101]. In a study using *L. (L.) mexicana* infected C3H mice, CPs appeared to play an immunomodulatory role in suppressing Th1 response and ensuring parasite persistence [102]. Additionally, there are evidence that CPs can interact with the host immune system subverting immune recognition by altering phagosome function and as a consequence inhibiting MHC-I cross-presentation [103]. Moreover, previous studies have shown that the activity of some *L. (L.) major* CPs is lower than in *L. (L.) mexicana* and this may account for different disease phenotypes in both species [102].

It is clear that *Leishmania* parasites have developed different adaptive strategies during the various stages of their life cycle in response to the contact between the surface of promastigotes and amastigotes with the tissues and host cells. Important virulence factors organized in large gene families reflect the genome complexity and evolutionary process that result in the parasite capacity of infection, subversion of the host immune system and parasite protection. More details about *Leishmania* spp. virulent factors and immune evasion are described in Chapter 7.

Trypanosoma Genus

Trypanosoma genus is a highly diverse taxon that includes parasites able to infect a broad range of vertebrate/invertebrate species. *T. cruzi* and *T. brucei* are the most studied species of this genus as they cause severe human diseases [21, 25, 104 - 106]. The organization of *Trypanosoma* genomes can be divided into two regions: 1- conserved and syntenic regions among different *Trypanosoma* species, that contain genes related to the parasite basal metabolism, and 2- species-specific gene clusters that encode surface proteins directly related to specific mechanisms of interaction with their hosts [54, 69]. These gene clusters are mainly composed by Variant Surface Glycoproteins (VSGs) and Expression Site Associated Genes (ESAGs) in *T. brucei*, and by *trans*-sialidases, TcMUCs and Mucin Associated Surface Proteins (MASPs) in *T. cruzi*, *T. marinkellei* and *T. rangeli*.

Trypanosoma brucei Sub-species Complex

In 2005, Berriman and co-workers sequenced the genome of *T. b. brucei* TREU 927/4 strain, producing the first *T. brucei* whole genome draft [56]. *T. b. brucei* TREU 927/4 genome was assembled into 11 Megabase-sized chromosomes totalizing ~26Mb [56]. The nuclear genome also contains small and medium-sized chromosomes, which vary in sizes from 30 to 700kb, totalizing ~10 Mb [56]. The core of the large chromosomes is mainly composed by housekeeping and hypothetical genes, which are highly conserved and syntenic with *T. cruzi* and *Leishmania* [56, 106]. The sub-telomeric regions of large chromosomes, as well as the majority of the small and medium-sized chromosomes are composed by *T. brucei* specific genes, such as VSGs and ESAGs, involved in immune evasion [60]. VSGs are *T. brucei*-specific highly antigenic surface proteins, comprising a hypervariable N-terminal domain that is exposed extracellularly and a conserved C-terminal region that is buried in the parasite's coat [107, 108]. In contrast to *T. cruzi* and *Leishmania* sp. that can invade vertebrate host cells, *T. brucei* develops exclusively extracellularly, and therefore is constantly exposed to the host humoral immune response. To evade antibody-mediated lysis, the parasite relies on antigenic variation, a mechanism based on monoallelic expression of VSGs. During its infection in the mammalian host, the surface of *T. brucei* is covered with ~10^7 copies of a single VSG variant, which shields other invariant antigens in the parasite surface [60, 109, 110]. As VSG is highly antigenic and the majority of the population expresses the same VSG variant, a strong immune response focused on the expressed VSG is established, greatly reducing the parasitemia. However, a small number of parasites in the same infecting population, express different variants, which are not promptly recognized by the host immune system, and therefore expand in the population, re-populating the host. Long-term infection is achieved by changing the expressed antigens, leading to successive waves of parasitemia and clearance, as novel antigenic determinants spread in the parasite population [56, 109, 110]. To allow monoallelic expression of a single VSG, the parasite relies on two structures, the Expression Site (ES) and the Expression Site Body (ESB) [111, 112]. The ES consists of specific sub-telomeric regions within the 11 Mb-sized *T. brucei* chromosomes, and can be further divided into Bloodstream Expression Site (BES) and Metacyclic Expression Site (MES), which are activated in the bloodstream or metacyclic trypomastigote forms, respectively [111, 113]. BES is composed by several ESAGs, a ~70pb repetitive sequence and a single downstream VSG gene (Fig. **7A**), whereas MES lacks the upstream ESAGs (Fig. **7B**) [111, 113]. BES and MES transcription is mediated by RNA polymerase I, however, their expression is controlled by different promoters, perhaps reflecting different regulatory mechanisms [114, 115]. While BES and MES are sequence-based structures, ESB is an extranucleolar region in *T. brucei* nucleus that contains the machinery to accommodate, expose and express the VSG gene in a given ES (Fig. **7C**) [113, 116]. As there is only one ESB in *T. brucei* nucleus, only the ESB-associated ES is expressed at a given time [113, 115, 116]. Changing the expressed VSG is achieved by gene conversion processes, which copy a silent VSG into the active ES, deleting the resident VSG, or by changing the ES into the ESB [111]. This change-rate of the expressed VSG was estimated as one event per 100 cell divisions [108, 117, 118]. To sustain a long-term infection, *T. brucei* relies on a large pool of VSG genes. From ~ 9,000 complete genes and pseudogenes identified in the *T. b. brucei* genome, approximately 800 encode VSGs, which are clustered into arrays of 3 to ~250 genes [59, 60]. However only 7% of these genes are fully functional, presenting all the typical features of a VSG, while 9% are atypical with folding limitations, 66% are full-length pseudogenes and 18% are gene fragments [60]. *T. brucei* can explore this arsenal of pseudogenes to increase

VSG variability by creating functional mosaic genes through gene conversion [111, 112, 119, 120]. As *T. b. brucei*, *T. b. gambiense* [106] as well as several isolates from the three subspecies share the majority of the VSG repertoire [62], the ratio of gene conversion is still to be determined.

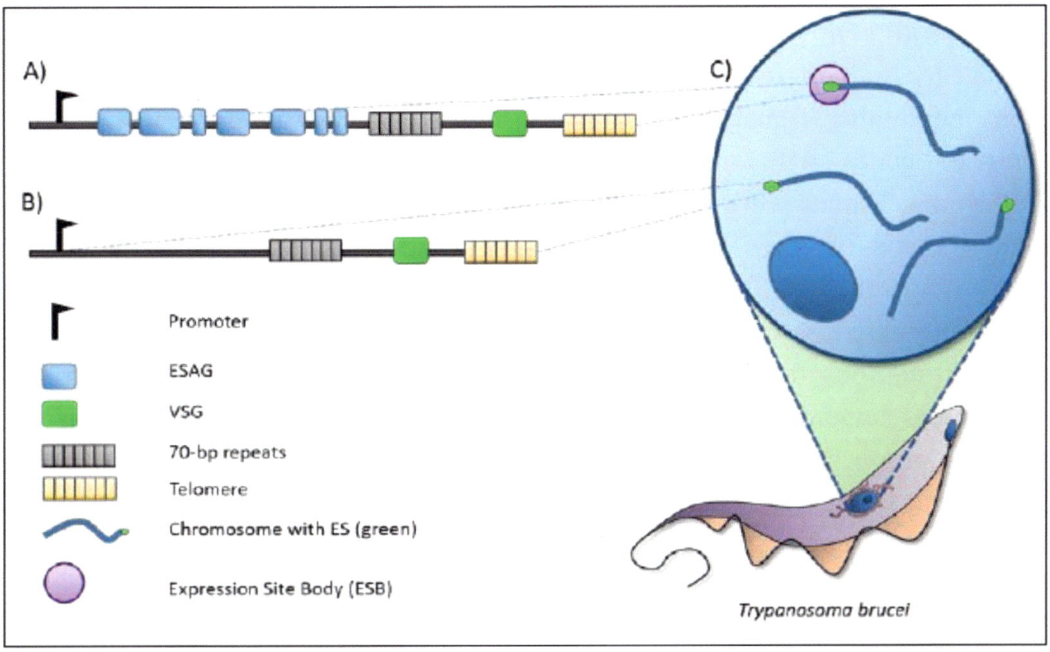

Fig. (7). *T. brucei* VSG Expression Sites. The structure of the BES (**A**), MES (**B**) and the ESB (**C**) are shown. Only the VSG in the ES that co-localize with the ESB is expressed.

The genome sequencing of *T. b. gambiense* MHOM/CI/86/DAL972 strain published in 2010 revealed large conservation in gene order and content with *T. b. brucei 927* strain. Regarding the chromosome core, no *T. b. gambiense* specific protein-encoding gene was identified [106]. *T. b. gambiense* presents the smallest genomes of *T. brucei* sub-species, having only 71-82% of the *T. b. brucei* DNA content [106, 121, 122]. This difference in genome size is mainly caused by differential CNV, with several cases of single copy genes in *T. b. gambiense* that are duplicated or expanded in *T. b. brucei* [106]. Usually, in *T. b. brucei* genome, these duplicated regions are identical, but in few cases they resulted in subspecies-specific isoforms with paralog genes presenting 30-40% of sequence divergence at protein level [106]. Analysis of *T. b. brucei* and *T. b. gambiense* ortholog genes showed that 86.4% of them vary less than 1%, with a mean nucleotide identity of 99.2%, while non-coding regions are more divergent with 95% sequence identity on average [106]. On the other hand, 69 orthologs showed evidence of positive selection, with identities varying from 60 to ~85%. The majority of these cases (35) correspond to VSGs, and 17 to uncharacterized proteins [106]. On rare occasions, the synteny between these two strains is disrupted by inversions, insertions and duplications, usually found in repetitive regions containing transposable elements, chromosomal internal VSG/ESAG islands or genes with unknown function [106]. In total, 17 VSG/ESAG chromosomic islands were identified in both genomes, and only six of them are unique to one subspecies, being the most dynamic features in the 11 megabase-sized chromosomes [106]. The subtelomeric regions of the large chromosomes on both genomes are also fairly similar in composition and gene order. These regions are mainly composed by VSGs (67.8% and 44%), ESAGs (13.4%, 15.8%) and transposable-element-related genes (7.6% and 13.5%) in *T. b. brucei* and *T. b. gambiense*, respectively [106]. Pulse field gel analysis shows that *T. b. gambiense* also have the fewest number of minichromosomes in the complex, which are also smaller in average size, 25kb, when compared with 100kb in *T. b. brucei* and *T. b. rhodesiense* [106, 121, 123]. This reduction in minichromosomes contributes to the smaller VSG repertoire in *T. b. gambiense* compared with the other *T. brucei* sub-species [106, 121, 124]. Sequence comparison between the 1,258 assembled VSG genes from *T. b. brucei* and unassembled reads from *T. b. gambiense* showed an average amino acid identity of 43.3%, suggesting positive selection or at least a relaxation of purifying selection [106]. Furthermore, 787 (63%) of the *T. b. brucei* VSG genes are closely related with each other than with *T. b. gambiense* reads, which suggests that gene duplication, gene conversion, and diversification occurred after the speciation event [106]. *T. b. brucei* VSGs do not cluster by chromosomal location or by developmental stage expression, indicating that gene conversion process frequently occurred among subtelomeric VSGs in a recent past regardless of genome positioning and parasite life stage [106]. These results reinforce that gene duplication or gene conversion are ongoing and active processes to generate VSG variants in *T. brucei*. Gene conversion processes also seem to be the evolutionary process behind the origin of the Serum Resistance-Associated gene (SRA) in *T. b. rhodesiense*, which is localized in one VSG ES

[125]. The SRA consists of a truncated VSG gene, which protein sequence interacts with the human trypanolytic factor apoL1 and neutralizes its effects [126, 127].

More recently, comparative analyses of 19 *T. b. brucei*, 3 *T. b. gambiense* and 13 *T. b. rhodesiense* genomes have further confirmed that the majority of *T. brucei* genome is under purifying selection [62]. However, SNP clustering analysis revealed several differences among the samples from the same sub-species, which are not necessary more related to each other; showing that the country of origin and host species are important factors in shaping the genomic diversity within *T. brucei* complex [62]. These results highlight the important role of geographic distance and host specificity in *T. brucei* evolution, and suggest that the actual subspecies classification could not completely reflect their phylogeny [62].

The three sub-species of the *T. brucei* complex, *T. b. brucei*, *T. b. gambiense* and *T. b. rhodesiense* present a highly conserved genome, however they show significant differences in host specificity and human infection outcomes [59, 62, 106]. Sequence polymorphism, copy number variation and distinct gene expression profile may contribute to distinct disease phenotypes caused by these three *T. brucei* subspecies [68, 106].

T. cruzi and Close Related Parasites

T. cruzi taxa are currently divided into six DTUs due to its high genotypic and phenotypic diversity [43, 44, 82, 128 - 131]. *T. cruzi* replication is usually clonal [45], however, genetic exchange including whole-genome hybridizations events is well characterized ([46, 48, 50], see also Chapter 5 for more details about trypanosome genetic exchange). Three distinct models based on nuclear and mitochondrial markers were proposed to explain *T. cruzi* evolution: The "Two-Hybridization Model" [50], the "Three Ancestor Model" [46], and more recently a "Two Ancestor Model" proposed by Burgos and co-workers [132] (Fig. **8**).

To date, the genome sequence of three *T. cruzi* isolates was sequenced and assembled: TcVI CL Brener clone, with an estimated genome size of 55Mb [54]; and two TcI genomes, Sylvio X10 and Dm28c strains, with genome sizes of 44Mb and 27Mb, respectively [55, 67, 70]. Recently, the genome sequencing of the Bat-parasite *T. c. marinkellei* (39Mb) [67], and the *T. cruzi* closed related parasite *T. rangeli* (24Mb) [61], further confirmed the extensive variation in genome sizes within the *Trypanosoma* genus.

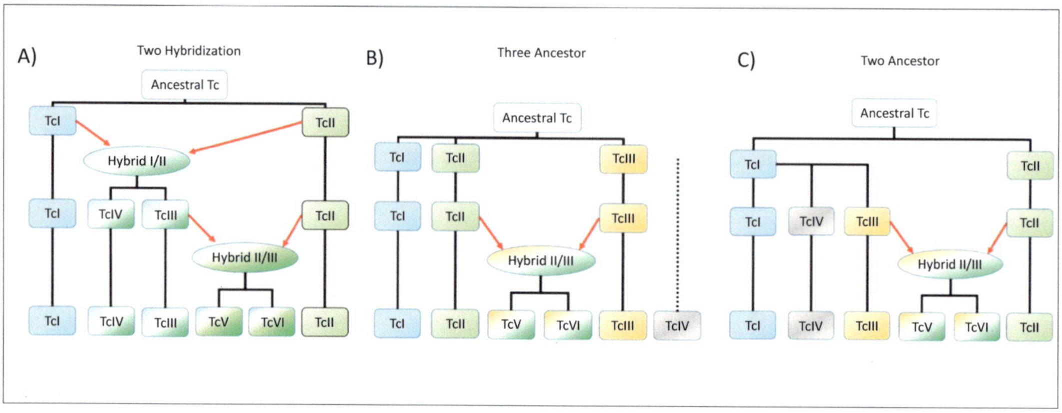

Fig. (8). *T. cruzi* evolution models. (**A**). In the Two-Hybridization model, a hybridization event between DTUs TcI and TcII originated TcIII and TcIV, while a second hybridization between the hybrid TcIII and the clonal TcII generated TcV and TcVI [50]. (**B**). In the Three Ancestor model, TcI, TcII and TcIII are ancestral DTUs, and a hybridization between TcII and TcIII generated TcV and TcVI [46]. (**C**). In the two-ancestor model, only TcI and TcII are ancestor DTUs, and TcI originates TcIII and TcIV, and a hybridization event between TcII and TcIII originated TcV and TcVI [132]. More details about the classification of *T. cruzi* in DTUs are described in Chapter 9.

CL Brener, a clone from the TcVI CL strain, was the first *T. cruzi* genome sequenced, revealing the hybrid nature of this DTU [46, 54]. Post-assembly comparisons of CL Brener contigs with reads from the *T. cruzi* Esmeraldo strain (TcII) led to the identification of two haplotypes within its genome: the Esmeraldo-like haplotype, derived from a TcII ancestor, and the non-Esmerald--like haplotype, derived from a TcIII ancestor [54]. These haplotypes display a high level of synteny and conservation, especially in the directional gene clusters encoding housekeeping genes, with a sequence similarity of 97.8% [54, 55]. In 2010, the sequencing of *T. cruzi* strain Sylvio X10 (TcI) showed that the core genomic regions of this isolate is highly conserved with CL Brener [55, 67]. Furthermore, Sylvio has a higher mean nucleotide similarity with the CL Brener Non-esmeraldo haplotype (98.2%) than with the Esmeraldo-like (97.5%) [55]. This closer similarity between TcI and TcIII were further supported by the preferential mapping of the whole genome reads from TcI Arequipa and

Colombiana strains to the non-Esmeraldo like haplotype compared to Esmeraldo-like [83]. These results suggest a closer evolutionary relationship between TcI and TcIII, by the Two Ancestor model, proposed by Burgos [132]. The similarity between *T. cruzi* DTUs is smaller than between *T. brucei* subspecies (99.2%) [106], but higher than between *Leishmania* species (94%), which still supports the classification of *T. cruzi* DTUs in the same species [52].

Both CL Brener haplotypes, Esmeraldo-like and non-Esmeraldo-like, are highly syntenic in regions encoding housekeeping genes. Synteny breaks occur in regions enriched in genes encoding surface proteins involved in host-parasite interactions, such as MASPs, Mucins, GP63, *Trans*-Sialidases and DGF-1 (see details about *T. cruzi* virulent factors in Chapter 8), as well as transposable elements and retrotransposon hot spot proteins (RHS) [54, 133 - 136]. In these genomic regions, genes of each multigene family are not clustered together. Instead, they alternate with other gene families in a non-ordered fashion, constantly changing the coding strand [54, 135, 137] (Fig. **9**). These clusters can be as large as 600kb, and vary in sequence and length among CL Brener haplotypes as well as among *T. cruzi* DTUs [54, 55, 138]. In fact, expansion of multigene families in CL Brener account for 5.9Mb of the genome size difference between this clone and Sylvio X10 strain [55]. In contrast to the usual subtelomeric localization of *T. brucei* VSG/ESAG clusters, in CL Brener, multigene family clusters are evenly distributed in subtelomeric and internal chromosomal regions (Fig. **9**) [54, 137]. Details about the organization of *T. cruzi* sub-telomeric regions are described in Chapter 3.

It is evident that very distinct evolutionary forces have shaped the evolution of *T. cruzi* genome resulting in conservation of genomic organization in regions encoding housekeeping genes and intense sequence rearrangement and diversification in regions containing genes encoding surface proteins.

T. cruzi has the largest expansion of species-specific genes among the trypanosomatids, which account for approximately 45% of the parasite genome, while in *T. brucei*, *T. rangeli* and *L. major* species-specific genes account for only 25%, 25% and 20%, respectively [52, 54, 59, 60, 139]. Although the mammalian non-pathogenic parasite *T. rangeli* has the same multigene families as *T. cruzi*, the number and proportion of MASP, Mucin, *trans*-sialidase and KMP-11 genes in these two genomes were quite different [54, 61] (Fig. **10**). *Trans*-sialidases catalyzes the transference of sialic acid from host glycoconjugates to β-galactopyranose acceptors in the parasite mucin surface proteins, generating a negatively charged surface that confers protection against alternative complement and antibody opsonization [133, 134, 140 - 143]. This family is also involved in cellular adhesion and invasion of epithelial [144], neural and glial cells [145], and promotes anti-apoptotic signals in infected cells [146, 147]. Mucins are the major *T. cruzi* surface proteins, covering the parasite outer membrane with approximately $2x10^6$ copies [134, 148]. *T. cruzi* ability to invade different cell types has also been associated with the highly polymorphic MASP gene family [135, 149]. The massive reduction of gene number from these three families in the *T. rangeli* genome is consistent with the inability of this parasite to establish a productive infection in mammals. More details about *T. cruzi* virulent factors and immune evasion are described in Chapter 8.

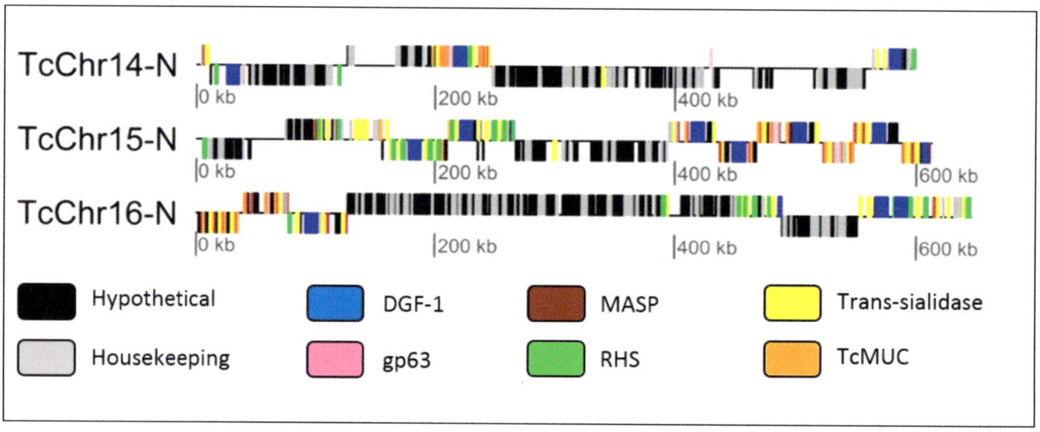

Fig. (9). *T. cruzi* CL Brener non-Esmeraldo-like chromosomal structure. The gene distribution of the Non-Esmeraldo-like chromosomes 14, 15 and 16 is shown. Multicopy gene families are represented as colorful boxes, while hypothetical and housekeeping genes are represented by black and grey boxes, respectively. Box-less regions represent intergenic and gap regions. The genes and the chromosome are drawn in proportion to their size, where genes above the chromosomal line are in the positive coding strand and genes below are in the negative coding strand.

KPM-11, the only multigene family expanded in *T. rangeli* compared to *T. cruzi,* encodes a 92-amino acid antigen present in a wide range of trypanosomatids, which elicits an immunoprotective anti-parasite immune response [150]. While the KPM-11 expansion in *T. rangeli* could provide some advantage in colonizing the insect vector [61], it could be detrimental to the mammalian host infection.

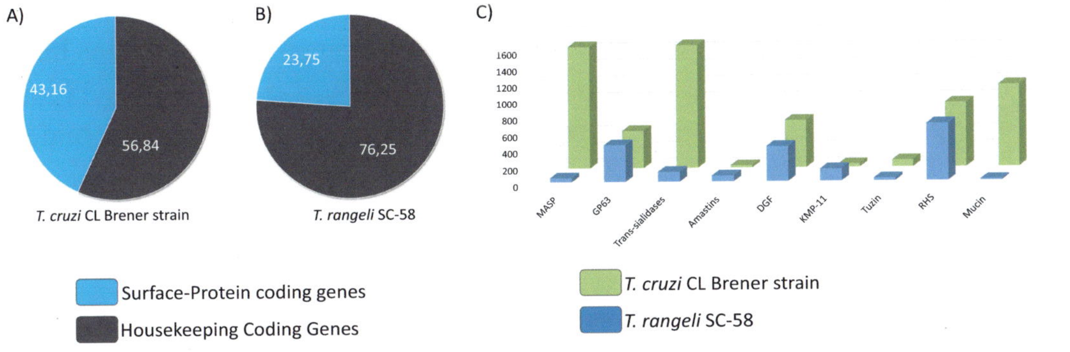

Fig. (10). *T. cruzi* CL Brener and *T. rangeli* SC-58 multigene families. The percentage of surface protein coding genes (cyan) among all the CDSs in the genome of (**A**) *T. cruzi* CL Brener and (**B**) *T. rangeli* SC-58 are shown. (**C**) The number of CDSs belonging to each gene family (MASP, GP63, *Trans*-sialidase, Amastin, DGF-1, KMP-11, Tuzin, RHS and Mucin) in the genome of *T. cruzi* CL Brener strain (green) or *T. rangeli* SC-58 (blue) is shown. The large reduction in gene counts from the families MASP, *Trans*-sialidase and Mucin in *T. rangeli* SC-58 could be related to the inability of this parasite to establish a productive infection in mammalian hosts.

The hybrid nature and large repetitive content of CL Brener genome hampered the genome assembly into chromosomes. In 2009, based on scaffold information, BAC-end sequences and synteny maps with *T. brucei* and *L. major*, the CL Brener contigs were assembled into 41 putative chromosomes [137], in accordance with previous pulse-field gel electrophoresis results [151]. The availability of a sequence representation of both Esmeraldo-like (TcII) and non-Esmeraldo (TcIII) chromosomes allowed a comparative analysis of *T. cruzi* strain ploidy using unassembled next generation sequence reads [83]. This analysis revealed several chromosomal copy number variations (CCNVs) among strains from different *T. cruzi* DTUs. Six *T. cruzi* strains were evaluated: Arequipa, Colombiana and Sylvio from TcI, Esmeraldo and Y from TcII and 231 from TcIII [83]. From the three DTUs evaluated, TcI has the fewest number of chromosomal expansion, which could be a contributing factor to the reduced heterozygosity and smaller DNA content per cell found in this DTU [55, 83, 129, 152]. On the other hand,

Esmeraldo and Y TcII strains present several whole chromosomal duplications, suggesting extensive chromosomal variation within this DTU. Interestingly, the chromosome 31 is the only one that was supernumerary in all six strains evaluated [83]. This chromosome has ~10 copies of the UDP-GlcNAc-dependent glycosyltransferase, an essential enzyme in *T. cruzi* mucins glycosylation [134]. Previous comparative genomic hybridization assays among *T. cruzi* strains have also shown gene CNV in another gene enrolled in *T. cruzi* mucin glycosylation, the beta-galactofuranosyl transferase [130, 153]. As mucins are highly glycosylated proteins that cover the parasite surface with ~$2x10^6$ copies, the expansion of genes related with the glycosylation of these genes could be a driving force for chromosome 31 duplication in *T. cruzi* [83].

CONCLUDING REMARKS

Over the last decade, several advances in the field of trypanosomatid genomics have contributed to a better understanding of evolutionary forces that may have shaped the genome architecture of these parasites. The whole genome sequencing of the model organisms *L. major* [53], *T. brucei* [56] and *T. cruzi* [54] as well as comparative genomic analysis of these three parasites [59], allowed the definition of highly syntenic genomic regions, interspaced by species-specific genes related to their parasitic lifestyle. However, there are some important aspects regarding the genome structure of these parasites that should be addressed in a near future. Although *T. cruzi* CL Brener genome was assembled in 41 putative chromosomes [137], there are still several hundred kilobase-long gaps in Esmeraldo and non-Esmeraldo-like predicted chromosomes, as well as a chromosomal-size content of sequences that was not incorporated into these putative chromosomes [137]. These unresolved genomic regions are mainly composed of highly repetitive sequences, including multigene families involved in key aspects of the *T. cruzi* biology, that could not be correctly assembled into chromosomes with Sanger or next-generation sequencing reads. Third-generation single molecule sequencing is required not only to close these gaps in the CL Brener genome, but also to generate reliable chromosomal sequences to different *T. cruzi* DTUs. More recently, the occurrence of intense CCNV in *Leishmania* [63, 64], as well as in *T. cruzi* [83] has revealed another level of genomic plasticity of these parasites. However, the occurrence and frequency of these events in a broader range of trypanosomatids including *T. brucei* subspecies, *T. rangeli* and *T. marinkellei* are currently unknown. Also, analysis of CCNV in natural populations as well as the functional impact of these events in gene expression could shed light on the role of CCNV in the evolution of genome structure of trypanosomatids and its implications for parasite pathogenicity.

CONFLICT OF INTEREST

The authors confirm that they have no conflict of interest to declare for this publication.

ACKNOWLEDGEMENTS

This work is funded by The National Institute of Science and Technology for Vaccines (Brazil) (MCT/CNPq, grant CNPq 573547/2008-4), Fundação de Amparo à Pesquisa do Estado de Minas Gerais (FAPEMIG, grant # APQ-0407--10, PPM-00219-13) and Coordenação de Aperfeiçoamento de Pessoal de Nivel Superior (CAPES, grant # 051/2013). DCB is a CNPq research fellow. JLRC and HOV received scholarships from CAPES.

ABBREVIATIONS

ApoL1	Apolipoprotein L1
BAC	Bacterial Artificial Chromosome
BES	Bloodstream Expression Site
CCNV	Chromosomal Copy Number Variation
CL	Cutaneous Leishmaniasis
CNV	Copy Number Variation
CP	Cysteine Protease
DCL	Diffuse Cutaneous Leishmaniasis
DGC	Directional Gene Clusters
DGF-1	Dispersed Gene Family -1
DNA	Deoxyribonucleic Acid
DTU	Discrete Typing Units
ES	Expression Site
ESAG	Expression Site Associated Genes
ESB	Expression Site Body

FISH	Fluorescence *in situ* hybridization
GP	Glycoprotein
HIV	Human Immunodeficiency Virus
KMP-11	Kinetoplastid membrane protein 11
MASPs	Mucin Associated Surface Proteins
Mb	Mega bases
MES	Metacyclic Expression Site
MHC	Major Histocompatibility Complex
ML	Mucosal Leishmaniasis
RHS	Retro-transposon hotspot proteins
RNAi	RNA interference
SNP	Single Nucleotide Polymorphism
SRA	Serum Resistance-Associated
TcMUC	*Trypanosoma cruzi* mucin
Th1	Type 1 T helper
TL	Tegumentary leishmaniasis
UTR	Untranslated Region
VL	Visceral Leishmaniasis
VSG	Variant Surface Glycoproteins
WHO	World Health Organization

REFERENCES

[1] Alvar J, Vélez ID, Bern C, *et al.* Leishmaniasis worldwide and global estimates of its incidence. PLoS One 2012; 7(5): e35671.
[http://dx.doi.org/10.1371/journal.pone.0035671] [PMID: 22693548]

[2] Desjeux P. Leishmaniasis: current situation and new perspectives. Comp Immunol Microbiol Infect Dis 2004; 27(5): 305-18.
[http://dx.doi.org/10.1016/j.cimid.2004.03.004] [PMID: 15225981]

[3] Murray HW, Berman JD, Davies CR, Saravia NG. Advances in leishmaniasis. Lancet 2005; 366(9496): 1561-77.
[http://dx.doi.org/10.1016/S0140-6736(05)67629-5] [PMID: 16257344]

[4] Kassi M, Kassi M, Afghan AK, Rehman R, Kasi PM. Marring leishmaniasis: the stigmatization and the impact of cutaneous leishmaniasis in Pakistan and Afghanistan. PLoS Negl Trop Dis 2008; 2(10): e259.
[http://dx.doi.org/10.1371/journal.pntd.0000259] [PMID: 18958168]

[5] Silveira FT, Lainson R, Corbett CE. Clinical and immunopathological spectrum of American cutaneous leishmaniasis with special reference to the disease in Amazonian Brazil: a review. Mem Inst

Oswaldo Cruz 2004; 99(3): 239-51.
[http://dx.doi.org/10.1590/S0074-02762004000300001] [PMID: 15273794]

[6] Guerra JA, Prestes SR, Silveira H, *et al.* Mucosal Leishmaniasis caused by *Leishmania* (*Viannia*) *braziliensis* and *Leishmania* (*Viannia*) *guyanensis* in the Brazilian Amazon. PLoS Negl Trop Dis 2011; 5(3): e980.
[http://dx.doi.org/10.1371/journal.pntd.0000980] [PMID: 21408116]

[7] Silveira FT, Lainson R, Corbett CE. Further observations on clinical, histopathological, and immunological features of borderline disseminated cutaneous leishmaniasis caused by *Leishmania* (*Leishmania*) *amazonensis*. Mem Inst Oswaldo Cruz 2005; 100(5): 525-34.
[http://dx.doi.org/10.1590/S0074-02762005000500013] [PMID: 16184231]

[8] Akuffo H, Maasho K, Blostedt M, Höjeberg B, Britton S, Bakhiet M. *Leishmania* aethiopica derived from diffuse leishmaniasis patients preferentially induce mRNA for interleukin-10 while those from localized leishmaniasis patients induce interferon-gamma. J Infect Dis 1997; 175(3): 737-41.
[http://dx.doi.org/10.1093/infdis/175.3.737] [PMID: 9041358]

[9] Sinha S, Fernández G, Kapila R, Lambert WC, Schwartz RA. Diffuse cutaneous leishmaniasis associated with the immune reconstitution inflammatory syndrome. Int J Dermatol 2008; 47(12): 1263-70.
[http://dx.doi.org/10.1111/j.1365-4632.2008.03804.x] [PMID: 19126013]

[10] Bañuls AL, Hide M, Prugnolle F. *Leishmania* and the leishmaniases: a parasite genetic update and advances in taxonomy, epidemiology and pathogenicity in humans. Adv Parasitol 2007; 64: 1-109.
[http://dx.doi.org/10.1016/S0065-308X(06)64001-3] [PMID: 17499100]

[11] Zijlstra EE, Musa AM, Khalil EA, el-Hassan IM, el-Hassan AM. Post-kala-azar dermal leishmaniasis. Lancet Infect Dis 2003; 3(2): 87-98.
[http://dx.doi.org/10.1016/S1473-3099(03)00517-6] [PMID: 12560194]

[12] Desjeux P. The increase in risk factors for leishmaniasis worldwide. Trans R Soc Trop Med Hyg 2001; 95(3): 239-43.
[http://dx.doi.org/10.1016/S0035-9203(01)90223-8] [PMID: 11490989]

[13] Salmanpour R, Handjani F, Nouhpisheh MK. Comparative study of the efficacy of oral ketoconazole with intra-lesional meglumine antimoniate (Glucantime) for the treatment of cutaneous leishmaniasis. J Dermatolog Treat 2001; 12(3): 159-62.
[http://dx.doi.org/10.1080/09546630152607899] [PMID: 12243707]

[14] Alkhawajah AM, Larbi E, al-Gindan Y, Abahussein A, Jain S. Treatment of cutaneous leishmaniasis with antimony: intramuscular *versus* intralesional administration. Ann Trop Med Parasitol 1997; 91(8): 899-905.
[http://dx.doi.org/10.1080/00034989760284] [PMID: 9579209]

[15] Arana BA, Navin TR, Arana FE, Berman JD, Rosenkaimer F. Efficacy of a short course (10 days) of high-dose meglumine antimonate with or without interferon-gamma in treating cutaneous leishmaniasis in Guatemala. Clin Infect Dis 1994; 18(3): 381-4.
[http://dx.doi.org/10.1093/clinids/18.3.381] [PMID: 8011819]

[16] Murray HW. Progress in the treatment of a neglected infectious disease: visceral leishmaniasis. Expert Rev Anti Infect Ther 2004; 2(2): 279-92.

[http://dx.doi.org/10.1586/14787210.2.2.279] [PMID: 15482193]

[17] Killingley B, Lamb LE, Davidson RN. Miltefosine to treat cutaneous leishmaniasis caused by *Leishmania tropica.* Ann Trop Med Parasitol 2009; 103(2): 171-5. [http://dx.doi.org/10.1179/136485909X398177] [PMID: 19208301]

[18] Soto J, Toledo JT. Oral miltefosine to treat new world cutaneous leishmaniasis. Lancet Infect Dis 2007; 7(1): 7. [http://dx.doi.org/10.1016/S1473-3099(06)70665-X] [PMID: 17182338]

[19] Berman J. Miltefosine to treat leishmaniasis. Expert Opin Pharmacother 2005; 6(8): 1381-8. [http://dx.doi.org/10.1517/14656566.6.8.1381] [PMID: 16013987]

[20] Sundar S, Chakravarty J, Agarwal D, Rai M, Murray HW. Single-dose liposomal amphotericin B for visceral leishmaniasis in India. N Engl J Med 2010; 362(6): 504-12. [http://dx.doi.org/10.1056/NEJMoa0903627] [PMID: 20147716]

[21] Rassi A Jr, Rassi A, Marcondes de Rezende J. American trypanosomiasis (Chagas disease). Infect Dis Clin North Am 2012; 26(2): 275-91. [http://dx.doi.org/10.1016/j.idc.2012.03.002] [PMID: 22632639]

[22] Organization WH. Chagas disease (American trypanosomiasis): Fact Sheet No 340. WHO, Geneva (June 2010). 2010.

[23] Rassi A Jr, Rassi A, Marin-Neto JA. Chagas disease. Lancet 2010; 375(9723): 1388-402. [http://dx.doi.org/10.1016/S0140-6736(10)60061-X] [PMID: 20399979]

[24] Bern C, Montgomery SP. An estimate of the burden of Chagas disease in the United States. Clin Infect Dis 2009; 49(5): e52-4. [http://dx.doi.org/10.1086/605091] [PMID: 19640226]

[25] Brun R, Blum J, Chappuis F, Burri C. Human African trypanosomiasis. Lancet 2010; 375(9709): 148-59. [http://dx.doi.org/10.1016/S0140-6736(09)60829-1] [PMID: 19833383]

[26] Malvy D, Chappuis F. Sleeping sickness. Clinical microbiology and infection: the official publication of the European Society of Clinical Microbiology and Infectious Diseases 2011; 17(7): 986-5. [http://dx.doi.org/10.1111/j.1469-0691.2011.03536.x]

[27] Deborggraeve S, Koffi M, Jamonneau V, *et al.* Molecular analysis of archived blood slides reveals an atypical human *Trypanosoma* infection. Diagn Microbiol Infect Dis 2008; 61(4): 428-33. [http://dx.doi.org/10.1016/j.diagmicrobio.2008.03.006] [PMID: 18455900]

[28] Odiit M, Kansiime F, Enyaru JC. Duration of symptoms and case fatality of sleeping sickness caused by *Trypanosoma brucei rhodesiense* in Tororo, Uganda. East Afr Med J 1997; 74(12): 792-5. [PMID: 9557424]

[29] Checchi F, Filipe JA, Haydon DT, Chandramohan D, Chappuis F. Estimates of the duration of the early and late stage of gambiense sleeping sickness. BMC Infect Dis 2008; 8: 16. [http://dx.doi.org/10.1186/1471-2334-8-16] [PMID: 18261232]

[30] Kennedy PG. Human African trypanosomiasis-neurological aspects. J Neurol 2006; 253(4): 411-6. [http://dx.doi.org/10.1007/s00415-006-0093-3] [PMID: 16541214]

[31] Burri C. Chemotherapy against human African trypanosomiasis: is there a road to success? Parasitology 2010; 137(14): 1987-94.
[http://dx.doi.org/10.1017/S0031182010001137] [PMID: 20961469]

[32] Kato H, Gomez EA, Cáceres AG, Uezato H, Mimori T, Hashiguchi Y. Molecular epidemiology for vector research on leishmaniasis. Int J Environ Res Public Health 2010; 7(3): 814-26.
[http://dx.doi.org/10.3390/ijerph7030814] [PMID: 20617005]

[33] Killick-Kendrick R. Phlebotomine vectors of the leishmaniases: a review. Med Vet Entomol 1990; 4(1): 1-24.
[http://dx.doi.org/10.1111/j.1365-2915.1990.tb00255.x] [PMID: 2132963]

[34] Brener Z. Life cycle of *Trypanosoma cruzi*. Rev Inst Med Trop Sao Paulo 1971; 13(3): 171-8.
[PMID: 5563126]

[35] Tyler KM, Engman DM. The life cycle of *Trypanosoma cruzi* revisited. Int J Parasitol 2001; 31(5-6): 472-81.
[http://dx.doi.org/10.1016/S0020-7519(01)00153-9] [PMID: 11334932]

[36] Maeda FY, Cortez C, Yoshida N. Cell signaling during *Trypanosoma cruzi* invasion. Front Immunol 2012; 3: 361.
[http://dx.doi.org/10.3389/fimmu.2012.00361] [PMID: 23230440]

[37] Vickerman K. Developmental cycles and biology of pathogenic trypanosomes. Br Med Bull 1985; 41(2): 105-14.
[PMID: 3928017]

[38] Lainson R, Shaw JJ, Peters W, Killick-Kendrick R. Evolution, classification and geographical distribution. Academic Press 1987.

[39] Lainson R, Shaw J. The role of animals in the epidemiology of South American leishmaniasis 1979.

[40] Taylor VM, Muñoz DL, Cedeño DL, Vélez ID, Jones MA, Robledo SM. *Leishmania tarentolae*: utility as an *in vitro* model for screening of antileishmanial agents. Exp Parasitol 2010; 126(4): 471-5.
[http://dx.doi.org/10.1016/j.exppara.2010.05.016] [PMID: 20685203]

[41] Balmer O, Beadell JS, Gibson W, Caccone A. Phylogeography and taxonomy of *Trypanosoma brucei*. PLoS Negl Trop Dis 2011; 5(2): e961.
[http://dx.doi.org/10.1371/journal.pntd.0000961] [PMID: 21347445]

[42] Lewis MD, Ma J, Yeo M, Carrasco HJ, Llewellyn MS, Miles MA. Genotyping of *Trypanosoma cruzi*: systematic selection of assays allowing rapid and accurate discrimination of all known lineages. Am J Trop Med Hyg 2009; 81(6): 1041-9.
[http://dx.doi.org/10.4269/ajtmh.2009.09-0305] [PMID: 19996435]

[43] Zingales B, Andrade SG, Briones MR, *et al.* A new consensus for *Trypanosoma cruzi* intraspecific nomenclature: second revision meeting recommends TcI to TcVI. Mem Inst Oswaldo Cruz 2009; 104(7): 1051-4.
[http://dx.doi.org/10.1590/S0074-02762009000700021] [PMID: 20027478]

[44] Zingales B, Miles MA, Campbell DA, *et al.* The revised *Trypanosoma cruzi* subspecific nomenclature: rationale, epidemiological relevance and research applications. Infection, genetics and evolution. J Molecul Epidemiol Evol Genet Infect Dise 2012; 12(2): 240-53.

[http://dx.doi.org/10.1016/j.meegid.2011.12.009]

[45] Tibayrenc M, Kjellberg F, Ayala FJ. A clonal theory of parasitic protozoa: the population structures of *Entamoeba, Giardia, Leishmania, Naegleria, Plasmodium, Trichomonas*, and *Trypanosoma* and their medical and taxonomical consequences. Proc Natl Acad Sci USA 1990; 87(7): 2414-8. [http://dx.doi.org/10.1073/pnas.87.7.2414] [PMID: 2320563]

[46] de Freitas JM, Augusto-Pinto L, Pimenta JR, *et al.* Ancestral genomes, sex, and the population structure of *Trypanosoma cruzi*. PLoS Pathog 2006; 2(3): e24. [http://dx.doi.org/10.1371/journal.ppat.0020024] [PMID: 16609729]

[47] Brisse S, Henriksson J, Barnabe C, *et al.* Evidence for genetic exchange and hybridization in *Trypanosoma cruzi* based on nucleotide sequences and molecular karyotype. Infection, genetics and evolution : journal of molecular epidemiology and evolutionary genetics in infectious diseases 2003; 2(3): 173-83. [http://dx.doi.org/10.1016/S1567-1348(02)00097-7]

[48] Machado CA, Ayala FJ. Nucleotide sequences provide evidence of genetic exchange among distantly related lineages of *Trypanosoma cruzi*. Proc Natl Acad Sci USA 2001; 98(13): 7396-401. [http://dx.doi.org/10.1073/pnas.121187198] [PMID: 11416213]

[49] Ferreira KA, Fajardo EF, Baptista RP, *et al.* Species-specific markers for the differential diagnosis of *Trypanosoma cruzi* and *Trypanosoma rangeli* and polymorphisms detection in *Trypanosoma rangeli*. Parasitol Res 2014; 113(6): 2199-207. [http://dx.doi.org/10.1007/s00436-014-3872-2] [PMID: 24728520]

[50] Westenberger SJ, Barnabé C, Campbell DA, Sturm NR. Two hybridization events define the population structure of *Trypanosoma cruzi*. Genetics 2005; 171(2): 527-43. [http://dx.doi.org/10.1534/genetics.104.038745] [PMID: 15998728]

[51] Burgos JM, Risso MG, Brenière SF, Barnabé C, Campetella O, Leguizamón MS. Differential distribution of genes encoding the virulence factor *trans*-sialidase along *Trypanosoma cruzi* Discrete typing units. PLoS One 2013; 8(3): e58967. [http://dx.doi.org/10.1371/journal.pone.0058967] [PMID: 23536842]

[52] Peacock CS, Seeger K, Harris D, *et al.* Comparative genomic analysis of three *Leishmania* species that cause diverse human disease. Nat Genet 2007; 39(7): 839-47. [http://dx.doi.org/10.1038/ng2053] [PMID: 17572675]

[53] Ivens AC, Peacock CS, Worthey EA, *et al.* The genome of the kinetoplastid parasite, *Leishmania major*. Science 2005; 309(5733): 436-42. [http://dx.doi.org/10.1126/science.1112680] [PMID: 16020728]

[54] El-Sayed NM, Myler PJ, Bartholomeu DC, *et al.* The genome sequence of *Trypanosoma cruzi*, etiologic agent of Chagas disease. Science 2005; 309(5733): 409-15. [http://dx.doi.org/10.1126/science.1112631] [PMID: 16020725]

[55] Franzén O, Ochaya S, Sherwood E, *et al.* Shotgun sequencing analysis of *Trypanosoma cruzi* I Sylvio X10/1 and comparison with *T. cruzi* VI CL Brener. PLoS Negl Trop Dis 2011; 5(3): e984. [http://dx.doi.org/10.1371/journal.pntd.0000984] [PMID: 21408126]

[56] Berriman M, Ghedin E, Hertz-Fowler C, *et al.* The genome of the African trypanosome *Trypanosoma*

brucei. Science 2005; 309(5733): 416-22.
[http://dx.doi.org/10.1126/science.1112642] [PMID: 16020726]

[57] Myler PJ, Audleman L, deVos T, *et al. Leishmania* major Friedlin chromosome 1 has an unusual distribution of protein-coding genes. Proc Natl Acad Sci USA 1999; 96(6): 2902-6.
[http://dx.doi.org/10.1073/pnas.96.6.2902] [PMID: 10077609]

[58] Haile S, Papadopoulou B. Developmental regulation of gene expression in trypanosomatid parasitic protozoa. Curr Opin Microbiol 2007; 10(6): 569-77.
[http://dx.doi.org/10.1016/j.mib.2007.10.001] [PMID: 18177626]

[59] El-Sayed NM, Myler PJ, Blandin G, *et al.* Comparative genomics of trypanosomatid parasitic protozoa. Science 2005; 309(5733): 404-9.
[http://dx.doi.org/10.1126/science.1112181] [PMID: 16020724]

[60] Berriman M, Ghedin E, Hertz-Fowler C, *et al.* The genome of the African trypanosome *Trypanosoma brucei*. Science 2005; 309(5733): 416-22.
[http://dx.doi.org/10.1126/science.1112642] [PMID: 16020726]

[61] Stoco PH, Wagner G, Talavera-Lopez C, *et al.* Genome of the avirulent human-infective trypanosome *Trypanosoma rangeli*. PLoS Negl Trop Dis 2014; 8(9): e3176.
[http://dx.doi.org/10.1371/journal.pntd.0003176] [PMID: 25233456]

[62] Sistrom M, Evans B, Bjornson R, *et al.* Comparative genomics reveals multiple genetic backgrounds of human pathogenicity in the *Trypanosoma brucei* complex. Genome Biol Evol 2014; 6(10): 2811-9.
[http://dx.doi.org/10.1093/gbe/evu222] [PMID: 25287146]

[63] Rogers MB, Hilley JD, Dickens NJ, *et al.* Chromosome and gene copy number variation allow major structural change between species and strains of *Leishmania*. Genome Res 2011; 21(12): 2129-42.
[http://dx.doi.org/10.1101/gr.122945.111] [PMID: 22038252]

[64] Valdivia HO, Reis-Cunha JL, Rodrigues-Luiz GF, *et al.* Comparative genomic analysis of *Leishmania* (*Viannia*) peruviana and *Leishmania* (*Viannia*) *braziliensis*. BMC Genomics 2015; 16: 715.
[http://dx.doi.org/10.1186/s12864-015-1928-z] [PMID: 26384787]

[65] Stevens JR, Noyes HA, Dover GA, Gibson WC. The ancient and divergent origins of the human pathogenic trypanosomes, *Trypanosoma brucei* and *T. cruzi*. Parasitology 1999; 118(Pt 1): 107-16.
[http://dx.doi.org/10.1017/S0031182098003473] [PMID: 10070668]

[66] Gibson W, Stevens J. Genetic exchange in the trypanosomatidae. Adv Parasitol 1999; 43: 1-46.
[http://dx.doi.org/10.1016/S0065-308X(08)60240-7] [PMID: 10214689]

[67] Franzén O, Talavera-López C, Ochaya S, *et al.* Comparative genomic analysis of human infective *Trypanosoma cruzi* lineages with the bat-restricted subspecies *T. cruzi* marinkellei. BMC Genomics 2012; 13: 531.
[http://dx.doi.org/10.1186/1471-2164-13-531] [PMID: 23035642]

[68] Teixeira SM, de Paiva RM, Kangussu-Marcolino MM, Darocha WD. Trypanosomatid comparative genomics: Contributions to the study of parasite biology and different parasitic diseases. Genet Mol Biol 2012; 35(1): 1-17.
[http://dx.doi.org/10.1590/S1415-47572012005000008] [PMID: 22481868]

[69] Bartholomeu DC, de Paiva RM, Mendes TA, DaRocha WD, Teixeira SM. Unveiling the intracellular

survival gene kit of trypanosomatid parasites. PLoS Pathog 2014; 10(12): e1004399. [http://dx.doi.org/10.1371/journal.ppat.1004399] [PMID: 25474314]

[70] Grisard EC, Teixeira SM, de Almeida LG, *et al.* *Trypanosoma cruzi* clone Dm28c draft genome sequence. Genome Announc 2014; 2(1): e01114-13. [http://dx.doi.org/10.1128/genomeA.01114-13] [PMID: 24482508]

[71] Franzén O, Talavera-López C, Ochaya S, *et al.* Comparative genomic analysis of human infective *Trypanosoma cruzi* lineages with the bat-restricted subspecies *T. cruzi* marinkellei. BMC Genomics 2012; 13: 531. [http://dx.doi.org/10.1186/1471-2164-13-531] [PMID: 23035642]

[72] Lye LF, Owens K, Shi H, *et al.* Retention and loss of RNA interference pathways in trypanosomatid protozoans. PLoS Pathog 2010; 6(10): e1001161. [http://dx.doi.org/10.1371/journal.ppat.1001161] [PMID: 21060810]

[73] Inoue N, Otsu K, Ferraro DM, Donelson JE. Tetracycline-regulated RNA interference in *Trypanosoma congolense.* Mol Biochem Parasitol 2002; 120(2): 309-13. [http://dx.doi.org/10.1016/S0166-6851(02)00015-4] [PMID: 11897138]

[74] DaRocha WD, Otsu K, Teixeira SM, Donelson JE. Tests of cytoplasmic RNA interference (RNAi) and construction of a tetracycline-inducible T7 promoter system in *Trypanosoma cruzi.* Mol Biochem Parasitol 2004; 133(2): 175-86. [http://dx.doi.org/10.1016/j.molbiopara.2003.10.005] [PMID: 14698430]

[75] Burleigh BA, Andrews NW. The mechanisms of *Trypanosoma cruzi* invasion of mammalian cells. Annu Rev Microbiol 1995; 49: 175-200. [http://dx.doi.org/10.1146/annurev.mi.49.100195.001135] [PMID: 8561458]

[76] Epting CL, Coates BM, Engman DM. Molecular mechanisms of host cell invasion by *Trypanosoma cruzi.* Exp Parasitol 2010; 126(3): 283-91. [http://dx.doi.org/10.1016/j.exppara.2010.06.023] [PMID: 20599990]

[77] Fernandes MC, Andrews NW. Host cell invasion by *Trypanosoma cruzi*: a unique strategy that promotes persistence. FEMS Microbiol Rev 2012; 36(3): 734-47. [http://dx.doi.org/10.1111/j.1574-6976.2012.00333.x] [PMID: 22339763]

[78] Flores-Lopez CA, Machado CA. Differences in inferred genome-wide signals of positive selection during the evolution of *Trypanosoma cruzi* and *Leishmania* spp. lineages: A result of disparities in host and tissue infection ranges? Infection, genetics and evolution: journal of molecular epidemiology and evolutionary genetics in infectious diseases 2015; 33: 37-46.

[79] Stranger BE, Forrest MS, Dunning M, *et al.* Relative impact of nucleotide and copy number variation on gene expression phenotypes. Science 2007; 315(5813): 848-53. [http://dx.doi.org/10.1126/science.1136678] [PMID: 17289997]

[80] Iskow RC, Gokcumen O, Lee C. Exploring the role of copy number variants in human adaptation. Trends Genet 2012; 28(6): 245-57. [http://dx.doi.org/10.1016/j.tig.2012.03.002] [PMID: 22483647]

[81] Martinez-Calvillo S, Vizuet-de-Rueda JC, Florencio-Martinez LE, Manning-Cela RG, Figueroa-Angulo EE. Gene expression in trypanosomatid parasites. Journal of biomedicine and biotechnology

2010; 2010: 525241.
[http://dx.doi.org/10.1155/2010/525241]

[82] Minning TA, Weatherly DB, Flibotte S, Tarleton RL. Widespread, focal copy number variations (CNV) and whole chromosome aneuploidies in *Trypanosoma cruzi* strains revealed by array comparative genomic hybridization. BMC Genomics 2011; 12: 139.
[http://dx.doi.org/10.1186/1471-2164-12-139] [PMID: 21385342]

[83] Reis-Cunha JL, Mendes TA, dc Almcida Lourdes R, *et al.* Genome-wide screening and identification of new *Trypanosoma cruzi* antigens with potential application for chronic Chagas disease diagnosis. PLoS One 2014; 9(9): e106304.
[http://dx.doi.org/10.1371/journal.pone.0106304] [PMID: 25225853]

[84] Abbey D, Hickman M, Gresham D, Berman J. High-Resolution SNP/CGH microarrays reveal the accumulation of loss of heterozygosity in commonly used *Candida albicans* strains. G3 (Bethesda) 2011; 1(7): 523-30.
[http://dx.doi.org/10.1534/g3.111.000885] [PMID: 22384363]

[85] Sheltzer JM, Blank HM, Pfau SJ, *et al.* Aneuploidy drives genomic instability in yeast. Science 2011; 333(6045): 1026-30.
[http://dx.doi.org/10.1126/science.1206412] [PMID: 21852501]

[86] Farrer RA, Henk DA, Garner TW, Balloux F, Woodhams DC, Fisher MC. Chromosomal copy number variation, selection and uneven rates of recombination reveal cryptic genome diversity linked to pathogenicity. PLoS Genet 2013; 9(8): e1003703.
[http://dx.doi.org/10.1371/journal.pgen.1003703] [PMID: 23966879]

[87] Real F, Vidal RO, Carazzolle MF, *et al.* The genome sequence of *Leishmania* (Leishmania) amazonensis: functional annotation and extended analysis of gene models. DNA Res 2013; 20(6): 567-81.

[88] Llanes A, Restrepo CM, Del Vecchio G, Anguizola FJ, Lleonart R. The genome of *Leishmania panamensis*: insights into genomics of the *L. (Viannia)* subgenus. Sci Rep 2015; 5: 8550.
[http://dx.doi.org/10.1038/srep08550] [PMID: 25707621]

[89] Aslett M, Aurrecoechea C, Berriman M, *et al.* TriTrypDB: a functional genomic resource for the Trypanosomatidae. Nucl Acids Res 2010; 38: D457-62.
[http://dx.doi.org/10.1093/nar/gkp851] [PMID: 19843604]

[90] Wincker P, Ravel C, Blaineau C, *et al.* The *Leishmania* genome comprises 36 chromosomes conserved across widely divergent human pathogenic species. Nucleic Acids Rcs 1996; 24(9): 1688-94.
[http://dx.doi.org/10.1093/nar/24.9.1688] [PMID: 8649987]

[91] Britto C, Ravel C, Bastien P, *et al.* Conserved linkage groups associated with large-scale chromosomal rearrangements between Old World and New World *Leishmania* genomes. Gene 1998; 222(1): 107-17.
[http://dx.doi.org/10.1016/S0378-1119(98)00472-7] [PMID: 9813266]

[92] Sterkers Y, Lachaud L, Bourgeois N, Crobu L, Bastien P, Pagès M. Novel insights into genome plasticity in Eukaryotes: mosaic aneuploidy in *Leishmania*. Mol Microbiol 2012; 86(1): 15-23.
[http://dx.doi.org/10.1111/j.1365-2958.2012.08185.x] [PMID: 22857263]

[93] Sterkers Y, Crobu L, Lachaud L, Pagès M, Bastien P. Parasexuality and mosaic aneuploidy in *Leishmania*: alternative genetics. Trends Parasitol 2014; 30(9): 429-35. [http://dx.doi.org/10.1016/j.pt.2014.07.002] [PMID: 25073852]

[94] Sterkers Y, Lachaud L, Crobu L, Bastien P, Pagès M. FISH analysis reveals aneuploidy and continual generation of chromosomal mosaicism in *Leishmania major*. Cell Microbiol 2011; 13(2): 274-83. [http://dx.doi.org/10.1111/j.1462-5822.2010.01534.x] [PMID: 20964798]

[95] Valdivia HO, Scholte LL, Oliveira G, Gabaldón T, Bartholomeu DC. The *Leishmania metaphylome*: a comprehensive survey of *Leishmania* protein phylogenetic relationships. BMC Genomics 2015; 16(1): 887. [http://dx.doi.org/10.1186/s12864-015-2091-2] [PMID: 26518129]

[96] Jackson AP. The evolution of amastin surface glycoproteins in trypanosomatid parasites. Mol Biol Evol 2010; 27(1): 33-45. [http://dx.doi.org/10.1093/molbev/msp214] [PMID: 19748930]

[97] Jecna L, Dostalova A, Wilson R, *et al.* The role of surface glycoconjugates in *Leishmania* midgut attachment examined by competitive binding assays and experimental development in sand flies. Parasitology 2013; 140(8): 1026-32. [http://dx.doi.org/10.1017/S0031182013000358] [PMID: 23611086]

[98] Brittingham A, Morrison CJ, McMaster WR, McGwire BS, Chang KP, Mosser DM. Role of the *Leishmania* surface protease gp63 in complement fixation, cell adhesion, and resistance to complement-mediated lysis. J Immunol 1995; 155(6): 3102-11. [PMID: 7673725]

[99] Gomez MA, Contreras I, Hallé M, Tremblay ML, McMaster RW, Olivier M. *Leishmania* GP63 alters host signaling through cleavage-activated protein tyrosine phosphatases. Sci Signal 2009; 2(90): ra58. [http://dx.doi.org/10.1126/scisignal.2000213] [PMID: 19797268]

[100] Corradin S, Ransijn A, Corradin G, *et al.* MARCKS-related protein (MRP) is a substrate for the *Leishmania* major surface protease leishmanolysin (gp63). J Biol Chem 1999; 274(36): 25411-8. [http://dx.doi.org/10.1074/jbc.274.36.25411] [PMID: 10464270]

[101] Judice WA, Manfredi MA, Souza GP, *et al.* Heparin modulates the endopeptidase activity of *Leishmania mexicana* cysteine protease cathepsin L-Like rCPB2.8. PLoS One 2013; 8(11): e80153. [http://dx.doi.org/10.1371/journal.pone.0080153] [PMID: 24278253]

[102] Buxbaum LU, Denise H, Coombs GH, Alexander J, Mottram JC, Scott P. Cysteine protease B of *Leishmania mexicana* inhibits host Th1 responses and protective immunity. J Immunol 2003; 171(7): 3711-7. [http://dx.doi.org/10.4049/jimmunol.171.7.3711] [PMID: 14500670]

[103] Matheoud D, Moradin N, Bellemare-Pelletier A, *et al. Leishmania* evades host immunity by inhibiting antigen cross-presentation through direct cleavage of the SNARE VAMP8. Cell Host Microbe 2013; 14(1): 15-25. [http://dx.doi.org/10.1016/j.chom.2013.06.003] [PMID: 23870310]

[104] Van Nieuwenhove S. Gambiense sleeping sickness: re-emerging and soon untreatable? Bull World Health Organ 2000; 78(11): 1283.

[PMID: 11143187]

[105] Coura JR, Dias JC. Epidemiology, control and surveillance of Chagas disease: 100 years after its discovery. Mem Inst Oswaldo Cruz 2009; 104 (Suppl. 1): 31-40. [http://dx.doi.org/10.1590/S0074-02762009000900006] [PMID: 19753455]

[106] Jackson AP, Sanders M, Berry A, *et al.* The genome sequence of *Trypanosoma brucei* gambiense, causative agent of chronic human african trypanosomiasis. PLoS Negl Trop Dis 2010; 4(4): e658. [http://dx.doi.org/10.1371/journal.pntd.0000658] [PMID: 20404998]

[107] Van der Ploeg LH, Valerio D, De Lange T, Bernards A, Borst P, Grosveld FG. An analysis of cosmid clones of nuclear DNA from *Trypanosoma brucei* shows that the genes for variant surface glycoproteins are clustered in the genome. Nucleic Acids Res 1982; 10(19): 5905-23. [http://dx.doi.org/10.1093/nar/10.19.5905] [PMID: 6292859]

[108] Turner CM, Barry JD. High frequency of antigenic variation in *Trypanosoma brucei* rhodesiense infections. Parasitology 1989; 99(Pt 1): 67-75. [http://dx.doi.org/10.1017/S0031182000061035] [PMID: 2797873]

[109] Borst P. Antigenic variation and allelic exclusion. Cell 2002; 109(1): 5-8. [http://dx.doi.org/10.1016/S0092-8674(02)00711-0] [PMID: 11955440]

[110] Pays E, Vanhamme L, Pérez-Morga D. Antigenic variation in Trypanosoma brucei: facts, challenges and mysteries. Curr Opin Microbiol 2004; 7(4): 369-74. [http://dx.doi.org/10.1016/j.mib.2004.05.001] [PMID: 15288623]

[111] Barry JD, McCulloch R. Antigenic variation in trypanosomes: enhanced phenotypic variation in a eukaryotic parasite. Adv Parasitol 2001; 49: 1-70. [http://dx.doi.org/10.1016/S0065-308X(01)49037-3] [PMID: 11461029]

[112] McCulloch R, Horn D. What has DNA sequencing revealed about the VSG expression sites of African trypanosomes? Trends Parasitol 2009; 25(8): 359-63. [http://dx.doi.org/10.1016/j.pt.2009.05.007] [PMID: 19632154]

[113] McCulloch R. Antigenic variation in African trypanosomes: monitoring progress. Trends Parasitol 2004; 20(3): 117-21. [http://dx.doi.org/10.1016/j.pt.2003.12.004] [PMID: 15036032]

[114] Günzl A, Bruderer T, Laufer G, *et al.* RNA polymerase I transcribes procyclin genes and variant surface glycoprotein gene expression sites in *Trypanosoma brucei.* Eukaryot Cell 2003; 2(3): 542-51. [http://dx.doi.org/10.1128/EC.2.3.542-551.2003] [PMID: 12796299]

[115] Figueiredo LM, Janzen CJ, Cross GA. A histone methyltransferase modulates antigenic variation in African trypanosomes. PLoS Biol 2008; 6(7): e161. [http://dx.doi.org/10.1371/journal.pbio.0060161] [PMID: 18597556]

[116] Navarro M, Gull K. A pol I transcriptional body associated with VSG mono-allelic expression in *Trypanosoma brucei.* Nature 2001; 414(6865): 759-63. [http://dx.doi.org/10.1038/414759a] [PMID: 11742402]

[117] Turner CM. The rate of antigenic variation in fly-transmitted and syringe-passaged infections of *Trypanosoma brucei.* FEMS Microbiol Lett 1997; 153(1): 227-31. [http://dx.doi.org/10.1111/j.1574-6968.1997.tb10486.x] [PMID: 9252591]

[118] Turner CM. Antigenic variation in *Trypanosoma brucei* infections: an holistic view. J Cell Sci 1999; 112(Pt 19): 3187-92.
[PMID: 10504324]

[119] Roth C, Bringaud F, Layden RE, Baltz T, Eisen H. Active late-appearing variable surface antigen genes in *Trypanosoma equiperdum* are constructed entirely from pseudogenes. Proc Natl Acad Sci USA 1989; 86(23): 9375-9.
[http://dx.doi.org/10.1073/pnas.86.23.9375] [PMID: 2574459]

[120] Marcello L, Barry JD. Analysis of the VSG gene silent archive in *Trypanosoma brucei* reveals that mosaic gene expression is prominent in antigenic variation and is favored by archive substructure. Genome Res 2007; 17(9): 1344-52.
[http://dx.doi.org/10.1101/gr.6421207] [PMID: 17652423]

[121] Dero B, Zampetti-Bosseler F, Pays E, Steinert M. The genome and the antigen gene repertoire of *Trypanosoma brucei* gambiense are smaller than those of T. b. brucei. Mol Biochem Parasitol 1987; 26(3): 247-56.
[http://dx.doi.org/10.1016/0166-6851(87)90077-6] [PMID: 3431572]

[122] Kanmogne GD, Bailey M, Gibson WC. Wide variation in DNA content among isolates of *Trypanosoma brucei* ssp. Acta Trop 1997; 63(2-3): 75-87.
[http://dx.doi.org/10.1016/S0001-706X(96)00600-6] [PMID: 9088421]

[123] Gibson WC. Will the real *Trypanosoma b. gambiense* please stand up. Parasitol Today (Regul Ed) 1986; 2(9): 255-7.
[http://dx.doi.org/10.1016/0169-4758(86)90011-6] [PMID: 15462856]

[124] Pays E, Dekerck P, Van Assel S, *et al.* Comparative analysis of a *Trypanosoma brucei gambiense* antigen gene family and its potential use in epidemiology of sleeping sickness. Mol Biochem Parasitol 1983; 7(1): 63-74.
[http://dx.doi.org/10.1016/0166-6851(83)90117-2] [PMID: 6843593]

[125] De Greef C, Hamers R. The serum resistance-associated (SRA) gene of *Trypanosoma brucei* rhodesiense encodes a variant surface glycoprotein-like protein. Mol Biochem Parasitol 1994; 68(2): 277-84.
[http://dx.doi.org/10.1016/0166-6851(94)90172-4] [PMID: 7739673]

[126] Pérez-Morga D, Vanhollebeke B, Paturiaux-Hanocq F, *et al.* Apolipoprotein L-I promotes trypanosome lysis by forming pores in lysosomal membranes. Science 2005; 309(5733): 469-72.
[http://dx.doi.org/10.1126/science.1114566] [PMID: 16020735]

[127] Pays E, Vanhollebeke B. Mutual self-defence: the trypanolytic factor story. Microbes and infection / Institut Pasteur 2008; 10(9): 985-.
[http://dx.doi.org/10.1016/j.micinf.2008.07.020]

[128] Vargas N, Pedroso A, Zingales B. Chromosomal polymorphism, gene synteny and genome size in *T. cruzi* I and *T. cruzi* II groups. Mol Biochem Parasitol 2004; 138(1): 131-41.
[http://dx.doi.org/10.1016/j.molbiopara.2004.08.005] [PMID: 15500924]

[129] Lewis MD, Llewellyn MS, Gaunt MW, Yeo M, Carrasco HJ, Miles MA. Flow cytometric analysis and microsatellite genotyping reveal extensive DNA content variation in *Trypanosoma cruzi* populations

and expose contrasts between natural and experimental hybrids. Int J Parasitol 2009; 39(12): 1305-17. [http://dx.doi.org/10.1016/j.ijpara.2009.04.001] [PMID: 19393242]

[130] Ackermann AA, Panunzi LG, Cosentino RO, Sánchez DO, Agüero F. A genomic scale map of genetic diversity in *Trypanosoma cruzi*. BMC Genomics 2012; 13: 736. [http://dx.doi.org/10.1186/1471-2164-13-736] [PMID: 23270511]

[131] Panunzi LG, Agüero F. A genome-wide analysis of genetic diversity in *Trypanosoma cruzi* intergenic regions. PLoS Negl Trop Dis 2014; 8(5): e2839. [http://dx.doi.org/10.1371/journal.pntd.0002839] [PMID: 24784238]

[132] Burgos JM, Risso MG, Brenière SF, Barnabé C, Campetella O, Leguizamón MS. Differential distribution of genes encoding the virulence factor *trans*-sialidase along *Trypanosoma cruzi* Discrete typing units. PLoS One 2013; 8(3): e58967. [http://dx.doi.org/10.1371/journal.pone.0058967] [PMID: 23536842]

[133] Schenkman S, Eichinger D, Pereira ME, Nussenzweig V. Structural and functional properties of Trypanosoma *trans*-sialidase. Annu Rev Microbiol 1994; 48: 499-523. [http://dx.doi.org/10.1146/annurev.mi.48.100194.002435] [PMID: 7826016]

[134] Buscaglia CA, Campo VA, Frasch AC, Di Noia JM. *Trypanosoma cruzi* surface mucins: host-dependent coat diversity. Nat Rev Microbiol 2006; 4(3): 229-36. [http://dx.doi.org/10.1038/nrmicro1351] [PMID: 16489349]

[135] Bartholomeu DC, Cerqueira GC, Leão AC, *et al.* Genomic organization and expression profile of the mucin-associated surface protein (masp) family of the human pathogen *Trypanosoma cruzi*. Nucleic Acids Res 2009; 37(10): 3407-17. [http://dx.doi.org/10.1093/nar/gkp172] [PMID: 19336417]

[136] De Pablos LM, Osuna A. Multigene families in *Trypanosoma cruzi* and their role in infectivity. Infect Immun 2012; 80(7): 2258-64. [http://dx.doi.org/10.1128/IAI.06225-11] [PMID: 22431647]

[137] Weatherly DB, Boehlke C, Tarleton RL. Chromosome level assembly of the hybrid *Trypanosoma cruzi* genome. BMC Genomics 2009; 10: 255. [http://dx.doi.org/10.1186/1471-2164-10-255] [PMID: 19486522]

[138] De Pablos LM, Osuna A. Multigene families in *Trypanosoma cruzi* and their role in infectivity. Infect Immun 2012; 80(7): 2258-64. [http://dx.doi.org/10.1128/IAI.06225-11] [PMID: 22431647]

[139] Acosta-Serrano A, Almeida IC, Freitas-Junior LH, Yoshida N, Schenkman S. The mucin-like glycoprotein super-family of *Trypanosoma cruzi*: structure and biological roles. Mol Biochem Parasitol 2001; 114(2): 143-50. [http://dx.doi.org/10.1016/S0166-6851(01)00245-6] [PMID: 11378194]

[140] Tomlinson S, Pontes de Carvalho LC, Vandekerckhove F, Nussenzweig V. Role of sialic acid in the resistance of *Trypanosoma cruzi* trypomastigotes to complement. J Immunol 1994; 153(7): 3141-7. [PMID: 8089492]

[141] Pereira-Chioccola VL, Acosta-Serrano A, Correia de Almeida I, *et al.* Mucin-like molecules form a negatively charged coat that protects *Trypanosoma cruzi* trypomastigotes from killing by human anti-

alpha-galactosyl antibodies. J Cell Sci 2000; 113(Pt 7): 1299-307.
[PMID: 10704380]

[142] Freire-de-Lima CG, Nunes MP, Corte-Real S, *et al.* Proapoptotic activity of a *Trypanosoma cruzi* ceramide-containing glycolipid turned on in host macrophages by IFN-gamma. J Immunol 1998; 161(9): 4909-16.
[PMID: 9794425]

[143] Freire-de-Lima L, Oliveira IA, Neves JL, *et al.* Sialic acid: a sweet swing between mammalian host and *Trypanosoma cruzi*. Front Immunol 2012; 3: 356.
[http://dx.doi.org/10.3389/fimmu.2012.00356] [PMID: 23230438]

[144] Butler CE, de Carvalho TM, Grisard EC, Field RA, Tyler KM. Trans-sialidase stimulates eat me response from epithelial cells. Traffic 2013; 14(7): 853-69.
[http://dx.doi.org/10.1111/tra.12078] [PMID: 23601193]

[145] de Melo-Jorge M, PereiraPerrin M. The Chagas disease parasite *Trypanosoma cruzi* exploits nerve growth factor receptor TrkA to infect mammalian hosts. Cell Host Microbe 2007; 1(4): 251-61.
[http://dx.doi.org/10.1016/j.chom.2007.05.006] [PMID: 18005706]

[146] Chuenkova MV, PereiraPerrin M. *Trypanosoma cruzi* targets Akt in host cells as an intracellular antiapoptotic strategy. Sci Signal 2009; 2(97): ra74.
[http://dx.doi.org/10.1126/scisignal.2000374] [PMID: 19920252]

[147] Chuenkova MV, Pereiraperrin M. *Trypanosoma cruzi*-Derived Neurotrophic Factor: Role in Neural Repair and Neuroprotection. J Neuroparasitology 2010; 1: 55-60.
[http://dx.doi.org/10.4303/jnp/N100507] [PMID: 21572925]

[148] Serrano AA, Schenkman S, Yoshida N, Mehlert A, Richardson JM, Ferguson MA. The lipid structure of the glycosylphosphatidylinositol-anchored mucin-like sialic acid acceptors of *Trypanosoma cruzi* changes during parasite differentiation from epimastigotes to infective metacyclic trypomastigote forms. J Biol Chem 1995; 270(45): 27244-53.
[http://dx.doi.org/10.1074/jbc.270.45.27244] [PMID: 7592983]

[149] dos Santos SL, Freitas LM, Lobo FP, Rodrigues-Luiz GF, Mendes TA, Oliveira AC, *et al.* The MASP family of *Trypanosoma cruzi*: changes in gene expression and antigenic profile during the acute phase of experimental infection. PLoS neglected tropical diseases 2012; 6(8): e1779.

[150] Diez H, López MC, Del Carmen Thomas M, *et al.* Evaluation of IFN-gamma production by CD8 T lymphocytes in response to the K1 peptide from KMP-11 protein in patients infected with *Trypanosoma cruzi*. Parasite Immunol 2006; 28(3): 101-5.
[http://dx.doi.org/10.1111/j.1365-3024.2005.00815.x] [PMID: 16441508]

[151] Branche C, Ochaya S, Aslund L, Andersson B. Comparative karyotyping as a tool for genome structure analysis of *Trypanosoma cruzi*. Mol Biochem Parasitol 2006; 147(1): 30-8.
[http://dx.doi.org/10.1016/j.molbiopara.2006.01.005] [PMID: 16481054]

[152] Llewellyn MS, Lewis MD, Acosta N, *et al. Trypanosoma cruzi* IIc: phylogenetic and phylogeographic insights from sequence and microsatellite analysis and potential impact on emergent Chagas disease. PLoS Negl Trop Dis 2009; 3(9): e510.
[http://dx.doi.org/10.1371/journal.pntd.0000510] [PMID: 19721699]

[153] Jones C, Todeschini AR, Agrellos OA, Previato JO, Mendonça-Previato L. Heterogeneity in the biosynthesis of mucin O-glycans from *Trypanosoma cruzi* tulahuen strain with the expression of novel galactofuranosyl-containing oligosaccharides. Biochemistry 2004; 43(37): 11889-97. [http://dx.doi.org/10.1021/bi048942u] [PMID: 15362875]

CHAPTER 3

Chromosomes Ends and Telomere Biology of Trypanosomatids

Miguel Angel Chiurillo[1,*], Cristiane Regina Antonio[2], Marjorie Mendes Marini[2], Renata Torres de Souza[2] and José Franco da Silveira[2,*]

[1] *Departament of Clinical Pathology, Faculdade de Ciências Médicas, Universidade de Campinas, UNICAMP, Campinas, Brazil*

[2] *Departament of Microbiology, Immunology and Parasitology, Escola Paulista de Medicina, Universidade Federal de São Paulo, UNIFESP, São Paulo, Brazil*

Abstract: In this chapter, we focus on the structure and function of telomeres and subtelomeres of human protozoan parasites *T. cruzi*, *T. rangeli* and *Leishmania* spp.. Beyond their role in maintaining the integrity of chromosomes, telomeres and subtelomeres are involved in the survival mechanisms of these single-celled parasites. The telomeric repeat $(5'\text{-TTAGGG-}3')_n$ is conserved among trypanosomatid species, but adjacent subtelomers vary between species and chromosomes within the same cell. The chromosome ends of *T. rangeli*, for example, exhibit a simple organization with short subtelomeres whereas *T. cruzi* subtelomeres are a complex mosaic of genomic fragments including gene/pseudogenes corresponding to large gene families of surface proteins and retrotransposons. Differences in the copy number and organization of these genes determine the variation in the size of subtelomeres on each *T. cruzi* chromosome. *Leishmania* subtelomeres, in contrast, lack genes encoding surface antigens; instead they carry conserved repeat sequences referred to as telomere-associated sequences. *T. cruzi* and *T. rangeli* chromosomes share a high level of synteny which is lost in the subtelomeric regions. It has been suggested that *T. cruzi* subtelomeres can serve as recombination hotspots and thus promoting the increase of the repertoire of surface antigens. Many pieces of evidence indicate that telomere maintenance in Kinetoplastids occurs primarily by a telomerase-mediated elongation.

[*] **Corresponding authors Miguel Angel Chiurillo:** Departament of Clinical Pathology, Faculdade de Ciências Médicas, Universidade de Campinas, UNICAMP, Campinas, Brazil; Tel/Fax: +55-19-35217370; Email: mchiurillo@ucla.edu.ve; **José Franco da Silveira:** Departamento de Microbiologia, Imunologia e Parasitologia, Escola Paulista de Medicina, Universidade Federal de São Paulo, São Paulo, Brazil; Tel/Fax: +55-11-55711095; Email: jose.franco@unifesp.br

Marcelo Santos da Silva & Maria Isabel N. Cano (Eds.)

The catalytic subunit of telomerase (TERT) is present in all sequenced trypanosomatid species, whereas the RNA component containing a template for telomere repeat extension has recently been identified in *T. brucei* and *Leishmania*. Further studies are needed to understanding the regulation of telomere homeostasis and the biology of subtelomeres of trypanosomatids.

Keywords: Base modification, Chromosomes ends, Contingency genes, Retrotransposons, Subtelomere, Surface protein genes, Telomerase, Telomere generation, Telomere junction, Telomere repeats.

INTRODUCTION

Telomeres are nucleoprotein complexes whose function is to stabilize the physical ends of a linear eukaryotic chromosome, protecting them from nucleases, end-to-end fusions and cellular recombination machinery, ensuring their stability and integrity. Telomeres may also be involved in establishing the spatial architecture of the interphase nucleus. This chapter is addressed to emphasize some recent findings to establish a comparison between telomere of three trypanosomatids species (*Trypanosoma cruzi*, *Trypanosoma rangeli* and *Leishmania*), taking into account the established knowledge from other eukaryotes. It does not contain a compilation of all available data.

OVERALL STRUCTURE OF TELOMERES

A general picture of the structure of trypanosomatid telomeres and subtelomeres is shown in Fig. (**1**). Telomeres consist of repetitive DNA associated with a variety of proteins. Tandem arrays of double-stranded 5'-TTAGGG-3' hexamers of variable size are found in *T. cruzi* (0.3-0.5 kb) and *T. brucei* (10-20 kb) telomeres [1 - 4]. Toward the chromosomal terminus, the G-rich strand is single-stranded ending in a common 5'-GGGTTAGGG-3' sequence in all *Trypanosoma* species. Moreover, overhangs of G-rich strand in *T. brucei* originate structures known as telomere loops (T-loops) because they may fold back and invade the double-stranded telomeric DNA [5] (Fig. **2**). Telomeres in *Leishmania* are constituted by tandem arrays of the same 5′-TTAGGG-3′ hexamer [4, 6, 7], although in *L. braziliensis* a different telomere repeat of 14 bp 5'-CCCTA-

-CCCGTGGA-3' is also found in some chromosome ends in addition to the conventional hexameric repeats [6, 8, 9]. Moreover, the telomere overhangs (size range: ≥ 9 nt in *L. major*, *L. donovani* and *Trypanosoma* species; ≤ 12 nt in *L. amazonensis*) differ among some *Leishmania* species and also between *Leishmania* and *Trypanosoma*. The sequence of *L. donovani* and *L. major* 3′G-overhang is 5'-GGTTAGGGT-3' [4, 7] and GTTAGGGTTAGG-3' in *L. amazonensis* [10] (Figs. **1** and **2**).

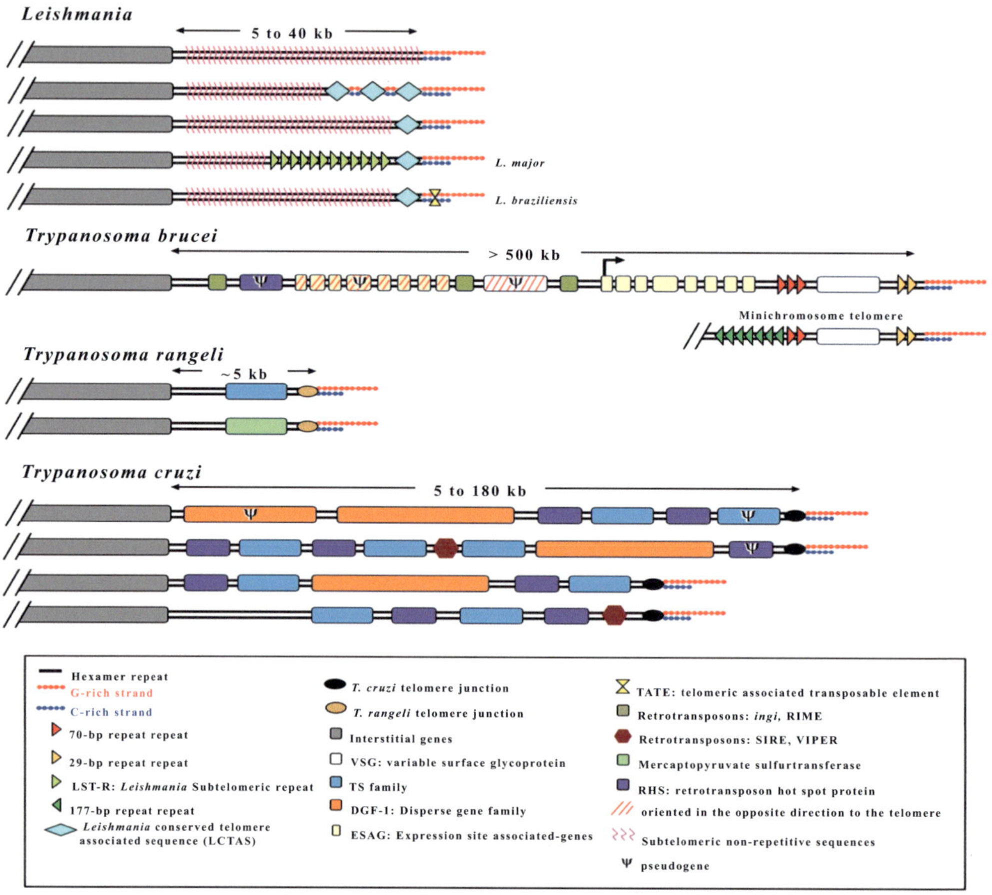

Fig. (1). Schematic representation of trypanosomatid telomeres. Two types of telomeres identified in *T. rangeli* are shown. The heterogeneity of *T. cruzi* subtelomeres is represented with the four most common organizations observed in chromosome ends of this parasite. The size of the subtelomeric region is indicated above each map. Boxes indicate genes and other sequence elements. The maps are not to scale. *T. brucei*, *T. cruzi* and *T. rangeli* maps were adapted from El-Sayed *et al.* [14], Barros *et al.* [15], Stoco *et al.* [11], respectively.

On the other side, moving away towards the center of the chromosome there is a telomere junction region which differs in its sequence between *T. cruzi* and *T. rangeli* telomeres [1, 4], therefore they can be considered as species-specific sequences. Telomeric junctions of *T. cruzi* (Tc189) and *T. rangeli* (SubTr) are 189 bp and 516–639 bp long, respectively. Moreover, the fact that Tc189 and SubTr related sequences can be found in intergenic regions of multigene families (*e.g. Trans*-sialidase) in *T. cruzi* and *T. rangeli* allowing us to suggest that these telomeric junctions could have originated from a common sequence [11].

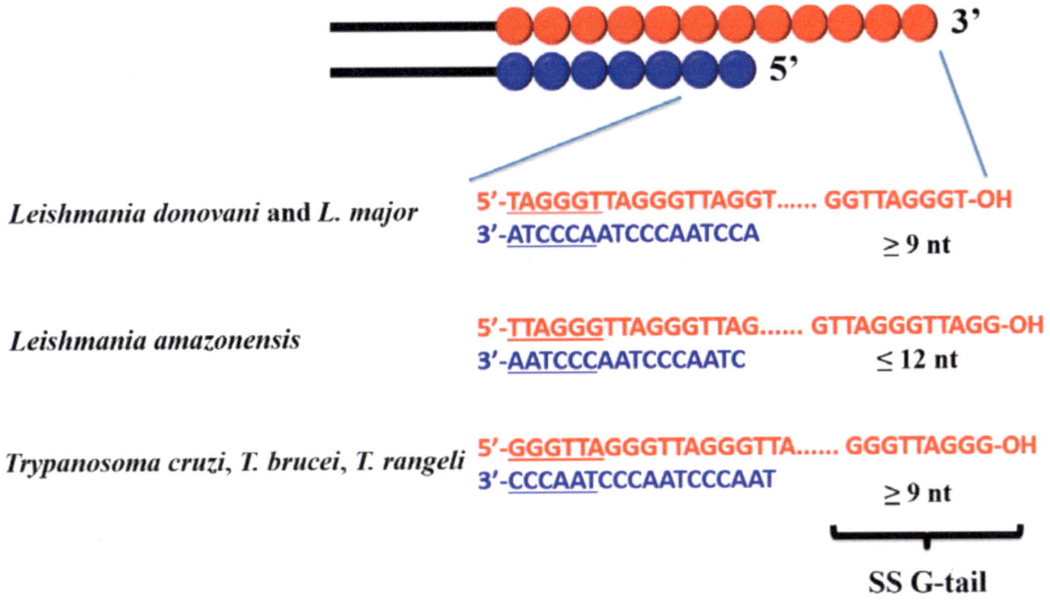

Fig. (2). Telomeric repeat and 3` G-overhang in trypanosomatid chromosome ends. Telomeric DNA termini of *Leishmania* and *Trypanosoma* species showing the G-rich and C-rich strands painted in red and blue, respectively.

The region extending from the hexamer repeats to the first internal (interstitial) chromosome-specific gene can be defined as subtelomere or subtelomeric region, which in trypanosomatids varies between species and chromosomes. The trypanosome subtelomeres contain several expanded multigene families that encode species-specific proteins involved in host and parasite interactions. *T. brucei* subtelomeres are extremely large (>500 kb) while in *T. rangeli* they are quite short (<5 kb) [11 - 16]. *In silico* analysis determined that the *T. cruzi* subtelomeric regions size vary in the range of 5 kb to 180 kb. Size variation could be the result of different genomic rearrangement events, such as breakage and healing, occurred recently at subtelomeric level [12, 15]. The subtelomeric regions of *L. major*, in contrast are quite short (~20 kb) [7, 14, 17].

THE STRUCTURE OF SUBTELOMERIC REGIONS

Size, Sequence Content and Organization of *T. cruzi* and *T. rangeli* Subtelomeres

The subtelomeric regions of *T. cruzi* are composed of a patchwork of genes and pseudogenes of multigene families encoding surface proteins and retrotransposons *Trans*-sialidases (TS); Dispersed Gene Family 1 (DGF-1), Retrotransposon Hot Spot Proteins (RHS), RNA-helicase, N-acetyltransferase and retrotransposon elements [12, 15] (Figs. **1** and **3**). However, the length of subtelomeres is highly variable due to the differences in the abundance and organization of these genes, therefore each chromosome end is singular. For instance, RHS, DGF-1 and TS are often duplicated. Recently, considering the gene content and organization of the subtelomeric region, *T. cruzi* chromosome ends were classified into 11 groups (I to XI) [15].

Fig. (**3**) shows the most representative chromosome ends (groups I and IV) in *T. cruzi* clone CL Brener and a chromosome end (group VI) in which a retrotransposon is the first gene found close the telomeric repeats towards the centromere. The organization most frequently found in the *T. cruzi* subtelomeric regions is two gene/pseudogenes RHS flanking one or more TS sequences. It is noteworthy the presence of complete coding sequences for known and hypothetical proteins, suggesting that *T. cruzi* subtelomeres are functional genomic structures that may be transcriptionally active, and do not merely nonfunctional DNA portions of chromosomes. *T. cruzi* chromosome ends are extensively polymorphic among them both in the size of subtelomeric regions as in the content and organization of genes encoding surface proteins (TS, DGF-1), RHS, retrotransposon elements, RNA-helicase and N-acetyltransferase (Fig. **1**).

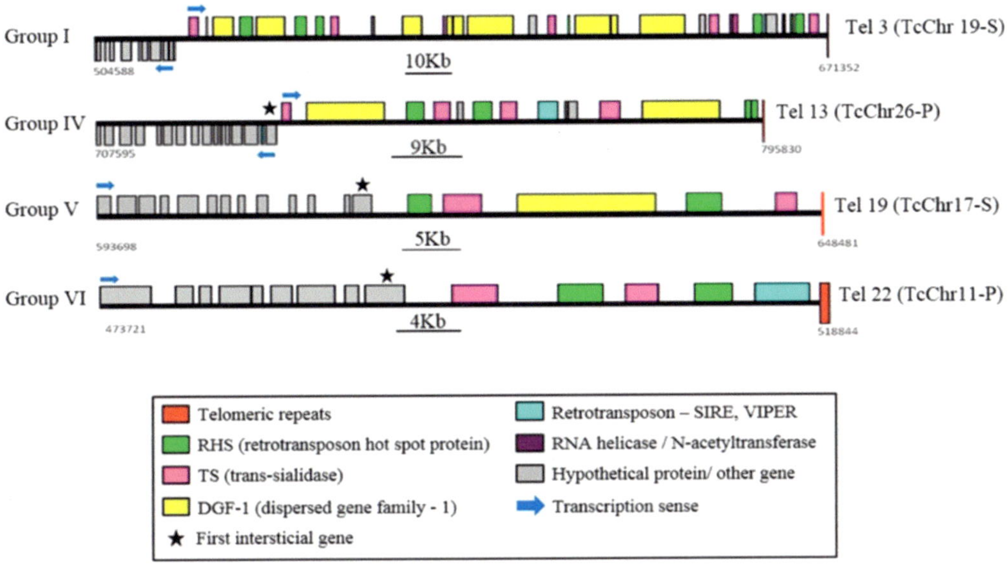

Fig. (3). Organization and gene content of *Trypanosoma cruzi* subtelomeres. Schematic representation of one representative member of *T. cruzi* chromosome ends (Tels) groups I, IV, V and VI. These groups were formed considering gene content in subtelomeric regions. The maps of chromosome ends display localization of genes and other genomic elements oriented 5' to 3' according to TriTrypDB annotation (http://tritrypdb.org/tritrypdb/): hexamer repeats (TTAGGG)n, subtelomeric genes/pseudogenes and interstitial genes including those encoding hypothetical proteins are represented by red, color and grey boxes, respectively. Maps are to scale. Start and end genomic coordinates are shown on each map. Transcription sense is indicated by blue arrows. Adapted from Moraes Barros RR, Marini MM, Antônio CR, Cortez DR, Miyake AM, Lima FM, Ruiz JC, Bartholomeu DC, Chiurillo MA, Ramirez JL, da Silveira JF. Anatomy and evolution of telomeric and subtelomeric regions in the human protozoan parasite *Trypanosoma cruzi*. BMC Genomics. 2012 Jun 8;13:229. doi: 10.1186/1471-2164-13-229. (Paper published under Creative Commons Attribution License 4.0).

Comparative analysis of homologous chromosome pairs showed that the high degree of synteny observed in the interstitial chromosomal regions can be broken at the subtelomeric regions [15] (Fig. **4**). Subtelomeres from homologous chromatids can be well conserved as observed in the chromosome pair TcChr13-P and TcChr13-S, however the synteny is found disrupted downstream an RHS gene by the insertion of a gene cluster containing RNA helicase, N-acetyltransferase and genes encoding hypothetical proteins (Fig. **4A**). On the other hand, subtelomeres from homologous chromatids can also diverge greatly in size and gene content such as in the homologous pairs TcChr 36-S and 36-P, and TcChr

39-S and 39-P (Fig. **4B** and **4C**). Comparison of homologous chromosomes showed that synteny breaks down the subtelomeric region, reinforcing the hypothesis that recombination events occurred between subtelomeric regions in this parasite. When comparing homologous chromosome ends of *T. cruzi*, the assumption that subtelomeric regions are prone to recombination events occurring between them is reinforced by the frequent breaks of synteny found nearby these regions.

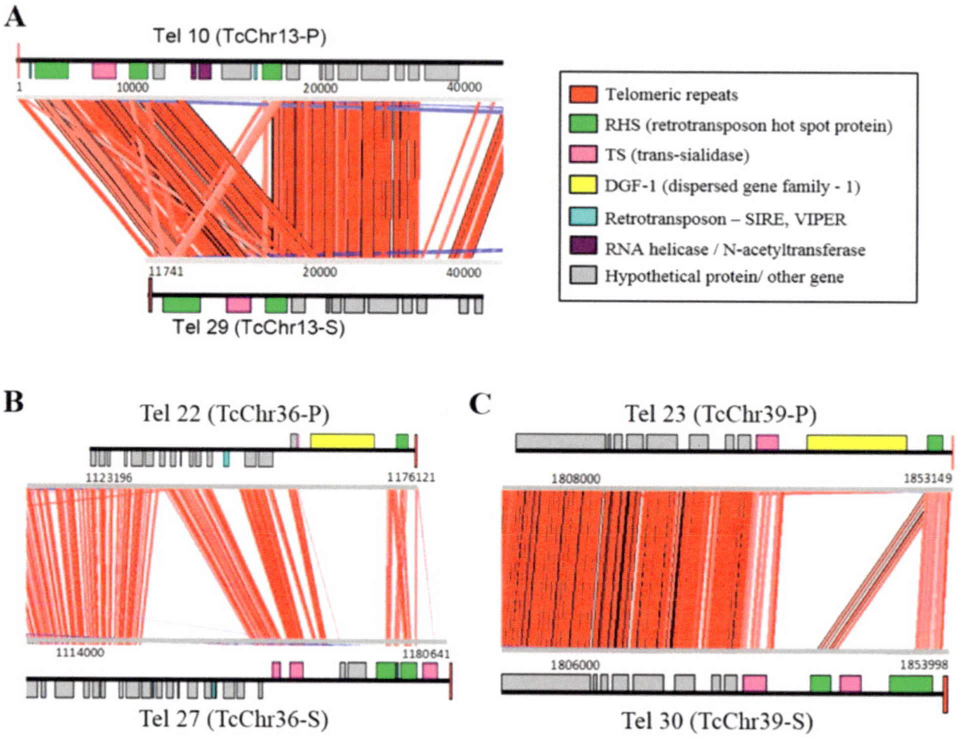

Fig. (4). Analysis of synteny between *T. cruzi* homologous chromosome ends. Homologous chromosome ends are compared: TcChr13-P and S (panel A); TcChr36-P and S (panel B) and TcChr39-P and S (panel C). Chromosome ends Tel 10 and Tel 29 are from the homologous chromosomes TcChr13-P and TcChr13-S; Tel 22 and Tel 27 in the homologous TcChr36-P and TcChr36-S pair and Tel 23 and Tel 30 in the homologous TcChr39-P and TcChr39-S. Colored boxes indicate genes. Homology between assemblies is indicated with red lines. Reproduced of Moraes Barros RR, Marini MM, Antônio CR, Cortez DR, Miyake AM, Lima FM, Ruiz JC, Bartholomeu DC, Chiurillo MA, Ramirez JL, da Silveira JF. Anatomy and evolution of telomeric and subtelomeric regions in the human protozoan parasite *Trypanosoma cruzi*. BMC Genomics. 2012 Jun 8;13:229. doi: 10.1186/1471-2164-13-229. (Paper published under Creative Commons Attribution License 4.0).

Within a single organism the subtelomeric regions of different chromosomes behave as dynamic elements or "recombinogenic regions" participating in ectopic recombination. DNA exchange events involving repetitive DNA families located at subtelomeres play a major role in the evolution of genomes. It has been proposed that *T. cruzi* subtelomeric regions may act as sites for DNA recombination and the generation of new variants of surface proteins. Therefore, the special abundance of genes encoding surface proteins implicated in host cell invasion at subtelomeres can be involved in homologous recombination events. As a consequence, for example, subtelomeric gene copies can be duplicated without affecting the functional interstitial copies [12]. In addition to the participation of telomere and subtelomere in chromosomal stability, they can play a role in the survival of this single-celled parasite. Moreover, the role of subtelomeric regions in the control of expression of genes encoding surface antigens and the generation of new antigenic variants has been widely demonstrated in protozoan parasites, such as *T. brucei* and *Plasmodium falciparum* [18, 19].

When compared with *T. cruzi* telomeres, the chromosome ends of *T. rangeli* exhibit shorter subtelomeric regions with a simpler and homogeneous organization [4, 11, 16]. They have been classified into two types according to the subtelomeric sequences localized upstream of the telomeric junction (SubTr) [11] (Fig. **1**). They differ in that SubTr telomere junction is followed (towards the centromere) by a gp85/*trans*-sialidase gene/pseudogene (type 1), or a copy of the mercaptopyruvate sulfurtransferase gene (type 2). Moreover, *T. cruzi* and *T. rangeli* chromosomes exhibit a high level of synteny between them, which is lost in the subtelomeric regions (Fig. **5**) because of the absence of large blocks of TS, DGF-1 and RHS genes/pseudogenes in *T. rangeli* chromosome ends. Differences in size and gene content between the subtelomeric regions of *T. rangeli* and *T. cruzi* are coherent with the lower number of repeated sequences from multigene families in the *T. rangeli* genome [11].

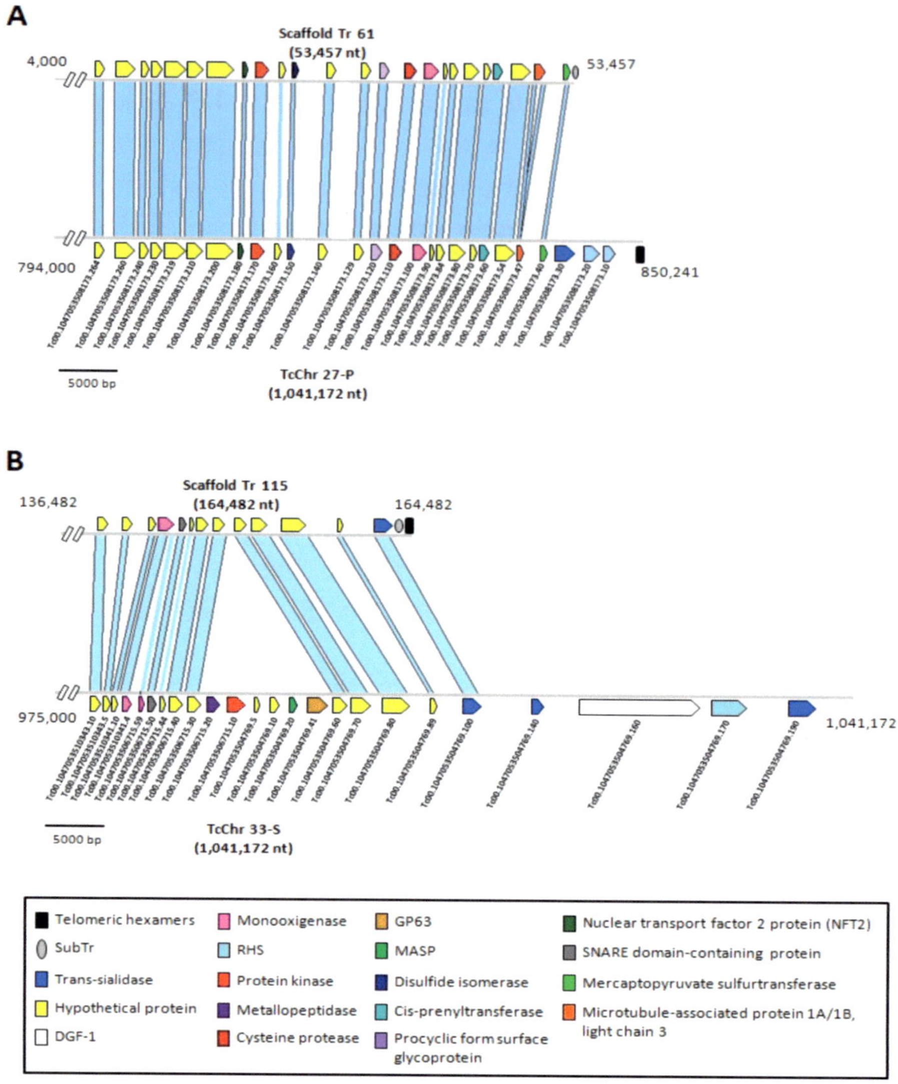

Fig. (5). Syntenic relationships between *Trypanosoma rangeli* and *T. cruzi* chromosome ends. Homologous sequences are connected with blue lines. Colored boxes indicated genes and other sequence features. Boxes pointing right indicate transcription sense. (**A**). Alignment of Scaffold Tr 61 (4,000-53,457 nt) and TcChr27-P (794,000-850,241 nt). (**B**). Alignment of Scaffold Tr 115 (136,482-164,482 nt) and TcChr33-S (975,000-1,041,172 nt). Contig ends were oriented 5' to 3' according to the TriTrypDB assemblies of *T. cruzi* scaffolds. TriTrypDB accession numbers of annotated sequences below *T. cruzi* scaffolds. Reproduced from Stoco PH *et al.* (2014). The genome of the avirulent human-infective trypanosome- *Trypanosoma rangeli*. PLoS Negl Trop Dis. 2014 Sep 18;8(9):e3176. (Paper published under Creative Commons Attribution (CC BY) license).

Organization of *Leishmania* Subtelomeres

The telomeres of *Leishmania* lack genes encoding surface antigens, instead adjacent to the telomeric repeats are conserved sequences referred to as *Leishmania* subtelomeric sequences (LSTS) or *Leishmania* conserved telomere-associated sequences (LCTAS) [4, 6 - 9]. The organization, number and level of conservation of LCTASs in telomeric regions differ greatly among *Leishmania* species or chromosomes. The composition and organization of LCTAS repeats usually occur as ~100 bp sequences with two conserved elements named CSB1 (conserved sequence block 1) and CSB2 (conserved sequence block 2). The separation between the two boxes varies from 1 to 51 bp, and between two LCTAS exists a variable number of 5′-TTAGGG-3′ repeats. Moreover, in *L. donovani*, the well-defined 100-bp LCTAS sequence reported in *L. major* and *L. mexicana* is composed of a variable number of octamer repeats and a single 62-bp sequence [7]. These differences were exploited to design a highly sensitive and specific PCR assay for detecting visceral leishmaniasis [17, 20]. Details about molecular diagnosis of trypanosomiasis are found in Chapter 9.

L. braziliensis telomeres also contain a family of DNA transposable elements inserted in the telomeric hexamer repeats, named "telomere-associated transposable elements" (TATEs), each including DNA and/or RNA polymerase domains, putative reverse transcriptase and phage integrase (site-specific recombinase) [21]. Several other repeat elements (*e.g.* 272-bp repeats) are located at subtelomeric regions of *Leishmania* and are responsible for the length polymorphism observed between homologous chromosomes [22]. The ~20 kb length variation between the two homologs of *major* chromosome 1 is mainly due to the difference in the number of LCTAS and other sequence repeats in their telomeres [22, 23].

Unlike what is found in *T. brucei*, *T. cruzi* and *T. rangeli* and other pathogenic protozoa, which show contingency genes at subtelomeric location principally genes encoding for surface molecules associated with parasite virulence, adjacent to *Leishmania* subtelomeric repeats usually is observed a barren or geneless region [21, 24]. However, up to 5 kb from the telomere, clusters of housekeeping genes are located at some *Leishmania* subtelomeric regions [25]. In *L. major* subtelomeres are found members of *SCG/L/R* gene family, which encode phosphoglycan scß-galactosyltransferases involved in modifications of lipophosphoglycan during the differentiation of promastigote into infective metacyclic forms in the sandfly midgut [26]. The exclusive telomeric location of *SCG*1–7 genes and evidence of gene conversion suggest that the telomeric "environment" could participate in gene expression regulation and in determining an increase of diversity of this gene family through recombination. Moreover, the gamma-glutamylcysteine synthetase gene (*GSH1*), which is the rate-limiting enzyme of glutathione synthesis in *Leishmania*, is also found at a subtelomeric position in these parasites [27].

In addition, *Leishmania* spp. telomere clusters shows different nuclear distribution changing from random to concentrated at the nuclear periphery, depending on the cell cycle phase and life cycle stage ([28], Dr. Maria Isabel N. Cano, personal communication). This fact has been explained by the existence of housekeeping genes at the subtelomeres of *Leishmania*. On the contrary, in *P. falciparum* clustering of chromosome ends at the nuclear periphery has been involved in promoting ectopic recombination as well as gene expression regulation of subtelomeric genes involved in antigenic variation like those of the *var* gene family [29, 30].

Telomeric DNA Modification, the Base J

Base J, ß-d-glucosyl-hydroxymethyluracil, is a modified DNA base replacing a portion of total thymine in the nuclear DNA of kinetoplastid parasites such as *T. brucei*, *T. cruzi*, *Crithidia fasciculata*, *Leishmania* sp. and *Trypanoplasma borreli*, and other closely related unicellular flagellates [31, 32]. Biosynthesis of base J involves two steps, in which a thymidine hydroxylase, JBP1 or JBP2, hydroxylates specific thymidine residues to form 5-hydroxymethyl-2′-deoxyuridine (5-HmdU), followed by the action of a glucosyltransferase that adds a glucose moiety to HMedU to form base J [33].

Base J is mainly a telomeric DNA modification found in the telomeric repeats of all kinetoplastid parasites analyzed, although it is also found in sequences between transcription units (internal J) participating in transcription initiation and termination by RNA polymerase II (RNAP II) [33 - 35]. In *T. brucei* genome, base J is found in the telomeric repeats (~50% of total base J) and in other repetitive sequences, such as the 70 and 177 bp subtelomeric repeats [36, 37]. Moreover, a considerable proportion of base J is found in non-telomeric sites in *T. equiperdum*, *T. cruzi*, and *T. borreli* [33, 39], in contrast to the small amount (only 1% of base J) located in regions other than the telomere repeats sequences of *Leishmania* sp. and *C. fasciculata* [38].

Base J is associated with genes encoding cell-surface antigens located at subtelomeric regions of *T. brucei* and *T. cruzi*, suggesting a role in regulating the expression of telomeric genes involved in pathogenesis [33, 39, 40]. However, currently available evidence suggests that base J does not play a direct role in gene silencing at telomeres, but instead, at least in *T. brucei* base J is involved in regulation of gene expression by transcriptional attenuation at specific genomic loci, whereas in *Leishmania* sp. base J function is restricted in regulating RNAP II termination genome-wide [41].

FORMATION OF NEW TELOMERES IN *T. CRUZI*

Chromosomes that lack telomeres are unstable which implicit the notion that formation of new telomeres is an important but relatively rare event in most organisms. However, for some organisms such as *Ascaris* and ciliated protozoa, the large-scale formation of new telomeres occurs at specific developmental stages [42 - 44]. The architecture of the telomere junction sequence of *T. cruzi* suggests an early broken-healing event in the evolution of the telomeres in this parasite [1, 12].

The *T. cruzi* telomere junction (Tc189) was initially observed only next to telomeric 5'-TTAGGG-3' repeats towards the center of the chromosome. However, accurate analyses of non-telomeric genomic sequences from this organism showed that they share a moderate level (~70%) of identity with sequences found in the intergenic regions of the *tandem* of TS genes. Taking into account this fact, the particular organization of Tc189 at telomeres and the "recombinogenic" nature of telomeres, allowed Kim *et al.* [12] to propose a hypothesis for the origin of *T. cruzi* telomeres [12] (Fig. **6**).

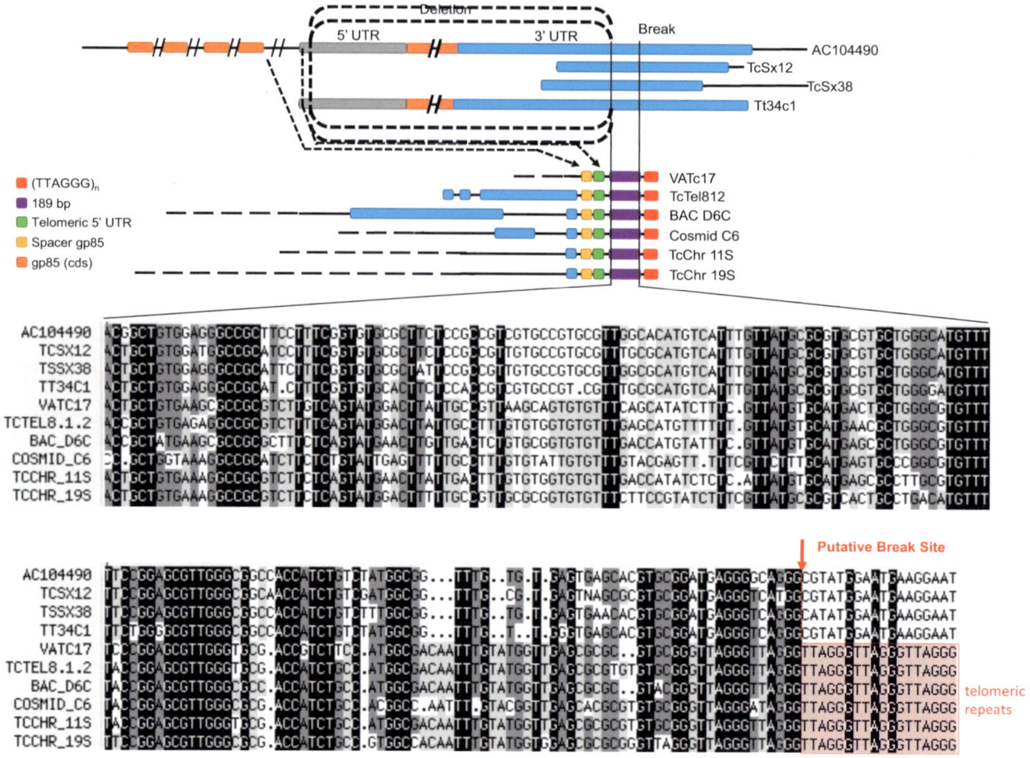

Fig. (6). Proposed model for the generation of *T. cruzi* telomeres. Schematic representation of different telomeric and non-telomeric *T. cruzi* sequences having the Tc189 homologous sequence. Upper panel: the proposed events that originated *T. cruzi* telomeres occurred at a tandem of *trans*-sialidase (gp85) genes, including an event deleting most of the 5' UTR, the coding region, and part of the 3' UTR of a gp85 gene, the remaining part of the UTRs sequences put together. Subsequently a second breakage event occurred and a putative telomerase-dependent mechanism repaired this DNA broken end. The bottom panel shows the comparison of nucleotide sequences from regions including the 189-bp sequence. It is indicated the site where the putative double-strand break occurred in telomeric-like repeats adjacent to the 189-bp homologous regions of non-telomeric sequences (red arrow). Enclosed in a red box are telomeric repeats. Sequences are follows: cosmid harboring an array of gp85 genes (GenBank AC104490); TcSx12 and TsSx38, cDNA clones from a reiterated family of transcribed oligo (A)-terminated interspersed DNA (GenBank AF510088 and AF510086); Tt34c1, gp85 cDNA (GenBank M64836); Tctel8.1.2 and VATc17, telomeric associated sequences (GenBank AF100646 and AF100652); BAC D6C and cosmid 6 are artificial bacterial chromosome and cosmid carrying large telomeric regions, respectively GenBank AY551440 and AY552588); TcChr 11S and TcChr 19S are chromosome-sized scaffolds assigned to the Esmeraldo-like haplotype of *T. cruzi* CL Brener strain (Tritrypdb.org). Modified from Kim D, Chiurillo MA, El-Sayed N, Jones K, Santos MRM, Porcile PE, *et al.* Telomere and subtelomere of *Trypanosoma cruzi* chromosomes are enriched in (pseudo)genes of retrotransposon hot spot and *trans*-sialidase-like gene families: the origins of *T. cruzi* telomeres. Gene (Internet). 2005 Feb 14 (cited 2014 Feb 8);346:153–61. Available from: http://www.sciencedirect.com/science/article/pii/S0378111904006389.

Sequence analysis suggested that an initial deletion led to join a segment of DNA containing the spacer sequence separating two TS genes and the 5'-UTR of one of them to the 3'-UTR of the same TS gene. Afterward, a break probably occurred at the 3'-UTR, resulting in a broken end, which could have been repaired by a telomerase-based mechanism. This structure was able to stabilize one telomere and later, perhaps by recombination it functioned as a seed, which was widespread and fixed at each *T. cruzi* telomere (Fig. **6**). As it is shown in Fig. (**6**) the alignment of telomeric and non-telomeric sequences indicates that originally the Tc189 junction was part of the 3'-UTR region of a TS gene. In agreement with this hypothesis, in the alignment it is also observed that next to the non-telomeric Tc189-bp homologous region there are telomeric-like repeats which could be an indication where the putative breakage occurred, to be subsequently healed by telomerase.

Similarly, in *P. falciparum* was described short telomere-like sequence motifs in fragile sites susceptible to double strand break. These broken ends might be healed by *P. falciparum* telomerase, thus creating new telomeres [45]. Moreover, in humans chromosome-healing events are also mediated by telomerase and occur in sequences that resemble telomeric repeats found at the end of the broken chromosome [46, 47].

Moreover, considering that TS superfamily is one of the most widely expanded gene families in *T. cruzi* genome [13, 14] and that similar organization of telomeric junction has been identified in almost all *T. cruzi* chromosome ends, the participation of TS genes in the composition of the telomere junction and the origin of *T. cruzi* telomeres suggests that these events took place during the expansion of the TS gene superfamily within the genome of this parasite.

TELOMERE MAINTENANCE IN TRYPANOSOMATIDS

DNA polymerases synthesize DNA unidirectionally in the 5'-3' direction and needs an RNA primer to initiate replication. Whereas the replication of leading strand generates blunt telomeric ends, the replication of lagging strand generates short telomeres. It was recently shown that at the end of lagging strand synthesis, the terminal RNA primer is retained and the same process that designated the final nucleotides on C-strand will remove it later. Curiously, the terminal Okazaki fragment is positioned about 70-100nt from the chromosome end, meaning that every DNA replication event leaves about 70-100nt at lagging strand unreplicated. In both leading and lagging telomeres, resection of the C-strand by specific nucleases create the mature G-rich overhang or mature telomeres [48].

Most eukaryotes solve the shortcomings of semiconservative DNA replication with the enzyme telomerase [49]. Telomerase is a ribonucleoprotein enzyme containing two essential components in its catalytically core: the telomerase reverse transcriptase protein (TERT) and the telomerase RNA (TER) subunit, that provides the template for copying the hexameric repeats to the telomeric overhang [50, 51]. Although, some mechanisms such as replication slippage and gene conversion have been proposed [6], evidence indicates that the mechanism responsible for maintaining telomere length in trypanosomatids is the telomerase activity.

The TERT component has the typical telomerase conserved domains such as TEN (Telomerase N-terminal), TRBD (Telomerase RNA Binding Domain) and RT domain (Reverse Transcriptase), located near the less conserved C-terminus of the protein. The RT domain, for example, contains all specific motifs conserved across different species: 1, 2, A, B', C, D and E (Fig. 7). On the contrary, the TER subunits from a large number of organisms vary greatly in sequence and size, ranging from 150 to 2,200 nt [52 - 54]. However, some degree of conservation has been observed in its secondary structure such as the presence of a template region, the pseudoknot and the template boundary element (TBE) [52, 53].

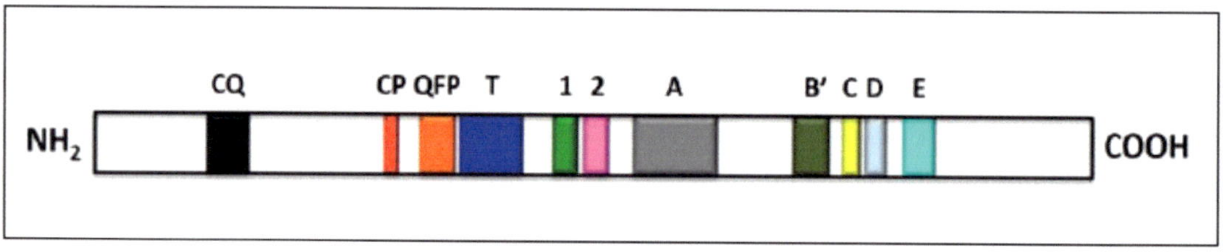

Fig. (7). Schematic representation of TERT protein motif. From N- to C-terminal ends of TERT protein are indicated the conserved TERT motifs. The GQ motif in TEN domain plays a critical role in binding telomerase to telomeric DNA. Telomerase RNA binding domain (TRBD) contains motifs CP, QFP and T and the RT-like domain contained motifs 1, 2, A, B', C, D and E.

All trypanosomatid TERTs already described share the conserved structural features displayed by other TERTs, although with some important amino acid substitutions [55, 56]. Differences in size and molecular mass are more evident when comparing TERTs from *Trypanosoma vs. Leishmania* (Table **1**) although all bear the conserved structural domains including: TEN domain, necessary for TER binding and telomerase activity, the TRBD domain containing the less-conserved T motif, the catalytic RT domain, and the less conserved C-terminal domain [57 - 59].

Table 1. Biochemical properties of putative telomerase reverse transcriptase proteins (TERT) from trypanosomatid organisms.

Organism	GenBank Accession Number	Length (amino acids)	Theoretical Isoelectric point	Predicted molecular weight (kDa)
L. major	XP_001686976	1,451	10.69	157,906
L. braziliensis	XP_001569036	1,452	10.44	157,638
L. infantum	XP_001569036	1,451	10.45	157,577
L. mexicana	XP_003874749	1,442	10.55	156,983
T. brucei	XP_829083	1,193	9.75	132,512
T. cruzi	XP_819198	1,207	9.89	134,135
T. rangeli	ESL08195	1,173	9.57	130,515

The evolutionary relationship between trypanosomatids TERTs and those of other organisms is shown in a phylogenetic tree constructed from a comparison of the amino acid sequence of TERT motifs using neighbor-joining method (Fig. **8**). As shown at the trypanosomatids branch, the TERT motif of *Leishmania* species displayed a higher sequence identity among them than with the telomerase of other eukaryotic organisms, including their trypanosome counterparts.

Telomerase activity has been detected in protein extracts from promastigotes of *L. amazonensis*, *L. major* and *L. tarentolae*, from amastigotes of *L. major* as well as from different *T. brucei* and *T. cruzi* replicative forms ([60-62]; Dr. Maria Isabel N. Cano, personal communication). From the biochemical point of view, the activity of *Leishmania* and *Trypanosoma* TERTs resembles other telomeres, in their ability to add hexamer repeats (TTAGGG) encoded by the intrinsic TER component to the 3' end of the G-rich overhang, as well as they show sensitivity to RNase A and RNase H digestion, and to protein degradation or denaturation. Moreover, it has been reported that *L. major* and *T. cruzi* telomerase activities obtained from nuclear lysates showed thermal stability, since their activities were enhanced by heat (37 °C), and they were very resistant to heat denaturation [56, 63]. In contrast, using DEAE-cellulose purified protein extracts; it has been recently observed that in *L. amazonensis*, telomerase activity can be detected at 27-28°C in promastigotes and 37°C in lesion-derived amastigotes. In this case, telomerase activity obtained from promastigotes was inactive at 37°C (Dr. Maria Isabel N. Cano, personal communication). Finally, trypanosomatid telomerase differ among them in processivity, since *T. brucei* telomerase is higher processive than *T. cruzi* and *Leishmania* TERTs [60 - 62].

Recently, the telomerase RNA component (TER) was identified and characterized in *T. brucei*, *L. major* and *L. amazonensis* [54, 64, 65]. These RNAs are around 900 and 2100 nucleotides long in *T. brucei* and *Leishmania*, respectively. Interestingly, both TERT and TER molecules in *Leishmania* (TERT: ~156 kDa; TER: ~2,100 nt) and *P. falciparum* (TERT: ~280; kDa; TER: ~2,200 nt) might be the largest telomerase components identified so far for protozoa. [53, 54, 66]. Also, *T. brucei* and *Leishmania* TERs are produced from larger transcripts subsequently processed by *trans*-splicing. These RNA molecules were co-purified with the TERT component, indicating their association *in vivo*. Moreover, similarly to TERs from different eukaryotes, trypanosomatid TERs display large differences in size and nucleotide sequence, but they share several conserved secondary structural features present in most TERs from eukaryotes, such as the template sequence and the TBE element [54, 64, 65].

The characterization of *T. brucei* TER confirms that the template sequence 5'-CCCTAACCCTA-3' is very similar to the previously predicted template (5'-CCCTAACCC-3') [60, 64, 65]. However, *T. cruzi* telomerase shows primer specificity since higher activity was observed using primer-template ended in 5'-GGG-3' sequence, which is included in the hexameric repeats and also is coincident with the 3' end sequence of the telomeric overhang found in this parasite [1, 56]. Moreover, the *Leishmania* putative template domain (5'-ACCCTAACCCTA-3') is identical in all *Leishmania* TER sequences [54]. Therefore, the fact that the telomeric terminal 3' G-overhangs of *L. amazonensis* and *L. donovani/L. major* [17, 67, 68] are different both in length and in the 3' terminus nucleotide sequence, suggests the existence of a species-specific mechanism involving a putative exonuclease activity that promotes chromosome 5' end resection to generate the overhangs that are required for telomerase elongation [69, 70].

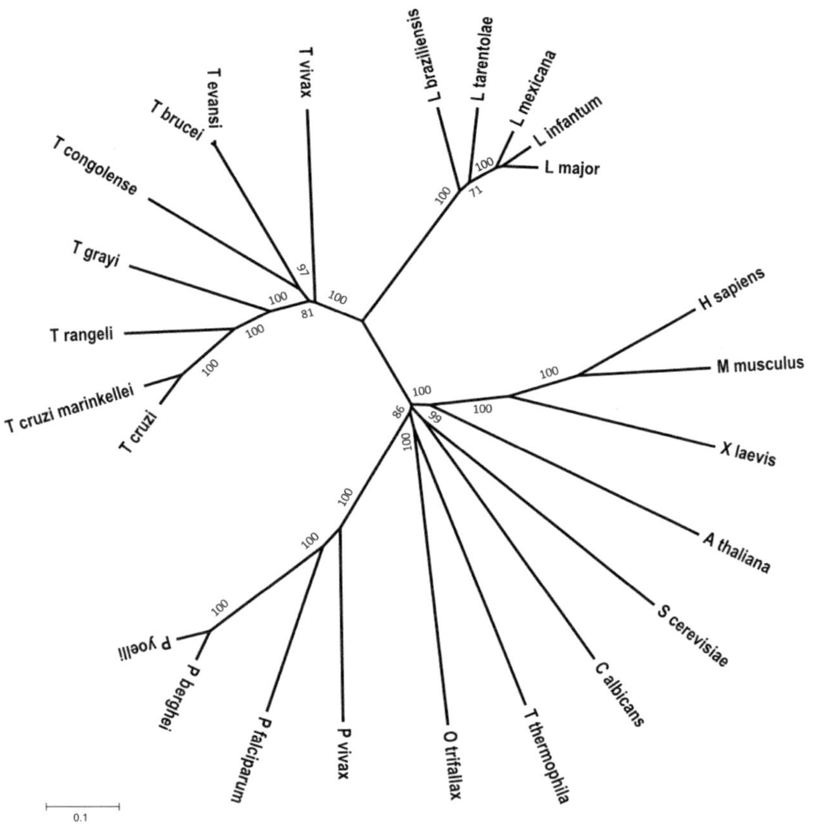

Fig. (8). Phylogenetic tree based on the amino acid sequences of putative TERT proteins. The tree was constructed by the neighbor-joining method from the aligned sequences of TERT telomerase component motif. The statistical support for each node is indicated as the percentage of 1,000 bootstrap replicates. GenBank accession numbers of sequences: *L. braziliensis*: XP_001569036; *L. infantum*: XP_001569036; *L. major*: XP_001686976; *L. mexicana*: XP_003874749; *T. brucei*: XP_829083; *T. congolense*: CCC95595; *T. cruzi*: XP_819198; *T. grayi*: XP_009310609; *T. vivax*: CCC53620; *T. cruzi marinkellei*: EKF29103; *T. evansi*: AGH70337; *T. rangeli*: ESL08195; *P. falciparum*: XP_001349878; *P. berghei*: CDS51551; *P. vivax*: XP_001617069; *P. yoelii*: CDZ14124; *H. sapiens*: NP_937983; *M. musculus*: NP_033380; X. *laevis*: NP_001079102; *A. thaliana*: NP_197187; *T. thermophila*: XP_001010651; *O. trifallax*: EJY64832; *S. cerevisiae*: AAB64520; *C. albicans*: XP_717626.

In *T. brucei*, the study of telomerase activity has gained importance in recent years due to the evidence indicating that telomere length influences the VSG switching. *T. brucei* TERT gene deletion results in shorter and fragile telomeres [71]. Moreover, contrasting to *T. brucei* strains with long telomeres, VSGs switching has been observed occurring at a higher rate (~10-fold) in telomerase-deficient strains bearing short telomeres (~1 kb) [72]. In addition, *T. brucei* strains that did not express the TER component displayed progressive telomere shortening [64, 65]. Therefore, telomerase-mediated telomere maintenance is essential for *T. brucei* pathogenesis.

Finally, as is has been observed that similarly to humans, in stress conditions *L. major* telomerase shows extra-telomeric activity and extranuclear localization, such as mitochondrial and cytoplasmic compartments [73 - 75]. It was observed that when *Leishmania* cells are exposed to H_2O_2, TERT changed from its nuclear localization and in relative abundance moving to the mitochondrion. Moreover, *Leishmania* overexpressing TERT has an increased cell growth rate and resistance to oxidative stress. Therefore, at least one of the extra-telomeric functions of *Leishmania* telomerase might be to participate in resistance to oxidative stress [75].

ARTIFICIAL CHROMOSOMES

An artificial chromosome is a useful tool for genetic manipulation that allows the expression of the gene of interest in the absence of selective pressure and avoiding detrimental effects at specific chromosomal locations. Telomeric sequences have been used to generate some vectors destined to genetic manipulation in different organisms. Notwithstanding the high level of synteny between *Trypanosoma* spp. and *Leishmania* genomes, there are a number of differences in gene content as well as in important functional features determining that several genetic manipulation tools can not be used interchangeably among trypanosomatids [14, 76]. Vectors developed for one trypanosomatid species do not work in other members of the family [77 - 82]. On the other hand, contrasting to the scenario found in *T. brucei*, the available options for genetic manipulation in *T. cruzi* and most *Leishmania* spp. are restricted, for example because the RNAi mechanism is absent [83, 84].

In *Leishmania* spp., linear constructs with added telomere hexameric repeats were conceived to obtain stable *Leishmania* artificial chromosomes used for protein over-expression or functional studies of genomic sequences [85, 86].

Moreover, stably inherited artificial linear minichromosomes for *T. brucei* were obtained by transfection of procyclic parasite forms with linear constructs containing a chloramphenicol acetyltransferase (CAT) gene under the control of the PARP gene promoter [87]. In this construct, downstream the CAT gene is found one or three tandemly linked hygromycin phosphotransferase (Hph) genes. These linear constructs end in subtelomeric sequences followed by telomere repeats, and also they include a hygromycin resistance gene for the selection and maintenance of eukaryotic cells. After transfection into *T. brucei* procyclic forms the constructs persisted as linear molecules with telomere sequences at both ends. However, the constructs increased in length (20-50 kb or 150 kb according to the vector) as a result of the incorporation of new genomic DNA sequences to the constructs [87, 88]. Artificial linear chromosomes were maintained stable in trypanosome cell lines without antibiotic pressure for at least 20 generations.

Similar constructs were made carrying *T. brucei* autonomously replicating genetic elements, neomycin phosphotransferase gene under control of a PARP promoter, signals for *trans*-splicing and polyadenylation, CAT gene and one or three Hph genes [88]. These chromosomes (1-2 copies per cell) are maintained stable under drug selection, however they are lost in approximately 7 generations when grown without antibiotic pressure. As observed in natural *T. brucei* chromosomes, telomere length increases by 6-9 bp per generation in artificial chromosomes.

The hexameric repeats and telomere junction (Tc189) from *T. cruzi* were also used to construct a series of linear vectors called pTACs for Trypanosome Artificial Chromosomes [89]. pTAC constructs harbor two cassettes in opposite orientations that confer Neomycin and Puromycin resistance. Two types of pTACs were made: i) pTAC-gfp includes a GFP gene with an internal EcoRI recognition sequence as the cloning site, and ii) pTAC-odc which harbors a transgene, ornithine decarboxylase (odc) from *C. fasciculata*, which allows the functional complementation of a natural auxotrophy of *T. cruzi*. After the electroporation of CL Brener epimastigotes, the generated molecules remained linear and stable for 60 or 150 generations in absence or presence of drug selection. pTAC segregated with efficiency and was stably maintained throughout the parasite's life cycle, as either 6-8 copies. Rearrangements or deletions of insert sequences were not detected. pTAC-odc also, can complement the auxotrophy of *T. cruzi* for polyamines. The pTAC system shows an efficient maintenance and faithful segregation as a linear episome through the different *T. cruzi* developmental forms, representing a useful tool for reverse functional genetics, to investigate mechanisms of genetic exchange and to identify elements involved in chromosome replication, maintenance and segregation in this parasite.

CONCLUDING REMARKS

In general, our current knowledge about trypanosomatid telomeres is mostly related to the mechanisms involved in antigenic variation in *T. brucei*. However, recent findings have also revealed important biological functions for telomeres in other trypanosomatids. Moreover, the architecture of *T. cruzi* telomeres, far from being monotonous, represents a fascinating field of study about the pathogenic mechanisms of this parasite. In this regard, a pending goal is to prove the role of *T. cruzi* subtelomeric regions for the generation of diversity in the multigene families and its influence in parasite pathogenicity.

Challenges to extend our knowledge about telomere biology of trypanosomatids include fully elucidation of the mechanisms underlying telomere maintenance. Moreover, although some telomeric proteins from trypanosomatids have already been identified, much more needs to be studied, especially in *T. cruzi*, in which there are no reports so far about the telomeric chromatin composition. Finally, a very interesting feature that is starting to be uncovered in trypanosomatids is the recently discovered of long noncoding RNAs that are transcribed from subtelomeric regions in *Leishmania* spp (Dr. Maria Isabel N. Cano, personal communication). These molecules, named telomeric repeat-containing RNA (TERRA), play key roles in telomere biology, including heterochromatin formation at chromosome ends and regulation of telomerase activity. We anticipate that the identification of TERRA in trypanosomatids will significantly improve the understanding of their telomere biology.

CONFLICT OF INTEREST

The authors confirm that they have no conflict of interest to declare for this publication.

ACKNOWLEDGEMENTS

This work was supported by grants from FAPESP (11/51475-3) and CNPq to JFS. MC receives a postdoctoral fellowship from FAPESP (2014/13148-9). CRA was awarded a doctoral fellowship by FAPESP (2013/04372-0) and MMM was awarded a postdoctoral fellowship by FAPESP (2011/06653-0). RTS receives a CNPq fellowship for young researchers (BJT, 401699/2013-7).

ABBREVIATIONS

BAC	Bacterial artificial chromosome
CAT	Chloramphenicol acetyltransferase
CSB1	Conserved sequence block 1
CSB2	Conserved sequence block 2
DEAE-cellulose	Diethylaminoethyl cellulose
DGF-1	Dispersed gene family 1
GFP	Green fluorescent protein
GSH1	Gamma-glutamylcysteine synthetase gene
Hph	Hygromycin phosphotransferase gene
LCTAS	*Leishmania* conserved telomere associated sequences
LSTS	*Leishmania* subtelomeric sequences
ODC	Ornithine decarboxylase
PARP	Poly ADP ribose polymerase
PCR	Polymerase chain reaction
pTAC	Trypanosome artificial chromosome
RHS	Retrotransposon Hot Spot Proteins
RNAPII	RNA polymerase II
RT	Reverse transcriptase domain
SubTr	Telomere junction of *Trypanosoma rangeli*
TATE	Telomere-associated transposable elements
TBE	Template boundary element
Tc189	Telomere junction of *Trypanosoma cruzi*

TcChr	*T. cruzi* chromosome scaffold
TEN	Telomerase N-terminal domain
TER	Telomerase RNA subunit
TERRA	Telomeric repeat-containing RNA
TERT	Telomerase reverse transcriptase protein
TRBD	Telomerase RNA Binding domain
TS	*Trans*-sialidases
UTR	Untranslated sequence
VSG	Variant surface glycoprotein

REFERENCES

[1] Chiurillo MA, Cano I, Da Silveira JF, Ramirez JL. Organization of telomeric and sub-telomeric regions of chromosomes from the protozoan parasite *Trypanosoma cruzi*. Mol Biochem Parasitol 1999; 100(2): 173-83.
[http://dx.doi.org/10.1016/S0166-6851(99)00047-X] [PMID: 10391379]

[2] Freitas-Junior LH, Porto RM, Pirrit LA, Schenkman S, Scherf A. Identification of the telomere in *Trypanosoma cruzi* reveals highly heterogeneous telomere lengths in different parasite strains. Nucleic Acids Res 1999; 27(12): 2451-6.
[http://dx.doi.org/10.1093/nar/27.12.2451] [PMID: 10352173]

[3] Stansel RM, de Lange T, Griffith JD. T-loop assembly *in vitro* involves binding of TRF2 near the 3 telomeric overhang. EMBO J 2001; 20(19): 5532-40.
[http://dx.doi.org/10.1093/emboj/20.19.5532] [PMID: 11574485]

[4] Chiurillo MA, Santos MR, Franco Da Silveira J, Ramírez JL. An improved general approach for cloning and characterizing telomeres: the protozoan parasite *Trypanosoma cruzi* as model organism. Gene 2002; 294(1-2): 197-204.
[http://dx.doi.org/10.1016/S0378-1119(02)00768-0] [PMID: 12234681]

[5] Muñoz-Jordán JL, Cross GA, de Lange T, Griffith JD. t-loops at trypanosome telomeres. EMBO J 2001; 20(3): 579-88.
[http://dx.doi.org/10.1093/emboj/20.3.579] [PMID: 11157764]

[6] Fu G, Barker DC. Characterisation of *Leishmania* telomeres reveals unusual telomeric repeats and conserved telomere-associated sequence. Nucleic Acids Res 1998; 26(9): 2161-7.
[http://dx.doi.org/10.1093/nar/26.9.2161] [PMID: 9547275]

[7] Chiurillo MA, Beck AE, Devos T, Myler PJ, Stuart K, Ramirez JL. Cloning and characterization of *Leishmania donovani* telomeres. Exp Parasitol 2000; 94(4): 248-58.
[http://dx.doi.org/10.1006/expr.2000.4499] [PMID: 10831393]

[8] Fu G, Perona-Wright G, Barker DC. *Leishmania braziliensis*: characterisation of a complex specific subtelomeric repeat sequence and its use in the detection of parasites. Exp Parasitol 1998; 90(3): 236-43.
[http://dx.doi.org/10.1006/expr.1998.4326] [PMID: 9806868]

[9] Fu G, Melville SE. Polymorphism in the subtelomeric regions of chromosomes of Kinetoplastida. Trans R Soc Trop Med Hyg 2002; 96 (Suppl. 1): S31-40. [http://dx.doi.org/10.1016/S0035-9203(02)90049-0] [PMID: 12055849]

[10] Conte FF, Cano MI. Genomic organization of telomeric and subtelomeric sequences of *Leishmania (Leishmania) amazonensis*. Int J Parasitol 2005; 35(13): 1435-43. [http://dx.doi.org/10.1016/j.ijpara.2005.05.011] [PMID: 16126212]

[11] Stoco PH, Wagner G, Talavera-Lopez C, *et al.* Genome of the avirulent human-infective trypanosome *Trypanosoma rangeli*. PLoS Negl Trop Dis 2014; 8(9): e3176. [http://dx.doi.org/10.1371/journal.pntd.0003176] [PMID: 25233456]

[12] Kim D, Chiurillo MA, El-Sayed N, *et al.* Telomere and subtelomere of *Trypanosoma cruzi* chromosomes are enriched in (pseudo)genes of retrotransposon hot spot and *trans*-sialidase-like gene families: the origins of *T. cruzi* telomeres. Gene 2005; 346: 153-61. [http://dx.doi.org/10.1016/j.gene.2004.10.014] [PMID: 15716016]

[13] El-Sayed NM, Myler PJ, Bartholomeu DC, *et al.* The genome sequence of *Trypanosoma cruzi*, etiologic agent of Chagas disease. Science 2005; 309(5733): 409-15. [http://dx.doi.org/10.1126/science.1112631] [PMID: 16020725]

[14] El-Sayed NM, Myler PJ, Blandin G, *et al.* Comparative genomics of trypanosomatid parasitic protozoa. Science 2005; 309(5733): 404-9. [http://dx.doi.org/10.1126/science.1112181] [PMID: 16020724]

[15] Moraes Barros RR, Marini MM, Antônio CR, *et al.* Anatomy and evolution of telomeric and subtelomeric regions in the human protozoan parasite *Trypanosoma cruzi*. BMC Genomics 2012; 13(1): 229. [http://dx.doi.org/10.1186/1471-2164-13-229] [PMID: 22681854]

[16] Cabrine-Santos M, Ramírez LE, Lages-Silva E, de Souza BF, Pedrosa AL. Sequencing and analysis of chromosomal extremities of *Trypanosoma rangeli* in comparison with *Trypanosoma cruzi* lineages. Parasitol Res 2011; 108(2): 459-66. [http://dx.doi.org/10.1007/s00436-010-2087-4] [PMID: 20886234]

[17] Chiurillo MA, Ramírez JL. Charaterization of *Leishmania* major Friedlin telomeric terminus. Mem Inst Oswaldo Cruz 2002; 97(3): 343-6. [http://dx.doi.org/10.1590/S0074-02762002000300011] [PMID: 12048562]

[18] Freitas-Junior LH, Bottius E, Pirrit LA, *et al.* Frequent ectopic recombination of virulence factor genes in telomeric chromosome clusters of *P. falciparum*. Nature 2000; 407(6807): 1018-22. [http://dx.doi.org/10.1038/35039531] [PMID: 11069183]

[19] Lira CB, Giardini MA, Neto JL, Conte FF, Cano MI. Telomere biology of trypanosomatids: beginning to answer some questions. Trends Parasitol 2007; 23(8): 357-62. [http://dx.doi.org/10.1016/j.pt.2007.06.005] [PMID: 17580124]

[20] Feliciangeli MD, Delgado O, Suarez B, Chiurillo MA. The burden of the *Leishmania* chagasi/infantum infection in a closed rural focus of visceral leishmaniasis in Lara state, west-central Venezuela. Trop Med Int Health 2005; 10(5): 444-9. [http://dx.doi.org/10.1111/j.1365-3156.2005.01408.x] [PMID: 15860091]

[21] Peacock CS, Seeger K, Harris D, *et al.* Comparative genomic analysis of three *Leishmania* species that cause diverse human disease. Nat Genet 2007; 39(7): 839-47.
[http://dx.doi.org/10.1038/ng2053] [PMID: 17572675]

[22] Sunkin SM, Kiser P, Myler PJ, Stuart K. The size difference between *Leishmania* major friedlin chromosome one homologues is localized to sub-telomeric repeats at one chromosomal end. Mol Biochem Parasitol 2000; 109(1): 1-15.
[http://dx.doi.org/10.1016/S0166-6851(00)00215-2] [PMID: 10924752]

[23] Myler PJ, Audleman L, deVos T, *et al. Leishmania major* Friedlin chromosome 1 has an unusual distribution of protein-coding genes. Proc Natl Acad Sci USA 1999; 96(6): 2902-6.
[http://dx.doi.org/10.1073/pnas.96.6.2902] [PMID: 10077609]

[24] Barry JD, Ginger ML, Burton P, McCulloch R. Why are parasite contingency genes often associated with telomeres? Int J Parasitol 2003; 33(1): 29-45.
[http://dx.doi.org/10.1016/S0020-7519(02)00247-3] [PMID: 12547344]

[25] Ivens AC, Peacock CS, Worthey EA, *et al.* The genome of the kinetoplastid parasite, *Leishmania major*. Science 2005; 309(5733): 436-42.
[http://dx.doi.org/10.1126/science.1112680] [PMID: 16020728]

[26] Dobson DE, Scholtes LD, Myler PJ, Turco SJ, Beverley SM. Genomic organization and expression of the expanded SCG/L/R gene family of *Leishmania major*: internal clusters and telomeric localization of SCGs mediating species-specific LPG modifications. Mol Biochem Parasitol 2006; 146(2): 231-41.
[http://dx.doi.org/10.1016/j.molbiopara.2005.12.012] [PMID: 16464509]

[27] Mukherjee A, Langston LD, Ouellette M. Intrachromosomal tandem duplication and repeat expansion during attempts to inactivate the subtelomeric essential gene GSH1 in *Leishmania*. Nucleic Acids Res 2011; 39(17): 7499-511.
[http://dx.doi.org/10.1093/nar/gkr494] [PMID: 21693561]

[28] Dossin FdeM, Dufour A, Dusch E, *et al.* Automated nuclear analysis of *Leishmania major* telomeric clusters reveals changes in their organization during the parasites life cycle. PLoS One 2008; 3(6): e2313.
[http://dx.doi.org/10.1371/journal.pone.0002313] [PMID: 18545650]

[29] Scherf A, Lopez-Rubio JJ, Riviere L. Antigenic variation in *Plasmodium falciparum*. Annu Rev Microbiol 2008; 62: 445-70.
[http://dx.doi.org/10.1146/annurev.micro.61.080706.093134] [PMID: 18785843]

[30] Figueiredo L, Scherf A. Plasmodium telomeres and telomerase: the usual actors in an unusual scenario. Chromosome Res 2005; 13(5): 517-24.
[http://dx.doi.org/10.1007/s10577-005-0996-3] [PMID: 16132816]

[31] van Leeuwen F, Taylor MC, Mondragon A, *et al.* beta-D-glucosyl-hydroxymethyluracil is a conserved DNA modification in kinetoplastid protozoans and is abundant in their telomeres. Proc Natl Acad Sci USA 1998; 95(5): 2366-71.
[http://dx.doi.org/10.1073/pnas.95.5.2366] [PMID: 9482891]

[32] Dooijes D, Chaves I, Kieft R, Dirks-Mulder A, Martin W, Borst P. Base J originally found in kinetoplastida is also a minor constituent of nuclear DNA of *Euglena gracilis*. Nucleic Acids Res

2000; 28(16): 3017-21.
[http://dx.doi.org/10.1093/nar/28.16.3017] [PMID: 10931915]

[33] Borst P, Sabatini R. Base J: discovery, biosynthesis, and possible functions. Annu Rev Microbiol 2008; 62: 235-51.
[http://dx.doi.org/10.1146/annurev.micro.62.081307.162750] [PMID: 18729733]

[34] Cliffe LJ, Siegel TN, Marshall M, Cross GA, Sabatini R. Two thymidine hydroxylases differentially regulate the formation of glucosylated DNA at regions flanking polymerase II polycistronic transcription units throughout the genome of *Trypanosoma brucei*. Nucleic Acids Res 2010; 38(12): 3923-35.
[http://dx.doi.org/10.1093/nar/gkq146] [PMID: 20215442]

[35] van Luenen HG, Farris C, Jan S, *et al.* Glucosylated hydroxymethyluracil, DNA base J, prevents transcriptional readthrough in *Leishmania*. Cell 2012; 150(5): 909-21.
[http://dx.doi.org/10.1016/j.cell.2012.07.030] [PMID: 22939620]

[36] van Leeuwen F, Wijsman ER, Kuyl-Yeheskiely E, van der Marel GA, van Boom JH, Borst P. The telomeric GGGTTA repeats of *Trypanosoma brucei* contain the hypermodified base J in both strands. Nucleic Acids Res 1996; 24(13): 2476-82.
[http://dx.doi.org/10.1093/nar/24.13.2476] [PMID: 8692684]

[37] van Leeuwen F, Kieft R, Cross M, Borst P. Tandemly repeated DNA is a target for the partial replacement of thymine by beta-D-glucosyl-hydroxymethyluracil in *Trypanosoma brucei*. Mol Biochem Parasitol 2000; 109(2): 133-45.
[http://dx.doi.org/10.1016/S0166-6851(00)00247-4] [PMID: 10960172]

[38] Genest P-A, Ter Riet B, Cijsouw T, van Luenen HG, Borst P. Telomeric localization of the modified DNA base J in the genome of the protozoan parasite *Leishmania*. Nucleic Acids Res 2007; 35(7): 2116-24.
[http://dx.doi.org/10.1093/nar/gkm050] [PMID: 17329373]

[39] Ekanayake DK, Cipriano MJ, Sabatini R. Telomeric co-localization of the modified base J and contingency genes in the protozoan parasite *Trypanosoma cruzi*. Nucleic Acids Res 2007; 35(19): 6367-77.
[http://dx.doi.org/10.1093/nar/gkm693] [PMID: 17881368]

[40] van Leeuwen F, Wijsman ER, Kieft R, van der Marel GA, van Boom JH, Borst P. Localization of the modified base J in telomeric VSG gene expression sites of *Trypanosoma brucei*. Genes Dev 1997; 11(23): 3232-41.
[http://dx.doi.org/10.1101/gad.11.23.3232] [PMID: 9389654]

[41] Reynolds D, Cliffe L, Förstner KU, Hon CC, Siegel TN, Sabatini R. Regulation of transcription termination by glucosylated hydroxymethyluracil, base J, in *Leishmania major* and *Trypanosoma brucei*. Nucleic Acids Res 2014; 42(15): 9717-29.
[http://dx.doi.org/10.1093/nar/gku714] [PMID: 25104019]

[42] Jentsch S, Tobler H, Müller F. New telomere formation during the process of chromatin diminution in *Ascaris suum*. Int J Dev Biol 2002; 46(1): 143-8.
[PMID: 11902675]

[43] Möllenbeck M, Klobutcher LA. De novo telomere addition to spacer sequences prior to their

developmental degradation in *Euplotes crassus*. Nucleic Acids Res 2002; 30(2): 523-31. [http://dx.doi.org/10.1093/nar/30.2.523] [PMID: 11788715]

[44] Cristofari G, Lingner J. Fingering the ends: how to make new telomeres. Cell 2003; 113(5): 552-4. [http://dx.doi.org/10.1016/S0092-8674(03)00397-0] [PMID: 12787493]

[45] Bottius E, Bakhsis N, Scherf A. Plasmodium falciparum telomerase: *de novo* telomere addition to telomeric and nontelomeric sequences and role in chromosome healing. Mol Cell Biol 1998; 18(2): 919-25.
[http://dx.doi.org/10.1128/MCB.18.2.919] [PMID: 9447988]

[46] Wilkie AO, Lamb J, Harris PC, Finney RD, Higgs DR. A truncated human chromosome 16 associated with α thalassaemia is stabilized by addition of telomeric repeat (TTAGGG)n. Nature 1990; 346(6287): 868-71.
[http://dx.doi.org/10.1038/346868a0] [PMID: 1975428]

[47] Flint J, Craddock CF, Villegas A, *et al.* Healing of broken human chromosomes by the addition of telomeric repeats. Am J Hum Genet 1994; 55(3): 505-12.
[PMID: 7521575]

[48] Lundblad V. Telomere end processing: unexpected complexity at the end game. Genes Dev 2012; 26(11): 1123-7.
[http://dx.doi.org/10.1101/gad.195339.112] [PMID: 22661225]

[49] Greider CW, Blackburn EH. A telomeric sequence in the RNA of Tetrahymena telomerase required for telomere repeat synthesis. Nature 1989; 337(6205): 331-7.
[http://dx.doi.org/10.1038/337331a0] [PMID: 2463488]

[50] Yu GL, Bradley JD, Attardi LD, Blackburn EH. *In vivo* alteration of telomere sequences and senescence caused by mutated Tetrahymena telomerase RNAs. Nature 1990; 344(6262): 126-32. [http://dx.doi.org/10.1038/344126a0] [PMID: 1689810]

[51] Feng J, Funk WD, Wang SS, *et al.* The RNA component of human telomerase. Science 1995; 269(5228): 1236-41.
[http://dx.doi.org/10.1126/science.7544491] [PMID: 7544491]

[52] Chen J-L, Greider CW. An emerging consensus for telomerase RNA structure. Proc Natl Acad Sci USA 2004; 101(41): 14683-4.
[http://dx.doi.org/10.1073/pnas.0406204101] [PMID: 15466703]

[53] Chakrabarti K, Pearson M, Grate L, *et al.* Structural RNAs of known and unknown function identified in malaria parasites by comparative genomics and RNA analysis. RNA 2007; 13(11): 1923-39. [http://dx.doi.org/10.1261/rna.751807] [PMID: 17901154]

[54] Vasconcelos EJ, Nunes VS, da Silva MS, *et al.* The putative *Leishmania* telomerase RNA (LeishTER) undergoes trans-splicing and contains a conserved template sequence. PLoS One 2014; 9(11): e112061.
[http://dx.doi.org/10.1371/journal.pone.0112061] [PMID: 25391020]

[55] Giardini MA, Lira CB, Conte FF, *et al.* The putative telomerase reverse transcriptase component of *Leishmania amazonensis*: gene cloning and characterization. Parasitol Res 2006; 98(5): 447-54. [http://dx.doi.org/10.1007/s00436-005-0036-4] [PMID: 16416120]

[56] Campelo R, Galindo MM, Ramirez JL. Characterization of *Trypanosoma cruzi* telomerase. Acta Trop 2011; 120(3): 173-8.
[http://dx.doi.org/10.1016/j.actatropica.2011.08.003] [PMID: 21893016]

[57] Lingner J, Hughes TR, Shevchenko A, Mann M, Lundblad V, Cech TR. Reverse transcriptase motifs in the catalytic subunit of telomerase. Science 1997; 276(5312): 561-7.
[http://dx.doi.org/10.1126/science.276.5312.561] [PMID: 9110970]

[58] Armbruster BN, Banik SS, Guo C, Smith AC, Counter CM. N-terminal domains of the human telomerase catalytic subunit required for enzyme activity *in vivo*. Mol Cell Biol 2001; 21(22): 7775-86.
[http://dx.doi.org/10.1128/MCB.21.22.7775-7786.2001] [PMID: 11604512]

[59] Lai CK, Mitchell JR, Collins K. RNA binding domain of telomerase reverse transcriptase. Mol Cell Biol 2001; 21(4): 990-1000.
[http://dx.doi.org/10.1128/MCB.21.4.990-1000.2001] [PMID: 11158287]

[60] Cano MI, Dungan JM, Agabian N, Blackburn EH. Telomerase in kinetoplastid parasitic protozoa. Proc Natl Acad Sci USA 1999; 96(7): 3616-21.
[http://dx.doi.org/10.1073/pnas.96.7.3616] [PMID: 10097086]

[61] Muñoz DP, Collins K. Biochemical properties of *Trypanosoma cruzi* telomerase. Nucleic Acids Res 2004; 32(17): 5214-22.
[http://dx.doi.org/10.1093/nar/gkh864] [PMID: 15459291]

[62] Giardini MA, Fernández MF, Lira CB, Cano MI. *Leishmania amazonensis*: partial purification and study of the biochemical properties of the telomerase reverse transcriptase activity from promastigote-stage. Exp Parasitol 2011; 127(1): 243-8.
[http://dx.doi.org/10.1016/j.exppara.2010.08.001] [PMID: 20692257]

[63] Galindo MM, Rodriguez E, Rojas MG, Figarella K, Campelo R, Ramírez JL. A heat-activated and thermoresistant telomerase activity in *Leishmania major* Friedlin. Acta Trop 2009; 111(1): 86-9.
[http://dx.doi.org/10.1016/j.actatropica.2009.02.002] [PMID: 19426669]

[64] Gupta SK, Kolet L, Doniger T, *et al.* The *Trypanosoma brucei* telomerase RNA (TER) homologue binds core proteins of the C/D snoRNA family. FEBS Lett 2013; 587(9): 1399-404.
[http://dx.doi.org/10.1016/j.febslet.2013.03.017] [PMID: 23523918]

[65] Sandhu R, Sanford S, Basu S, *et al.* A *trans*-spliced telomerase RNA dictates telomere synthesis in *Trypanosoma brucei*. Cell Res 2013; 23(4): 537-51.
[http://dx.doi.org/10.1038/cr.2013.35] [PMID: 23478302]

[66] Figueiredo LM, Rocha EP, Mancio-Silva L, Prevost C, Hernandez-Verdun D, Scherf A. The unusually large *Plasmodium* telomerase reverse-transcriptase localizes in a discrete compartment associated with the nucleolus. Nucleic Acids Res 2005; 33(3): 1111-22.
[http://dx.doi.org/10.1093/nar/gki260] [PMID: 15722485]

[67] Chiurillo MA, Beck AE, Devos T, Myler PJ, Stuart K, Ramirez JL. Cloning and characterization of *Leishmania donovani* telomeres. Exp Parasitol 2000; 94(4): 248-58.
[http://dx.doi.org/10.1006/expr.2000.4499] [PMID: 10831393]

[68] Conte FF, Cano MI. Genomic organization of telomeric and subtelomeric sequences of *Leishmania*

(*Leishmania*) *amazonensis*. Int J Parasitol 2005; 35(13): 1435-43.
[http://dx.doi.org/10.1016/j.ijpara.2005.05.011] [PMID: 16126212]

[69] Li B, Reddy S, Comai L. Sequence-specific processing of telomeric 3 overhangs by the Werner syndrome protein exonuclease activity. Aging (Albany, NY) 2009; 1(3): 289-302.
[http://dx.doi.org/10.18632/aging.100032] [PMID: 20157518]

[70] Longhese MP, Bonetti D, Manfrini N, Clerici M. Mechanisms and regulation of DNA end resection. EMBO J 2010; 29(17): 2864-74.
[http://dx.doi.org/10.1038/emboj.2010.165] [PMID: 20647996]

[71] Dreesen O, Li B, Cross GA. Telomere structure and shortening in telomerase-deficient *Trypanosoma brucei*. Nucleic Acids Res 2005; 33(14): 4536-43.
[http://dx.doi.org/10.1093/nar/gki769] [PMID: 16091631]

[72] Hovel-Miner GA, Boothroyd CE, Mugnier M, Dreesen O, Cross GA, Papavasiliou FN. Telomere length affects the frequency and mechanism of antigenic variation in *Trypanosoma brucei*. PLoS Pathog 2012; 8(8): e1002900.
[http://dx.doi.org/10.1371/journal.ppat.1002900] [PMID: 22952449]

[73] Santos JH, Meyer JN, Van Houten B. Mitochondrial localization of telomerase as a determinant for hydrogen peroxide-induced mitochondrial DNA damage and apoptosis. Hum Mol Genet 2006; 15(11): 1757-68.
[http://dx.doi.org/10.1093/hmg/ddl098] [PMID: 16613901]

[74] Ahmed S, Passos JF, Birket MJ, *et al.* Telomerase does not counteract telomere shortening but protects mitochondrial function under oxidative stress. J Cell Sci 2008; 121(Pt 7): 1046-53.
[http://dx.doi.org/10.1242/jcs.019372] [PMID: 18334557]

[75] Campelo R, Díaz Lozano I, Figarella K, Osuna A, Ramírez JL. *Leishmania major* telomerase TERT protein has a nuclear/mitochondrial eclipsed distribution that is affected by oxidative stress. Infect Immun 2015; 83(1): 57-66.
[http://dx.doi.org/10.1128/IAI.02269-14] [PMID: 25312950]

[76] Obado SO, Bot C, Nilsson D, Andersson B, Kelly JM. Repetitive DNA is associated with centromeric domains in *Trypanosoma brucei* but not *Trypanosoma cruzi*. Genome Biol 2007; 8(3): R37.
[http://dx.doi.org/10.1186/gb-2007-8-3-r37] [PMID: 17352808]

[77] Kelly JM, Ward HM, Miles MA, Kendall G. A shuttle vector which facilitates the expression of transfected genes in *Trypanosoma cruzi* and *Leishmania*. Nucleic Acids Res 1992; 20(15): 3963-9.
[http://dx.doi.org/10.1093/nar/20.15.3963] [PMID: 1324472]

[78] Martínez-Calvillo S, López I, Hernández R. pRIBOTEX expression vector: a pTEX derivative for a rapid selection of *Trypanosoma cruzi* transfectants. Gene 1997; 199(1-2): 71-6.
[http://dx.doi.org/10.1016/S0378-1119(97)00348-X] [PMID: 9358041]

[79] Vazquez MP, Levin MJ. Functional analysis of the intergenic regions of TcP2beta gene loci allowed the construction of an improved *Trypanosoma cruzi* expression vector. Gene 1999; 239(2): 217-25.
[http://dx.doi.org/10.1016/S0378-1119(99)00386-8] [PMID: 10548722]

[80] Tetaud E, Lecuix I, Sheldrake T, Baltz T, Fairlamb AH. A new expression vector for *Crithidia fasciculata* and *Leishmania*. Mol Biochem Parasitol 2002; 120(2): 195-204.

[http://dx.doi.org/10.1016/S0166-6851(02)00002-6] [PMID: 11897125]

[81] Wirtz E, Leal S, Ochatt C, Cross GA. A tightly regulated inducible expression system for conditional gene knock-outs and dominant-negative genetics in *Trypanosoma brucei*. Mol Biochem Parasitol 1999; 99(1): 89-101.
[http://dx.doi.org/10.1016/S0166-6851(99)00002-X] [PMID: 10215027]

[82] Taylor MC, Kelly JM. pTcINDEX: a stable tetracycline-regulated expression vector for *Trypanosoma cruzi*. BMC Biotechnol 2006; 6: 32.
[http://dx.doi.org/10.1186/1472-6750-6-32] [PMID: 16824206]

[83] DaRocha WD, Otsu K, Teixeira SM, Donelson JE. Tests of cytoplasmic RNA interference (RNAi) and construction of a tetracycline-inducible T7 promoter system in *Trypanosoma cruzi*. Mol Biochem Parasitol 2004; 133(2): 175-86.
[http://dx.doi.org/10.1016/j.molbiopara.2003.10.005] [PMID: 14698430]

[84] Lye LF, Owens K, Shi H, *et al.* Retention and loss of RNA interference pathways in trypanosomatid protozoans. PLoS Pathog 2010; 6(10): e1001161.
[http://dx.doi.org/10.1371/journal.ppat.1001161] [PMID: 21060810]

[85] Casagrande L, Ruiz JC, Beverley SM, Cruz AK. Identification of a DNA fragment that increases mitotic stability of episomal linear DNAs in *Leishmania* major. Int J Parasitol 2005; 35(9): 973-80.
[http://dx.doi.org/10.1016/j.ijpara.2005.04.004] [PMID: 15996670]

[86] Kushnir S, Cirstea IC, Basiliya L, Lupilova N, Breitling R, Alexandrov K. Artificial linear episome-based protein expression system for protozoon *Leishmania* tarentolae. Mol Biochem Parasitol 2011; 176(2): 69-79.
[http://dx.doi.org/10.1016/j.molbiopara.2010.12.002] [PMID: 21167214]

[87] Lee MG, e Y, Axelrod N. Construction of trypanosome artificial mini-chromosomes. Nucleic Acids Res 1995; 23(23): 4893-9.
[http://dx.doi.org/10.1093/nar/23.23.4893] [PMID: 8532534]

[88] Patnaik PK, Axelrod N, Van der Ploeg LH, Cross GA. Artificial linear mini-chromosomes for *Trypanosoma brucei*. Nucleic Acids Res 1996; 24(4): 668-75.
[http://dx.doi.org/10.1093/nar/24.4.668] [PMID: 8604308]

[89] Curto MdeL, Lorenzi HA, Moraes Barros RR, *et al.* Cloning and expression of transgenes using linear vectors in *Trypanosoma cruzi*. Int J Parasitol 2014; 44(7): 447-56.
[http://dx.doi.org/10.1016/j.ijpara.2014.03.009] [PMID: 24759431]

CHAPTER 4

Nuclear and Kinetoplast DNA Replication in Trypanosomatids

Marcelo S. da Silva[1], **Maria Alejandra Viviescas**[2], **Raphael Souza Pavani**[1], **Edna Gicela Ortiz**[2], **Camila B. Storti**[2] and **Maria Isabel N. Cano**[2,*]

[1] *Special Laboratory of Cell Cycle, Butantan Institute, São Paulo, Brazil; Center of Toxins, Immune Response and Cell Signaling-CeTICS, Butantan Institute, São Paulo, Brazil*

[2] *Departament of Genetics, Biosciences Institute, Universidade Estadual Paulista "Júlio de Mesquita Filho" (UNESP), Botucatu, São Paulo, Brazil*

Abstract: Trypanosomatids are among the most primitive eukaryotes and therefore exhibit both conserved and unique non-conserved features in the DNA replication machinery. In eukaryotes, nuclear DNA replication is preceded by the assembly of the pre-replication complex (pre-RC), which is coordinated by the six-subunit origin recognition complex (ORC), which together with the Cdc6 and Cdt1 proteins play a central role in the loading of the hetero-hexamer Mcm_{2-7}. In the domain Archaea there are no Cdt1 protein homologs, Mcm is a homo-hexamer, which is recruited by a protein that shows homology with ORC, and Cdc6 (called Orc/Cdc6). Curiously, trypanosomatid pre-RC differs from others eukaryotes in this context, and it appears more similar to that of Archaea, presenting a homolog of protein Orc/Cdc6 and no homologs of Cdt1, in addition to present Mcm as a hetero-hexamer complex. The completion of DNA replication, at trypanosomatid telomeres, apparently is similar to other eukaryotes, although the processing of the leading and lagging telomeres required to generate the 3' overhangs, which serves as telomerase substrate, remains unknown. With the generation of overhangs at the ends of the chromosomes, telomeres are frequently extended by the action of telomerase, whose control also remains unknown. It is worth mentioning that DNA replication in trypanosomatids initiates almost simultaneously in the nucleus and the kinetoplast, suggesting that regulation of DNA synthesis in the two DNA-containing organelles may be coordinated. The kinetoplast DNA (kDNA) consists of mini- and maxicircles, which are replicated by many proteins

* **Corresponding author Maria Isabel N. Cano:** Genetics Departament - IB, São Paulo State University "Júlio de Mesquita Filho" (UNESP), Botucatu, São Paulo, Brazil; Tel/Fax: (+5514) 38800 388; E-mail: micano@ibb.unesp.br

Marcelo Santos da Silva & Maria Isabel N. Cano (Eds.)

whose mechanisms of action remain unclear. This chapter aims to review and discuss the complex DNA replication mechanisms that act independently in the kinetoplast and the nucleus, as well as some fascinating peculiarities exclusive to trypanosomatids protozoa group.

Keywords: DNA polymerases, Kinetoplast DNA replication, Nuclear DNA replication, Origin Firing, Origin Licensing, Origin recognition complex, Replication forks, Replisome, Telomere replication, Topoisomerases.

INTRODUCTION

For most eukaryotes the cell cycle can be considered, in general, as two distinct events: DNA replication, or synthesis (the S phase), and mitosis (the M phase), separated by two gap phases (G1 and G2). In trypanosomatids, the cell cycle is a fundamental process, essential for efficient proliferation, even in inhospitable environments, ensuring a correct duplication and segregation of single-copy specialized organelles, such as flagellum, nucleus, and mitochondrion. The duplication of these single-copy organelles requires precise temporal and spatial segregation of certain specialized structures, which trigger a very clear morphological pattern that changes according to the cell cycle. Chapter 1 contains descriptions of trypanosomatids life stages and their morphological changes within their specific hosts. Also, detailed morphological descriptions of *T. cruzi*, *T. brucei* and *L. amazonensis*, show that these morphological patterns are found in each phase of the cell cycle and vary among trypanosomatid species [1 - 3]. In addition, in these parasites DNA replication initiates almost simultaneously in the nucleus and the kinetoplast, suggesting that DNA synthesis is coördinately regulated [3 - 5].

DNA replication is an important biological process due to its importance in the maintenance and storage of genetic information, which is responsible (directly or indirectly) for cellular metabolic control. Moreover, as mentioned in the previous chapter, trypanosomatids alternate between replicative and non-replicative forms during their developmental life cycle, suggesting that the control of replication may be directly involved in the infective ability of these organisms. As these protozoa diverged early in the eukaryotic lineage, unique aspects of their DNA synthesis can explain how DNA replication mechanisms have evolved.

In this chapter, we will discuss various aspects of DNA replication in the two organelles that contain DNA: nucleus and mitochondrion (kinetoplast). Also, we will explain some of the main functions of DNA polymerases in these protozoans as these are essential enzymes adapted to perform a wide variety of DNA transactions [6, 7].

DNA POLYMERASES

In 1955, Arthur Kornberg began elucidating the fascinating DNA synthesis process. Over two years, Kornberg and his students measured DNA synthesis in *E. coli* extracts, using ^{14}C-labeled thymidine, a known constituent of DNA [8]. Afterward he isolated the first DNA polymerizing enzyme and named it DNA polymerase. In 1959, Kornberg shared the Nobel Prize in Physiology or Medicine with Severo Ochoa, who had demonstrated the mechanisms in the biological synthesis of ribonucleic acid (RNA) [8]. Kornenberg studies paved the way for subsequent studies seeking to elucidate DNA replication mechanisms in Bacteria, Archaea and Eukarya.

DNA polymerases can replicate DNA in both strands, by adding nucleotides continuously on the leading strand and discontinuously on the lagging strand. They play an essential role in DNA replication [9, 10] and are involved in several processes from DNA chain elongation, to recombination, replication, and repair processes [11, 12]. DNA polymerases are divided into seven families according to their sequence similarities (Table **1**). Polymerases γ, θ and Pol I are members of A family. The B family is represented by α, ε, δ and Pol II and Pol III is the sole member of C family. Some members of D family remains uncovered, but here are include some enzymes found in the Archaea. Tdt, β, λ, μ and σ members represent the X family. Y family includes Pol I, η, κ, Pol IV and Pol V. RT family is the last group and includes the reverse transcriptases (RT) in retroviruses, and the eukaryotic RNA-dependent DNA polymerases; both can do the reverse synthesis of RNA into DNA (Table **1**). In most eukaryotes, the RT family is usually restricted to telomerase [13].

Table 1. Distribution of DNA polymerases family among the domains of life.

Family	Types	Function	Domain of life
A	γ, θ and Pol I	Replicative and repair polymerase	Eukarya (including trypanosomatids) and Bacteria
B	α, ε, δ and Pol II	Replicative and repair polymerases	Eukarya (including trypanosomatids) and Archaea
C	Pol III	Chromosomal replicase leading and lagging strands simultaneously	Bacteria
D	uncertain	Likely chromosomal replicase, but still uncertain	Archaea
X	Tdt, β, λ, μ and σ	DNA repair	Eukarya (including trypanosomatids)
Y	ι, η, κ, Pol IV and Pol V	Translation synthesis on damaged DNA	Eukaryotic (including trypanosomatids) and Bacteria
RT	RT	Transcriptase Reverse	Retroviruses and Eukarya (including trypanosomatids)

The most representative family of DNA polymerases in eukaryotic organisms is the B family, which is found in Archaea, Bacteria, and large DNA viruses. Members of the B family have conserved domains in these organisms, suggesting that they may share a common ancestor. According to some authors, the evolution of eukaryotic DNA polymerases may have involved previously unnoticed complex events [14, 15]. This idea is based on the fact that the archaeal ancestor of eukaryotes encoded three DNA polymerases (two B family polymerase members and a D family polymerase member) all of which contributed to the evolution of the eukaryotic replication machinery. These authors also suggest that the essential zinc (Zn) finger domain, present in most eukaryotic polymerases, might have been acquired from the archaeal D family by polymerases in the B family. This may have given rise to Pol ε either before to or during the eukaryogenesis. Pol ε, in turn, should have been captured by the ancestor of other eukaryotic B family polymerases [5, 14]. Another recent hypothesis suggests that DNA polymerase genes evolved through horizontal gene transfer between nucleo-cytoplasmic large DNA viruses and archaeal-eukaryotic lineages [15].

Trypanosomatids DNA Polymerases

Although distinct DNA polymerases have been described in many eukaryote species [13, 16], the number of DNA polymerases in trypanosomatids and their relationship with homologous enzymes are not well established. There are some interesting descriptions about mitochondrial DNA polymerases, but nuclear DNA polymerases are poorly described. Nevertheless, a more detailed description of the four DNA polymerase families found in trypanosomes (A, B, X and Y) and its peculiarities comparing with polymerases from other eukaryotes, will be shown below.

A Family: The A-Family contains the Pol A domain in the C-terminal region considered a conserved signature. Members of this family play a variety of roles such as gap filling and primer removal, replication of mitochondrial and phage DNA and DNA repair. The bacterial DNA polymerase I is included in this family and curiously, multiple mitochondrial enzymes related to Pol I (Pol IA, IB, IC and ID) were found in *T. brucei* [17]. Pol IB, IC and ID are required to maintain the integrity of the kDNA complex network during the life cycle of both procyclic and bloodstream developmental forms [17 - 20]. These enzymes appear to be essential since the others do not compensate the loss of one member of the family. In contrast, Pol IA does not seem essential [17, 18]. As mentioned before, these polymerases are predominantly mitochondrial in *T. brucei*, with Pol ID showing a dynamic localization, from the mitochondrial matrix fraction to the antipodal sites [21].

Although most studies involving A family in trypanosomatids have been performed in *T. brucei*, orthologues of Pol IA, IB, IC, and ID were found in *T. cruzi* and *L. major* [17]. This evidence suggests that Pol I is conserved among trypanosomatids.

B Family: As previously mentioned, this family is subdivided into α, ε, δ and Pol II. Until now, only Pol α was described in some trypanosomatids: *T. brucei*, *L. mexicana* and *L. donovani* [22 - 24].

In mot eukaryotes, Pol α contains six conserved regions, which are essential for its catalytic function and lagging strand DNA synthesis [22, 25, 26]. Curiously, in Kinetoplastids there is a seventh region localized in the middle of the active site of the enzyme, which is unique to this group [22].

X Family: The main characteristic of the X family of polymerases is the presence of 26 essential amino acids in the catalytic core of the enzyme. This family is associated with the DNA synthesis reaction and dRP lyase activity, acting in the mitochondrial base excision repair (BER) process [27]. Pol β is the main representative member of the X family, and in various eukaryotes it is involved in DNA repair [28, 29], DNA replication, recombination and base excision repair [28].

In trypanosomatids, in contrast, Pol β-like proteins were identified in *Chrithidia fasciculata* and *T. brucei* [17, 26, 30]. In both organisms, the proteins are localized in the mitochondria and are involved in kinetoplast DNA replication [17, 31]. *T. brucei* presents two distinct Pol β, Pol β and Pol β-pak, which are active as DNA polymerases and deoxyribose phosphate lyases, but are structurally and biochemically distinct [31].

In *Leishmania infantum*, in contrast, Pol β is nuclear and plays a role in BER [32]. Its expression is cell cycle dependent and appears to be higher at the infective stationary phase [32]. In *L. mexicana*, Pol β was found in both promastigote and amastigote forms [26].

Pol β and Pol β-pak were also found in *T. cruzi* kinetoplasts, showing intrinsic dRP lyase and replication *in vitro* activity [33, 34]. Moreover, recent studies have shown that in *T. cruzi* Pol β gene is highly expressed in both epimastigote and trypomastigote forms. This evidence suggests that, in addition to kDNA replication, in *T. cruzi* Pol β may have another function, such as an in DNA repair, since trypomastigotes are the non-replicative forms of the parasite [34].

Y Family: As mentioned above, polymerases κ and η are members of the Y family, and carry out translesion synthesis on damaged DNA [35, 36]. Pol κ can replicate DNA across massive distortions formed by multiple lesions [37 - 42], although has lower fidelity and processivity than the replicative polymerases [43]. It is worth noting that some Pol κ are required for specific DNA repair and they may have had a fundamental role in evolution and adaptation of a particular environment [44].

Trypanosomatids present tandem copies of the Pol κ-encoding gene: *T. cruzi* has two copies, *L. major* has three copies and *T. brucei* has ten copies [44]. *T. cruzi* Pol κ has an intrinsic DNA-binding activity, and its localization is restricted to the kDNA, without any detection in the nucleus [44]. Its presence in the kinetoplast suggests a role in kinetoplast DNA maintenance and recombination [44]. Moreover, Pol η activity was also described in *T. cruzi*. This enzyme can synthesize DNA *in vitro* using different templates [45].

Although most trypanosomatid nuclear DNA polymerases are poorly described, gene sequences encoding for some of them, such as Pol α, Pol β and Pol δ, have been reported in the public TriTrypDB database. With this information, *in silico* analysis were performed and functional domains, such as the catalytic domains, were identified in each polymerase [5]. Interestingly, the percentage of identity/similarity between the trypanosomatid sequences and their mammalian counterparts was very low, suggesting that although apparently conserved, these enzymes have specific features, which are probably related to the peculiar functions restricted to the trypanosomatids replication machinery.

NUCLEAR DNA REPLICATION

Nuclear DNA replication is a complex event that occurs inside the nucleus of eukaryotic cells allowing the complete DNA duplication before cells enter in the division. This process needs to be precisely regulated since it can be directly affected by the extracellular and intracellular environments. To be replicated, double-stranded DNA must be opened allowing the assembly of the components of the replication machinery that will provide access of DNA polymerases to copy each DNA strand. The assembly and interaction of the replication machinery components are didactically subdivided into small events, which will be explained in more detail below.

Recognition of Origins and Assembly/Activation of Pre-replication Complex

DNA replication starts at specific sites in the DNA molecule called replication origins, which are present in every chromosome, although their number per chromosome varies among organisms. Bacterial chromosomes for example, generally have a single origin (named *Ori*C) with ~260bp in size, composed of conserved repeats, necessary for initiator recognition, and an AT-rich sequence where the replication bubble originates [46]. In *E. coli*, the ATP-dependent initiator protein DnaA recognizes a 9 bp sequence called the DnaA box that is repeated many times and can be present in both DNA strands. Several DnaA molecules bind cooperatively to the DnaA box, forming a multimeric complex that distorts the double helix, resulting in rupture of the hydrogen bond of the adjacent AT-rich sequence, allowing DnaB helicase loading and replisome assembly [46 - 48]. In contrast, eukaryotic chromosomes have multiple origins of replication and exhibit several differences compared to their bacterial counterparts. Replication origins in eukaryotes are set by a three step process: recognition of origins, assembly of pre-replication complex (pre-RC), and activation of pre-RC [49]. It is worth noting that most knowledge about replication comes from mammalian and yeast models.

Recognition of Origins in Eukaryotes: The budding yeast *S. cerevisiae* is the only known eukaryote that possesses the best characterized and conserved replication origins called the Autonomous Replication Sequence (ARS). ARS are recognized by the Origin Recognition Complex (ORC), the eukaryotic protein complex responsible for initiation of replication [50, 51]. In all other eukaryotes, replication origins appear to be defined locally by DNA structure rather than by DNA sequence conservation [49, 52].

Assembly of Pre-RC: The pre-RC is formed during the M/G1 transition when CDK activity is low. In most eukaryotes, pre-RC is composed of ORC_{1-6}, Cdc6, Cdt1, and Mcm_{2-7}, which are conserved among taxa ranging from yeast to human [53]. ORC was first described in *S. cerevisiae* and is composed of six subunits Orc1-Orc6 [51, 54]. Orc1, Orc4, and Orc6 possess the family domain AAA+ that confers ATPase activity essential for origin recognition. Orc_{1-5} contains a Winged-Helix Domain (WHD) responsible for DNA binding activity [55, 56]. ORC is the first protein that recognizes the replication origin and it recruits two other proteins, Cell division cycle 6 (Cdc6) and Cdc dependent transcript 1 (Cdt1), both of which are required to load Mini-Chromosome Maintenance (Mcm). Mcm is a hexameric ring-shaped complex composed of six proteins Mcm2-7 and exhibits helicase activity that can unwind double-stranded DNA [57 - 59]. All six subunits have the same AAA+ ATPase domain essential for its activity [60]. Once Mcm is loaded to the pre-RC assembles and remains bonded to DNA throughout G1 phase, it is activated only at the beginning of S-phase [61].

Activation of Pre-RC: The loading of Mcm onto DNA is referred to as DNA replication licensing. Although Mcm_{2-7} serves as the primary motor for unwinding DNA during replication, it does not exhibit robust activity until additional factors join with it. In other words, conversion of the pre-RC to a replication fork complex (called replisome), requires the association of additional factors, such as Cdc45 and the GINS complex (GINS is the abbreviation for Go-Ichi-Ni-San, Japanese words that represents the numbers 5, 1, 2, 3 for the proteins Sld5-Psf--Psf2-Psf3 that forms the complex) [62]. At the G1/S phase transition, two protein kinases, Cdc7-Dbf4 kinase (DDK) and the cyclin-dependent kinase (CDK), trigger a chain reaction that results in the phosphorylation of the Mcm complex [62]. The Mcm_{2-7} serves as a platform for GINS and Cdc45 recruitment, forming the Cdc45/Mcm/GINS (CMG) complex, which then act as a replicative helicase competent to unwind the double-stranded DNA [62, 63]. It is worth noting that DDK and CDKs phosphorylate several other replication factors (*e.g.* Mcm10, CDC45, ATP-dependent DNA helicase Q4 (RECQL4), treslin, GINS, DNA topoisomerase 2-binding protein 1 (TOPBP1) and DNA polymerase ε) to promote their loading to the origins [64].

Origins Recognition, Licensing and Firing in Trypanosomatids

Unlike other eukaryotes, little is known about the DNA replication process in trypanosomatids, protozoa parasites that diverged early in eukaryotic evolution. Curiously, the pre-RC of these organisms differs from pre-RC of other eukaryotes and appears to be similar to that of *Archaea*. Trypanosomatids and *Archaea* lack homologs of ORC, Cdc6, and Cdt1, but possess a single chimeric molecule called Orc1/Cdc6 that shares similarities with both Orc1 and Cdc6 [65, 66]. Orc1/Cdc6 from *Archaea* is a member of AAA+ ATPase superfamily and a component of pre-RC that recognizes replication origins and recruits the Mcm complex [65]. In trypanosomatids, it was shown that Orc1/Cdc6 from *T. cruzi* and *T. brucei* have ATPase activity and localize in the nucleus throughout the cell cycle [66]. Interestingly, Orc1/Cdc6 complement yeast Cdc6 but do not complement yeast Orc1, probably because yeast Orc1 recognizes specifically ARS elements (see "Recognition of origins" topic) present in *S. cerevisiae* chromosomes. Another hypothesis is that the smaller size and structure of Orc1/Cdc6 compared to yeast Orc1 do not support ORC complex formation [66].

In *T. brucei*, the absence of Orc1/Cdc6 results in parasites without a nucleus and a decrease in the number of cells in G2/M phases, indicating that DNA replication is compromised [66, 67]. Recent studies identified a novel Orc1-like protein (named Orc1b). It was shown that both Orc1/Cdc6 and Orc1b associate with each other and with Mcm complex *via* interactions with Mcm3 [68]. Orc1b possess the motifs Walker A (involved in ATP binding) and Walker B (functions to orient the nucleophilic water molecule) of a typical AAA-type ATPase domain, although it has not presented any detectable ATPase activity *in vitro* [68]. It is worth mentioning that Orc1 proteins possess an arginine finger motif that exhibits a key arginine residue which may interact with the γ-phosphate of the adenine nucleotide, an interaction that was shown to be important in coordinating ATP hydrolysis with a conformational change, a critical feature of protein-protein interactions. This arginine finger motif is absent in Orc1b [68 - 70]. Immuno-precipitation assays followed by mass spectrometry revealed three other ORC-interacting factors in *T. brucei*. Two of them display sequence features that suggest they may be ORC-related proteins, but they so diverged that their orthology is uncertain. The other, TbOrc4, is related to eukaryotes Orc4 and shows homology with Walker A and B boxes, and a sensor 1 motif, found within AAA+ ATPase proteins. However, it is still unclear whether TbOrc4 is a functional NTPase, since it lacks critical conserved residues, although the arginine finger appears to be conserved in this protein [71]. It is proposed though that Orc1b and Orc4 may be part of the pre-replication complex in Trypanosomatids. Their roles in replication are still unclear, and we believe that more experimental data are needed to determine their functions (Fig. **1A**).

In contrast with *Archaea*, which contain only one Mcm protein that forms a homohexameric ring, trypanosomatids encode six distinct Mcm proteins (Mcm_{2-7}) that can form a heterohexamer ring as in higher eukaryotes [68, 72, 73]. Since there is no Cdt1 in trypanosomatids, the mechanism by which the Mcm complex is recruited to the replication origin remains unknown. Some authors speculate that the Mcm complex is recruited only by Orc1/Cdc6, as in Archaea, or by Mcm9, which has been proposed to be a "co-licenser" of the Mcm complex present in trypanosomatid genomes [5, 74]. Another possibility is that an unknown protein could recruit Mcm to the replication origin, in a manner analogous to Cdt1, although further studies are necessary to confirm this hypothesis (Fig. **1B** and **1C**). These examples of proteins that compose pre-RC support the idea that trypanosomatids possess intermediate characteristics between Archaea and higher eukaryotes. Fig. (**1**) summarizes the possible processes of origin recognition, licensing and firing in trypanosomatids.

Replication Fork Progress

Replication fork progression is a very conserved process throughout eukaryotes and involves many proteins and protein interactions. In this section, we will focus on some of the core proteins involved in this essential process, on their functions, and on the putative orthologues in trypanosomatids.

During the bidirectional synthesis of new DNA strands, double-stranded DNA is unwound by DNA helicases to form two replication forks containing two single-stranded templates [75], providing access of DNA polymerase α (Pol α)-primase, replicative polymerase δ (Pol δ), and polymerase ε (Pol ε) to both DNA strands. Pol α-primase complex assembles the RNA-DNA primers required by the processive Pol δ and Pol ε synthesizes of bulk DNA on the lagging and leading strand, respectively. During primer synthesis, the primase subunits synthesize *de novo* the RNA primer (an oligomer of 7-12 ribonucleotides), which undergoes limited extension with deoxyribonucleotides by Pol α [76].

To strengthen the interaction between template and polymerase, DNA sliding clamps have evolved to promote the processivity of replicative polymerases. In eukaryotes, the clamp (called Proliferating Cell Nuclear Antigen - PCNA) is recruited by the clamp loader, the Replication Factor C (RFC) complex [77]. Therefore, the Pol δ function must be considered in the context of these two factors (PCNA and RFC), which together build the moving platform for Pol δ. This moving platform also provides an essential framework for Pol δ, which may play roles such as recruitment, binding to the primer terminus, processivity, prevention of non-productive binding, release after DNA synthesis, and bridging to other replication proteins [78 - 80]. Also, RFC and PCNA regulate the switch between Pol α-primase and the replicative polymerases, and act as a loading platform for FEN-1 (Flap Endonuclease-1) and DNA ligase I, which process and seal Okazaki fragments [81, 82].

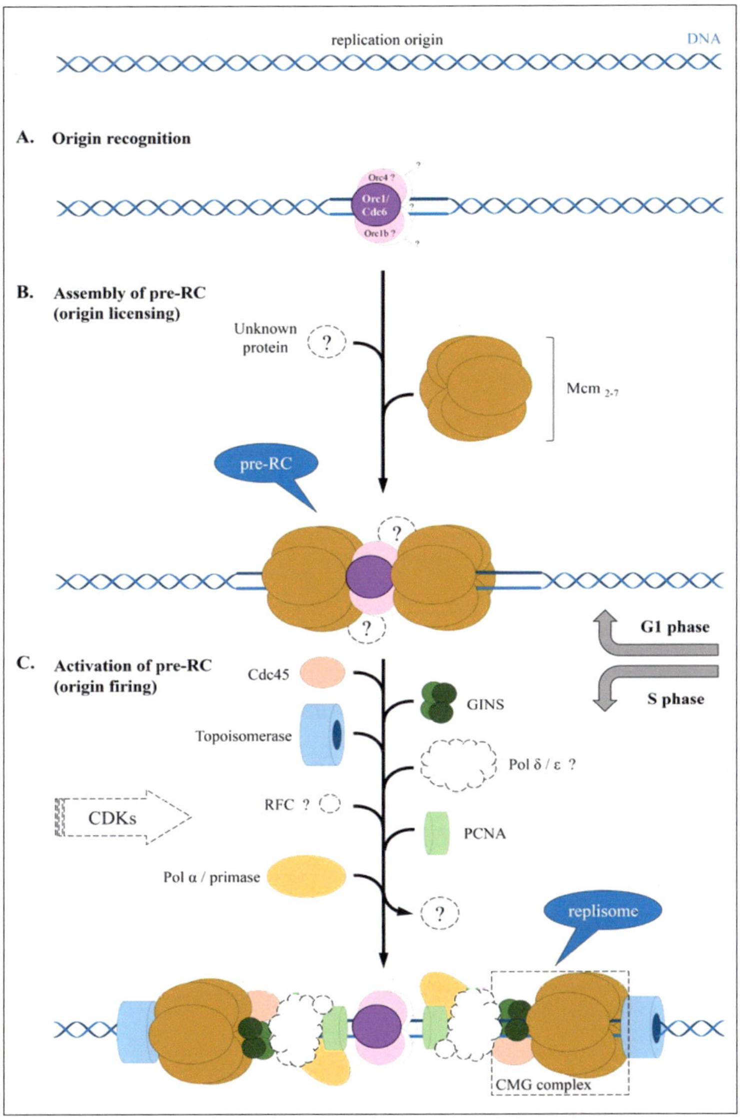

Fig. (1). Origin recognition, licensing and firing in trypanosomatids. **(A).** Origin recognition on DNA is done by Origin Recognition Complex - ORC (Orc1/Cdc6 in trypanosomatids). Although it is proposed that Orc1b and Orc4 may be part of the pre-replication complex in Trypanosomatids, their roles in replication are still unclear, as well as whether Orc1/Cdc6 is a heterohexamer or act as a homodimer, similarly to its homologs in Archaea. **(B).** Assembly of pre-RC (also called origin licensing) is restricted to the G1 phase of the cell cycle. To assemble the pre-RC, the recruitment of Orc1/Cdc6 to replication origins is followed by the loading of

Mcm helicase complex, which contains six subunits Mcm_{2-7} in trypanosomatids. The potential Orc1/Cdc6 interactor responsible for Mcm recruitment remains unknown, as well as whether Orc1/Cdc6 can fulfill this role alone. **(C).** Activation of pre-RC (also called origin firing) involves assembly and activation of a functional replisome, which occurs in the S phase. Probably CDKs phosphorylate several replication factors (of which CDC45, GINS, DNA topoisomerase, potential DNA polymerase ε / δ are the most important) to promote their loading on origins. In some point between origin licensing and firing, the Mcm_{2-7} double hexamer divides into two hexamers that function at the two replication forks emanating from the replication origin. This probably induces the recruitment of the clamp loader (replication factor C - RFC), the clamp (proliferating cell nuclear antigen - PCNA), and other DNA polymerases (*e.g.* DNA Pol α) establishing two functional replication forks that move in opposite directions from the activated origin, with the functional replisome at each replication fork. It is not yet clear whether one or two Orc1/Cdc6 remain on the duplicated origin after origin firing, and whether the unknown protein, responsible for Mcm recruitment, is unbound before replisome assembly. The functional helicase at the forks is the CMG complex (which consists of Cdc45, Mcm_{2-7} and the GINS complex).

Pol ε is considered a component of the preloading complex (pre-LC), which contains GINS, Sld2, and Dpb11, suggesting that Pol ε may also play an active role in replication initiation [83]. Pol ε and Pol δ differ in their biochemical interactions with PCNA. In the presence of PCNA, RFC, and Replication Protein A (RPA), budding yeast Pol ε processivity is stimulated ~6-fold, which is less than observed for Pol δ [84].

Topoisomerases are also important in DNA replication. These proteins catalyze addition or removal of supercoils. Topoisomerase type I removes supercoils by nicking and closing one strand of double-stranded DNA. This enzyme is a nuclear protein mostly associated with the nucleolus and chromatin regions that are transcribing genes [85]. Topoisomerase type II can cut both strands of a double-stranded DNA molecule, pass another portion of the duplex through the cut, and reseal the cut in a process that utilizes ATP. Topoisomerases II are essential for DNA replication and chromosome condensation [85], in the movement of the growing fork and in resolving finished chromosomes after DNA duplication [86].

Another important protein in DNA replication is the Replication Protein A (RPA). This protein is the major single-stranded binding protein of eukaryotes that participates in DNA replication, repair, damage response, and telomere maintenance [87]. In most eukaryotes RPA is a heterotrimeric complex, composed of RPA-1, RPA-2, and RPA-3. It was first identified and purified as a protein factor required for both initiation and elongation phases of SV40 DNA replication [88].

RPA binds single-stranded DNA regions exposed at the replication fork to stabilize this region and protect against nucleases. RPA binds DNA *via* oligonucleotide/oligosaccharide binding (OB-fold) domains distributed along all RPA heterotrimeric components, which are called DNA binding domains (DBD) A-E. Here it is worth mentioning that RPA-1 is the most critical subunit of the RPA complex because it is the only subunit that interacts directly with the single-stranded DNA [89, 90]. Following DNA unwinding, RPA helps in the recruitment of Pol α to replication origins where RNA priming is required for initiation [91, 92]. RPA stimulates Pol δ, Pol ε and also Pol α activity and processivity, reducing its misincorporation [90, 93 - 95].

Although little is known about the progress of the replication fork or 'joint action' of the replication fork proteins in trypanosomatids, some components of the replication fork have already been characterized in most of these organisms, as previously described for DNA polymerases and will be discussed later in this chapter.

The DNA replication clamp PCNA was already described in *T. cruzi* and *Leishmania donovani*. TcPCNA is localized in the nuclear periphery during S phase, in agreement with previous studies that showed that replication occurs in the nuclear periphery at least in *T. cruzi* [96, 97]. During all other cell cycle phases, LdPCNA and TcPCNA are visible in a dispersed pattern throughout the nucleus. In contrast, PCNA from *T. brucei* (TbPCNA) are not preferentially located at the nuclear periphery. TbPCNA also appears to be degraded when cells enter in G2 phase, whereas LdPCNA and TcPCNA are present in the nucleus throughout the cell cycle, suggesting that the regulation of DNA replication in these organisms should be different [96 - 98]. Homologs for RFC, FEN-1 and DNA ligase I are also found in the trypanosomatids genomes (TriTrypDB), but functional studies are necessary to understand their participation in the DNA replication machinery [99 - 101].

Topoisomerases I and II have also been characterized in several trypanosomatids. In *Trypanosoma* and *Leishmania* genera, topoisomerase I is composed of two subunits encoded by two separate genes. It is an abundant nuclear enzyme mostly associated with the nucleolus and chromatin regions undergoing transcription [85,

102]. Treatment of the parasites with the topoisomerase I inhibitor camptothecin compromises the incorporation and proliferation of nucleotides, suggesting that this enzyme can influence DNA replication [102].

Topoisomerase II activity is essential for nuclear and kinetoplast DNA replication and segregation. Topoisomerase II purified from *T. cruzi* and *L. donovani* has ATP-dependent and ATP-independent decatenating activities, although the ATP-independent activity may be an artifact [103]. *L. donovani* topoisomerase II (LdTOP2) appears in the nucleus and the kinetoplast, and *T. cruzi* topoisomerase II is only found in the nucleus [103, 104]. Most trypanosomatids topoisomerases II share high sequence identity and work almost exclusively as mitochondrial enzymes. Recently, two nuclear type II topoisomerases (TbTOP2α and TbTOP2β) were described in *T. brucei*. Phylogenetically, they share a common node with other nuclear topoisomerases from higher eukaryotes. TbTOP2α encodes an ATP-dependent topoisomerase whereas the role played by TbTOP2β is still unclear. Although transcribed in trypanosomes, TbTOP2β is not detected by a β-specific serum and RNAi silencing results in no obvious phenotype [105].

The potential of DNA topoisomerases as drug targets has led to increased interest in their study among pharmacology and clinical medicine researchers [106]. The identification of DNA topoisomerases as promising drug targets and recent studies showing that Kinetoplastida topoisomerases are markedly distinct from their human counterparts supports the idea that topoisomerase-based antiparasitic drugs would be of great importance in future therapies [107].

The RPA complex is also present in trypanosomatid genomes, although little is known about this protein in these organisms. The best-characterized component of RPA heterotrimer in trypanosomatids is RPA-1, which lacks the 70N domain present in mammalian RPA-1, responsible for the interaction with checkpoint and repair proteins. In addition, the linkers between its three OBfold domains are more flexible compared with yeast and human RPA-1, conferring better flexibility of this protein [108, 109]. As in other eukaryotes, trypanosomatid RPA is composed of subunits RPA-1, RPA-2, and RPA-3 and, in *T. cruzi*, RPA-1 and RPA-2 co-localize with EdU foci after five minutes of incorporation, confirming its role in DNA replication (Dr. Maria Carolina Elias, personal communication). Recent studies have shown that despite the absence of the 70N domain, RPA-1 from *L. amazonensis* probably participates in DNA double-stranded damage response generated by phleomycin treatment, since it interacts with LaRAD51, a protein involved in homologous recombination [108]. Furthermore, *L. major* RPA is also enriched in the chromatin fraction after damage and co-localizes with LmHus1 (a protein component of the 911 complex), although their direct interaction could not be confirmed [110].

In another recent study, Calderano and collaborators [113] calculated the replication rate in *T. brucei* and found that replication is about 15% faster in the bloodstream form (4,4 ± 0,1kb/min) compared to the procyclic form (3,7 ± 0,1kb/min). The authors concluded that these rates are very similar to what was observed in yeast (2kb/min) and mammal cells (2kb-3kb/min) [49, 111 - 113]. These data suggest that, although there are differences in the replication machinery components between higher eukaryotes and trypanosomatids, the DNA replication as a whole is a well-conserved process among eukaryotes and probably has experienced few modifications during evolution.

Telomere Replication and Elongation

Telomeres are the physical ends of eukaryotic linear chromosomes that function mainly to protect the termini from nucleolytic attack. This protective function is called 'capping' and allows the cell to distinguish between telomeres, and DNA double-strand breaks [114 - 116]. With few exceptions, telomeres are composed primarily of repetitive DNA associated with proteins that interact specifically with double- or single-stranded telomeric DNA or with each other, forming highly ordered and dynamic DNA/protein complexes involved in their maintenance and length regulation [117 - 119]. Single- and double-stranded telomeric DNA contain binding sites for isolated proteins and/or protein complexes responsible for telomeric maintenance and homeostasis. These sites are considerably more complex in higher eukaryotes, although a common feature shared by all telomeres is the presence of telomeric proteins that interact directly with the DNA or with other proteins and whose functions appear to be related to the telomeric region to which they interact [119, 120]. These protein complexes are dynamic and vary according to the cell cycle phase, external stimuli, and cell aging [121 - 123].

Also, telomeres are essential for the conclusion of DNA replication, being associated with cell division control, transcription regulation, genome integrity, and maintenance of nuclear architecture [120, 124, 125].

The single-stranded telomeric DNA that protrudes towards the end of the chromosome is G-rich and known as 3' G-overhang or G-overhang (Fig. **2**). This structure was first described in *Tetrahymena* and *Oxytrichia* and after that in many other eukaryotes [126, 127]. Among the trypanosomatids, only *T. brucei* and few *Leishmania* species have their 3′G-overhangs examined by native in-gel hybridization coupled or not to ligation-mediated primer extension. Both parasites show very short 3′G-overhangs ranging in size from 10-12nt long and at least in *T. brucei* no telomere C-overhangs were detected since in-gel hybridization with a telomeric G-strand probe does not yield any signal [128 - 130].

The telomere repeats are very conserved across a wide range of eukaryotic organisms, especially mammals, and the 3′ G-overhang is the substrate for telomere extension by telomerase, the enzyme responsible for replicating the ends of the chromosomes and thus, to maintain chromosome length. The enzyme was first characterized in the ciliate protozoan *Tetrahymena thermophila* and consists in a specialized reverse transcriptase ribonucleoprotein composed of two main components: a telomerase reverse transcriptase protein (known as TERT, telomerase reverse transcriptase) and a long non-coding RNA component (known as TER, telomerase RNA). Both are integral components of the enzyme [131 - 133]. TER contains a short template sequence complementary to the telomeric repeat sequence represented by the 3' G-overhang, which is used as a template by telomerase during telomere extension (Fig. **2**) [134, 135]. The requirement for a 3' G-overhang in telomere extension was first demonstrated 11 years after the discovery of telomerase [136]. It is worth noting that without telomerase, cells lose genetic material at their chromosomal ends at each replication cycle [137, 138].

Sequence analysis of the telomerase reverse transcriptase protein component in *Euplotes aediculatus*, *Saccharomyces cerevisiae*, *Schizosaccharomyces pombe,* and human shows an evolutionary relationship among telomerase and retrotransposons, used by some insects to maintain telomere integrity as an alternative for the absence of telomerase [139 - 143]. Curiously, some authors have shown that the success of telomerase in the eukaryotic world can be caused not only by the efficient synthesis of telomeric repeats but also by its extratelomeric functions, such as an increase in cellular viability independently of telomere maintenance [142, 144 - 146].

The regulation and control of telomerase activity are highly complex and involve multiple layers of control. While the main target of enzyme regulation is the catalytic subunit TERT, the TER expression levels are also implicated in telomerase control. Moreover, expression of enzymatic subcomponents, interactions with telomeric DNA, with telomeric and non-telomeric proteins, the subcellular localization and variations in the size range of 3′ G-overhangs is somehow responsible for controlling telomerase activity [5, 147 - 149]. The 3′ G-overhang can also regulate telomerase recruitment to the telomeres by forming special structures that prevent enzymes from binding and modifying the single strand. The two most relevant conformations have been described in a variety of organisms: t-loops, which also occur in trypanosomatids and single strand G-quadruplexes (also known as G-tetrads or G4-DNA) [118, 123, 124, 150 - 152].

The recruitment of telomerase to the telomeres, which provides the enzyme access to the telomeric DNA, is dependent on protein complexes that vary in several model organisms. In mammals, a group of six proteins (TRF1, TRF2, RAP1, TEN1, TPP1, and POT1), known collectively as shelterin, forms a high order complex that dynamically interacts with both single and double-stranded telomeric DNA. TRF1 and TRF2 bind directly to the double-stranded telomeric DNA, and POT1 binds to single-stranded DNA, while the others (RAP1, TEN1, and TPP1) interact with each other to form a bridge between the TRF1 and TRF2 bound to the double strand DNA and POT1. Shelterin is responsible for telomere protection, t-loop formation and maintenance and the regulation of telomerase access in a telomere length-dependent manner [153 - 155]. Protein complexes similar to shelterin are also present in yeast and plants. Furthermore, budding yeast presents a protein complex known as CST (Cdc13/Stn1/Tin1), whose main actor is the telomere end-binding protein Cdc13, that interacts with RPA and the telomerase component EST1 to control DNA replication at telomeres by recruiting both telomerase and DNA Pol α for C-strand synthesis [154, 156].

Furthermore, components of DNA repair machinery (KU70/80, Mre11 and RAD50), and homologs of DNA damage checkpoint kinases (Tel1 and Mec1) also participate in the recruitment of telomeric proteins for telomerase extension [157]. On the other hand, CST-like telomeric complexes are also present in higher eukaryotes (mammals and plants) with proteins that share structural similarities but not the same functions [155]. These findings indicate that the current concept of how telomerase gains access to short telomeres is very complex, involving many mechanisms that evolved toward a balance among multiprotein complexes that share many similarities to the DNA damage response machinery [147, 148, 158 - 162].

It can be concluded that telomerase requires the presence of a 3' overhang to extend telomeres, and it was recently shown that these 3'overhangs could potentially be generated at both chromosome ends by an additional processing (known as resection) after the completion of DNA replication. It is worth noting that expected leading-strand DNA synthesis generates a blunt end telomere whereas lagging-strand synthesis generates a shorter telomere with a long overhang due to the terminal Okazaki fragment is positioned about 70-100 nt from the end. Thus, both termini require resection to give rise to a functional telomere terminus. It is unclear, however, how this 3' overhang is generated in some eukaryotes, although in mice and yeasts, 3'overhang synthesis has been partially elucidated. In mice, for example, resection of the 5' end occurs at leading telomeres, but not at lagging strand, by the action of the nuclease Apollo bound to TRF2, a shelterin subunit; POT1b, also a shelterin subunit, controls Apollo hyper resection. The nuclease Exo1, in contrast, generates long overhangs in both DNA strands, and its action is blocked by the interaction between POT1b and CST, which recruits Polα/primase for C-strand fill-in synthesis. Thus, the generation of 3' overhangs, although necessary for telomere synthesis, should be regulated since excessive resection may cause telomere shortening [162 - 165]. Although much is known about telomere end processing in eukaryotes [166], relatively little is known of how primitive eukaryotes, such as trypanosomatids, generate telomeric overhangs.

Before describing some aspects of telomeric processing in trypanosomatids, it is worth mentioning general characteristics of these chromosome regions. Telomeres in trypanosomatids show the standard conformation of tandem repeats found in other eukaryotes (5'-TTAGGG-3')$_n$, although these parasites exhibit a peculiarity in their telomeric composition: a modified base β-D-glucosyl-hydroxymethyluracil (known as base J) is enriched in telomeric and subtelomeric regions [5, 167, 168]. Some evidence suggests that this modified base may participate in the regulation of gene expression in subtelomeric regions of trypanosomes [169 - 171]. Moreover, recent studies show that base J prevents transcriptional readthrough in *Leishmania* species [172], although we have evidence for telomere transcription in these organisms ([173,174], Dr. Maria Isabel N. Cano, personal communication).

Trypanosome subtelomeric regions vary among species. *T. cruzi* subtelomeric regions present a small number of hexameric repeats, genes related to parasite virulence, conserved elements, and retrotransposons that may participate in transposition, duplication, and recombination of nearby genes [5, 173 - 177]. The most well-studied subtelomeric regions in trypanosomatids are from *T. brucei*, which are classified into groups that contain a variable number of copies of variant surface glycoprotein (VSG), genes, and pseudogenes known to be involved in parasite evasion of the mammalian immune system [5]. This division in VSG groups occurs due to the position of functional DNA elements in relation to VSG genes and pseudogenes. Some authors claim that the positioning of VSGs adjacent to the telomere facilitates switching, increasing rates of recombination involving duplicate gene conversion and reciprocal telomere translocation [5, 178]. Subtelomeric regions have also been extensively investigated in *Leishmania* species, where they tend to be relatively short in comparison to their trypanosome counterparts. Although variations may occur in different species and even among chromosomes, they consist mainly of 100 bp repeats, known as LCTAS (*Leishmania* Conserved Telomere-Associated Sequence), interspersed with CCCTAA repeats [129, 179 - 182]. In some species of *Leishmania*, subtelomeric regions also contain a Non-Repeated Telomeric Associated Sequence (NRTAS), as well as developmentally-transcribed noncoding RNA elements that are exclusively expressed in amastigotes and vary in size among individual chromosomes [183 - 185]. Details about the organization of the subtelomeric regions of trypanosomatids are found in Chapter 3 of this book.

Currently, few telomeric proteins have been identified and characterized in trypanosomatids protozoa. *T. brucei* telomeres contain functional homologs to the higher eukaryote proteins TRF, Tin2, Tpp1, and Rap1 [168, 185, 186]. *L. amazonensis, L. major*, and *T. cruzi* show only TRF homologs and, together with *L. tarentolae* and *T. brucei*, have been shown to have telomerase activity [187 - 190]. The gene encoding the telomerase protein component (TERT) from four *Leishmania* species has been cloned and characterized. Besides being one of the largest telomerase described to date, trypanosomatid TERTs show specific amino acid substitutions within the conserved telomerase motifs and present a peculiar T motif (telomerase specific motif) [191]. Recently, the telomerase RNA component (TER) was identified and characterized in *L. amazonensis* and *T. brucei*. Both TER also present unusual secondary structures, but contain conserved structural elements such as the template sequence and the TBE motif [192, 193].

To date, the best-characterized trypanosomatid telomeric complex is that of *L. amazonensis*. This complex consists basically of the proteins LaRBP38 and LaRPA-1 which bind *in vitro* and *in vivo* the 3′G overhang. LaRBP38 is also present in the kinetoplast and bind double-stranded telomeric DNA [194]. In other trypanosomatids, homologs of LaRPB38 are named: Tc38 in *T. cruzi* and p38 in *T. brucei*, are exclusively mitochondrial playing roles in RNA stabilization and minicircle replication [195, 196]. It is worth mentioning that LaRPA-1 and LaRbp38 exhibit some structural and functional similarities with the eukaryote CST-like components such as specific subcellular localizations, similar DNA binding mode and binding domain and mutual interaction [5, 109, 196, 197]. Other telomeric proteins identified in *L. amazonensis*, such as LaTBP1 and LaTRF, bind double-stranded telomeric DNA using a Myb-like DNA binding domain. LaTBP1 Myb-like domain is closely related to RAP1 and TFIIIB, whereas LaTRF is a chimeric protein that shares structural and functional similarities to both mammalian TRF1 and TRF2 proteins [189, 198, 199]. Fig. (**2**) summarizes the current knowledge about the fork progression through telomeres, as well as the telomeric chromatin composition in trypanosomatids.

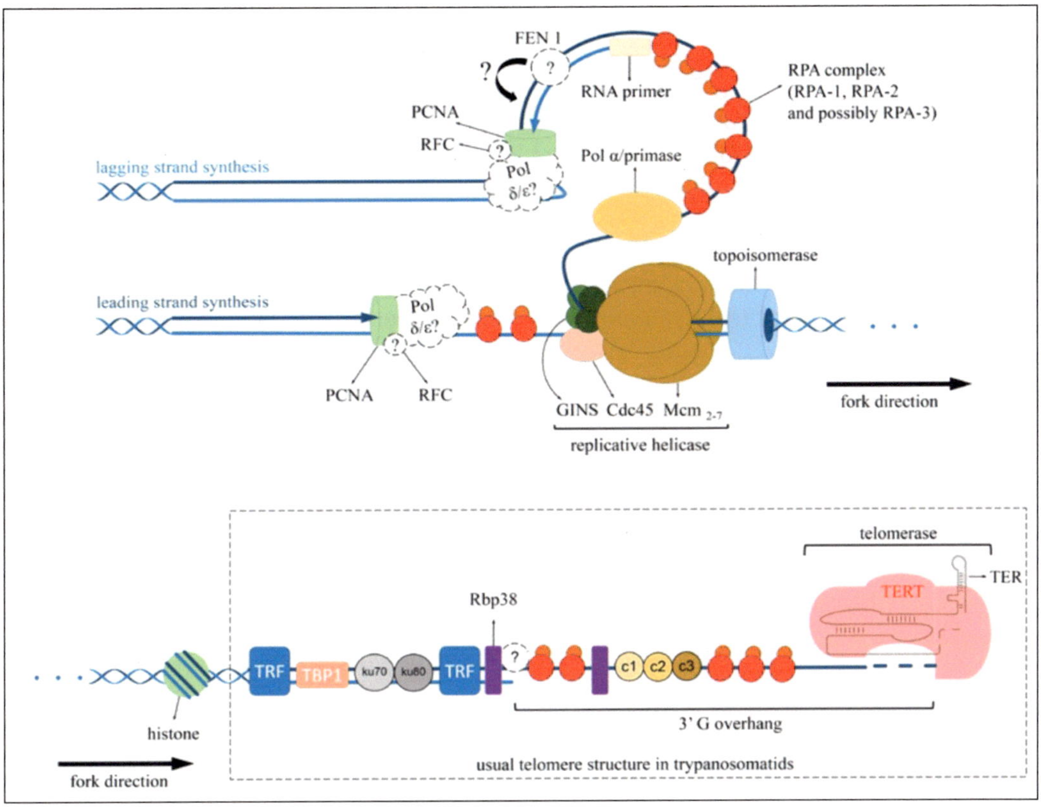

Fig. (2). Scheme for fork progression through telomeres in trypanosomatids. This scheme describes the progression of replisome towards telomeres in trypanosomatids. All the colored proteins belonging to the replisome (topoisomerases, Mcm_{2-7}, Cdc45, GINS, RPA, PCNA and Pol α/primase) have already been described in trypanosomatids [68, 200 - 203]. The proteins with dashed border (DNA polymerase δ and ε (Pol δ and ε), the clamp loader Replication Factor C (RFC) and the 5' flap endonuclease (FEN1)) are present in the TriTrypDB genome databank, but they were not yet characterized. The usual telomere structure in trypanosomatids includes proteins that bind double-stranded telomeric DNA (TRF, TBP1, Ku70, Ku80 and Rbp38) and proteins that bind the G-rich single-stranded telomeric DNA (RPA, c1, c2, c3 and Rbp38) [194, 201, 204]. Although Rbp38 homologs were characterized in other trypanosomatids (Tc38 in *T. cruzi* [195] and p38 in *T. brucei* [196]), only in *L. amazonensis* this protein was found interacting with telomeres [194]. It is still unclear whether RPA interacts with Rbp38 or with other proteins that may bridge double-stranded to single-stranded telomeric proteins (Dr. Maria Isabel Cano, personal communication). The telomeric single-stranded DNA or 3' G-overhang is used as a substrate by telomerase (composed by TER and TERT) to promote telomere elongation. Small 3' G-overhangs can be potentially originated at lagging-strand telomeric DNA synthesis; however, leading-strand telomeric DNA synthesis is expected to generate a blunt-end telomere, which requires an additional processing at its 5' end to generate a 3' overhang and thus be also elongated by telomerase (see main text).

Telomere processing and generation of 3'G-overhangs in trypanosomatids have not yet been described, although, *T. brucei* and few *Leishmania* species have homologs to some of the proteins involved in this process found in TritrypDB, *e.g.* Exo1, MRN, RPA and XRS2/NBS1 [101, 205]. XRS2/NBS1, for example, participate in 5' end resection during homologous recombination to solve DNA Double Strand Breaks, a process similar to lagging-telomere resection that generates the 3' G-overhang [101, 205]. Here is worth mention that the generation of 3′G-overhangs is not only necessary for telomere length maintenance, but it is also related to cellular aging since telomere attrition rates determine the replicative lifespan of most eukaryotic cells [206].

A better understanding of replication, elongation, and telomeric processing in trypanosomatids may facilitate the discovery of new parasitic targets for the development of future strategies for the eradication of the diseases caused by these organisms, since telomeres homeostasis is directly related to genome maintenance and cell viability.

Control and Regulation of Nuclear DNA Replication

DNA replication is an essential process in eukaryotes, including trypanosomatids that ensure high fidelity copy of the entire genome. It is also highly regulated involving the actions of many licensing and replication factors. Although this mechanism is only recently beginning to be uncovered in trypanosomatids, it is widely believed that similarly to other eukaryotes, parasites have evolved an intricate origin replication licensing and firing system to avoid DNA re-replication within the same cell cycle. The two steps comprising the origin replication licensing and firing is temporally separated and are coordinated by changes in CDKs (Cyclin Dependent Kinases) activity. High CDK activity triggers origin firing and marks the start of DNA synthesis in S phase. CDKs also inhibit assembly of pre-RCs and, consequently, licensing. Once an origin has fired, it cannot be re-licensed until the cell has exited mitosis and CDK activity decrease, ensuring origins fire just once in each cell cycle [207].

Normally, CDK phosphorylates and inhibits many of pre-RC components in several organisms. In yeast, CDK phosphorylates Cdc6, which undergoes ubiquitin-mediated proteolysis by the Cdc4/34/53 pathway [207, 208]. CDK also causes phosphorylation and inactivation of some proteins belonging to the ORC complex and nuclear export of the Mcm complex [207, 209]. In mammals, pre-RC components phosphorylated by CDK can block licensing [207]. Phosphorylated Cdt1 experiences ubiquitin-mediated proteolysis [210] and phosphorylation. Ubiquitylation of Orc1 can cause its degradation or nuclear export during S and G2 phase [211, 212]. Chromatin-unbound Cdc6 appears to be exported when phosphorylated by CDK during S phase. However, chromatin-bound Cdc6 is unchanged throughout S phase, and it remains unclear how this process inhibits pre-RC assembly [213 - 215].

Briefly, after initiation of DNA replication, CDKs prevent DNA re-replication mainly through three mechanisms: phosphorylation of the origin recognition complex (ORC), downregulation of Cdc6 activity, and nuclear exclusion of the Mcm2-7 complex. The control of DNA replication in Trypanosomatids appears to differ from that of other eukaryotes. As previously mentioned, trypanosomes express a single Orc1/Cdc6-like protein, which shows similarities to the eukaryote homologs Orc1 and Cdc6. It associates with chromatin throughout the cell cycle and is essential for DNA replication [66]. In *T. brucei*, an Orc1-like protein (TbOrc1b) was found associated with TbOrc1/Cdc6 both *in vitro* and *in vivo* [68]. Three other proteins were also identified as potential partners of TbOrc1/Cdc6. One of these, TbOrc4, is distantly related to Orc4 while the other two have low homology to ORC complex proteins and have not yet been characterized [71].

Despite uncertainties regarding pre-RC composition in trypanosomatids, these protozoa apparently contain a well-conserved CMG complex that interacts with TbOrc1/Cdc6 and TbOrc1b *via* Mcm3 [68]. Curiously, it is unclear how the CMG complex is loaded onto the replication origins since homologs of Cdt1 are absent in the trypanosome genome. The interactions between Mcm3 and both TbOrc1/Cdc6 and TbOrc1b appear to be enough to recruit CMG complex to the replication origins [216].

Unlike in yeast [217], the Mcm complex in trypanosomes remains in the nucleus throughout the cell cycle [68], whereas Cdc45, a member of the CMG complex described previously, appears to be exported after DNA replication [68, 216].

Since trypanosomatids diverged very early in the eukaryotic lineage, they have apparently evolved distinct mechanisms to prevent DNA re-replication and promote DNA replication initiation. Further studies are necessary to clarify the control and regulation of nuclear DNA replication in trypanosomatids.

KINETOPLAST DNA REPLICATION

Protozoa belonging to the Kinetoplastida order are characterized by their network of circular DNA present inside the unique large mitochondrion, known as kinetoplast. Kinetoplast DNA (kDNA) consists of a vast number of relaxed circular DNA molecules interlocked with each other to form a DNA network that resembles medieval armor. This complex and unique structure requires equally complex replication machinery to maintain both the structure and the sequences of the network. This section will provide an overview of kDNA components and the replication mechanisms of kDNA network. Early studies on these processes were performed mainly in *Crithidia fasciculata,* an insect parasite that is especially easy to culture *in vitro*. Recent studies were done mainly using *T. brucei* as a model, due to the numerous genetic tools available for this organism. Many excellent reviews on this subject are available [218 - 221].

Components of kDNA: Mini- and Maxicircles

kDNA contains two types of DNA molecules, mini- and maxicircles, which differ in size, function, and replication mechanism. Maxicircles comprise the mitochondrial genome of Kinetoplastids, which encode few genes whose products present local functions such as rRNA genes and protein genes that are involved in the energy transduction pathway. The kDNA network contains from 25 to 50 maxicircles that range from 20-40 kb in size. Maxicircles are identical in sequence to one another and contain both conserved and variable regions [222]. The genes transcribed by the maxicircles have to undergo an editing process consisting of uridines insertions and deletions to create functional mRNAs. The edited residues sometimes represent half of the mature mRNA and, since these insertions are made in the middle of the sequence, a high degree of accuracy is essential. RNA edition is achieved through the action of guide RNAs (gRNAs), encoded by the second component of the kinetoplast network, the minicircles [223 - 225].

C. fasciculata kinetoplast network is comprised of several thousand minicircles of about 2.5 kb each, which vary in sequence since they code for various gRNAs. The more editing means more sequences present in a given organism; although, there are usually predominant sequences among the minicircles [223]. When they are not undergoing replication minicircles are fully closed and relaxed [226], linked together to form the fishnet structure of the kinetoplast network. The fact that they are relaxed is odd since most of the other circular DNA molecules are negatively supercoiled. However, the relaxed state of minicircles might be a requirement of network formation, since a supercoiled circular DNA would not have enough space to be intertwined with other molecules. In *C. fasciculata*, all the minicircles are linked together forming an arrangement in which each minicircle has other three minicircles attached to it (minicircle valence = 3) [227]. Other organisms might have a different valence, which implies only slight variations to the network topology. Maxicircles also form catenates but instead of forming a net they form rosette-like structures and are therefore more densely packed in the center [222].

In the classical model, the maxicircle catenate is intertwined with the minicircle catenate, forming a network within a network. The kDNA network is compacted together to form a disk-shaped structure about 1 μm in diameter and 0.35 μm thick, which is localized in the mitochondrial matrix near the flagellum. This structure is visible by electron microscopy and is called kinetoplast disk [218].

The previously described network-within-a-network model has mainly been studied in *C. fasciculata* but is also present in trypanosomatids, such as *Leishmania* spp. and *Trypanosoma* spp. This has led to the idea that all Kinetoplastids share the same kDNA structure with only minor variations. However, recent studies in other groups like the Bodonida order have shown an amazing array of different kDNA arrangements, including rudimentary networks, monomeric minicircles (both relaxed and supercoiled), and even the complete absence of minicircles [228]. This presents interesting questions regarding the evolution of the kDNA disk. However, the remaining part of the section will focus on the replication mechanisms of network-arranged kDNA.

kDNA Replication: Mechanisms and Proteins Involved

Minicircle replication: Minicircles are not replicated while attached to the network. To undergo replication they must first be detached by a type II topoisomerase and released into the kinetoflagelar zone (KFZ), a region near the flagellum where free minicircles are replicated [229]. Minicircle sequences are highly heterogenic, even within the same species; however, they all share a conserved region of about 100-200bp [230, 231]. This region contains two sequences present in almost every minicircle studied to date. The first (5' GGGTTGGTGTA 3' – 12-mer) is present in the lagging strand and has been termed the 'universal minicircle sequence' (UMS) [230], which is recognized by the replication initiation machinery, and is where the first complementary primer of the leading strand is inserted. The number of 12-mer sequences present in a single minicircle varies depending on the species: *T. brucei* and *L. tarantola* have one, *C. fasciculata* have two, and *Trypanosoma cruzi* have four. In species with multiple replication origin sequences, only one is used, and this is randomly determined [232, 233]. The second sequence present in almost every minicircle is a less conserved sequence (5' ACGCCC 3' - hexamer) and is located 70-90 nucleotides downstream from the origin sequence in the leading strand. The first Okazaki fragment starts complementary to that sequence. It is worth mentioning that Okazaki fragments in minicircles have approximately 100 bp in length [234, 235].

Minicircles are replicated unidirectionally with the formation of theta structures intermediates [224, 225]. Once replication is completed, the minicircles migrate to the antipodal sites of the kDNA network. In the antipodal sites, there are two protein complexes, located 180° about each other, which contain proteins that aid the kDNA replication process. These enzymes remove the RNA primers and reattach minicircles to the network.

Since every minicircle can only be replicated once, a control mechanism to know which minicircles were already replicated is needed. When the enzymes at the antipodal sites replace the RNA primers for DNA, they always leave at least one primer space without DNA, so the minicircle is reattached to the network containing at least one gap [224, 225]. The last gaps to be closed correspond to the origins sequences (12-mer and hexamer) [232]. Thereby, the replication machinery can differentiate minicircles because the unreplicated ones are covalently closed while the replicated have gaps. Once all the minicircles have been replicated, all the gaps are filled, and then minicircles are covalently closed.

Maxicircle replication: The mechanism underlining maxicircles replication is far less understood than that of minicircles. Maxicircles also replicate unidirectionally to form theta structures intermediates but, unlike minicircles, they do so while they are still attached to the network [236]. Maxicircle replication starts in the variable region and proceeds clockwise from there [222, 236]. Two copies of the sequence (5' GGGGTTGGTGT 3') were found in the variable region of *T. brucei*. This sequence is very similar to the 12-mer UMS from minicircles, and this sequence similarity raises the possibility that both mini- and maxicircles may use similar replication machinery.

Mechanisms of kDNA Replication

The kDNA arrangement in a network-like structure implies that a very complex replication process is necessary to maintain not only the individual components but also the overall structure. The reason for such an intricate structure has arisen, and its preservation of the Kinetoplastida order is yet to be found. However, one of the most accepted hypotheses is that the network arrangement permits control of minicircle migration which helps prevent the loss of sequences by random segregation, guarantying the presence of all essential gRNAs in the kinetoplast progeny [221].

As discussed above, minicircles are detached from the network before being reattached to it. Consequently, at the end of minicircles replication, the net has twice as many minicircles than at the beginning. Since the space occupied by the network is the same, the replicated network is more densely packed, and the minicircle has higher valence. At the end of the replication, the kinetoplast membrane enlarges, and more space is available. The densely packed network is rearranged, increasing the network area correspondingly. At the end of this first rearrangement, the network is divided into two daughter networks whose shape suggests that the division is done by a straight cut along the short axis of the network. Finally, the newly divided networks go through a final remodeling which gives them their characteristic oval shape (Fig. **3**) [237].

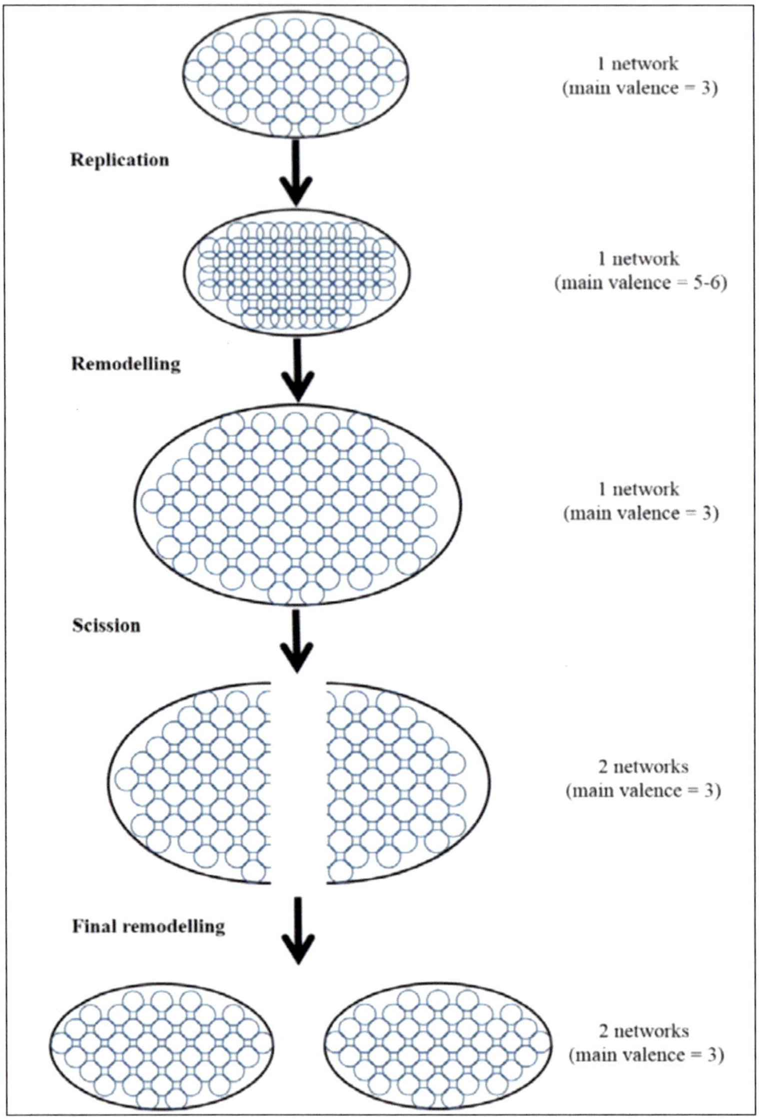

Fig. (3). Network changes during kDNA replication. In the beginning, there is a single network with minicircle valence of 3. After minicircle replication, the network has the same size but twice as many minicircles, thus the main valence increases to 5-6. Then, during remodeling the network is twice as big, and the minicircle valence goes back to 3. After scission, there are two networks with the original minicircle number, but without the characteristic oval shape. Then, final remodeling generates 2 daughters networks identical to the parental in minicircle number, valence, and shape.

To date, two mechanisms of minicircle replication and reattachment have been described: the Ring mechanism, and the Polar mechanism [238].

Ring mechanism: This is the kDNA replication mechanism found in *C. fasciculta, T. cruzi, L. major* and *L. tarentolae*. Pulse experiments and electron microscopy have shown that, at the beginning of replication, the newly replicated minicircles are aggregated at two points 180° about each other. Since the replicated minicircles are reattached to the network by the protein complexes located at the antipodal sites of the kinetoplast disk, the experiments suggest that the protein complexes rotate in relation to the kinetoplast network. It is still unknown whether it is the protein complex or the DNA that are moving (Fig. **4A**) [237 - 239].

Polar mechanism: In this mechanism there is a movement of the network relative to the antipodal sites. The newly replicated minicircles are aggregated at the polar sites of the network. Along the replication, these polar regions enlarge until only maxicircles remain in the central region (Fig. **4B**) [238, 239].

It is not yet clear how individual minicircles are not lost during replication. Different hypotheses have been proposed: It may be that every time a minicircle is replicated, each new molecule goes to separate antipodal sites. It has also been suggested that the leading chain minicircle is incorporated at a higher rate than the lagging chain minicircle [221]. Finally, it has also been suggested that the same mechanism used to exchange genetic material that takes place during *T. brucei* cell cycle also helps to maintain a varied amount of minicircles sequences. Even though, the relative amount of a specific sequence can vary with time [235], which shows that there is an intrinsic randomness in the minicircle distribution model and that the presence of many minicircles that share the same sequence is necessary.

Proteins Involved in kDNA Replication

It is thought that the replication of kDNA involves more than 100 different proteins while other organisms use only one protein to carry out similar functions in mitochondrial DNA replication. Most of the enzymes involved in kDNA replication have not yet been characterized, in part because trypanosomatids use many similar enzymes for different functions. Fig. (**5**) shows the regions in the kinetoplast where the proteins involved in kDNA replication are located. To further understand the mechanism of kDNA replication, some of the essential proteins and their function will be reviewed bellow.

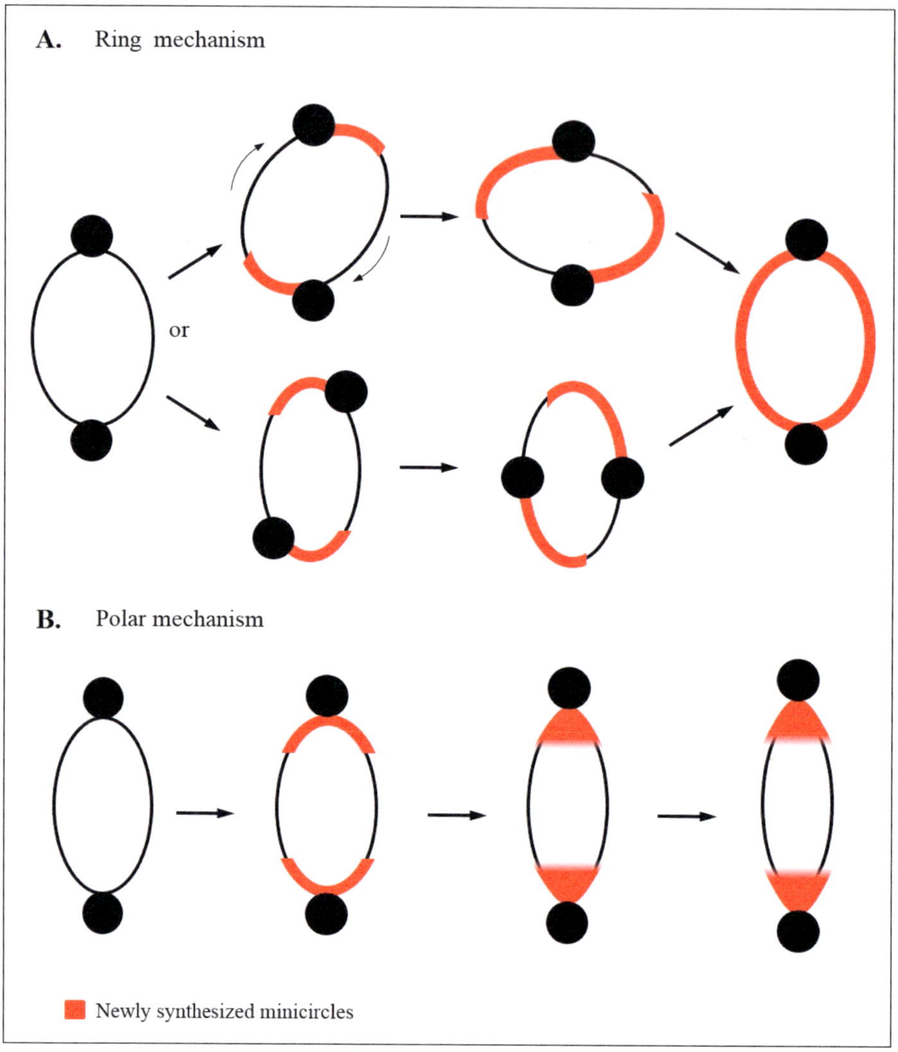

Fig. (4). Comparison between the ring and the polar mechanisms of kDNA replication. **(A)**. In the ring mechanism protein complexes (black filled circle) rotate in relation to the minicircle network, so the newly synthesized minicircles (red lane) are distributed over the network periphery. It is worth mentioning that it is still unknown whether it is the protein complex or the DNA that are moving. **(B)**. In the polar mechanism, there is no rotation of the protein complex, thereby newly synthesized minicircles (red) are aggregated at the polar sites of the kDNA network.

Universal minicircle sequence binding protein (UMSBP): USMBP binds to the universal minicircle sequence and is thought to mediate the initiation of minicircle replication. This protein is located in the KFZ, where the minicircle replication starts. This makes sense because USMBP is thought to assemble the proteins necessary for replication [240]. Recently, a protein from *T. brucei* (p38) was shown to localize and bind to kDNA replication origin. It might be that this protein interacts with USMBP during replication initiation, although the relationship between them remains to be determined [241].

Mitochondrial Topo II: Trypanosomes Type II topoisomerases localizes at the antipodal sites and are critical in the kDNA replication process since they decatenate, catenate, and rearrange the kinetoplast network [242].

Mitochondrial DNA Polymerases: *T. brucei* has at least five mitochondrial polymerases. The first to be described was a DNA polymerase similar to Pol β, which is located at the antipodal sites. This polymerase is more characteristic of a repair enzyme and its role, more than to replicate kDNA, is to perform gap filling in the late S phase [243].

Attempts to find kinetoplast polymerases similar to Pol γ (the single replicative mitochondrial polymerase found in most eukaryotes) have shown no results [17]. Instead, four other mitochondrial polymerases related to bacterial Pol I were found in *T. brucei*. At least three of them (Pol IB, C and D), are essential for parasite survival [20], implying that they are not redundant to each other, but instead each plays a specific role in kDNA replication process.

Mitochondrial Primases: Two primases have been characterized in *T. brucei* mitochondria, and homologs have been found in other trypanosomatids [244]. PRI2 is located at the antipodal sites and is thought to be the minicircle primase [245]. Whereas PRI1, appears to be a maxicircle primase [244], raising the possibility that minicircles and maxicircles use different sets of enzymes to replicate.

As mentioned before, other proteins are involved in the minicircle replication, but this phenomenon is commonly summarized as follow: minicircles are released from the network into the KFZ, where the first stages of minicircle replication take place. Minicircles initiate replication as theta structures by the action of UMSBP, Pol 1B and Pol 1C and other proteins. Then, replicated minicircles are thought to migrate to the antipodal sites where the late stages of minicircle replication take place. SSE-1 and PIF5 proteins are thought to perform primer removal; Pol β provides gap filling, and a specific DNA ligase (named DNA ligase kβ) seal the nicks. Then, minicircles are attached to the network periphery by topo II and those containing at least one gap are repaired in the network by DNA ligase kα, Pol β-PAK and other proteins located at kDNA disk [221]. Fig. (**5**) summarizes this hypothesis.

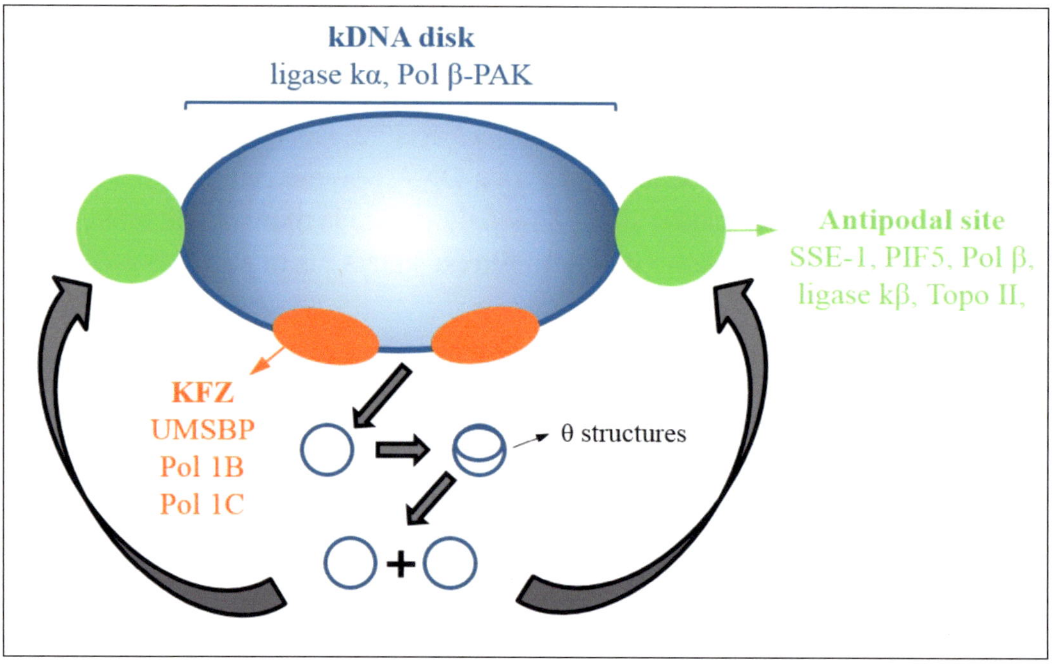

Fig. (5). The actual hypothesis of minicircle replication. The diagram shows the places for kDNA replication and the location of the principal proteins involved in this process. Minicircles are released from the network into the kinetoflagellar zone (KFZ), where the first stages of minicircle replication take place. Minicircles initiate replication as theta (θ) structures by universal minicircle sequence binding protein (UMSBP), DNA polymerase 1B and 1C (Pol 1B and Pol 1C) and other proteins. Then, replicated minicircles are thought to migrate to the antipodal sites where the later steps of minicircle replication take place. The proteins SSE-1 and PIF5 perform primer removal, DNA polymerase β (Pol β) provides gap filling and DNA ligase kβ seal the nicks. Then, minicircles are attached to the network periphery by topoisomerase II (topo II). At the end of replication, minicircles containing at least one gap are repaired in the network by DNA ligase kα, DNA polymerase β-PAK (Pol β-PAK) and other proteins located at kDNA disk.

Control and Regulation of kDNA Replication

A unique feature of kDNA replication is that it takes place at the same time as cell division. In contrast to other mitochondrial DNA that replicates throughout the whole cell cycle, kDNA replicates only at some point near the S phase in all trypanosomatids [3, 245 - 249]. The studies about control mechanisms of kDNA replication have only recently started to be available, although we can highlight two mechanisms in particular.

Redox Regulation of UMSBP Binding

As mentioned before, the protein UMSBP binds the 12-mer universal origin sequence of the minicircles to initiate minicircle replication. The binding of this protein to the minicircles might serve as a control for kDNA replication process. This protein contains five zinc finger domains that mediate binding to DNA. The zinc finger domains are very sensitive to changes in the redox environment, *e.g.*, when oxidized, they can no longer bind DNA. It has been shown that the redox state of UMSBP changes during the cell cycle [250, 251].

Expression of mRNA

It has been proposed that the levels of mRNA encoding proteins related to kDNA replication are higher at the beginning of S phase and lower during M and G2 phases. Since kDNA is replicated almost at the same time as the nuclear DNA, expression of the replication proteins may be regulated in a similar manner. This would explain how these events are synchronized in time. This idea is supported by the fact that both nuclear and mitochondrial genes of Trypanosomes appear to be under the control of the same transcription factors [4].

CONCLUDING REMARKS

During the past several years, significant advances in our understanding of the molecular mechanisms of trypanosomatid DNA replication have been achieved. All this has been made possible by the availability of genomic sequences of these organisms in public databases and also by significant advances in molecular tools and techniques. An excellent example is the emergence of the most advanced genome editing tool ever developed, Clustered Regularly Interspaced Short

Palindromic Repeats (CRISPR) / CRISPR-associated genes (Cas), and its early implementation in some trypanosomatids [252 - 254]. The results obtained established a powerful new tool for the analysis of gene functions in trypanosomatids, enabling the study and analysis of essential genes involved in DNA replication and other important metabolic process. Compared to higher eukaryotes, these techniques provide a huge advantage, since many trypanosomatids that have a substantial impact on human health (*e.g. T. cruzi*, *L. major*, *L. amazonensis*, and *L. donovani*), perform polycistronic transcription and lack a functional RNAi pathway [255 - 257].

Although significant advances have emerged, our knowledge of most events in trypanosomatid DNA replication remains limited. Little is known about the activity of nuclear DNA polymerases and proteins involved in kDNA replication, even though their genes are annotated in the genome databases and the putative proteins show conserved structural domains. Furthermore, as previously mentioned, the processes involved in the control and regulation of kDNA replication are not well established. A particularly important question that remains to be answered is this: How can kDNA, unlike the mitochondrial DNA present in metazoan, replicate only once and almost at the same time as nuclear DNA? Although speculations have been advanced, further studies are necessary to begin to understand how the unique 'large specialized mitochondria' have developed during the evolution of this phylogenetically primitive organism.

Issues related to replicative initiation sites also remain unanswered: The initiation sites in trypanosomatids function similarly to those in higher eukaryotes, *i.e.* are specific but not in the sense applied to bacterial DNA replication origins. The specificity of each origin is much more relaxed, and initiation sites are selected from the active pre-RC sites. The decision to activate pre-RC sites may occur in the preceding M phase, or it may be determined during G1 phase by an unknown mechanism. Curiously, at M and G1 phases, the majority of potential origins are rendered inactive, although Orc1/Cdc6 remains bounded to DNA throughout the cell cycle. Alternatively, changes in chromatin structures may be formed to differentiate the initiation sites, but it is still unclear what kind of chromatin mark is left or what kind of chromatin structures are set up to differentiate the initiation sites at the beginning of S phase.

Also, despite extensive telomere characterization in *T. brucei* and *L. amazonensis*, much remains to be understood regarding telomere maintenance and replication in trypanosomatids. Undoubtedly, searching for differences in protein structure, protein interactions, telomeric complex, telomerase composition and regulation of DNA replication, would provide the discovery of specific parasitic targets that will be useful for the development of novel antiparasite drugs and alternative therapies with minimal side effects for the host.

CONFLICT OF INTEREST

The authors confirm that they have no conflict of interest to declare for this publication.

ACKNOWLEDGEMENTS

The authors thank Dr. Eric C. K. Gren for assisting with English translation and writing syntax during the preparation of the manuscript. This work was supported by M.I.N. Cano grants from Fundação de Amparo à Pesquisa do Estado de São Paulo (FAPESP 2012/50263-5), and Conselho Nacional de Pesquisa (CNPq), Brazil. M.S. da Silva and R.S. Pavani are respectively, post doctor and Ph.D. fellows from FAPESP; M.A. Viviescas and C.B. Storti are respectively Ph.D. and undergraduate fellows (FAPESP 2014/24170-5 and FAPESP 2014/02978-0) from CNPq, and *E.G.*O. Morea is a Ph.D. fellow from CAPES (Coordenação de Aperfeiçoamento de Pessoal de Ensino Superior).

ABBREVIATIONS

^{14}C	Carbon-14 / radiocarbon
911 complex:	Rad9-Hus1-Rad1 complex
AAA+:	Family domain that confers ATPase activity
Apollo	DNA nuclease
ARS:	Autonomous replication sequence
ATP:	Adenosine triphosphate
AT-rich	Adenine/Thymidine rich
base J	β-D-glucosyl-hydroxymethyluracil
BER:	Base excision repair
bp:	Base pair

Cas:	CRISPR-associated genes
Cdc45	Cell division cycle subunit 45
Cdc6	Cell division cycle subunit 6
CDK:	Cyclin-dependent kinase
Cdt1:	Cdc dependent transcript 1
CMG:	Cdc45/Mcm/GINS
C-rich	Cytosine rich
CRISPR	Clustered Regularly Interspaced Short Palindromic Repeats
CST	Cdc13/Stn1/Tin2
C-strand:	Telomeric DNA strand C-rich
C-terminal	Carboxy-terminal
DBD	DNA binding domains
DDK:	Cdc7-Dbf4 kinase
DNA:	Deoxyribonucleic Acid
DnaA:	Protein that activates initiation of DNA replication (in Bacteria)
DnaB helicase:	Enzyme that opens the replication fork (in Bacteria)
Dpb11	DNA Polymerase B subunit 2 (in *Saccharomyces*)
dRP:	5'-deoxyribose-5-phosphate
EdU:	5-ethynyl-2′-deoxyuridine
EST1:	Ever shorter telomeres protein 1
Exo1:	Exonuclease 1
FEN-1:	Flap Endonuclease-1
G1 phase:	Gap 1 phase
G2 phase:	Gap 2 phase
GINS: abbreviation for Go-Ichi--i-San	Japanese words that represent the numbers 5, 1, 2, 3 for the proteins Sld5-Psf1-Psf2-Psf3.
G-rich	Guanine-rich
gRNA:	Guide RNAs
G-strand:	Telomeric DNA strand G-rich
G-tetrads or G4-DNA	Telomeric single strand DNA G-quadruplexes
Hus1	Hydroxyurea sensitive subunit 1
kb:	Kilo base pair

kDNA:	kinetoplast Deoxyribonucleic Acid
KFZ:	Kinetoflagelar zone
Ku70	ATP-dependent DNA helicase II 70 kDa subunit
Ku80	ATP-dependent DNA helicase II 80 kDa subunit
L. :	*Leishmania*
LaRAD51:	*Leishmania amazonensis*RAD51
LaRBP38	*Leishmania amazonensis*DNA binding protein 38 kDa subunit
LaRPA-1	*Leishmania amazonensis* replication protein A subunit 1
LaTBP1	*Leishmania amazonensis* telomere binding protein 1
LaTRF	Leishmania amazonensis TTAGG repeat-binding factor
LCTAS:	*Leishmania* Conserved Telomere-Associated Sequence
LdPCNA	*Leishmania donovani* Proliferating Cell Nuclear Antigen
LdTOP2:	*Leishmania donovani* topoisomerase II
Ligase κα:	DNA ligase κα
Ligase κβ:	DNA ligase κβ
LmHus1:	*Leishmania major* Hus1
M phase	Mitosis phase
Mcm	Mini-chromosome maintenance
Mcm2-7	Minichromosome maintenance subunits 2-7
Mcm9	Minichromosome maintenance subunit 9
Mec1:	Mitosis entry checkpoint protein 1
mer:	part (in Greek)
Mre11	Meiotic recombination subunit 11
MRN	Mre11-Rad50-Nbs1
mRNA:	messenger RNA
NRTAS	Non-Repeated Telomeric Associated Sequence
nt:	nucleotides
OB-fold	oligonucleotide/oligosaccharide binding domain
ORC:	Origin recognition complex
ORC1b:	Origin recognition complex subunit 1b
OriC:	DNA replication origin (in Bacteria)
p38:	mitochondrial protein (in *Trypanosoma brucei*)
PCNA:	Proliferating Cell Nuclear Antigen
PIF5:	DNA repair and recombination helicase PIF5
Polμ:	DNA polymerase μ

Pol A:	domain present in DNA polymerases of A-family
Pol I:	DNA polymerase I
Pol IA:	DNA polymerase I subtype A
Pol IB	DNA polymerase I – subtype B
Pol IC:	DNA polymerase I subtype C
Pol ID:	DNA polymerase I subtype D
Pol II	DNA polymerase II
Pol III	DNA polymerase III
Pol IV	DNA polymerase IV
Pol V:	DNA polymerase V
Pol α:	DNA polymerase α
Pol β:	DNA polymerase β
Pol β-pak	DNA polymerase β pak
Pol γ:	DNA polymerase γ
Pol δ:	DNA polymerase δ
Pol ε:	DNA polymerase ε
Polη:	DNA polymerase η
Pol θ:	DNA polymerase θ
Polι:	DNA polymerase ι
Polκ:	DNA polymerase κ
Polλ:	DNA polymerase λ
Polσ:	DNA polymerase σ
Pot1	Protection of telomeres protein 1
pre-RC:	pre-Replication Complex
PRI1:	Primase subunit 1
PRI2:	Primase subunit 2
Rad50:	Radiation sensitive subtype 50 (DNA repair protein Rad50)
Rad51:	Radiation sensitive subtype 51 (DNA repair protein Rad51)
Rap1:	Repressor/activator site-binding protein
RECQL4:	ATP-dependent DNA helicase Q4
RFC:	Replication Factor C
RNA:	Ribonucleic Acid
RNAi	Ribonucleic Acid Interference
RPA:	Replication Protein A
RPA1	Replication Protein A subunit 1

RPA2	Replication Protein A subunit 2
RPA3	Replication Protein A subunit 3
rRNA:	ribosomal RNA
RT:	Reverse Transcriptase
S phase	Synthesis phase
S. :	*Saccharomyces*
Sld2:	Synthetically lethal with Dpb11 subunit 2
SSE1:	Heat shock protein homologue SSE1
Stn1:	Suppressor of cdc thirteen
SV40:	Simian Virus subunit 40
T motif:	telomerase-specific motif
T. :	*Trypanosoma*
TbORC:	*Trypanosoma brucei*Origin Recognition Complex
TbOrc1/Cdc6:	*Trypanosoma brucei*ORC1/Cdc6
TbORC1b:	*Trypanosoma brucei*Origin Recognition Complex subunit 1b
TbOrc4	*Trypanosoma brucei*Origin Recognition Complex subunit 4
TbPCNA:	*Trypanosoma brucei* Proliferating Cell Nuclear Antigen
TbTOP2α	*Trypanosoma brucei* topoisomerase II subtype α
TbTOP2β:	*Trypanosoma brucei* topoisomerase II subtype β
Tc38:	*Trypanosoma cruzi* protein 38 kDa
TcPCNA	*Trypanosoma cruzi* Proliferating Cell Nuclear Antigen
Tdt:	Terminal deoxynucleotidyl transferase
Tel1:	Telomere maintenance
Ten1:	Telomeric pathways with Stn1
TER:	telomerase RNA
TERT:	Telomerase reverse transcriptase
TFIIIB	Subunit of RNA polymerase III transcription factor
Tin2:	TRF1-interacting nuclear protein 2
TOPBP1	DNA topoisomerase 2-binding protein 1
TPP1	Tripeptidyl-peptidase (protein associated with POT1)
TRF1:	TTAGGG repeat-binding factor subunit 1
TRF2	TTAGGG repeat-binding factor subunit 2
Tri-TrypDB	Kinetoplastid Genomics Resource website
UMS:	Universal minicircle sequence
UMSBP:	Universal minicircle sequence binding protein

VSG: Variant surface glycoprotein

WHD Winged-Helix domain

XRS2/NBS1 Proteins required for DNA repair and telomere maintenance

REFERENCES

[1] Elias MC, da Cunha JP, de Faria FP, Mortara RA, Freymüller E, Schenkman S. Morphological events during the *Trypanosoma cruzi* cell cycle. Protist 2007; 158(2): 147-57. [http://dx.doi.org/10.1016/j.protis.2006.10.002] [PMID: 17185034]

[2] Vaughan S, Gull K. The structural mechanics of cell division in *Trypanosoma brucei.* Biochem Soc Trans 2008; 36(Pt 3): 421-4. [http://dx.doi.org/10.1042/BST0360421] [PMID: 18481972]

[3] da Silva MS, Monteiro JP, Nunes VS, *et al. Leishmania amazonensis* promastigotes present two distinct modes of nucleus and kinetoplast segregation during cell cycle. PLoS One 2013; 8(11): e81397. [http://dx.doi.org/10.1371/journal.pone.0081397] [PMID: 24278433]

[4] Pasion SG, Brown GW, Brown LM, Ray DS. Periodic expression of nuclear and mitochondrial DNA replication genes during the trypanosomatid cell cycle. J Cell Sci 1994; 107(Pt 12): 3515-20. [PMID: 7706402]

[5] da Silva MS, da Silveira RC, Perez AM, *et al.* Nuclear DNA replication in trypanosomatid protozoa. In: Leitner RP, Ed. DNA replication and mutation. 1st ed., United Kingdom: Nova Science Publishers 2012.

[6] Pavlov YI, Shcherbakova PV. DNA polymerases at the eukaryotic fork-20 years later. Mutat Res 2010; 685(1-2): 45-53. [http://dx.doi.org/10.1016/j.mrfmmm.2009.08.002] [PMID: 19682465]

[7] Makarova KS, Krupovic M, Koonin EV. Evolution of replicative DNA polymerases in archaea and their contributions to the eukaryotic replication machinery. Front Microbiol 2014; 5: 354. [http://dx.doi.org/10.3389/fmicb.2014.00354] [PMID: 25101062]

[8] Lehman IR. Historical perspective: Arthur Kornberg, a giant of 20th century biochemistry. Trends Biochem Sci 2008; 33(6): 291-6. [http://dx.doi.org/10.1016/j.tibs.2008.03.002] [PMID: 18467101]

[9] Kornberg A, Baker TA, Eds. DNA Replication. 2nd ed., New York, NY: University Science Books 1992.

[10] Johnson A, ODonnell M. Cellular DNA replicases: components and dynamics at the replication fork. Annu Rev Biochem 2005; 74: 283-315. [http://dx.doi.org/10.1146/annurev.biochem.73.011303.073859] [PMID: 15952889]

[11] Goodman MF, Tippin B. The expanding polymerase universe. Nat Rev Mol Cell Biol 2000; 1(2): 101-9. [http://dx.doi.org/10.1038/35040051] [PMID: 11253362]

[12] Pavlov YI, Shcherbakova PV, Rogozin IB. Roles of DNA polymerases in replication, repair, and

recombination in eukaryotes. Int Rev Cytol 2006; 255: 41-132.
[http://dx.doi.org/10.1016/S0074-7696(06)55002-8] [PMID: 17178465]

[13] Burgers PM. Eukaryotic DNA polymerases in DNA replication and DNA repair. Chromosoma 1998; 107(4): 218-27.
[http://dx.doi.org/10.1007/s004120050300] [PMID: 9745046]

[14] Tahirov TH, Makarova KS, Rogozin IB, Pavlov YI, Koonin EV. Evolution of DNA polymerases: an inactivated polymerase-exonuclease module in Pol epsilon and a chimeric origin of eukaryotic polymerases from two classes of archaeal ancestors. Biol Direct 2009; 4: 11.
[http://dx.doi.org/10.1186/1745-6150-4-11] [PMID: 19296856]

[15] Takemura M, Yokobori S, Ogata H. Evolution of Eukaryotic DNA Polymerases *via* interaction between cells and large DNA viruses. J Mol Evol 2015; 81(1-2): 24-33.
[http://dx.doi.org/10.1007/s00239-015-9690-z] [PMID: 26177821]

[16] Hübscher U, Maga G, Spadari S. Eukaryotic DNA polymerases. Annu Rev Biochem 2002; 71: 133-63.
[http://dx.doi.org/10.1146/annurev.biochem.71.090501.150041] [PMID: 12045093]

[17] Klingbeil MM, Motyka SA, Englund PT. Multiple mitochondrial DNA polymerases in *Trypanosoma brucei.* Mol Cell 2002; 10(1): 175-86.
[http://dx.doi.org/10.1016/S1097-2765(02)00571-3] [PMID: 12150917]

[18] Chandler J, Vandoros AV, Mozeleski B, Klingbeil MM. Stem-loop silencing reveals that a third mitochondrial DNA polymerase, POLID, is required for kinetoplast DNA replication in trypanosomes. Eukaryot Cell 2008; 7(12): 2141-6.
[http://dx.doi.org/10.1128/EC.00199-08] [PMID: 18849470]

[19] Bruhn DF, Mozeleski B, Falkin L, Klingbeil MM. Mitochondrial DNA polymerase POLIB is essential for minicircle DNA replication in African trypanosomes. Mol Microbiol 2010; 75(6): 1414-25.
[http://dx.doi.org/10.1111/j.1365-2958.2010.07061.x] [PMID: 20132449]

[20] Bruhn DF, Sammartino MP, Klingbeil MM. Three mitochondrial DNA polymerases are essential for kinetoplast DNA replication and survival of bloodstream form *Trypanosoma brucei.* Eukaryot Cell 2011; 10(6): 734-43.
[http://dx.doi.org/10.1128/EC.05008-11] [PMID: 21531873]

[21] Concepción-Acevedo J, Luo J, Klingbeil MM. Dynamic localization of *Trypanosoma brucei* mitochondrial DNA polymerase ID. Eukaryot Cell 2012; 11(7): 844-55.
[http://dx.doi.org/10.1128/EC.05291-11] [PMID: 22286095]

[22] Leegwater PA, Strating M, Murphy NB, Kooy RF, van der Vliet PC, Overdulve JP. The *Trypanosoma brucei* DNA polymerase alpha core subunit gene is developmentally regulated and linked to a constitutively expressed open reading frame. Nucleic Acids Res 1991; 19(23): 6441-7.
[http://dx.doi.org/10.1093/nar/19.23.6441] [PMID: 1754381]

[23] Nolan LL, Rivera JH. The DNA polymerases of *Leishmania mexicana.* FEMS Microbiol Lett 1992; 74(1): 71-5.
[http://dx.doi.org/10.1111/j.1574-6968.1992.tb05344.x] [PMID: 1516809]

[24] Luton K, Johnson AM. Cloning and sequence analysis of the DNA polymerase alpha gene of *Leishmania donovani*: comparison with the human homologue. Biochem Biophys Res Commun 1997;

234(1): 95-100.
[http://dx.doi.org/10.1006/bbrc.1997.6583] [PMID: 9168968]

[25] Wang TS. Eukaryotic DNA polymerases. Annu Rev Biochem 1991; 60: 513-52.
[http://dx.doi.org/10.1146/annurev.bi.60.070191.002501] [PMID: 1883203]

[26] Chang LM, Cheriathundam E, Mahoney EM, Cerami A. DNA polymerases in parasitic protozoans differ from host enzymes. Science 1980; 208(4443): 510-1.
[http://dx.doi.org/10.1126/science.7367875] [PMID: 7367875]

[27] Sawaya MR, Pelletier H, Kumar A, Wilson SH, Kraut J. Crystal structure of rat DNA polymerase beta: evidence for a common polymerase mechanism. Science 1994; 264(5167): 1930-5.
[http://dx.doi.org/10.1126/science.7516581] [PMID: 7516581]

[28] Wilson SH. Mammalian base excision repair and DNA polymerase beta. Mutat Res 1998; 407(3): 203-15.
[http://dx.doi.org/10.1016/S0921-8777(98)00002-0] [PMID: 9653447]

[29] Dianov GL, Prasad R, Wilson SH, Bohr VA. Role of DNA polymerase beta in the excision step of long patch mammalian base excision repair. J Biol Chem 1999; 274(20): 13741-3.
[http://dx.doi.org/10.1074/jbc.274.20.13741] [PMID: 10318775]

[30] Ivens AC, Peacock CS, Worthey EA, *et al.* The genome of the kinetoplastid parasite, *Leishmania major.* Science 2005; 309(5733): 436-42.
[http://dx.doi.org/10.1126/science.1112680] [PMID: 16020728]

[31] Saxowsky TT, Choudhary G, Klingbeil MM, Englund PT. *Trypanosoma brucei* has two distinct mitochondrial DNA polymerase beta enzymes. J Biol Chem 2003; 278(49): 49095-101.
[http://dx.doi.org/10.1074/jbc.M308565200] [PMID: 12966090]

[32] Taladriz S, Hanke T, Ramiro MJ, *et al.* Nuclear DNA polymerase beta from *Leishmania infantum.* Cloning, molecular analysis and developmental regulation. Nucleic Acids Res 2001; 29(18): 3822-34.
[http://dx.doi.org/10.1093/nar/29.18.3822] [PMID: 11557814]

[33] Lopes DdeO, Schamber-Reis BL, Regis-da-Silva CG, *et al.* Biochemical studies with DNA polymerase beta and DNA polymerase beta-PAK of *Trypanosoma cruzi* suggest the involvement of these proteins in mitochondrial DNA maintenance. DNA Repair (Amst) 2008; 7(11): 1882-92.
[http://dx.doi.org/10.1016/j.dnarep.2008.07.018] [PMID: 18761429]

[34] Venegas JA, Aslund L, Solari A. Cloning and characterization of a DNA polymerase beta gene from *Trypanosoma cruzi.* Parasitol Int 2009; 58(2): 187-92.
[http://dx.doi.org/10.1016/j.parint.2009.01.007] [PMID: 19567232]

[35] Johnson RE, Washington MT, Prakash S, Prakash L. Bridging the gap: a family of novel DNA polymerases that replicate faulty DNA. Proc Natl Acad Sci USA 1999; 96(22): 12224-6.
[http://dx.doi.org/10.1073/pnas.96.22.12224] [PMID: 10535901]

[36] Santiago MJ, Alejandre-Durán E, Ruiz-Rubio M. Analysis of UV-induced mutation spectra in *Escherichia coli* by DNA polymerase eta from *Arabidopsis thaliana.* Mutat Res 2006; 601(1-2): 51-60.
[http://dx.doi.org/10.1016/j.mrfmmm.2006.05.036] [PMID: 16857217]

[37] Johnson RE, Prakash S, Prakash L. The human DINB1 gene encodes the DNA polymerase Poltheta.

Proc Natl Acad Sci USA 2000; 97(8): 3838-43.
[http://dx.doi.org/10.1073/pnas.97.8.3838] [PMID: 10760255]

[38] Zhang Y, Yuan F, Wu X, *et al.* Error-free and error-prone lesion bypass by human DNA polymerase kappa *in vitro*. Nucleic Acids Res 2000; 28(21): 4138-46.
[http://dx.doi.org/10.1093/nar/28.21.4138] [PMID: 11058110]

[39] Gerlach VL, Feaver WJ, Fischhaber PL, Friedberg EC. Purification and characterization of pol kappa, a DNA polymerase encoded by the human DINB1 gene. J Biol Chem 2001; 276(1): 92-8.
[http://dx.doi.org/10.1074/jbc.M004413200] [PMID: 11024016]

[40] Suzuki N, Ohashi E, Hayashi K, Ohmori H, Grollman AP, Shibutani S. Translesional synthesis past acetylaminofluorene-derived DNA adducts catalyzed by human DNA polymerase kappa and *Escherichia coli* DNA polymerase IV. Biochemistry 2001; 40(50): 15176-83.
[http://dx.doi.org/10.1021/bi010702g] [PMID: 11735400]

[41] Suzuki N, Ohashi E, Kolbanovskiy A, *et al.* Translesion synthesis by human DNA polymerase kappa on a DNA template containing a single stereoisomer of dG-(+)- or dG-(-)-anti-N(2)-BPDE (7,8-dihydroxy-anti-9,10-epoxy-7,8,9,10-tetrahydrobenzo[a]pyrene). Biochemistry 2002; 41(19): 6100-6.
[http://dx.doi.org/10.1021/bi020049c] [PMID: 11994005]

[42] Fischhaber PL, Gerlach VL, Feaver WJ, Hatahet Z, Wallace SS, Friedberg EC. Human DNA polymerase kappa bypasses and extends beyond thymine glycols during translesion synthesis *in vitro*, preferentially incorporating correct nucleotides. J Biol Chem 2002; 277(40): 37604-11.
[http://dx.doi.org/10.1074/jbc.M206027200] [PMID: 12145297]

[43] Prakash S, Johnson RE, Prakash L. Eukaryotic translesion synthesis DNA polymerases: specificity of structure and function. Annu Rev Biochem 2005; 74: 317-53.
[http://dx.doi.org/10.1146/annurev.biochem.74.082803.133250] [PMID: 15952890]

[44] Rajão MA, Passos-Silva DG, DaRocha WD, *et al.* DNA polymerase kappa from *Trypanosoma cruzi* localizes to the mitochondria, bypasses 8-oxoguanine lesions and performs DNA synthesis in a recombination intermediate. Mol Microbiol 2009; 71(1): 185-97.
[http://dx.doi.org/10.1111/j.1365-2958.2008.06521.x] [PMID: 19007414]

[45] de Moura MB, Schamber-Reis BL, Passos Silva DG, *et al.* Cloning and characterization of DNA polymerase eta from *Trypanosoma cruzi*: roles for translesion bypass of oxidative damage. Environ Mol Mutagen 2009; 50(5): 375-86.
[http://dx.doi.org/10.1002/em.20450] [PMID: 19229999]

[46] Kaguni JM. DnaA: controlling the initiation of bacterial DNA replication and more. Annu Rev Microbiol 2006; 60: 351-75.
[http://dx.doi.org/10.1146/annurev.micro.60.080805.142111] [PMID: 16753031]

[47] Kaguni JM. *Escherichia coli* DnaA protein: the replication initiator. Mol Cells 1997; 7(2): 145-57.
[PMID: 9163725]

[48] Messer W, Blaesing F, Jakimowicz D, *et al.* Bacterial replication initiator DnaA. Rules for DnaA binding and roles of DnaA in origin unwinding and helicase loading. Biochimie 2001; 83(1): 5-12.
[http://dx.doi.org/10.1016/S0300-9084(00)01216-5] [PMID: 11254968]

[49] Méchali M. Eukaryotic DNA replication origins: many choices for appropriate answers. Nat Rev Mol

Cell Biol 2010; 11(10): 728-38.
[http://dx.doi.org/10.1038/nrm2976] [PMID: 20861881]

[50] Stinchcomb DT, Struhl K, Davis RW. Isolation and characterisation of a yeast chromosomal replicator. Nature 1979; 282(5734): 39-43.
[http://dx.doi.org/10.1038/282039a0] [PMID: 388229]

[51] Bell SP, Stillman B. ATP-dependent recognition of eukaryotic origins of DNA replication by a multiprotein complex. Nature 1992; 357(6374): 128-34.
[http://dx.doi.org/10.1038/357128a0] [PMID: 1579162]

[52] Costa A, Hood IV, Berger JM. Mechanisms for initiating cellular DNA replication. Annu Rev Biochem 2013; 82: 25-54.
[http://dx.doi.org/10.1146/annurev-biochem-052610-094414] [PMID: 23746253]

[53] Nguyen VQ, Co C, Li JJ. Cyclin-dependent kinases prevent DNA re-replication through multiple mechanisms. Nature 2001; 411(6841): 1068-73.
[http://dx.doi.org/10.1038/35082600] [PMID: 11429609]

[54] Diffley JF, Cocker JH. Protein-DNA interactions at a yeast replication origin. Nature 1992; 357(6374): 169-72.
[http://dx.doi.org/10.1038/357169a0] [PMID: 1579168]

[55] Iyer LM, Leipe DD, Koonin EV, Aravind L. Evolutionary history and higher order classification of AAA+ ATPases. J Struct Biol 2004; 146(1-2): 11-31.
[http://dx.doi.org/10.1016/j.jsb.2003.10.010] [PMID: 15037234]

[56] Clarey MG, Botchan M, Nogales E. Single particle EM studies of the *Drosophila melanogaster* origin recognition complex and evidence for DNA wrapping. J Struct Biol 2008; 164(3): 241-9.
[http://dx.doi.org/10.1016/j.jsb.2008.08.006] [PMID: 18824234]

[57] Cocker JH, Piatti S, Santocanale C, Nasmyth K, Diffley JF. An essential role for the Cdc6 protein in forming the pre-replicative complexes of budding yeast. Nature 1996; 379(6561): 180-2.
[http://dx.doi.org/10.1038/379180a0] [PMID: 8538771]

[58] Nishitani H, Lygerou Z, Nishimoto T, Nurse P. The Cdt1 protein is required to license DNA for replication in fission yeast. Nature 2000; 404(6778): 625-8.
[http://dx.doi.org/10.1038/35007110] [PMID: 10766248]

[59] Tanaka S, Diffley JF. Interdependent nuclear accumulation of budding yeast Cdt1 and Mcm27 during G1 phase. Nat Cell Biol 2002; 4(3): 198-207.
[http://dx.doi.org/10.1038/ncb757] [PMID: 11836525]

[60] Da-Silva LF, Duncker BP. ORC function in late G1: maintaining the license for DNA replication. Cell Cycle 2007; 6(2): 128-30.
[http://dx.doi.org/10.4161/cc.6.2.3743] [PMID: 17314509]

[61] Bell SP, Dutta A. DNA replication in eukaryotic cells. Annu Rev Biochem 2002; 71: 333-74.
[http://dx.doi.org/10.1146/annurev.biochem.71.110601.135425] [PMID: 12045100]

[62] Lei M, Tye BK. Initiating DNA synthesis: from recruiting to activating the MCM complex. J Cell Sci 2001; 114(Pt 8): 1447-54.
[PMID: 11282021]

[63] Petojevic T, Pesavento JJ, Costa A, *et al.* Cdc45 (cell division cycle protein 45) guards the gate of the Eukaryote Replisome helicase stabilizing leading strand engagement. Proc Natl Acad Sci USA 2015; 112(3): E249-58.
[http://dx.doi.org/10.1073/pnas.1422003112] [PMID: 25561522]

[64] Fragkos M, Ganier O, Coulombe P, Méchali M. DNA replication origin activation in space and time. Nat Rev Mol Cell Biol 2015; 16(6): 360-74.
[http://dx.doi.org/10.1038/nrm4002] [PMID: 25999062]

[65] Tada S, Kundu LR, Enomoto T. Insight into initiator-DNA interactions: a lesson from the archaeal ORC. BioEssays 2008; 30(3): 208-11.
[http://dx.doi.org/10.1002/bies.20726] [PMID: 18293358]

[66] Godoy PD, Nogueira-Junior LA, Paes LS, *et al.* Trypanosome prereplication machinery contains a single functional orc1/cdc6 protein, which is typical of archaea. Eukaryot Cell 2009; 8(10): 1592-603.
[http://dx.doi.org/10.1128/EC.00161-09] [PMID: 19717742]

[67] Benmerzouga I, Concepción-Acevedo J, Kim HS, *et al. Trypanosoma brucei* Orc1 is essential for nuclear DNA replication and affects both VSG silencing and VSG switching. Mol Microbiol 2013; 87(1): 196-210.
[http://dx.doi.org/10.1111/mmi.12093] [PMID: 23216794]

[68] Dang HQ, Li Z. The Cdc45·Mcm27·GINS protein complex in trypanosomes regulates DNA replication and interacts with two Orc1-like proteins in the origin recognition complex. J Biol Chem 2011; 286(37): 32424-35.
[http://dx.doi.org/10.1074/jbc.M111.240143] [PMID: 21799014]

[69] Neuwald AF, Aravind L, Spouge JL, Koonin EV. AAA+: A class of chaperone-like ATPases associated with the assembly, operation, and disassembly of protein complexes. Genome Res 1999; 9(1): 27-43.
[PMID: 9927482]

[70] Erzberger JP, Pirruccello MM, Berger JM. The structure of bacterial DnaA: implications for general mechanisms underlying DNA replication initiation. EMBO J 2002; 21(18): 4763-73.
[http://dx.doi.org/10.1093/emboj/cdf496] [PMID: 12234917]

[71] Tiengwe C, Marcello L, Farr H, *et al.* Identification of ORC1/CDC6-interacting factors in *Trypanosoma brucei* reveals critical features of origin recognition complex architecture. PLoS One 2012; 7(3): e32674.
[http://dx.doi.org/10.1371/journal.pone.0032674] [PMID: 22412905]

[72] Pape T, Meka H, Chen S, Vicentini G, van Heel M, Onesti S. Hexameric ring structure of the full-length archaeal MCM protein complex. EMBO Rep 2003; 4(11): 1079-83.
[http://dx.doi.org/10.1038/sj.embor.7400010] [PMID: 14566326]

[73] Miller JM, Enemark EJ. Archaeal MCM proteins as an analog for the eukaryotic Mcm2–7 helicase to reveal essential features of structure and function. Archaea 2015; 2015: 305497. [in press.].
[http://dx.doi.org/10.1155/2015/305497] [PMID: 26539061]

[74] Lutzmann M, Méchali M. MCM9 binds Cdt1 and is required for the assembly of prereplication complexes. Mol Cell 2008; 31(2): 190-200.

[http://dx.doi.org/10.1016/j.molcel.2008.07.001] [PMID: 18657502]

[75] Leman AR, Noguchi E. The replication fork: understanding the eukaryotic replication machinery and the challenges to genome duplication. Genes (Basel) 2013; 4(1): 1-32.
[http://dx.doi.org/10.3390/genes4010001] [PMID: 23599899]

[76] Pellegrini L. The Pol α-primase complex. In: MacNeill S, Ed. The Eukaryotic Replisome: a Guide to Protein Structure and Function (Subcellular Biochemistry). 1st ed. Netherlands: Springer 2012; pp. 157-69.
[http://dx.doi.org/10.1007/978-94-007-4572-8_9]

[77] Branzei D, Foiani M. Maintaining genome stability at the replication fork. Nat Rev Mol Cell Biol 2010; 11(3): 208-19.
[http://dx.doi.org/10.1038/nrm2852] [PMID: 20177396]

[78] Prelich G, Tan CK, Kostura M, *et al.* Functional identity of proliferating cell nuclear antigen and a DNA polymerase-delta auxiliary protein. Nature 1987; 326(6112): 517-20.
[http://dx.doi.org/10.1038/326517a0] [PMID: 2882424]

[79] Hindges R, Hübscher U. DNA polymerase delta, an essential enzyme for DNA transactions. Biol Chem 1997; 378(5): 345-62.
[PMID: 9191022]

[80] Langston LD, ODonnell M. DNA polymerase delta is highly processive with proliferating cell nuclear antigen and undergoes collision release upon completing DNA. J Biol Chem 2008; 283(43): 29522-31.
[http://dx.doi.org/10.1074/jbc.M804488200] [PMID: 18635534]

[81] Stukenberg PT, Turner J, ODonnell M. An explanation for lagging strand replication: polymerase hopping among DNA sliding clamps. Cell 1994; 78(5): 877-87.
[http://dx.doi.org/10.1016/S0092-8674(94)90662-9] [PMID: 8087854]

[82] Jones RM, Petermann E. Replication fork dynamics and the DNA damage response. Biochem J 2012; 443(1): 13-26.
[http://dx.doi.org/10.1042/BJ20112100] [PMID: 22417748]

[83] Muramatsu S, Hirai K, Tak YS, Kamimura Y, Araki H. CDK-dependent complex formation between replication proteins Dpb11, Sld2, Pol (ε, and GINS in budding yeast. Genes Dev 2010; 24(6): 602-12.
[http://dx.doi.org/10.1101/gad.1883410] [PMID: 20231317]

[84] Chilkova O, Stenlund P, Isoz I, *et al.* The eukaryotic leading and lagging strand DNA polymerases are loaded onto primer-ends *via* separate mechanisms but have comparable processivity in the presence of PCNA. Nucleic Acids Res 2007; 35(19): 6588-97.
[http://dx.doi.org/10.1093/nar/gkm741] [PMID: 17905813]

[85] Muller MT, Pfund WP, Mehta VB, Trask DK. Eukaryotic type I topoisomerase is enriched in the nucleolus and catalytically active on ribosomal DNA. EMBO J 1985; 4(5): 1237-43.
[PMID: 2988941]

[86] Lodish H, Berk A, Zipursky SL, *et al.* The Role of Topoisomerases in DNA Replication. In: Lodish HF, Ed. Molecular Cell Biology. 4th ed., New York: WH Freeman 2000.

[87] Sakaguchi K, Ishibashi T, Uchiyama Y, Iwabata K. The multi-replication protein A (RPA) systema new perspective. FEBS J 2009; 276(4): 943-63.

[http://dx.doi.org/10.1111/j.1742-4658.2008.06841.x] [PMID: 19154342]

[88] Wold MS, Kelly T. Purification and characterization of replication protein A, a cellular protein required for *in vitro* replication of simian virus 40 DNA. Proc Natl Acad Sci USA 1988; 85(8): 2523-7.
[http://dx.doi.org/10.1073/pnas.85.8.2523] [PMID: 2833742]

[89] Fanning E, Klimovich V, Nager AR. A dynamic model for replication protein A (RPA) function in DNA processing pathways. Nucleic Acids Res 2006; 34(15): 4126-37.
[http://dx.doi.org/10.1093/nar/gkl550] [PMID: 16935876]

[90] Oakley GG, Patrick SM. Replication protein A: directing traffic at the intersection of replication and repair. Front Biosci (Landmark Ed) 2010; 15: 883-900.
[http://dx.doi.org/10.2741/3652] [PMID: 20515732]

[91] Tanaka T, Nasmyth K. Association of RPA with chromosomal replication origins requires an Mcm protein, and is regulated by Rad53, and cyclin- and Dbf4-dependent kinases. EMBO J 1998; 17(17): 5182-91.
[http://dx.doi.org/10.1093/emboj/17.17.5182] [PMID: 9724654]

[92] Zou L, Stillman B. Assembly of a complex containing Cdc45p, replication protein A, and Mcm2p at replication origins controlled by S-phase cyclin-dependent kinases and Cdc7p-Dbf4p kinase. Mol Cell Biol 2000; 20(9): 3086-96.
[http://dx.doi.org/10.1128/MCB.20.9.3086-3096.2000] [PMID: 10757793]

[93] Waga S, Stillman B. Anatomy of a DNA replication fork revealed by reconstitution of SV40 DNA replication *in vitro*. Nature 1994; 369(6477): 207-12.
[http://dx.doi.org/10.1038/369207a0] [PMID: 7910375]

[94] Waga S, Bauer G, Stillman B. Reconstitution of complete SV40 DNA replication with purified replication factors. J Biol Chem 1994; 269(14): 10923-34.
[PMID: 8144677]

[95] Braun KA, Lao Y, He Z, Ingles CJ, Wold MS. Role of protein-protein interactions in the function of replication protein A (RPA): RPA modulates the activity of DNA polymerase α by multiple mechanisms. Biochemistry 1997; 36(28): 8443-54.
[http://dx.doi.org/10.1021/bi970473r] [PMID: 9214288]

[96] Elias MC, Faria M, Mortara RA, *et al.* Chromosome localization changes in the *Trypanosoma cruzi* nucleus. Eukaryot Cell 2002; 1(6): 944-53.
[http://dx.doi.org/10.1128/EC.1.6.944-953.2002] [PMID: 12477795]

[97] Kumar D, Minocha N, Rajanala K, Saha S. The distribution pattern of proliferating cell nuclear antigen in the nuclei of *Leishmania donovani.* Microbiology 2009; 155(Pt 11): 3748-57.
[http://dx.doi.org/10.1099/mic.0.033217-0] [PMID: 19729406]

[98] Calderano SG, de Melo Godoy PD, Motta MC, Mortara RA, Schenkman S, Elias MC. *Trypanosoma cruzi* DNA replication includes the sequential recruitment of pre-replication and replication machineries close to nuclear periphery. Nucleus 2011; 2(2): 136-45.
[http://dx.doi.org/10.4161/nucl.2.2.15134] [PMID: 21738836]

[99] Downey N, Hines JC, Sinha KM, Ray DS. Mitochondrial DNA ligases of *Trypanosoma brucei.*

Eukaryot Cell 2005; 4(4): 765-74.
[http://dx.doi.org/10.1128/EC.4.4.765-774.2005] [PMID: 15821136]

[100] Jackson AP, Sanders M, Berry A, *et al.* The genome sequence of *Trypanosoma brucei gambiense*, causative agent of chronic human african trypanosomiasis. PLoS Negl Trop Dis 2010; 4(4): e658.
[http://dx.doi.org/10.1371/journal.pntd.0000658] [PMID: 20404998]

[101] Genois MM, Paquet ER, Laffitte MC, *et al.* DNA repair pathways in trypanosomatids: from DNA repair to drug resistance. Microbiol Mol Biol Rev 2014; 78(1): 40-73.
[http://dx.doi.org/10.1128/MMBR.00045-13] [PMID: 24600040]

[102] Bodley AL, Shapiro TA. Molecular and cytotoxic effects of camptothecin, a topoisomerase I inhibitor, on trypanosomes and *Leishmania.* Proc Natl Acad Sci USA 1995; 92(9): 3726-30.
[http://dx.doi.org/10.1073/pnas.92.9.3726] [PMID: 7731973]

[103] Das BB, Sen N, Ganguly A, Majumder HK. Reconstitution and functional characterization of the unusual bi-subunit type I DNA topoisomerase from *Leishmania donovani.* FEBS Lett 2004; 565(1-3): 81-8.
[http://dx.doi.org/10.1016/j.febslet.2004.03.078] [PMID: 15135057]

[104] Fragoso SP, Mattei D, Hines JC, Ray D, Goldenberg S. Expression and cellular localization of *Trypanosoma cruzi* type II DNA topoisomerase. Mol Biochem Parasitol 1998; 94(2): 197-204.
[http://dx.doi.org/10.1016/S0166-6851(98)00066-8] [PMID: 9747970]

[105] Kulikowicz T, Shapiro TA. Distinct genes encode type II Topoisomerases for the nucleus and mitochondrion in the protozoan parasite *Trypanosoma brucei.* J Biol Chem 2006; 281(6): 3048-56.
[http://dx.doi.org/10.1074/jbc.M505977200] [PMID: 16316982]

[106] Topoisomerases RJ. Advances Gen Biol 1998; 5B: 463-85.

[107] Das BB, Sen N, Dasgupta SB, Ganguly A, Das R, Majumder HK. Topoisomerase research of kinetoplastid parasite *Leishmania*, with special reference to development of therapeutics. Indian J Med Res 2006; 123(3): 221-32.
[PMID: 16778306]

[108] Da Silveira RdeC, Da Silva MS, Nunes VS, Perez AM, Cano MI. The natural absence of RPA1N domain did not impair *Leishmania amazonensis* RPA-1 participation in DNA damage response and telomere protection. Parasitology 2013; 140(4): 547-59.
[http://dx.doi.org/10.1017/S0031182012002028] [PMID: 23388105]

[109] Pavani RS, Fernandes C, Perez AM, *et al.* RPA-1 from *Leishmania amazonensis* (LaRPA-1) structurally differs from other eukaryote RPA-1 and interacts with telomeric DNA *via* its N-terminal OB-fold domain. FEBS Lett 2014; 588(24): 4740-8.
[http://dx.doi.org/10.1016/j.febslet.2014.11.005] [PMID: 25451229]

[110] Damasceno JD, Nunes VS, Tosi LR. LmHus1 is required for the DNA damage response in *Leishmania major* and forms a complex with an unusual Rad9 homologue. Mol Microbiol 2013; 90(5): 1074-87.
[http://dx.doi.org/10.1111/mmi.12418] [PMID: 24118609]

[111] Brown EH, Iqbal MA, Stuart S, Hatton KS, Valinsky J, Schildkraut CL. Rate of replication of the murine immunoglobulin heavy-chain locus: evidence that the region is part of a single replicon. Mol Cell Biol 1987; 7(1): 450-7.

[http://dx.doi.org/10.1128/MCB.7.1.450] [PMID: 3031474]

[112] Hodgson B, Calzada A, Labib K. Mrc1 and Tof1 regulate DNA replication forks in different ways during normal S phase. Mol Biol Cell 2007; 18(10): 3894-902.
[http://dx.doi.org/10.1091/mbc.E07-05-0500] [PMID: 17652453]

[113] Calderano SG, Drosopoulos WC, Quaresma MM, *et al.* Single molecule analysis of *Trypanosoma brucei* DNA replication dynamics. Nucleic Acids Res 2015; 43(5): 2655-65.
[http://dx.doi.org/10.1093/nar/gku1389] [PMID: 25690894]

[114] Day JP, Marder BA, Morgan WF. Telomeres and their possible role in chromosome stabilization. Environ Mol Mutagen 1993; 22(4): 245-9.
[http://dx.doi.org/10.1002/em.2850220411] [PMID: 8223505]

[115] Lundblad V. DNA ends: maintenance of chromosome termini *versus* repair of double strand breaks. Mutat Res 2000; 451(1-2): 227-40.
[http://dx.doi.org/10.1016/S0027-5107(00)00052-X] [PMID: 10915875]

[116] Ferreira MG, Miller KM, Cooper JP. Indecent exposure: when telomeres become uncapped. Mol Cell 2004; 13(1): 7-18.
[http://dx.doi.org/10.1016/S1097-2765(03)00531-8] [PMID: 14731390]

[117] Blackburn EH, Gall JG. A tandemly repeated sequence at the termini of the extrachromosomal ribosomal RNA genes in *Tetrahymena.* J Mol Biol 1978; 120(1): 33-53.
[http://dx.doi.org/10.1016/0022-2836(78)90294-2] [PMID: 642006]

[118] Teixeira MT, Gilson E. Telomere maintenance, function and evolution: the yeast paradigm. Chromosome Res 2005; 13(5): 535-48.
[http://dx.doi.org/10.1007/s10577-005-0999-0] [PMID: 16132818]

[119] Giardini MA, Segatto M, da Silva MS, Nunes VS, Cano MI. Telomere and telomerase biology. Prog Mol Biol Transl Sci 2014; 125: 1-40.
[http://dx.doi.org/10.1016/B978-0-12-397898-1.00001-3] [PMID: 24993696]

[120] Dmitriev PV, Petrov AV, Dontsova OA. Yeast telosome complex: components and their functions. Biochemistry (Mosc) 2003; 68(7): 718-34.
[http://dx.doi.org/10.1023/A:1025022630840] [PMID: 12946253]

[121] Laroche T, Martin SG, Tsai-Pflugfelder M, Gasser SM. The dynamics of yeast telomeres and silencing proteins through the cell cycle. J Struct Biol 2000; 129(2-3): 159-74.
[http://dx.doi.org/10.1006/jsbi.2000.4240] [PMID: 10806066]

[122] Stewart SA, Weinberg RA. Senescence: does it all happen at the ends? Oncogene 2002; 21(4): 627-30.
[http://dx.doi.org/10.1038/sj.onc.1205062] [PMID: 11850788]

[123] Smogorzewska A, de Lange T. Regulation of telomerase by telomeric proteins. Annu Rev Biochem 2004; 73: 177-208.
[http://dx.doi.org/10.1146/annurev.biochem.73.071403.160049] [PMID: 15189140]

[124] Chan SR, Blackburn EH. Telomeres and telomerase. Philos Trans R Soc Lond B Biol Sci 2004; 359(1441): 109-21.
[http://dx.doi.org/10.1098/rstb.2003.1370] [PMID: 15065663]

[125] Oberdoerffer P, Sinclair DA. The role of nuclear architecture in genomic instability and ageing. Nat Rev Mol Cell Biol 2007; 8(9): 692-702.
[http://dx.doi.org/10.1038/nrm2238] [PMID: 17700626]

[126] Klobutcher LA, Swanton MT, Donini P, Prescott DM. All gene-sized DNA molecules in four species of hypotrichs have the same terminal sequence and an unusual 3 terminus. Proc Natl Acad Sci USA 1981; 78(5): 3015-9.
[http://dx.doi.org/10.1073/pnas.78.5.3015] [PMID: 6265931]

[127] Henderson ER, Blackburn EH. An overhanging 3 terminus is a conserved feature of telomeres. Mol Cell Biol 1989; 9(1): 345-8.
[http://dx.doi.org/10.1128/MCB.9.1.345] [PMID: 2927395]

[128] Chiurillo MA, Beck AE, Devos T, Myler PJ, Stuart K, Ramirez JL. Cloning and characterization of *Leishmania donovani* telomeres. Exp Parasitol 2000; 94(4): 248-58.
[http://dx.doi.org/10.1006/expr.2000.4499] [PMID: 10831393]

[129] Conte FF, Cano MI. Genomic organization of telomeric and subtelomeric sequences of *Leishmania (Leishmania) amazonensis.* Int J Parasitol 2005; 35(13): 1435-43.
[http://dx.doi.org/10.1016/j.ijpara.2005.05.011] [PMID: 16126212]

[130] Sandhu R, Li B. Examination of the telomere G-overhang structure in *Trypanosoma brucei.* J Vis Exp 2011; (47): 1959.
[PMID: 21307825]

[131] Greider CW, Blackburn EH. Identification of a specific telomere terminal transferase activity in *Tetrahymena* extracts Cell 1985; 43(2-1): 405-13.

[132] Greider CW, Blackburn EH. The telomere terminal transferase of *Tetrahymena* is a ribonucleoprotein enzyme with two kinds of primer specificity. Cell 1987; 51(6): 887-98.
[http://dx.doi.org/10.1016/0092-8674(87)90576-9] [PMID: 3319189]

[133] Chiang YJ, Hemann MT, Hathcock KS, *et al.* Expression of telomerase RNA template, but not telomerase reverse transcriptase, is limiting for telomere length maintenance *in vivo*. Mol Cell Biol 2004; 24(16): 7024-31.
[http://dx.doi.org/10.1128/MCB.24.16.7024-7031.2004] [PMID: 15282303]

[134] Shippen-Lentz D, Blackburn EH. Functional evidence for an RNA template in telomerase. Science 1990; 247(4942): 546-52.
[http://dx.doi.org/10.1126/science.1689074] [PMID: 1689074]

[135] Londoño-Vallejo JA, Wellinger RJ. Telomeres and telomerase dance to the rhythm of the cell cycle. Trends Biochem Sci 2012; 37(9): 391-9.
[http://dx.doi.org/10.1016/j.tibs.2012.05.004] [PMID: 22727244]

[136] Lingner J, Cech TR. Purification of telomerase from *Euplotes aediculatus*: requirement of a primer 3 overhang. Proc Natl Acad Sci USA 1996; 93(20): 10712-7.
[http://dx.doi.org/10.1073/pnas.93.20.10712] [PMID: 8855245]

[137] Harley CB, Futcher AB, Greider CW. Telomeres shorten during ageing of human fibroblasts. Nature 1990; 345(6274): 458-60.
[http://dx.doi.org/10.1038/345458a0] [PMID: 2342578]

[138] Counter CM, Avilion AA, LeFeuvre CE, *et al.* Telomere shortening associated with chromosome instability is arrested in immortal cells which express telomerase activity. EMBO J 1992; 11(5): 1921-9.
[PMID: 1582420]

[139] Eickbush TH. Telomerase and retrotransposons: which came first? Science 1997; 277(5328): 911-2. [http://dx.doi.org/10.1126/science.277.5328.911] [PMID: 9281073]

[140] Nakamura TM, Morin GB, Chapman KB, *et al.* Telomerase catalytic subunit homologs from fission yeast and human. Science 1997; 277(5328): 955-9.
[http://dx.doi.org/10.1126/science.277.5328.955] [PMID: 9252327]

[141] Pardue ML, DeBaryshe PG. Retrotransposons provide an evolutionarily robust non-telomerase mechanism to maintain telomeres. Annu Rev Genet 2003; 37: 485-511.
[http://dx.doi.org/10.1146/annurev.genet.38.072902.093115] [PMID: 14616071]

[142] Nosek J, Kosa P, Tomaska L. On the origin of telomeres: a glimpse at the pre-telomerase world. BioEssays 2006; 28(2): 182-90.
[http://dx.doi.org/10.1002/bies.20355] [PMID: 16435294]

[143] Belfort M, Curcio MJ, Lue NF. Telomerase and retrotransposons: reverse transcriptases that shaped genomes. Proc Natl Acad Sci USA 2011; 108(51): 20304-10.
[http://dx.doi.org/10.1073/pnas.1100269109] [PMID: 22187457]

[144] Chang S, DePinho RA. Telomerase extracurricular activities. Proc Natl Acad Sci USA 2002; 99(20): 12520-2.
[http://dx.doi.org/10.1073/pnas.212514699] [PMID: 12271146]

[145] Shay JW, Wright WE. Does telomerase moonlight? Scientist 2005; 19(4): 18-9.

[146] Jaiswal RK, Kumar P, Yadava PK. Telomerase and its extracurricular activities. Cell Mol Biol Lett 2013; 18(4): 538-54.
[http://dx.doi.org/10.2478/s11658-013-0105-0] [PMID: 24048710]

[147] Collins K. The biogenesis and regulation of telomerase holoenzymes. Nat Rev Mol Cell Biol 2006; 7(7): 484-94.
[http://dx.doi.org/10.1038/nrm1961] [PMID: 16829980]

[148] Lydall D. Taming the tiger by the tail: modulation of DNA damage responses by telomeres. EMBO J 2009; 28(15): 2174-87.
[http://dx.doi.org/10.1038/emboj.2009.176] [PMID: 19629039]

[149] Cifuentes-Rojas C, Shippen DE. Telomerase regulation. Mutat Res 2012; 730(1-2): 20-7.
[http://dx.doi.org/10.1016/j.mrfmmm.2011.10.003] [PMID: 22032831]

[150] Griffith JD, Comeau L, Rosenfield S, *et al.* Mammalian telomeres end in a large duplex loop. Cell 1999; 97(4): 503-14.
[http://dx.doi.org/10.1016/S0092-8674(00)80760-6] [PMID: 10338214]

[151] Muñoz-Jordán JL, Cross GA, de Lange T, Griffith JD. t-loops at trypanosome telomeres. EMBO J 2001; 20(3): 579-88.
[http://dx.doi.org/10.1093/emboj/20.3.579] [PMID: 11157764]

[152] He Q, Zeng P, Tan JH, *et al.* G-quadruplex-mediated regulation of telomere binding protein POT1 gene expression. Biochim Biophys Acta 2014; 1840(7): 2222-33. [http://dx.doi.org/10.1016/j.bbagen.2014.03.001] [PMID: 24631651]

[153] Teixeira MT, Arneric M, Sperisen P, Lingner J. Telomere length homeostasis is achieved *via* a switch between telomerase- extendible and -nonextendible states. Cell 2004; 117(3): 323-35. [http://dx.doi.org/10.1016/S0092-8674(04)00334-4] [PMID: 15109493]

[154] Wellinger RJ. The CST complex and telomere maintenance: the exception becomes the rule. Mol Cell 2009; 36(2): 168-9. [http://dx.doi.org/10.1016/j.molcel.2009.10.001] [PMID: 19854124]

[155] Price CM, Boltz KA, Chaiken MF, Stewart JA, Beilstein MA, Shippen DE. Evolution of CST function in telomere maintenance. Cell Cycle 2010; 9(16): 3157-65. [http://dx.doi.org/10.4161/cc.9.16.12547] [PMID: 20697207]

[156] Gao H, Cervantes RB, Mandell EK, Otero JH, Lundblad V. RPA-like proteins mediate yeast telomere function. Nat Struct Mol Biol 2007; 14(3): 208-14. [http://dx.doi.org/10.1038/nsmb1205] [PMID: 17293872]

[157] Longhese MP. DNA damage response at functional and dysfunctional telomeres. Genes Dev 2008; 22(2): 125-40. [http://dx.doi.org/10.1101/gad.1626908] [PMID: 18198332]

[158] Gilson E, Géli V. How telomeres are replicated. Nat Rev Mol Cell Biol 2007; 8(10): 825-38. [http://dx.doi.org/10.1038/nrm2259] [PMID: 17885666]

[159] Palm W, de Lange T. How shelterin protects mammalian telomeres. Annu Rev Genet 2008; 42: 301-34. [http://dx.doi.org/10.1146/annurev.genet.41.110306.130350] [PMID: 18680434]

[160] DeZwaan DC, Toogun OA, Echtenkamp FJ, Freeman BC. The Hsp82 molecular chaperone promotes a switch between unextendable and extendable telomere states. Nat Struct Mol Biol 2009; 16(7): 711-6. [http://dx.doi.org/10.1038/nsmb.1616] [PMID: 19525972]

[161] Giraud-Panis MJ, Teixeira MT, Géli V, Gilson E. CST meets shelterin to keep telomeres in check. Mol Cell 2010; 39(5): 665-76. [http://dx.doi.org/10.1016/j.molcel.2010.08.024] [PMID: 20832719]

[162] Soudet J, Jolivet P, Teixeira MT. Elucidation of the DNA end-replication problem in *Saccharomyces cerevisiae*. Mol Cell 2014; 53(6): 954-64. [http://dx.doi.org/10.1016/j.molcel.2014.02.030] [PMID: 24656131]

[163] Lundblad V. Telomere end processing: unexpected complexity at the end game. Genes Dev 2012; 26(11): 1123-7. [http://dx.doi.org/10.1101/gad.195339.112] [PMID: 22661225]

[164] Wu P, Takai H, de Lange T. Telomeric 3 overhangs derive from resection by Exo1 and Apollo and fill-in by POT1b-associated CST. Cell 2012; 150(1): 39-52. [http://dx.doi.org/10.1016/j.cell.2012.05.026] [PMID: 22748632]

[165] de Lange T. How telomeres solve the end-protection problem. Science 2009; 326(5955): 948-52. [http://dx.doi.org/10.1126/science.1170633] [PMID: 19965504]

[166] Longhese MP, Bonetti D, Manfrini N, Clerici M. Mechanisms and regulation of DNA end resection. EMBO J 2010; 29(17): 2864-74.
[http://dx.doi.org/10.1038/emboj.2010.165] [PMID: 20647996]

[167] Borst P, van Leeuwen F. β-D-glucosyl-hydroxymethyluracil, a novel base in African trypanosomes and other Kinetoplastida. Mol Biochem Parasitol 1997; 90(1): 1-8.
[http://dx.doi.org/10.1016/S0166-6851(97)00170-9] [PMID: 9497027]

[168] Dreesen O, Li B, Cross GA. Telomere structure and function in trypanosomes: a proposal. Nat Rev Microbiol 2007; 5(1): 70-5.
[http://dx.doi.org/10.1038/nrmicro1577] [PMID: 17160000]

[169] van Leeuwen F, Wijsman ER, Kieft R, van der Marel GA, van Boom JH, Borst P. Localization of the modified base J in telomeric VSG gene expression sites of *Trypanosoma brucei.* Genes Dev 1997; 11(23): 3232-41.
[http://dx.doi.org/10.1101/gad.11.23.3232] [PMID: 9389654]

[170] Ekanayake DK, Cipriano MJ, Sabatini R. Telomeric co-localization of the modified base J and contingency genes in the protozoan parasite *Trypanosoma cruzi.* Nucleic Acids Res 2007; 35(19): 6367-77.
[http://dx.doi.org/10.1093/nar/gkm693] [PMID: 17881368]

[171] Genest PA, Ter Riet B, Cijsouw T, van Luenen HG, Borst P. Telomeric localization of the modified DNA base J in the genome of the protozoan parasite *Leishmania.* Nucleic Acids Res 2007; 35(7): 2116-24.
[http://dx.doi.org/10.1093/nar/gkm050] [PMID: 17329373]

[172] van Luenen HG, Farris C, Jan S, *et al.* Glucosylated hydroxymethyluracil, DNA base J, prevents transcriptional readthrough in *Leishmania.* Cell 2012; 150(5): 909-21.
[http://dx.doi.org/10.1016/j.cell.2012.07.030] [PMID: 22939620]

[173] Rudenko G, Van der Ploeg LH. Transcription of telomere repeats in protozoa. EMBO J 1989; 8(9): 2633-8.
[PMID: 2511008]

[174] Dumas C, Chow C, Müller M, Papadopoulou B. A novel class of developmentally regulated noncoding RNAs in *Leishmania.* Eukaryot Cell 2006; 5(12): 2033-46.
[http://dx.doi.org/10.1128/EC.00147-06] [PMID: 17071827]

[175] Chiurillo MA, Cano I, Da Silveira JF, Ramirez JL. Organization of telomeric and sub-telomeric regions of chromosomes from the protozoan parasite *Trypanosoma* cruzi. Mol Biochem Parasitol 1999; 100(2): 173-83.
[http://dx.doi.org/10.1016/S0166-6851(99)00047-X] [PMID: 10391379]

[176] Freitas-Junior LH, Porto RM, Pirrit LA, Schenkman S, Scherf A. Identification of the telomere in *Trypanosoma cruzi* reveals highly heterogeneous telomere lengths in different parasite strains. Nucleic Acids Res 1999; 27(12): 2451-6.
[http://dx.doi.org/10.1093/nar/27.12.2451] [PMID: 10352173]

[177] Kim D, Chiurillo MA, El-Sayed N, *et al.* Telomere and subtelomere of *Trypanosoma cruzi* chromosomes are enriched in (pseudo)genes of retrotransposon hot spot and trans-sialidase-like gene

families: the origins of *T. cruzi* telomeres. Gene 2005; 346: 153-61.
[http://dx.doi.org/10.1016/j.gene.2004.10.014] [PMID: 15716016]

[178] Navarro M, Peñate X, Landeira D. Nuclear architecture underlying gene expression in *Trypanosoma brucei.* Trends Microbiol 2007; 15(6): 263-70.
[http://dx.doi.org/10.1016/j.tim.2007.04.004] [PMID: 17481901]

[179] Fu G, Barker DC. Characterisation of *Leishmania* telomeres reveals unusual telomeric repeats and conserved telomere-associated sequence. Nucleic Acids Res 1998; 26(9): 2161-7.
[http://dx.doi.org/10.1093/nar/26.9.2161] [PMID: 9547275]

[180] Fu G, Perona-Wright G, Barker DC. Leishmania braziliensis: characterisation of a complex specific subtelomeric repeat sequence and its use in the detection of parasites. Exp Parasitol 1998; 90(3): 236-43.
[http://dx.doi.org/10.1006/expr.1998.4326] [PMID: 9806868]

[181] Chiurillo MA, Beck AE, Devos T, Myler PJ, Stuart K, Ramirez JL. Cloning and characterization of *Leishmania donovani* telomeres. Exp Parasitol 2000; 94(4): 248-58.
[http://dx.doi.org/10.1006/expr.2000.4499] [PMID: 10831393]

[182] Lira CB, Giardini MA, Neto JL, Conte FF, Cano MI. Telomere biology of trypanosomatids: beginning to answer some questions. Trends Parasitol 2007; 23(8): 357-62.
[http://dx.doi.org/10.1016/j.pt.2007.06.005] [PMID: 17580124]

[183] Chiurillo MA, Ramírez JL. Charaterization of *Leishmania major* Friedlin telomeric terminus. Mem Inst Oswaldo Cruz 2002; 97(3): 343-6.
[http://dx.doi.org/10.1590/S0074-02762002000300011] [PMID: 12048562]

[184] Dumas C, Chow C, Müller M, Papadopoulou B. A novel class of developmentally regulated noncoding RNAs in *Leishmania.* Eukaryot Cell 2006; 5(12): 2033-46.
[http://dx.doi.org/10.1128/EC.00147-06] [PMID: 17071827]

[185] Lowell JE, Cross GA. A variant histone H3 is enriched at telomeres in *Trypanosoma brucei.* J Cell Sci 2004; 117(Pt 24): 5937-47.
[http://dx.doi.org/10.1242/jcs.01515] [PMID: 15522895]

[186] Yang X, Figueiredo LM, Espinal A, Okubo E, Li B. RAP1 is essential for silencing telomeric variant surface glycoprotein genes in *Trypanosoma brucei.* Cell 2009; 137(1): 99-109.
[http://dx.doi.org/10.1016/j.cell.2009.01.037] [PMID: 19345190]

[187] Cano MI, Dungan JM, Agabian N, Blackburn EH. Telomerase in kinetoplastid parasitic protozoa. Proc Natl Acad Sci USA 1999; 96(7): 3616-21.
[http://dx.doi.org/10.1073/pnas.96.7.3616] [PMID: 10097086]

[188] Muñoz DP, Collins K. Biochemical properties of *Trypanosoma cruzi* telomerase. Nucleic Acids Res 2004; 32(17): 5214-22.
[http://dx.doi.org/10.1093/nar/gkh864] [PMID: 15459291]

[189] da Silva MS, Perez AM, da Silveira RdeC, *et al.* The *Leishmania amazonensis* TRF (TTAGGG repeat-binding factor) homologue binds and co-localizes with telomeres. BMC Microbiol 2010; 10: 136.
[http://dx.doi.org/10.1186/1471-2180-10-136] [PMID: 20459667]

[190] Giardini MA, Fernández MF, Lira CB, Cano MI. *Leishmania amazonensis*: partial purification and study of the biochemical properties of the telomerase reverse transcriptase activity from promastigote-stage. Exp Parasitol 2011; 127(1): 243-8.
[http://dx.doi.org/10.1016/j.exppara.2010.08.001] [PMID: 20692257]

[191] Giardini MA, Lira CB, Conte FF, *et al.* The putative telomerase reverse transcriptase component of *Leishmania amazonensis*: gene cloning and characterization. Parasitol Res 2006; 98(5): 447-54.
[http://dx.doi.org/10.1007/s00436-005-0036-4] [PMID: 16416120]

[192] Gupta SK, Kolet L, Doniger T, *et al.* The *Trypanosoma brucei* telomerase RNA (TER) homologue binds core proteins of the C/D snoRNA family. FEBS Lett 2013; 587(9): 1399-404.
[http://dx.doi.org/10.1016/j.febslet.2013.03.017] [PMID: 23523918]

[193] Vasconcelos EJ, Nunes VS, da Silva MS, Segatto M, Myler PJ, Cano MI. The putative *Leishmania* telomerase RNA (LeishTER) undergoes trans-splicing and contains a conserved template sequence. PLoS One 2014; 9(11): e112061.
[http://dx.doi.org/10.1371/journal.pone.0112061] [PMID: 25391020]

[194] Lira CB, Siqueira Neto JL, Giardini MA, Winck FV, Ramos CH, Cano MI. LaRbp38: a *Leishmania amazonensis* protein that binds nuclear and kinetoplast DNAs. Biochem Biophys Res Commun 2007; 358(3): 854-60.
[http://dx.doi.org/10.1016/j.bbrc.2007.05.005] [PMID: 17506988]

[195] Duhagon MA, Pastro L, Sotelo-Silveira JR, *et al.* The Trypanosoma cruzi nucleic acid binding protein Tc38 presents changes in the intramitochondrial distribution during the cell cycle. BMC Microbial 2009; 9(1): 34.

[196] Liu B, Molina H, Kalume D, Pandey A, Griffith JD, Englund PT. Role of p38 in replication of *Trypanosoma brucei* kinetoplast DNA. Mol Cell Biol 2006; 26(14): 5382-93.
[http://dx.doi.org/10.1128/MCB.00369-06] [PMID: 16809774]

[197] Chen LY, Zhang Y, Zhang Q, *et al.* Mitochondrial localization of telomeric protein TIN2 links telomere regulation to metabolic control. Mol Cell 2012; 47(6): 839-50.
[http://dx.doi.org/10.1016/j.molcel.2012.07.002] [PMID: 22885005]

[198] Fernández MF, Castellari RR, Conte FF, *et al.* Identification of three proteins that associate *in vitro* with the *Leishmania* (*Leishmania*) *amazonensis* G-rich telomeric strand. Eur J Biochem 2004; 271(14): 3050-63.
[http://dx.doi.org/10.1111/j.1432-1033.2004.04237.x] [PMID: 15233802]

[199] Lira CB, de Siqueira Neto JL, Khater L, *et al.* LaTBP1: a Leishmania amazonensis DNA-binding protein that associates in vivo with telomeres and GT-rich DNA using a Myb-like domain. Arch Biochem Biophys 2007; 465(2): 399-409.
[http://dx.doi.org/10.1016/j.abb.2007.06.020] [PMID: 17678615]

[200] Riou G, Gabillot M, Douc-Rasy S, Kayser A. Partial purification and characterization of nuclear DNA-topoisomerases from trypanosomes (authors transl). C R Seances Acad Sci III 1982; 294(11): 439-42.
[PMID: 6284321]

[201] Neto JL, Lira CB, Giardini MA, *et al.* *Leishmania* replication protein A-1 *binds in vivo* single-stranded

telomeric DNA. Biochem Biophys Res Commun 2007; 358(2): 417-23.
[http://dx.doi.org/10.1016/j.bbrc.2007.04.144] [PMID: 17498665]

[202] Kaufmann D, Gassen A, Maiser A, Leonhardt H, Janzen CJ. Regulation and spatial organization of PCNA in *Trypanosoma brucei*. Biochem Biophys Res Commun 2012; 419(4): 698-702.
[http://dx.doi.org/10.1016/j.bbrc.2012.02.082] [PMID: 22387477]

[203] Leegwater PA, Strating M, Murphy NB, Kooy RF, van der Vliet PC, Overdulve JP. The *Trypanosoma brucei* DNA polymerase alpha core subunit gene is developmentally regulated and linked to a constitutively expressed open reading frame. Nucleic Acids Res 1991; 19(23): 6441-7.
[http://dx.doi.org/10.1093/nar/19.23.6441] [PMID: 1754381]

[204] Cano MI, Blake JJ, Blackburn EH, Agabian N. A *Trypanosoma brucei* protein complex that binds G-overhangs and co-purifies with telomerase activity. J Biol Chem 2002; 277(2): 896-906.
[http://dx.doi.org/10.1074/jbc.M104111200] [PMID: 11673453]

[205] Klingbeil MM, Burton P, Barnes R, McCulloch R. The three Rs of the Trypanosomatids genomes: Replication, recombination and repair. In: Barry D, McCulloch R, Mottram J, Acosta-Serrano A, Eds. Trypanosomes, after the genome. 1st ed., United Kingdom: Horizon Biosciense 2007.

[206] Stewart SA, Ben-Porath I, Carey VJ, OConnor BF, Hahn WC, Weinberg RA. Erosion of the telomeric single-strand overhang at replicative senescence. Nat Genet 2003; 33(4): 492-6.
[http://dx.doi.org/10.1038/ng1127] [PMID: 12652299]

[207] Hills SA, Diffley JF. DNA replication and oncogene-induced replicative stress. Curr Biol 2014; 24(10): R435-44.
[http://dx.doi.org/10.1016/j.cub.2014.04.012] [PMID: 24845676]

[208] Drury LS, Perkins G, Diffley JF. The Cdc4/34/53 pathway targets Cdc6p for proteolysis in budding yeast. EMBO J 1997; 16(19): 5966-76.
[http://dx.doi.org/10.1093/emboj/16.19.5966] [PMID: 9312054]

[209] Nguyen VQ, Co C, Irie K, Li JJ. Clb/Cdc28 kinases promote nuclear export of the replication initiator proteins Mcm27. Curr Biol 2000; 10(4): 195-205.
[http://dx.doi.org/10.1016/S0960-9822(00)00337-7] [PMID: 10704410]

[210] Liu E, Li X, Yan F, Zhao Q, Wu X. Cyclin-dependent kinases phosphorylate human Cdt1 and induce its degradation. J Biol Chem 2004; 279(17): 17283-8.
[http://dx.doi.org/10.1074/jbc.C300549200] [PMID: 15004027]

[211] Méndez J, Zou-Yang XH, Kim SY, Hidaka M, Tansey WP, Stillman B. Human origin recognition complex large subunit is degraded by ubiquitin-mediated proteolysis after initiation of DNA replication. Mol Cell 2002; 9(3): 481-91.
[http://dx.doi.org/10.1016/S1097-2765(02)00467-7] [PMID: 11931757]

[212] Saha T, Ghosh S, Vassilev A, DePamphilis ML. Ubiquitylation, phosphorylation and Orc2 modulate the subcellular location of Orc1 and prevent it from inducing apoptosis. J Cell Sci 2006; 119(Pt 7): 1371-82.
[http://dx.doi.org/10.1242/jcs.02851] [PMID: 16537645]

[213] Fujita M, Yamada C, Goto H, *et al.* Cell cycle regulation of human CDC6 protein. Intracellular localization, interaction with the human mcm complex, and CDC2 kinase-mediated hyper-

phosphorylation. J Biol Chem 1999; 274(36): 25927-32. [http://dx.doi.org/10.1074/jbc.274.36.25927] [PMID: 10464337]

[214] Méndez J, Stillman B. Chromatin association of human origin recognition complex, cdc6, and minichromosome maintenance proteins during the cell cycle: assembly of prereplication complexes in late mitosis. Mol Cell Biol 2000; 20(22): 8602-12. [http://dx.doi.org/10.1128/MCB.20.22.8602-8612.2000] [PMID: 11046155]

[215] Alexandrow MG, Hamlin JL. Cdc6 chromatin affinity is unaffected by serine-54 phosphorylation, S-phase progression, and overexpression of cyclin A. Mol Cell Biol 2004; 24(4): 1614-27. [http://dx.doi.org/10.1128/MCB.24.4.1614-1627.2004] [PMID: 14749377]

[216] Li Z. Regulation of the cell division cycle in *Trypanosoma brucei.* Eukaryot Cell 2012; 11(10): 1180-90. [http://dx.doi.org/10.1128/EC.00145-12] [PMID: 22865501]

[217] Blow JJ, Dutta A. Preventing re-replication of chromosomal DNA. Nat Rev Mol Cell Biol 2005; 6(6): 476-86. [http://dx.doi.org/10.1038/nrm1663] [PMID: 15928711]

[218] Shapiro TA, Englund PT. The structure and replication of kinetoplast DNA. Annu Rev Microbiol 1995; 49: 117-43. [http://dx.doi.org/10.1146/annurev.mi.49.100195.001001] [PMID: 8561456]

[219] Morris JC, Drew ME, Klingbeil MM, *et al.* Replication of kinetoplast DNA: an update for the new millennium. Int J Parasitol 2001; 31(5-6): 453-8. [http://dx.doi.org/10.1016/S0020-7519(01)00156-4] [PMID: 11334929]

[220] Liu B, Liu Y, Motyka SA, Agbo EE, Englund PT. Fellowship of the rings: the replication of kinetoplast DNA. Trends Parasitol 2005; 21(8): 363-9. [http://dx.doi.org/10.1016/j.pt.2005.06.008] [PMID: 15967722]

[221] Jensen RE, Englund PT. Network news: the replication of kinetoplast DNA. Annu Rev Microbiol 2012; 66: 473-91. [http://dx.doi.org/10.1146/annurev-micro-092611-150057] [PMID: 22994497]

[222] Shapiro TA. Kinetoplast DNA maxicircles: networks within networks. Proc Natl Acad Sci USA 1993; 90(16): 7809-13. [http://dx.doi.org/10.1073/pnas.90.16.7809] [PMID: 8395055]

[223] Hong M, Simpson L. Genomic organization of *Trypanosoma brucei* kinetoplast DNA minicircles. Protist 2003; 154(2): 265-79. [http://dx.doi.org/10.1078/143446103322166554] [PMID: 13677453]

[224] Ryan KA, Englund PT. Synthesis and processing of kinetoplast DNA minicircles in *Trypanosoma equiperdum.* Mol Cell Biol 1989; 9(8): 3212-7. [http://dx.doi.org/10.1128/MCB.9.8.3212] [PMID: 2552285]

[225] Ryan KA, Englund PT. Replication of kinetoplast DNA in *Trypanosoma equiperdum.* Minicircle H strand fragments which map at specific locations. J Biol Chem 1989; 264(2): 823-30. [PMID: 2536025]

[226] Rauch CA, Perez-Morga D, Cozzarelli NR, Englund PT. The absence of supercoiling in kinetoplast

DNA minicircles. EMBO J 1993; 12(2): 403-11.
[PMID: 8382606]

[227] Chen J, Rauch CA, White JH, Englund PT, Cozzarelli NR. The topology of the kinetoplast DNA network. Cell 1995; 80(1): 61-9.
[http://dx.doi.org/10.1016/0092-8674(95)90451-4] [PMID: 7813018]

[228] Lukes J, Guilbride DL, Votýpka J, Zíková A, Benne R, Englund PT. Kinetoplast DNA network: evolution of an improbable structure. Eukaryot Cell 2002; 1(4): 495-502.
[http://dx.doi.org/10.1128/EC.1.4.495-502.2002] [PMID: 12455998]

[229] Drew ME, Englund PT. Intramitochondrial location and dynamics of *Crithidia fasciculata* kinetoplast minicircle replication intermediates. J Cell Biol 2001; 153(4): 735-44.
[http://dx.doi.org/10.1083/jcb.153.4.735] [PMID: 11352935]

[230] Ray DS. Conserved sequence blocks in kinetoplast minicircles from diverse species of trypanosomes. Mol Cell Biol 1989; 9(3): 1365-7.
[http://dx.doi.org/10.1128/MCB.9.3.1365] [PMID: 2542768]

[231] Kidane GZ, Hughes D, Simpson L. Sequence heterogeneity and anomalous electrophoretic mobility of kinetoplast minicircle DNA from *Leishmania tarentolae.* Gene 1984; 27(3): 265-77.
[http://dx.doi.org/10.1016/0378-1119(84)90071-4] [PMID: 6329906]

[232] Birkenmeyer L, Sugisaki H, Ray DS. Structural characterization of site-specific discontinuities associated with replication origins of minicircle DNA from *Crithidia fasciculata.* J Biol Chem 1987; 262(5): 2384-92.
[PMID: 3029077]

[233] Myler PJ, Glick D, Feagin JE, Morales TH, Stuart KD. Structural organization of the maxicircle variable region of *Trypanosoma brucei*: identification of potential replication origins and topoisomerase II binding sites. Nucleic Acids Res 1993; 21(3): 687-94.
[http://dx.doi.org/10.1093/nar/21.3.687] [PMID: 8382804]

[234] Ryan KA, Shapiro TA, Rauch CA, Englund PT. Replication of kinetoplast DNA in trypanosomes. Annu Rev Microbiol 1988; 42: 339-58.
[http://dx.doi.org/10.1146/annurev.mi.42.100188.002011] [PMID: 2849370]

[235] Shlomai J. The structure and replication of kinetoplast DNA. Curr Mol Med 2004; 4(6): 623-47.
[http://dx.doi.org/10.2174/1566524043360096] [PMID: 15357213]

[236] Carpenter LR, Englund PT. Kinetoplast maxicircle DNA replication in *Crithidia fasciculata* and *Trypanosoma brucei.* Mol Cell Biol 1995; 15(12): 6794-803.
[http://dx.doi.org/10.1128/MCB.15.12.6794] [PMID: 8524245]

[237] Pérez-Morga DL, Englund PT. The attachment of minicircles to kinetoplast DNA networks during replication. Cell 1993; 74(4): 703-11.
[http://dx.doi.org/10.1016/0092-8674(93)90517-T] [PMID: 8395351]

[238] Guilbride DL, Englund PT. The replication mechanism of kinetoplast DNA networks in several trypanosomatid species. J Cell Sci 1998; 111(Pt 6): 675-9.
[PMID: 9471996]

[239] Liu Y, Englund PT. The rotational dynamics of kinetoplast DNA replication. Mol Microbiol 2007;

64(3): 676-90.
[http://dx.doi.org/10.1111/j.1365-2958.2007.05686.x] [PMID: 17462016]

[240] Abu-Elneel K, Robinson DR, Drew ME, Englund PT, Shlomai J. Intramitochondrial localization of universal minicircle sequence-binding protein, a trypanosomatid protein that binds kinetoplast minicircle replication origins. J Cell Biol 2001; 153(4): 725-34.
[http://dx.doi.org/10.1083/jcb.153.4.725] [PMID: 11352934]

[241] Melendy T, Sheline C, Ray DS. Localization of a type II DNA topoisomerase to two sites at the periphery of the kinetoplast DNA of *Crithidia fasciculata.* Cell 1988; 55(6): 1083-8.
[http://dx.doi.org/10.1016/0092-8674(88)90252-8] [PMID: 2849507]

[242] Hines JC, Engel ML, Zhao H, Ray DS. RNA primer removal and gap filling on a model minicircle replication intermediate. Mol Biochem Parasitol 2001; 115(1): 63-7.
[http://dx.doi.org/10.1016/S0166-6851(01)00272-9] [PMID: 11377740]

[243] Hines JC, Ray DS. A mitochondrial DNA primase is essential for cell growth and kinetoplast DNA replication in *Trypanosoma brucei.* Mol Cell Biol 2010; 30(6): 1319-28.
[http://dx.doi.org/10.1128/MCB.01231-09] [PMID: 20065037]

[244] Hines JC, Ray DS. A second mitochondrial DNA primase is essential for cell growth and kinetoplast minicircle DNA replication in *Trypanosoma brucei.* Eukaryot Cell 2011; 10(3): 445-54.
[http://dx.doi.org/10.1128/EC.00308-10] [PMID: 21257796]

[245] Gull K, Birkett C, Gerke-Bonet R, *et al.* The cell cycle and cytoskeletal morphogenesis in *Trypanosoma brucei.* Biochem Soc Trans 1990; 18(5): 720-2.
[http://dx.doi.org/10.1042/bst0180720] [PMID: 2083657]

[246] Woodward R, Gull K. Timing of nuclear and kinetoplast DNA replication and early morphological events in the cell cycle of *Trypanosoma brucei.* J Cell Sci 1990; 95(Pt 1): 49-57.
[PMID: 2190996]

[247] Minocha N, Kumar D, Rajanala K, Saha S. Kinetoplast morphology and segregation pattern as a marker for cell cycle progression in *Leishmania donovani.* J Eukaryot Microbiol 2011; 58(3): 249-53.
[http://dx.doi.org/10.1111/j.1550-7408.2011.00539.x] [PMID: 21401783]

[248] Wheeler RJ, Gluenz E, Gull K. The cell cycle of *Leishmania*: morphogenetic events and their implications for parasite biology. Mol Microbiol 2011; 79(3): 647-62.
[http://dx.doi.org/10.1111/j.1365-2958.2010.07479.x] [PMID: 21255109]

[249] Ambit A, Woods KL, Cull B, Coombs GH, Mottram JC. Morphological events during the cell cycle of *Leishmania major.* Eukaryot Cell 2011; 10(11): 1429-38.
[http://dx.doi.org/10.1128/EC.05118-11] [PMID: 21926331]

[250] Sela D, Yaffe N, Shlomai J. Enzymatic mechanism controls redox-mediated protein-DNA interactions at the replication origin of kinetoplast DNA minicircles. J Biol Chem 2008; 283(46): 32034-44.
[http://dx.doi.org/10.1074/jbc.M804417200] [PMID: 18799461]

[251] Sela D, Shlomai J. Regulation of UMSBP activities through redox-sensitive protein domains. Nucleic Acids Res 2009; 37(1): 279-88.
[http://dx.doi.org/10.1093/nar/gkn927] [PMID: 19039000]

[252] Cong L, Zhang F. Genome engineering using CRISPR-Cas9 system. Methods Mol Biol 2015; 1239:

197-217.
[http://dx.doi.org/10.1007/978-1-4939-1862-1_10] [PMID: 25408407]

[253] Sollelis L, Ghorbal M, MacPherson CR, *et al.* First efficient CRISPR-Cas9-mediated genome editing in *Leishmania* parasites. Cell Microbiol 2015; 17(10): 1405-12.
[http://dx.doi.org/10.1111/cmi.12456] [PMID: 25939677]

[254] Peng D, Kurup SP, Yao PY, Minning TA, Tarleton RL. CRISPR-Cas9-mediated single-gene and gene family disruption in *Trypanosoma cruzi.* MBio 2014; 6(1): e02097-14.
[http://dx.doi.org/10.1128/mBio.02097-14] [PMID: 25550322]

[255] Robinson KA, Beverley SM. Improvements in transfection efficiency and tests of RNA interference (RNAi) approaches in the protozoan parasite *Leishmania.* Mol Biochem Parasitol 2003; 128(2): 217-28.
[http://dx.doi.org/10.1016/S0166-6851(03)00079-3] [PMID: 12742588]

[256] DaRocha WD, Otsu K, Teixeira SM, Donelson JE. Tests of cytoplasmic RNA interference (RNAi) and construction of a tetracycline-inducible T7 promoter system in *Trypanosoma cruzi.* Mol Biochem Parasitol 2004; 133(2): 175-86.
[http://dx.doi.org/10.1016/j.molbiopara.2003.10.005] [PMID: 14698430]

[257] Barnes RL, Shi H, Kolev NG, Tschudi C, Ullu E. Comparative genomics reveals two novel RNAi factors in *Trypanosoma brucei* and provides insight into the core machinery. PLoS Pathog 2012; 8(5): e1002678.
[http://dx.doi.org/10.1371/journal.ppat.1002678] [PMID: 22654659]

CHAPTER 5

Genome Maintenance in Trypanosomatids

Gonzalo Cabrera[1], **Viviane G. Silva**[2], **Isabela C. Mendes**[2], **Carlos R. Machado**[2,*] and **Richard McCulloch**[3,*]

[1] *Programa de Biología Celular y Molecular, Instituto de Ciencias Biomédicas, Facultad de Medicina, Universidad de Chile, Santiago, Chile*

[2] *Departamento de Bioquímica e Imunologia, Instituto de Ciências Biológicas, Universidade Federal de Minas Gerais, Belo Horizonte, Minas Gerais, Brazil*

[3] *The Wellcome Trust Centre for Molecular Parasitology, Institute of Infection, Immunity and Inflammation, University of Glasgow, Sir Graeme Davis Building, 120 University Place, Glasgow, G12 8TA, UK*

Abstract: The genome is the source of life, providing the information needed to direct all aspects of organismal function. Propagation of life requires copying of the genome and faithful transmission from parent to offspring. Many challenges confront genome propagation, including ensuring the accurate and complete copying of the DNA, circumventing impediments to DNA replication, and maintaining genome integrity in the face of myriad insults and during periods of cellular quiescence. Just as importantly, the genome must be allowed to change, either incrementally through small mutations in sequence or by large-scale rearrangements. Such changes not only drive evolution, but can be integral components of an organism's life cycle. In this chapter we consider the rapidly growing body of knowledge on how the genomes of kinetoplastid parasites are maintained, by describing the range of genome repair and damage tolerance pathways that operate. We focus on *Trypanosoma brucei*, *Trypanosoma cruzi* and *Leishmania*, three important human and animal pathogens, but we believe the lessons learned from the study of genome maintenance in these genetically tractable parasites are applicable widely, not only to other parasites but throughout biology.

* **Corresponding authors Richard McCulloch:** University of Glasgow, Scotland, UK; Tel: 0044(0)1413305946; Fax: 0044(0)1413305422; E-mail: richard.mcculloch@glasgow.ac.uk; **Carlos R. Machado:** Universidade Federal de Minas Gerais, Belo Horizonte, Minas Gerais, Brazil; E-mail: crmachad@icb.ufmg.br

Marcelo Santos da Silva & Maria Isabel N. Cano (Eds.)

Keywords: Base excision repair, DNA repair, DRP lyase activities, Genome integrity, Homologous recombination, Microhomology-mediated end-joining, Mismatch repair, Non-homologous end-joining, Nucleotide excision repair, Translesion DNA synthesis.

INTRODUCTION

Propagation of life requires transmission of the genome from parent to offspring. A high level of fidelity in transmission allows the retention of selected genetic characteristics. For this reason, complex machineries for highly accurate genome duplication have evolved and are identifiably conserved throughout extant life. However, in addition to replication, accurate genome transmission requires processes that identify and respond to threats to the integrity and content of the genetic information. All genomes are subject to constant attacks, which can arise exogenously, such as from environmental radiation or chemicals, or endogenously, such as due to free radicals generated during metabolism. In cellular organisms, such attacks can lead to DNA damage in both the nuclear and organellar genomes, potentially resulting in genetic alterations, which may be beneficial but are frequently detrimental and may cause death. The forms of damage to DNA genomes are wide-ranging, including modified or lost bases, base mismatches, incorrect intra- or inter-strand chemical bonds and single- or double-strand breaks (SSBs and DSBs). The primary response to damage is mediated by DNA repair processes, which survey the genome for all these forms of damage and, having identified them, correct the structure. Normally, DNA repair ensures that the genetic content is maintained, but occasionally genome sequence or organisation change is an unavoidable by- product of the reactions. Sometimes the repair is not possible and, in these circumstances, damage bypass may be invoked, potentially resulting in an elevated rate of genome change. Indeed, in some circumstances organisms deliberately invoke genome alterations for continued survival, including by harnessing the normally protective functions of the repair processes.

In this chapter we discuss our growing understanding of the diverse array of genome maintenance processes that are active in kinetoplastid parasites, reflecting the wide range of genome damage that the organisms face. In addition, we discuss growing evidence for the absence of at least one key repair process, as well as kinetoplastid adaptations in what are normally considered highly conserved reactions. Finally, we highlight areas of limited understanding, which may become the focus of future research. The interested reader is referred to recent reviews that provide further insight into these topics [1 - 3].

The Genomes of Kinetoplastids

Kinetoplastids represent an evolutionary grouping of eukaryotic microbes marked by shared, diverged features of core eukaryotic biology. One such divergence, for which the grouping is named, is the mitochondrial genome (the kinetoplast), which is composed of an interlinked lattice of ~1,000 minicircles and ~20 maxicircles [4, 5]. Divergence is not limited to the kinetoplast, however, since nuclear genome sequencing has revealed a highly unusual genetic landscape in which virtually all genes are organised in a relatively small number (~150-200) of so-called directional gene clusters (DGCs), some of which can encompass hundreds of genes [6, 7]. This arrangement reflects diverged RNA Polymerase (Pol) II transcription, in that virtually every gene is expressed from a multigene transcription unit (which, again, can encompass hundreds of genes). Thus, transcription of each gene within a DGC initiates from a single, shared transcription start site and ends at a shared termination site, with coupled trans-splicing and polyadenylation used to generate mature mRNAs from a multigene primary transcript. Initiation and termination sites have been termed strand-switch regions (SSRs), represent only a fraction (~1%) of the number of nuclear genes and are marked by the accumulation of variant histones, modified histones and a modified base (named J) [8 - 16]. It is increasingly clear that the diverged organisation of transcription in kinetoplastids is reflected in gene expression novelties, most notably highly diverged RNA Pol II promoters and the devolution of gene expression controls from transcription initiation to post-transcriptional processes [17, 18]. However, the impact of multigenic transcription is unlikely to be limited to gene expression controls, with wider (and, as yet, less well explored) consequences for genome stability.

Central to the potential impact of multigenic transcription on genome stability is replication of the nuclear genome, a reaction whose machinery and dynamics have only recently begun to be examined [19]. The interaction between transcription and replication exists on at least two levels, as revealed by genome-wide mapping of the initiator factor ORC1/CDC6 [20 - 22] and the pattern of replication [23]. A profound overlap between transcription and replication is found at the SSR boundaries of the multigene transcription units: ORC1/CDC6 appears to bind to potentially all SSRs, though mediates DNA replication initiation at only a fraction; in addition, loss of ORC1/CDC6 not only impedes replication, but leads to altered transcription of the genes around the SSRs [23]. In any organism, collisions between the DNA replisome and RNA Pol at genes transcribed during S phase are associated with genomic instability, which is limited by the evolution of replication and transcription co-directionality in some highly expressed genes in bacteria and humans [24 - 26]. The extent of the kinetoplastid genome that is transcribed, and the distances that RNA Pol II must traverse in a single direction suggest that such collisions must be pronounced in these genomes [27]. Though replication origin mapping in *T. brucei* [23] and *Leishmania* [28] has shown that replication initiation occurs at an only subset of the available SSRs, these sites are not selected to prevent head-on collisions between multigenic transcription and the replication forks: multiple sites of such collisions are seen and cause detectable slowing of replication rate [23]. The fact that replication-transcription collisions are present shows that they are not lethal, suggesting that the conflicts at these sites must be resolved. However, how kinetoplastids tackle such replication stress is unknown.

Sequencing the nuclear genomes of *T. brucei* [29], *T. cruzi* [30] and multiple *Leishmania* [31 - 33] species revealed very high levels of gene synteny [6], perhaps due to the constraints of multigenic transcription. Despite this, the parasites display considerable variation in genome architecture, ploidy stability and the dynamics of nuclear DNA replication. Indeed, a picture appears to be emerging of a commonality in genome biology between *T. cruzi* and *Leishmania*, which differs radically from *T. brucei*. Genome sequencing and karyotype analysis suggest that the genomes of *T. cruzi* and *Leishmania* are composed of ~3-4 times more chromosomes (~35-40) than are seen in *T. brucei* (11), despite broadly similar overall haploid genome sizes (~26-40 Mb) [6, 34]. As a consequence of this differing organisation, most *T. cruzi* and *Leishmania* chromosomes are smaller (size range ~0.1-3.0 Mb) than the *T. brucei* chromosomes (size range ~1.0-4.5 Mb). The genomes of *T. cruzi* and *Leishmania* are also marked by highly variable ploidy, seen as chromosome copy number variation during growth and between cells in the population [35 - 39]. In contrast, the megabase chromosomes of *T. brucei* appear stably diploid, with aneuploidy limited to Variant Surface Glycoprotein (*VSG*) genes found in the megabase chromosome subtelomeres [40], as well as in further chromosomes, termed the intermediate and minichromosomes, that appear to have evolved in *T. brucei* to provide *VSG*s for immune evasion [41]. Gene amplification and loss is pervasive throughout the *Leishmania* genome and provides a route for adaptation during growth (such as in response to drugs) [42]; in contrast, in *T. brucei* such gene copy number variation may be limited to the *VSG*s, which rearrange to mediate immune evasion during mammal infections [43]. Finally, genome-wide mapping suggests that replication dynamics differ markedly between *Leishmania* and *T. brucei*: where multiple mapped origins are seen in each chromosome in *T. brucei* [23], only a single origin can be mapped in each *Leishmania* chromosome [28]; in addition, while origins display a temporal order of activation in *T. brucei*, all origins activate at the same time in *Leishmania*.

The above description provides an oversight into the genetic landscape in which DNA maintenance mechanisms operate in the kinetoplastids, including some of the challenges that are faced. More details about genome organization and ploidy in trypanosomatids are described in Chapter 2. Below we consider how DNA replication, gene transcription and genome maintenance by repair and lesion bypass intersect and shape the genomes of the kinetoplastids.

NUCLEOTIDE EXCISION REPAIR: MECHANISTIC STREAMLINING AND NEOFUNCTIONALISATION

Three forms of excision repair are found widely in nature. In each of these repair pathways, altered DNA composition is recognized, a stretch of single-stranded DNA containing the lesion is then excised by nuclease activity and the sequence that has been removed is resynthesized. Each of these pathways, termed nucleotide excision repair, mismatch repair and base excision repair, operate in kinetoplastids. However, to a greater or lesser extent the mode of action of the three pathways has been modified to suit the specific demands of the cell biology of these organisms. Each repair pathway is considered in turn.

Nucleotide Excision Repair (NER) is an evolutionarily conserved repair mechanism that acts in the removal of a range helix-distorting DNA lesions [44, 45]. Examples of NER substrates include cyclobutane pyrimidine dimers and pyrimidine-6-4-pyrimidine photoproducts, both induced by UV radiation, and base adducts created by exogenous chemical agents such as cisplatin. These DNA alterations can impede the progress of DNA or RNA Pols, resulting in replication fork collapse or transcription stalling. In humans, NER defects are associated with the cancer-prone syndrome xeroderma pigmentosum (XP), with the developmental disorders Cockayne syndrome (CS) and trichothiodystrophy and with a combined CS/XP syndrome, making this pathway the object of extensive study. In mammals, the NER response involves more than 30 proteins [46]. Though not all these factors are conserved in other organisms, NER is considered always to operate in two subpathways in bacteria [47], eukaryotes [46] and, most likely, in archaea [48]. One subpathway, termed global genome repair (GGR), removes lesions throughout the genome and is considered a slower transcription-independent process. The second, faster subpathway is called transcription-coupled repair (TCR) and is activated when a lesion prevents a gene from continuing to be transcribed. The two subpathways of NER differ at the step of lesion recognition and then utilize common downstream machinery to complete the repair process.

In eukaryotic GGR, a complex of the XP group C (XPC) and hHR23B proteins is responsible for the damage sensing and repair factor recruitment step, binding the strand opposite the lesion and not the chemical adduct itself [49, 50]. This step can be aided by a dimer of DNA damage-binding protein 1 (DDB1) – XPE (DDB2), the latter of which specifically recognizes UV-induced lesions [51, 52]. In eukaryotic TCR the signal for initiation of NER is blockage of transcription elongation by RNA Pol II and, in some circumstances, RNA Pol I [53]. The NER machinery is recruited to the stalled RNA Pol by the CSA and CSB (Cockayne syndrome A and B proteins) proteins, which, like the GGR lesion recognition factors, facilitate the recruitment of downstream NER factors [54, 55]. As mentioned above, the repair steps and machinery following lesion recognition are the same in GGR and TCR. The first common step is the recruitment of the transcription factor II H (TFIIH) complex. TFIIH contains ten subunits, including two helicases, XPB and XPD, five associated proteins (p62, p52, p44, p34 and p8), and the cyclin-dependent kinase (CDK)-activating kinase (CAK) sub-complex, containing MaT1, CDK7 and cyclin H. Through the activity of the helicase subunits, TFIIH opens the DNA double strands around the lesion, creating a 'bubble' for recruitment of XPA and replication protein A (RPA). XPA promotes the release of the CAK subcomplex, facilitating the recruitment of two endonucleases: the XPF-excision repair cross-complementing 1 (ERCC1) complex, and XPG. XPF-ERCC1 and XPG are responsible for incisions 5' and 3', respectively, of the DNA damage. This dual incision leads to the removal of an oligonucleotide of approximately 30 nucleotides, resulting in a single strand gap that is filled by DNA Pols δ and k in non-dividing cells, and DNA Pol ε in replicative cells, reactions that are aided by replication factor C (RFC) and the proliferating cellular nuclear antigen (PCNA). Finally, the nicks at the ends of the filled gap are sealed, either by the X-ray repair cross-complementing protein 1 (XRCC1)-DNA ligase III (LIG3) complex or a flap endonuclease 1 (FEN1)-DNA ligase I (LIG1) complex in dividing cells, and in non-dividing cells by XRCC1-LIG3.

Genome sequencing allowed the cataloguing of putative NER factors encoded by the kinetoplastids. These analyses suggested broad conservation of all factors, with only apparently minor divergences [3, 30, 56]. In the GGR recognition step, orthologues of XPC, RAD23 and DDB1 proteins could be identified, though the smaller subunit of the UV-DDB dimer, XPE (DDB2), was not seen. In *T. cruzi*, the putative DDB1-encoding gene appears duplicated, a situation not seen in the other kinetoplastids, though whether this has functional implications for NER has not been tested. Only CSB is recognizably present for lesion recognition during TCR, suggesting kinetoplastids may lack CSA. In the downstream, common NER steps, tandem affinity purification allowed identification of a *T. brucei* TFIIH complex containing XPB and XPD along with their regulators TFB1 (p62), TFB2 (p52), TFB4 (p34), TFB5 and SSL1 (p44) [57]. However, *T. brucei* TFIIH lacks the CAK complex and is instead associated with two kinetoplastid-specific proteins, TSP1 and TSP2 [58]. In addition, a variant form of XPB, termed XPBz, was found [57]. Expression of two XPB-like proteins is not unique to kinetoplastids, but is also seen in plants [59] and some archaea [60]. However, whereas the two plant XPBs are closely sequence related, the two XPB-like proteins in kinetoplastids and archaea display substantial sequence and size divergence. Downstream of TFIIH, no gene encoding XPA could be identified, whereas the predicted NER endonucleases could: both subunits of the XPF/ERCC1 complex are well conserved, while XPG appears to be substantially smaller than orthologues in *H. sapiens* and *A. thaliana* due to the absence of residues within the predicted 'spacer' region that mediates TFIIH interaction [61].

The above bioinformatic predictions suggested broad conservation of NER, with both the GGR and TCR pathways in operation. However, two functional studies in *T. brucei* reveal that these gene predictions mask a pronounced divergence in NER activity relative to other eukaryotes (and, indeed, all organisms thus far examined) Fig. (**1**) summarises the available data. The first study [62] focused on XPBz and showed that the factor is non-essential and plays no role in transcription, suggesting it is not a major component of the TFIIH complex, if at all. In addition, by analyzing the sensitivity of the mutants to UV and cisplatin, it was possible to show that XPBz acts in repair. In the second study [63], RNAi was used to examine the functions of most of the predicted *T. brucei* NER genes, revealing that whereas loss of CSB, XPBz and XPG does not impair growth and sensitizes the cells to UV and cisplatin, RNAi of XPC, DDB1, XPB or XPD is lethal but does not sensitise the cells to UV or cisplatin. In fact, there was no evidence that XPB or XPD act in the repair of any damage thus far tested [62, 63], whereas RNAi of XPC and DDB1 sensitises *T. brucei* cells to cyclophosphamide, suggesting a role in cross-link repair and not NER [63]. Preliminary work in *T. cruzi*, using single allele knockouts or overexpression of CSB or XPC support the proposed divergence in functions seen in *T. brucei* (Mendes and Machado, unpublished): changing the expression levels of XPC causes no detectable variation in UV or cisplatin sensitivity, whereas changes in CSB levels cause clear changes in sensitivity. As these responses mirror the findings observed in *T. brucei* following RNAi, it seems likely that the alterations in GGR and TCR function are common to all kinetoplastids. Collectively, the above data suggest two things. First, it appears that TCR is the predominant (and perhaps sole) NER pathway that is active, but operates without engaging at least the XPB and XPD components of TFIIH, and uses a variant XPB instead. Thus, TCR is a streamlined NER reaction. Second, GGR does not act detectably in NER and instead the activities of XPC and DDB1 have been diverted to a distinct and essential genome maintenance function. The implications of these predictions are still to be explored fully, but the simplest explanation for why such changes have evolved is the predominance of multigenic transcription in the kinetoplastid genomes.

One further facet of the analysis of CSB, XPBz and XPG gene function in *T. brucei* and *T. cruzi* is that there is some evidence for the tolerance of at least some level of UV-induced lesions that is not seen with cisplatin, as manifest in pronounced differences in repair kinetics [63]. It has been proposed that this can be explained by RNA Pol II bypass of some lesions in some circumstances, a reaction that is well established in other eukaryotes [64, 65] but not so far described in kinetoplastids. In an organism with multigenic transcription, translesion RNA synthesis may be preferred to enacting repair in some circumstances, since it can avoid prolonged accumulation of stalled RNA Pol II (and perhaps Pol I) and ensure the continued movement of transcription across the multigene units. In yeast, CSB can act independently of NER to promote translesion transcription [66], but in kinetoplastids it is possible that CSB may remain more closely tied to the NER machinery to act in deciding, after RNA Pol blockage, whether to allow recruitment of the downstream TCR machinery or allowing lesion bypass. In bacteria and eukaryotes an increasing number of factors are being identified that act on stalled RNA Pol and influence its behavior, including during NER [67, 68]. In this regard, the potential absence of CSA in kinetoplastids may be significant. Recent evidence indicates that CSA is involved in CSB ubiquitination and degradation following UV irradiation [68], which would restore transcription at a normal rate after the repair. The potential absence of CSA again implies that the regulation of TCR may differ in kinetoplastids, though what the absence of this component might mean, or if an alternative protein replaces the action of CSA on CSB, is still to be determined. More widely, new approaches will need to be adopted to test the effect of UV damage and loss of the NER factors on transcription dynamics in the kinetoplastids.

Further questions about the operation of TCR and the function of the GGR lesion sensors also remain. To date, only CSB, XPBz and XPG have been shown to act in NER. Is it possible that this collection of factors is alone sufficient to execute the major steps in the reaction? Given the regulatory role of CSA (above), the conservation of residues in kinetoplastid CSBs that mark the protein as a member of the superfamily 2 of nucleic acid translocases/helicases [69], suggests it is possible that CSB alone may be sufficient to initiate NER, but functional studies are needed. The available data in *T. brucei* indicate that NER occurs without the engagement of at least the XPB and XPD components of the TFIIH complex, due to the presence of the repair-specific helicase, TbXPBz [57, 62, 63]. XPBz orthologues in the kinetoplastids retain all conserved domains related to DNA helicase function, suggesting that XPB duplication emerged to provide activities for repair that replace, or assume greater importance than, XPB-TFIIH. This is intriguing, since in other eukaryotes XPB is a weak helicase and impeding this activity does not harm NER, while loss of XPD helicase activity is harmful [25], indicating that XPD has the major role in strand lesion extrusion [70]. It will be important to determine if XPBz acts alone or associates with another helicase, perhaps in a variant of the TFIIH complex, or even acts in an unrelated complex. The absence of a CAK complex in kinetoplastid TFIIH [58], allied with the potential absence of XPA and the variant structure of XPG (lacking a TFIIH interaction domain) [63], provide further evidence of streamlining. CAK disassociation from TFIIH is considered a critical step for recruitment of XPF/ERCC1 and XPG to the lesion during NER [71], and XPA's role is in stabilizing TFIIH association with the NER machinery [72]. Thus, the absence of CAK and XPA, and truncation of XPG may suggest that XPG recruitment during kinetoplastid NER bypasses the need for TFIIH and the remodeling provided by XPA. Retention of XPG for kinetoplastid NER is consistent with the evidence that the endonuclease acts in TCR and, indeed, can interact with CSB during early steps in this NER subpathway [73]. To date, testing whether XPG acts alone during incision or whether XPF/ERCC1 also contributes to kinetoplastid NER has been hampered by technical issues during RNAi [63]. RNAi suggests, surprisingly, that XPC and DDB1 are essential genes, but the reasons for this crucial role are unclear, as the only phenotypic consequence of their loss to date has been increased sensitivity to one interstrand crosslinking agent [63]. Further work will therefore be needed to examine what roles these factors play, which may be illuminating given the increasing evidence that XPC can provide a range of functions, including in non-NER repair, transcription and genome stability [50]. Fig. (**1**) summarises streamlining and diversification of NER in *T. brucei*.

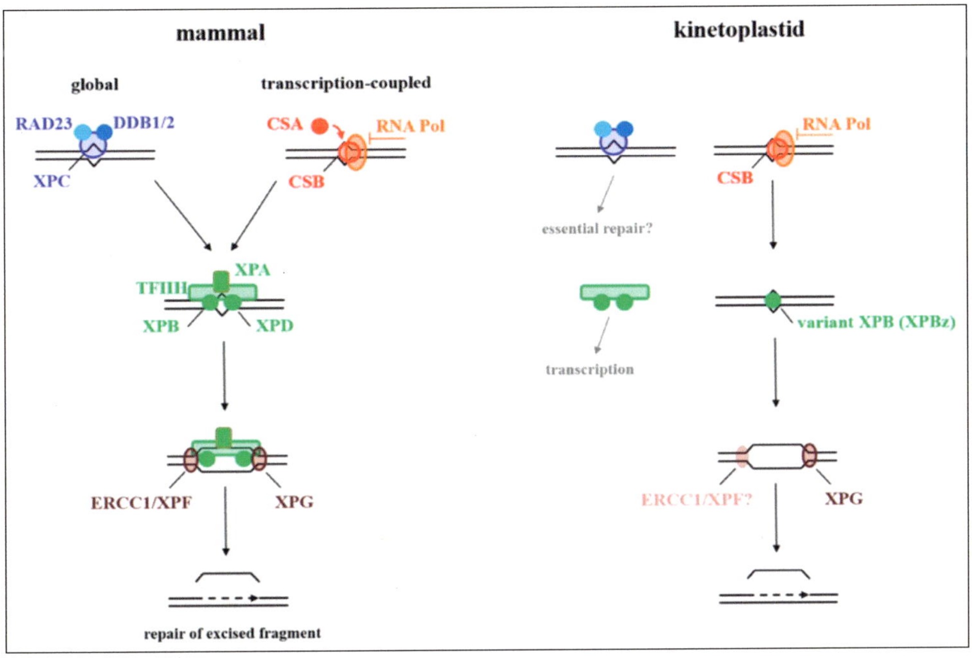

Fig. (1). Streamlining and diversification of *T. brucei* nucleotide excision repair. The diagram to the left provides a simplified outline of the reaction steps and machinery of nucleotide excision repair (NER) in mammals. Two subpathways operate in tandem: in global genome NER (global), lesions are recognised by XPC acting with RAD23 and DDB1/2, initiating downstream lesion excision, whereas in transcription-coupled NER repair is triggered when movement of RNA polymerase (Pol) is blocked, leading to recruitment of CSB, whose activity is influenced by CSA. Downstream steps are common and begin by recruitment of the TFIIH transcription complex to the lesion, allowing the helicases XPB and XPB to open the DNA strands. XPA alters the conformation of TFIIH, releasing some components (not shown) and allowing recruitment of two endonucleases: XPG and a complex of XPF and ERCC1. The endonucleases excise an oligonucleotide around the lesion and repair is completed by DNA resynthesis (dotted arrow) and ligation (proteins involved are not detailed). The diagram to the right details known divergence in *T. brucei* NER, which appears to be common to *T. cruzi* and perhaps all kinetoplastids. RNAi indicates that XPC and DDB are essential but do not contribute to NER, suggesting diversion of their activity to an unknown feature of genome maintenance (RAD23 is presumed to act in the same way, but its activity has not been tested). XPB and XPD do not engage in NER and XPA has not been detected, suggesting that TFIIH acts only in transcription. *T. brucei* NER appears to be exclusively or predominantly executed by CSB-directed transcription-coupled repair; to date, CSA has not been detected and might be absent. A diverged paralogue of XPB, termed XPBz, is the only helicase so far detected to act in NER. XPG acts as a nuclease in NER, though has a predicted variant structure (not shown); the role of XPF/ERCC1 has not to date been tested.

THE DUAL ROLE OF MISMATCH REPAIR IN DNA REPAIR AND IN THE OXIDATIVE STRESS RESPONSE

During DNA replication some errors can escape the proofreading activity of replicative DNA Pols, generating base mismatches or insertions and deletions that can generate loops in the DNA strand. The main system to repair such errors is the Mismatch Repair (MMR) pathway. In addition to repairing post-replicative base mismatches and insertion/deletion loops (IDLs), MMR also scans for mismatches during homologous recombination between non-identical molecules and prevents exchange. More widely, MMR and MMR-related proteins can also be involved in meiotic recombination, triplet repeat expansion, class switch recombination, somatic hypermutation and DNA damage signaling [74 - 76].

The MMR machinery is well conserved between prokaryotes and eukaryotes [77]. The first step of error recognition in eukaryotes is performed by homologues of the bacterial protein MutS, named MSH (MutS homologues), and numbered according to the order they were cloned and characterised. MutS and MSH proteins have numerous conserved domains: domain I is the mismatch binding domain that interacts with DNA; domains II and III are connectors; and domains IV and V represent the ATPase element of the proteins [78]. In eukaryotic cells, the recognition of lesions is carried out by heterodimers formed between two MutS homologues, forming complexes called MutSα or MutSβ. MutSα (MSH2-MSH6) recognizes base mismatches and 1 to 2 unpaired nucleotides, whereas MutSβ (MSH2-MSH3) recognizes longer IDLs with 1 to 15 unpaired nucleotides [78]. Even small distortions in DNA caused by single-base mismatches or IDLs can trigger the repair cascade, with MutSα or MutSβ binding leading to recruitment of downstream factors, followed by excision of incorrect bases and synthesis of a new DNA strand containing the correct base. MSH2-MSH6 wraps around DNA, forming a 'sliding clamp', and scans for errors through close contact with the phosphate backbone [79]. After recognition of mismatches or IDLs, an ATP-dependent conformational change of the MSH heterodimers occurs, promoting interaction with the eukaryotic MutL homologue (MLH) heterodimers at the site of the error. PCNA also has a role in the MMR response by activating the MLH heterodimers and so initiating excision of the daughter DNA strand, meaning that an oligonucleotide is excised on the strand that contains mismatch errors [80]. In some bacteria, MutH catalyses the excision step. However, this protein is not found in all bacteria and homologues have not been described in any eukaryote. Instead, the excision step is catalysed by endonuclease activity found within a heterodimer formed by the proteins MLH1 and PMS1 [81]. In addition to these factors, some eukaryotes encode multiple further MutL homologues. It appears that pre-existing nicks in the newly synthesised DNA are sufficient to activate the MMR machinery in eukaryotic cells [82, 83]. Because MMR increases the fidelity of DNA replication [84], defects in the pathway result in an increased mutation rate, including in *T. brucei* [85] and *T. cruzi* [86]. In addition, or a direct result of such mutations, defective MMR can result in cancer in humans, including hereditary colon cancer and a variety of sporadic cancers [87].

Studies of the MMR pathway in *T. brucei* and *T. cruzi* began before the sequencing of the parasite genomes [85, 86, 88, 89]. However, with the sequencing of the genomes it was possible to identify kinetoplastid orthologous genes encoding multiple MutS and MutL homologues [90 - 92]. Genes encoding MSH2, MSH3, MSH4, MSH5 and MSH6 (originally called MSH8) [85] are found, as well as a smaller number of MutL homologues: only PMS1 and MLH1 were identified, suggesting PMS2, MLH3 and MLH2 are absent [56, 91]. No homologue of MSH1 was identified in any of the parasites. In *Saccharomyces cerevisiae*, MSH1 is involved in mitochondrial DNA repair, including repair of oxidative damage [93 - 95] when acting in conjunction with Base Excision Repair (BER; below). In addition, MSH1 might act in homologous recombination of DNA DSBs that arise from oxidative damage, preventing rearrangements of mitochondrial DNA [94]. To date, the greatest analysis of MMR has been focused on *T. brucei* and *T. cruzi*, with fewer studies in *Leishmania*. Nonetheless, the low efficiency of recombination between diverged sequences in *Leishmania* [96], and the increase in efficiency of such diverged recombination upon inactivation of MSH2 [97], suggests these parasites have a functional MMR pathway.

The first functional analysis of kinetoplastid MMR was in *T. brucei*, when Bell and colleagues [85, 88] described single copy genes encoding MSH2, MSH3, MSH6, MLH1 and PMS1. Sequence comparisons of *T. brucei* MSH6 with orthologues in other eukaryotic homologs showed that the parasite protein is truncated at the N-terminus, for reasons that have not been so far tested. MSH2 or MLH1 knockout mutants were generated in bloodstream form *T. brucei* cells and shown to have increased tolerance to DNA damage caused by exposure to the methylating agent MNNG (N-methyl-N'-nitro-N-nitrosoguanidine) and increased microsatellite instability, phenotypes characteristic of MMR–impaired cells [85]. The MMR pathway can act in response to DNA damage after treatment with genotoxic drugs, including alkylating agents such as MNNG, which methylate guanines and gives rise to O^6-methylguanine (O^6-MeG). Such alkylation damage can be removed by a methyltransferase but, if not repaired, can lead to the incorrect base pairing of guanine with thymine, thus activating the MMR machinery [98]. Cells that are MMR-deficient are more tolerant to the DNA damage caused by alkylating agents [99] because, in this case, MMR removes not the damaged base but the correct base, generating a futile cycle of repair and error recognition [100]. Recently, this response has been exploited to develop new anti-trypanosomal compounds, which comprise methyltriazenyl purine prodrugs that are selectively taken up trypanosome nucleobase transporters and are then processed to yield methyldiazonium cations, which kill by MMR-mediated DNA damage [101]. The destabilization of microsatellites, which comprise tandem arrays of small sequences, in MMR mutants is an indicator of the efficiency with which a cell is able to repair mismatches that arise due to DNA Pol slippages during genome replication [102, 103]. In addition to these phenotypes, Bell and colleagues showed that MMR plays a role in controlling homologous recombination (HR) in *T. brucei*, since MSH2 and MLH1 knockouts in bloodstream form cells show increased efficiency of the integration of transformed constructs, in particular when the constructs have sequence differences with the genome target [88]. A later refinement of this approach showed HR efficiency in *T. brucei* is dependent on substrate length as well as homology, though a potentially MMR-independent HR pathway also exists that assumes greater importance on relatively short substrates [104]. It is possible that the use of such an MMR-independent HR reaction during antigenic variation (see below) may explain the finding that loss of MSH2 or MLH1 does not detectably alter VSG switching rate (see below) [88]. More widely, the influence of MMR on the acquisition of small- and large-scale sequence changes, which may give rise to drug resistance, has not to date been explored in the kinetoplastids, despite evidence that MMR status (and that of other repair reactions) may be important in these processes in *Plasmodium* [105 - 107]. In fact, the increasing evidence for MMR variation in *T. cruzi* strains and the action of some parts of the pathway in other repair processes suggest that such a link is likely.

In *T. cruzi* MMR characterization has largely focused on MSH2, due to its central role in the repair process. MSH2 is expressed in all life cycle stages of the parasite [86], but analysis of gene sequences from various *T. cruzi* strains revealed single nucleotide polymorphisms (SNPs) that lead to a variation in the encoded polypeptides. Proteins displaying the amino acid substitutions were labelled TcMSH2a, TcMSH2b and TcMSH2c isoforms, and different strains of *T. cruzi* expressing each isoform were classified into three distinct haplogroups, named A, B and C [89]. The division into haplogroups A, B and C using MSH2 as a molecular marker correlates with the most recent division of the strains in DTUs I-VI [108], representing *T. cruzi* lineages I, III and II, respectively. Hybrid strains express two isoforms of MSH2, such as MSH2b and MSH2c in CL Brener [89, 108, 109]. Analysis of microsatellite instability in the presence of hydrogen peroxide (H_2O_2) and levels of survival after treatment with cisplatin suggest that strains belonging to different MSH2 haplogroups present variation in the efficiency of MMR [89, 110]. Since it is well established that less efficient MMR can lead to increased mutation rate, with a consequent increase in genetic diversity of the population [98, 111], one can speculate that *T. cruzi* strains with greatest genetic diversity would most successfully infect humans, since it would have more genetic resources to evade the immune system. On the other hand, a more efficient MMR would lead to a lower mutation rate, with consequent decrease of genetic diversity [1]. It was observed that strains belonging to *T. cruzi* I lineage are more frequently associated with the parasite sylvatic cycle, whereas strains belonging to group II are generally associated with the domestic cycle of the disease and more often found in isolates from patients with Chagas disease [109, 112]. Interestingly, *T. cruzi* strains belonging to the lineage II (haplogroup MSH2c) possess less efficient MMR and greater genetic variability, whereas strains belonging to lineage I (haplogroup MSH2a) have more efficient MMR and lower genetic variability [1, 110, 113].

Recently, another role has been attributed to MMR proteins in *T. brucei* and *T.*

cruzi: response to oxidative stress. Although oxidative damage to DNA is not usually a lesion that constitutes a barrier to transcription, when bound to MMR proteins transcription is reduced [114]. In most cells, NER and BER (below) are the pathways responsible for repairing DNA damage induced by Reactive Oxygen Species (ROS), but MMR also plays a fundamental role in the prevention of mutations associated with oxidative DNA damage [115]. In other organisms, MMR has been observed to be involved in the removal of 7,8-dihydro-8-oxoguanine (also known as 8-oxoguanine or 8-oxoG), which arises as a result of oxidation of guanine at position 8 [116 - 118]. In human fibroblasts it was suggested that this role involves MSH2 and MSH6 acting independently of MMR. The complex formed by these proteins is associated with the mono-ubiquitination of PCNA in response to oxidative DNA lesions. According to this hypothesis, in oxidative stress conditions clusters of damaged DNA refractory to the action of DNA glycosylases (see below) could be formed. This type of injury is then recognized by MSH2-MSH6, which recruits an exonuclease like *Exo*1. PCNA can bind to a DNA fragment containing clustered lesions or may interact directly with MSH2-MSH6, since MSH6 has a PCNA binding motif. MSH2-MSH6 is indispensable for PCNA monoubiquination, and monoubiquitinated PCNA will form a complex with DNA Pol η, which is probably associated with DNA synthesis in the final stages of repair [118]. Initial studies on the role of MMR in the oxidative stress response in kinetoplastids were performed in *T. brucei* bloodstream form cells. Here, MSH2 null mutants showed increased sensitivity to treatment with H_2O_2, suggesting MSH2 is involved in the oxidative stress response in this parasite [119]. Increased H_2O_2 sensitivity of the mutants could be reversed by expression of *T. cruzi* MSH2. However, analysis of microsatellite instability and resistance to MNNG showed that *T. cruzi* MSH2 could not support the full functionality of the MMR pathway in *T. brucei*. Furthermore, MLH1 mutants in *T. brucei* had no change in sensitivity to H_2O_2, suggesting that MSH2 acts independently of the downstream MMR components. *T. cruzi* I strains are more tolerant to treatment with H_2O_2 and accumulate less 8-oxoguanine in their genome after exposure to H_2O_2 when compared to *T. cruzi* II strains [110], suggesting a link between the oxidative stress response and MMR also in this trypanosome. To examine this link further, MSH2 null mutants have now been generated in *T. cruzi* epimastigote cells, and MSH2 and MLH1 null mutants generated in *T. brucei* procyclic form cells [120]. In *T. cruzi*, whereas a single allele mutant shows increased susceptibility to H_2O_2 treatment [110], MSH2 null mutants show increased tolerance [120]. In addition, a larger number of intracellular amastigotes can be generated in infected macrophages with the MSH2 null mutants relative to wild type parasites. In *T. brucei*, procyclic form MSH2 null mutants also display increased tolerance to H_2O_2, whereas MLH1 null mutants do not. Moreover, re-expression of MSH2 in the null mutants had a different outcome in bloodstream or procyclic form of *T. brucei*: whereas in both life cycle forms the response to genotoxic agents and microsatellite instability could be reverted, suggesting that the MMR function can be restored, the altered sensitivity to H_2O_2 could only be reverted in bloodstream form cells [119, 120]. Taken as a whole, these data reveal a complex interplay between MMR and the oxidative stress response, with an adaptation process occurring in the fly stages of both parasites after the generation of MSH2 (but not MLH1) null mutants. Such an adaptation, specifically in the fly, most likely results from the added oxidative stress due to increased aerobic metabolism relative to the mammal stages.

Further evidence of a role for MSH2 in response to the oxidative stress is seen in the analysis of benznidazole (BZ) resistance, suggesting again that MSH2 is able to recognise 8-oxoG lesions in DNA. BZ is commonly used for the treatment of Chaga's disease and has a mechanism of action based on the production of free radicals and electrophilic metabolites [121]. Treatment with BZ is capable of increasing the amount of 8-oxoG present in the nucleotide pool, which makes cells over-expressing MutT more resistant to this drug. On the other hand, BZ is not able to induce the formation of high levels of 8-oxoG in genomic DNA, since cells over-expressing OGG1 (see below) and MutY do not show any phenotype related to the treatment with BZ. However, when one allele of MSH2 is deleted in *T. cruzi* an increased resistance to BZ is observed, suggesting 8-oxoG from the nucleotide pool that is not degraded by MutT is incorporated into DNA and further repaired by MSH2.

A number of questions remain about the mode of action of kinetoplastid MSH2 in the response to oxidative stress. Most prominent is how much of the MMR machinery is involved: MLH1 appears not to be, but does the role require MSH2 interaction with MSH6 (or MSH3)? More broadly, are there any other repair factors, if any, involved? A more complex aspect of the process is that increased 8oxoG, which is indicative of DNA damage, is not limited to the nuclear genome in the MSH2 mutants, but is also seen in the kDNA; indeed, loss of MSH2 in bloodstream form *T. brucei* results in increased numbers of cells that lack detectable kDNA [110, 120]. Whether this indicates a mitochondrial function for MSH2, or whether these effects are due to undescribed signalling processes, is unknown. However, understanding these roles may be important in considering the mode of action of some drugs and the acquisition of resistance.

BASE EXCISION REPAIR

As intracellular pathogens, *T. cruzi* and *L. major* must overcome reactive oxygen and nitrogen species (ROS/RNS) generated not only from endogenous sources but also by host mammalian cell defense mechanisms, mainly inside host macrophages [122, 123]. Although all cellular macromolecules are susceptible to damage by ROS/RNS, the primary deleterious consequences of oxidative stress arise from damage to DNA [124]. DNA damage induced by ROS/RNS includes SSBs and DSBs, base and sugar modifications, and DNA depurination or depirymidation, among others [125]. In response to such damage, several cellular response pathways are activated, such as DNA repair mechanisms, stimulation of cell cycle checkpoints and triggering of apoptosis or other mechanisms of programmed cell death [125]. Base excision repair (BER) acts as a key response, meaning it plays an important role in how kinetoplastids address hostile environments that potentially produce DNA damage.

BER is an evolutionarily conserved pathway that acts to maintain both nuclear and mitochondrial genomic integrity [126, 127], and is capable of eliminating DNA damage including oxidation, alkylation, deamination and depurination, as well as single-strand breaks. BER is a multistep pathway that requires four enzymes in the basic reaction steps to carry out repair of DNA containing apurinic/apyrimidinic (AP) sites or base damage: (i) a DNA glycosylase, (ii) an AP endonuclease, (iii) a DNA Pol, and (iv) a DNA ligase. Initiation of BER occurs by recognition of a damaged DNA base by a DNA glycosylase, which catalyzes the hydrolysis of the N-glycosidic bond of the damaged deoxynucleotide, creating an AP site. In a second step, the AP endonuclease cleaves the AP site generating 3′ OH and 5′ deoxyribose phosphate (dRP) termini. Afterwards, a DNA pol fills in the single nucleotide gap produced by the base removal. In mammalian cells, the Pol β repair protein has an intrinsic dRP lyase activity, which cleaves the dRP residue, generating a 5′ phosphate; the resulting nick, after single correct nucleotide incorporation, is sealed by DNA Lig III in the final step [128].

In mammals, DNA repair *via* BER uses two possible sub-pathways: i) a short-patch (SP-BER) reaction, in which a single nucleotide is incorporated into DNA by Pol β and the DNA Lig III/XRCC1 complex seals both single DNA strands; and ii) a long-patch reaction (LP-BER), in which the newly synthesized DNA oligonucleotide (2-20 nucleotides), generated by DNA Pols delta (Pol δ) or epsilon (Pol ε), that displaces the old DNA strand into a flap structure that is removed by the flap endonuclease 1 (FEN1). The resulting nick is sealed by DNA Lig I. LP-BER also requires the presence of PCNA, which coordinates the enzymatic activities of DNA Pols and FEN1. On the other hand, the presence of poly (ADP-ribose) polymerase (PARP; see below) facilitates the access and the recruitment of BER enzymes directly to the sites of DNA damage. In bacteria, LP-BER is a minor backup pathway [129]. In mammals, though SP-BER is proposed as the major repair pathway, LP-BER still accounts for some DNA repair. In contrast, in *S. cerevisiae* [130] and *P. falciparum* [131] SP-BER appears to be absent, and LP-BER is the major, if not sole, BER pathway. Up until 2010 there was only one report indicating the presence of the SP-BER in *T. cruzi* [132]. To date information about LP-BER in these parasites is absent. Given the importance of BER in tackling host insults, examination of this repair reaction is important, at least to evaluate whether it may provide possible targets for anti-parasite therapy.

DNA glycosylases are crucial in BER because they are the initial enzymes that recognize damaged DNA bases and catalyze the cleavage of an N-glycosidic bond, leading to AP sites. The range of DNA glycosylases encoded by an organism dictates the range of threats they can tackle and are exposed to. DNA glycosylases can be monofunctional, meaning they only remove a specific modified base from DNA, or bifunctional, which means they also possess AP lyase activity that generate a SSB. To date 11 different mammalian DNA glycosylases have been characterized [128]. Genes for many DNA glycosylases have been predicted in kinetoplastid genomes, but few have been examined functionally. The enzymatic activity of a uracil-DNA glycosylase (UNG), which excises uracils inserted in DNA, was first demonstrated in *T. cruzi* [133]. Moreover, the presence of an AP endonuclease from *L. major* increases this activity, suggesting a functional interaction between those enzymes. The expression of *T. cruzi* UNG in *E. coli ung* mutant cells restored the wild-type BER, similar to human UNG, indicating a functional homology between the *E. coli*, *H. sapiens* and *T. cruzi* enzymes [132]. The *T. brucei* genome also presents a uracil-DNA glycosylase responsible for removal of uracil from DNA. No excision activity is detected on U:G, U: strand uracil-containing DNA in uracil-DNA glycosylase null mutant *T. brucei* extracts [134], indicating the absence of backup uracil excision activities. In addition, the lack in uracil DNA glycosylase renders *T. brucei* cells hypersensitive to antifolates and hydrogen peroxide, with reduced infectivity *in vivo,* demonstrating that efficient uracil elimination is important for survival within the mammalian host. There are six known DNA glycosylases related to oxidative DNA damage repair in humans (OGG1, NTH1, NEIL1, NEIL2, NEIL3 and MYH) [135]. The *T. brucei* genome appears to possess only three of these DNA glycosylases: putative homologues of OGG1, NTH1 and MYH. However, in the *T. cruzi* genome it is has also been possible to identify the coding gene for NEIL3. ROS generate mainly the oxidation of guanine, adenine and thymine, but the most stable product is 8-oxoG, which is highly mutagenic because it mispairs with cytosine and adenine with an almost equal efficiency during DNA replication [136]. The human bifunctional DNA glycosylase OGG1 presents an enzymatic activity to directly remove 8-oxoG from the DNA. Recently, Furtado *et al.* showed that ectopic expression of the putative *T. cruzi* OGG1 orthologue complemented an OGG1-defective *S. cerevisiae* strain [137], decreasing the spontaneous mutation rate. However, despite *T. cruzi* epimastigote parasites that overexpress OGG1 having reduced levels of 8-oxoG, they are more sensitive to oxidative stress. This paradoxical phenotype may be explained by uncoupling of the coordination of the excision step from downstream steps, and may relate to OGG1 in *T. cruzi* being both nuclear and mitochondrial [137].

In *T. cruzi* AP1 and AP2, both putative AP endonucleases (orthologues of *H.*

sapiens APE1 and APE2, respectively) [138], have been recently characterized, providing insight into the later steps in BER [139]. Both enzymes are expressed in the two replicative forms of the parasite (epimastigotes and amastigotes) as well as in the non-replicative, infective trypomastigotes, and are localized in the nucleus, with no evidence of kinetoplast targeting. These data expand upon and confirm the nuclear localization of one DNA AP endonuclease in each of *L. major* [140] and *T. brucei* [141] (LMAP and TBAPE1, respectively). Interestingly, overexpression of *T. cruzi* AP1, but not of AP2, increases epimastigote survival when exposed to acute and sustained oxidative stress, suggesting that AP1 is the major AP endonuclease in *T. cruzi*. On the other hand, methoxyamine (MX), a compound that binds to AP sites and thereby impedes the catalytic activity of the AP endonucleases, inhibits the recombinant activity of both AP1 and AP2 and increases the cytotoxic effect induced by hydrogen peroxide and sodium peroxynitrite in *T. cruzi* epimastigote and trypomastigote cellular forms [139, 142]. Other enzymatic studies, by Perez *et al.* [143] and Charret *et al.* [141], reported that recombinant *L. major* and *T. brucei* AP proteins exhibit AP endonuclease and 3' phosphodiesterase activity *in vitro*, and can remove 3′-phosphoglycolate groups from 3′-ends. Finally, *T. brucei* APE1-deficient bloodstream form cells have been generated, showing this AP endonuclease is not essential for parasite viability but results in a significant increase in abasic DNA damage, rendering cells hypersensitive to methotrexate, MX, phleomycin and nitric oxide [141].

Different structural DNA Pols that could participate in the BER pathway are described in mammalian cells, but DNA Pol β is the main gap-filling enzyme in eukaryotes [135]. *Crithidia fasciculata* was the first kinetoplastid in which DNA Pol β was characterized. However, contrary to mammalian DNA Pol β, whose expression has been restricted to the nucleus, *C. fasciculata* Pol β is a mitochondrial enzyme [144, 145]. *T. brucei* and *T. cruzi* each have two distinct DNA Pol β enzymes (TbPol β and TbPolβPAK), both of which localize in the parasite's mitochondrion and have dRP lyase activities [146, 147]. The lack of nuclear DNA Pol β in trypanosomes remains inexplicable, and perhaps indicates that other DNA Pols can play a similar role in nuclear BER, although it is at odds with the reported nuclear localisation of DNA Pol β in *L. infantum* [148]. One other Pol contributes to BER, but has very different enzymatic activity: Poly (ADP-ribose) Polymerase (PARP). PARP is frequently encoded in multiple variants, which are implicated in many cellular processes, such as DNA replication, transcription regulation, apoptosis, necrosis and DNA repair [149]. PARP action contributes to SSB repair, which overlaps with BER, as described below.

HOMOLOGOUS RECOMBINATION: ADAPTATIONS FOR GENOME VARIATION?

DNA breaks involve loss of integrity of the phosphodiester backbone. SSBs probably arise more frequently than DSBs, since many routes cause their formation, including base oxidation, as a by-product of BER, incomplete topoisomerase I reactions, and mistaken incorporation of ribonucleotides into DNA [150, 151]. However, SSBs appear less toxic, due to retention of a template strand for repair, which occurs by using some of the enzymatic machinery of BER. However, SSB sensing can differ depending on how the lesion forms. SSB repair clearly operates in kinetoplastids, since PARP activity has been described [152, 153] and, indeed, has been shown to be important for the survival of *T. cruzi* [154] and *Leishmania* [155] in the mammal. Exposure of *T. cruzi* epimastigotes to H_2O_2 or β-lapachone leads to an increase in the levels of poly (ADP-ribose) polymer synthesis in the parasite nucleus [152]. Moreover, after exposure to different DNA-damaging agents (H_2O_2 or UV-C radiation) PARP is activated and translocated from the cytosol to the parasite nucleus [154]. SSB repair after RNase removal of ribonucleotides also appears active, both in the nucleus [156] and, intriguingly, in the mitochondrion [157].

DSBs are a highly genotoxic form of damage since both strands of the DNA helix are compromised. DSBs can be induced by radiation or other agents, can arise when the DNA replication fork encounters impediments to progress and may be generated deliberately to initiate locus-directed sequence change. Two enzymatically and mechanistically distinct pathways for DSB repair exist: homologous recombination (HR), and non-homologous DNA end-joining (NHEJ). This section considers what is known about HR in kinetoplastids, whereas NHEJ is discussed in the next section.

HR repairs DSBs by identifying homologous DNA sequence in an unbroken DNA duplex and using that sequence to regenerate the broken molecule by copying the intact information [158, 159]. Homology-directed DNA strand exchange is catalysed by RecA in bacteria, and by its homologues Rad51 and RadA in eukaryotes and archaea, respectively [158]. Indeed, RecA/Rad51/RadA appear to be present in all genomes, consistent with the universality of HR repair [159]. Given this, it is unsurprising that HR is readily detected in all kinetoplastids, such as during the generation of gene knockout mutants and other genetic modifications after transfection [160]. However, increasing evidence suggests that in each of *T. brucei*, *T. cruzi* and *Leishmania*, HR may have been adapted or co-opted to provide for lineage-specific adaptive genome change (Fig. **2**). If the mechanisms that underlie these processes can be fully understood it remains possible that therapeutic interventions that target kinetoplastid parasite HR, including Rad51 [161, 162], might be considered, perhaps based on approaches that have been developed for cancer intervention [163].

Antigenic variation in *T. brucei* occurs by periodic, stochastic switches in the identity of the Variant Surface Glycoprotein (VSG) expressed on the cell surface [164 - 166]. At any one time an individual *T. brucei* cell in the mammal expresses only one *VSG* gene, which encodes a dense VSG 'coat' that is thought to shield invariant antigens from immune recognition. The mammal-expressed *VSG* gene is sited in one of ~15 *VSG* transcription sites, termed bloodstream expression sites (BES) and the primary route of VSG switching occurs by recombination reactions that displace the BES-resident *VSG* and introduce a new *VSG*, which is selected from huge genome archive of ~2,000 *VSG*s [167, 168]. Recombination is a very widespread strategy for antigenic variation in diverse pathogens [169], most likely because it generates antigen diversity, which prolongs infection and facilitates transmission [170 - 172]. The *T. brucei VSG* archive is spread across all three chromosome classes that comprise the nuclear genome, with ~90% found in arrays in the megabase chromosomes subtelomeres. Most array *VSG*s are pseudogenes or partial genes [29, 167], but act as substrates for gene conversion reactions that can segmental combine multiple *VSG* ORFs to generate novel, functional 'mosaic' *VSG*s [170, 171, 173]. Segmental gene conversion of impaired *VSG*s predominates late in *T. brucei* infections, whereas the complete gene conversion of intact *VSG*s based on flanking sequence homology predominates early in infections [167, 174 - 176]. Genetic evidence strongly indicates that switching of intact *VSG*s is catalyzed by HR, a finding first suggested by experiments that showed mutation of RAD51 impairs (though does not abolish) *VSG* gene conversion [177]. Subsequently, mutation of factors that aid RAD51 function - BRCA2 [178] and at least one of four RAD51 paralogues [179] - was shown to have the same outcome. Finally, mutation of TOP3α or RMI1, two factors that probably interact with a RecQ helicase (Devlin, McCulloch *et al.*, submitted), was shown to increase the rate of VSG switching, a phenotype that is RAD51-dependent [180, 181]. Catalysis of antigenic variation by HR is not limited to *T. brucei*, as genetic studies implicate many HR factors in *Neisseria gonorrhoeae* pilin switching [182]. However, the observation that *T. brucei* VSG gene conversion can continue in the absence of RAD51 indicates that an as yet unidentified recombination reaction (or reactions) also operates, which may have parallels with RecA-independent antigenic variation in at least one species of *Borrelia* [183]. Indeed, there is accumulating evidence for recombination of *B. burgdorferi vlsE* genes being catalyzed by only a subset of predicted HR factors [184 - 186]. Thus, it remains possible that segmental *VSG* gene conversion in *T. brucei* shares some of this adaptation, since no studies have yet tested the genetic requirements of this process [187].

Analysis of the known machinery and mechanism of VSG switching by recombination reveals more concrete evidence for adapting HR for the needs of immune evasion. As noted above, BRCA2 is a central enzyme in HR, where it promotes the activity of Rad51 on single-stranded DNA ends formed at a DSB. BRCA2 directs Rad51 to DNA damage *in vivo* [178] and enhances Rad51-mediated strand exchange *in vitro* [188, 189]. These activities are the result of BRCA2 interaction with Rad51, which occurs both through a poorly conserved C-terminal domain of BRCA2 and conserved, BRCA2-specific BRC repeat motifs towards the N-terminus [190]. The number of such BRC repeats varies between organisms, though a dramatic expansion in number is seen in *T. brucei* relative to most single-celled eukaryotes, including its close relatives *T. cruzi* and *Leishmania* [178, 191]. In fact, the organization of the BRC repeats in *T. brucei* BRCA2 is also unusual, being arranged as a tandem array of identical sequences.

Though this BRC repeat organization was initially predicted to be an adaptation for the mechanistic requirements of RAD51-directed VSG switching, *T. brucei* cells manipulated to express a BRCA2 variant with only a single BRC repeat are competent for switching of intact VSGs, despite being impaired in general HR [178]. Further analysis revealed that BRC repeat number dictates the efficiency of RAD51 subnuclear relocalization into repair foci, though BRCA2 itself rarely colocalizes with RAD51 [43]. Because loss of BRCA2 leads to impaired VSG switching as well as attrition of the *VSG* arrays in mammal-infective cells (but not in procyclic form, tsetse infective cells), it is clear that BRCA2 contributes to VSG activation and maintenance [43, 178]. However, the purpose of the BRC repeat adaptation remains somewhat unclear, with the current best model being that BRCA2 provides a dual role: promoting RAD51-directed recombination, as well as a potentially RAD51-independent activity in the *VSG* array [43]. It remains possible that this latter role contributes to activation or reassortment of impaired *VSG*s during segmental gene conversion, but a direct test of this hypothesis has not been conducted to date. Whatever the precise purpose, if the potentially novel interaction between RAD51 and the BRC repeats of BRCA2 in *T. brucei* can be characterized and disrupted [192], this may impede crucial elements of *VSG* function.

A mechanistic adaptation of HR during antigenic variation may be necessary due to the fact that the VSG switching reaction can occur at rates substantially higher than spontaneous, 'background' recombination [193, 194]. It is also possible that VSG switching is focused on, or at least more frequent in, the active BES. These features lead to comparisons with other locus-directed frequent recombination reactions such as immunoglobulin gene switching in mammals [195] and mating type switching in yeast [196], where targeted regions drive high rates of recombination. How VSG switching is initiated remains unclear, but an association with DSBs has recently been proposed. Two lines of evidence support this association: targeting of ISceI endonuclease cleavage to the vicinity of the *VSG* in the active BES leads to increased VSG switching [197, 198]; and specialized PCR suggests DNA breaks are found in the BES, though with conflicting evidence as to whether the breaks localize and whether or not they are limited to the active site or are also found in the silent BES [197 - 199]. If DSBs initiate VSG switching, an important question is how they arise. Several models have been proposed [200], and none can be simply discounted: an uncharacterised endonuclease might target the BES [201], telomere homeostasis might lead to BES damage [199, 202, 203], and transcription or replication of the BES might lead to damage [180, 181]. Critically testing both the association between DSBs and VSG switching and the various models is the next step in understanding the activation of at least intact *VSG*s.

Increasing evidence suggests that HR in *Leishmania* and *T. cruzi* may contribute to more widespread genome variation than that seen in *T. brucei*. As noted above, the genomes of *T. cruzi* and *Leishmania* display more widespread plasticity than is found in *T. brucei*. Deviation from diploidy is found in multiple chromosomes both *T. cruzi* and *Leishmania*, an effect that affects different chromosomes in different strains or species [35, 36]. Moreover, at least in *Leishmania*, such chromosome copy number variation fluctuates during growth of individual cells and between cells in the population [37 - 39, 204]. How this aneuploidy is generated and tolerated remains unclear [205 - 207], though a potential explanation lies in the difficulty of completing replication of each chromosome from a single origin [28]. Gene copy number variation is also pervasive throughout the *Leishmania* genome [36, 42] and provides a route for adaptation, including the treatment with drugs [208 - 212]. Frequently, gene amplification is seen as extrachromosomal circular or linear molecules that form as a result of flanking repeats [42], but such changes may also occur within the chromosome and can be gene losses as well as gains [213]. Recent studies have revealed that HR contributes to at least the extrachromosomal gene amplification events, since mutation of RAD51 [42] and RAD51-4 (one of the three *Leishmania* RAD51 paralogues) [214] leads to reduced levels of circular amplicons, while mutation of MRE11 results in fewer linear amplicons [215]. Whether these factors contribute only to extrachromosomal amplification, or provide related functions in intra-chromosomal and whole chromosome copy number variation events remains to be seen. Indeed, the emerging evidence that BRCA2 [216] and components of the 9-1-1 complex [217, 218] contribute to *Leishmania* chromosome stability and the response to replicative stress means that growing tools are available to unravel the potential interplay between HR, repair and replication in *Leishmania* genome variation and adaptation. Like for *T. brucei*, if these reactions can be understood then the potential for therapeutic intervention against *Leishmaniasis* with anti-HR compounds will be clarified [162, 163, 219].

Evidence that HR may have been adapted to play a widespread role in *T. cruzi* genome maintenance has emerged from studies that sought to explain the pronounced resistance the parasite displays to irradiation. The foundation for these studies was attempts to use irradiation to attenuate *T. cruzi*, which revealed unusually high resistance to gamma radiation [220], with epimastigote cells capable of surviving doses of up to ~1.5 kGy, much greater than most eukaryotic cells. Treatment of *T. cruzi* with gamma radiation results in growth arrest and causes the chromosomes to shatter [221], indicating pronounced DNA damage. Recovery occurs first by reassembly of the chromosomes and then by a return to growth, and in both responses HR plays a role, since intact chromosomes become detectable more quickly when RAD51 is overexpressed and growth recovery is more rapid [221]. Indeed, slower chromosome repair and growth recovery is seen when just one allele of the *T. cruzi RAD51* gene is deleted (Passos-Silva *et al.*, submitted). Transcriptomic [222] and proteomic [223] analyses have sought to understand this response in more detail, but are complicated due to the predominance of post-transcriptional expression controls in *T. cruzi* and the growing realization that a major response of cells to gamma irradiation, perhaps as a result of oxidative damage, is a process that protect the proteome [223]. Understanding the mechanism of the remarkable chromosome reassembly will shed light on which aspects HR are involved and, crucially, will determine how accurate the process is, including how comparable the reaction is to genome reformation reactions in bacteria [224 - 226]. It is currently unclear what feature of *T. cruzi* biology is reflected by irradiation resistance and repair but it is worth asking whether it may be related to chromosome copy number variation [35], or it may be a result of genome-wide adaptive variation reactions that are similar to those seen in *Leishmania*. Fig. (**2**) summarises the adaptive genome changes directed by HR in trypanosomatids.

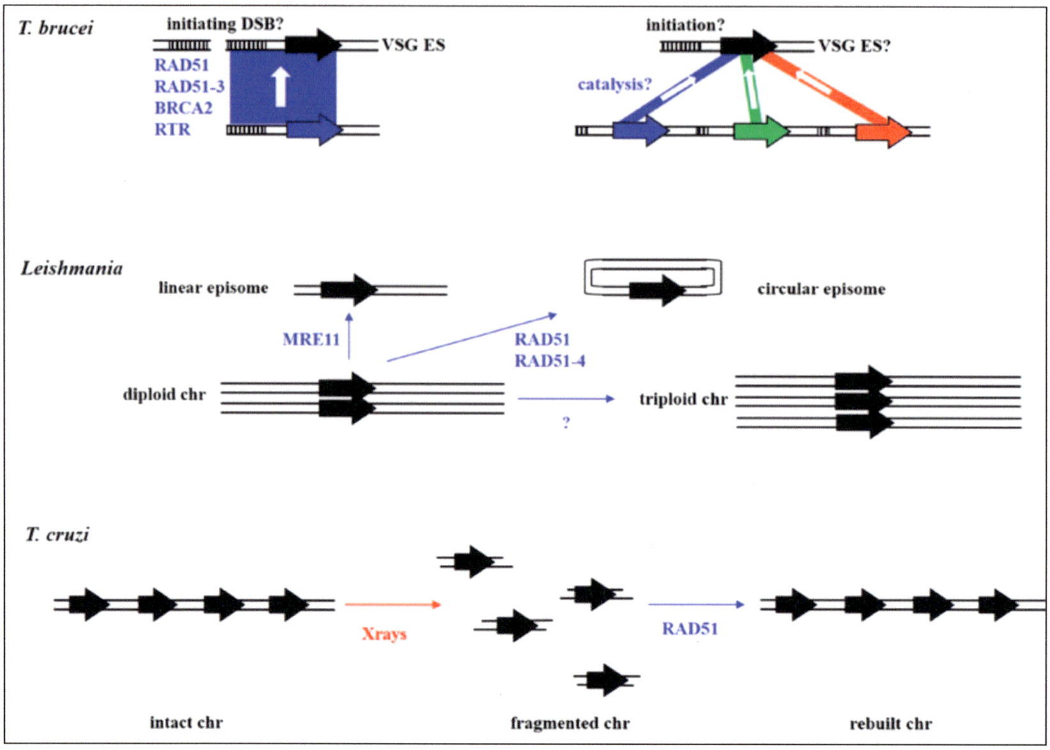

Fig. (2). Adaptive genome changes directed by homologues recombination in trypanosomatids. Antigenic variation in *T. brucei* (top left) can occur by gene conversion (extent of sequence copied and direction indicated by white arrow and surrounding blue box) of a functional silent Variant Surface Glycoprotein (VSG) gene (blue arrow) into the active VSG expression site (ES), displaying the resident VSG (black arrow). This form of VSG switching by recombination is catalysed by the homologous recombination factors indicated (RTR denotes a putative orthologous complex of RecQ-Top3-Rmi1), and it has been suggested to be initiated by a DNA double strand break (DSB) in the ES, here shown within the 70 bp repeats (hatched box) that are found upstream of most VSGs. Antigenic variation also occurs by segmental gene conversion (top right), where an intra-VSG sequence from multiple silent VSGs can be copied to generate a novel mosaic VSG. How this reaction is initiated, whether it occurs within the ES and the machinery that catalyses the process are unknown. *Leishmania* chromosomes have been shown to undergo extensive change (middle diagram). Intra-chromosome (chr) sequences, sometimes including a specific gene (black arrow) can be copied to generate extra-chromosomal linear or circular episomes; the former reaction has been shown involve MRE11, while RAD51 and RAD51-4 act in the latter. In addition, the ploidy of a whole chromosome (shown as conventionally diploid) can change (shown increasing to triploidy, but can be greater numbers, or less than diploid also); the pathway(s) that leads to such change and how it is maintained is unknown. Gamma irradiation of *T. cruzi* has been shown to result in fragmentation of whole chromosomes (bottom diagram), which can then be rebuilt by homologous recombination involving RAD51; the extent to which the rebuilt chromosomes (and indeed the whole genome) match or differ from those prior to fragmentation awaits detailed analysis.

Homologous Recombination During Sexual Genetic Exchange

Exchange of genetic information between cells appears to be exceptionally widespread in biology. Though the strategies and frequency of genetic exchange vary widely, genetic information transfer can be seen, along with mutation, to have shaped the genomes of nearly all living organisms. Sexual reproduction, the fusion of haploid cells generated by meiotic cell divisions, appears not only to be a specific feature of eukaryotes but to have been present in the ancestor of all extant eukaryotic organisms [227]. Though it has proved difficult to detect sexual recombination in some organisms, including some fungi, *Giardia* and *Trichomonas vaginalis*, the existence of conserved, meiosis-specific factors in eukaryotes previously thought to be asexual hinted at the potential for sex [228, 229]. Indeed, more detailed analysis of the organisms' lifestyles revealed that sex can occur, if infrequently or only in specific circumstances [230, 231].

Homologous recombination is central to eukaryotic meiosis [232] and is initiated by DSBs generated by the enzyme Spo11, which is a meiosis-specific topoisomerase IV homologue whose activity in mammals is guided by PRDM9 [233]. HR directs repair of the SpoII DSBs, and strand exchange is catalysed not merely by Rad51 but, normally, also by a closely related meiosis-specific homologue called Dmc1 [234]. Two other proteins, Hop2 and Mnd1, interact and stimulate Rad51 and Dmc1 [235]. Meiosis differs markedly from mitosis in that, in addition to sister chromatids remaining attached following DNA replication, homologous chromosomes are brought together also. HR is crucial in synapsis of the homologues: though these can be held together by a structure called the synaptonemal complex, recombination-driven exchanges generate chiasmata that connects the homologues until separation and allows genetic exchange between maternal and paternal chromosomes. Multiple HR pathways occur during meiotic recombination and can lead to gene conversion with and without crossing over between homologous chromosomes. A key intermediate in some of these strand exchange reactions is a Holliday junction intermediate, on which two meiosis-specific MutS homologues, called Msh4 and Msh5, can act together to promote crossing over [236]. A growing list of DNA endonucleases can also act on Holliday junctions, though which, if any, are meiosis-specific is unclear [237].

Population studies have long provided evidence for genetic exchange in *T. brucei*, *T. cruzi* and *L. major*, even though the population structures suggest that much of the parasites' reproduction is asexual [238]. Compelling evidence now exists for sexual exchange in all three kinetoplastids [239], since it has been possible to observe hybrids as a result of crosses between strains, subspecies or species of *T. brucei* [240 - 243], *T. cruzi* [244] and *L. major* [245 - 247] in the laboratory. In all these experiments, mating appears not to be obligatory, but the ability to recover hybrids shows that each parasite retains the capacity for sexual recombination. In addition, genetic exchange occurs at a specific part of the life cycle: during fly transmission for *T. brucei* and *Leishmania*, and following co-infection of mammalian cells in *T. cruzi*. Gene inventories reveal that each of *T. brucei*, *T. cruzi* and *Leishmania* contain homologues of all the meiosis-specific HR genes [248], including both MSH4 and MSH5 (Barnes R, 2006, PhD thesis, University of Glasgow). In fact, and for reasons that remain unclear, the parasites contain two putative Spo11 homologues [249]. Retention of these genes is consistent with mating occurring by meiosis, though to date this has been most fully tested in *T. brucei*. Mutation of DMC1 has been shown to be possible in bloodstream form cells, but has no discernible phenotypic consequence [250]. More importantly, expression of DMC1, MND1 and HOP1 (a synaptonemal complex component homologue) is limited to the nucleus of previously unidentified *T. brucei* cell types in the tsetse salivary glands, where mating was predicted to occur [251]. Further analysis of these cells revealed them to be haploid and capable of flagellum-mediated interaction and fusion, consistent with being gamete-like life cycle stages [252]. Taken together, the evidence for meiosis appears overwhelming, though many questions remain. Do *T. brucei* gamete cells always form during the life cycle, and are similar forms found in the other kinetoplastids? How and where does DSB formation occur in the unusual genetic landscape of the kinetoplastid genomes? How does meiosis cope with the pronounced differences in size of chromosome homologues in kinetoplastids? Does meiosis occur in *T. vivax* and *T. equiperdum*, which have variant tsetse fly life cycles relative to *T. brucei*? And, finally, how important is sexual recombination for the spread of important genetic traits such as drug resistance and human infectivity? [243]. More widely, it will be interesting to see how comparable meiosis is between *T. brucei* and its kinetoplastid cousins.

HAS LOSS OF NON-HOMOLOGOUS END-JOINING IN KINETOPLASTIDS ALLOWED INCREASED USE OF MICROHOMOLOGY-MEDIATED END-JOINING?

NHEJ is a DSB repair pathway that is found in bacteria, archaea and eukaryotes [253 - 255]. In eukaryotes NHEJ is the major pathway of DSB repair in some settings; in mammals, for instance, NHEJ predominates in the G1 phase of the cell cycle, when sister chromatids are not available as repair substrates [256]. In contrast with HR, NHEJ can be imprecise [257] as a result of the fact that the reaction does not employ a template to guide repair, but instead brings the two ends of the DSB together and re-ligates them. If DSB ends are incompatible for ligation some sequence can be lost by nuclease degradation or added by DNA polymerization; indeed, at least in prokaryotes, ribonucleotides can be incorporated during NHEJ [255]. In addition, ends may be joined from two DSBs, leading to translocations [258]. The core components of eukaryotic NHEJ are the Ku70/80 heterodimer (Ku) and DNA Lig IV, which acts in concert with XRCC4/Lif1 (Fig. **3**); in some eukaryotes, these factors are aided by the DNA-dependent protein kinase catalytic subunit (DNA-PKcs), Artemis, NEJ1 and specific DNA Pols. Mechanistically, NHEJ occurs by DSB recognition, bridging and processing of the ends, and repair of the broken molecule by ligation. The choice between enacting NHEJ or HR appears to be largely dictated by the extent of DSB end resection, which is catalysed by the Mre11-Rad50-Nbs1 complex and other factors [259, 260]: Rad51 filaments form on relatively long single-stranded DNA ends, which are inhibitory to NHEJ. Ku plays a key role in NHEJ, since it not only binds the ends of DSBs (thus preventing further resection and diverting them from HR) but can bridge and process the ends, and acts as a platform for recruitment of downstream NHEJ factors. NHEJ in prokaryotes resembles the process in eukaryotes, but uses diverged, often streamlined machinery. Instead of a heterodimer, prokaryotic Ku is a homodimer and, in several species of bacteria, downstream reactions are catalysed by DNA ligase, DNA polymerase and phosphoesterase activities that are housed in a single polypeptide [254].

The demonstration of end-joining activity, catalysed by related machinery in bacteria, archaea and eukaryotes suggests that NHEJ is an ancient repair process [261]. However, there is increasing evidence that NHEJ may not be active in kinetoplastids and, in fact, may also be absent from a number of other eukaryotes. Despite poor Ku protein conservation, KU70 and KU80-encoding genes were identified in *T. brucei* and mutation has shown that their loss impairs telomere maintenance, demonstrating the genes are functional [262, 263]. Orthologues of both Ku subunits are present in all kinetoplastids, and in *Leishmania* KU70 may be a target for CRK3 protein kinase-mediated phosphorylation, though for what function is unknown [264]. However, sequence-based searches have failed to identify genes encoding any other NHEJ-specific factor. The most notable absences from such searches were identifiable homologues of DNA Lig IV and XRCC4/Lif1 [265]. In fact, searches of the genomes with any ligase, including bacterial NHEJ-specific ATP-dependent DNA ligases, recovered only homologues of DNA Lig I and three putative mitochondrial ligases, two of which act in maintenance of the kinetoplast [266]. These bioinformatics predictions suggest a potential absence of NHEJ, a prediction that was supported by two initial studies that sought to detect NHEJ activity in *T. brucei.* In one study, end-joining activity was detected using artificial substrates and cell extracts, but was mediated by 5-15 bp stretches of microhomology (see below) and was equally detectable in extracts from wild-type cells and Ku mutants [265]. In another study, plasmid constructs were introduced into *T. brucei* cells that lacked any terminal sequence homology for HR-guided genome integration; though transformed clones could be recovered, in all cases the constructs had integrated using 5-15 microhomology, reminiscent of the end-joining seen in cell extracts and longer than the very limited homology that can act in NHEJ [267]. Since these studies were published a widening range of assays have failed to detect NHEJ in any kinetoplastid. In *T. brucei*, targeted formation of DSBs after expression of ISceI has been used to evaluate repair in a number of loci and with varying levels of available sequence homology. In the first of these studies both HR and microhomology-mediated end-joining (MMEJ) repair events were detected [268]; later, and more revealingly, an experimental set up in which HR was excluded revealed only MMEJ repair of a DSB [269]. More recently, CRISPR-cas9 has been developed for genome manipulation of *L. donovani* [270], *L. major* [271] and *T. cruzi* [272, 273] and in all cases repair of the induced DSBs have been documented by HR and MMEJ, with no examples of NHEJ found even when a homologous substrate was excluded.

Collectively, the above data suggest either NHEJ is absent from kinetoplastids or operates as the least efficient DSB repair pathway, subservient to both HR and MMEJ. *T. brucei KU70* and *KU80* mutants display no increased sensitivity to DNA damaging agents relative to wild type cells, providing no evidence that the factors contribute to repair [262]. Nonetheless, NHEJ may still operate at very low efficiency, perhaps as a result of the absence of DNA Lig IV and XRCC4/Lif1: since this ligase complex is dedicated for NHEJ in other eukaryotes, Ku would need then to act with another ligase to enact repair. It is also possible that the assays to date have failed to detect NHEJ because they have been conducted in the wrong cell types, since NHEJ in other organisms appears to predominate in haploid cells, in the G1 cell cycle stage or in non-replicative cells. However, perhaps the simplest explanation is that NHEJ is truly absent in kinetoplastids. This conclusion has been reached in several other eukaryotes, including *Encephalitozoon cuniculi* [274], *Giardia lamblia* [275] and *Plasmodium falciparum* [276], arguing that the reaction can be discarded and has a sporadic distribution in both eukaryotes and bacteria. One notable difference between kinetoplastids and other putative NHEJ-deficient eukaryotes is that whereas normally absence of Ku is associated with the same absence of Lig IV and XRCC4 (such as, notably, within the apicomplexans) [277], kinetoplastids have retained both Ku subunits. This may be because of important features of telomere function in kinetoplatids [262, 263], though testing whether kinetoplastid Ku retains the capacity to bind DSBs, and thereby influences repair dynamics, is necessary. Why organisms might discard NHEJ is unknown, but one obvious potential consequence is that MMEJ assumes greater prominence. Long thought merely a repair pathway of last resort, MMEJ is now known to play an important role in some settings, though the detailed enzymatic machinery remains unclear (Fig. **3**) [278, 279]. Sequence signatures of MMEJ are widely associated with large scale genome variations [280, 281], including replication-associated events that can lead to copy number variation [282]. Indeed, MMEJ can play a pronounced role in class switch recombination in immunoglobulin genes [283, 284]. Perhaps, then, MMEJ may play an as yet untested role in kinetoplastid adaptive genome changes, including VSG array stability and activation in *T. brucei*, copy number variation in *Leishmania* and chromosome reassembly and multigene array stability in *T. cruzi*. Indeed, there may be overlap in such putative functions with mechanisms of acquisition of drug resistance and evolution of subtelomeric gene families in *P. falciparum* [285 - 288] and other pathogens [289]. It seems likely that the time to dissect MMEJ function in kinetoplatids has arrived. Fig. (**3**) summarises what is known about non-HR DSB repair pathways in trypanosomatids.

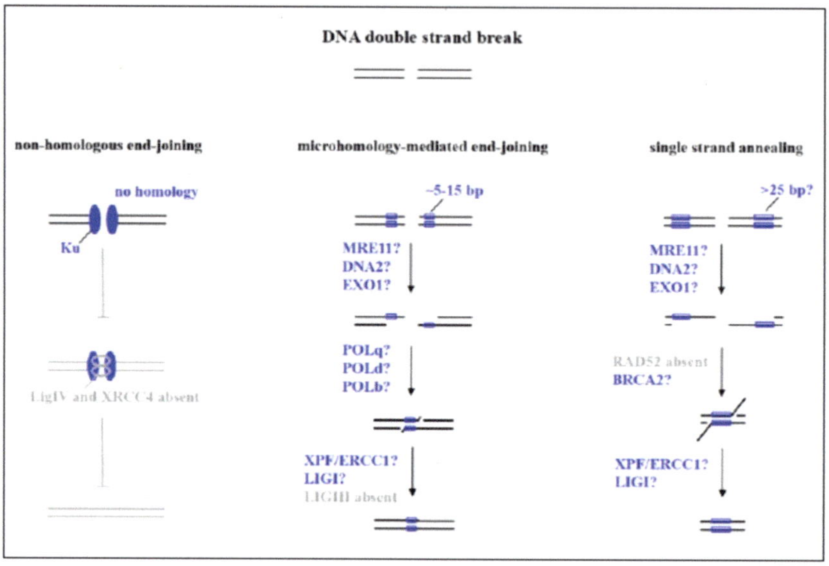

Fig. (3). DNA double-strand break repair by end-joining in trypanosomatids. Pathways for the repair of DNA double-strand break (DSBs) by end-joining, and some of the potential factors involved, are diagrammed. In most eukaryotes, the most prevalent form of DNA end-joining is non-homologous end-joining (NHEJ; left pathway), which can occur without any sequence homology around the break, though can involve very short stretches of homology. In NHEJ, DSB ends are bound by the Ku heterodimer, which recruits downstream factors, the most conserved of which are a complex between DNA ligase IV (LigIV) and XRCC4. Though both Ku subunits are present in kinetoplastids, LigIV-XRCC4 cannot be found and experiments suggest that NHEJ does not occur; indeed, whether or not Ku in kinetoplastids binds DSBs (as shown) is untested. In the absence of NHEJ, end-joining that relies on ~5-15 bp stretches of sequence homology flanking the DSB has been widely documented in kinetoplastids. Such microhomology-mediated end-joining (MMEJ) is presumed to require DSB end-processing by redundant nucleases to reveal the homology, with several DNA Polymerases (Pols) implicated in guiding base pairing, and repair thought to be completed by endonuclease processing of unpaired strands and ligation; none of the factors listed in the above steps have been examined for a role in MMEJ in any kinetoplastid, though homologues of LigIII have not been detected. Single-strand annealing (SSA) is similar to MMEJ in being initiated by similar DSB processing and completion steps, but uses more extensive flanking homology (perhaps >25 bp), with RAD52 and BRCA2 having being implicated in guiding base-pairing. Though SSA has been documented in the tubulin array in *T. brucei* after IScel cleavage, RAD52 appears to be absent from kinetoplastids and the role of BRCA2 in the reaction has not been tested.

WHAT DO WE KNOW ABOUT LESION BYPASS?

Normally cells attempt to limit the levels of genetic change from generation to generation, which is achieved by the high fidelity of replication by the replisome as well as DNA repair processes recognising and correcting damage and errors, most notably the post-replicative repair function of MMR. However, some forms of lesions cannot be repaired and in these circumstances cells resort to translesion DNA synthesis (TLS). During TLS, specialised translesion DNA Pols, which are found in all domains of life, are used to bypass the damage, often by inserting incorrect sequence opposite the lesion [290, 291]. TLS is not, therefore, a repair pathway but a damage tolerance pathway, since the damage is not immediately removed and the cost of completing genome replication can be elevated levels of sequence change at the damage site. Nonetheless, these costs are tolerated because the alternative is a failure to transmit the genome and, potentially, death.

Genome indications suggest that kinetoplastids encode a number of potential translesion DNA Pols [30], of which only a small number have been studied to date, meaning the roles played by TLS is only partly understood. DNA Pol H (eta; η) is found in all kinetoplastids, but has only so far been examined functionally in *T. cruzi* [292]. Overexpression of Pol H was shown not to improve survival after UV exposure, despite evidence from other eukaryotes that Pol H homologues can bypass cyclobutane pyrimidine dimers. Overexpression of Pol H in *T. cruzi* also does not improve survival after exposure to cisplatin, perhaps consistent with most repair of transcription-blocking lesions being tackled by variant NER. In fact, the role of Pol H remains somewhat unclear, since overexpression sensitised the cells to gamma irradiation and zeocin exposure, even though reports have implicated Pol H in DSB repair by HR [293, 294] and in immunoglobulin gene diversification [295]. In contrast, Pol H overexpression improved survival in oxidative stress conditions [292]. DNA Pol K (Kappa; κ) is capable of bypassing a number of lesions and is intriguing in kinetoplastids because, uniquely amongst the translesion DNA Pols, it is present in multicopy [30], with evidence for sequence diversification of the individual genes [296]. One of the *T. cruzi* Pol K genes has been analysed and the encoded protein shown to localise to the kDNA, suggesting a role in mitochondrial genome maintenance. Purified Pol K was shown to be capable of bypassing 8oxoG and able to promote the extension of D-loop substrates, suggesting a potential role in HR. The putative HR role of Pol K may be in the kinetoplast, but the fact that overexpression of Pol K improved recovery of *T. cruzi* cells after gamma irradiation [296], which causes the nuclear genome to shatter, may also indicate nuclear functions are yet to be uncovered. Two further putative translesion Pols have recently been described in *T. brucei*, each of which belong to the archaeo-eukaryotic primase superfamily [297], some members of which have been implicated in DNA repair and replication [298]. The two putative primase-polymerases in *T. brucei* are named PPL1 and PPL2 and RNAi indicates that while the latter is essential in bloodstream cells, the former is dispensable [299]. Loss of PPL2 results in accumulation of cells in late S/G2 phase and causes accumulation of repair factors in nuclear foci, suggesting elevated damage. Purification of PPL2 revealed that it is a translesion Pol, as is PPL1 (which is also capable of RNA or DNA primer synthesis) [299]. Collectively, these data indicate that PPL2 plays a critical role in ensuring the completion of *T. brucei* nuclear DNA replication, though what feature of the genome necessitates its activity remains unclear. Clearly, we have much to learn about the range of activities played by the growing number of translesion Pols in kinetoplastids. Pursuing these activities, in particular those that are critical or diverged from the host, may be valuable as translesion Pols are increasingly a focus for disease therapy [300].

WHAT REPAIR ACTIVITIES ARE USED IN THE KINETOPLAST?

The mitochondrion is a crucial source of energy in at least some life cycle stages of kinetoplastids. However, the mechanism of ATP generation leads to the generation of ROS, which can damage DNA. The proximity of the mitochondrial genome to the membrane-bound respiratory chain suggests it may be more immediately a target for damage than the nuclear genome, which would necessitate mitochondrial repair machinery. It is increasingly clear that such mitochondrial repair activities are present in eukaryotes and, indeed, may provide a wider range of functions than merely the response to ROS damage [301 - 303]. To date, however, our understanding of the extent to which such repair activities are also present in the kinetoplast is somewhat fragmentary, and it remains an open question whether the highly unusual mitochondrial genome of kinetoplastids has necessitated a greater or lesser range of repair pathways. However, the fact that the mitochondrion is single copy may suggest genome maintenance is more critical than in eukaryotic cells that possess many mitochondria.

BER is the best established repair activity in the mitochondrion, perhaps reflecting the damage caused by reactive oxygen, including DNA SSBs. Many of the factors that direct BER have been shown to display mitochondrial localization in mammals [303], suggesting the full pathway may be active. To date the two DNA Pol β homologues, Pol β and PolβPAK, have been described in the mitochondrion of *T. brucei* [146] and *T. cruzi* [147], and *T. cruzi* OGG1 is found in both the nucleus and kinetoplast [137]. Though the presence of these enzymes would be consistent with mitochondrial BER activity, the differing localization of the two DNA Pol β proteins to the antipodal sites and the kDNA disc has been interpreted as the DNA Pols providing key replication, rather than repair, roles [146]. Nonetheless, Pol β is found in the mitochondrion of non-replicating *T. cruzi* cells [304, 305]. Moreover, overexpression of Pol β reduces 8oxoG in the kDNA of *T. cruzi*, and the addition of H_2O_2 results in Pol β localisation into a putative repair focus that is separate from the predominant kDNA antipodal localisation [304]. Taken together, it is possible that DNA Pol β provides both replication and repair functions in the mitochondrion, but whether the same flexibility in seen with PolβPAK and the plethora of bacterial DNA Pol 1-like Pols that also localize to the organelle [306] remains untested.

Beyond BER, there is evidence for both MMR and break repair activities in the mitochondria of eukaryotes, though the extent of the machinery that acts in these pathways is somewhat less clear. In yeast, a mitochondrion-specific MutS-homologue (MSH1) counteracts oxidative lesions [307], and plants, corals and *Toxoplasma* each express highly variant mitochondrial MutS homologues that possess endonuclease activity [308 - 310]. In mammals, no such MSH proteins are found, but a form of MMR in the mitochondrion is facilitated by the Y-box binding protein [311, 312]. Genome sequencing has not revealed homologues of these proteins in kinetoplastids, but whether some of the nuclear MMR factors might also localise to the kinetoplast has not been fully explored [120]. To what extent break repair by end-joining occurs in mitochondria is somewhat unclear, but extensive evidence is emerging for HR. Bacterial RecA homologues are targeted to both the mitochondrion and chloroplast in plants [313, 314], where evidence for recombination can be found [315]. Such RecA homologues have not been described in other eukaryotes, but a Rad52-related protein is widely conserved and acts in mitochondrial maintenance in yeast [316]; in addition, plants express multiple Rad52 variants, some of which are targeted to the organelles [317], and increasing evidence suggests that some of the cellular pool of mammalian Rad51 can be found in the mitochondria and contribute to maintenance [318, 319]. Given the predicted role of *T. cruzi* DNA PolβPAK in HR and its mitochondrial localization, it may be valuable to assess HR activities more widely in kinetoplastids. More generally, it is clear that a wide range of DNA helicases, including Pif- and RecQ- related helicases, are localised to the mitochondrion in mammals [320]. Kinetoplastids encode a large repertoire of helicases [321], amongst which a spectacular expansion of Pif-related helicases is seen, most of which appear to be targeted to the kDNA [322, 323]. Whether some of these helicases play DNA repair roles awaits analysis.

CONCLUSION

The last 15-20 years have seen substantial progress in understanding the genome maintenance reactions that operate in kinetoplastids, including novelties, losses and adaptations that reflect the wide range of novel features of genome biology in this organismal grouping. Many of these reactions are crucial for kinetoplastid survival, including during adapting to life in their vectors and hosts, not least in surviving the hosts' attempts to eradicate the pathogens. However, many gaps remain in our knowledge, which should be filled by the range of genetic tools available and the capacity to perform genome-wide analysis of gene and protein expression. In addition, little work has to date attempted to exploit the growing knowledge of genome maintenance pathways to develop disease therapies, but the novelties being described bring this possibility ever closer.

CONFLICT OF INTEREST

The authors confirm that they have no conflict of interest to declare for this publication.

ACKNOWLEDGEMENTS

Work in Richard McCulloch's lab has recently been supported by the Wellcome Trust (089172), the BBSRC (BB/K006495/10 and the Royal Society; the Wellcome Trust Centre for Molecular Parasitology is supported by core funding from the Wellcome Trust (104111). Carlos Machado lab has recently been supported by CNPq, and FAPEMIG. Richard McCulloch and Carlos Machado are joint recipients of an RCUK-CONFAP Partnership award (BB/M028909/1), and thank all lab members, past and present, for their experimental and intellectual efforts.

ABBREVIATIONS

8oxoG	7,8-dihydro-8-oxoguanine
AP	Apurinic/apyrimidinic
ATP	Adenosine triphosphate
BER	Base excision repair
BES	Bloodstream expression sites
BZ	Benznidazole
CS	Cockayne syndrome
DDB	Damage DNA binding
DGC	Directional gene cluster
DSB	Double strand break
ERCC1	Excision repair cross complementing 1
FEN	Flap endonuclease
GGR	Global genome repair
HR	Homologous recombination
H_2O_2	Hydrogen peroxide
IDLs	Insertion/deletion loops
LIG	Ligase
LP	Long patch
MLH	MutL homologue
MMEJ	Microhomology-mediated end-joining
MMR	Mismatch repair
MNNG	N-methyl-N'-nitro-N-nitrosoguanidine
MSH	MutS homologue

NER Nucleotide excision repair
NHEJ Non-homologous DNA end-joining
PARP Poly (ADP-ribose) polymerase
PCNA Proliferating cellular nuclear antigen
Pol Polymerase
PPL Primase polymerase
RFC Replication factor C
ROS Reactive oxygen species
SP Short patch
SSB Single strand break
SSR Strand switch region
TCR Transcription coupled repair
TFIIH Transcription factor II H
TLS Translesion synthesis
UNG Uracil-DNA glycosylase
UV Ultraviolet
VSG Variant Surface Glycoprotein
XP Xeroderma pigmentosum
XRCC1 X-ray repair cross-complementing protein 1

REFERENCES

[1] Machado CR, Augusto-Pinto L, McCulloch R, Teixeira SM. DNA metabolism and genetic diversity in Trypanosomes. Mutat Res 2006; 612(1): 40-57. [http://dx.doi.org/10.1016/j.mrrev.2005.05.001] [PMID: 16040270]

[2] Passos-Silva DG, Rajao MA, Nascimento de Aguiar PH, Vieira-da-Rocha JP, Machado CR, Furtado C. Overview of DNA Repair in *Trypanosoma cruzi*, *Trypanosoma brucei*, and *Leishmania major*. J Nucleic Acids 2010; 2010: 840768.

[3] Genois MM, Paquet ER, Laffitte MC, *et al.* DNA repair pathways in trypanosomatids: from DNA repair to drug resistance. Microbiol Mol Biol Rev 2014; 78(1): 40-73. [http://dx.doi.org/10.1128/MMBR.00045-13] [PMID: 24600040]

[4] Povelones ML. Beyond replication: division and segregation of mitochondrial DNA in kinetoplastids. Mol Biochem Parasitol 2014; 196(1): 53-60. [http://dx.doi.org/10.1016/j.molbiopara.2014.03.008] [PMID: 24704441]

[5] Jensen RE, Englund PT. Network news: the replication of kinetoplast DNA. Annu Rev Microbiol 2012; 66: 473-91. [http://dx.doi.org/10.1146/annurev-micro-092611-150057] [PMID: 22994497]

[6] El-Sayed NM, Myler PJ, Blandin G, *et al.* Comparative genomics of trypanosomatid parasitic protozoa. Science 2005; 309(5733): 404-9.
[http://dx.doi.org/10.1126/science.1112181] [PMID: 16020724]

[7] Daniels JP, Gull K, Wickstead B. Cell biology of the trypanosome genome. Microbiol Mol Biol Rev 2010; 74(4): 552-69.
[http://dx.doi.org/10.1128/MMBR.00024-10] [PMID: 21119017]

[8] Moretti NS, Schenkman S. Chromatin modifications in trypanosomes due to stress. Cell Microbiol 2013; 15(5): 709-17.
[http://dx.doi.org/10.1111/cmi.12111] [PMID: 23336291]

[9] Siegel TN, Hekstra DR, Kemp LE, *et al.* Four histone variants mark the boundaries of polycistronic transcription units in *Trypanosoma brucei*. Genes Dev 2009; 23(9): 1063-76.
[http://dx.doi.org/10.1101/gad.1790409] [PMID: 19369410]

[10] Wright JR, Siegel TN, Cross GA. Histone H3 trimethylated at lysine 4 is enriched at probable transcription start sites in *Trypanosoma brucei*. Mol Biochem Parasitol 2010; 172(2): 141-4.
[http://dx.doi.org/10.1016/j.molbiopara.2010.03.013] [PMID: 20347883]

[11] Respuela P, Ferella M, Rada-Iglesias A, Aslund L. Histone acetylation and methylation at sites initiating divergent polycistronic transcription in *Trypanosoma cruzi*. J Biol Chem 2008; 283(23): 15884-92.
[http://dx.doi.org/10.1074/jbc.M802081200] [PMID: 18400752]

[12] Anderson BA, Wong IL, Baugh L, Ramasamy G, Myler PJ, Beverley SM. Kinetoplastid-specific histone variant functions are conserved in *Leishmania major*. Mol Biochem Parasitol 2013; 191(2): 53-7.
[http://dx.doi.org/10.1016/j.molbiopara.2013.09.005] [PMID: 24080031]

[13] Thomas S, Green A, Sturm NR, Campbell DA, Myler PJ. Histone acetylations mark origins of polycistronic transcription in *Leishmania major*. BMC Genomics 2009; 10: 152.
[http://dx.doi.org/10.1186/1471-2164-10-152] [PMID: 19356248]

[14] Ekanayake DK, Minning T, Weatherly B, *et al.* Epigenetic regulation of transcription and virulence in *Trypanosoma cruzi* by O-linked thymine glucosylation of DNA. Mol Cell Biol 2011; 31(8): 1690-700.
[http://dx.doi.org/10.1128/MCB.01277-10] [PMID: 21321080]

[15] Reynolds D, Cliffe L, Förstner KU, Hon CC, Siegel TN, Sabatini R. Regulation of transcription termination by glucosylated hydroxymethyluracil, base J, in *Leishmania major* and *Trypanosoma brucei*. Nucleic Acids Res 2014; 42(15): 9717-29.
[http://dx.doi.org/10.1093/nar/gku714] [PMID: 25104019]

[16] van Luenen HG, Farris C, Jan S, *et al.* Glucosylated hydroxymethyluracil, DNA base J, prevents transcriptional readthrough in *Leishmania*. Cell 2012; 150(5): 909-21.
[http://dx.doi.org/10.1016/j.cell.2012.07.030] [PMID: 22939620]

[17] Clayton CE. Networks of gene expression regulation in *Trypanosoma brucei*. Mol Biochem Parasitol 2014; 195(2): 96-106.
[http://dx.doi.org/10.1016/j.molbiopara.2014.06.005] [PMID: 24995711]

[18] Kolev NG, Ullu E, Tschudi C. The emerging role of RNA-binding proteins in the life cycle of

Trypanosoma brucei. Cell Microbiol 2014; 16(4): 482-9.
[http://dx.doi.org/10.1111/cmi.12268] [PMID: 24438230]

[19] Tiengwe C, Marques CA, McCulloch R. Nuclear DNA replication initiation in kinetoplastid parasites: new insights into an ancient process. Trends Parasitol 2014; 30(1): 27-36.
[http://dx.doi.org/10.1016/j.pt.2013.10.009] [PMID: 24287149]

[20] Calderano S, Godoy P, Soares D, SantAnna OA, Schenkman S, Elias MC. ORC1/CDC6 and MCM7 distinct associate with chromatin through *Trypanosoma cruzi* life cycle. Mol Biochem Parasitol 2014; 193(2): 110-3.
[http://dx.doi.org/10.1016/j.molbiopara.2014.03.004] [PMID: 24681203]

[21] Godoy PD, Nogueira-Junior LA, Paes LS, *et al.* Trypanosome prereplication machinery contains a single functional orc1/cdc6 protein, which is typical of archaea. Eukaryot Cell 2009; 8(10): 1592-603.
[http://dx.doi.org/10.1128/EC.00161-09] [PMID: 19717742]

[22] Tiengwe C, Marcello L, Farr H, *et al.* Identification of ORC1/CDC6-interacting factors in *Trypanosoma brucei* reveals critical features of origin recognition complex architecture. PLoS One 2012; 7(3): e32674.
[http://dx.doi.org/10.1371/journal.pone.0032674] [PMID: 22412905]

[23] Tiengwe C, Marcello L, Farr H, *et al.* Genome-wide analysis reveals extensive functional interaction between DNA replication initiation and transcription in the genome of *Trypanosoma brucei*. Cell Reports 2012; 2(1): 185-97.
[http://dx.doi.org/10.1016/j.celrep.2012.06.007] [PMID: 22840408]

[24] Brambati A, Colosio A, Zardoni L, Galanti L, Liberi G. Replication and transcription on a collision course: eukaryotic regulation mechanisms and implications for DNA stability. Front Genet 2015; 6: 166.
[http://dx.doi.org/10.3389/fgene.2015.00166] [PMID: 25972894]

[25] Jinks-Robertson S, Bhagwat AS. Transcription-associated mutagenesis. Annu Rev Genet 2014; 48: 341-59.
[http://dx.doi.org/10.1146/annurev-genet-120213-092015] [PMID: 25251854]

[26] Aguilera A, García-Muse T. Causes of genome instability. Annu Rev Genet 2013; 47: 1-32.
[http://dx.doi.org/10.1146/annurev-genet-111212-133232] [PMID: 23909437]

[27] Wilson TE, Arlt MF, Park SH, *et al.* Large transcription units unify copy number variants and common fragile sites arising under replication stress. Genome Res 2015; 25(2): 189-200.
[http://dx.doi.org/10.1101/gr.177121.114] [PMID: 25373142]

[28] Marques CA, Dickens NJ, Paape D, Campbell SJ, McCulloch R. Genome-wide mapping reveals single-origin chromosome replication in Leishmania, a eukaryotic microbe. Genome Biol 2015; 16(1): 230.
[http://dx.doi.org/10.1186/s13059-015-0788-9] [PMID: 26481451]

[29] Berriman M, Ghedin E, Hertz-Fowler C, *et al.* The genome of the African trypanosome *Trypanosoma brucei*. Science 2005; 309(5733): 416-22.
[http://dx.doi.org/10.1126/science.1112642] [PMID: 16020726]

[30] El-Sayed NM, Myler PJ, Bartholomeu DC, *et al.* The genome sequence of *Trypanosoma cruzi*,

etiologic agent of Chagas disease. Science 2005; 309(5733): 409-15.
[http://dx.doi.org/10.1126/science.1112631] [PMID: 16020725]

[31] Ivens AC, Peacock CS, Worthey EA, *et al.* The genome of the kinetoplastid parasite, *Leishmania major*. Science 2005; 309(5733): 436-42.
[http://dx.doi.org/10.1126/science.1112680] [PMID: 16020728]

[32] Parsons M, Worthey EA, Ward PN, Mottram JC. Comparative analysis of the kinomes of three pathogenic trypanosomatids: *Leishmania major*, *Trypanosoma brucei* and *Trypanosoma cruzi*. BMC Genomics 2005; 6(1): 127.
[http://dx.doi.org/10.1186/1471-2164-6-127] [PMID: 16164760]

[33] Peacock CS, Seeger K, Harris D, *et al.* Comparative genomic analysis of three *Leishmania* species that cause diverse human disease. Nat Genet 2007; 39(7): 839-47.
[http://dx.doi.org/10.1038/ng2053] [PMID: 17572675]

[34] Weatherly DB, Boehlke C, Tarleton RL. Chromosome level assembly of the hybrid *Trypanosoma cruzi* genome. BMC Genomics 2009; 10(1): 255.
[http://dx.doi.org/10.1186/1471-2164-10-255] [PMID: 19486522]

[35] Reis-Cunha JL, Rodrigues-Luiz GF, Valdivia HO, *et al.* Chromosomal copy number variation reveals differential levels of genomic plasticity in distinct *Trypanosoma cruzi* strains. BMC Genomics 2015; 16: 499.
[http://dx.doi.org/10.1186/s12864-015-1680-4] [PMID: 26141959]

[36] Rogers MB, Hilley JD, Dickens NJ, *et al.* Chromosome and gene copy number variation allow major structural change between species and strains of *Leishmania*. Genome Res 2011; 21(12): 2129-42.
[http://dx.doi.org/10.1101/gr.122945.111] [PMID: 22038252]

[37] Sterkers Y, Lachaud L, Crobu L, Bastien P, Pagès M. FISH analysis reveals aneuploidy and continual generation of chromosomal mosaicism in *Leishmania major*. Cell Microbiol 2011; 13(2): 274-83.
[http://dx.doi.org/10.1111/j.1462-5822.2010.01534.x] [PMID: 20964798]

[38] Valdivia HO, Reis-Cunha JL, Rodrigues-Luiz GF, *et al.* Comparative genomic analysis of *Leishmania* (*Viannia*) peruviana and *Leishmania* (*Viannia*) braziliensis. BMC Genomics 2015; 16(1): 715.
[http://dx.doi.org/10.1186/s12864-015-1928-z] [PMID: 26384787]

[39] Downing T, Imamura H, Decuypere S, *et al.* Whole genome sequencing of multiple *Leishmania donovani* clinical isolates provides insights into population structure and mechanisms of drug resistance. Genome Res 2011; 21(12): 2143-56.
[http://dx.doi.org/10.1101/gr.123430.111] [PMID: 22038251]

[40] Callejas S, Leech V, Reitter C, Melville S. Hemizygous subtelomeres of an African trypanosome chromosome may account for over 75% of chromosome length. Genome Res 2006; 16(9): 1109-18.
[http://dx.doi.org/10.1101/gr.5147406] [PMID: 16899654]

[41] Wickstead B, Ersfeld K, Gull K. The small chromosomes of *Trypanosoma brucei* involved in antigenic variation are constructed around repetitive palindromes. Genome Res 2004; 14(6): 1014-24.
[http://dx.doi.org/10.1101/gr.2227704] [PMID: 15173109]

[42] Ubeda JM, Raymond F, Mukherjee A, *et al.* Genome-wide stochastic adaptive DNA amplification at direct and inverted DNA repeats in the parasite *Leishmania*. PLoS Biol 2014; 12(5): e1001868.

[http://dx.doi.org/10.1371/journal.pbio.1001868] [PMID: 24844805]

[43] Trenaman A, Hartley C, Prorocic M, *et al. Trypanosoma brucei* BRCA2 acts in a life cycle-specific genome stability process and dictates BRC repeat number-dependent RAD51 subnuclear dynamics. Nucleic Acids Res 2013; 41(2): 943-60.
[http://dx.doi.org/10.1093/nar/gks1192] [PMID: 23222131]

[44] Nouspikel T. DNA repair in mammalian cells : Nucleotide excision repair: variations on versatility. Cell Mol Life Sci 2009; 66(6): 994-1009.
[http://dx.doi.org/10.1007/s00018-009-8737-y] [PMID: 19153657]

[45] Naegeli H, Sugasawa K. The xeroderma pigmentosum pathway: decision tree analysis of DNA quality. DNA Repair (Amst) 2011; 10(7): 673-83.
[http://dx.doi.org/10.1016/j.dnarep.2011.04.019] [PMID: 21684221]

[46] Schärer OD. Nucleotide excision repair in eukaryotes. Cold Spring Harb Perspect Biol 2013; 5(10): a012609.
[http://dx.doi.org/10.1101/cshperspect.a012609] [PMID: 24086042]

[47] Van Houten B, Kad N. Investigation of bacterial nucleotide excision repair using single-molecule techniques. DNA Repair (Amst) 2014; 20: 41-8.
[http://dx.doi.org/10.1016/j.dnarep.2013.10.012] [PMID: 24472181]

[48] Grogan DW. Understanding DNA Repair in Hyperthermophilic Archaea: Persistent Gaps and Other Reasons to Focus on the Fork. Archaea 2015; 2015: 942605.

[49] Min JH, Pavletich NP. Recognition of DNA damage by the Rad4 nucleotide excision repair protein. Nature 2007; 449(7162): 570-5.
[http://dx.doi.org/10.1038/nature06155] [PMID: 17882165]

[50] Nemzow L, Lubin A, Zhang L, Gong F. XPC: Going where no DNA damage sensor has gone before. DNA Repair (Amst) 2015; 36: 19-27.

[51] Feldberg RS, Grossman L. A DNA binding protein from human placenta specific for ultraviolet damaged DNA. Biochemistry 1976; 15(11): 2402-8.
[http://dx.doi.org/10.1021/bi00656a024] [PMID: 1276148]

[52] Chu G, Chang E. Xeroderma pigmentosum group E cells lack a nuclear factor that binds to damaged DNA. Science 1988; 242(4878): 564-7.
[http://dx.doi.org/10.1126/science.3175673] [PMID: 3175673]

[53] Charton R, Guintini L, Peyresaubes F, Conconi A. Repair of UV induced DNA lesions in ribosomal gene chromatin and the role of "Odd" RNA polymerases (I and III). DNA Repair (Amst) 2015.

[54] Hanawalt PC, Spivak G. Transcription-coupled DNA repair: two decades of progress and surprises. Nat Rev Mol Cell Biol 2008; 9(12): 958-70.
[http://dx.doi.org/10.1038/nrm2549] [PMID: 19023283]

[55] Spivak G, Ganesan AK. The complex choreography of transcription-coupled repair. DNA Repair (Amst) 2014; 19: 64-70.
[http://dx.doi.org/10.1016/j.dnarep.2014.03.025] [PMID: 24751236]

[56] Passos-Silva DG, Rajão MA, Nascimento de Aguiar PH, Vieira-da-Rocha JP, Machado CR, Furtado C. Overview of DNA Repair in *Trypanosoma cruzi*, *Trypanosoma brucei*, and *Leishmania major* J

Nucleic Acids 2010; 2010: 840768.

[57] Lecordier L, Devaux S, Uzureau P, *et al.* Characterization of a TFIIH homologue from *Trypanosoma brucei*. Mol Microbiol 2007; 64(5): 1164-81. [http://dx.doi.org/10.1111/j.1365-2958.2007.05725.x] [PMID: 17542913]

[58] Lee JH, Jung HS, Günzl A. Transcriptionally active TFIIH of the early-diverged eukaryote *Trypanosoma brucei* harbors two novel core subunits but not a cyclin-activating kinase complex. Nucleic Acids Res 2009; 37(11): 3811-20. [http://dx.doi.org/10.1093/nar/gkp236] [PMID: 19386623]

[59] Morgante PG, Berra CM, Nakabashi M, Costa RM, Menck CF, Van Sluys MA. Functional XPB/RAD25 redundancy in Arabidopsis genome: characterization of AtXPB2 and expression analysis. Gene 2005; 344: 93-103. [http://dx.doi.org/10.1016/j.gene.2004.10.006] [PMID: 15656976]

[60] Richards JD, Cubeddu L, Roberts J, Liu H, White MF. The archaeal XPB protein is a ssDNA-dependent ATPase with a novel partner. J Mol Biol 2008; 376(3): 634-44. [http://dx.doi.org/10.1016/j.jmb.2007.12.019] [PMID: 18177890]

[61] Dunand-Sauthier I, Hohl M, Thorel F, Jaquier-Gubler P, Clarkson SG, Schärer OD. The spacer region of XPG mediates recruitment to nucleotide excision repair complexes and determines substrate specificity. J Biol Chem 2005; 280(8): 7030-7. [http://dx.doi.org/10.1074/jbc.M412228200] [PMID: 15590680]

[62] Badjatia N, Nguyen TN, Lee JH, Günzl A. Trypanosoma brucei harbours a divergent XPB helicase paralogue that is specialized in nucleotide excision repair and conserved among kinetoplastid organisms. Mol Microbiol 2013; 90(6): 1293-308. [http://dx.doi.org/10.1111/mmi.12435] [PMID: 24134817]

[63] Machado CR, Vieira-da-Rocha JP, Mendes IC, *et al.* Nucleotide excision repair in *Trypanosoma brucei*: specialization of transcription-coupled repair due to multigenic transcription. Mol Microbiol 2014; 92(4): 756-76. [http://dx.doi.org/10.1111/mmi.12589] [PMID: 24661334]

[64] Saxowsky TT, Doetsch PW. RNA polymerase encounters with DNA damage: transcription-coupled repair or transcriptional mutagenesis? Chem Rev 2006; 106(2): 474-88. [http://dx.doi.org/10.1021/cr040466q] [PMID: 16464015]

[65] Brégeon D, Doetsch PW. Transcriptional mutagenesis: causes and involvement in tumour development. Nat Rev Cancer 2011; 11(3): 218-27. [http://dx.doi.org/10.1038/nrc3006] [PMID: 21346784]

[66] Lee SK, Yu SL, Prakash L, Prakash S. Yeast RAD26, a homolog of the human CSB gene, functions independently of nucleotide excision repair and base excision repair in promoting transcription through damaged bases. Mol Cell Biol 2002; 22(12): 4383-9. [http://dx.doi.org/10.1128/MCB.22.12.4383-4389.2002] [PMID: 12024048]

[67] Kamarthapu V, Nudler E. Rethinking transcription coupled DNA repair. Curr Opin Microbiol 2015; 24: 15-20. [http://dx.doi.org/10.1016/j.mib.2014.12.005] [PMID: 25596348]

[68] Mullenders L. DNA damage mediated transcription arrest: Step back to go forward. DNA Repair (Amst) 2015; 36: 28-35.

[69] Berquist BR, Wilson DM III. Nucleic acid binding activity of human Cockayne syndrome B protein and identification of Ca(2+) as a novel metal cofactor. J Mol Biol 2009; 391(5): 820-32. [http://dx.doi.org/10.1016/j.jmb.2009.06.078] [PMID: 19580815]

[70] Oksenych V, Coin F. The long unwinding road: XPB and XPD helicases in damaged DNA opening. Cell Cycle 2010; 9(1): 90-6. [http://dx.doi.org/10.4161/cc.9.1.10267] [PMID: 20016270]

[71] Coin F, Oksenych V, Mocquet V, Groh S, Blattner C, Egly JM. Nucleotide excision repair driven by the dissociation of CAK from TFIIH. Mol Cell 2008; 31(1): 9-20. [http://dx.doi.org/10.1016/j.molcel.2008.04.024] [PMID: 18614043]

[72] Krasikova YS, Rechkunova NI, Maltseva EA, Petruseva IO, Lavrik OI. Localization of *Xeroderma pigmentosum* group A protein and replication protein A on damaged DNA in nucleotide excision repair. Nucleic Acids Res 2010; 38(22): 8083-94. [http://dx.doi.org/10.1093/nar/gkq649] [PMID: 20693538]

[73] Sarker AH, Tsutakawa SE, Kostek S, *et al.* Recognition of RNA polymerase II and transcription bubbles by XPG, CSB, and TFIIH: insights for transcription-coupled repair and Cockayne Syndrome. Mol Cell 2005; 20(2): 187-98. [http://dx.doi.org/10.1016/j.molcel.2005.09.022] [PMID: 16246722]

[74] Peña-Diaz J, Jiricny J. Mammalian mismatch repair: error-free or error-prone? Trends Biochem Sci 2012; 37(5): 206-14. [http://dx.doi.org/10.1016/j.tibs.2012.03.001] [PMID: 22475811]

[75] Jiricny J. The multifaceted mismatch-repair system. Nat Rev Mol Cell Biol 2006; 7(5): 335-46. [http://dx.doi.org/10.1038/nrm1907] [PMID: 16612326]

[76] Li GM. Mechanisms and functions of DNA mismatch repair. Cell Res 2008; 18(1): 85-98. [http://dx.doi.org/10.1038/cr.2007.115] [PMID: 18157157]

[77] Fukui K. DNA mismatch repair in eukaryotes and bacteria. J Nucleic Acids 2010; 2010

[78] Sharma M, Predeus AV, Kovacs N, Feig M. Differential mismatch recognition specificities of eukaryotic MutS homologs, MutSα and MutSβ. Biophys J 2014; 106(11): 2483-92. [http://dx.doi.org/10.1016/j.bpj.2014.04.026] [PMID: 24896128]

[79] Gorman J, Plys AJ, Visnapuu ML, Alani E, Greene EC. Visualizing one-dimensional diffusion of eukaryotic DNA repair factors along a chromatin lattice. Nat Struct Mol Biol 2010; 17(8): 932-8. [http://dx.doi.org/10.1038/nsmb.1858] [PMID: 20657586]

[80] Erie DA, Weninger KR. Single molecule studies of DNA mismatch repair. DNA Repair (Amst) 2014; 20: 71-81. [http://dx.doi.org/10.1016/j.dnarep.2014.03.007] [PMID: 24746644]

[81] Kadyrov FA, Dzantiev L, Constantin N, Modrich P. Endonucleolytic function of MutLα in human mismatch repair. Cell 2006; 126(2): 297-308. [http://dx.doi.org/10.1016/j.cell.2006.05.039] [PMID: 16873062]

[82] Pillon MC, Miller JH, Guarné A. The endonuclease domain of MutL interacts with the β sliding clamp. DNA Repair (Amst) 2011; 10(1): 87-93.
[http://dx.doi.org/10.1016/j.dnarep.2010.10.003] [PMID: 21050827]

[83] Smith CE, Bowen N, Graham WJ V, Goellner EM, Srivatsan A, Kolodner RD. Activation of *Saccharomyces cerevisiae* Mlh1-Pms1 endonuclease in a reconstituted mismatch repair system. J Biol Chem 2015; 290(35): 21580-90.
[http://dx.doi.org/10.1074/jbc.M115.662189] [PMID: 26170454]

[84] Iyama T, Wilson DM III. DNA repair mechanisms in dividing and non-dividing cells. DNA Repair (Amst) 2013; 12(8): 620-36.
[http://dx.doi.org/10.1016/j.dnarep.2013.04.015] [PMID: 23684800]

[85] Bell JS, Harvey TI, Sims AM, McCulloch R. Characterization of components of the mismatch repair machinery in *Trypanosoma brucei*. Mol Microbiol 2004; 51(1): 159-73.
[http://dx.doi.org/10.1046/j.1365-2958.2003.03804.x] [PMID: 14651619]

[86] Augusto-Pinto L, Bartholomeu DC, Teixeira SM, Pena SD, Machado CR. Molecular cloning and characterization of the DNA mismatch repair gene class 2 from the *Trypanosoma cruzi*. Gene 2001; 272(1-2): 323-33.
[http://dx.doi.org/10.1016/S0378-1119(01)00549-2] [PMID: 11470539]

[87] Stadler ZK. Diagnosis and management of DNA mismatch repair-deficient colorectal cancer. Hematol Oncol Clin North Am 2015; 29(1): 29-41.
[http://dx.doi.org/10.1016/j.hoc.2014.09.008] [PMID: 25475571]

[88] Bell JS, McCulloch R. Mismatch repair regulates homologous recombination, but has little influence on antigenic variation, in *Trypanosoma brucei*. J Biol Chem 2003; 278(46): 45182-8.
[http://dx.doi.org/10.1074/jbc.M308123200] [PMID: 12933800]

[89] Augusto-Pinto L, Teixeira SM, Pena SD, Machado CR. Single-nucleotide polymorphisms of the *Trypanosoma cruzi* MSH2 gene support the existence of three phylogenetic lineages presenting differences in mismatch-repair efficiency. Genetics 2003; 164(1): 117-26.
[PMID: 12750325]

[90] Berriman M, Ghedin E, Hertz-Fowler C, *et al.* The genome of the African trypanosome *Trypanosoma brucei*. Science 2005; 309(5733): 416-22.
[http://dx.doi.org/10.1126/science.1112642] [PMID: 16020726]

[91] El-Sayed NM, Myler PJ, Bartholomeu DC, *et al.* The genome sequence of *Trypanosoma cruzi*, etiologic agent of Chagas disease. Science 2005; 309(5733): 409-15.
[http://dx.doi.org/10.1126/science.1112631] [PMID: 16020725]

[92] Ivens AC, Peacock CS, Worthey EA, *et al.* The genome of the kinetoplastid parasite, *Leishmania major*. Science 2005; 309(5733): 436-42.
[http://dx.doi.org/10.1126/science.1112680] [PMID: 16020728]

[93] Reenan RA, Kolodner RD. Characterization of insertion mutations in the *Saccharomyces cerevisiae* MSH1 and MSH2 genes: evidence for separate mitochondrial and nuclear functions. Genetics 1992; 132(4): 975-85.
[PMID: 1334021]

[94] Kaniak A, Dzierzbicki P, Rogowska AT, Malc E, Fikus M, Ciesla Z. Msh1p counteracts oxidative lesion-induced instability of mtDNA and stimulates mitochondrial recombination in *Saccharomyces cerevisiae*. DNA Repair (Amst) 2009; 8(3): 318-29. [http://dx.doi.org/10.1016/j.dnarep.2008.11.004] [PMID: 19056520]

[95] Pogorzala L, Mookerjee S, Sia EA. Evidence that msh1p plays multiple roles in mitochondrial base excision repair. Genetics 2009; 182(3): 699-709. [http://dx.doi.org/10.1534/genetics.109.103796] [PMID: 19398768]

[96] Papadopoulou B, Dumas C. Parameters controlling the rate of gene targeting frequency in the protozoan parasite *Leishmania*. Nucleic Acids Res 1997; 25(21): 4278-86. [http://dx.doi.org/10.1093/nar/25.21.4278] [PMID: 9336458]

[97] Coelho AC, Leprohon P, Ouellette M. Generation of *Leishmania* hybrids by whole genomic DNA transformation. PLoS Negl Trop Dis 2012; 6(9): e1817.

[98] Schofield MJ, Hsieh P. DNA mismatch repair: molecular mechanisms and biological function. Annu Rev Microbiol 2003; 57: 579-608. [http://dx.doi.org/10.1146/annurev.micro.57.030502.090847] [PMID: 14527292]

[99] de Wind N, Dekker M, Berns A, Radman M, te Riele H. Inactivation of the mouse Msh2 gene results in mismatch repair deficiency, methylation tolerance, hyperrecombination, and predisposition to cancer. Cell 1995; 82(2): 321-30. [http://dx.doi.org/10.1016/0092-8674(95)90319-4] [PMID: 7628020]

[100] Noonan EM, Shah D, Yaffe MB, Lauffenburger DA, Samson LD. O6-Methylguanine DNA lesions induce an intra-S-phase arrest from which cells exit into apoptosis governed by early and late multi-pathway signaling network activation. Integr Biol (Camb) 2012; 4(10): 1237-55. [http://dx.doi.org/10.1039/c2ib20091k] [PMID: 22892544]

[101] Rodenko B, Wanner MJ, Alkhaldi AA, *et al.* Targeting the parasites DNA with methyltriazenyl purine analogs is a safe, selective, and efficacious antitrypanosomal strategy. Antimicrob Agents Chemother 2015; 59(11): 6708-16. [http://dx.doi.org/10.1128/AAC.00596-15] [PMID: 26282430]

[102] Sia EA, Dominska M, Stefanovic L, Petes TD. Isolation and characterization of point mutations in mismatch repair genes that destabilize microsatellites in yeast. Mol Cell Biol 2001; 21(23): 8157-67. [http://dx.doi.org/10.1128/MCB.21.23.8157-8167.2001] [PMID: 11689704]

[103] Yamamoto H, Imai K. Microsatellite instability: an update. Arch Toxicol 2015; 89(6): 899-921. [http://dx.doi.org/10.1007/s00204-015-1474-0] [PMID: 25701956]

[104] Barnes RL, McCulloch R. *Trypanosoma brucei* homologous recombination is dependent on substrate length and homology, though displays a differential dependence on mismatch repair as substrate length decreases. Nucleic Acids Res 2007; 35(10): 3478-93. [http://dx.doi.org/10.1093/nar/gkm249] [PMID: 17478508]

[105] Miotto O, Almagro-Garcia J, Manske M, *et al.* Multiple populations of artemisinin-resistant *Plasmodium falciparum* in Cambodia. Nat Genet 2013; 45(6): 648-55. [http://dx.doi.org/10.1038/ng.2624] [PMID: 23624527]

[106] Takala-Harrison S, Clark TG, Jacob CG, *et al.* Genetic loci associated with delayed clearance of

Plasmodium falciparum following artemisinin treatment in Southeast Asia. Proc Natl Acad Sci USA 2013; 110(1): 240-5.
[http://dx.doi.org/10.1073/pnas.1211205110] [PMID: 23248304]

[107] Rathod PK, McErlean T, Lee PC. Variations in frequencies of drug resistance in *Plasmodium falciparum*. Proc Natl Acad Sci USA 1997; 94(17): 9389-93.
[http://dx.doi.org/10.1073/pnas.94.17.9389] [PMID: 9256492]

[108] Zingales B, Andrade SG, Briones MR, *et al.* A new consensus for *Trypanosoma cruzi* intraspecific nomenclature: second revision meeting recommends TcI to TcVI. Mem Inst Oswaldo Cruz 2009; 104(7): 1051-4.
[http://dx.doi.org/10.1590/S0074-02762009000700021] [PMID: 20027478]

[109] de Freitas JM, Augusto-Pinto L, Pimenta JR, *et al.* Ancestral genomes, sex, and the population structure of *Trypanosoma cruzi*. PLoS Pathog 2006; 2(3): e24.
[http://dx.doi.org/10.1371/journal.ppat.0020024] [PMID: 16609729]

[110] Campos PC, Silva VG, Furtado C, *et al. Trypanosoma cruzi* MSH2: Functional analyses on different parasite strains provide evidences for a role on the oxidative stress response. Mol Biochem Parasitol 2011; 176(1): 8-16.
[http://dx.doi.org/10.1016/j.molbiopara.2010.11.001] [PMID: 21073906]

[111] Iyer RR, Pluciennik A, Burdett V, Modrich PL. DNA mismatch repair: functions and mechanisms. Chem Rev 2006; 106(2): 302-23.
[http://dx.doi.org/10.1021/cr0404794] [PMID: 16464007]

[112] Zingales B, Souto RP, Mangia RH, *et al.* Molecular epidemiology of American trypanosomiasis in Brazil based on dimorphisms of rRNA and mini-exon gene sequences. Int J Parasitol 1998; 28(1): 105-12.
[http://dx.doi.org/10.1016/S0020-7519(97)00178-1] [PMID: 9504338]

[113] Augusto-Pinto L, Teixeira SM, Pena SD, Machado CR. Single-nucleotide polymorphisms of the *Trypanosoma cruzi* MSH2 gene support the existence of three phylogenetic lineages presenting differences in mismatch-repair efficiency. Genetics 2003; 164(1): 117-26.
[PMID: 12750325]

[114] Yanamadala S, Ljungman M. Potential role of MLH1 in the induction of p53 and apoptosis by blocking transcription on damaged DNA templates. Mol Cancer Res 2003; 1(10): 747-54.
[PMID: 12939400]

[115] Brierley DJ, Martin SA. Oxidative stress and the DNA mismatch repair pathway. Antioxid Redox Signal 2013; 18(18): 2420-8.
[http://dx.doi.org/10.1089/ars.2012.4994] [PMID: 23121537]

[116] Ni TT, Marsischky GT, Kolodner RD. MSH2 and MSH6 are required for removal of adenine misincorporated opposite 8-oxo-guanine in S. cerevisiae. Mol Cell 1999; 4(3): 439-44.
[http://dx.doi.org/10.1016/S1097-2765(00)80346-9] [PMID: 10518225]

[117] Wyrzykowski J, Volkert MR. The Escherichia coli methyl-directed mismatch repair system repairs base pairs containing oxidative lesions. J Bacteriol 2003; 185(5): 1701-4.
[http://dx.doi.org/10.1128/JB.185.5.1701-1704.2003] [PMID: 12591888]

[118] Zlatanou A, Despras E, Braz-Petta T, *et al.* The hMsh2-hMsh6 complex acts in concert with monoubiquitinated PCNA and Pol η in response to oxidative DNA damage in human cells. Mol Cell 2011; 43(4): 649-62.
[http://dx.doi.org/10.1016/j.molcel.2011.06.023] [PMID: 21855803]

[119] Machado-Silva A, Teixeira SM, Franco GR, *et al.* Mismatch repair in *Trypanosoma brucei*: heterologous expression of MSH2 from *Trypanosoma cruzi* provides new insights into the response to oxidative damage. Gene 2008; 411(1-2): 19-26.
[http://dx.doi.org/10.1016/j.gene.2007.12.021] [PMID: 18262734]

[120] Grazielle-Silva V, Zeb TF, Bolderson J, *et al.* Distinct phenotypes caused by mutation of MSH2 in trypanosome insect and mammalian life cycle forms are associated with parasite adaptation to oxidative stress. PLoS Negl Trop Dis 2015; 9(6): e0003870.
[http://dx.doi.org/10.1371/journal.pntd.0003870] [PMID: 26083967]

[121] Rajão MA, Furtado C, Alves CL, *et al.* Unveiling benznidazoles mechanism of action through overexpression of DNA repair proteins in *Trypanosoma cruzi*. Environ Mol Mutagen 2014; 55(4): 309-21.
[http://dx.doi.org/10.1002/em.21839] [PMID: 24347026]

[122] Piacenza L, Alvarez MN, Peluffo G, Radi R. Fighting the oxidative assault: the *Trypanosoma cruzi* journey to infection. Curr Opin Microbiol 2009; 12(4): 415-21.
[http://dx.doi.org/10.1016/j.mib.2009.06.011] [PMID: 19616990]

[123] Van Assche T, Deschacht M, da Luz RA, Maes L, Cos P. *Leishmania*-macrophage interactions: insights into the redox biology. Free Radic Biol Med 2011; 51(2): 337-51.
[http://dx.doi.org/10.1016/j.freeradbiomed.2011.05.011] [PMID: 21620959]

[124] Wang D, Kreutzer DA, Essigmann JM. Mutagenicity and repair of oxidative DNA damage: insights from studies using defined lesions. Mutat Res 1998; 400(1-2): 99-115.
[http://dx.doi.org/10.1016/S0027-5107(98)00066-9] [PMID: 9685598]

[125] Shackelford RE, Kaufmann WK, Paules RS. Oxidative stress and cell cycle checkpoint function. Free Radic Biol Med 2000; 28(9): 1387-404.
[http://dx.doi.org/10.1016/S0891-5849(00)00224-0] [PMID: 10924858]

[126] Mandavilli BS, Santos JH, Van Houten B. Mitochondrial DNA repair and aging. Mutat Res 2002; 509(1-2): 127-51.
[http://dx.doi.org/10.1016/S0027-5107(02)00220-8] [PMID: 12427535]

[127] Hegde ML, Hazra TK, Mitra S. Early steps in the DNA base excision/single-strand interruption repair pathway in mammalian cells. Cell Res 2008; 18(1): 27-47.
[http://dx.doi.org/10.1038/cr.2008.8] [PMID: 18166975]

[128] Robertson AB, Klungland A, Rognes T, Leiros I. DNA repair in mammalian cells: Base excision repair: the long and short of it. Cell Mol Life Sci 2009; 66(6): 981-93.
[http://dx.doi.org/10.1007/s00018-009-8736-z] [PMID: 19153658]

[129] Dianov G, Bischoff C, Piotrowski J, Bohr VA. Repair pathways for processing of 8-oxoguanine in DNA by mammalian cell extracts. J Biol Chem 1998; 273(50): 33811-6.
[http://dx.doi.org/10.1074/jbc.273.50.33811] [PMID: 9837971]

[130] Blank A, Kim B, Loeb LA. DNA polymerase delta is required for base excision repair of DNA methylation damage in *Saccharomyces cerevisiae*. Proc Natl Acad Sci USA 1994; 91(19): 9047-51. [http://dx.doi.org/10.1073/pnas.91.19.9047] [PMID: 8090767]

[131] Haltiwanger BM, Matsumoto Y, Nicolas E, Dianov GL, Bohr VA, Taraschi TF. DNA base excision repair in human malaria parasites is predominantly by a long-patch pathway. Biochemistry 2000; 39(4): 763-72.
[http://dx.doi.org/10.1021/bi9923151] [PMID: 10651642]

[132] Peña-Diaz J, Akbari M, Sundheim O, *et al.* Trypanosoma cruzi contains a single detectable uracil-DNA glycosylase and repairs uracil exclusively *via* short patch base excision repair. J Mol Biol 2004; 342(3): 787-99.
[http://dx.doi.org/10.1016/j.jmb.2004.07.043] [PMID: 15342237]

[133] Fárez-Vidal ME, Gallego C, Ruiz-Pérez LM, González-Pacanowska D. Characterization of uracil-DNA glycosylase activity from *Trypanosoma cruzi* and its stimulation by AP endonuclease. Nucleic Acids Res 2001; 29(7): 1549-55.
[http://dx.doi.org/10.1093/nar/29.7.1549] [PMID: 11266557]

[134] Castillo-Acosta VM, Aguilar-Pereyra F, Vidal AE, Navarro M, Ruiz-Pérez LM, González-Pacanowska D. Trypanosomes lacking uracil-DNA glycosylase are hypersensitive to antifolates and present a mutator phenotype. Int J Biochem Cell Biol 2012; 44(9): 1555-68.
[http://dx.doi.org/10.1016/j.biocel.2012.06.014] [PMID: 22728162]

[135] Zharkov DO. Base excision DNA repair. Cell Mol Life Sci 2008; 65(10): 1544-65.
[http://dx.doi.org/10.1007/s00018-008-7543-2] [PMID: 18259689]

[136] Shibutani S, Takeshita M, Grollman AP. Insertion of specific bases during DNA synthesis past the oxidation-damaged base 8-oxodG. Nature 1991; 349(6308): 431-4.
[http://dx.doi.org/10.1038/349431a0] [PMID: 1992344]

[137] Furtado C, Kunrath-Lima M, Rajão MA, *et al.* Functional characterization of 8-oxoguanine DNA glycosylase of *Trypanosoma cruzi*. PLoS One 2012; 7(8): e42484.
[http://dx.doi.org/10.1371/journal.pone.0042484] [PMID: 22876325]

[138] Vidal AE, Harkiolaki M, Gallego C, *et al.* Crystal structure and DNA repair activities of the AP endonuclease from *Leishmania major*. J Mol Biol 2007; 373(4): 827-38.
[http://dx.doi.org/10.1016/j.jmb.2007.08.001] [PMID: 17870086]

[139] Sepúlveda S, Valenzuela L, Ponce I, *et al.* Expression, functionality, and localization of apurinic/apyrimidinic endonucleases in replicative and non-replicative forms of *Trypanosoma cruzi*. J Cell Biochem 2014; 115(2): 397-409.
[http://dx.doi.org/10.1002/jcb.24675] [PMID: 24114998]

[140] Castillo-Acosta VM, Ruiz-Pérez LM, Yang W, González-Pacanowska D, Vidal AE. Identification of a residue critical for the excision of 3-blocking ends in apurinic/apyrimidinic endonucleases of the Xth family. Nucleic Acids Res 2009; 37(6): 1829-42.
[http://dx.doi.org/10.1093/nar/gkp021] [PMID: 19181704]

[141] Charret KS, Requena CE, Castillo-Acosta VM, Ruiz-Pérez LM, González-Pacanowska D, Vidal AE. *Trypanosoma brucei* AP endonuclease 1 has a major role in the repair of abasic sites and protection

against DNA-damaging agents. DNA Repair (Amst) 2012; 11(1): 53-64.
[http://dx.doi.org/10.1016/j.dnarep.2011.10.006] [PMID: 22041024]

[142] Cabrera G, Barría C, Fernández C, *et al.* DNA repair BER pathway inhibition increases cell death caused by oxidative DNA damage in *Trypanosoma cruzi*. J Cell Biochem 2011; 112(8): 2189-99.
[http://dx.doi.org/10.1002/jcb.23138] [PMID: 21480362]

[143] Pérez J, Gallego C, Bernier-Villamor V, Camacho A, González-Pacanowska D, Ruiz-Pérez LM. Apurinic/apyrimidinic endonuclease genes from the trypanosomatidae *Leishmania major* and *Trypanosoma cruzi* confer resistance to oxidizing agents in DNA repair-deficient Escherichia coli. Nucleic Acids Res 1999; 27(3): 771-7.
[http://dx.doi.org/10.1093/nar/27.3.771] [PMID: 9889272]

[144] Torri AF, Englund PT. A DNA polymerase beta in the mitochondrion of the trypanosomatid *Crithidia fasciculata*. J Biol Chem 1995; 270(8): 3495-7.
[http://dx.doi.org/10.1074/jbc.270.8.3495] [PMID: 7876082]

[145] Torri AF, Kunkel TA, Englund PT. A beta-like DNA polymerase from the mitochondrion of the trypanosomatid *Crithidia fasciculata*. J Biol Chem 1994; 269(11): 8165-71.
[PMID: 8132542]

[146] Saxowsky TT, Choudhary G, Klingbeil MM, Englund PT. *Trypanosoma brucei* has two distinct mitochondrial DNA polymerase beta enzymes. J Biol Chem 2003; 278(49): 49095-101.
[http://dx.doi.org/10.1074/jbc.M308565200] [PMID: 12966090]

[147] Lopes DdeO, Schamber-Reis BL, Regis-da-Silva CG, *et al.* Biochemical studies with DNA polymerase beta and DNA polymerase beta-PAK of *Trypanosoma cruzi* suggest the involvement of these proteins in mitochondrial DNA maintenance. DNA Repair (Amst) 2008; 7(11): 1882-92.
[http://dx.doi.org/10.1016/j.dnarep.2008.07.018] [PMID: 18761429]

[148] Taladriz S, Hanke T, Ramiro MJ, *et al.* Nuclear DNA polymerase beta from *Leishmania infantum*. Cloning, molecular analysis and developmental regulation. Nucleic Acids Res 2001; 29(18): 3822-34.
[http://dx.doi.org/10.1093/nar/29.18.3822] [PMID: 11557814]

[149] Bai P. Biology of poly(ADP-Ribose) polymerases: the factotums of cell maintenance. Mol Cell 2015; 58(6): 947-58.
[http://dx.doi.org/10.1016/j.molcel.2015.01.034] [PMID: 26091343]

[150] Caldecott KW. DNA single-strand break repair. Exp Cell Res 2014; 329(1): 2-8.
[http://dx.doi.org/10.1016/j.yexcr.2014.08.027] [PMID: 25176342]

[151] Caldecott KW. Molecular biology. Ribose an internal threat to DNA. Science 2014; 343(6168): 260-1.
[http://dx.doi.org/10.1126/science.1248234] [PMID: 24436412]

[152] Fernández Villamil SH, Baltanás R, Alonso GD, Vilchez Larrea SC, Torres HN, Flawiá MM. TcPARP: A DNA damage-dependent poly(ADP-ribose) polymerase from *Trypanosoma cruzi*. Int J Parasitol 2008; 38(3-4): 277-87.
[http://dx.doi.org/10.1016/j.ijpara.2007.08.003] [PMID: 17936287]

[153] Farzaneh F, Shall S, Michels P, Borst P. ADP-ribosyl transferase activity in *Trypanosoma brucei*. Mol Biochem Parasitol 1985; 14(3): 251-9.
[http://dx.doi.org/10.1016/0166-6851(85)90053-2] [PMID: 2985984]

[154] Vilchez Larrea SC, Haikarainen T, Narwal M, *et al.* Inhibition of poly(ADP-ribose) polymerase interferes with *Trypanosoma cruzi* infection and proliferation of the parasite. PLoS One 2012; 7(9): e46063.
[http://dx.doi.org/10.1371/journal.pone.0046063] [PMID: 23049934]

[155] Khademvatan S, Gharavi MJ, Saki J. Miltefosine induces metacaspase and PARP genes expression in *Leishmania infantum*. The Brazilian J Infectious Dis: an official publication of the Brazilian Society of Infectious Diseases 2011; 15(5): 442-8.
[http://dx.doi.org/10.1016/S1413-8670(11)70225-2]

[156] Kobil JH, Campbell AG. *Trypanosoma brucei* RNase HI requires its divergent spacer subdomain for enzymatic function and its conserved RNA binding motif for nuclear localization. Mol Biochem Parasitol 2000; 107(1): 135-42.
[http://dx.doi.org/10.1016/S0166-6851(00)00182-1] [PMID: 10717310]

[157] Misra S, Bennett J, Friew YN, *et al.* A type II ribonuclease H from *Leishmania* mitochondria: an enzyme essential for the growth of the parasite. Mol Biochem Parasitol 2005; 143(2): 135-45.
[http://dx.doi.org/10.1016/j.molbiopara.2005.05.009] [PMID: 15978682]

[158] Daley JM, Gaines WA, Kwon Y, Sung P. Regulation of DNA pairing in homologous recombination. Cold Spring Harb Perspect Biol 2014; 6(11): a017954.
[http://dx.doi.org/10.1101/cshperspect.a017954] [PMID: 25190078]

[159] Heyer WD. Regulation of recombination and genomic maintenance. Cold Spring Harb Perspect Biol 2015; 7(8): a016501.
[http://dx.doi.org/10.1101/cshperspect.a016501] [PMID: 26238353]

[160] McCulloch R, Vassella E, Burton P, Boshart M, Barry JD. Transformation of monomorphic and pleomorphic *Trypanosoma brucei*. Methods Mol Biol 2004; 262: 53-86.
[PMID: 14769956]

[161] Huang F, Mazin AV. Targeting the homologous recombination pathway by small molecule modulators. Bioorg Med Chem Lett 2014; 24(14): 3006-13.
[http://dx.doi.org/10.1016/j.bmcl.2014.04.088] [PMID: 24856061]

[162] Huang F, Mazina OM, Zentner IJ, Cocklin S, Mazin AV. Inhibition of homologous recombination in human cells by targeting RAD51 recombinase. J Med Chem 2012; 55(7): 3011-20.
[http://dx.doi.org/10.1021/jm201173g] [PMID: 22380680]

[163] Carvalho JF, Kanaar R. Targeting homologous recombination-mediated DNA repair in cancer. Expert Opin Ther Targets 2014; 18(4): 427-58.
[http://dx.doi.org/10.1517/14728222.2014.882900] [PMID: 24491188]

[164] Matthews KR, McCulloch R, Morrison LJ. The within-host dynamics of African trypanosome infections. Philos Trans R Soc Lond B Biol Sci 2015; 370(1675): 370.
[http://dx.doi.org/10.1098/rstb.2014.0288] [PMID: 26150654]

[165] McCulloch R, Morrison LJ, Hall JP. DNA recombination strategies during antigenic variation in the african trypanosome. Microbiology spectrum 2015; 3(2): MDNA3-0016-2014.
[http://dx.doi.org/10.1128/microbiolspec.MDNA3-0016-2014]

[166] Horn D. Antigenic variation in African trypanosomes. Mol Biochem Parasitol 2014; 195(2): 123-9.

[http://dx.doi.org/10.1016/j.molbiopara.2014.05.001] [PMID: 24859277]

[167] Marcello L, Barry JD. Analysis of the VSG gene silent archive in *Trypanosoma brucei* reveals that mosaic gene expression is prominent in antigenic variation and is favored by archive substructure. Genome Res 2007; 17(9): 1344-52. [http://dx.doi.org/10.1101/gr.6421207] [PMID: 17652423]

[168] Cross GA, Kim HS, Wickstead B. Capturing the variant surface glycoprotein repertoire (the VSGnome) of *Trypanosoma brucei* Lister 427. Mol Biochem Parasitol 2014; 195(1): 59-73. [http://dx.doi.org/10.1016/j.molbiopara.2014.06.004] [PMID: 24992042]

[169] Palmer GH, Brayton KA. Gene conversion is a convergent strategy for pathogen antigenic variation. Trends Parasitol 2007; 23(9): 408-13. [http://dx.doi.org/10.1016/j.pt.2007.07.008] [PMID: 17662656]

[170] Hall JP, Wang H, Barry JD. Mosaic VSGs and the scale of *Trypanosoma brucei* antigenic variation. PLoS Pathog 2013; 9(7): e1003502. [http://dx.doi.org/10.1371/journal.ppat.1003502] [PMID: 23853603]

[171] Mugnier MR, Cross GA, Papavasiliou FN. The *in vivo* dynamics of antigenic variation in *Trypanosoma brucei.* Science 2015; 347(6229): 1470-3. [http://dx.doi.org/10.1126/science.aaa4502] [PMID: 25814582]

[172] McCulloch R, Field MC. Quantitative sequencing confirms VSG diversity as central to immune evasion by *Trypanosoma brucei.* Trends Parasitol 2015; 31(8): 346-9. [http://dx.doi.org/10.1016/j.pt.2015.05.001] [PMID: 25999027]

[173] Barbet AF, Kamper SM. The importance of mosaic genes to trypanosome survival. Parasitol Today (Regul Ed) 1993; 9(2): 63-6. [http://dx.doi.org/10.1016/0169-4758(93)90039-I] [PMID: 15463711]

[174] Morrison LJ, Majiwa P, Read AF, Barry JD. Probabilistic order in antigenic variation of *Trypanosoma brucei.* Int J Parasitol 2005; 35(9): 961-72. [http://dx.doi.org/10.1016/j.ijpara.2005.05.004] [PMID: 16000200]

[175] Capbern A, Giroud C, Baltz T, Mattern P. *Trypanosoma equiperdum*: antigenic variations in experimental trypanosomiasis of rabbits. Exp Parasitol 1977; 42(1): 6-13. [http://dx.doi.org/10.1016/0014-4894(77)90055-8] [PMID: 67961]

[176] Thon G, Baltz T, Giroud C, Eisen H. Trypanosome variable surface glycoproteins: composite genes and order of expression. Genes Dev 1990; 4(8): 1374-83. [http://dx.doi.org/10.1101/gad.4.8.1374] [PMID: 2227415]

[177] McCulloch R, Barry JD. A role for RAD51 and homologous recombination in *Trypanosoma brucei* antigenic variation. Genes Dev 1999; 13(21): 2875-88. [http://dx.doi.org/10.1101/gad.13.21.2875] [PMID: 10557214]

[178] Hartley CL, McCulloch R. *Trypanosoma brucei* BRCA2 acts in antigenic variation and has undergone a recent expansion in BRC repeat number that is important during homologous recombination. Mol Microbiol 2008; 68(5): 1237-51. [http://dx.doi.org/10.1111/j.1365-2958.2008.06230.x] [PMID: 18430140]

[179] Dobson R, Stockdale C, Lapsley C, Wilkes J, McCulloch R. Interactions among *Trypanosoma brucei*

RAD51 paralogues in DNA repair and antigenic variation. Mol Microbiol 2011; 81(2): 434-56. [http://dx.doi.org/10.1111/j.1365-2958.2011.07703.x] [PMID: 21615552]

[180] Kim HS, Cross GA. TOPO3alpha influences antigenic variation by monitoring expression-sit--associated VSG switching in *Trypanosoma brucei*. PLoS Pathog 2010; 6(7): e1000992. [http://dx.doi.org/10.1371/journal.ppat.1000992] [PMID: 20628569]

[181] Kim HS, Cross GA. Identification of *Trypanosoma brucei* RMI1/BLAP75 homologue and its roles in antigenic variation. PLoS One 2011; 6(9): e25313. [http://dx.doi.org/10.1371/journal.pone.0025313] [PMID: 21980422]

[182] Cahoon LA, Seifert HS. Focusing homologous recombination: pilin antigenic variation in the pathogenic *Neisseria*. Mol Microbiol 2011; 81(5): 1136-43. [http://dx.doi.org/10.1111/j.1365-2958.2011.07773.x] [PMID: 21812841]

[183] Liveris D, Mulay V, Sandigursky S, Schwartz I. *Borrelia burgdorferi* vlsE antigenic variation is not mediated by RecA. Infect Immun 2008; 76(9): 4009-18. [http://dx.doi.org/10.1128/IAI.00027-08] [PMID: 18606826]

[184] Dresser AR, Hardy PO, Chaconas G. Investigation of the genes involved in antigenic switching at the vlsE locus in *Borrelia burgdorferi*: an essential role for the RuvAB branch migrase. PLoS Pathog 2009; 5(12): e1000680. [http://dx.doi.org/10.1371/journal.ppat.1000680] [PMID: 19997508]

[185] Lin T, Gao L, Edmondson DG, Jacobs MB, Philipp MT, Norris SJ. Central role of the Holliday junction helicase RuvAB in vlsE recombination and infectivity of *Borrelia burgdorferi*. PLoS Pathog 2009; 5(12): e1000679. [http://dx.doi.org/10.1371/journal.ppat.1000679] [PMID: 19997622]

[186] Norris SJ. vls antigenic variation systems of lyme disease borrelia: eluding host immunity through both random, segmental gene conversion and framework heterogeneity. Microbiol Spectr 2014; 2(6) [http://dx.doi.org/10.1128/microbiolspec.MDNA3-0038-2014] [PMID: 26104445]

[187] Morrison LJ, Marcello L, McCulloch R. Antigenic variation in the African trypanosome: molecular mechanisms and phenotypic complexity. Cell Microbiol 2009; 11(12): 1724-34. [http://dx.doi.org/10.1111/j.1462-5822.2009.01383.x] [PMID: 19751359]

[188] Jensen RB, Carreira A, Kowalczykowski SC. Purified human BRCA2 stimulates RAD51-mediated recombination. Nature 2010; 467(7316): 678-83. [http://dx.doi.org/10.1038/nature09399] [PMID: 20729832]

[189] Liu J, Doty T, Gibson B, Heyer WD. Human BRCA2 protein promotes RAD51 filament formation on RPA-covered single-stranded DNA. Nat Struct Mol Biol 2010; 17(10): 1260-2. [http://dx.doi.org/10.1038/nsmb.1904] [PMID: 20729859]

[190] Shahid T, Soroka J, Kong EH, *et al.* Structure and mechanism of action of the BRCA2 breast cancer tumor suppressor. Nat Struct Mol Biol 2014; 21(11): 962-8. [http://dx.doi.org/10.1038/nsmb.2899] [PMID: 25282148]

[191] Oyola SO, Bringaud F, Melville SE. A kinetoplastid BRCA2 interacts with DNA replication protein CDC45. Int J Parasitol 2009; 39(1): 59-69. [http://dx.doi.org/10.1016/j.ijpara.2008.07.002] [PMID: 18723021]

[192] Hall M III, Misra S, Chaudhuri M, Chaudhuri G. Peptide aptamer mimicking RAD51-binding domain of BRCA2 inhibits DNA damage repair and survival in *Trypanosoma brucei*. Microb Pathog 2011; 50(5): 252-62.
[http://dx.doi.org/10.1016/j.micpath.2010.11.007] [PMID: 21296653]

[193] Turner CM. The rate of antigenic variation in fly-transmitted and syringe-passaged infections of *Trypanosoma brucei*. FEMS Microbiol Lett 1997; 153(1): 227-31.
[http://dx.doi.org/10.1111/j.1574-6968.1997.tb10486.x] [PMID: 9252591]

[194] Barry JD, McCulloch R. Antigenic variation in trypanosomes: enhanced phenotypic variation in a eukaryotic parasite. Adv Parasitol 2001; 49: 1-70.
[http://dx.doi.org/10.1016/S0065-308X(01)49037-3] [PMID: 11461029]

[195] Roth DB. V(D)J recombination: mechanism, errors, and fidelity. Microbiol Spectr 2014; 2(6)
[http://dx.doi.org/10.1128/microbiolspec.MDNA3-0041-2014] [PMID: 26104458]

[196] Lee CS, Haber JE. Mating-type Gene Switching in *Saccharomyces cerevisiae*. Microbiology Spectrum 2015; 3(2): MDNA3-0013-2014..
[http://dx.doi.org/10.1128/microbiolspec.MDNA3-0013-2014]

[197] Boothroyd CE, Dreesen O, Leonova T, *et al.* A yeast-endonuclease-generated DNA break induces antigenic switching in *Trypanosoma brucei*. Nature 2009; 459(7244): 278-81.
[http://dx.doi.org/10.1038/nature07982] [PMID: 19369939]

[198] Glover L, Alsford S, Horn D. DNA break site at fragile subtelomeres determines probability and mechanism of antigenic variation in African trypanosomes. PLoS Pathog 2013; 9(3): e1003260.
[http://dx.doi.org/10.1371/journal.ppat.1003260] [PMID: 23555264]

[199] Jehi SE, Wu F, Li B. *Trypanosoma brucei* TIF2 suppresses VSG switching by maintaining subtelomere integrity. Cell Res 2014; 24(7): 870-85.
[http://dx.doi.org/10.1038/cr.2014.60] [PMID: 24810301]

[200] Barry D, McCulloch R. Molecular microbiology: A key event in survival. Nature 2009; 459(7244): 172-3.
[http://dx.doi.org/10.1038/459172a] [PMID: 19444197]

[201] Barry JD. The relative significance of mechanisms of antigenic variation in African trypanosomes. Parasitol Today (Regul Ed) 1997; 13(6): 212-8.
[http://dx.doi.org/10.1016/S0169-4758(97)01039-9] [PMID: 15275073]

[202] Dreesen O, Li B, Cross GA. Telomere structure and function in trypanosomes: a proposal. Nat Rev Microbiol 2007; 5(1): 70-5.
[http://dx.doi.org/10.1038/nrmicro1577] [PMID: 17160000]

[203] Hovel-Miner GA, Boothroyd CE, Mugnier M, Dreesen O, Cross GA, Papavasiliou FN. Telomere length affects the frequency and mechanism of antigenic variation in *Trypanosoma brucei*. PLoS Pathog 2012; 8(8): e1002900.
[http://dx.doi.org/10.1371/journal.ppat.1002900] [PMID: 22952449]

[204] Lachaud L, Bourgeois N, Kuk N, *et al.* Constitutive mosaic aneuploidy is a unique genetic feature widespread in the *Leishmania* genus. Microbes and infection/Institut Pasteur 2014; 16(1): 61-6.
[http://dx.doi.org/10.1016/j.micinf.2013.09.005]

[205] Sterkers Y, Crobu L, Lachaud L, Pagès M, Bastien P. Parasexuality and mosaic aneuploidy in *Leishmania*: alternative genetics. Trends Parasitol 2014; 30(9): 429-35. [http://dx.doi.org/10.1016/j.pt.2014.07.002] [PMID: 25073852]

[206] Dujardin JC, Mannaert A, Durrant C, Cotton JA. Mosaic aneuploidy in Leishmania: the perspective of whole genome sequencing. Trends Parasitol 2014; 30(12): 554-5. [http://dx.doi.org/10.1016/j.pt.2014.09.004] [PMID: 25458156]

[207] Mannaert A, Downing T, Imamura H, Dujardin JC. Adaptive mechanisms in pathogens: universal aneuploidy in *Leishmania*. Trends Parasitol 2012; 28(9): 370-6. [http://dx.doi.org/10.1016/j.pt.2012.06.003] [PMID: 22789456]

[208] Beverley SM, Coderre JA, Santi DV, Schimke RT. Unstable DNA amplifications in methotrexate-resistant *Leishmania* consist of extrachromosomal circles which relocalize during stabilization. Cell 1984; 38(2): 431-9. [http://dx.doi.org/10.1016/0092-8674(84)90498-7] [PMID: 6467372]

[209] Grondin K, Kündig C, Roy G, Ouellette M. Linear amplicons as precursors of amplified circles in methotrexate-resistant *Leishmania tarentolae*. Nucleic Acids Res 1998; 26(14): 3372-8. [http://dx.doi.org/10.1093/nar/26.14.3372] [PMID: 9649621]

[210] Grondin K, Papadopoulou B, Ouellette M. Homologous recombination between direct repeat sequences yields P-glycoprotein containing amplicons in arsenite resistant *Leishmania*. Nucleic Acids Res 1993; 21(8): 1895-901. [http://dx.doi.org/10.1093/nar/21.8.1895] [PMID: 8098523]

[211] White TC, Fase-Fowler F, van Luenen H, Calafat J, Borst P. The H circles of *Leishmania* tarentolae are a unique amplifiable system of oligomeric DNAs associated with drug resistance. J Biol Chem 1988; 263(32): 16977-83. [PMID: 3182827]

[212] Haimeur A, Brochu C, Genest P, Papadopoulou B, Ouellette M. Amplification of the ABC transporter gene PGPA and increased trypanothione levels in potassium antimonyl tartrate (SbIII) resistant *Leishmania tarentolae*. Mol Biochem Parasitol 2000; 108(1): 131-5. [http://dx.doi.org/10.1016/S0166-6851(00)00187-0] [PMID: 10802326]

[213] Monte-Neto R, Laffitte MC, Leprohon P, Reis P, Frézard F, Ouellette M. Intrachromosomal amplification, locus deletion and point mutation in the aquaglyceroporin AQP1 gene in antimony resistant *Leishmania* (*Viannia*) *guyanensis*. PLoS Negl Trop Dis 2015; 9(2): e0003476. [http://dx.doi.org/10.1371/journal.pntd.0003476] [PMID: 25679388]

[214] Genois MM, Plourde M, Éthier C, *et al.* Roles of Rad51 paralogs for promoting homologous recombination in *Leishmania infantum*. Nucleic Acids Res 2015; 43(5): 2701-15. [http://dx.doi.org/10.1093/nar/gkv118] [PMID: 25712090]

[215] Laffitte MC, Genois MM, Mukherjee A, Légaré D, Masson JY, Ouellette M. Formation of linear amplicons with inverted duplications in *Leishmania* requires the MRE11 nuclease. PLoS Genet 2014; 10(12): e1004805. [http://dx.doi.org/10.1371/journal.pgen.1004805] [PMID: 25474106]

[216] Genois MM, Mukherjee A, Ubeda JM, *et al.* Interactions between BRCA2 and RAD51 for promoting

homologous recombination in *Leishmania infantum*. Nucleic Acids Res 2012; 40(14): 6570-84. [http://dx.doi.org/10.1093/nar/gks306] [PMID: 22505581]

[217] Damasceno JD, Nunes VS, Tosi LR. LmHus1 is required for the DNA damage response in *Leishmania major* and forms a complex with an unusual Rad9 homologue. Mol Microbiol 2013; 90(5): 1074-87. [http://dx.doi.org/10.1111/mmi.12418] [PMID: 24118609]

[218] Nunes VS, Damasceno JD, Freire R, Tosi LR. The Hus1 homologue of *Leishmania major* encodes a nuclear protein that participates in DNA damage response. Mol Biochem Parasitol 2011; 177(1): 65-9. [http://dx.doi.org/10.1016/j.molbiopara.2011.01.011] [PMID: 21291918]

[219] Garner KM, Pletnev AA, Eastman A. Corrected structure of mirin, a small-molecule inhibitor of the Mre11-Rad50-Nbs1 complex. Nat Chem Biol 2009; 5(3): 129-30. [http://dx.doi.org/10.1038/nchembio0309-129] [PMID: 19219009]

[220] Takeda GK, Campos R, Kieffer J, *et al.* Effect of gamma rays on blood forms of *Trypanosoma cruzi*. Experimental study in mice. Rev Inst Med Trop Sao Paulo 1986; 28(1): 15-8. [http://dx.doi.org/10.1590/S0036-46651986000100004] [PMID: 3094131]

[221] Regis-da-Silva CG, Freitas JM, Passos-Silva DG, *et al.* Characterization of the *Trypanosoma cruzi* Rad51 gene and its role in recombination events associated with the parasite resistance to ionizing radiation. Mol Biochem Parasitol 2006; 149(2): 191-200. [http://dx.doi.org/10.1016/j.molbiopara.2006.05.012] [PMID: 16828179]

[222] Grynberg P, Passos-Silva DG, Mourão MdeM, *et al. Trypanosoma cruzi* gene expression in response to gamma radiation. PLoS One 2012; 7(1): e29596. [http://dx.doi.org/10.1371/journal.pone.0029596] [PMID: 22247781]

[223] Vieira HG, Grynberg P, Bitar M, *et al.* Proteomic analysis of *Trypanosoma cruzi* response to ionizing radiation stress. PLoS One 2014; 9(5): e97526. [http://dx.doi.org/10.1371/journal.pone.0097526] [PMID: 24842666]

[224] Bentchikou E, Servant P, Coste G, Sommer S. A major role of the RecFOR pathway in DNA double-strand-break repair through ESDSA in *Deinococcus radiodurans*. PLoS Genet 2010; 6(1): e1000774. [http://dx.doi.org/10.1371/journal.pgen.1000774] [PMID: 20090937]

[225] Repar J, Cvjetan S, Slade D, Radman M, Zahradka D, Zahradka K. RecA protein assures fidelity of DNA repair and genome stability in *Deinococcus radiodurans*. DNA Repair (Amst) 2010; 9(11): 1151-61. [http://dx.doi.org/10.1016/j.dnarep.2010.08.003] [PMID: 20817622]

[226] Zahradka K, Slade D, Bailone A, *et al.* Reassembly of shattered chromosomes in *Deinococcus radiodurans*. Nature 2006; 443(7111): 569-73. [PMID: 17006450]

[227] Goodenough U, Heitman J. Origins of eukaryotic sexual reproduction. Cold Spring Harb Perspect Biol 2014; 6(3): a016154. [http://dx.doi.org/10.1101/cshperspect.a016154] [PMID: 24591519]

[228] Malik SB, Pightling AW, Stefaniak LM, Schurko AM, Logsdon JM Jr. An expanded inventory of conserved meiotic genes provides evidence for sex in *Trichomonas vaginalis*. PLoS One 2007; 3(8): e2879.

[http://dx.doi.org/10.1371/journal.pone.0002879] [PMID: 18663385]

[229] Schurko AM, Logsdon JM Jr. Using a meiosis detection toolkit to investigate ancient asexual scandals and the evolution of sex. BioEssays 2008; 30(6): 579-89. [http://dx.doi.org/10.1002/bies.20764] [PMID: 18478537]

[230] Heitman J. Evolution of eukaryotic microbial pathogens *via* covert sexual reproduction. Cell Host Microbe 2010; 8(1): 86-99. [http://dx.doi.org/10.1016/j.chom.2010.06.011] [PMID: 20638645]

[231] Poxleitner MK, Carpenter ML, Mancuso JJ, Wang CJ, Dawson SC, Cande WZ. Evidence for karyogamy and exchange of genetic material in the binucleate intestinal parasite *Giardia intestinalis*. Science 2008; 319(5869): 1530-3. [http://dx.doi.org/10.1126/science.1153752] [PMID: 18339940]

[232] Keeney S, Lange J, Mohibullah N. Self-organization of meiotic recombination initiation: general principles and molecular pathways. Annu Rev Genet 2014; 48: 187-214. [http://dx.doi.org/10.1146/annurev-genet-120213-092304] [PMID: 25421598]

[233] Borde V, de Massy B. Programmed induction of DNA double strand breaks during meiosis: setting up communication between DNA and the chromosome structure. Curr Opin Genet Dev 2013; 23(2): 147-55. [http://dx.doi.org/10.1016/j.gde.2012.12.002] [PMID: 23313097]

[234] Brown MS, Bishop DK. DNA strand exchange and RecA homologs in meiosis. Cold Spring Harb Perspect Biol 2014; 7(1): a016659. [http://dx.doi.org/10.1101/cshperspect.a016659] [PMID: 25475089]

[235] Sansam CL, Pezza RJ. Connecting by breaking and repairing: mechanisms of DNA strand exchange in meiotic recombination. FEBS J 2015; 282(13): 2444-57. [http://dx.doi.org/10.1111/febs.13317] [PMID: 25953379]

[236] Lynn A, Soucek R, Borner GV. ZMM proteins during meiosis: crossover artists at work. Chromosome Res 2007; 15(5): 591-605. [http://dx.doi.org/10.1007/s10577-007-1150-1]

[237] Schwartz EK, Heyer WD. Processing of joint molecule intermediates by structure-selective endonucleases during homologous recombination in eukaryotes. Chromosoma 2011; 120(2): 109-27. [http://dx.doi.org/10.1007/s00412-010-0304-7] [PMID: 21369956]

[238] Tomasini N, Lauthier JJ, Ayala FJ, Tibayrenc M, Diosque P. How often do they have sex? A comparative analysis of the population structure of seven eukaryotic microbial pathogens. PLoS One 2014; 9(7): e103131. [http://dx.doi.org/10.1371/journal.pone.0103131] [PMID: 25054834]

[239] Gibson W. Liaisons dangereuses: sexual recombination among pathogenic trypanosomes. Res Microbiol 2015; 166(6): 459-66. [http://dx.doi.org/10.1016/j.resmic.2015.05.005] [PMID: 26027775]

[240] Jenni L, Marti S, Schweizer J, *et al.* Hybrid formation between African trypanosomes during cyclical transmission. Nature 1986; 322(6075): 173-5. [http://dx.doi.org/10.1038/322173a0] [PMID: 3724860]

[241] Sternberg J, Tait A, Haley S, *et al.* Gene exchange in African trypanosomes: characterisation of a new hybrid genotype. Mol Biochem Parasitol 1988; 27(2-3): 191-200.
[http://dx.doi.org/10.1016/0166-6851(88)90038-2] [PMID: 2893975]

[242] Gibson W, Peacock L, Ferris V, Williams K, Bailey M. The use of yellow fluorescent hybrids to indicate mating in *Trypanosoma brucei*. Parasit Vectors 2008; 1(1): 4.
[http://dx.doi.org/10.1186/1756-3305-1-4] [PMID: 18298832]

[243] Gibson W, Peacock L, Ferris V, *et al.* Genetic recombination between human and animal parasites creates novel strains of human pathogen. PLoS Negl Trop Dis 2015; 9(3): e0003665.
[http://dx.doi.org/10.1371/journal.pntd.0003665] [PMID: 25816228]

[244] Gaunt MW, Yeo M, Frame IA, *et al.* Mechanism of genetic exchange in American trypanosomes. Nature 2003; 421(6926): 936-9.
[http://dx.doi.org/10.1038/nature01438] [PMID: 12606999]

[245] Akopyants NS, Kimblin N, Secundino N, *et al.* Demonstration of genetic exchange during cyclical development of *Leishmania* in the sand fly vector. Science 2009; 324(5924): 265-8.
[http://dx.doi.org/10.1126/science.1169464] [PMID: 19359589]

[246] Inbar E, Akopyants NS, Charmoy M, *et al.* The mating competence of geographically diverse *Leishmania major* strains in their natural and unnatural sand fly vectors. PLoS Genet 2013; 9(7): e1003672.
[http://dx.doi.org/10.1371/journal.pgen.1003672] [PMID: 23935521]

[247] Romano A, Inbar E, Debrabant A, *et al.* Cross-species genetic exchange between visceral and cutaneous strains of *Leishmania* in the sand fly vector. Proc Natl Acad Sci USA 2014; 111(47): 16808-13.
[http://dx.doi.org/10.1073/pnas.1415109111] [PMID: 25385616]

[248] Ramesh MA, Malik SB, Logsdon JM Jr. A phylogenomic inventory of meiotic genes; evidence for sex in Giardia and an early eukaryotic origin of meiosis. Curr Biol 2005; 15(2): 185-91.
[PMID: 15668177]

[249] Malik SB, Ramesh MA, Hulstrand AM, Logsdon JM Jr. Protist homologs of the meiotic Spo11 gene and topoisomerase VI reveal an evolutionary history of gene duplication and lineage-specific loss. Mol Biol Evol 2007; 24(12): 2827-41.
[http://dx.doi.org/10.1093/molbev/msm217] [PMID: 17921483]

[250] Proudfoot C, McCulloch R. Trypanosoma brucei DMC1 does not act in DNA recombination, repair or antigenic variation in bloodstream stage cells. Mol Biochem Parasitol 2006; 145(2): 245-53.
[http://dx.doi.org/10.1016/j.molbiopara.2005.10.007] [PMID: 16289356]

[251] Peacock L, Ferris V, Sharma R, *et al.* Identification of the meiotic life cycle stage of *Trypanosoma brucei* in the tsetse fly. Proc Natl Acad Sci USA 2011; 108(9): 3671-6.
[http://dx.doi.org/10.1073/pnas.1019423108] [PMID: 21321215]

[252] Peacock L, Bailey M, Carrington M, Gibson W. Meiosis and haploid gametes in the pathogen *Trypanosoma brucei*. Curr Biol 2014; 24(2): 181-6.
[http://dx.doi.org/10.1016/j.cub.2013.11.044] [PMID: 24388851]

[253] Shuman S, Glickman MS. Bacterial DNA repair by non-homologous end joining. Nat Rev Microbiol

2007; 5(11): 852-61.
[http://dx.doi.org/10.1038/nrmicro1768] [PMID: 17938628]

[254] Brissett NC, Doherty AJ. Repairing DNA double-strand breaks by the prokaryotic non-homologous end-joining pathway. Biochem Soc Trans 2009; 37(Pt 3): 539-45.
[http://dx.doi.org/10.1042/BST0370539] [PMID: 19442248]

[255] Bartlett EJ, Brissett NC, Plocinski P, Carlberg T, Doherty AJ. Molecular basis for DNA strand displacement by NHEJ repair polymerases. Nucleic Acids Res 2016; (5)44: 2173-86.
[PMID: 26405198]

[256] Lieber MR. The mechanism of double-strand DNA break repair by the nonhomologous DNA end-joining pathway. Annu Rev Biochem 2010; 79: 181-211.
[http://dx.doi.org/10.1146/annurev.biochem.052308.093131] [PMID: 20192759]

[257] Bétermier M, Bertrand P, Lopez BS. Is non-homologous end-joining really an inherently error-prone process? PLoS Genet 2014; 10(1): e1004086.
[http://dx.doi.org/10.1371/journal.pgen.1004086] [PMID: 24453986]

[258] Bunting SF, Nussenzweig A. End-joining, translocations and cancer. Nat Rev Cancer 2013; 13(7): 443-54.
[http://dx.doi.org/10.1038/nrc3537] [PMID: 23760025]

[259] Symington LS. End resection at double-strand breaks: mechanism and regulation. Cold Spring Harb Perspect Biol 2014; 6(8): a016436.
[http://dx.doi.org/10.1101/cshperspect.a016436] [PMID: 25085909]

[260] Symington LS, Gautier J. Double-strand break end resection and repair pathway choice. Annu Rev Genet 2011; 45: 247-71.
[http://dx.doi.org/10.1146/annurev-genet-110410-132435] [PMID: 21910633]

[261] Aravind L, Koonin EV. Prokaryotic homologs of the eukaryotic DNA-end-binding protein Ku, novel domains in the Ku protein and prediction of a prokaryotic double-strand break repair system. Genome Res 2001; 11(8): 1365-74.
[http://dx.doi.org/10.1101/gr.181001] [PMID: 11483577]

[262] Conway C, McCulloch R, Ginger ML, Robinson NP, Browitt A, Barry JD. Ku is important for telomere maintenance, but not for differential expression of telomeric VSG genes, in African trypanosomes. J Biol Chem 2002; 277(24): 21269-77.
[http://dx.doi.org/10.1074/jbc.M200550200] [PMID: 11919193]

[263] Janzen CJ, Lander F, Dreesen O, Cross GA. Telomere length regulation and transcriptional silencing in KU80-deficient *Trypanosoma brucei*. Nucleic Acids Res 2004; 32(22): 6575-84.
[http://dx.doi.org/10.1093/nar/gkh991] [PMID: 15602000]

[264] Maity AK, Goswami A, Saha P. Identification of substrates of an S-phase cell cycle kinase from *Leishmania donovani*. FEBS Lett 2011; 585(17): 2635-9.
[http://dx.doi.org/10.1016/j.febslet.2011.06.017] [PMID: 21708149]

[265] Burton P, McBride DJ, Wilkes JM, Barry JD, McCulloch R. Ku heterodimer-independent end joining in *Trypanosoma brucei* cell extracts relies upon sequence microhomology. Eukaryot Cell 2007; 6(10): 1773-81.

[http://dx.doi.org/10.1128/EC.00212-07] [PMID: 17693593]

[266] Downey N, Hines JC, Sinha KM, Ray DS. Mitochondrial DNA ligases of *Trypanosoma brucei*. Eukaryot Cell 2005; 4(4): 765-74.
[http://dx.doi.org/10.1128/EC.4.4.765-774.2005] [PMID: 15821136]

[267] Conway C, Proudfoot C, Burton P, Barry JD, McCulloch R. Two pathways of homologous recombination in *Trypanosoma brucei*. Mol Microbiol 2002; 45(6): 1687-700.
[http://dx.doi.org/10.1046/j.1365-2958.2002.03122.x] [PMID: 12354234]

[268] Glover L, McCulloch R, Horn D. Sequence homology and microhomology dominate chromosomal double-strand break repair in African trypanosomes. Nucleic Acids Res 2008; 36(8): 2608-18.
[http://dx.doi.org/10.1093/nar/gkn104] [PMID: 18334531]

[269] Glover L, Jun J, Horn D. Microhomology-mediated deletion and gene conversion in African trypanosomes. Nucleic Acids Res 2011; 39(4): 1372-80.
[http://dx.doi.org/10.1093/nar/gkq981] [PMID: 20965968]

[270] Zhang WW, Matlashewski G. CRISPR-cas9-mediated genome editing in *Leishmania donovani*. MBio 2015; 6(4): e00861.
[http://dx.doi.org/10.1128/mBio.00861-15] [PMID: 26199327]

[271] Sollelis L, Ghorbal M, MacPherson CR, *et al.* First efficient CRISPR-Cas9-mediated genome editing in *Leishmania* parasites. Cell Microbiol 2015; 17(10): 1405-12.
[http://dx.doi.org/10.1111/cmi.12456] [PMID: 25939677]

[272] Lander N, Li ZH, Niyogi S, Docampo R. CRISPR/cas9-induced disruption of paraflagellar rod protein 1 and 2 genes in *Trypanosoma cruzi* reveals their role in flagellar attachment. MBio 2015; 6(4): e01012.
[http://dx.doi.org/10.1128/mBio.01012-15] [PMID: 26199333]

[273] Peng D, Kurup SP, Yao PY, Minning TA, Tarleton RL. CRISPR-Cas9-mediated single-gene and gene family disruption in *Trypanosoma cruzi*. MBio 2014; 6(1): e02097-14.
[http://dx.doi.org/10.1128/mBio.02097-14] [PMID: 25550322]

[274] Gill EE, Fast NM. Stripped-down DNA repair in a highly reduced parasite. BMC Mol Biol 2007; 8(1): 24.
[http://dx.doi.org/10.1186/1471-2199-8-24] [PMID: 17374165]

[275] Manning G, Reiner DS, Lauwaet T, *et al.* The minimal kinome of Giardia lamblia illuminates early kinase evolution and unique parasite biology. Genome Biol 2011; 12(7): R66.
[http://dx.doi.org/10.1186/gb-2011-12-7-r66] [PMID: 21787419]

[276] Lee AH, Symington LS, Fidock DA. DNA repair mechanisms and their biological roles in the malaria parasite *Plasmodium falciparum*. Microbiol Mol Biol Rev 2014; 78(3): 469-86.
[http://dx.doi.org/10.1128/MMBR.00059-13] [PMID: 25184562]

[277] Kirkman LA, Deitsch KW. Recombination and diversification of the variant antigen encoding genes in the malaria parasite *Plasmodium falciparum*. Microbiol Spectr 2014; 2(6)
[http://dx.doi.org/10.1128/microbiolspec.MDNA3-0022-2014] [PMID: 26104446]

[278] McVey M, Lee SE. MMEJ repair of double-strand breaks (directors cut): deleted sequences and alternative endings. Trends Genet 2008; 24(11): 529-38.

[http://dx.doi.org/10.1016/j.tig.2008.08.007] [PMID: 18809224]

[279] Sfeir A, Symington LS. Microhomology-mediated end joining: a back-up survival mechanism or dedicated pathway? Trends Biochem Sci 2015; 40(11): 701-14. [http://dx.doi.org/10.1016/j.tibs.2015.08.006] [PMID: 26439531]

[280] Ottaviani D, LeCain M, Sheer D. The role of microhomology in genomic structural variation. Trends Genet 2014; 30(3): 85-94. [http://dx.doi.org/10.1016/j.tig.2014.01.001] [PMID: 24503142]

[281] van Schendel R, Tijsterman M. Microhomology-mediated intron loss during metazoan evolution. Genome Biol Evol 2013; 5(6): 1212-9. [http://dx.doi.org/10.1093/gbe/evt088] [PMID: 23737326]

[282] Anand RP, Tsaponina O, Greenwell PW, *et al.* Chromosome rearrangements *via* template switching between diverged repeated sequences. Genes Dev 2014; 28(21): 2394-406. [http://dx.doi.org/10.1101/gad.250258.114] [PMID: 25367035]

[283] Cortizas EM, Zahn A, Hajjar ME, Patenaude AM, Di Noia JM, Verdun RE. Alternative end-joining and classical nonhomologous end-joining pathways repair different types of double-strand breaks during class-switch recombination. J Immunol 2013; 191(11): 5751-63. [http://dx.doi.org/10.4049/jimmunol.1301300] [PMID: 24146042]

[284] Yan CT, Boboila C, Souza EK, *et al.* IgH class switching and translocations use a robust non-classical end-joining pathway. Nature 2007; 449(7161): 478-82. [http://dx.doi.org/10.1038/nature06020] [PMID: 17713479]

[285] Bopp SE, Manary MJ, Bright AT, *et al.* Mitotic evolution of *Plasmodium falciparum* shows a stable core genome but recombination in antigen families. PLoS Genet 2013; 9(2): e1003293. [http://dx.doi.org/10.1371/journal.pgen.1003293] [PMID: 23408914]

[286] Claessens A, Hamilton WL, Kekre M, *et al.* Generation of antigenic diversity in *Plasmodium falciparum* by structured rearrangement of Var genes during mitosis. PLoS Genet 2014; 10(12): e1004812. [http://dx.doi.org/10.1371/journal.pgen.1004812] [PMID: 25521112]

[287] Heinberg A, Siu E, Stern C, *et al.* Direct evidence for the adaptive role of copy number variation on antifolate susceptibility in *Plasmodium falciparum*. Mol Microbiol 2013; 88(4): 702-12. [http://dx.doi.org/10.1111/mmi.12162] [PMID: 23347134]

[288] Kirkman LA, Lawrence EA, Deitsch KW. Malaria parasites utilize both homologous recombination and alternative end joining pathways to maintain genome integrity. Nucleic Acids Res 2014; 42(1): 370-9. [http://dx.doi.org/10.1093/nar/gkt881] [PMID: 24089143]

[289] Barry JD, Ginger ML, Burton P, McCulloch R. Why are parasite contingency genes often associated with telomeres? Int J Parasitol 2003; 33(1): 29-45. [http://dx.doi.org/10.1016/S0020-7519(02)00247-3] [PMID: 12547344]

[290] Goodman MF, Woodgate R. Translesion DNA polymerases. Cold Spring Harb Perspect Biol 2013; 5(10): a010363. [http://dx.doi.org/10.1101/cshperspect.a010363] [PMID: 23838442]

[291] Sale JE. Translesion DNA synthesis and mutagenesis in eukaryotes. Cold Spring Harb Perspect Biol 2013; 5(3): a012708.
[http://dx.doi.org/10.1101/cshperspect.a012708] [PMID: 23457261]

[292] de Moura MB, Schamber-Reis BL, Passos Silva DG, *et al.* Cloning and characterization of DNA polymerase eta from *Trypanosoma cruzi*: roles for translesion bypass of oxidative damage. Environ Mol Mutagen 2009; 50(5): 375-86.
[http://dx.doi.org/10.1002/em.20450] [PMID: 19229999]

[293] McIlwraith MJ, Vaisman A, Liu Y, Fanning E, Woodgate R, West SC. Human DNA polymerase eta promotes DNA synthesis from strand invasion intermediates of homologous recombination. Mol Cell 2005; 20(5): 783-92.
[http://dx.doi.org/10.1016/j.molcel.2005.10.001] [PMID: 16337601]

[294] Kawamoto T, Araki K, Sonoda E, *et al.* Dual roles for DNA polymerase eta in homologous DNA recombination and translesion DNA synthesis. Mol Cell 2005; 20(5): 793-9.
[http://dx.doi.org/10.1016/j.molcel.2005.10.016] [PMID: 16337602]

[295] Martomo SA, Gearhart PJ. Somatic hypermutation: subverted DNA repair. Curr Opin Immunol 2006; 18(3): 243-8.
[http://dx.doi.org/10.1016/j.coi.2006.03.007] [PMID: 16616477]

[296] Rajão MA, Passos-Silva DG, DaRocha WD, *et al.* DNA polymerase kappa from *Trypanosoma cruzi* localizes to the mitochondria, bypasses 8-oxoguanine lesions and performs DNA synthesis in a recombination intermediate. Mol Microbiol 2009; 71(1): 185-97.
[http://dx.doi.org/10.1111/j.1365-2958.2008.06521.x] [PMID: 19007414]

[297] Iyer LM, Koonin EV, Leipe DD, Aravind L. Origin and evolution of the archaeo-eukaryotic primase superfamily and related palm-domain proteins: structural insights and new members. Nucleic Acids Res 2005; 33(12): 3875-96.
[http://dx.doi.org/10.1093/nar/gki702] [PMID: 16027112]

[298] Guilliam TA, Keen BA, Brissett NC, Doherty AJ. Primase-polymerases are a functionally diverse superfamily of replication and repair enzymes. Nucleic Acids Res 2015; 43(14): 6651-64.
[http://dx.doi.org/10.1093/nar/gkv625] [PMID: 26109351]

[299] Rudd SG, Glover L, Jozwiakowski SK, Horn D, Doherty AJ. PPL2 translesion polymerase is essential for the completion of chromosomal DNA replication in the African trypanosome. Mol Cell 2013; 52(4): 554-65.
[http://dx.doi.org/10.1016/j.molcel.2013.10.034] [PMID: 24267450]

[300] Tomasso A, Casari G, Maga G. What makes y family pols potential candidates for molecular targeted therapies and novel biotechnological applications. Curr Mol Med 2014; 14(1): 96-114.
[http://dx.doi.org/10.2174/15665240113136660080] [PMID: 24160487]

[301] Holt IJ. Mitochondrial DNA replication and repair: all a flap. Trends Biochem Sci 2009; 34(7): 358-65.
[http://dx.doi.org/10.1016/j.tibs.2009.03.007] [PMID: 19559620]

[302] Alexeyev M, Shokolenko I, Wilson G, LeDoux S. The maintenance of mitochondrial DNA integrity critical analysis and update. Cold Spring Harb Perspect Biol 2013; 5(5): a012641.

[http://dx.doi.org/10.1101/cshperspect.a012641] [PMID: 23637283]

[303] Scheibye-Knudsen M, Fang EF, Croteau DL, Wilson DM III, Bohr VA. Protecting the mitochondrial powerhouse. Trends Cell Biol 2015; 25(3): 158-70. [http://dx.doi.org/10.1016/j.tcb.2014.11.002] [PMID: 25499735]

[304] Schamber-Reis BL, Nardelli S, Régis-Silva CG, *et al.* DNA polymerase beta from *Trypanosoma cruzi* is involved in kinetoplast DNA replication and repair of oxidative lesions. Mol Biochem Parasitol 2012; 183(2): 122-31. [http://dx.doi.org/10.1016/j.molbiopara.2012.02.007] [PMID: 22369885]

[305] Venegas JA, Aslund L, Solari A. Cloning and characterization of a DNA polymerase beta gene from *Trypanosoma cruzi*. Parasitol Int 2009; 58(2): 187-92. [http://dx.doi.org/10.1016/j.parint.2009.01.007] [PMID: 19567232]

[306] Klingbeil MM, Motyka SA, Englund PT. Multiple mitochondrial DNA polymerases in *Trypanosoma brucei*. Mol Cell 2002; 10(1): 175-86. [http://dx.doi.org/10.1016/S1097-2765(02)00571-3] [PMID: 12150917]

[307] Kaniak A, Dzierzbicki P, Rogowska AT, Malc E, Fikus M, Ciesla Z. Msh1p counteracts oxidative lesion-induced instability of mtDNA and stimulates mitochondrial recombination in *Saccharomyces cerevisiae*. DNA Repair (Amst) 2009; 8(3): 318-29. [http://dx.doi.org/10.1016/j.dnarep.2008.11.004] [PMID: 19056520]

[308] Shedge V, Arrieta-Montiel M, Christensen AC, Mackenzie SA. Plant mitochondrial recombination surveillance requires unusual RecA and MutS homologs. Plant Cell 2007; 19(4): 1251-64. [http://dx.doi.org/10.1105/tpc.106.048355] [PMID: 17468263]

[309] Garrison EM, Arrizabalaga G. Disruption of a mitochondrial MutS DNA repair enzyme homologue confers drug resistance in the parasite *Toxoplasma gondii*. Mol Microbiol 2009; 72(2): 425-41. [http://dx.doi.org/10.1111/j.1365-2958.2009.06655.x] [PMID: 19291232]

[310] Xu YZ, Arrieta-Montiel MP, Virdi KS, *et al.* MutS HOMOLOG1 is a nucleoid protein that alters mitochondrial and plastid properties and plant response to high light. Plant Cell 2011; 23(9): 3428-41. [http://dx.doi.org/10.1105/tpc.111.089136] [PMID: 21934144]

[311] Chang YW, Mai RT, Fang WH, Lin CC, Chiu CC, Wu Lee YH. YB-1 disrupts mismatch repair complex formation, interferes with MutSα recruitment on mismatch and inhibits mismatch repair through interacting with PCNA. Oncogene 2014; 33(43): 5065-77. [http://dx.doi.org/10.1038/onc.2013.450] [PMID: 24141788]

[312] de Souza-Pinto NC, Mason PA, Hashıguchı K, *et al.* Novel DNA mismatch-repair activity involving YB-1 in human mitochondria. DNA Repair (Amst) 2009; 8(6): 704-19. [http://dx.doi.org/10.1016/j.dnarep.2009.01.021] [PMID: 19272840]

[313] Jeon H, Jin YM, Choi MH, Lee H, Kim M. Chloroplast-targeted bacterial RecA proteins confer tolerance to chloroplast DNA damage by methyl viologen or UV-C radiation in tobacco (Nicotiana tabacum) plants. Physiol Plant 2013; 147(2): 218-33. [http://dx.doi.org/10.1111/j.1399-3054.2012.01658.x] [PMID: 22651245]

[314] Miller-Messmer M, Kühn K, Bichara M, Le Ret M, Imbault P, Gualberto JM. RecA-dependent DNA repair results in increased heteroplasmy of the *Arabidopsis* mitochondrial genome. Plant Physiol 2012;

159(1): 211-26.
[http://dx.doi.org/10.1104/pp.112.194720] [PMID: 22415515]

[315] Zampini É, Lepage É, Tremblay-Belzile S, Truche S, Brisson N. Organelle DNA rearrangement mapping reveals U-turn-like inversions as a major source of genomic instability in *Arabidopsis* and humans. Genome Res 2015; 25(5): 645-54.
[http://dx.doi.org/10.1101/gr.188573.114] [PMID: 25800675]

[316] Chen XJ. Mechanism of homologous recombination and implications for aging-related deletions in mitochondrial DNA. Microbiol Mol Biol Rev 2013; 77(3): 476-96.
[http://dx.doi.org/10.1128/MMBR.00007-13] [PMID: 24006472]

[317] Samach A, Melamed-Bessudo C, Avivi-Ragolski N, Pietrokovski S, Levy AA. Identification of plant RAD52 homologs and characterization of the *Arabidopsis thaliana* RAD52-like genes. Plant Cell 2011; 23(12): 4266-79.
[http://dx.doi.org/10.1105/tpc.111.091744] [PMID: 22202891]

[318] Sage JM, Gildemeister OS, Knight KL. Discovery of a novel function for human Rad51: maintenance of the mitochondrial genome. J Biol Chem 2010; 285(25): 18984-90.
[http://dx.doi.org/10.1074/jbc.M109.099846] [PMID: 20413593]

[319] Sage JM, Knight KL. Human Rad51 promotes mitochondrial DNA synthesis under conditions of increased replication stress. Mitochondrion 2013; 13(4): 350-6.
[http://dx.doi.org/10.1016/j.mito.2013.04.004] [PMID: 23591384]

[320] Ding L, Liu Y. Borrowing nuclear DNA helicases to protect mitochondrial DNA. Int J Mol Sci 2015; 16(5): 10870-87.
[http://dx.doi.org/10.3390/ijms160510870] [PMID: 25984607]

[321] Gargantini PR, Lujan HD, Pereira CA. *In silico* analysis of trypanosomatids helicases. FEMS Microbiol Lett 2012; 335(2): 123-9.
[http://dx.doi.org/10.1111/j.1574-6968.2012.02644.x] [PMID: 22835260]

[322] Liu B, Wang J, Yaffe N, *et al.* Trypanosomes have six mitochondrial DNA helicases with one controlling kinetoplast maxicircle replication. Mol Cell 2009; 35(4): 490-501.
[http://dx.doi.org/10.1016/j.molcel.2009.07.004] [PMID: 19646907]

[323] Liu B, Yildirir G, Wang J, Tolun G, Griffith JD, Englund PT. TbPIF1, a *Trypanosoma brucei* mitochondrial DNA helicase, is essential for kinetoplast minicircle replication. J Biol Chem 2010; 285(10): 7056-66.
[http://dx.doi.org/10.1074/jbc.M109.084038] [PMID: 20042610]

CHAPTER 6

Mechanisms Controlling Gene Expression in Trypanosomatids

Santuza M. R. Teixeira* and **Bruna M. Valente**

Departamento de Bioquímica e Imunologia - ICB, Universidade Federal de Minas Gerais, Belo Horizonte, Minas Gerais, Brazil

Abstract: As members of a highly divergent group of eukaryotes, trypanosomatids present peculiar mechanisms of gene expression. These protozoan parasites have transcription and processing machineries that constitutively transcribe clusters of non-related genes into polycistronic pre-mRNAs, which are subsequently *trans*-spliced into monocistronic transcripts. Because of this, control of gene expression relies mainly on post-transcriptional mechanisms that are, for the most part, mediated by RNA binding proteins that control steady-state levels of mRNAs and/or their translation rates. Using primarily *Trypanosoma brucei* as a model, several groups have begun to elucidate the basic regulatory mechanisms and to define the cellular factors controlling transcription, processing, degradation and translation of mRNAs in trypanosomatids. This chapter describes studies that have been focused on a subset of genes that are differentially expressed during the life cycle of *T. brucei, T. cruzi* and few species of *Leishmania.* Although a predominance of regulatory pathways acting at a post-transcriptional level is found for most genes from all three parasites, it is also evident that the regulatory strategies chosen by different trypanosomatid species are not similar. Because of their complex and diversified gene regulatory machinery, *T. brucei, T. cruzi* and *Leishmania* spp. are able to respond rapidly to the drastic environmental changes they face during their life cycle, particularly when they move between their different hosts.

Keywords: Epigenetics, Gene expression, Polyadenylation, Post-transcription, RNA polymerase, RNA promoter, *Trans*-splicing, Transcription, *U*ntranslated region, VSG.

* **Corresponding author Santuza M. R. Teixeira:** Departamento de Bioquímica e Imunologia – ICB, Universidade Federal de Minas Gerais, Belo Horizonte, Minas Gerais, Brazil; Tel: 55 31 3409 2665; E-mail: santuzat@ufmg.br

Marcelo Santos da Silva & Maria Isabel N. Cano (Eds.)

INTRODUCTION

Trypanosomatids have Unusual Mechanisms of Gene Expression

Within the *Trypanosomatidae* family, the genera *Trypanosoma* and *Leishmania* consist of several species of uniflagellated protozoa parasites that are the causative agents of several tropical diseases, such as sleeping sickness (African trypanosomiasis), caused by *Trypanosoma brucei gambiensis* and *Trypanosoma brucei rhodesiense*; Chagas disease (American trypanosomiasis), caused by *Trypanosoma cruzi*; and different forms of leishmaniasis, caused by various species of *Leishmania*. According to recent World Health Organization surveys, it is estimated that approximately 20 million people are infected with either one of these parasites and 550 million people in the developing world are at risk of contracting leishmaniasis, Chagas disease or sleeping sickness (The World Health Report, 2014, http://www.who.ch). These digenetic parasites have distinct life cycles that present multiple differentiation forms alternating between various types of invertebrate and vertebrate hosts. Details about *Trypanosoma* and *Leishmania* life cycles and their hosts are described in Chapter 1. Therefore, the parasites must rely on multiple regulatory mechanisms to rapidly respond to the drastic environmental changes they face every time they move between their different hosts. *T. brucei, T. cruzi* and a few species of *Leishmania* are the model systems used in the studies herein described. These species belong to an ancient group of unicellular eukaryotes, which, although we tend to consider as closely related organisms, are in fact highly divergent evolutionarily [1]. Because there are still no effective drugs to treat or vaccines to prevent diseases caused by the trypanosomatids, there continues to be an urgent need for basic research that can reveal new molecular targets and parasite-specific pathways with the potential of creating new prophylactic tools and more efficient drug therapies. Details about vaccine candidates against all Tritryps and new therapeutic tools are available, respectively, on Chapters 10 and 9.

In addition to their medical relevance, trypanosomatids have been used as models for gene expression studies because of many eccentricities found in their biology. Unlike most eukaryotic genes, which are transcribed into pre-mRNAs containing exons (coding sequences) and introns (mostly non-coding sequences), trypanosome protein-coding genes are intronless and are transcribed into polycistronic pre-mRNAs that are processed into mature mRNAs through "*trans*-splicing" reactions (reviewed in [2]). RNA polymerase II (RNA pol II) is the enzyme responsible for the transcription of protein-coding genes in eukaryotes. In *T. brucei*, however, two groups of protein-coding genes are transcribed by RNA polymerase I (RNA pol I), which, in most eukaryotes, exclusively transcribes ribosomal RNA genes. The co-transcriptional capping of RNAs that occurs in most eukaryotes is mediated by interactions between RNA pol II and capping enzymes. Because in trypanosomatids the addition of an m7G (cap) to the 5′ end of the primary transcript occurs during *trans*-splicing, RNA pol II-dependent transcription of protein-coding genes can be bypassed in trypanosomes. RNA pol I-dependent transcription occurs not only for a few endogenous trypanosome protein-coding genes but also for most foreign genes that can be expressed in the three groups of parasites using vectors containing RNA polymerase I promoters [3]. Indeed, except for SL genes (see below), no RNA pol II promoters have been identified in trypanosomatids. In addition to these differences regarding transcription of nuclear genes, other bizarre aspects of trypanosome biology include the unusual structure of the mitochondrial genome, called kinetoplast DNA (kDNA), and the extensive post-transcriptional modification of mitochondrial RNAs known as RNA editing, which is required to produce functional mitochondrial mRNAs [4].

Direct evidence for polycistronic transcription at a genomic level was derived from the first complete sequence analysis of chromosome 1 from *L. major*, which contains 79 genes, 29 of them encoded on one DNA strand and the remaining 50 on the opposite strand [5]. Nuclear run-on assays with strand-specific probes showed that, although a low level of nonspecific transcription likely takes place over the entire chromosome, RNA pol II-mediated transcription initiates within the strand-switch region (between the two divergent polycistronic transcription units, or PTUs) and proceeds bidirectionally towards the telomeres [6] (Fig. **1**). Sequencing of the complete *T. brucei*, *T. cruzi* and *L. major* genomes, known as the TriTryp genomes [7 - 9], revealed similar architectures in all three parasite genomes. More recent studies based on genome-wide RNA-seq analyses of the *T. brucei* genome confirmed that transcription initiation is not restricted to regions at the beginning of gene clusters but also occurs at internal sites, albeit at lower levels [10]. However, even though a number of sequences corresponding to strand-switch regions (SSR) have been thoroughly investigated, no typical RNA pol II promoters were identified. Although the mechanism by which trypanosomes initiate transcription remains unclear, chromatin immunoprecipitation analyses indicate that transcription initiation is associated with changes in the acetylation state of histones located at the SSRs [6]. In line with these findings, genome analyses showed that, although homologs of most RNA polymerase II subunits are present in trypanosomatids, trypanosome genes encoding general transcription factors present very divergent sequences (for a review, see [11]).

In trypanosomatids, RNAP II termination must occur at a defined location between two convergent PTUs, called convergent strand-switch regions (cSSRs), to prevent the synthesis of antisense RNAs. Transcription also terminates and initiates at head-tail (HT) sites, where transcription of an upstream PTU terminates and transcription of a downstream PTU on the same strand initiates [6]. Recently, it was described that β-d-glucosyl-hydroxymethyluracil, or base J, an epigenetic modification of thymine found exclusively in Kinetoplastid [12, 13], plays a role in transcription. In *L. major*, reduction of base J resulted in genome-wide transcriptional readthrough at cSSRs and HT sites, as shown by strand-specific RT-PCR. Surprisingly, genome-wide base J regulation of RNA pol II transcription termination at cSSRs is restricted to *Leishmania* spp [14]. In *T. brucei*, base J regulates termination and gene expression at specific sites within polycistronic gene clusters, whereas, in *T. cruzi*, reduction of base J results in the formation of more active chromatin at transcription initiation sites, increased recruitment of RNA pol II and global changes in gene expression [13].

Following synthesis of polycistronic transcripts of protein-coding genes by trypanosome RNA polymerase II, processing reactions are required to generate mature mRNAs, which are monocistronic, capped and polyadenylated. Processing of mRNA through *trans*-splicing is a process that has been observed only in trypanosomatids, *Euglena* and in nematode and trematode worms [15]. Because of these differences in gene expression between trypanosomatids and higher eukaryotes, we shall first observe how the general processes of gene transcription and mRNA maturation take place before studying how gene regulation occurs. Similarly to most eukaryotes, for efficient translation by ribosomes, the addition of a methylated G nucleotide or cap at the 5'end and the poly(A) tail at the 3'end of each mRNA is essential. However, in contrast to the general process of eukaryote transcription, in which the enzymes required for the *capping* process are associated with the C-terminal domain of the large subunit of RNA polymerase II, in trypanosomatids, transcription and mRNA processing are not interdependent events. Moreover, no consensus signal sequence for poly-adenylation has been identified in trypanosomatids, and the cap has a unique, highly methylated structure [2]. In fact, the 5'cap is part of the 39 nucleotide miniexon (or spliced leader, SL) that is joined to the 5'end of every mRNA in a transesterification reaction called *trans*-splicing. Coupling between *trans*-splicing and polyadenylation was first described by LeBowitz and collaborators [16], who demonstrated that poly(A) selection is governed by the location of the splice acceptor site of the downstream gene in the polycistronic primary transcript of the DHFR-TS locus of *L. major*. Soon after that, other groups showed that this was also the case for several *T. brucei* genes [17, 18]. Because there is no consensus polyadenylation signal found in trypanosome genes, the model proposing the existence of a conserved distance between splice acceptor site and the site for polyadenylation for the gene located upstream has become largely accepted. Several experiments using deletion mutants have tested this hypothesis and shown that, in addition to the correct distance, the presence of a polypyrimidine-rich motif is also crucial because only AG dinucleotides situated downstream from a polypyrimidine tract (PPT) are used as splice acceptor site (Fig. **1**) [19].

After the completion of the TriTryp genomes, genome-wide analyses of mRNA in *T. brucei* [10, 20, 21] and *L. major* [22] precisely identified the 5' and 3' boundaries for almost all *T. brucei* and *Leishmania* genes, confirming the rules that govern *trans*-splicing and polyadenylation. These studies also allowed for the discovery of a large number of novel coding and noncoding transcripts from previously unannotated genes and, more importantly, the discovery of an unprecedented heterogeneity of pre-mRNA-processing sites. It has been determined that in *T. brucei* the PPT can be located up to 200 nt (median is 43) upstream of the splice site and has a median length of 18 nt, with a range of 7 to

79 nt [10]. In *L. major*, PPTs range from 7 to 123 nt, with a median value of 21 nt, and are located at a median distance of 64 nt from the downstream SL addition site [22]. Using a more limited dataset, it was determined that, in *T. cruzi*, the median length of PPT is 13 nt and its median distance from an SL addition site is 18 nt [23]. These distances are in agreement with the comparative analysis of the TriTryp genomes [24], which indicates that the *T. cruzi* genome is more compact, with a smaller average size of coding regions and inter-coding regions compared with *T. brucei* and *L. major*. Importantly, these analyses also showed that the high-throughput RNA sequencing studies performed with *T. brucei* and *L. major* RNAs revealed the existence of widespread alternative *trans*-splicing and polyadenylation [21, 22]. Although Dillon and collaborators [22] did not find a correlation between alternative processing and differences in mRNA levels of genes that are differentially regulated between distinct life-cycle stages of *L. major*, a major role of alternative *trans*-splicing was revealed in the study of the *T. brucei* transcriptome, in which a number of alternative *trans*-spliced mRNAs giving rise to proteins with distinct N-terminal sequences were identified [21].

Because of polycistronic transcription and the absence of conserved signals known to be involved in transcriptional regulation, studies on gene regulation in trypanosomatids led to the concept of post-transcriptional regulons, *i.e.*, networks of gene expression consisting of different genes that are coordinately regulated at particular stages of the life cycle, during the cell cycle, or in response to external signals [25, 26]. Transcripts belonging to a regulon are targets of a group of sequence-specific RNA-binding proteins (RBPs) that recognize cis-regulatory elements in their untranslated regions and mediate the synchronized control of their fate [25, 26]. Based on this assumption, several groups have performed genome-wide analyses to identify conserved regulatory elements in the untranslated regions of *T. brucei*, *T. cruzi* and *Leishmania* genes. In parallel with the identification of cis-acting elements, *trans*-acting protein factors containing RNA-binding motifs that act as co-regulating groups of genes are beginning to be characterized. Notwithstanding, it should be mentioned that recent studies are also starting to reveal a significant role for chromatin structure and post-translation modification of histones as regulatory elements, that are also involved in parasite differentiation and adaptation to changes in the environment [27].

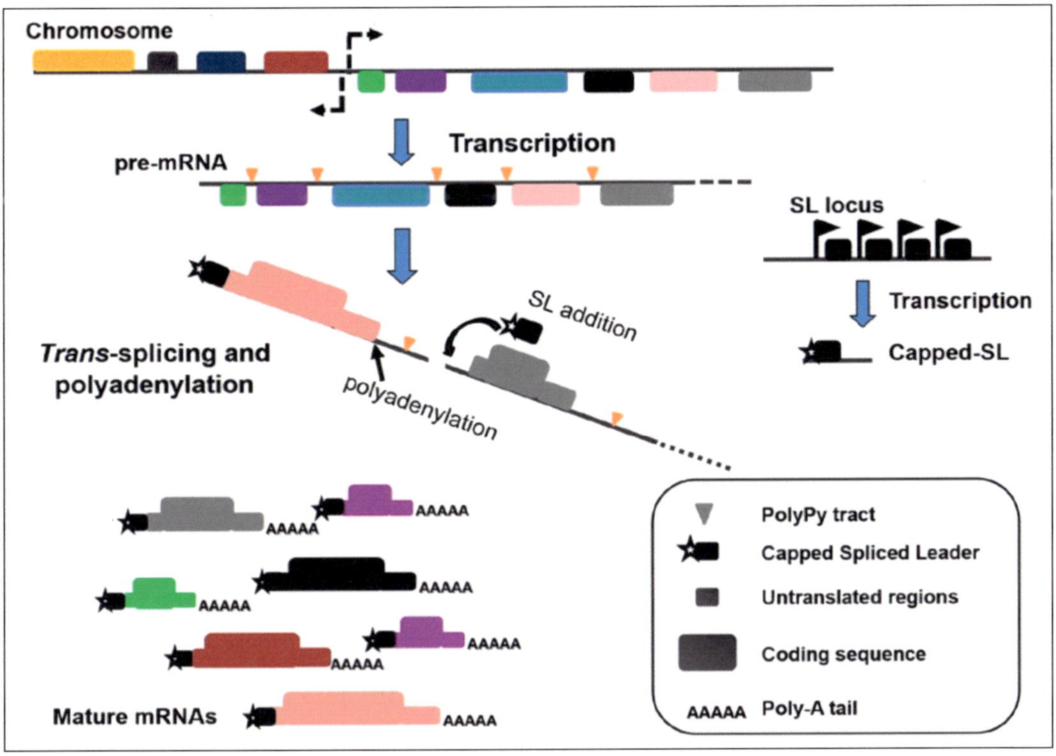

Fig. (1). The general process of mRNA synthesis in trypanosomatids. Trypanosomatid genes are organized in tandem repeats, as indicated by the boxes representing clusters of genes encoded on one DNA strand followed by a cluster of genes encoded on the opposite strand. Sequences corresponding to strand-switch regions (SSR) are represented by the two broken arrows between the two divergent gene clusters or polycistronic transcription units (PTUs). In a separate chromosomal location, there are several hundred tandem direct repeats are encoding spliced leader (SL) RNAs (small boxes), with each repeat bearing a transcriptional promoter (small flags). After transcription, the polycistronic pre-mRNA is processed by *trans*-splicing and polyadenylation reactions, which require signals present within the intergenic regions: a polypyrimidine-rich sequence (small triangles), the SL addition site and a polyadenylation addition site (arrows). Mature mRNAs composed of a coding region, untranslated regions, the capped SL sequence and the poly(A) tail are represented in the diagram at the bottom of the figure.

VSG GENES AND TRANSCRIPTIONAL CONTROL OF GENE EXPRESSION IN *T. BRUCEI*

A large number of studies aiming at revealing the molecular mechanisms involved in the control of gene expression in trypanosomatids have been performed with two groups of *T. brucei* genes encoding GPI-anchored surface proteins, variant surface glycoproteins (VSGs), which cover the bloodstream form of the parasite, and procyclins, the major surface protein of insect forms. VSG genes are responsible for antigenic variation, perhaps one of the most interesting discoveries in the field of parasitology and a powerful survival strategy devised by African trypanosomes and few other pathogens that allows them to escape from the immunological response built by their hosts [28]. Unlike *T. cruzi* and *Leishmania*, *T. brucei* does not have an intracellular stage in its mammalian host. To survive and multiply in the mammalian bloodstream, *T. brucei* bloodstream parasites rely on the constant change of VSGs expressed on their surface. In addition to a bloodstream form (BF), *T. brucei* has another developmental stage, the procyclic form (PF), which multiplies in the insect vector, the tsetse fly. When the parasite differentiates from the bloodstream to PF in the insect midgut, the VSG coat is replaced by two insect-specific surface proteins, named GPEET and EP procyclins [29]. Thus, besides switching on and off different VSG genes during the bloodstream stage, the parasite must completely turn off VSG expression in the procyclic stage and turn on EP and GPEET procyclins in an orderly manner. Early in tsetse fly transmission, both procyclic genes are expressed, but when the trypanosomes colonize the ectoperitrophic space, GPEET is down-regulated, and EP becomes the major surface protein of late PF [29].

Initial evidence for the phenomenon of antigenic variation in *T. brucei*, described in 1910 by Ross and Thompson [30], was based on observations of peaks of parasitemia every one to two weeks in patients suffering from sleeping sickness. In 1975, the molecular basis of antigenic variation began to be uncovered through the studies by George Cross, who isolated a surface protein that was found to be the single major protein present in the BF of *T. brucei* and named it VSG for variant surface glycoprotein [31]. Analyses of the N-terminal sequences of VSGs isolated from four cloned trypanosome populations derived from a single rabbit infected with *T. brucei* showed that each peak of parasitemia corresponded to a trypanosome population expressing only one VSG on their surface [32]. It soon became clear that, by switching their dense protective coat of VSG, African trypanosomes escape the host adaptive immune system. Each time a VSG switch occurs, with estimated switch rates of up to 10^{-2} cells per generation, 10^{7} identical new proteins are added to the surface of bloodstream parasites, allowing these variants to survive the host antibody response [28]. Two main questions regarding the molecular mechanisms underlying antigenic variation are still under investigation. One important question is how trypanosomes control VSG monoallelic expression, *i.e.*, switches from the expression of VSG gene A to B while keeping all other VSG genes silent. A second question is why VSG expression occurs exclusively in BF. With the completion of the *T. brucei* genome [7], it was possible to determine that a total of 1,600 VSG genes are scattered among the 12 pairs of large chromosomes at telomeric and internal locations, as well as throughout hundreds of African-trypanosome-specific minichromosomes. The active VSG is invariably encoded in a polycistronic pre-mRNA derived from a telomeric expression site (ES) such that only one among 10-20 telomeric ES is transcribed at a time [28]. Evidence of rearrangements of VSG genes in the parasite genome was shown by early studies using restriction fragment analyses [33], and soon thereafter, it was demonstrated that the expressed VSG gene is always located near a chromosome end, approximately 1-3 kb from the repetitive telomeric hexamer sequence. Interestingly, VSG genes are not alone, instead, they are transcribed together with a gene cluster that can extend for 50-60 kb and contains approximately 10 other genes [34]. In every *T. brucei* telomere, there is a resident VSG gene. VSG genes are also found internally in chromosomes and, due to DNA rearrangement, internal VSG genes can be activated after moving to a telomeric ES location. *In situ* gene activation/inactivation is another mechanism known to assure that only one telomeric VSG is expressed in one cell. Notably, transcription of the VSG located in an ES is mediated by an RNA pol I that presents an extranucleolar location identified as a nuclear structure called an ES body (ESB) [35]. Some studies have produced a model to explain the activation/inactivation of one particular ES in a bloodstream trypanosome, comparable to silencing mechanisms described in yeast. Active VSG expression sites in *T. brucei* are known to be depleted of nucleosomes; therefore, histone-modifying enzymes must be involved [36 - 38]. Recently, a role in telomeric VSG gene silencing has been ascribed for the histone acetyltransferase (HAT3) and SIR2rp1, the only NAD-dependent histone deacetylase in *T. brucei*, as well as for other histone-modifying enzymes and chromatin chaperones, [27, 39]. Although not yet fully understood, silencing mechanisms resulting in hypoacetylated, hypomethylated compact chromatin, regulate interactions between the RNA polymerase I and VSG genes. Moreover, DNA base modification by β-D-glucosyl-hydroxymethyluracil, known as base J, also appears to be involved in telomeric gene silencing. In *T. brucei*, 50% of base J is found within telomeric repeats. Although significant levels of base J can also be found in other genomic locations in the *T. brucei* genome, its presence at silent VSG expression sites has led to the hypothesis that base J also plays a role in the control of antigenic variation. In addition, because of its absence during the tsetse life-cycle stages that do not transcribe the telomeric VSGs, base J may also contribute to shutting down all VSGs in PF [40, 41].

According to the current model, transcription of the single active VSG gene by RNA pol I, rather than RNA pol II, is a key factor that allows for monoallelic VSG expression based on its physical association with the ESB within the parasite nucleus. This unique ability of *T. brucei* to transcribe a protein-coding gene using RNA polymerase I is only possible because, as previously mentioned, trypanosome maturation of the 5'end (and the addition of cap) is independent of the RNA polymerase because, in trypanosomes, the *trans*-splicing provides the SL sequence already containing a *cap* structure at its 5'end. Moreover, because the ESB is present exclusively in bloodstream cells and has never been observed in the insect, procyclic stage, the ESB model controlling monoallelic VSG expression also provides a mechanism for repression of VSG after the bloodstream parasites differentiate into procyclic.

POST-TRANSCRIPTIONAL ELEMENTS CONTROLLING GENE EXPRESSION IN TRYPANOSOMATIDS

In the mammalian bloodstream, *T. brucei* proliferates in morphologically "slender" forms, which are able to evade host immunity by antigenic variation. Proliferating slender forms undergo morphological and molecular transformation to stumpy forms through a differentiation process that appears to be parasite density-dependent, similar to quorum-sensing systems commonly found in bacteria [42]. Once in the tsetse fly, stumpy forms, which are non-dividing and show partial adaptation to conditions in the fly midgut, differentiate into PF that are no longer infectious to mammals but can multiply in the insect vector. Thus, extensive changes in gene expression must occur when the mammalian bloodstream form differentiates into the insect PF. Recent genome-wide analyses comparing transcript levels in *T. brucei* bloodstream and PF showed that almost 6% of all genes are life-cycle regulated by 2-fold or more [20]. However, the proteomic analysis showed that over 45% of the parasite protein levels changed between the two life cycles [43]. Procyclin genes, which encode highly abundant surface glycoproteins, are among the genes with up-regulated expression in *T. brucei* PF and that have been used as a model to study the mechanisms behind stage-specific gene expression in the parasite. Soon after bloodstream trypanosomes are taken up by a tsetse fly and differentiate into procyclic, the VSG coat is rapidly replaced by a new surface coat composed of two classes of small acidic proteins named procyclins, which are distinguished by the presence of an internal dipeptide (EP procyclins) or pentapeptide (GPEET procyclins) repeats [44]. EP and GPEET procyclins also have distinct patterns of expression, whereas early procyclic parasites transiently express GPEET procyclins, EP procyclins continue to be expressed throughout development in the insect host [44]. In addition to investigating the mechanisms behind the expression activation of GPEET and EP procyclin genes, the group led by Isabel Roditi has been investigating the factors involved in GPEET repression, which coincides with the parasite's migration to the ectoperitrophic space of the fly. Procyclin genes, initially named PARP genes, for procyclic acidic repetitive protein, are located on chromosomes VI and X and are followed by one or more procyclin-associated genes (PAGs) [45]. Similarly to VSG, procyclins are transcribed by RNA polymerase I, but, unexpectedly, procyclin and VSG promoters bear little resemblance to each other or to the classical RNA polymerase I rRNA promoter [46, 47].

Stage-specific control of procyclin gene expression occurs at several levels, including at the transcription initiation as well as transcription elongation rates of procyclic genes, *trans*-splicing efficiency, mRNA degradation and translation. Although procyclin mRNAs are not detectable in bloodstream trypanosomes, they are transcribed at low levels in bloodstream parasites, indicating that post-transcription mechanisms occur to prevent the accumulation of procyclin mRNA in BF [48]. Moreover, although procyclin and PAG genes are derived from the same polycistronic transcription unit, there is a 100-fold difference in the steady-state levels when procyclin and PAG mRNAs are compared [46]. Analysis of the sequence of events that occurs when early PF differentiate to late PF revealed several mechanisms that could explain the differences in the accumulation of EP and GPEET procyclin mRNAs, including the role of regulatory sequences in the 3' untranslated regions (UTRs) in determining the steady-state levels of these mRNAs. In a recent study, procyclin expression during synchronous differentiation of BF to PF, triggered *in vitro* by the addition of cis-aconitate and by dropping the temperature, was carefully analyzed by measuring steady-state levels and mRNA half-lives as well as by mapping the 3' ends of the different mRNAs. It was shown that GPEET mRNA has a half-life of more than 6 h in early differentiating PF, which is markedly reduced to 3.8 h in late PF. In contrast, EP procyclin mRNA has a half-life of 2.3 h in early PF that is increased to 6 h in late PFs [49]. These studies also indicate that differential processing of GPEET transcripts may also contribute to their repression by preventing the generation of mature mRNA in late procyclic. To identify cis-acting regulatory elements, sequences derived from the 3' UTR has been extensively investigated using transient and stable transfection assays with reporter genes [34 - 36]. Distinct elements, including a 16-mer stem-loop that stimulates translation and a 26-mer stem-loop that causes RNA instability in the BF, have been characterized. In addition, GPEET procyclin mRNA possesses a negative regulatory element known as the glycerol responsive element (GRE), which also forms a stem-loop structure and destabilizes GPEET mRNA in late PF [50, 51]. It appears that, by employing several layers of regulation, these parasites can ensure that the rapid changes associated with transmission between the insect vector and the mammalian host are followed by an instant reprogramming of genetic expression.

Studies aimed at understanding the mechanisms controlling gene expression in other trypanosomatids have shown that a significant number of genes are regulated at the mRNA stability level. In contrast to *T. brucei* VSG genes, which are transcribed by RNA pol I, in *Leishmania* and *T. cruzi* there are no reports describing transcriptional regulation affecting the expression of protein-coding genes. Recently, the first direct evidence for epigenetic regulation of RNA pol II transcription in *T. cruzi* was provided by R. Sabatini's group, who described changes in chromatin structure and RNA pol II recruitment to divergent SSRs [13]. By deleting the genes involved in base J synthesis, these authors showed a global increase in RNA pol II transcription, which was associated with increased histone acetylation and decreased nucleosome abundance. Most of the work describing gene regulation in *T. cruzi* and *Leishmania* has focused on a limited number of gene families that are developmentally regulated during the parasite life cycle including tubulin genes [52], the surface metalloprotease gp63 or *msp* genes [53], the *trans*-sialidase gene family [54] and the amastigote-specific genes encoding A2 [55] and amastin surface proteins [56].

Unlike *T. brucei*, *T. cruzi* and *Leishmania* spp. require an intracellular stage in the mammalian host. *Leishmania* spp. proliferates as promastigotes in the midgut of phlebotomine sand flies and is transmitted to several species of mammals as metacyclic promastigotes when the fly takes a blood meal. In the mammalian host, promastigotes are phagocytized by macrophages, and, once in the phagolysosome, they are converted into amastigotes, which multiply numerous times before being released during cell lysis. *T. cruzi* replicates as epimastigotes in the midgut of different species of reduviid bugs and develops into infective metacyclic trypomastigotes once it reaches the rectum, it is then excreted with the insect feces. Unlike *Leishmania* spp., which multiplies only inside host macrophages, *T. cruzi* trypomastigotes invade essentially any nucleated cell type by a mechanism involving lysosomal recruitment at the parasite invasion site. Also in contrast to *Leishmania*, *T. cruzi* trypomastigotes are able to escape from the phagolysosome into the cytosol, where they differentiate into amastigotes. After several rounds of cell division, amastigotes differentiate again into trypomastigotes and are released from the infected cell and can reach the bloodstream or invade neighboring cells [57]. Details about *Leishmania* and *T. cruzi* life cycles are described in Chapter 1.

Regulatory Elements in the 3' UTR of *T. cruzi* and *Leishmania* Genes

Amastin genes have been a model for several studies aiming at identifying cis and *trans*-acting factors involved in post-transcriptional regulation in *T. cruzi* and *Leishmania*. Initially characterized as *T. cruzi* and *Leishmania* genes whose expression is upregulated in amastigotes, amastin genes have been identified in the genomes of other trypanosomatids, including the insect parasites *Leptomonas seymouri*, *Angomonas deanei*, and *Strigomonas culicis* [57, 58]. Although their function remains unknown, because of the predicted structure and surface localization in the intracellular stage of *T. cruzi* and *Leishmania* spp., it has been proposed that amastins may play a role in host-parasite interactions within the mammalian cell involved in the transport of ions and nutrients across the membrane or in cell signaling events that trigger parasite differentiation. Phylogenetic analyses of amastin orthologs from various members of the Trypanosomatidae family resulted in a classification of all amastin sequences into four subfamilies, named α-, β-, γ-, and δ-amastins [59]. Importantly, through analyses of *Leishmania* spp., the δ-amastin sub-family has been widely expanded. With 45 copies identified in the genomes of *L. major* and *L. infantum* [60], amastins constitute the largest gene family in the *Leishmania* genus, whose members exhibit regulated expression during the life cycle of the parasite. *T. cruzi*, on the other hand, has only members of the β- and δ-amastins. Similarly to all *Leishmania* amastin genes, *T. cruzi* δ-amastin genes are upregulated in amastigotes compared to those in the epimastigote or trypomastigote stages; however, in contrast to *Leishmania* orthologs, *T. cruzi* β-amastin genes are up-regulated in epimastigotes [61].

The presence of elements in the 3' UTRs plus intergenic regions controlling the expression of amastin genes were examined using transient and stable trans-fections assays of *T. cruzi* and *Leishmania* with luciferase as a reporter gene [58, 62]. The presence of the δ-amastin 3' UTR plus downstream processing signals of luciferase results in a 36-fold increase in luciferase mRNA levels in *T. cruzi* amastigotes compared to epimastigotes, a difference that corresponds to the 50-fold increase in steady-state levels of amastin mRNA in amastigotes. It was previously shown that this large difference in steady-state amastin mRNA levels between amastigotes and epimastigotes is caused by, at least in part, to differences in amastin mRNA stability because the amastin mRNA is 7-fold more stable in amastigotes than in epimastigotes. Analysis of amastin mRNA in *L. infantum* showed that it is 4-fold more abundant in axenic amastigotes than in promastigotes. The decay kinetics of mRNAs in both stages of the parasite lifecycle indicates the amastin mRNA decays with an overall half-life of aprox. ~40 min in promastigotes *versus* >120 min in amastigotes. Reporter gene assays revealed the presence of a 100 nt long U-rich sequence that contributes significantly to the stage-specific accumulation and degradation of the amastin mRNA [63]. The 1.8 kb 3′UTR of *L. infantum* amastin mRNA also contains additional cis-acting regulatory sequences related to the *Leishmania*-specific SIDER retroelements [64] and a 300 nt sequence which is highly conserved among a large number of *Leishmania* mRNAs, that confers stage-specific gene expression by a mechanism affecting mRNA translation [63].

Several groups have investigated the presence of regulatory elements in the 3' UTR of other *Leishmania* genes besides amastin genes. The list includes the *Leishmania* aquaglyceroporin AQP1 mRNA [65], HSP70 [66], alfa-tubulin [67], HSP83 [68], histone [69], A2 [55], gp46 and gp63 [70]. A large number of *Leishmania* transcripts harbors in their 3' UTRs two phylogenetically distinct subfamilies of extinct short interspersed degenerated retroposons (SIDER1 and SIDER2) that are involved in post-transcriptional regulation of gene expression [64]. It has been recently shown that members of the SIDER2 subfamily promote mRNA destabilization and that, in contrast to most eukaryotic genes, in which mRNA degradation initiated by poly(A) tail shortening, degradation of SIDER2-containing mRNAs is triggered by site-specific endonucleolytic cleavage within SIDER2 signature sequence, without prior shortening of the poly(A) tail [71].

By conducting stable transfection experiments, we and others have described the role of sequences in the 3’ UTR from various stage-specific and constitutive *T. cruzi* genes that include, in addition to amastin genes, genes encoding α-tubulin and the surface proteins GP72 and GP85 [72, 73]. Using an *in silico* approach, Li and collaborators [74] characterized a 43 nt U-rich RNA element located in the 3' UTR of a large number of *T. cruzi* mRNAs, including *trans*-sialidase, mucin-associated protein (MASP), mucin, surface protease GP63, and protein kinases, that are upregulated in amastigotes. Immunoprecipitation analyses associated with *in silico* genome-wide screening, identified conserved structural RNA motifs, statistically enriched in the 3’ UTR of a group of genes encoding metabolically related transcripts, which are targets for two *T. cruzi* U-rich RNA-binding proteins named TcUBP1 and TcUBP2 [75]. This work not only uncovered signature motifs and conserved structural elements in the 3’ UTR of groups of functionally related mRNAs but also showed that groups of transcripts can be preferentially associated with a given RBP.

RNA-BINDING PROTEINS AS MAJOR PLAYERS OF STAGE-SPECIFIC GENE EXPRESSION IN TRYPANOSOMATIDS

Most studies aimed to understand the mechanisms determining mRNA half-lives and translation efficiency points towards the essential role of RNA-binding proteins (RBPs) and their dynamic interactions with the various subpopulations of mRNA in the different stages of the life cycles of trypanosomatid parasites. Different approaches have been used to identify these RBPs and their mRNA targets. In early studies, after characterizing a cis-acting regulatory element in the 3' UTR, scientists searched for a specific RBP able to recognize this element and influence the expression of the corresponding mRNA using RNA band-shift assays and affinity chromatography protocols. Using this initial gene-based approach, a limited number of RBPs have been characterized in *T. brucei*, *T. cruzi* and *L. infantum*. With the availability of complete genome sequences and, more recently, with new sequencing technologies, not only have much more *cis*-acting RNA regulatory elements been uncovered but many more sequences encoding RBPs have also been identified. Altogether, these studies lead to the description of RNA regulons or RBP-mediated regulatory maps that will significantly improve our understanding of how trypanosome post-transcription regulatory machinery operates. In conjunction with a variety of techniques that include nucleotide UV-cross-linking and RBP-immunoprecipitation coupled to deep sequencing (RIP-seq), high-throughput analyses allow for the characterization of not only many more RBPs containing different RNA binding domains (RBDs) but also detailed analyses of the effects caused by the binding of these regulatory *trans*-acting factors to their target transcripts [76].

Recognition by RBPs of their RNA targets is typically mediated by one or more RBDs, such as the RNA recognition motif (RRM), Cys3His (CCCH) zinc finger, Cold-shock, PIWI and Pumilio-like (PUF) domains (for a recent review, see [77]). Soon after the TriTryp genomes were completed, comparative analyses of predicted RBPs containing the RRM motif identified a total of 139 RRM-containing proteins in *T. cruzi*, 75 in *T. brucei*, and 80 in *L. major* [78]. It should be noted that, because of its hybrid nature, the *T. cruzi* CL Brener genome was assembled and annotated as two redundant datasets of sequences, referred to as "Esmeraldo-like" or "non-Esmeraldo-like" sequences, each corresponding to one haplotype, which were identified based on the sequences of the CL Brener parental strains [8]. Therefore, a larger number of sequences were identified in the CL Brener genome perhaps because, for various genes, two polymorphic alleles were counted. Interestingly, among the RRM-containing proteins described in all kinetoplastids, only 12 are conserved from kinetoplastids to yeast or mammals, indicating the presence of a large number of trypanosome-specific RRM proteins that may have coevolved with their RNA targets [78]. Furthermore, of the 75 RRM proteins identified in the *T. brucei* genome, only 18 have their function assigned, which include proteins involved in RNA splicing, ribosome biogenesis, translation, and control of mRNA turnover.

Because of the development of an inducible RNAi system, which has enormously facilitated gene function analyses in this parasite, a variety of *T. brucei* RBPs have been functionally characterized. In contrast to *T. cruzi* and *L. major*, which do not have functional RNAi machinery, in *T. brucei*, the efficacy of RNAi knockdown as a tool for systematic analysis of gene function has allowed not only gene-by-gene studies but also high-throughput approaches to studying RBP function (for a recent review, see [79]. Using RNAi screening, C. Clayton's group analyzed the effect of targeting 37 of the 72 RRM-containing proteins of *T. brucei*, showing that knocking down eight of these genes caused clear growth inhibition in BF trypanosomes. Further analysis of one of them, TbRBP3, which was found to be associated with 10 *T. brucei* mRNAs, showed that its depletion did not affect the parasite transcriptome, suggesting a role in translation [80].

Altering the expression levels of RBPs can produce valuable information regarding their function. By overexpressing a single RNA-binding protein, TbRBP6, in cultured noninfectious PF resulted in the appearance of developmental stages observed in the tsetse fly, including metacyclic trypomastigotes expressing a VSG coat [81]. Overexpression or RNAi knockdown of the *T. brucei* RBP named DRBD13 resulted in changes in the expression level of mRNA encoding various membranes associated proteins [82]. Additionally, RIP-seq data suggested that TbRBP6 transcript binds to DRBD13 protein, indicating the involvement of DRBD13 in the differentiation process that occurs in the insect vector and leads to the transformation of PF into the infectious metacyclic trypomastigotes.The first example of an RBP involved in life-cycle stage-specific regulation of the global trypanosome transcriptome is RBP10, a small RRM-containing protein whose expression is upregulated in BF [83]. Whereas RNAi knockdown of RBP10 in BF resulted in the downregulation of a large number of mRNAs normally found elevated in the bloodstream stage, overexpression of the protein in PF led to an increase in BF-specific mRNAs. Another group of RBPs that have been investigated as factors involved in *T. brucei* development is the ALBA family of proteins (acetylation lowers binding affinity), which are known to be involved in the developmental programs of various microorganisms. *T. brucei* genome encodes four ALBA proteins [84]. Depletion of ALBA3/4 proteins by RNAi in the cultured procyclic stage produces cell modifications mimicking several morphogenetic aspects of trypanosome differentiation that usually occur in the fly midgut [85]. *T. brucei* ALBA proteins have also been characterized as RNA-binding proteins that interact with translation machinery. Initially identified as proteins that interact with elements in the 3' UTR of procyclins, further studies showed that RNAi knockdown of all four Alba proteins specifically reduced translation of a reporter transcript flanked by the GPEET procyclin 3' UTR. Moreover, pull-down assays using tagged Alba proteins revealed interactions with poly(A) binding proteins, ribosomal protein P0 and, in the case of Alba3, the cap-binding protein eIF4E4, further confirming their role in translation control [84].

Recently, ALBA proteins have also been described as *trans*-acting factors regulating amastin developmental gene expression in *Leishmania*. Using RNA affinity chromatography with a 300 nt regulatory region within the amastin 3' UTR as bait, B. Papadopoulou's group identified the homolog of *T. brucei* ALBA3, named LiAlba20, as a specific amastin mRNA-binding factor [86]. Supporting the role of LiAlba20 in amastin gene regulation, gene knockout of LiAlba20 resulted in amastin mRNA destabilization in amastigotes. Further studies have shown that an arginine methyltransferase named LmjPRMT7 is associated with several RBPs, including LmjAlba20. LmjAlba20 is mono-methylated only in LmjPRMT7-expressing promastigote stages, and its protein levels are significantly altered in LmjPRMT7 knockout mutants [87]. In addition to ALBA proteins, only a few other RBPs have been characterized in *Leishmania*. Although proteins containing a PIWI domain such as Argonaute, involved in RNAi machinery, are not found in Old World *Leishmania*, a Piwi-like protein homolog has been identified in *L. infantum* and *L. major*. Knockout of the gene encoding this Piwi-like protein leads to changes in the expression of several developmental regulated *Leishmania* genes, suggesting that these proteins may play a role in post-transcriptional regulation that is independent of the RNAi pathway [88].

Among the RBPs that have been characterized in *T. cruzi* are several proteins that have been found to be involved in mRNA stabilization and translation. Using polyT beads as bait, S. Goldenberg's group isolated protein complexes bound to poly(A)+ mRNAs from polysomal and post-polysomal fractions of epimastigotes that were in the exponential phase of growth and from epimastigotes under nutritional stress. Mass spectrometry (LC-MS/MS) analyses resulted in the identification of 542 protein components of the mRNP complexes associated with mRNAs [89]. In addition to a number of ribosomal proteins, several RBPs were identified by the authors, including the CCR4-CAF1-NOT proteins implicated in mRNA turnover; the TIA-1-like RBP, known to repress the translation of the mRNAs in response to stress; and Dhh1, which belongs to a highly conserved subfamily of DEAD box helicases known to be involved in global repression of mRNA translation in yeast. These authors also identified TcUBP1, a U-rich RBP containing a single RRM motif, that was previously characterized as an mRNA destabilizing factor that binds to AU-rich elements present in the 3' UTR of mucin mRNAs [90]. Although other studies based on ribosome profiling have shown that control of mRNA translation is more extensive than previously predicted solely by the analyses of total transcript levels [91], the mechanisms involved in this regulation are still far from being understood.

TcUBP1 is the first RBP identified in trypanosomatids as a *trans*-acting factor involved in post-transcriptional control. Together with TcUBP2 and poly-adenylation binding protein, TcUBP1 is part of a ribonucleoprotein (mRNP) complex that recognizes U-rich motifs and is associated with mRNA granules [90]. The analysis of RNA-protein interactions by mRNP immunoprecipitation with specific antibodies against TcUPB1 and TcUPB3 allowed for the identification of 24 mRNAs encoding known proteins, as well as 10 hypothetical genes. The mRNA associated with TcUBP1 was found to encode mucin, amastin and proteases, such as cruzipain and GP63, as well as members of the TS family. In contrast, the transcripts that interacted with TcUBP3 were found to encode mainly ribosomal proteins [75]. The authors identified a stem-loop structure of 30-35 bases present in the majority of TcUBP-associated mRNAs, which was more frequently represented in the 3' UTRs of mRNAs [75].

RBPs bearing the CCCH zinc finger motif have also been characterized in a *T. cruzi* strain named Dm28c, which has been extensively studied as a model for parasite differentiation [92]. Goldenberg's group showed that the expression of TcZFP1 and TcZFP2 is increased in polysomal fractions from parasites submitted to nutritional stress. TcZFP2 is expressed in amastigotes and trypomastigotes, but its protein levels are reduced in *T. cruzi* metacyclic trypomastigotes. After identifying the TcZFN2 RNA targets using RIP-CHIP, these authors showed that the expression of most genes encoding the TcZFP2-associated mRNAs are upregulated in the metacyclic trypomastigotes, thus suggesting a role for TcZFP2 in metacyclogenesis [93]. TcPUF6 and a few other members of the PUF RNA-binding protein family have also been investigated as potential regulators of metacyclogenesis [94]. Other *T. cruzi* RBPs that have been characterized include two proteins containing an RRM motif: TcRBP19, which was only detected in the amastigote stage [95], and TcRBP40, which is a *T. cruzi*-specific RBP whose mRNA targets mainly encode putative *T. cruzi* transmembrane proteins [96].

Clearly, the number of RBPs known to be involved in post-transcriptional regulation in trypanosomatids is growing at a faster rate than it was a few years ago. However, the exact mechanism employed by these parasites to control the expression of their genes, including the ones coding for these RBPs, are far from being understood. Although almost all the main trypanosome homologs of yeast/mammalian mRNA degradation enzymes (such as components of the exosome), decapping and deadenylating enzymes, as well as exoribonucleases have been characterized and major pathways identified, there is still much work that must be done to understand how these activities are regulated. Because of the emerging role of non-coding RNA, which is now clearly established in various eukaryotes, several groups have also investigated whether small non-coding RNAs also participate in the regulatory processes associated with parasite differentiation [97]. However, although various components of the small RNome have been characterized in all three groups of parasites, to date, no clear evidence exists for the involvement of these molecules in gene regulation processes. With the powerful genome-wide approaches that have recently become available, the main gene regulatory networks, together with the RNA and protein components involved in these networks, will soon be fully characterized. Because they are essential components of the survival arsenal employed by these parasites, a better understating of these regulatory machines will ultimately reveal more adequate targets for pharmacological interventions that can block parasite development and lead to disease control.

CONFLICT OF INTEREST

The authors confirm that they have no conflict of interest to declare for this publication.

ACKNOWLEDGEMENTS

The work in our laboratory was supported by grants from Conselho Nacional de Desenvolvimento Científico e Tecnológico (CNPq), Fundação de Amparo à Pesquisa do Estado de Minas Gerais (FAPEMIG) and from the Instituto Nacional de Ciência e Tecnologia de Vacinas (INCTV).

ABBREVIATIONS

BF	Bloodstream form
ES	Expression site
ESB	Expression site body
kDNA	Kinetoplast DNA
PF	Procyclic form
PPT	Polypyrimidine tract
PTU	Polycistronic transcription unit
PUF	Pumilio-like domain
RBD	RNA binding domain
RBP	RNA binding protein
RNAi	RNA interference
RNA pol I	RNA polymerase I

RNA Pol II RNA polymerase II

RRM RNA recognition motif

SL Spliced leader

UTR Untranslated region

VSG Variant surface glycoprotein

REFERENCES

[1] Fernandes AP, Nelson K, Beverley SM. Evolution of nuclear ribosomal RNAs in kinetoplastid protozoa: perspectives on the age and origins of parasitism. Proc Natl Acad Sci USA 1993; 90(24): 11608-12.
[http://dx.doi.org/10.1073/pnas.90.24.11608] [PMID: 8265597]

[2] Michaeli S. Trans-splicing in trypanosomes: machinery and its impact on the parasite transcriptome. Future Microbiol 2011; 6(4): 459-74.
[http://dx.doi.org/10.2217/fmb.11.20] [PMID: 21526946]

[3] Teixeira SM, daRocha WD. Control of gene expression and genetic manipulation in the Trypanosomatidae. Genet Mol Res 2003; 2(1): 148-58.
[PMID: 12917811]

[4] Lukes J, Hashimi H, Zíková A. Unexplained complexity of the mitochondrial genome and transcriptome in kinetoplastid flagellates. Curr Genet 2005; 48(5): 277-99.
[http://dx.doi.org/10.1007/s00294-005-0027-0] [PMID: 16215758]

[5] Myler PJ, Audleman L, deVos T, *et al. Leishmania major* Friedlin chromosome 1 has an unusual distribution of protein-coding genes. Proc Natl Acad Sci USA 1999; 96(6): 2902-6.
[http://dx.doi.org/10.1073/pnas.96.6.2902] [PMID: 10077609]

[6] Thomas S, Green A, Sturm NR, Campbell DA, Myler PJ. Histone acetylations mark origins of polycistronic transcription in *Leishmania major*. BMC Genomics 2009; 10: 152.
[http://dx.doi.org/10.1186/1471-2164-10-152] [PMID: 19356248]

[7] Berriman M, Ghedin E, Hertz-Fowler C, *et al.* The genome of the African trypanosome *Trypanosoma brucei.* Science 2005; 309(5733): 416-22.
[http://dx.doi.org/10.1126/science.1112642] [PMID: 16020726]

[8] El-Sayed NM, Myler PJ, Bartholomeu DC, *et al.* The genome sequence of *Trypanosoma cruzi*, etiologic agent of Chagas disease. Science 2005; 309(5733): 409-15.
[http://dx.doi.org/10.1126/science.1112631] [PMID: 16020725]

[9] Ivens AC, Peacock CS, Worthey EA, *et al.* The genome of the kinetoplastid parasite, *Leishmania major.* Science 2005; 309(5733): 436-42.
[http://dx.doi.org/10.1126/science.1112680] [PMID: 16020728]

[10] Kolev NG, Franklin JB, Carmi S, Shi H, Michaeli S, Tschudi C. The transcriptome of the human pathogen *Trypanosoma brucei* at single-nucleotide resolution. PLoS Pathog 2010; 6(9): e1001090.
[http://dx.doi.org/10.1371/journal.ppat.1001090] [PMID: 20838601]

[11] Martinez-Calvillo S, Vizuet-de-Rueda JC, Florencio-Martinez LE, *et al.* Gene expression in trypanosomatid parasites. J Biomed Biotechnol 2010; 2010: 525241.

[12] Cliffe LJ, Siegel TN, Marshall M, Cross GA, Sabatini R. Two thymidine hydroxylases differentially regulate the formation of glucosylated DNA at regions flanking polymerase II polycistronic transcription units throughout the genome of *Trypanosoma brucei.* Nucleic Acids Res 2010; 38(12): 3923-35.
[http://dx.doi.org/10.1093/nar/gkq146] [PMID: 20215442]

[13] Ekanayake D, Sabatini R. Epigenetic regulation of polymerase II transcription initiation in Trypanosoma cruzi: modulation of nucleosome abundance, histone modification, and polymerase occupancy by O-linked thymine DNA glucosylation. Eukaryot Cell 2011; 10(11): 1465-72.
[http://dx.doi.org/10.1128/EC.05185-11] [PMID: 21926332]

[14] van Luenen HG, Farris C, Jan S, *et al.* Glucosylated hydroxymethyluracil, DNA base J, prevents transcriptional readthrough in *Leishmania.* Cell 2012; 150(5): 909-21.
[http://dx.doi.org/10.1016/j.cell.2012.07.030] [PMID: 22939620]

[15] Donelson JE, Zeng W. A comparison of *trans*-RNA splicing in trypanosomes and nematodes. Parasitol Today (Regul Ed) 1990; 6(10): 327-34.
[http://dx.doi.org/10.1016/0169-4758(90)90177-6] [PMID: 15463258]

[16] LeBowitz JH, Smith HQ, Rusche L, Beverley SM. Coupling of poly(A) site selection and *trans*-splicing in *Leishmania.* Genes Dev 1993; 7(6): 996-1007.
[http://dx.doi.org/10.1101/gad.7.6.996] [PMID: 8504937]

[17] Matthews KR, Tschudi C, Ullu E. A common pyrimidine-rich motif governs *trans*-splicing and polyadenylation of tubulin polycistronic pre-mRNA in trypanosomes. Genes Dev 1994; 8(4): 491-501.
[http://dx.doi.org/10.1101/gad.8.4.491] [PMID: 7907303]

[18] Schürch N, Hehl A, Vassella E, Braun R, Roditi I. Accurate polyadenylation of procyclin mRNAs in *Trypanosoma brucei* is determined by pyrimidine-rich elements in the intergenic regions. Mol Cell Biol 1994; 14(6): 3668-75.
[http://dx.doi.org/10.1128/MCB.14.6.3668] [PMID: 7910942]

[19] Liang XH, Haritan A, Uliel S, Michaeli S. trans and cis splicing in trypanosomatids: mechanism, factors, and regulation. Eukaryot Cell 2003; 2(5): 830-40.
[http://dx.doi.org/10.1128/EC.2.5.830-840.2003] [PMID: 14555465]

[20] Siegel TN, Hekstra DR, Wang X, Dewell S, Cross GA. Genome-wide analysis of mRNA abundance in two life-cycle stages of *Trypanosoma brucei* and identification of splicing and polyadenylation sites. Nucleic Acids Res 2010; 38(15): 4946-57.
[http://dx.doi.org/10.1093/nar/gkq237] [PMID: 20385579]

[21] Nilsson D, Gunasekera K, Mani J, *et al.* Spliced leader trapping reveals widespread alternative splicing patterns in the highly dynamic transcriptome of *Trypanosoma brucei.* PLoS Pathog 2010; 6(8): e1001037.
[http://dx.doi.org/10.1371/journal.ppat.1001037] [PMID: 20700444]

[22] Dillon LA, Okrah K, Hughitt VK, *et al.* Transcriptomic profiling of gene expression and RNA processing during *Leishmania major* differentiation. Nucleic Acids Res 2015; 43(14): 6799-813.

[http://dx.doi.org/10.1093/nar/gkv656] [PMID: 26150419]

[23] Campos PC, Bartholomeu DC, DaRocha WD, Cerqueira GC, Teixeira SM. Sequences involved in mRNA processing in *Trypanosoma cruzi.* Int J Parasitol 2008; 38(12): 1383-9. [http://dx.doi.org/10.1016/j.ijpara.2008.07.001] [PMID: 18700146]

[24] El-Sayed NM, Myler PJ, Blandin G, *et al.* Comparative genomics of trypanosomatid parasitic protozoa. Science 2005; 309(5733): 404-9. [http://dx.doi.org/10.1126/science.1112181] [PMID: 16020724]

[25] Clayton CE. Networks of gene expression regulation in *Trypanosoma brucei.* Mol Biochem Parasitol 2014; 195(2): 96-106. [http://dx.doi.org/10.1016/j.molbiopara.2014.06.005] [PMID: 24995711]

[26] De Gaudenzi JG, Carmona SJ, Agüero F, Frasch AC. Genome-wide analysis of 3-untranslated regions supports the existence of post-transcriptional regulons controlling gene expression in trypanosomes. Peer J 2013; 1: e118. [http://dx.doi.org/10.7717/peerj.118] [PMID: 23904995]

[27] Moretti NS, Schenkman S. Chromatin modifications in trypanosomes due to stress. Cell Microbiol 2013; 15(5): 709-17. [http://dx.doi.org/10.1111/cmi.12111] [PMID: 23336291]

[28] Horn D, McCulloch R. Molecular mechanisms underlying the control of antigenic variation in African trypanosomes. Curr Opin Microbiol 2010; 13(6): 700-5. [http://dx.doi.org/10.1016/j.mib.2010.08.009] [PMID: 20884281]

[29] Vassella E, Acosta-Serrano A, Studer E, Lee SH, Englund PT, Roditi I. Multiple procyclin isoforms are expressed differentially during the development of insect forms of *Trypanosoma brucei.* J Mol Biol 2001; 312(4): 597-607. [http://dx.doi.org/10.1006/jmbi.2001.5004] [PMID: 11575917]

[30] Ross R, Thomson D. A case of sleeping sickness showing regular periodical increase of the parasites disclosed. BMJ 1910; 1(2582): 1544-5. [http://dx.doi.org/10.1136/bmj.1.2582.1544] [PMID: 20765166]

[31] Cross GA. Identification, purification and properties of clone-specific glycoprotein antigens constituting the surface coat of *Trypanosoma brucei.* Parasitology 1975; 71(3): 393-417. [http://dx.doi.org/10.1017/S003118200004717X] [PMID: 645]

[32] Bridgen PJ, Cross GA, Bridgen J. N-terminal amino acid sequences of variant-specific surface antigens from *Trypanosoma brucei.* Nature 1976; 263(5578): 613-4. [http://dx.doi.org/10.1038/263613a0] [PMID: 980109]

[33] Williams RO, Young JR, Majiwa PA. Genomic rearrangements correlated with antigenic variation in *Trypanosoma brucei.* Nature 1979; 282(5741): 847-9. [http://dx.doi.org/10.1038/282847a0] [PMID: 514362]

[34] Kooter JM, van der Spek HJ, Wagter R, *et al.* The anatomy and transcription of a telomeric expression site for variant-specific surface antigens in *T. brucei.* Cell 1987; 51(2): 261-72. [http://dx.doi.org/10.1016/0092-8674(87)90153-X] [PMID: 2444341]

[35] Navarro M, Gull K. A pol I transcriptional body associated with VSG mono-allelic expression in

Trypanosoma brucei. Nature 2001; 414(6865): 759-63.
[http://dx.doi.org/10.1038/414759a] [PMID: 11742402]

[36] Povelones ML, Gluenz E, Dembek M, Gull K, Rudenko G. Histone H1 plays a role in heterochromatin formation and VSG expression site silencing in *Trypanosoma brucei.* PLoS Pathog 2012; 8(11): e1003010.
[http://dx.doi.org/10.1371/journal.ppat.1003010] [PMID: 23133390]

[37] Wang QP, Kawahara T, Horn D. Histone deacetylases play distinct roles in telomeric VSG expression site silencing in African trypanosomes. Mol Microbiol 2010; 77(5): 1237-45.
[http://dx.doi.org/10.1111/j.1365-2958.2010.07284.x] [PMID: 20624217]

[38] Günzl A, Kirkham JK, Nguyen TN, Badjatia N, Park SH. Mono-allelic VSG expression by RNA polymerase I in *Trypanosoma brucei*: expression site control from both ends? Gene 2015; 556(1): 68-73.
[http://dx.doi.org/10.1016/j.gene.2014.09.047] [PMID: 25261847]

[39] Glover L, Horn D. Locus-specific control of DNA resection and suppression of subtelomeric VSG recombination by HAT3 in the African trypanosome. Nucleic Acids Res 2014; 42(20): 12600-13.
[http://dx.doi.org/10.1093/nar/gku900] [PMID: 25300492]

[40] Glover L, Hutchinson S, Alsford S, McCulloch R, Field MC, Horn D. Antigenic variation in African trypanosomes: the importance of chromosomal and nuclear context in VSG expression control. Cell Microbiol 2013; 15(12): 1984-93.
[http://dx.doi.org/10.1111/cmi.12215] [PMID: 24047558]

[41] van Leeuwen F, Taylor MC, Mondragon A, *et al.* beta-D-glucosyl-hydroxymethyluracil is a conserved DNA modification in kinetoplastid protozoans and is abundant in their telomeres. Proc Natl Acad Sci USA 1998; 95(5): 2366-71.
[http://dx.doi.org/10.1073/pnas.95.5.2366] [PMID: 9482891]

[42] Mony BM, MacGregor P, Ivens A, *et al.* Genome-wide dissection of the quorum sensing signalling pathway in *Trypanosoma brucei.* Nature 2014; 505(7485): 681-5.
[http://dx.doi.org/10.1038/nature12864] [PMID: 24336212]

[43] Cirovic O, Ochsenreiter T. Whole proteome analysis of the protozoan parasite *Trypanosoma brucei* using stable isotope labeling by amino acids in cell culture and mass spectrometry. Methods Mol Biol 2014; 1188: 47-55.
[http://dx.doi.org/10.1007/978-1-4939-1142-4_4] [PMID: 25059603]

[44] Roditi I. The VSG-procyclin switch. Parasitol Today (Regul Ed) 1996; 12(2): 47-9.
[http://dx.doi.org/10.1016/0169-4758(96)80651-X] [PMID: 15275251]

[45] Roditi I, Furger A, Ruepp S, Schürch N, Bütikofer P. Unravelling the procyclin coat of *Trypanosoma brucei.* Mol Biochem Parasitol 1998; 91(1): 117-30.
[http://dx.doi.org/10.1016/S0166-6851(97)00195-3] [PMID: 9574930]

[46] Haenni S, Studer E, Burkard GS, Roditi I. Bidirectional silencing of RNA polymerase I transcription by a strand switch region in *Trypanosoma brucei.* Nucleic Acids Res 2009; 37(15): 5007-18.
[http://dx.doi.org/10.1093/nar/gkp513] [PMID: 19531741]

[47] Rudenko G, Le Blancq S, Smith J, Lee MG, Rattray A, Van der Ploeg LH. Procyclic acidic repetitive

protein (PARP) genes located in an unusually small alpha-amanitin-resistant transcription unit: PARP promoter activity assayed by transient DNA transfection of *Trypanosoma brucei.* Mol Cell Biol 1990; 10(7): 3492-504.
[http://dx.doi.org/10.1128/MCB.10.7.3492] [PMID: 1694012]

[48] Pays E, Coquelet H, Tebabi P, *et al. Trypanosoma brucei*: constitutive activity of the VSG and procyclin gene promoters. EMBO J 1990; 9(10): 3145-51.
[PMID: 1698609]

[49] Knüsel S, Roditi I. Insights into the regulation of GPEET procyclin during differentiation from early to late procyclic forms of *Trypanosoma brucei.* Mol Biochem Parasitol 2013; 191(2): 66-74.
[http://dx.doi.org/10.1016/j.molbiopara.2013.09.004] [PMID: 24076427]

[50] Furger A, Schürch N, Kurath U, Roditi I. Elements in the 3 untranslated region of procyclin mRNA regulate expression in insect forms of *Trypanosoma brucei* by modulating RNA stability and translation. Mol Cell Biol 1997; 17(8): 4372-80.
[http://dx.doi.org/10.1128/MCB.17.8.4372] [PMID: 9234695]

[51] Hehl A, Vassella E, Braun R, Roditi I. A conserved stem-loop structure in the 3 untranslated region of procyclin mRNAs regulates expression in *Trypanosoma brucei.* Proc Natl Acad Sci USA 1994; 91(1): 370-4.
[http://dx.doi.org/10.1073/pnas.91.1.370] [PMID: 8278396]

[52] Coulson RM, Connor V, Chen JC, Ajioka JW. Differential expression of *Leishmania major* beta-tubulin genes during the acquisition of promastigote infectivity. Mol Biochem Parasitol 1996; 82(2): 227-36.
[http://dx.doi.org/10.1016/0166-6851(96)02739-9] [PMID: 8946388]

[53] Ramamoorthy R, Swihart KG, McCoy JJ, Wilson ME, Donelson JE. Intergenic regions between tandem gp63 genes influence the differential expression of gp63 RNAs in *Leishmania chagasi* promastigotes. J Biol Chem 1995; 270(20): 12133-9.
[http://dx.doi.org/10.1074/jbc.270.20.12133] [PMID: 7744862]

[54] Jäger AV, Muiá RP, Campetella O. Stage-specific expression of *Trypanosoma cruzi trans*-sialidase involves highly conserved 3 untranslated regions. FEMS Microbiol Lett 2008; 283(2): 182-8.
[http://dx.doi.org/10.1111/j.1574-6968.2008.01170.x] [PMID: 18422619]

[55] Charest H, Zhang WW, Matlashewski G. The developmental expression of *Leishmania donovani* A2 amastigote-specific genes is post-transcriptionally mediated and involves elements located in the 3-untranslated region. J Biol Chem 1996; 271(29): 17081-90.
[http://dx.doi.org/10.1074/jbc.271.29.17081] [PMID: 8663340]

[56] Teixeira SM, Russell DG, Kirchhoff LV, Donelson JE. A differentially expressed gene family encoding amastin, a surface protein of *Trypanosoma cruzi* amastigotes. J Biol Chem 1994; 269(32): 20509-16.
[PMID: 8051148]

[57] Bartholomeu DC, de Paiva RM, Mendes TA, DaRocha WD, Teixeira SM. Unveiling the intracellular survival gene kit of trypanosomatid parasites. PLoS Pathog 2014; 10(12): e1004399.
[http://dx.doi.org/10.1371/journal.ppat.1004399] [PMID: 25474314]

[58] Teixeira SM, Kirchhoff LV, Donelson JE. Post-transcriptional elements regulating expression of

mRNAs from the amastin/tuzin gene cluster of *Trypanosoma cruzi.* J Biol Chem 1995; 270(38): 22586-94.
[http://dx.doi.org/10.1074/jbc.270.38.22586] [PMID: 7673251]

[59] Jackson AP. The evolution of amastin surface glycoproteins in trypanosomatid parasites. Mol Biol Evol 2010; 27(1): 33-45.
[http://dx.doi.org/10.1093/molbev/msp214] [PMID: 19748930]

[60] Rochette A, McNicoll F, Girard J, *et al.* Characterization and developmental gene regulation of a large gene family encoding amastin surface proteins in *Leishmania* spp. Mol Biochem Parasitol 2005; 140(2): 205-20.
[http://dx.doi.org/10.1016/j.molbiopara.2005.01.006] [PMID: 15760660]

[61] Kangussu-Marcolino MM, de Paiva RM, Araújo PR, *et al.* Distinct genomic organization, mRNA expression and cellular localization of members of two amastin sub-families present in *Trypanosoma cruzi.* BMC Microbiol 2013; 13: 10.
[http://dx.doi.org/10.1186/1471-2180-13-10] [PMID: 23327097]

[62] Coughlin BC, Teixeira SM, Kirchhoff LV, Donelson JE. Amastin mRNA abundance in *Trypanosoma cruzi* is controlled by a 3-untranslated region position-dependent cis-element and an untranslated region-binding protein. J Biol Chem 2000; 275(16): 12051-60.
[http://dx.doi.org/10.1074/jbc.275.16.12051] [PMID: 10766837]

[63] Boucher N, Wu Y, Dumas C, *et al.* A common mechanism of stage-regulated gene expression in *Leishmania* mediated by a conserved 3-untranslated region element. J Biol Chem 2002; 277(22): 19511-20.
[http://dx.doi.org/10.1074/jbc.M200500200] [PMID: 11912202]

[64] Bringaud F, Müller M, Cerqueira GC, *et al.* Members of a large retroposon family are determinants of post-transcriptional gene expression in *Leishmania.* PLoS Pathog 2007; 3(9): 1291-307.
[http://dx.doi.org/10.1371/journal.ppat.0030136] [PMID: 17907803]

[65] Mandal G, Mandal S, Sharma M, *et al.* Species-specific antimonial sensitivity in *Leishmania* is driven by post-transcriptional regulation of AQP1. PLoS Negl Trop Dis 2015; 9(2): e0003500.
[http://dx.doi.org/10.1371/journal.pntd.0003500] [PMID: 25714343]

[66] Requena JM, Chicharro C, García L, Parrado R, Puerta CJ, Cañavate C. Sequence analysis of the 3-untranslated region of HSP70 (type I) genes in the genus *Leishmania*: its usefulness as a molecular marker for species identification. Parasit Vectors 2012; 5: 87.
[http://dx.doi.org/10.1186/1756-3305-5-87] [PMID: 22541251]

[67] Ramírez CA, Requena JM, Puerta CJ. Alpha tubulin genes from *Leishmania braziliensis*: genomic organization, gene structure and insights on their expression. BMC Genomics 2013; 14: 454.
[http://dx.doi.org/10.1186/1471-2164-14-454] [PMID: 23829570]

[68] David M, Gabdank I, Ben-David M, *et al.* Preferential translation of Hsp83 in *Leishmania* requires a thermosensitive polypyrimidine-rich element in the 3 UTR and involves scanning of the 5 UTR. RNA 2010; 16(2): 364-74.
[http://dx.doi.org/10.1261/rna.1874710] [PMID: 20040590]

[69] Abanades DR, Ramírez L, Iborra S, *et al.* Key role of the 3 untranslated region in the cell cycle regulated expression of the *Leishmania infantum* histone H2A genes: minor synergistic effect of the 5

untranslated region. BMC Mol Biol 2009; 10: 48.
[http://dx.doi.org/10.1186/1471-2199-10-48] [PMID: 19460148]

[70] Myung KS, Beetham JK, Wilson ME, Donelson JE. Comparison of the post-transcriptional regulation of the mRNAs for the surface proteins PSA (GP46) and MSP (GP63) of *Leishmania chagasi.* J Biol Chem 2002; 277(19): 16489-97.
[http://dx.doi.org/10.1074/jbc.M200174200] [PMID: 11856749]

[71] Haile S, Dupé A, Papadopoulou B. Deadenylation-independent stage-specific mRNA degradation in *Leishmania.* Nucleic Acids Res 2008; 36(5): 1634-44.
[http://dx.doi.org/10.1093/nar/gkn019] [PMID: 18250085]

[72] Araújo PR, Teixeira SM. Regulatory elements involved in the post-transcriptional control of stage-specific gene expression in *Trypanosoma cruzi*: a review. Mem Inst Oswaldo Cruz 2011; 106(3): 257-66.
[http://dx.doi.org/10.1590/S0074-02762011000300002] [PMID: 21655811]

[73] Nozaki T, Cross GA. Effects of 3 untranslated and intergenic regions on gene expression in *Trypanosoma cruzi.* Mol Biochem Parasitol 1995; 75(1): 55-67.
[http://dx.doi.org/10.1016/0166-6851(95)02512-X] [PMID: 8720175]

[74] Li ZH, De Gaudenzi JG, Alvarez VE, *et al.* A 43-nucleotide U-rich element in 3-untranslated region of large number of *Trypanosoma cruzi* transcripts is important for mRNA abundance in intracellular amastigotes. J Biol Chem 2012; 287(23): 19058-69.
[http://dx.doi.org/10.1074/jbc.M111.338699] [PMID: 22500021]

[75] Noé G, De Gaudenzi JG, Frasch AC. Functionally related transcripts have common RNA motifs for specific RNA-binding proteins in trypanosomes. BMC Mol Biol 2008; 9: 107.
[http://dx.doi.org/10.1186/1471-2199-9-107] [PMID: 19063746]

[76] DOrso I, De Gaudenzi JG, Frasch AC. RNA-binding proteins and mRNA turnover in trypanosomes. Trends Parasitol 2003; 19(4): 151-5.
[http://dx.doi.org/10.1016/S1471-4922(03)00035-7] [PMID: 12689640]

[77] Sanchez-Diaz P, Penalva LO. Post-transcription meets post-genomic: the saga of RNA binding proteins in a new era. RNA Biol 2006; 3(3): 101-9.
[http://dx.doi.org/10.4161/rna.3.3.3373] [PMID: 17114948]

[78] De Gaudenzi J, Frasch AC, Clayton C. RNA-binding domain proteins in Kinetoplastids: a comparative analysis. Eukaryot Cell 2005; 4(12): 2106-14.
[http://dx.doi.org/10.1128/EC.4.12.2106-2114.2005] [PMID: 16339728]

[79] Kolev NG, Ullu E, Tschudi C. The emerging role of RNA-binding proteins in the life cycle of *Trypanosoma brucei.* Cell Microbiol 2014; 16(4): 482-9.
[http://dx.doi.org/10.1111/cmi.12268] [PMID: 24438230]

[80] Wurst M, Robles A, Po J, *et al.* An RNAi screen of the RRM-domain proteins of *Trypanosoma brucei.* Mol Biochem Parasitol 2009; 163(1): 61-5.
[http://dx.doi.org/10.1016/j.molbiopara.2008.09.001] [PMID: 18840477]

[81] Kolev NG, Ramey-Butler K, Cross GA, Ullu E, Tschudi C. Developmental progression to infectivity in *Trypanosoma brucei* triggered by an RNA-binding protein. Science 2012; 338(6112): 1352-3.

[http://dx.doi.org/10.1126/science.1229641] [PMID: 23224556]

[82] Jha BA, Gazestani VH, Yip CW, Salavati R. The DRBD13 RNA binding protein is involved in the insect-stage differentiation process of *Trypanosoma brucei.* FEBS Lett 2015; 589(15): 1966-74. [http://dx.doi.org/10.1016/j.febslet.2015.05.036] [PMID: 26028502]

[83] Wurst M, Seliger B, Jha BA, Klein C, Queiroz R, Clayton C. Expression of the RNA recognition motif protein RBP10 promotes a bloodstream-form transcript pattern in *Trypanosoma brucei*. Mol Microbiol 2012; 83(5): 1048-63. [http://dx.doi.org/10.1111/j.1365-2958.2012.07988.x] [PMID: 22296558]

[84] Mani J, Güttinger A, Schimanski B, *et al.* Alba-domain proteins of *Trypanosoma brucei* are cytoplasmic RNA-binding proteins that interact with the translation machinery. PLoS One 2011; 6(7): e22463. [http://dx.doi.org/10.1371/journal.pone.0022463] [PMID: 21811616]

[85] Subota I, Rotureau B, Blisnick T, *et al.* ALBA proteins are stage regulated during trypanosome development in the tsetse fly and participate in differentiation. Mol Biol Cell 2011; 22(22): 4205-19. [http://dx.doi.org/10.1091/mbc.E11-06-0511] [PMID: 21965287]

[86] Dupé A, Dumas C, Papadopoulou B. An Alba-domain protein contributes to the stage-regulated stability of amastin transcripts in *Leishmania.* Mol Microbiol 2014; 91(3): 548-61. [http://dx.doi.org/10.1111/mmi.12478] [PMID: 24620725]

[87] Ferreira TR, Alves-Ferreira EV, Defina TP, Walrad P, Papadopoulou B, Cruz AK. Altered expression of an RBP-associated arginine methyltransferase 7 in *Leishmania major* affects parasite infection. Mol Microbiol 2014. [http://dx.doi.org/10.1111/mmi.12819] [PMID: 25294169]

[88] Padmanabhan PK, Dumas C, Samant M, Rochette A, Simard MJ, Papadopoulou B. Novel features of a PIWI-like protein homolog in the parasitic protozoan *Leishmania.* PLoS One 2012; 7(12): e52612. [http://dx.doi.org/10.1371/journal.pone.0052612] [PMID: 23285111]

[89] Alves LR, Avila AR, Correa A, *et al.* Proteomic analysis reveals the dynamic association of proteins with translated mRNAs in *Trypanosoma cruzi.* Gene 2010; 452(2): 72-8. [http://dx.doi.org/10.1016/j.gene.2009.12.009] [PMID: 20060445]

[90] DOrso I, Frasch AC. TcUBP-1, an mRNA destabilizing factor from trypanosomes, homodimerizes and interacts with novel AU-rich element- and Poly(A)-binding proteins forming a ribonucleoprotein complex. J Biol Chem 2002; 277(52): 50520-8. [http://dx.doi.org/10.1074/jbc.M209092200] [PMID: 12403777]

[91] Jensen BC, Ramasamy G, Vasconcelos EJ, Ingolia NT, Myler PJ, Parsons M. Extensive stage-regulation of translation revealed by ribosome profiling of *Trypanosoma brucei.* BMC Genomics 2014; 15: 911. [http://dx.doi.org/10.1186/1471-2164-15-911] [PMID: 25331479]

[92] Alves LR, Oliveira C, Mörking PA, *et al.* The mRNAs associated to a zinc finger protein from *Trypanosoma cruzi* shift during stress conditions. RNA Biol 2014; 11(7): 921-33. [http://dx.doi.org/10.4161/rna.29622] [PMID: 25180711]

[93] Mörking PA, Rampazzo RdeC, Walrad P, *et al.* The zinc finger protein TcZFP2 binds target mRNAs

enriched during *Trypanosoma cruzi* metacyclogenesis. Mem Inst Oswaldo Cruz 2012; 107(6): 790-9. [http://dx.doi.org/10.1590/S0074-02762012000600014] [PMID: 22990970]

[94] Dallagiovanna B, Pérez L, Sotelo-Silveira J, Smircich P, Duhagon MA, Garat B. *Trypanosoma cruzi*: molecular characterization of TcPUF6, a Pumilio protein. Exp Parasitol 2005; 109(4): 260-4. [http://dx.doi.org/10.1016/j.exppara.2005.01.003] [PMID: 15755425]

[95] Pérez-Díaz L, Correa A, Moretão MP, Goldenberg S, Dallagiovanna B, Garat B. The overexpression of the trypanosomatid-exclusive TcRBP19 RNA-binding protein affects cellular infection by *Trypanosoma cruzi.* Mem Inst Oswaldo Cruz 2012; 107(8): 1076-9. [http://dx.doi.org/10.1590/S0074-02762012000800021] [PMID: 23295764]

[96] Guerra-Slompo EP, Probst CM, Pavoni DP, Goldenberg S, Krieger MA, Dallagiovanna B. Molecular characterization of the *Trypanosoma cruzi* specific RNA binding protein TcRBP40 and its associated mRNAs. Biochem Biophys Res Commun 2012; 420(2): 302-7. [http://dx.doi.org/10.1016/j.bbrc.2012.02.154] [PMID: 22425988]

[97] Michaeli S. Non-coding RNA and the complex regulation of the trypanosome life cycle. Curr Opin Microbiol 2014; 20: 146-52. [http://dx.doi.org/10.1016/j.mib.2014.06.006] [PMID: 25063970]

CHAPTER 7

Virulence Factors and Immune Evasion in *Leishmania* spp.

Jose M. Requena* and **Manuel Soto**

Centro de Biología Molecular Severo Ochoa (CSIC-UAM), Universidad Autónoma de Madrid, 28049 Madrid, Spain

Abstract: Protozoan parasites of the genus *Leishmania* cause a group of diseases, known as leishmaniasis, affecting humans and also household pets, mainly canids. In the human host, different pathological outcomes ranging from self-healing cutaneous lesions to systemic visceral leishmaniasis are produced by these parasites; these diseases affect millions of people worldwide. Similar to a virus, bacteria and other parasites, *Leishmania* need to evade immune destruction with the aim of completing their life cycle in their mammalian hosts. Moreover, the long co-evolutionary history between parasites of the genus *Leishmania* and their hosts for several millions of years has led to a balanced relationship. To avoid the powerful immune system of mammals, the parasite has developed a set of sophisticated mechanisms to persist, replicate, and spread.

Keywords: Complement system, Exosomes, Glycosylinositolphospholipids (GIPLs), IFN-γ, Immune response, Lipophosphoglycan (LPG), Macrophage, Neutrophil, Phagolysosome, Virulence factor.

INTRODUCTION

The diverse clinical manifestations of leishmaniasis, an infectious disease caused by species of the genus *Leishmania*, are dependent both on the virulence of the infecting species and on the immunological status of the host. This disease, in its cutaneous form, has been known and described since antiquity. Thus, human

* **Corresponding author Jose M. Requena:** Centro de Biologia Molecular Severo Ochoa (CSIC-UAM), Universidad Autonoma de Madrid, Madrid, Spain; Tel/Fax: +34 911964617; E-mail: jmrequena@cbm.csic.es

Marcelo Santos da Silva & Maria Isabel N. Cano (Eds.)

representations showing faces afflicted with cutaneous leishmaniasis (CL) have been found in Peru and Ecuador and dated 400-900 AD. The first description of the infectious agent was done by David D. Cuningham in 1885 when analyzing samples from lesions of "Delhi boil" in India. In 1903, independent investigations carried out by William Leishman and Charles Donovan led to the identification of the parasite in the spleen of visceral leishmaniasis (VL) patients. Soon after, Sir Ronald Ross named the parasite *Leishmania donovani* (see [1] and references therein).

Leishmaniasis cases have been reported in 98 countries and three territories [2]. Around 0.2 to 0.4 million and 0.7 to 1.2 million of cases of visceral and cutaneous leishmaniasis, respectively, occur each year. More than 90% of global VL cases occur in India, Bangladesh, Sudan, Ethiopia and Brazil. CL is more widely distributed, but 70 to 75% of cases concentrate in ten countries: Afghanistan, Algeria, Brazil, Colombia, Costa Rica, Ethiopia, Iran, North Sudan, Peru and Syria. It is likely that the actual incidence of leishmaniasis is underestimated, due to misdiagnosis and under-declaration. The relevance of this parasitic disease is further stressed out by the rise of *Leishmania*/HIV co-infection in many parts of the world, including European countries where up to 9% of the AIDS patients, before the introduction of HAART therapy, also experienced VL [3].

The long evolutionary history between *Leishmania* and their mammalian hosts has modulated both the strength of the immune response against the parasite and the virulence of the pathogen, leading often to a peaceful cohabitation (asymptomatic infection). Thus, the understanding of this balanced relationship will be useful for understanding disease progression and, therefore, paramount for developing immunotherapies for controlling leishmaniasis.

At present, there is no effective vaccine to prevent leishmaniasis in humans, despite effective vaccination against cutaneous leishmaniasis has been practiced for centuries by a procedure known as leishmanization, which consisted in the deliberate inoculation of virulent organisms from the exudates of active lesions [4]. The major hurdles for developing immunotherapies to either prevent or treat leishmaniasis are both the complexity of the host-pathogen interactions and our limited understanding of the precise immune mechanisms required for controlling parasite growth without causing disease. Details about *Leishmania* spp. vaccine candidates and prophylactic protocols are available in Chapter 11. Current treatments for leishmaniasis patients rely on chemotherapy, but drug treatment is undermined by unwanted toxicity problems in patients and the appearance of drug-resistant parasites [5].

Table 1. ***Leishmania*** **virulence factors involved in immune evasion.**

Leishmania **molecule**	**Effects on immune response and pathology**	**References**
A2	Associate with visceral tropism	[18]
Cysteine peptidase B (CPB)	Promotes Th2-type responses	[222, 224]
Elongation translation factor-1α (EF-1α)	Blockage of iNOS expression in response to macrophage treatment with IFN-γ	[195]
Fructose-1,6-bisphosphate aldolase	Interferes with the induction of nitric-oxide synthase in response to IFN-γ treatment	[197]
GP63 (leishmanolysin)	Resistance to complement-mediated lysis, inhibition of natural killer cellular functions, resistance to antimicrobial peptides	[39]
Lipophosphoglycan (LPG)	Blocks NADPH oxidase assembly at the phagosome membrane; suppresses IL-12 production by host macrophages	[81, 221]
Glycosylinositolphospholipids (GIPLs)	Inhibition of nitric oxide synthase; inhibit production of TNF-α	[35, 36]
Amastigote proteophosphoglycan (aPPG)	Ectopic activation of complement, avoiding parasite lysis by complement	[34]
Promastigote surface antigen (PSA)	Binding to CR3 of macrophages, protects *Leishmania* from complement-mediated lysis	[49] [48]
Ribosomal protein S3a	Inhibits T-cell proliferation and downregulates secretion of IFN-γ, IL-2 and IL-12	[220]
Leishmania RNA virus	Its presence is associated with metastasizing ability of *L. guyanensis*	[27]

This chapter is focused on characterized virulence factors (Table **1**), either membrane-associated or secreted by *Leishmania* that are involved in subverting host immune responses, favoring the immune evasion and persistence of the parasite. For getting a greater understanding of the evasion mechanisms, a comprehensive overview of our current knowledge on the host immune responses during *Leishmania* infection and the immune cells hosting the parasite is also presented.

LIFE CYCLE

As already detailed in previous chapters, parasites of the genus *Leishmania* have a digenetic life cycle: they proliferate as flagellated promastigotes in the gut of sandflies and as intracellular amastigotes inside mammalian macrophages, although some other cell types may host the parasite. In addition to humans, animals such as dogs and rodents are also susceptible to *Leishmania* infections [6]. The natural vectors of all forms of leishmaniasis are the female sand flies belonging to the subfamily Phlebotominae [7]. In the Old World the parasite is transmitted by phlebotomine insects of the genus *Phlebotomus*, and in the New World it is transmitted primarily by *Lutzomyia* and rarely by species of the genus *Psychodopygus*.

In the gut of the insect vector, two developmental stages (procyclic and metacyclic forms) are adopted by the parasite. Procyclic forms multiply attached to the epithelial cells of the midgut and, under nutritional stress, differentiate into the infective metacyclic stage, which is the transmission form of *Leishmania* [8]. Accordingly, metacyclic promastigotes are prepared to resist the human complement system. These vector-transmitted forms enter the host skin when the infected sandfly bites for taking its blood meal. Apart from the gut environmental conditions of the sandfly, necessary for parasite thriving and differentiation, an important reason why sandflies, and not another arthropods, are the vector for *Leishmania* is found in components of sandfly saliva that counteract the local immunological response and favor parasite multiplication, enhancing the development of leishmanial lesions.

After inoculation into the mammalian host, promastigotes are phagocytosed by different cells: neutrophils, dendritic cells and macrophages. The parasite proliferates inside the phagolysosome compartment, where it differentiates into the obligate intracellular amastigote stage. A number of surface molecules, including the abundant lipophosphoglycan (LPG) and several GPI-anchored glycoproteins, have been shown to be essential for promastigote survival immediately after transmission and during initial stages of infection and cell invasion (see below). The amastigotes multiply within one or more phagolysosomal vacuoles (PV). The morphologies of mature PVs vary depending upon the *Leishmania* species: large communal PVs (*Leishmania amazonensis* and *Leishmania mexicana*) and tight individual PVs (*Leishmania major and L. donovani*) are observed [9]. At some stage, the infected cell, when a large number of amastigotes are present, bursts and releases free amastigotes, which may infect other cells. Thus, amastigotes are the disseminating form, and it is commonly admitted that, after their release from infected macrophages, they can be phagocytosed by adjacent macrophages. However, an alternative view of parasite egress has been proposed: amastigotes would be released in a synchronized fashion, through an exocytosis-like process [10]. More recently, evidence has been found that amastigotes are transferred from cell-to-cell during cell death of the infected cell. Thus, enclosed within host cell membranes, amastigotes are moved to a new cell without exposure to the extracellular milieu, representing a strategy to evade host immune system [11].

Another relevant question to understand is how *Leishmania* survives under the harsh conditions existing in the acidic phagolysosomes of macrophage cells. During phagocytosis, the enzyme NADPH oxidase provokes an oxidative burst, which is responsible for the dramatic increase in oxygen consumption that is experienced by the macrophage. Thus, activation of macrophages and the production of the cytokines IFN-γ and TNF-α lead to an increase in the NADPH oxidase activity with the subsequent production of reactive oxygen species (ROS), mainly superoxide radical (O_2^-). In turn, O_2^- combines with other elements, either spontaneously or enzymatically, leading to the formation of hydrogen peroxide (H_2O_2), hydroxyl radical (HO^-), hypochlorite (OCl^-), and peroxynitrite ($ONOO^-$). Additionally, the activated macrophages, through the inducible nitric oxide synthase (iNOS or NOS2), produce high levels of nitric oxide (NO) and NO-derived metabolites. NO is generated by the conversion of L-arginine to L-citrulline; the molecule is unstable and converts spontaneously to nitrite and nitrate. From the moment the *Leishmania* parasite enters the macrophage, the parasite must express antioxidant enzymes and proteins, as well as actively to inhibit the production of reactive species such as O_2^- and NO in the macrophage (reviewed in [12]). Thus, parasite persistence within the macrophages is determined by a balance between the ability of the immune response to activate

Leishmania-infected macrophages and the capacity of the parasite to resist the toxic products resulting from macrophage activation.

Although the life cycles are similar in the different *Leishmania* species, an important question to resolve is why some species, like *L. donovani*, are able to disseminate into human internal organs, whereas other species, like *L. major*, remain in the skin and produce only cutaneous affections [13]. The features that control this tropism are thought to include both host and parasite genetics as well as the status of immunity. In part, these conclusions arise from experimental subcutaneous needle infection of mice, in which the tropism showed by these *Leishmania* species is not maintained. Thus, infection in the footpad of BALB/c mice with *L. major* leads to parasite multiplication and lesion development at the site of injection, but parasite dissemination to visceral organs is also observed [14]. In contrast, dissemination of *L. major* to the visceral organs does not occur in C57BL/6 mice, which are naturally resistant, indicating an important role for the host genetics and immune response in the control of visceralization [14]. However, an unexpected outcome occurs when BALB/c mice are subcutaneously infected with *L. donovani*, a viscerotropic species in humans; thus, this species causes minimal swelling at the site of injection and no dissemination to the viscera occurs in the mice. Nevertheless, intravenous inoculation resulted in hepatosplenomegaly and an increasing visceral parasite burden [15].

On the other hand, experimental data suggest that viscerotropic *Leishmania* species withstand *in vitro* higher temperatures than dermotropic ones [16]. Furthermore, some *L. donovani/Leishmania infantum* genes, like the genes A2, when introduced into *L. major* have been found to enhance *L. major* ability to visceralize in BALB/c mice [17], and remarkably the expression of A2 in *L. major* also inhibits its ability to establish a cutaneous lesion in mice, a phenotype typical of *L. donovani* [18]. Moreover, expression of A2 in *Leishmania tarentolae*, a lizard parasite which is nonpathogenic to mammals, resulted in increased survival of the A2-expressing line in mouse visceral organs [19]. Genes A2 encode a family of proteins (42-100 kDa), made up almost entirely of copies of a repetitive amino acid sequence, that are mainly expressed in amastigotes of *L. donovani* and *L. infantum*, whereas *L. major* possesses only a truncated A2 pseudogene [20]. More recently, it was found to be stress-inducible and to protect against heat shock and oxidative stress [21], although the exact mechanism remains to be understood.

CLINICAL MANIFESTATIONS OF LEISHMANIASIS

A variety of syndromes, ranging from self-healing skin ulcers to severe systemic disease, is grouped under the term leishmaniasis [22]. The type of disease that evolves depends both on the species of the parasites and on the host immune status at the time of infection. Thus, species like *L. major*, *L. mexicana*, and *Leishmania braziliensis* primarily cause cutaneous lesions, whereas *L. donovani* and *L. infantum* cause visceral leishmaniasis [23]. Nevertheless, this is not a fixed role as, depending on the host immune response, a given *Leishmania* species can cause different pathologies, including either subclinical infections, self-resolving lesions or systemic disease. The ongoing battle between the robust immune response mounted by the mammalian host and the evasion strategies, used by the parasite to counteract it, will ultimately decide the outcome of the infection. Likewise, the clinical spectrum of leishmaniasis may be envisioned as the phenotypic manifestation of the combat between *Leishmania* and the host immune response. Thus, spontaneously healing lesions are associated with antigen-specific T-cell responsiveness, diffuse cutaneous and visceral disease with T-cell non-responsiveness, and mucocutaneous disease with T-cell hyper-responsiveness [24]. Furthermore, many people in endemic regions have positive skin tests without having shown external signs of disease, suggesting that susceptible and resistant phenotypes also exist within human populations.

In humans, many *Leishmania* species produce localized cutaneous leishmaniasis (CL), which are the most benign form of leishmaniasis. At the site of the sandfly bite, a papule develops and enlarges until 2-5 cm in diameter, hereafter the lesion ulcerates and most often it heals spontaneously. However, even self-limiting forms of CL require weeks or months to heal, and ultimately leave fibrotic scars [22].

Visceral leishmaniasis (VL), also known as Kala-azar in the Indian subcontinent, is a systemic disease caused by the dissemination of *L. donovani* or *L. infantum* throughout the reticuloendothelial system (spleen, liver, and bone marrow). In VL patients, it is observed a loss of antigen-specific cell-mediated immunity, which is demonstrated by the lack of proliferation against leishmanial antigens in peripheral blood mononuclear cells (see [25] and references therein). It is the most severe form of leishmaniasis, having high rates of mortality and morbidity; in the absence of drug treatment, the disease is fatal. The emergence of *Leishmania*-HIV co-infection and increased prevalence of drug-resistant strains are worsening the impact of the disease. Moreover, acquired immunosuppression induced either during organ transplantation or associated with rheumatology, hematology and oncology disorders is another important factor contributing to the VL immunopathology [26]. In summary, the outcome of the infection with a viscerotropic species (*L. donovani* or *L. infantum*), whether the individual remains asymptomatic or progresses toward VL, depends on many factors related to parasite virulence, host genetic features and environmental conditions. These factors govern the quality and magnitude of the host immune response and ultimately the severity of the infection and disease.

In addition to the most frequent forms of leishmaniasis, CL and VL, there are other types of clinical manifestations [24]. Thus, diffuse cutaneous leishmaniasis (DCL) is considered to be an anergic variant of CL in which lesions are disseminated; sometimes, symptoms may be confused with lepromatous processes. Post-kala-azar dermal leishmaniasis (PKDL) is primarily caused by *L. donovani* and is endemic in East Africa and India. PKDL is a dermotropic form of the disease that may occur several months or years after kala-azar is cured. The disease is characterized by the presence of hypopigmented macules and erythematous eruptions.

Metastatic complications, even though can arise in immunocompromised patients infected with different *Leishmania* species, are particularly prevalent and aggressive in the New World, mainly associated with the infection of species of the subgenus *Viannia*; *L. braziliensis* is the most common species causing mucocutaneous leishmaniasis (MCL), also known as espundia. This destructive disease appears in some persons months or years after healing of a localized cutaneous ulcer. These *Leishmania* species show an especial tropism for the delicate nasopharynx tissues leading to an inordinately potent inflammation that contrast with the low number (sometimes undetectable) of viable parasites. These findings emphasize a major role of the immune response in disease pathology [23].

For *Leishmania guyanensis*, another species of the subgenus *Viannia*, the ability of the parasite to metastasize seems to be associated with the presence in its cytoplasm of an RNA virus, named *Leishmania* RNA virus (LRV). Evidence that the presence of LRV in the *Leishmania* parasite can influence the course of leishmaniasis has been experimental observed. Thus, in a hamster model of infection, LRV-bearing strains of *L. guyanensis* were found to metastasize repeatedly in contrast to their LRV-negative counterparts [23]. An in-depth characterization of the immunological response showed that the viral double-stranded RNA (dsRNA) provoked a potent inflammatory response after interacting with endosomal Toll-like receptor 3 (TLR3), resulting in the production of high levels of IFN-β [27]. LRV has since been found in various strains of *L. guyanensis* and *L. braziliensis* isolated from patients with MCL [28].

MEMBRANE COMPONENTS AND RELATED MOLECULES

Lipophosphoglycan (LPG) is the major glycoconjugate on the surface of *Leishmania* promastigotes (Fig. **1**). There are about 1.25×10^6 molecules of LPG per cell, covering the entire surface, including the flagellum, and it forms a densely packed filamentous glycocalyx. The LPG molecule consists of a backbone of multiple repeats of the phosphodisaccharide –6-Gal-β(1,4)-Man-α(1)-PO^-_4, which is linked *via* a hexasaccharide glycan core to a 1-O–alkyl-2-l-so-phosphatidylinositol lipid anchor, and it is capped by a nonphosphorylated oligosaccharide [7, 29]. During metacyclogenesis LPG undergoes structural alterations, being elongated by the addition of phosphorylated disaccharide units and the molecule may double its length in some *Leishmania* species. LPG is considered to be a major determinant of vector competence for *Leishmania* species. This molecule mediates promastigote attachment to the mid-gut epithelium and prevents parasite loss during blood-meal excretion; the insect galectin PpGalec has been identified as the receptor for LPG [30]. Also, LPG seems to be relevant during the initial steps in the mammalian host, as the continuous surface glycocalyx of LPG would be protecting the promastigote stage from complement-mediated lysis and macrophage-induced oxidative stress during phagocytosis (see below). However, although LPG has been identified as an important virulence factor for *L. major*, it is not so for *L. mexicana* [31].

LPG is considered to be a multifunctional virulence factor required for parasite infectivity in sandflies and mammals. Most of its biological activity has been associated with the phosphoglycan repeats, cap structures and side-chain modifications. However, these structures are not restricted to LPG, but also occur in a family of *Leishmania* glycoproteins, known as proteophosphoglycans or PPGs [32]. Moreover, *Leishmania* amastigotes, which do not synthesize LPG, secrete amastigote PPG (aPPG) when they thrive into the parasitophorous vacuole of macrophages. This molecule is present in high concentrations in the lesion tissue of *L. mexicana*-infected mice and might accumulate to the mg/ml level in the parasite-harboring parasitophorous vacuoles [33]. Furthermore, the soluble molecule aPPG possesses the remarkable property of activating the complement *via* the mannose-binding lectin pathway. Thus, intravenous injection of only 10 μg of aPPG into a mouse depletes complement to such an extent that the alternative pathway is nonfunctional for more than 24 h [34]. Therefore, it has been postulated that, in active lesions caused by *L. mexicana*, the release of aPPG by ruptured host cells or vesicular traffic might exhaust C3 locally, and thereby help the parasites to avoid complement lysis [32].

Leishmania glycosylinositolphospholipids (GIPLs) are present in high copy numbers and cover most of the amastigote cell surface. These molecules belong to the glycosylphosphatidylinositol (GPI) family and are structurally analogous to the LPG and GP63 anchors (Fig. **1**). GIPLs consist of a Man-α(1-4)-GlcN core linked to an alkyl-acylglycerol through a phosphatidylinositol (PI) molecule; distinct groups of GIPLs exist that differ in the sugar-chains bound to the mannose residue [29]. These molecules are endowed with immunoregulatory properties; for example, the *L. mexicana* GIPLs inhibits TNF-α expression, down-regulating macrophage responses to infection [35]. Additionally, it has been described that *Leishmania* GIPLs inhibit the synthesis of nitric oxide, the most efficient leishmanicidal factor synthesized by IFN-γ activated macrophages [36]. Finally, glycoconjugates have been found to interfere with the correct functioning of the macrophage transduction pathways, inhibiting IL-12 secretion in the infected macrophages [37, 38].

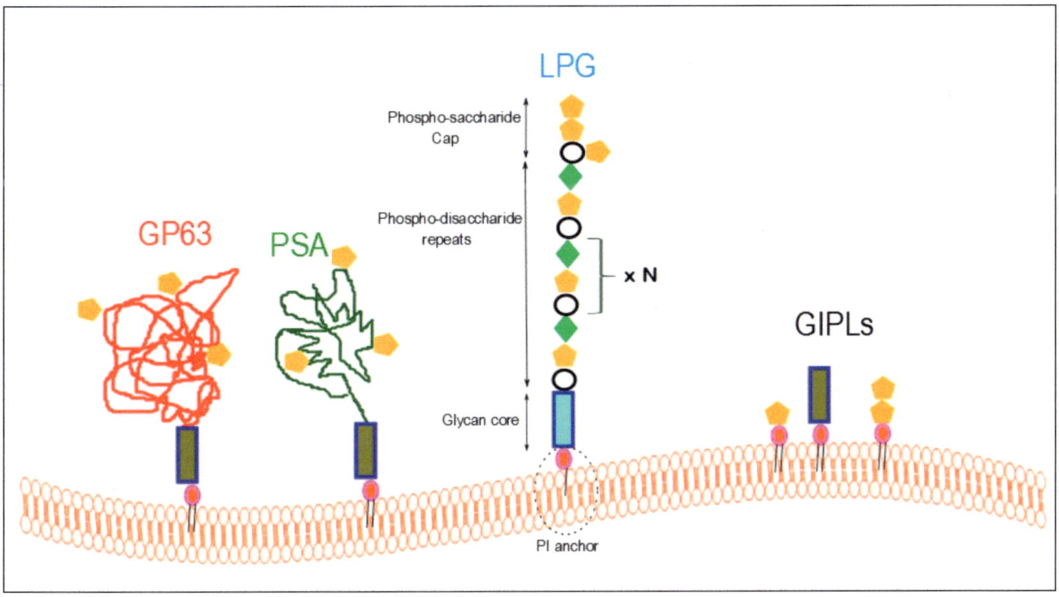

Fig. (1). Major molecules anchored to the membrane of *Leishmania*. Surface glycoproteins GP63 and glycolipids GIPLs are present in both parasites stages, but glycoprotein PSA and lipophosphoglycan (LPG) are specific for promastigotes. Most of these structures are anchored to the membrane by a GPI anchor ($Gal_3(Glc\text{-}1\text{-}PO_4)Man_2GlcN$ linked to the PI anchor). The LPG has four domains, the PI anchor (phosphatidylinositol linked to lyso-alkylglycerolipid anchor), the conserved hexasaccharide core, the phosphoglycan repeats (Gal-β(1,4)-Man-α(1)-PO^-_4) and the cap (generally mannose-rich). The number of phosphoglycan repeats varies and depends on the stage of differentiation and the species of *Leishmania*.

Leishmania is a zinc-dependent metalloprotease of about 63-kDa, designated GP63 or leishmanolysin. GP63 belongs to the metzincin class (EC 3.4.24.36) and is attached to the membrane *via* a GPI-anchor (reviewed in [39]). It is the most abundant surface glycoprotein of the parasite; it has been estimated that GP63 accounts for 1% of the total protein content of *Leishmania* promastigotes. Although GP63 of different *Leishmania* species shows high sequence conservation, substrate specificity variations have been reported [40]. This class of metalloproteases possesses a conserved zinc binding/catalytic signature motif HEXXHXXGXXH and an N-terminal pro-peptide that serves to maintain the pro-enzyme inactive during translation; the pro-peptide is removed upon protein maturation and activation. The mature GP63 contains three domains: the catalytic domain located at the N-terminal moiety, which is separated from the C-terminal domain containing the site of glycosylphosphatidylinositol (GPI) anchor addition by a central domain, which contains multiple sites of N-linked glycosylation.

Apart from its membrane location, there is a constant shedding of GP63 to the extracellular space by cleavage of the GPI anchor *via* phospholipase C. In addition, it has been suggested that GP63 may be also liberated from the parasite by a mechanism dependent on autoproteolysis [41]. Also, intracellular pools of GP63 exist that might be released upon certain extracellular triggers [42]. The protein is encoded in a multigene array; the different GP63 genes have differences in sequence as well as in expression patterns, even though the roles played by the different genes/proteins are not fully characterized [43].

Several functions in the pathogenesis of leishmaniasis have been ascribed to GP63, including evasion of complement-mediated lysis, facilitation of the entry of promastigotes in macrophages, interaction with the extracellular matrix, inhibition of natural killer cellular functions, resistance to antimicrobial peptides, and survival of intracellular amastigotes [39]. Thus, GP63, like LPG, promotes the entry into the phagocyte through the interaction with the mannose/fucose receptor on the host cell surface. In addition, it was experimentally demonstrated that GP63 contributes to parasite adhesion by interacting with human fibronectin receptors [44]. Moreover, GP63 was found to be also involved in the defense against antimicrobial peptides (AMPs), as *gp63*-deficient promastigotes were efficiently killed in a dose-dependent manner by AMPs [45]. Remarkably, to interact and survive within its host, *Leishmania* GP63 is able to gain access to the macrophage cytoplasm and cause cleavage of several host proteins, interfering with prominent signaling pathways (see below). In contrast, targeted deletion of the *L. major gp63* genes did not, however, significantly alter the growth and development of the parasite within the phlebotomine vector [46].

The promastigote surface protein (PSA) belongs to a family of *Leishmania* glycoproteins, including Gp46/M2 proteins, that contains a large number of leucine-rich repeats (LRR) in their structure; these proteins may be either anchored to the membrane by a glycosylphosphatidylinositol (GPI) anchor or excreted as soluble forms [47]. A member of the family, PSA-2, mediates the entry in macrophages by binding to the receptor CR3 [48]. In addition, PSA has been found to protect *L. infantum* from complement-mediated lysis [49].

On the other hand, membrane alterations associated with parasite cell death have been found relevant for modulating host immune response and favoring infection progression. In particular, it was reported that virulent inocula of *Leishmania* promastigotes contain a high ratio of annexin A5-binding parasites, which shown apoptotic features [50]. After depleting the apoptotic parasites from the population, the remaining promastigotes are not able to survive in phagocytes *in vitro* and lose their ability to induce disease in mice. Apoptotic parasites elicit production of TGF-β, which in turn blocks macrophage activation, allowing the survival and multiplication of *Leishmania*. In fact, it is a belief that the parasite would be exploiting a property of the immune system, *i.e.* the immune-silencing phagocytotic process occurring during up-taking of apoptotic cells by macrophages [51]. In general, the recognition of phosphatidylserine (PS) on the membrane of apoptotic cells by immune cells leads to the release of anti-inflammatory cytokines (such as TGF-β and IL-10) and down-regulation of the proinflammatory cytokine TNF-α. In this context, phagocyte functions are suppressed. Thus, it is not surprising that parasites, and even viruses, take advantage of the system used for the removal of apoptotic cells, as a strategy for minimizing the host inflammatory response. This process is known as apoptotic mimicry, by which the pathogenic organism, without death as the necessary outcome, is able to mislead phagocytic cells [52].

PHAGOCYTES ARE HOST CELLS FOR *LEISHMANIA*

Leishmania and the Neutrophil

Leishmania parasites are obligate intracellular pathogens and phagocytes are their preferred host cells. Although classically macrophages are mentioned as the host cells for *Leishmania*, it is now known that they are neither the first nor the only to be recruited to the site of inoculation ([53] and references therein). Using dynamic intravital microscopy and flow cytometry, it was observed a rapid and sustained infiltration of neutrophils (polymorphonuclear neutrophil granulocytes or PMN) at the site of sand fly bite [54]. Furthermore, it was observed that the invading neutrophils efficiently capture *L. major* parasites, but phagocytosed parasites remain viable. In fact, neutrophil depletion reduced the ability of parasites to establish productive infections, suggesting that initial infection of PMN is a key step in the infection process. It is not immediately intuitive how neutrophils may enhance the capacity of *Leishmania* to establish infection in the mammalian host, but some clues exist. Thus, co-incubation of PMN with *L. major* promastigotes results in a substantial increased survival time of neutrophils, which in another way is very short due to spontaneous apoptosis [55]. Other studies have shown that *Leishmania* infection leads to enhanced expression of the anti-apoptotic proteins Bcl-2 and Bfl-1 in neutrophils [56]. The enhanced life span of host cells is a common goal for intracellular parasites. Nevertheless, it remains to be uncovered how *Leishmania* dismantle the microbicide arsenal of a cell born to kill infectious agents [57]. It should be taken into account that neutrophils are key players for early defense against infections; they are prepared to kill obligate intracellular pathogens *via* ROS, neutrophil elastase (NE), and neutrophil extracellular traps (NETs) [57]. Although it was described that *Leishmania* promastigotes induces release of NETs and are killed by them [58, 59], being mammalian histones especially toxic for the parasite [60], evidence exists that promastigotes are able to escape NETs mediated killing throughout the action of some surface components such as the LPG or active nuclease enzymes (Fig. **2**) [59, 61].

Neutrophils have the capacity to phagocytose 80-90% of the infective inoculum [62]. It is thought that contact occurs through the interaction of neutrophil CR3 with C3bi deposited on the surface of the parasite, but also by the direct interaction of parasite with neutrophil pattern recognition receptors (PRRs) [63]. However, neutrophils are unable to lyse the internalized promastigotes [54]. In contrast, parasites have the ability to modulate the expression of some chemokines and cytokines in the infected neutrophil [64]. In particular, increased production of TGF-β combined with a decrease in TNF-α production in the infected neutrophil contributes to generating a permissive immunological environment allowing parasite survival [65]. However, internalized promastigotes have not been observed to differentiate into amastigotes or undergo intracellular proliferation within neutrophils [63], suggesting that these cells are not appropriate for parasite multiplication. In this scenario, neutrophils behave as parasite reservoirs until they are ingested by macrophages.

Although delayed by the *Leishmania* infection, PMNs finally die by apoptosis and this feature results important for the parasite. Macrophages readily phagocytose infected apoptotic PMN, but *Leishmania* internalized by this indirect way survive and multiply inside of these phagocytes [66]. As apoptotic neutrophils are phagocytosed through receptor-mediated pathways that fail to trigger macrophage defense responses, their cargoes of promastigotes are not destructed in the macrophage phagosome. Moreover, ingestion of apoptotic infected PMN resulted in the release of the anti-inflammatory cytokine TGF-β by the macrophages. Thus, it has been suggested that *Leishmania* can misuse PMNs as a "Trojan horse" to enter their preferred host cells (macrophages) in an immunological silent manner [66, 67]. Moreover, certain molecules such as high mobility group protein B1 (HMGB1) and IL-1β (dubbed alarmins), which are released from ruptured neutrophils, act as both chemoattractants to the site of infection and activators of monocyte-derived DCs (moDCs). It has been suggested that those infected moDCs would facilitate parasite traffic to the draining lymph node (reviewed in [68]). Alternatively, free promastigotes that are not phagocytosed by neutrophils or liberated by them after neutrophil lysis [54] could infect macrophages in an anti-inflammatory environment associated with the presence of apoptotic neutrophils [69, 70]. Both processes are illustrated in Fig. (**2**).

Leishmania and the Macrophage

Although parasites can be readily found in neutrophils, it is within macrophages when a new step of the infective process begins: parasites undergo transformation to the amastigote form and replicate. Substantial evidence exists that *Leishmania* might increase the lifespan of infected macrophages; life prolongation of the host is a common strategy used for many parasites to ensure completion of its infectious life cycle. Thus, it was found that infection by *L. donovani* promastigotes of bone marrow-derived macrophages (BMMs) inhibits apoptosis in the macrophages and, remarkably, the same effect was observed after treatment of BMMs with a preparation of lipophosphoglycan [71]. More recently, it was described the ability of the *L. major* protein Lm1740MIF, which shows significant structural homology with human macrophage migration inhibitory factor (MIF), to inhibit apoptosis of macrophages [72]. This *Leishmania* MIF-like protein shows significant binding interaction with the human MIF receptor, and, like its mammalian counterpart, Lm1740MIF induces ERK1/2 MAP kinase activation in a CD74-dependent manner, inhibiting the activation-induced apoptosis of macrophages [72].

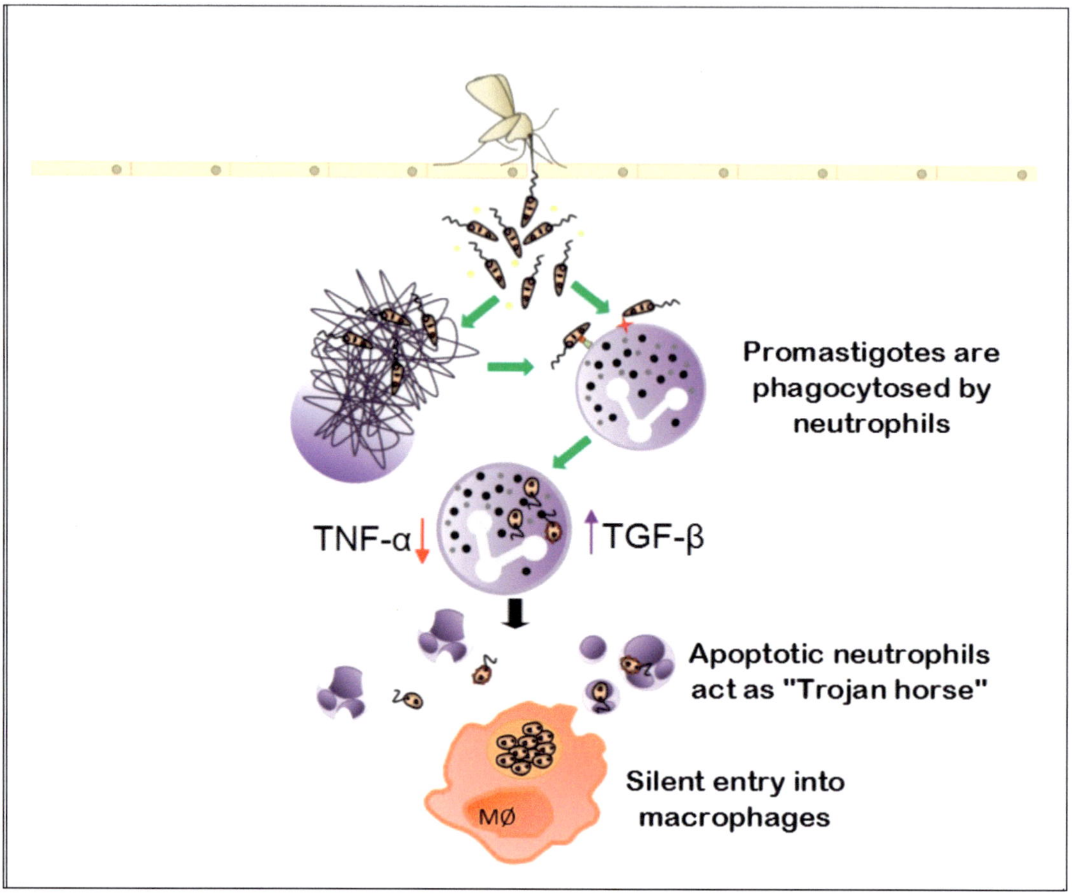

Fig. (2). Dual role of the neutrophils in the early events occurring after parasite entry in the host. On one side, the contents of neutrophil extracellular traps (NETs) can kill some promastigotes while others are resistant because of their membrane components. On the other side, phagocytosed promastigotes are not destroyed by the neutrophil and enter the macrophages inside the neutrophils showing apoptotic markers (as a Trojan horse) or by direct contact to macrophages attracted for the phagocytosis of apoptotic neutrophils. (Image generously provided by Laura Ramirez).

Regarding the molecules mediating parasite-host cell adhesion, the two major membrane components of the parasite (GP63 and LPG) and a variety of host cell receptors (mannose receptor and fibronectin receptor, among others) have been involved [73]. Thus, *Leishmania* GP63 interacts with cellular receptors for fibronectin, which are involved in the binding of both promastigote and amastigote forms [44]. Although the mannose-fucose receptor (MR) has been implicated in the entry of *Leishmania* parasites [74], its relevance in the entry of the parasite has been questioned and seems to depend on the species and life-cycle stage used in the studies [75]. The macrophage Fc receptor (FcR) is also implicated in parasite entry, since it is able to bind the immunoglobulins present on the parasite surface [76]. Finally, as indicated in the previous section, some promastigotes can infect the macrophage when this cell takes up apoptotic infected neutrophils. Moreover, complement receptors (CRs) have been found to play a major role in the uptake of *Leishmania* [77]. There are three receptors that could potentially mediate C3-dependent phagocytosis, CR1 (CD35) and CR3 (Mac-1; CD11b/CD18; aMb2) that recognize C3b and C3bi, respectively, and CR4 (p150, 95). The use of CR1 and CR3 may provide a survival advantage to the parasites, since engagement of these receptors promote phagocytosis without prompting an oxidative burst. Another benefit for the parasite, when uses the CR3 receptor, is that an inhibition of IL-12 induced cell-mediated immunity occurs [78].

In macrophages, it is generally accepted that the endosomal vacuole containing the parasite undergoes a maturation process to acquire lysosomal properties, but promastigotes seem to have the ability to regulate this process. A delayed phagosomal maturation was first demonstrated with *L. donovani* [79]; this study suggested that LPG is responsible for the inhibition of phagolysosomal biogenesis and, therefore, this would represent an intramacrophage survival strategy used by promastigotes to establish infection. Additional studies indicated that the acidification of endosomes, typically occurring during their transformation into phagolysosomes, results inhibited due to the integration of LPG into lipid microdomains of the phagosome membrane. The presence of LPG leads to exclusion or loss of synaptotagmin V, which is responsible for the recruitment of the vesicular proton ATPase (V-ATPase) that in turn drives phagosome acidification [80]. In addition, it was observed that during the first hour of phagocytosis, F-actin accumulates around phagosomes containing *L. donovani* promastigotes. This effect was observed after phagocytosis of yeasts coated with purified LPG but not after phagocytosis of LPG- defective mutants; thus, these experiments allowed to attribute this effect to the presence of LPG. Likewise, LPG causes the exclusion of cytosolic components of the NADPH oxidase from the phagosome membrane, avoiding production of reactive oxygen species (ROS) inside the parasitophorous vacuole [81].

In *L. amazonensis* infected macrophages, the parasite-containing phagosomes are enlarged, and it has been suggested that this may be another strategy used by the parasite to dilute the leishmanicidal effects, like acidification and nitric oxide production [82]. Thus, in the large parasitophorous vacuoles, in which the parasites are hosted, the hydrolytic content of the lysosomes, upon fusion, is diluted to a level below their effectiveness, allowing promastigotes to differentiate without any requirements of fusion delay.

Once inside the parasitophorous vacuole, parasites must neutralize the effects of ROS and reactive nitrogen species (RNS) that are generated by the membrane-bound NADPH-dependent oxidases and by the nitric oxide synthase (NOS), respectively. *Leishmania* parasites possess several antioxidant systems, including superoxide dismutases, peroxidases, and low-molecular weight thiols [12, 83]. The redox homeostasis in *Leishmania* relies on the dithiol trypanothione, whereas in nearly all living organisms maintenance of an intracellular reducing milieu is based on the sulphur-containing tripeptide, glutathione. Thus, the *Leishmania* trypanothione reductase has been shown to be essential for the fight against ROS and RNS, as parasites with reduced enzymatic activity are susceptible to intracellular killing by macrophages [84]. Other enzymes associated with protection against these reactive species are the *L. infantum* peroxiredoxins LicTXNPx and LinTXNPx [85], the *L. major* pteridin reductase [86], and the *L. donovani* superoxide dismutase [87]. Direct evidence of the role played by TXNPx in defense against NOS has been obtained by ectopic over-expression of the molecule in *L. amazonensis* promastigotes, which showed a significant enhanced resistance against peroxynitrite ($ONOO^-$) cytotoxicity. Moreover, TXNPx-overexpressing parasites exhibited greater intra-macrophage survival, and increased parasite load and lesion development in a murine model of leishmaniasis [88].

Another hurdle that *Leishmania* must face inside the macrophage parasitophorous vacuole is the limited availability of iron, which the parasite needs for growth and replication (reviewed in [89]). In macrophages, the efflux pump Nramp1 actively translocates Fe^{2+} from the phagolysosome into the cytosol, as a microbicide arm that restricts the iron availability to the phagocytosed parasite. *Leishmania* counteracts this effect by the activation of its own iron transporters, which effectively compete with the host iron transporter [90]. Arginine is also a matter of dispute between the host and the parasite, this amino acid is essential for the synthesis of polyamines by intracellular *Leishmania* amastigotes, but is also needed by macrophages for the production of the microbicidal molecule nitric oxide [91]. Accordingly, using deficient gene mutants of *Leishmania*, it was found that the arginase enzyme not only facilitates parasite growth by supplying substrate for the parasite polyamine pathway, but also interferes with the iNOS-dependent microbicidal mechanism of infected macrophages by competing for intracellular arginine [92].

Leishmania is Internalized by Other Mammalian Cells

Dendritic cells (DCs), bone marrow-derived antigen presenting cells (APCs) that play a central role in initiating both innate and adaptive immune responses, have been found to be relevant cells for hosting *Leishmania*. The outcome of the parasite-DC interaction varies widely; the magnitude and profile of DC activation depend on the *Leishmania* species/strains, developmental stages (promastigote/amastigote), DC subsets, serum opsonization, and exogenous DC stimuli, among others [93]. On the other hand, the Langerhans cells (a type of DC present in the epidermis) also internalize *Leishmania* and transport them to lymph nodes for priming of *Leishmania*-specific T cells [94]. Interestingly, mice depleted of epidermal DCs develop smaller lesions and have a reduced number of disease promoting T regulatory cells, suggesting that Langerhans cells are negative regulators of the anti-*Leishmania* response [95].

Skin fibroblasts have been described as host cells for both promastigotes and amastigotes. Compared with macrophages, fibroblasts had a reduced ability to express type 2 nitric oxide synthase and to kill intracellular *L. major* parasites. It has been suggested that fibroblasts might be important host cells for *Leishmania* as they can serve as safe cells during the chronic phase of infection [96]. Other cell types that have been postulated to function as "safe targets" during long-term persistence of *Leishmania* are: immature myeloid precursor cells and monocytes, sialoadhesin-positive stromal macrophages of the bone marrow and hepatocytes ([97] and references therein).

IMMUNOLOGY OF LEISHMANIASIS

The current knowledge of the responses involved in the resistance and/or susceptibility to leishmaniasis has been based on studies with leishmaniasis patients as well as using experimental infection models, especially murine models [98, 99]. Although there are differences in the responses generated following infection by different species of *Leishmania*, and in different strains of mice, some general conclusions can be extracted. These correlate, partially at least, to the immunological aspects associated with clinical features following natural infection in humans and dogs. In general, protective immunity is associated with the development of controlled inflammatory immune responses, mediated mainly by $CD4^+$ Th1 lymphocytes, but also by $CD8^+$ T cells, through IFN-γ secretion. This stimulation by IFN-γ enhances the antigen presentation process, the synthesis of TNF-α (cytokine which acts in coordination with the IFN-γ to maintain the activated status of the macrophage) and the production of nitric oxide (NO), a molecule with an elevated leishmanicidal capacity [100, 101]. This general mechanism for the destruction of intercellular parasites is belief to occur also in human and canine leishmaniasis [102 - 104]. On the other hand, Th2 responses and the production of antibodies by B lymphocytes, as well as immuno-suppressive responses mediated by IL-10, are related to susceptibility to infection and the development of the pathology [105 - 107]. Fig. (**3**) summarizes the immune response associated with resistance and susceptibility to *Leishmania* infection.

Regarding the initiation of the acquired immune response after infection, DCs are responsible for capturing the parasites and, after maturation, for migrating to the draining lymph nodes, where they present *Leishmania* antigens to $CD4^+$ and $CD8^+$ T cells *via* MHC-II and MHC-I, respectively [108]. At the early stages of infection, uptake of the parasites occurs mainly by the interaction with the DC Fc-receptor, given the presence of natural antibodies deposited on the parasite surface. Fc-receptor mediated phagocytosis stimulates the secretion of IL-12, a cytokine produced by different DC subtypes that result crucial for the priming and maintenance of local CD4$^+$ Th1 cells producing IFN-γ [108 - 110]. In addition, DCs are also able to activate CD8$^+$ T cell responses by cross-presentation, especially for the *Leishmania* antigens that are secreted-excreted [111]. In contrast, phagocytosis of neutrophils carrying *Leishmania* parasites by DCs results in the inhibition of pro-inflammatory cytokine production and a down-expression of costimulatory molecules [112]. Additionally, parasite-induced production of IL-10 by regulatory B cells during antigen presentation has been implicated in the induction of CD4$^+$ Th2 cell responses against *Leishmania* antigens, due to the down-regulation of IL-12 secretion by the DCs involved in the differentiation of naive T cells [113]. In humans, *L. major* promastigotes also prime human DCs for the production of CD40L-triggered IL-12, and these cells drive a Th1 response in autologous T cells [114]. However, this stimulation was found to be species-specific: contrary to *L. major*, neither *L. donovani* nor *L. tropica* induced IL-12 production after uptaking by DCs [115].

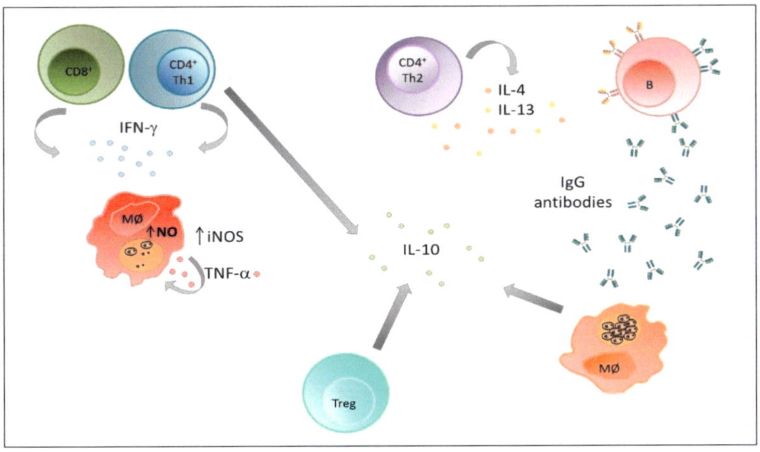

Fig. (3). Cell types and cytokines associated with the resistance and susceptibility to *Leishmania* infection. Resistance to infection is related to the induction of IFN-γ mediated responses. This cytokine is mainly produced by CD4$^+$ and CD8$^+$ T cells. The coordinated effects of IFN-γ and TNF-α, mainly produced by IFN-γ stimulated macrophages, upregulate the expression of genes implicated in nitric oxide synthesis, creating a leishmanicidal state. The production of IL-10 by different cells interferes with macrophage activation, allowing parasite replication. Patients affected by the visceral forms of the disease have a CD4$^+$ Th2 dependent high humoral response against parasite proteins. (Image generously provided by Laura Ramirez).

The profiles of cytokines present during the early interaction between T cells and APCs appear to be also important determinants of susceptibility or resistance to infection. For example, following infection in BALB/c mice by *L. major*, a progressive cutaneous lesion is produced and the parasite is disseminated to the internal organs, concurrent with the generation of a powerful humoral response mediated by $CD4^+$ Th2 lymphocytes, with the production of parasite-specific IL-4 and IL-13 [116]. The production of both cytokines is associated with the progression of the infection. Thus, IL-4 knockout BALB/c mice are resistant to infection by some strains of *L. major,* although the degree of resistance increases when they are deficient in the IL-4 receptor, which is shared by both cytokines [117]. Moreover, over-expression of IL-13 in C57BL/6 mice makes them susceptible to infection by *L. major* [118]. The Th2 lymphocytes collaborate in the activation of B lymphocytes to produce a large quantity of IgG1 isotype antibodies [119], aimed specifically against parasite proteins (Fig. **3**).

In humans, $CD4^+$ Th1 cells are also considered to be critical for the control of *Leishmania* infections, owing to their ability to make IFN-γ, which activates macrophages and DCs, leading to the death of the intracellular hosted parasite. However, an excessive Th1 response may be detrimental for the host. Thus, for example, human CL caused by *L. braziliensis* is characterized by a strong Th1 response with the production of high levels of IFN-γ and TNF-α. Moreover, the strength of this exaggerated Th1 response is correlated with the development of lesions and the severity of the disease [120].

In contrast, the relevance of $CD8^+$ T cells in controlling leishmaniasis remains controversial (reviewed in [68, 121]). Thus, early studies suggested that $CD8^+$ T cells may play a role in the healing process in CL [122]. However, in other studies, the recruitment of $CD8^+$ T cells that express the granule-associated serine protease granzyme B has been correlated with lesion progression in patients infected with *L. braziliensis* [123, 124].

IL-10 is produced by different cell types (B lymphocytes, infected macrophages and several classes of T lymphocytes [125]) and plays a leading role during infection (Fig. **3**). This cytokine is recognized for its ability to inhibit activation and effector functions of T cells and, therefore, it is related to susceptibility. Thus,

BALB/c mice treated with antibodies against the IL-10 receptor are highly resistant to infection by *L. major* [126, 127]. In the *L. infantum* murine infection models, splenic chronification depends on the synthesis of IL-10, since it has been verified that BALB/c or C57BL/6 mice treated with an IL-10 receptor blocking antibody, or transgenic strains that are genetically deficient for IL-10, have the ability to destroy the parasites in the spleen [128 - 130]. In addition, the immune complexes, formed by the interaction between parasite antigens and the abundant antibodies generated in the host during the infection, stimulate synthesis of IL-10 (and suppress production of IL-12) by the macrophages [131]. The relationship existing between the production of IL-10 and susceptibility in murine models is also observed in human leishmaniasis. Thus, VL patients present high serum IL-10 levels and an elevated expression of IL-10 mRNAs in the affected tissues [105, 132]. In these patients, IL-10 is produced by adaptive regulatory T lymphocytes (Tr1: $CD4^{+}Foxp3^{-}$) [105] and even by *Leishmania*-antigen specific Th1 lymphocytes [133]. All these data clearly relate IL-10 and its anti-inflammatory role in susceptibility to infection by *Leishmania*. Nevertheless, there is evidence indicating that this cytokine may also play a major role in resistance to re-infection. In the model *L. major*-C57BL/6 the production of IL-10 by natural regulatory T cells ($CD4^{+}$ $Foxp3^{+}$) has been associated with the control of the Th1 effector response at the end of the resolution phase. Such control, which occurs before the complete sterilization of the affected tissues, leads to the chronification of the infection [134, 135]. Natural $CD4^{+}$ $CD25^{+}$ T-regulatory (Treg) cells, which constitute 5–10% of peripheral CD4+ T cells in mice and humans, has been involved in the persistence of *L. major* in the skin after lesion-healing in resistant C57BL/6 mice [135]. In humans, functional Tregs are present in skin lesions of patients with mucocutaneous leishmaniasis (infected with *L. braziliensis*) and they produce large amounts of IL-10 and TGF-β [136]. Remarkably, chronic infections have been related to protection against re-infection, fundamentally by means of maintaining effector T lymphocytes against the parasite proteins [135, 137 - 139]. Thus, the persistence of parasites can be envisaged as the result of the co-evolution between the host and the parasite: the presence of the parasite provides a continuous antigenic stimulation of the host immune system, required for maintaining protective memory cells [138].

Finally, it should be mentioned that salivary gland components have been found to affect the immune response significantly and contribute to increasing *Leishmania* infectivity [140]. Thus, saliva from either *Lutzomyia longipalpis* or *Phlebotomus papatasi* exerts immunomodulatory activities on T cells and macrophages by inhibiting the expression of Th1 type cytokines and inducing the expression of Th2 cytokines by activated macrophages [141, 142]. Thus, co-administration of parasites with sandfly saliva greatly increases the infectivity of cutaneotropic species in mice [143]. Sand fly saliva components can inhibit antigen presentation by DCs and up-regulate cytokines related to the induction of Th2 responses, inhibiting pro-inflammatory responses [140, 144].

Animal Models of Leishmaniasis

Murine experimental leishmaniasis is a useful model to study disease pathogenesis, which in its various aspects has given valuable insights into the understanding of human leishmaniasis [106, 145]. In particular, the cutaneous leishmaniasis model based on the inoculation of *L. major* in mice represents one of the best models of chronic infection, and it served to obtain the first evidence of the *in vivo* relevance of the balance Th1/Th2 in the immune response [98, 146]. Activated *Leishmania*-specific $CD4^+$ T cells are detected in draining lymph nodes as early as 16 h post-infection in mice [147]. Remarkably, these lymphocytes rapidly expand and differentiate into cytokine-secreting cells that exhibit different phenotypes depending on the mouse strain. *L. major* infection in most mouse strains (C57BL/6, C3H, CBA) causes skin lesions similar to those seen in humans [148]. These lesions heal spontaneously, and animals become resistant to reinfection. As depicted above, control of infection is mainly mediated by the activation of *Leishmania*-specific $CD4^+$ T cells that differentiate in T-helper type 1 (Th1) cells, which secrete IFN-γ and efficiently contribute to the parasite elimination through the activation of macrophages and $CD8^+$ cells [149]. This strong cellular response is accompanied by a mild humoral response dominated by the production of IgG2a isotype antibodies against different parasite antigens. In contrast, in genetically susceptible BALB/c mice, the cellular response is dominated by Th2 cells that secrete IL-4, IL-5, and IL-13 [98]. DCs play a critical role on the type of immune response elicited, they not only process and present parasite peptides to $CD4^+$ T cells, but also secrete cytokines such as IL-12 that promote the polarization of naive T cells into Th1 effector lymphocytes [150].

However, this dichotomy is not observed in other *Leishmania* species. Thus, infection with *L. amazonensis*, another cutaneotropic species for humans, causes progressive cutaneous lesions in most inbred strains of mice [151]. A deficient activation of early immune responses, in part due to the inhibition of DCs migration [152] and in part as a consequence of the parasite-specific IL-10-mediated responses, contributes to the pathogenesis of leishmaniasis [151]. On the other hand, BALB/c mice infected by *L. braziliensis* are capable of controlling the infection after the development of Th1/Th2 mediated response [153].

The relationship between Th1-resistance and Th2-susceptibility may be correlated with the immune response observed in natural human and canine infections. Thus, resistant persons and patients recovered from leishmaniasis display a positive delayed-type hypersensitivity (DTH) reaction and possessed $CD4^+$ Th1 lymphocytes producing IFN-γ against the parasite proteins [154 - 156]. In contrast, patients presenting clinical signs of disease are usually DTH-negative [157] and show strong humoral responses against parasite proteins [158]. Moreover, the antibodies generated against the parasite proteins can form immune complexes that are responsible for the pathology in different tissues because of their deposition on filtration barriers such as the central nervous system [159] and the kidneys [160, 161]. However, patients with active VL express elevated levels of pro-inflammatory cytokines both systemically and at the site of infection, suggesting that an absolute Th2 polarization cannot by itself explain VL.

Regarding experimental infections with viscerotropic species, the most common model consist in the intravenous inoculation of high parasite doses [128]. After the challenge, at earliest phases, intracellular amastigotes are observed in the liver, spleen and bone marrow in mice. Four to 8 weeks later (depending on the number of parasites used in the challenge [162]) an effective immune response is mounted, leading to parasite clearance in the liver. This response is mainly mediated by IL-12, which favors the expansion of IFN-γ producing Th1 cells. The control of parasite replication in the liver ultimately depends on the upregulation of the inducible nitric oxide synthase gene and the production of NO [15]. Nevertheless, the infection at spleen and bone marrow becomes chronic. The lack of control of parasite multiplication in the spleen depends on different factors, namely: disruption of the microarchitecture of the splenic marginal zone due to the production of high levels of TNF-α affecting the interactions between APCs and T lymphocytes, the induction of parasite-specific IL-10 production by different cells, and the inhibition of $CD8^+$ T cell responses [68, 163, 164]. A similar evolution of the disease has been observed in mice infected with viscerotropic species using the intradermal or subcutaneous routes, an administration that in part mimics the natural infection [165, 166]. In addition, although low titers of parasite-specific antibodies are found in the sera of mice suffering VL, a polyclonal B-cell activation occurs at the initial stages of infection, as occur in human patients with VL [167]. However, although the mouse model seems to be a good model of VL during earliest stages of infection, it is not appropriate for the pathogenic stages because the lack of progressive disease [168].

VL pathogenesis is better reproduced in the golden hamster (*Mesocricetus auratus*), when the parasite is administered intravenously. The animals develop a progressive disease with close similarities to human and canine VL pathologies: progressive increase of parasite load in viscera, cachexia, hepatomegaly, splenomegaly, pancytopenia, hypergammaglobulinemia, and finally death of the animal [169, 170]. More recently, the hamster model has been refined, using intradermal inoculation of the parasite combined with vector saliva [171]. The dog is another susceptible model for VL [102, 172, 173], and studies performed in this model are of great interest because they are naturally infected by *L. infantum/L. chagasi* and develop a serious form of the disease. The intravenous challenge with the parasite reproduces the same clinical forms of leishmaniasis found in the field, such as asymptomatic, oligosymptomatic and symptomatic animals. The resistance is associated with cell response (cellular proliferative responses and DTH) and absence of circulating antibodies against parasite antigens. In contrast, susceptible animals exhibit high antibody titers and no cellular responses against parasite antigens [174]. As occur in other mammalian models, a combination of parasites and sandfly saliva for the experimental intradermic infections in dogs have also been established for testing new vaccines against the disease [175]. Development of experimental models that more closely mimics natural infection are on development by exposing laboratory species (mice, dogs and even non-human primates) to the bite of the infected sandflies [54, 176, 177]. Optimization of these models will generate considerable progress in the close development of protective vaccines and more accurate treatments that will help to control the expansion of this neglected disease.

STRATEGIES EVOLVED BY *LEISHMANIA* FOR EVASION OF THE IMMUNE RESPONSE

Leishmania parasites have evolved actively to subvert many aspects of the host immune response, ranging from preventing the production of microbicidal molecules and protective cytokines to interfering with effective antigen presentation and induction of an adaptive immune response. In addition to surface-exposed virulence factors, which may serve to establish a direct contact with the host cell receptors, *Leishmania* is able actively to secrete a large repertoire of molecules [178]. Additional studies on the secretion pathways have led to the discovery that *Leishmania* uses an exosome-based secretion mechanism to release effector molecules either to the extracellular environment or into the parasitophorous vacuole [179].

Avoiding Lysis by the Complement System

The complement system is an essential and efficient component of the immunological response, owing to its ability to recognize rapidly and eliminate microorganisms. As a consequence, most of the pathogenic organisms have developed different mechanisms to escape the attack of complement lytic factors [180].

Regarding *Leishmania*, it is often said that metacyclic promastigotes are resistant to complement-mediated lysis. This statement is mainly based on the article by Puentes *et al.* [181], in which it was documented that *L. major* metacyclic promastigotes, purified from stationary stage culture, have an increased resistance to complement lysis regarding logarithmic growth forms. After analyzing deposition of different complement components, it was concluded that some developmental modification of the parasite membrane during metacyclogenesis blocks insertion of the C5b-9 lytic complex into the promastigote membrane. It was suggested that the elongation of LPG on the surface of metacyclic promastigotes may protect the parasite from complement-mediated lysis. However, an absolute resistance to complement lysis does not seem a feature for *Leishmania* infections. Thus, classical studies from Sacks' laboratory indicated that survival of stationary phase promastigotes in undiluted human serum varies among species: all *L. major* and *L. amazonensis* were killed, while up to 30% of *Leishmania panamensis* and 10% of *L. donovani* survived [182]. Similarly, using near-physiological serum concentration and temperature conditions, it was shown that approximately 90% of axenic- and sand fly-derived *L. infantum* promastigotes were rapidly killed by complement [183].

Complement resistance is associated with the major components of the promastigote glycocalyx: LPG and GP63 [37, 184]. Thus, it was found that *L. major* GP63-deficient promastigotes, even though develop normally inside the sandfly vector, either recovered from sandflies or grown in culture show an increased sensitivity to complement-mediated lysis and a delay in lesion formation in BALB/c mice [185]. Similarly, an *L. major lpg1*$^{-}$ mutant, lacking the galactofuranosyltransferase necessary for the synthesis of the LPG glycan core, was highly susceptible to human complement [186].

On the other hand, the interaction of natural antibodies (IgM or IgG) with promastigotes activates the classical pathway of the complement, and it was suggested to be an efficient way for the destruction of the parasites inoculated by the vector [187]. Nevertheless, there is evidence demonstrating that the interaction with the circulating natural antibodies favors the immunoadherence of the parasite to human erythrocytes (or platelets in the case of non-primate mammals). This interaction constitutes an evasive mechanism since, in less time than it takes for the complement to destroy the parasite, *Leishmania* is transferred to the professional APCs facilitating host colonization [188 - 190]. Similarly, it has been proved that the lectin pathway can be activated because of the interaction of the host mannose-binding lectin (MBL) with sugars present on the parasite surface; thus, binding of the parasite to MBL promotes its phagocytosis by APCs, allowing disease progression [191].

Nevertheless, the presence of complement inhibitors in the sand fly saliva co-inoculated with the parasites in natural infection [192] has ruled out a relevant role of complement-mediated lysis as an early mechanism of resistance to infection. Moreover, promastigotes would be taking advantage of deposition of complement components (opsonization) to modulate the responses elicited by APCs, including DCs, and promoting the entry into macrophages [193].

In summary, even though some of the parasites may be destroyed by the complement system, an important fraction of the promastigotes are not only able to evade the lytic action of this system, but use it to facilitate the interaction and entry into the host macrophages.

Subverting Intracellular Signaling and Exosomes

Leishmania parasites are able to secure their survival and propagation within their host by altering signaling pathways involved in regulating the ability of macrophages to kill pathogens or to engage adaptive immune system (reviewed in [194]).

A rapid activation of host tyrosine phosphatases, in particular SHP-1, by *Leishmania* seems to be an important evasion strategy whereby the parasites can negatively regulate several key macrophage pathways and render them unresponsive to activating stimuli such as IFN-γ and, by doing so, *Leishmania* can interfere with several molecules involved in NO production including JAK2, Erk1/2, and the transcription factors NF-κB and AP-1 [194]. In this regards, *Leishmania* elongation factor-1α (EF-1α) has been found to interact with and activate the SHP-1 protein, promoting a deactivated phenotype in the infected macrophages [195]. This finding was somewhat surprising, given that EF-1α is a highly conserved protein, known by its activity as a regulator of the protein translation process. Nevertheless, the SHP-1-binding and -activating properties of *L. donovani* EF-1α were found to be highly specific, as mammalian EF-1α orthologue is devoid of such properties. Moreover, *Leishmania* EF-1α has been found among the proteins secreted in exosomes, which are delivered to macrophage cytosol, where its interaction with host SHP-1 occurs [196].

Another *Leishmania* protein, fructose-1,6-bisphosphate aldolase was also identified as a ligand for human SHP-1 [197]. The presence of the *L. donovani* aldolase in the cytosolic fractions of infected macrophages indicates that the protein aldolase is exported from phagolysosomes into the cytoplasm of infected cells. Co-immunoprecipitation experiments have shown the association of *Leishmania* aldolase with SHP-1. Moreover, *Leishmania* aldolase-expressing macrophages were unable to induce expression of nitric-oxide synthase in response to IFN-γ treatment [197]. Therefore, these data suggest that *Leishmania* aldolase is an SHP-1 binding and activating protein that contributes to macrophage dysfunction.

On the other hand, Gomez and co-workers [198] showed that SHP-1 and other protein tyrosine phosphatases are cleaved by GP63, which is able to gain access to the macrophage cytoplasm probably through secretion of exosomes by *Leishmania* [199]. Also, it has been described that *Leishmania*-derived GP63 enters the nucleus of the host macrophage to cleave some subunits of the factor AP-1(Activated Protein-1) rendering it inactive. AP-1 regulates pro-inflammatory cytokines, chemokines, and NO production. In particular, c-Jun, the central component of AP-1, results cleaved by *Leishmania* GP63. Furthermore, immunoprecipitation experiments showed that GP63 interacts, and cleaves c-Jun at the perinuclear area, shortly after infection [200]. Another cellular target of GP63 is the mammalian/mechanistic target of rapamycin (mTOR), a serine/threonine kinase that regulates the translational repressor 4E-BP1. As a consequence of mTOR cleavage, mTORC1 complex results inhibited, and the concomitant activation of 4E-BP1 promotes *Leishmania* proliferation [201]. Translational control by mTORC1 is a key component of the innate immune response.

Toll-like receptors (TLRs) are a family of membrane-associated molecules specialized in the recognition of a variety of pathogen-associated molecular patterns (PAMPs); they are essential for microbial recognition and are important mediators of innate immune responses to infection [202]. Upon engagement, TLRs mediate the activation of different transcription factors, such as NF-kB, leading to the production of inflammatory cytokines and microbicidal effectors [203]. Several TLRs such as TLR2, TLR3, TLR4, TLR7 and TLR9 have been shown to contribute to the innate sensing and recognition of *Leishmania* by various innate immune cells ([204] and references therein). To counteract this innate immune response, parasites have evolved strategies to interfere with TLR-signaling cascades, subverting the inflammatory responses. In this regards, it is noticeable the role played by ectoin-like molecules identified in *L. major* that are able to prevent the activation of TLR4 during the *Leishmania*–macrophage interaction, exerting a beneficial effect on intracellular amastigote survival and growth in macrophages, due to the inhibition of ROS production [205].

Substantial evidence supports that exosomes, produced by *Leishmania* in its intracellular niche, contain molecules able to modulate and/or subvert signaling processes of the host cell [68, 179, 206, 207]. Exosomes are 40–100 nm vesicles that bud from the plasma membrane and their existence has been documented in a variety of mammalian cells but also in parasitic pathogens [208], including *Leishmania* [209]. Thus, exosomes would facilitate the entry of *Leishmania* virulence factors into the host cytoplasm, through fusion with the mammalian cell plasma membrane (if externally located) or with the phagolysosome membrane (in the intracellular stage). *Leishmania* exosomes have been found to suppress the immune response, and the heat-shock protein 100 (HSP100) has been found to play a major role in the selection of packed proteins into exosomes. Thus, the absence of HSP100, in mutant parasites, leads to the formation of exosomes with a different cargo that exerts a pro-inflammatory effect on immune cells [206]. Similarly, by performing comparative proteomic analyses of wild-type exosomes and exosomes derived from *L. major gp63*$^{-/-}$, it has been shown that GP63 may be playing a key role in exosomal protein sorting [210].

Leishmania Interferes with Antigen Presentation by Professional Cells

Effective protection from infections with intracellular microorganisms requires the induction of pathogen-specific cell-mediated immune responses. Th cells and CTL need to be activated by professional APCs to gain effector functions and to respond in an antigen-specific manner [211]. APCs, either by phagocytosis of microorganisms or by intracellular parasite degradation processes, present antigen fragments in the context of MHC molecules type I and II [212].

As depicted before, *Leishmania* amastigotes reside and multiply in macrophages, within phagolysosomal compartments called parasitophorous vacuoles. These structures have properties similar to those described for the MHC class II compartments of APCs, where peptide-class II molecule complexes are formed before being trafficked to the cell surface. Thus, it has been described that *L. amazonensis* or *L. mexicana* intracellular amastigotes endocytose and degrade class II molecules, as a possible strategy evolved by these parasites to escape the potentially hazardous antigen presentation process [213].

The maturation of phagosomes into phagolysosomes, a process referred to as phagolysosome biogenesis, is essential for the killing of phagocytosed microbes. During the process, transfer of sets of hydrolases, acquisition of the oxidative machinery, involved in the killing of microorganisms, and the processing of some of their proteins for antigen presentation on MHC class II molecules occur. Additionally, interactions between phagosomes and subregions of the endoplasmic reticulum contribute to cross-presentation, the process by which exogenous microbial antigens are presented on MHC class I molecules [214]. Matheoud and co-workers [215] showed that *Leishmania* evades immune recognition of the infected cell by inhibiting phagolysosome biogenesis. They found that the metalloprotease GP63 cleaves a subset of SNARE proteins, including VAMP8. Thus, inactivation of VAMP8 would prevent the assembly of the NADPH oxidase complex on phagosomes, altering the presentation of exogenous *Leishmania* antigens on MHC class I molecules.

Interfering with Cytokine Production

By interfering with host cell signaling pathways, the different species of *Leishmania* have evolved the ability to modulate the profile of cytokines expressed by infected host cells to favor dissemination of the parasite and prevent its eradication [216]. As shown in Fig. (**3**), IFN-γ, released by Th1 cells, results key for triggering the leishmanicidal activity of macrophages *via* expression of the inducible NO synthase, whose activity is paramount for killing of intracellular parasites [217]. Therefore, it is not surprising that some *Leishmania* molecules are able to interfere with IFN-γ production. Thus, GP63 has been involved in the reduction of IFN-γ producing cells in BALB/c mice infected with *L. amazonensis* [218]. In addition, GP63, by a process involving degradation of synaptotagmin XI, induces the release of TNF-α and IL-6 *in vivo*; these pro-inflammatory cytokines may be responsible for the influx of neutrophils and inflammatory monocytes to the *Leishmania* inoculation site [219].

Cordeiro-da-Silva and co-workers [220] reported that the *Leishmania* LmS3arp protein, homologous to mammalian ribosomal protein S3a, inhibits T-cell proliferation and downregulates secretion of IFN-γ, IL-2 and IL-12.

LPG has been described to be able to block protein kinase C (PKC) activity [38]. PKC signaling is known to play a vital role in the regulation of macrophage functions, activating the expression of cytokines such as IFN-γ and TNF-α. LPG glycoconjugates also have a profound effect on the survival of *Leishmania* parasites through their ability to modulate the expression of the inducible nitric oxide synthase in macrophages [36]. Additionally, *Leishmania* lipophosphoglycans inhibit macrophage IL-12 production by stimulating the MAP kinase ERK [221]. On the other hand, infection with lesion-derived *L. mexicana* amastigotes leads to the inhibition of LPS-induced IL-12 production by mouse bone marrow-derived macrophages. This inhibitory effect was associated with expression of cysteine peptidase B (CPB) because inhibition is not observed either when infections are done with CPB-deletion mutants or when cells are preincubated with a CPB inhibitor. The results of this study suggest that this amastigote-specific cysteine peptidase inhibits IL-12 production by modulating the NF-κB signaling pathway [222]. The effects of CPB on the murine host immune system has also been tested by subcutaneously injecting an enzymatically active recombinant CPB protein into the foot-pads of BALB/c mice. The protein elicits increases in IL-4 and IL-5 expression (Th2-related cytokines) and in the levels of circulating IgE [223]. Therefore, CPB seems to favor the establishment of a Th2-type response for the benefit of *Leishmania* [224]. Another cysteine, CPA (cys1) from *L. infantum*, might also be a virulence factor as a *ΔLicpa* mutant was found to have low infectivity for mammalian hosts *in vivo* and for cells *in vitro* [225].

Another important function exerted by LPG is the suppression of the pro-inflammatory cytokine IL-1β expression, by inhibiting transcription of its coding gene, in human monocytes, which contributes to the macrophage deactivation, and defective T-cell responses observed in patients with leishmaniasis [226].

Finally, it has been shown that exosomes from *L. donovani* modulate human monocyte cytokine responses to IFN-γ in a bimodal fashion, by promoting IL-10 production and inhibiting that of TNF-α [206]. The characterization of the particular molecule(s) responsible for this immunomodulation remains to be done.

CONCLUDING REMARKS AND FUTURE TRENDS

Leishmania has evolved many molecular mechanisms to evade or subvert host immune responses to establish chronic infection. A range of effector molecules, either surface-anchored or secreted by *Leishmania*, play key roles in this process. Elucidating the intricate cellular and molecular events that orchestrate the parasite biology and the mechanisms by which *Leishmania* alter the function of the host cells is essential for the design of novel and efficient approaches to prevent or treat leishmaniasis. In pursuing these objectives, the application of "omics" methodologies is very useful for identifying *Leishmania* species-specific genes [227] and analyzing its contribution to the host tropism and pathology [228]. Thus, the application of 'omics' analyses, to both the parasite and the host cell during infection, will certainly lead to an in-depth understanding of the molecules responsible for the pathogenesis of the different forms of leishmaniasis. Additionally, a better knowledge of the various types of T cells, and other immune cells, responsible for protection (as well as pathology) results pivotal.

There is a general agreement that development of vaccines against leishmaniasis seems achievable since patients recovered from leishmaniasis develop a protective immune response against subsequent infections. Unfortunately, although several approaches have been taken to develop a vaccine for leishmaniasis, to date none has been successful in humans. Details about *Leishmania* vaccines and vaccine candidates are found in Chapter 11.

Persistence of parasites in host cells seems to be necessary for acquiring resistance to re-infection. The presence of low numbers of parasites, even after apparent clinical cure, would activate continuously the immune system in a way akin to concomitant immunity. Substantial progress has been made towards the characterization of T cell subsets that contribute to such immunity [68]. The question of how to generate these cells by vaccination remains a challenge, as does the question of which T cells may provide the best protection.

CONFLICT OF INTEREST

The authors confirm that they have no conflict of interest to declare for this publication.

ACKNOWLEDGEMENTS

Work at the JMR's lab is funded by Ministerio de Economía y Competitividad-FEDER (grant SAF2013-47556-R), the Network of Tropical Diseases Research RICET-ISCIII-FEDER (grant RD12/0018/0009), the Comunidad Autónoma de Madrid (grant S2010/BMD-2361), and the European Community Seventh Framework Programme (grant agreement No. 603181, project MuLeVaClin). Work at the MS's lab is funded by the Spanish grant from Ministerio de Economía y Competitividad-FEDER (FISPI14/00366 from the Instituto de Salud Carlos III) and by the Brazilian grant from CNPq (Ciencia sem Fronteiras-PVE 300174/2014-4). An institutional grant from Ramon Areces is also acknowledged. JMR's ORCID ID: 0000-0003-3410-9488.

ABBREVIATIONS

AMPs	Antimicrobial peptides
AP-1	Activated Protein-1
APCs	Antigen presenting cells
aPPG	Amastigote proteophosphoglycan
BMMs	Bone marrow-derived macrophages
CL	Cutaneous leishmaniasis
CPB	Cysteine Peptidase B
CR	Complement receptor
DC	Dendritic cell
DCL	Diffuse cutaneous leishmaniasis
DTH	Delayed-type hypersensitivity
EF-1α	Elongation translation factor-1α
GIPLs	Glycosylinositolphospholipids
GPI	Glycosylphosphatidylinositol
iNOS	Inducible nitric oxide synthase

LPG	Lipophosphoglycan
LRV	*Leishmania* RNA virus
MBL	Mannose-binding lectin
MCL	Mucocutaneous leishmaniasis
MIF	Migration inhibitory factor
moDC	Monocyte-derived DC
MR	Mannose-fucose receptor
mTOR	Mammalian/mechanistic target of rapamycin
NETs	Neutrophil extracellular traps
NO	Nitric oxide
NOS2	See iNOS
PAMPs	Pathogen-associated molecular patterns
PI	Phosphatidylinositol
PKC	Protein kinase C
PKDL	Post-kala-azar dermal leishmaniasis
PMN	Polymorphonuclear neutrophil
PRRs	Pattern recognition receptors
PSA	Promastigote surface antigen
PV	Phagolysosomal vacuoles
RNS	Reactive nitrogen species
ROS	Reactive oxygen species
Th1	T-helper type 1 cells
TLR	Toll-like receptor
VL	Visceral leishmaniasis

REFERENCES

[1] Grevelink SA, Lerner EA. Leishmaniasis. J Am Acad Dermatol 1996; 34(2 Pt 1): 257-72. [http://dx.doi.org/10.1016/S0190-9622(96)80121-6] [PMID: 8642091]

[2] Alvar J, Vélez ID, Bern C, *et al.* Leishmaniasis worldwide and global estimates of its incidence. PLoS One 2012; 7(5): e35671. [http://dx.doi.org/10.1371/journal.pone.0035671] [PMID: 22693548]

[3] Alvar J, Aparicio P, Aseffa A, *et al.* The relationship between leishmaniasis and AIDS: the second 10 years. Clin Microbiol Rev 2008; 21(2): 334-59. [http://dx.doi.org/10.1128/CMR.00061-07] [PMID: 18400800]

[4] Kumar R, Engwerda C. Vaccines to prevent leishmaniasis. Clin Transl Immunology 2014; 3(3): e13. [http://dx.doi.org/10.1038/cti.2014.4] [PMID: 25505961]

[5] Croft SL, Sundar S, Fairlamb AH. Drug resistance in leishmaniasis. Clin Microbiol Rev 2006; 19(1): 111-26.
[http://dx.doi.org/10.1128/CMR.19.1.111-126.2006] [PMID: 16418526]

[6] Gramiccia M, Gradoni L. The current status of zoonotic leishmaniases and approaches to disease control. Int J Parasitol 2005; 35(11-12): 1169-80.
[http://dx.doi.org/10.1016/j.ijpara.2005.07.001] [PMID: 16162348]

[7] Sacks D, Kamhawi S. Molecular aspects of parasite-vector and vector-host interactions in leishmaniasis. Annu Rev Microbiol 2001; 55: 453-83.
[http://dx.doi.org/10.1146/annurev.micro.55.1.453] [PMID: 11544364]

[8] da Silva R, Sacks DL. Metacyclogenesis is a major determinant of *Leishmania* promastigote virulence and attenuation. Infect Immun 1987; 55(11): 2802-6.
[PMID: 3666964]

[9] Courret N, Fréhel C, Gouhier N, *et al.* Biogenesis of *Leishmania*-harbouring parasitophorous vacuoles following phagocytosis of the metacyclic promastigote or amastigote stages of the parasites. J Cell Sci 2002; 115(Pt 11): 2303-16.
[PMID: 12006615]

[10] Rittig MG, Bogdan C. *Leishmania*-host-cell interaction: complexities and alternative views. Parasitol Today (Regul Ed) 2000; 16(7): 292-7.
[http://dx.doi.org/10.1016/S0169-4758(00)01692-6] [PMID: 10858648]

[11] Real F, Florentino PT, Reis LC, *et al.* Cell-to-cell transfer of *Leishmania amazonensis* amastigotes is mediated by immunomodulatory LAMP-rich parasitophorous extrusions. Cell Microbiol 2014; 16(10): 1549-64.
[http://dx.doi.org/10.1111/cmi.12311] [PMID: 24824158]

[12] Van Assche T, Deschacht M, da Luz RA, Maes L, Cos P. *Leishmania*-macrophage interactions: insights into the redox biology. Free Radic Biol Med 2011; 51(2): 337-51.
[http://dx.doi.org/10.1016/j.freeradbiomed.2011.05.011] [PMID: 21620959]

[13] McCall LI, Zhang WW, Matlashewski G. Determinants for the development of visceral leishmaniasis disease. PLoS Pathog 2013; 9(1): e1003053.
[http://dx.doi.org/10.1371/journal.ppat.1003053] [PMID: 23300451]

[14] Laskay T, Diefenbach A, Röllinghoff M, Solbach W. Early parasite containment is decisive for resistance to *Leishmania major* infection. Eur J Immunol 1995; 25(8): 2220-7.
[http://dx.doi.org/10.1002/eji.1830250816] [PMID: 7664785]

[15] Melby PC, Yang YZ, Cheng J, Zhao W. Regional differences in the cellular immune response to experimental cutaneous or visceral infection with *Leishmania donovani.* Infect Immun 1998; 66(1): 18-27.
[PMID: 9423834]

[16] Callahan HL, Portal IF, Bensinger SJ, Grogl M. *Leishmania* spp: temperature sensitivity of promastigotes *in vitro* as a model for tropism *in vivo.* Exp Parasitol 1996; 84(3): 400-9.
[http://dx.doi.org/10.1006/expr.1996.0128] [PMID: 8948329]

[17] Zhang WW, Matlashewski G. Characterization of the A2-A2rel gene cluster in *Leishmania donovani*:

involvement of A2 in visceralization during infection. Mol Microbiol 2001; 39(4): 935-48. [http://dx.doi.org/10.1046/j.1365-2958.2001.02286.x] [PMID: 11251814]

[18] Zhang WW, Mendez S, Ghosh A, *et al.* Comparison of the A2 gene locus in *Leishmania donovani* and *Leishmania major* and its control over cutaneous infection. J Biol Chem 2003; 278(37): 35508-15. [http://dx.doi.org/10.1074/jbc.M305030200] [PMID: 12829719]

[19] Mizbani A, Taslimi Y, Zahedifard F, Taheri T, Rafati S. Effect of A2 gene on infectivity of the nonpathogenic parasite *Leishmania tarentolae.* Parasitol Res 2011; 109(3): 793-9. [http://dx.doi.org/10.1007/s00436-011-2325-4] [PMID: 21442256]

[20] Matlashewski G. *Leishmania* infection and virulence. Med Microbiol Immunol (Berl) 2001; 190(1-2): 37-42. [http://dx.doi.org/10.1007/s004300100076] [PMID: 11770107]

[21] McCall LI, Matlashewski G. Involvement of the *Leishmania donovani* virulence factor A2 in protection against heat and oxidative stress. Exp Parasitol 2012; 132(2): 109-15. [http://dx.doi.org/10.1016/j.exppara.2012.06.001] [PMID: 22691540]

[22] Pearson RD, Sousa AQ. Clinical spectrum of Leishmaniasis. Clin Infect Dis 1996; 22(1): 1-13. [http://dx.doi.org/10.1093/clinids/22.1.1] [PMID: 8824958]

[23] Hartley MA, Drexler S, Ronet C, Beverley SM, Fasel N. The immunological, environmental, and phylogenetic perpetrators of metastatic leishmaniasis. Trends Parasitol 2014; 30(8): 412-22. [http://dx.doi.org/10.1016/j.pt.2014.05.006] [PMID: 24954794]

[24] Murray HW, Berman JD, Davies CR, Saravia NG. Advances in leishmaniasis. Lancet 2005; 366(9496): 1561-77. [http://dx.doi.org/10.1016/S0140-6736(05)67629-5] [PMID: 16257344]

[25] Flora R, Aghazadeh-Dibavar S, Bandyopadhyay M, Dasgupta S. Immunosuppression during *Leishmania donovani* infection: a potential target for the development of therapy. Ann Parasitol 2014; 60(4): 239-45. [PMID: 25706420]

[26] van Griensven J, Carrillo E, López-Vélez R, Lynen L, Moreno J. Leishmaniasis in immunosuppressed individuals. Clin Microbiol Infect 2014; 20(4): 286-99. [http://dx.doi.org/10.1111/1469-0691.12556] [PMID: 24450618]

[27] Ives A, Ronet C, Prevel F, *et al. Leishmania* RNA virus controls the severity of mucocutaneous leishmaniasis. Science 2011; 331(6018): 775-8. [http://dx.doi.org/10.1126/science.1199326] [PMID: 21311023]

[28] Zangger H, Ronet C, Desponds C, *et al.* Detection of *Leishmania* RNA virus in *Leishmania* parasites. PLoS neglected tropical diseases 2013; 7(1): e2006.

[29] de Assis RR, Ibraim IC, Nogueira PM, Soares RP, Turco SJ. Glycoconjugates in new world species of *Leishmania*: polymorphisms in lipophosphoglycan and glycoinositolphospholipids and interaction with hosts. Biochim Biophys Acta 2012; 1820(9): 1354-65. [http://dx.doi.org/10.1016/j.bbagen.2011.11.001] [PMID: 22093608]

[30] Kamhawi S. Phlebotomine sand flies and *Leishmania* parasites: friends or foes? Trends Parasitol 2006; 22(9): 439-45.

[http://dx.doi.org/10.1016/j.pt.2006.06.012] [PMID: 16843727]

[31] Ilg T. Lipophosphoglycan is not required for infection of macrophages or mice by *Leishmania mexicana.* EMBO J 2000; 19(9): 1953-62.
[http://dx.doi.org/10.1093/emboj/19.9.1953] [PMID: 10790362]

[32] Ilg T. Proteophosphoglycans of *Leishmania.* Parasitol Today 2000; 16(11): 489-97.
[http://dx.doi.org/10.1016/S0169-4758(00)01791-9] [PMID: 11063860]

[33] Aebischer T, Harbecke D, Ilg T. Proteophosphoglycan, a major secreted product of intracellular *Leishmania mexicana* amastigotes, is a poor B-cell antigen and does not elicit a specific conventional CD4+ T-cell response. Infect Immun 1999; 67(10): 5379-85.
[PMID: 10496920]

[34] Peters C, Kawakami M, Kaul M, Ilg T, Overath P, Aebischer T. Secreted proteophosphoglycan of *Leishmania mexicana* amastigotes activates complement by triggering the mannan binding lectin pathway. Eur J Immunol 1997; 27(10): 2666-72.
[http://dx.doi.org/10.1002/eji.1830271028] [PMID: 9368624]

[35] Tachado SD, Gerold P, Schwarz R, Novakovic S, McConville M, Schofield L. Signal transduction in macrophages by glycosylphosphatidylinositols of *Plasmodium, Trypanosoma*, and *Leishmania*: activation of protein tyrosine kinases and protein kinase C by inositolglycan and diacylglycerol moieties. Proc Natl Acad Sci USA 1997; 94(8): 4022-7.
[http://dx.doi.org/10.1073/pnas.94.8.4022] [PMID: 9108098]

[36] Proudfoot L, Nikolaev AV, Feng GJ, *et al.* Regulation of the expression of nitric oxide synthase and leishmanicidal activity by glycoconjugates of *Leishmania* lipophosphoglycan in murine macrophages. Proc Natl Acad Sci USA 1996; 93(20): 10984-9.
[http://dx.doi.org/10.1073/pnas.93.20.10984] [PMID: 8855295]

[37] Sacks D, Sher A. Evasion of innate immunity by parasitic protozoa. Nat Immunol 2002; 3(11): 1041-7.
[http://dx.doi.org/10.1038/ni1102-1041] [PMID: 12407413]

[38] Descoteaux A, Turco SJ. Glycoconjugates in *Leishmania* infectivity. Biochim Biophys Acta 1999; 1455(2-3): 341-52.
[http://dx.doi.org/10.1016/S0925-4439(99)00065-4] [PMID: 10571023]

[39] Yao C. Major surface protease of trypanosomatids: one size fits all? Infect Immun 2010; 78(1): 22-31.
[http://dx.doi.org/10.1128/IAI.00776-09] [PMID: 19858295]

[40] Bouvier J, Schneider P, Etges R. Leishmanolysin: surface metalloproteinase of *Leishmania.* Methods Enzymol 1995; 248: 614-33.
[http://dx.doi.org/10.1016/0076-6879(95)48039-0] [PMID: 7674949]

[41] McGwire BS, OConnell WA, Chang KP, Engman DM. Extracellular release of the glycosylphosphatidylinositol (GPI)-linked *Leishmania* surface metalloprotease, gp63, is independent of GPI phospholipolysis: implications for parasite virulence. J Biol Chem 2002; 277(11): 8802-9.
[http://dx.doi.org/10.1074/jbc.M109072200] [PMID: 11777912]

[42] Yao C, Luo J, Hsiao C, Donelson JE, Wilson ME. Internal and surface subpopulations of the major surface protease (MSP) of *Leishmania chagasi.* Mol Biochem Parasitol 2005; 139(2): 173-83.
[http://dx.doi.org/10.1016/j.molbiopara.2004.11.005] [PMID: 15664652]

[43] Medina-Acosta E, Karess RE, Russell DG. Structurally distinct genes for the surface protease of *Leishmania mexicana* are developmentally regulated. Mol Biochem Parasitol 1993; 57(1): 31-45. [http://dx.doi.org/10.1016/0166-6851(93)90241-O] [PMID: 8426614]

[44] Brittingham A, Chen G, McGwire BS, Chang KP, Mosser DM. Interaction of *Leishmania* gp63 with cellular receptors for fibronectin. Infect Immun 1999; 67(9): 4477-84. [PMID: 10456889]

[45] Kulkarni MM, McMaster WR, Kamysz E, Kamysz W, Engman DM, McGwire BS. The major surface-metalloprotease of the parasitic protozoan, *Leishmania*, protects against antimicrobial peptide-induced apoptotic killing. Mol Microbiol 2006; 62(5): 1484-97. [http://dx.doi.org/10.1111/j.1365-2958.2006.05459.x] [PMID: 17074074]

[46] Joshi PB, Sacks DL, Modi G, McMaster WR. Targeted gene deletion of *Leishmania major* genes encoding developmental stage-specific leishmanolysin (GP63). Mol Microbiol 1998; 27(3): 519-30. [http://dx.doi.org/10.1046/j.1365-2958.1998.00689.x] [PMID: 9489664]

[47] Jiménez-Ruiz A, Boceta C, Bonay P, Requena JM, Alonso C. Cloning, sequencing, and expression of the PSA genes from *Leishmania infantum.* Eur J Biochem 1998; 251(1-2): 389-97. [http://dx.doi.org/10.1046/j.1432-1327.1998.2510389.x] [PMID: 9492309]

[48] Kedzierski L, Montgomery J, Bullen D, *et al.* A leucine-rich repeat motif of *Leishmania* parasite surface antigen 2 binds to macrophages through the complement receptor 3. J Immunol 2004; 172(8): 4902-6. [http://dx.doi.org/10.4049/jimmunol.172.8.4902] [PMID: 15067069]

[49] Lincoln LM, Ozaki M, Donelson JE, Beetham JK. Genetic complementation of *Leishmania* deficient in PSA (GP46) restores their resistance to lysis by complement. Mol Biochem Parasitol 2004; 137(1): 185-9. [http://dx.doi.org/10.1016/j.molbiopara.2004.05.004] [PMID: 15279966]

[50] van Zandbergen G, Bollinger A, Wenzel A, *et al. Leishmania* disease development depends on the presence of apoptotic promastigotes in the virulent inoculum. Proc Natl Acad Sci USA 2006; 103(37): 13837-42. [http://dx.doi.org/10.1073/pnas.0600843103] [PMID: 16945916]

[51] Henson PM, Bratton DL, Fadok VA. Apoptotic cell removal. Curr Biol 2001; 11(19): R795-805. [http://dx.doi.org/10.1016/S0960-9822(01)00474-2] [PMID: 11591341]

[52] Wanderley JL, Barcinski MA. Apoptosis and apoptotic mimicry: the *Leishmania* connection. Cell Mol Life Sci 2010; 67(10): 1653-9. [http://dx.doi.org/10.1007/s00018-010-0291-0] [PMID: 20140747]

[53] Cecílio P, Pérez-Cabezas B, Santarém N, Maciel J, Rodrigues V, Cordeiro da Silva A. Deception and manipulation: the arms of *leishmania*, a successful parasite. Front Immunol 2014; 5: 480. [PMID: 25368612]

[54] Peters NC, Egen JG, Secundino N, *et al. In vivo* imaging reveals an essential role for neutrophils in leishmaniasis transmitted by sand flies. Science 2008; 321(5891): 970-4. [http://dx.doi.org/10.1126/science.1159194] [PMID: 18703742]

[55] Aga E, Katschinski DM, van Zandbergen G, *et al.* Inhibition of the spontaneous apoptosis of

neutrophil granulocytes by the intracellular parasite *Leishmania major*. J Immunol 2002; 169(2): 898-905.
[http://dx.doi.org/10.4049/jimmunol.169.2.898] [PMID: 12097394]

[56] Sarkar A, Aga E, Bussmeyer U, *et al.* Infection of neutrophil granulocytes with *Leishmania major* activates ERK 1/2 and modulates multiple apoptotic pathways to inhibit apoptosis. Med Microbiol Immunol (Berl) 2013; 202(1): 25-35.
[http://dx.doi.org/10.1007/s00430-012-0246-1] [PMID: 22661217]

[57] Spaan AN, Surewaard BG, Nijland R, van Strijp JA. Neutrophils *versus Staphylococcus aureus*: a biological tug of war. Annu Rev Microbiol 2013; 67: 629-50.
[http://dx.doi.org/10.1146/annurev-micro-092412-155746] [PMID: 23834243]

[58] Guimarães-Costa AB, Nascimento MT, Froment GS, *et al. Leishmania amazonensis* promastigotes induce and are killed by neutrophil extracellular traps. Proc Natl Acad Sci USA 2009; 106(16): 6748-53.
[http://dx.doi.org/10.1073/pnas.0900226106] [PMID: 19346483]

[59] Gabriel C, McMaster WR, Girard D, Descoteaux A. *Leishmania donovani* promastigotes evade the antimicrobial activity of neutrophil extracellular traps. J Immunol 2010; 185(7): 4319-27.
[http://dx.doi.org/10.4049/jimmunol.1000893] [PMID: 20826753]

[60] Wang Y, Chen Y, Xin L, *et al.* Differential microbicidal effects of human histone proteins H2A and H2B on *Leishmania* promastigotes and amastigotes. Infect Immun 2011; 79(3): 1124-33.
[http://dx.doi.org/10.1128/IAI.00658-10] [PMID: 21189319]

[61] Guimarães-Costa AB, DeSouza-Vieira TS, Paletta-Silva R, Freitas-Mesquita AL, Meyer-Fernandes JR, Saraiva EM. 3-nucleotidase/nuclease activity allows *Leishmania* parasites to escape killing by neutrophil extracellular traps. Infect Immun 2014; 82(4): 1732-40.
[http://dx.doi.org/10.1128/IAI.01232-13] [PMID: 24516114]

[62] Peters NC, Sacks DL. The impact of vector-mediated neutrophil recruitment on cutaneous leishmaniasis. Cell Microbiol 2009; 11(9): 1290-6.
[http://dx.doi.org/10.1111/j.1462-5822.2009.01348.x] [PMID: 19545276]

[63] Laufs H, Müller K, Fleischer J, *et al.* Intracellular survival of *Leishmania major* in neutrophil granulocytes after uptake in the absence of heat-labile serum factors. Infect Immun 2002; 70(2): 826-35.
[http://dx.doi.org/10.1128/IAI.70.2.826-835.2002] [PMID: 11796617]

[64] Marques CS, Passero LF, Vale-Gato I, *et al.* New insights into neutrophil and *Leishmania infantum in vitro* immune interactions. Comp Immunol Microbiol Infect Dis 2015; 40: 19-29.
[http://dx.doi.org/10.1016/j.cimid.2015.03.003] [PMID: 25857442]

[65] Laskay T, van Zandbergen G, Solbach W. Neutrophil granulocytes as host cells and transport vehicles for intracellular pathogens: apoptosis as infection-promoting factor. Immunobiology 2008; 213(3-4): 183-91.
[http://dx.doi.org/10.1016/j.imbio.2007.11.010] [PMID: 18406366]

[66] van Zandbergen G, Klinger M, Mueller A, *et al.* Cutting edge: neutrophil granulocyte serves as a vector for *Leishmania* entry into macrophages. J Immunol 2004; 173(11): 6521-5.
[http://dx.doi.org/10.4049/jimmunol.173.11.6521] [PMID: 15557140]

[67] Laskay T, van Zandbergen G, Solbach W. Neutrophil granulocytesTrojan horses for *Leishmania major* and other intracellular microbes? Trends Microbiol 2003; 11(5): 210-4.
[http://dx.doi.org/10.1016/S0966-842X(03)00075-1] [PMID: 12781523]

[68] Kaye P, Scott P. Leishmaniasis: complexity at the host-pathogen interface. Nat Rev Microbiol 2011; 9(8): 604-15.
[http://dx.doi.org/10.1038/nrmicro2608] [PMID: 21747391]

[69] Peters N, Sacks D. Immune privilege in sites of chronic infection: *Leishmania* and regulatory T cells. Immunol Rev 2006; 213: 159-79.
[http://dx.doi.org/10.1111/j.1600-065X.2006.00432.x] [PMID: 16972903]

[70] Ritter U, Frischknecht F, van Zandbergen G. Are neutrophils important host cells for *Leishmania* parasites? Trends Parasitol 2009; 25(11): 505-10.
[http://dx.doi.org/10.1016/j.pt.2009.08.003] [PMID: 19762280]

[71] Moore KJ, Matlashewski G. Intracellular infection by *Leishmania donovani* inhibits macrophage apoptosis. J Immunol 1994; 152(6): 2930-7.
[PMID: 8144893]

[72] Kamir D, Zierow S, Leng L, *et al.* A *Leishmania* ortholog of macrophage migration inhibitory factor modulates host macrophage responses. J Immunol 2008; 180(12): 8250-61.
[http://dx.doi.org/10.4049/jimmunol.180.12.8250] [PMID: 18523291]

[73] Ueno N, Wilson ME. Receptor-mediated phagocytosis of *Leishmania*: implications for intracellular survival. Trends Parasitol 2012; 28(8): 335-44.
[http://dx.doi.org/10.1016/j.pt.2012.05.002] [PMID: 22726697]

[74] Russell DG, Talamas-Rohana P. *Leishmania* and the macrophage: a marriage of inconvenience. Immunol Today 1989; 10(10): 328-33.
[http://dx.doi.org/10.1016/0167-5699(89)90188-6] [PMID: 2679629]

[75] Akilov OE, Kasuboski RE, Carter CR, McDowell MA. The role of mannose receptor during experimental leishmaniasis. J Leukoc Biol 2007; 81(5): 1188-96.
[http://dx.doi.org/10.1189/jlb.0706439] [PMID: 17261547]

[76] Kima PE, Constant SL, Hannum L, *et al.* Internalization of *Leishmania mexicana* complex amastigotes *via* the Fc receptor is required to sustain infection in murine cutaneous leishmaniasis. J Exp Med 2000; 191(6): 1063-8.
[http://dx.doi.org/10.1084/jem.191.6.1063] [PMID: 10727468]

[77] Mosser DM, Springer TA, Diamond MS. *Leishmania* promastigotes require opsonic complement to bind to the human leukocyte integrin Mac-1 (CD11b/CD18). J Cell Biol 1992; 116(2): 511-20.
[http://dx.doi.org/10.1083/jcb.116.2.511] [PMID: 1730765]

[78] Cunningham AC. Parasitic adaptive mechanisms in infection by *leishmania.* Exp Mol Pathol 2002; 72(2): 132-41.
[http://dx.doi.org/10.1006/exmp.2002.2418] [PMID: 11890722]

[79] Desjardins M, Descoteaux A. Inhibition of phagolysosomal biogenesis by the *Leishmania* lipophosphoglycan. J Exp Med 1997; 185(12): 2061-8.
[http://dx.doi.org/10.1084/jem.185.12.2061] [PMID: 9182677]

[80] Vinet AF, Fukuda M, Turco SJ, Descoteaux A. The *Leishmania donovani* lipophosphoglycan excludes the vesicular proton-ATPase from phagosomes by impairing the recruitment of synaptotagmin V. PLoS Pathog 2009; 5(10): e1000628.
[http://dx.doi.org/10.1371/journal.ppat.1000628] [PMID: 19834555]

[81] Lodge R, Diallo TO, Descoteaux A. *Leishmania donovani* lipophosphoglycan blocks NADPH oxidase assembly at the phagosome membrane. Cell Microbiol 2006; 8(12): 1922-31.
[http://dx.doi.org/10.1111/j.1462-5822.2006.00758.x] [PMID: 16848789]

[82] Wilson J, Huynh C, Kennedy KA, *et al.* Control of parasitophorous vacuole expansion by LYST/Beige restricts the intracellular growth of *Leishmania amazonensis.* PLoS Pathog 2008; 4(10): e1000179.
[http://dx.doi.org/10.1371/journal.ppat.1000179] [PMID: 18927622]

[83] Requena JM. The stressful life of pathogenic *Leishmania* species. In: Requena JM, Ed. Stress Response in Microbiology Caister Academic Press. 2012.

[84] Tovar J, Cunningham ML, Smith AC, Croft SL, Fairlamb AH. Down-regulation of *Leishmania donovani* trypanothione reductase by heterologous expression of a trans-dominant mutant homologue: effect on parasite intracellular survival. Proc Natl Acad Sci USA 1998; 95(9): 5311-6.
[http://dx.doi.org/10.1073/pnas.95.9.5311] [PMID: 9560272]

[85] Castro H, Sousa C, Santos M, Cordeiro-da-Silva A, Flohé L, Tomás AM. Complementary antioxidant defense by cytoplasmic and mitochondrial peroxiredoxins in *Leishmania infantum.* Free Radic Biol Med 2002; 33(11): 1552-62.
[http://dx.doi.org/10.1016/S0891-5849(02)01089-4] [PMID: 12446213]

[86] Nare B, Garraway LA, Vickers TJ, Beverley SM. PTR1-dependent synthesis of tetrahydrobiopterin contributes to oxidant susceptibility in the trypanosomatid protozoan parasite *Leishmania major.* Curr Genet 2009; 55(3): 287-99.
[http://dx.doi.org/10.1007/s00294-009-0244-z] [PMID: 19396443]

[87] Ghosh S, Goswami S, Adhya S. Role of superoxide dismutase in survival of *Leishmania* within the macrophage. Biochem J 2003; 369(3): 447-52
[http://dx.doi.org/10.1042/BJ20021684] [PMID: 12459037]

[88] Henard CA, Carlsen ED, Hay C, Kima PE, Soong L. *Leishmania amazonensis* amastigotes highly express a tryparedoxin peroxidase isoform that increases parasite resistance to macrophage antimicrobial defenses and fosters parasite virulence. PLoS Negl Trop Dis 2014; 8(7): e3000.
[http://dx.doi.org/10.1371/journal.pntd.0003000] [PMID: 25033301]

[89] Flannery AR, Renberg RL, Andrews NW. Pathways of iron acquisition and utilization in *Leishmania.* Curr Opin Microbiol 2013; 16(6): 716-21.
[http://dx.doi.org/10.1016/j.mib.2013.07.018] [PMID: 23962817]

[90] Flannery AR, Huynh C, Mittra B, Mortara RA, Andrews NW. LFR1 ferric iron reductase of *Leishmania amazonensis* is essential for the generation of infective parasite forms. J Biol Chem 2011; 286(26): 23266-79.
[http://dx.doi.org/10.1074/jbc.M111.229674] [PMID: 21558274]

[91] Iniesta V, Gómez-Nieto LC, Corraliza I. The inhibition of arginase by N(omega)-hydroxy-l-arginine controls the growth of *Leishmania* inside macrophages. J Exp Med 2001; 193(6): 777-84.

[http://dx.doi.org/10.1084/jem.193.6.777] [PMID: 11257143]

[92] Reguera RM, Balaña-Fouce R, Showalter M, Hickerson S, Beverley SM. *Leishmania major* lacking arginase (ARG) are auxotrophic for polyamines but retain infectivity to susceptible BALB/c mice. Mol Biochem Parasitol 2009; 165(1): 48-56.
[http://dx.doi.org/10.1016/j.molbiopara.2009.01.001] [PMID: 19393161]

[93] Soong L. Modulation of dendritic cell function by *Leishmania* parasites. J Immunol 2008; 180(7): 4355-60.
[http://dx.doi.org/10.4049/jimmunol.180.7.4355] [PMID: 18354154]

[94] Moll H, Fuchs H, Blank C, Röllinghoff M. Langerhans cells transport *Leishmania major* from the infected skin to the draining lymph node for presentation to antigen-specific T cells. Eur J Immunol 1993; 23(7): 1595-601.
[http://dx.doi.org/10.1002/eji.1830230730] [PMID: 8325337]

[95] Kautz-Neu K, Noordegraaf M, Dinges S, *et al.* Langerhans cells are negative regulators of the anti-*Leishmania* response. J Exp Med 2011; 208(5): 885-91.
[http://dx.doi.org/10.1084/jem.20102318] [PMID: 21536741]

[96] Bogdan C, Donhauser N, Döring R, Röllinghoff M, Diefenbach A, Rittig MG. Fibroblasts as host cells in latent leishmaniosis. J Exp Med 2000; 191(12): 2121-30.
[http://dx.doi.org/10.1084/jem.191.12.2121] [PMID: 10859337]

[97] Bogdan C. Mechanisms and consequences of persistence of intracellular pathogens: leishmaniasis as an example. Cell Microbiol 2008; 10(6): 1221-34.
[http://dx.doi.org/10.1111/j.1462-5822.2008.01146.x] [PMID: 18363880]

[98] Sacks D, Noben-Trauth N. The immunology of susceptibility and resistance to *Leishmania major* in mice. Nat Rev Immunol 2002; 2(11): 845-58.
[http://dx.doi.org/10.1038/nri933] [PMID: 12415308]

[99] Kedzierski L, Evans KJ. Immune responses during cutaneous and visceral leishmaniasis. Parasitology 2014; 141: 1-19.
[http://dx.doi.org/10.1017/S003118201400095X] [PMID: 25075460]

[100] Bogdan C, Gessner A, Solbach W, Röllinghoff M. Invasion, control and persistence of *Leishmania* parasites. Curr Opin Immunol 1996; 8(4): 517-25.
[http://dx.doi.org/10.1016/S0952-7915(96)80040-9] [PMID: 8794010]

[101] Bogdan C, Röllinghoff M. The immune response to *Leishmania*: mechanisms of parasite control and evasion. Int J Parasitol 1998; 28(1): 121-34.
[http://dx.doi.org/10.1016/S0020-7519(97)00169-0] [PMID: 9504340]

[102] Baneth G, Koutinas AF, Solano-Gallego L, Bourdeau P, Ferrer L. Canine leishmaniosis - new concepts and insights on an expanding zoonosis: part one. Trends Parasitol 2008; 24(7): 324-30.
[http://dx.doi.org/10.1016/j.pt.2008.04.001] [PMID: 18514028]

[103] Reis AB, Giunchetti RC, Carrillo E, Martins-Filho OA, Moreno J. Immunity to *Leishmania* and the rational search for vaccines against canine leishmaniasis. Trends Parasitol 2010; 26(7): 341-9.
[http://dx.doi.org/10.1016/j.pt.2010.04.005] [PMID: 20488751]

[104] Castellano LR, Filho DC, Argiro L, *et al.* Th1/Th2 immune responses are associated with active

cutaneous leishmaniasis and clinical cure is associated with strong interferon-gamma production. Hum Immunol 2009; 70(6): 383-90.
[http://dx.doi.org/10.1016/j.humimm.2009.01.007] [PMID: 19480861]

[105] Nylén S, Sacks D. Interleukin-10 and the pathogenesis of human visceral leishmaniasis. Trends Immunol 2007; 28(9): 378-84.
[http://dx.doi.org/10.1016/j.it.2007.07.004] [PMID: 17689290]

[106] Faleiro RJ, Kumar R, Hafner LM, Engwerda CR. Immune regulation during chronic visceral leishmaniasis. PLoS Negl Trop Dis 2014; 8(7): e2914.
[http://dx.doi.org/10.1371/journal.pntd.0002914] [PMID: 25010815]

[107] Gradoni L. An update on antileishmanial vaccine candidates and prospects for a canine *Leishmania* vaccine. Vet Parasitol 2001; 100(1-2): 87-103.
[http://dx.doi.org/10.1016/S0304-4017(01)00486-1] [PMID: 11522409]

[108] Kautz-Neu K, Schwonberg K, Fischer MR, Schermann AI, von Stebut E. Dendritic cells in *Leishmania major* infections: mechanisms of parasite uptake, cell activation and evidence for physiological relevance. Med Microbiol Immunol (Berl) 2012; 201(4): 581-92.
[http://dx.doi.org/10.1007/s00430-012-0261-2] [PMID: 22983754]

[109] Ashok D, Schuster S, Ronet C, *et al.* Cross-presenting dendritic cells are required for control of *Leishmania* major infection. Eur J Immunol 2014; 44(5): 1422-32.
[http://dx.doi.org/10.1002/eji.201344242] [PMID: 24643576]

[110] Ashok D, Acha-Orbea H. Timing is everything: dendritic cell subsets in murine *Leishmania* infection. Trends Parasitol 2014; 30(10): 499-507.
[http://dx.doi.org/10.1016/j.pt.2014.08.001] [PMID: 25190685]

[111] Bertholet S, Debrabant A, Afrin F, *et al.* Antigen requirements for efficient priming of CD8+ T cells by *Leishmania major*-infected dendritic cells. Infect Immun 2005; 73(10): 6620-8.
[http://dx.doi.org/10.1128/IAI.73.10.6620-6628.2005] [PMID: 16177338]

[112] Ribeiro-Gomes FL, Peters NC, Debrabant A, Sacks DL. Efficient capture of infected neutrophils by dendritic cells in the skin inhibits the early anti-leishmania response. PLoS Pathog 2012; 8(2): e1002536.
[http://dx.doi.org/10.1371/journal.ppat.1002536] [PMID: 22359507]

[113] Ronet C, Voigt H, Himmelrich H, *et al. Leishmania major*-specific B cells are necessary for Th2 cell development and susceptibility to L. major LV39 in BALB/c mice. J Immunol 2008; 180(7): 4825-35.
[http://dx.doi.org/10.4049/jimmunol.180.7.4825] [PMID: 18354206]

[114] Marovich MA, McDowell MA, Thomas EK, Nutman TB. IL-12p70 production by *Leishmania major*-harboring human dendritic cells is a CD40/CD40 ligand-dependent process. J Immunol 2000; 164(11): 5858-65.
[http://dx.doi.org/10.4049/jimmunol.164.11.5858] [PMID: 10820265]

[115] McDowell MA, Marovich M, Lira R, Braun M, Sacks D. *Leishmania* priming of human dendritic cells for CD40 ligand-induced interleukin-12p70 secretion is strain and species dependent. Infect Immun 2002; 70(8): 3994-4001.
[http://dx.doi.org/10.1128/IAI.70.8.3994-4001.2002] [PMID: 12117904]

[116] Reiner SL, Locksley RM. The regulation of immunity to *Leishmania major*. Annu Rev Immunol 1995; 13: 151-77.
[http://dx.doi.org/10.1146/annurev.iy.13.040195.001055] [PMID: 7612219]

[117] Noben-Trauth N, Paul WE, Sacks DL. IL-4- and IL-4 receptor-deficient BALB/c mice reveal differences in susceptibility to *Leishmania major* parasite substrains. J Immunol 1999; 162(10): 6132-40.
[PMID: 10229856]

[118] Matthews DJ, Emson CL, McKenzie GJ, Jolin HE, Blackwell JM, McKenzie AN. IL-13 is a susceptibility factor for *Leishmania major* infection. J Immunol 2000; 164(3): 1458-62.
[http://dx.doi.org/10.4049/jimmunol.164.3.1458] [PMID: 10640762]

[119] Coffman RL, Seymour BW, Lebman DA, *et al.* The role of helper T cell products in mouse B cell differentiation and isotype regulation. Immunol Rev 1988; 102: 5-28.
[http://dx.doi.org/10.1111/j.1600-065X.1988.tb00739.x] [PMID: 2966762]

[120] Bacellar O, Lessa H, Schriefer A, *et al.* Up-regulation of Th1-type responses in mucosal leishmaniasis patients. Infect Immun 2002; 70(12): 6734-40.
[http://dx.doi.org/10.1128/IAI.70.12.6734-6740.2002] [PMID: 12438348]

[121] Novais FO, Scott P. CD8+ T cells in cutaneous leishmaniasis: the good, the bad, and the ugly. Semin Immunopathol 2015; 37(3): 251-9.
[http://dx.doi.org/10.1007/s00281-015-0475-7] [PMID: 25800274]

[122] Da-Cruz AM, Conceição-Silva F, Bertho AL, Coutinho SG. *Leishmania*-reactive CD4+ and CD8+ T cells associated with cure of human cutaneous leishmaniasis. Infect Immun 1994; 62(6): 2614-8.
[PMID: 7910596]

[123] Faria DR, Souza PE, Durães FV, *et al.* Recruitment of CD8(+) T cells expressing granzyme A is associated with lesion progression in human cutaneous leishmaniasis. Parasite Immunol 2009; 31(8): 432-9.
[http://dx.doi.org/10.1111/j.1365-3024.2009.01125.x] [PMID: 19646207]

[124] Cardoso TM, Machado Á, Costa DL, *et al.* Protective and pathological functions of CD8+ T cells in *Leishmania braziliensis* infection. Infect Immun 2015; 83(3): 898-906.
[http://dx.doi.org/10.1128/IAI.02404-14] [PMID: 25534940]

[125] Moore KW, de Waal Malefyt R, Coffman RL, OGarra A. Interleukin-10 and the interleukin-10 receptor. Annu Rev Immunol 2001; 19: 683-765.
[http://dx.doi.org/10.1146/annurev.immunol.19.1.683] [PMID: 11244051]

[126] Kane MM, Mosser DM. The role of IL-10 in promoting disease progression in leishmaniasis. J Immunol 2001; 166(2): 1141-7.
[http://dx.doi.org/10.4049/jimmunol.166.2.1141] [PMID: 11145695]

[127] Noben-Trauth N, Lira R, Nagase H, Paul WE, Sacks DL. The relative contribution of IL-4 receptor signaling and IL-10 to susceptibility to *Leishmania major*. J Immunol 2003; 170(10): 5152-8.
[http://dx.doi.org/10.4049/jimmunol.170.10.5152] [PMID: 12734362]

[128] Garg R, Dube A. Animal models for vaccine studies for visceral leishmaniasis. Indian J Med Res 2006; 123(3): 439-54.

[PMID: 16778322]

[129] Murphy ML, Wille U, Villegas EN, Hunter CA, Farrell JP. IL-10 mediates susceptibility to *Leishmania donovani* infection. Eur J Immunol 2001; 31(10): 2848-56. [http://dx.doi.org/10.1002/1521-4141(2001010)31:10<2848::AID-IMMU2848>3.0.CO;2-T] [PMID: 11592059]

[130] Murray HW, Lu CM, Mauze S, *et al.* Interleukin-10 (IL-10) in experimental visceral leishmaniasis and IL-10 receptor blockade as immunotherapy. Infect Immun 2002; 70(11): 6284-93. [http://dx.doi.org/10.1128/IAI.70.11.6284-6293.2002] [PMID: 12379707]

[131] Miles SA, Conrad SM, Alves RG, Jeronimo SM, Mosser DM. A role for IgG immune complexes during infection with the intracellular pathogen *Leishmania.* J Exp Med 2005; 201(5): 747-54. [http://dx.doi.org/10.1084/jem.20041470] [PMID: 15753208]

[132] Caldas A, Favali C, Aquino D, *et al.* Balance of IL-10 and interferon-gamma plasma levels in human visceral leishmaniasis: implications in the pathogenesis. BMC Infect Dis 2005; 5: 113. [http://dx.doi.org/10.1186/1471-2334-5-113] [PMID: 16364177]

[133] Anderson CF, Oukka M, Kuchroo VJ, Sacks D. CD4(+)CD25(-)Foxp3(-) Th1 cells are the source of IL-10-mediated immune suppression in chronic cutaneous leishmaniasis. J Exp Med 2007; 204(2): 285-97. [http://dx.doi.org/10.1084/jem.20061886] [PMID: 17283207]

[134] Belkaid Y, Hoffmann KF, Mendez S, *et al.* The role of interleukin (IL)-10 in the persistence of *Leishmania major* in the skin after healing and the therapeutic potential of anti-IL-10 receptor antibody for sterile cure. J Exp Med 2001; 194(10): 1497-506. [http://dx.doi.org/10.1084/jem.194.10.1497] [PMID: 11714756]

[135] Belkaid Y, Piccirillo CA, Mendez S, Shevach EM, Sacks DL. CD4+CD25+ regulatory T cells control *Leishmania major* persistence and immunity. Nature 2002; 420(6915): 502-7. [http://dx.doi.org/10.1038/nature01152] [PMID: 12466842]

[136] Campanelli AP, Roselino AM, Cavassani KA, *et al.* CD4+CD25+ T cells in skin lesions of patients with cutaneous leishmaniasis exhibit phenotypic and functional characteristics of natural regulatory T cells. J Infect Dis 2006; 193(9): 1313-22. [http://dx.doi.org/10.1086/502980] [PMID: 16586370]

[137] Uzonna JE, Wei G, Yurkowski D, Bretscher P. Immune elimination of *Leishmania major* in mice: implications for immune memory, vaccination, and reactivation disease. J Immunol 2001; 167(12): 6967-74. [http://dx.doi.org/10.4049/jimmunol.167.12.6967] [PMID: 11739516]

[138] Okwor I, Uzonna J. Persistent parasites and immunologic memory in cutaneous leishmaniasis: implications for vaccine designs and vaccination strategies. Immunol Res 2008; 41(2): 123-36. [http://dx.doi.org/10.1007/s12026-008-8016-2] [PMID: 18389179]

[139] Sacks DL. Vaccines against tropical parasitic diseases: a persisting answer to a persisting problem. Nat Immunol 2014; 15(5): 403-5. [http://dx.doi.org/10.1038/ni.2853] [PMID: 24747701]

[140] Abdeladhim M, Kamhawi S, Valenzuela JG. Whats behind a sand fly bite? The profound effect of

sand fly saliva on host hemostasis, inflammation and immunity. Infect Genet Evol 2014; 28: 691-703. [http://dx.doi.org/10.1016/j.meegid.2014.07.028] [PMID: 25117872]

[141] Hall LR, Titus RG. Sand fly vector saliva selectively modulates macrophage functions that inhibit killing of *Leishmania major* and nitric oxide production. J Immunol 1995; 155(7): 3501-6. [PMID: 7561045]

[142] Mbow ML, Bleyenberg JA, Hall LR, Titus RG. *Phlebotomus papatasi* sand fly salivary gland lysate down-regulates a Th1, but up-regulates a Th2, response in mice infected with *Leishmania major.* J Immunol 1998; 161(10): 5571-7. [PMID: 9820534]

[143] Belkaid Y, Kamhawi S, Modi G, *et al.* Development of a natural model of cutaneous leishmaniasis: powerful effects of vector saliva and saliva preexposure on the long-term outcome of *Leishmania major* infection in the mouse ear dermis. J Exp Med 1998; 188(10): 1941-53. [http://dx.doi.org/10.1084/jem.188.10.1941] [PMID: 9815271]

[144] Gomes R, Oliveira F. The immune response to sand fly salivary proteins and its influence on *Leishmania* immunity. Front Immunol 2012; 3: 110. [http://dx.doi.org/10.3389/fimmu.2012.00110] [PMID: 22593758]

[145] Solbach W, Laskay T. The host response to *Leishmania* infection. Adv Immunol 2000; 74: 275-317. [http://dx.doi.org/10.1016/S0065-2776(08)60912-8] [PMID: 10605608]

[146] Scott P, Artis D, Uzonna J, Zaph C. The development of effector and memory T cells in cutaneous leishmaniasis: the implications for vaccine development. Immunol Rev 2004; 201: 318-38. [http://dx.doi.org/10.1111/j.0105-2896.2004.00198.x] [PMID: 15361250]

[147] Malherbe L, Filippi C, Julia V, *et al.* Selective activation and expansion of high-affinity CD4+ T cells in resistant mice upon infection with *Leishmania major.* Immunity 2000; 13(6): 771-82. [http://dx.doi.org/10.1016/S1074-7613(00)00075-3] [PMID: 11163193]

[148] Belkaid Y, Mendez S, Lira R, Kadambi N, Milon G, Sacks D. A natural model of *Leishmania major* infection reveals a prolonged silent phase of parasite amplification in the skin before the onset of lesion formation and immunity. J Immunol 2000; 165(2): 969-77. [http://dx.doi.org/10.4049/jimmunol.165.2.969] [PMID: 10878373]

[149] Belkaid Y, Von Stebut E, Mendez S, *et al.* CD8+ T cells are required for primary immunity in C57BL/6 mice following low-dose, intradermal challenge with *Leishmania major.* J Immunol 2002; 168(8): 3992-4000. [http://dx.doi.org/10.4049/jimmunol.168.8.3992] [PMID: 11937556]

[150] Sypek JP, Chung CL, Mayor SE, *et al.* Resolution of cutaneous leishmaniasis: interleukin 12 initiates a protective T helper type 1 immune response. J Exp Med 1993; 177(6): 1797-802. [http://dx.doi.org/10.1084/jem.177.6.1797] [PMID: 8098733]

[151] Ji J, Sun J, Soong L. Impaired expression of inflammatory cytokines and chemokines at early stages of infection with *Leishmania amazonensis.* Infect Immun 2003; 71(8): 4278-88. [http://dx.doi.org/10.1128/IAI.71.8.4278-4288.2003] [PMID: 12874303]

[152] Hermida MD, Doria PG, Taguchi AM, Mengel JO, dos-Santos W. *Leishmania amazonensis* infection impairs dendritic cell migration from the inflammatory site to the draining lymph node. BMC Infect

Dis 2014; 14: 450.
[http://dx.doi.org/10.1186/1471-2334-14-450] [PMID: 25142021]

[153] DeKrey GK, Lima HC, Titus RG. Analysis of the immune responses of mice to infection with *Leishmania braziliensis.* Infect Immun 1998; 66(2): 827-9.
[PMID: 9453649]

[154] Kemp K, Kemp M, Kharazmi A, *et al.* *Leishmania*-specific T cells expressing interferon-gamma (IFN-gamma) and IL-10 upon activation are expanded in individuals cured of visceral leishmaniasis. Clin Exp Immunol 1999; 116(3): 500-4.
[http://dx.doi.org/10.1046/j.1365-2249.1999.00918.x] [PMID: 10361241]

[155] Khalil EA, Ayed NB, Musa AM, *et al.* Dichotomy of protective cellular immune responses to human visceral leishmaniasis. Clin Exp Immunol 2005; 140(2): 349-53.
[http://dx.doi.org/10.1111/j.1365-2249.2005.02768.x] [PMID: 15807861]

[156] Probst P, Stromberg E, Ghalib HW, *et al.* Identification and characterization of T cell-stimulating antigens from *Leishmania* by CD4 T cell expression cloning. J Immunol 2001; 166(1): 498-505.
[http://dx.doi.org/10.4049/jimmunol.166.1.498] [PMID: 11123329]

[157] Crescente JA, Silveira FT, Lainson R, Gomes CM, Laurenti MD, Corbett CE. A cross-sectional study on the clinical and immunological spectrum of human *Leishmania* (*L.*) *infantum chagasi* infection in the Brazilian Amazon region. Trans R Soc Trop Med Hyg 2009; 103(12): 1250-6.
[http://dx.doi.org/10.1016/j.trstmh.2009.06.010] [PMID: 19615710]

[158] Kar K. Serodiagnosis of leishmaniasis. Crit Rev Microbiol 1995; 21(2): 123-52.
[http://dx.doi.org/10.3109/10408419509113537] [PMID: 7639932]

[159] Garcia-Alonso M, Nieto CG, Blanco A, Requena JM, Alonso C, Navarrete I. Presence of antibodies in the aqueous humour and cerebrospinal fluid during *Leishmania* infections in dogs. Pathological features at the central nervous system. Parasite Immunol 1996; 18(11): 539-46.
[http://dx.doi.org/10.1046/j.1365-3024.1996.d01-28.x] [PMID: 9226692]

[160] Nieto CG, Barrera R, Habela MA, *et al.* Changes in the plasma concentrations of lipids and lipoprotein fractions in dogs infected with *Leishmania infantum.* Vet Parasitol 1992; 44(3-4): 175-82.
[http://dx.doi.org/10.1016/0304-4017(92)90115-P] [PMID: 1466128]

[161] Nieto CG, Navarrete I, Habela MA, Serrano F, Redondo E. Pathological changes in kidneys of dogs with natural *Leishmania* infection. Vet Parasitol 1992; 45(1-2): 33-47.
[http://dx.doi.org/10.1016/0304-4017(92)90025-5] [PMID: 1485420]

[162] Carrión J, Nieto A, Iborra S, *et al.* Immunohistological features of visceral leishmaniasis in BALB/c mice. Parasite Immunol 2006; 28(5): 173-83.
[http://dx.doi.org/10.1111/j.1365-3024.2006.00817.x] [PMID: 16629702]

[163] Engwerda CR, Kaye PM. Organ-specific immune responses associated with infectious disease. Immunol Today 2000; 21(2): 73-8.
[http://dx.doi.org/10.1016/S0167-5699(99)01549-2] [PMID: 10652464]

[164] Stäger S, Joshi T, Bankoti R. Immune evasive mechanisms contributing to persistent *Leishmania donovani* infection. Immunol Res 2010; 47(1-3): 14-24.
[http://dx.doi.org/10.1007/s12026-009-8135-4] [PMID: 20087685]

[165] Ahmed S, Colmenares M, Soong L, *et al.* Intradermal infection model for pathogenesis and vaccine studies of murine visceral leishmaniasis. Infect Immun 2003; 71(1): 401-10. [http://dx.doi.org/10.1128/IAI.71.1.401-410.2003] [PMID: 12496190]

[166] Oliveira DM, Costa MA, Chavez-Fumagalli MA, *et al.* Evaluation of parasitological and immunological parameters of *Leishmania chagasi* infection in BALB/c mice using different doses and routes of inoculation of parasites. Parasitol Res 2012; 110(3): 1277-85. [http://dx.doi.org/10.1007/s00436-011-2628-5] [PMID: 21915627]

[167] Deak E, Jayakumar A, Cho KW, *et al.* Murine visceral leishmaniasis: IgM and polyclonal B-cell activation lead to disease exacerbation. Eur J Immunol 2010; 40(5): 1355-68. [http://dx.doi.org/10.1002/eji.200939455] [PMID: 20213734]

[168] Handman E. Leishmaniasis: current status of vaccine development. Clin Microbiol Rev 2001; 14(2): 229-43. [http://dx.doi.org/10.1128/CMR.14.2.229-243.2001] [PMID: 11292637]

[169] Requena JM, Soto M, Doria MD, Alonso C. Immune and clinical parameters associated with *Leishmania infantum* infection in the golden hamster model. Vet Immunol Immunopathol 2000; 76(3-4): 269-81. [http://dx.doi.org/10.1016/S0165-2427(00)00221-X] [PMID: 11044559]

[170] Melby PC, Chandrasekar B, Zhao W, Coe JE. The hamster as a model of human visceral leishmaniasis: progressive disease and impaired generation of nitric oxide in the face of a prominent Th1-like cytokine response. J Immunol 2001; 166(3): 1912-20. [http://dx.doi.org/10.4049/jimmunol.166.3.1912] [PMID: 11160239]

[171] Gomes R, Teixeira C, Teixeira MJ, *et al.* Immunity to a salivary protein of a sand fly vector protects against the fatal outcome of visceral leishmaniasis in a hamster model. Proc Natl Acad Sci USA 2008; 105(22): 7845-50. [http://dx.doi.org/10.1073/pnas.0712153105] [PMID: 18509051]

[172] Solano-Gallego L, Koutinas A, Miró G, *et al.* Directions for the diagnosis, clinical staging, treatment and prevention of canine leishmaniosis. Vet Parasitol 2009; 165(1-2): 1-18. [http://dx.doi.org/10.1016/j.vetpar.2009.05.022] [PMID: 19559536]

[173] Miró G, Cardoso L, Pennisi MG, Oliva G, Baneth G. Canine leishmaniosis new concepts and insights on an expanding zoonosis: part two. Trends Parasitol 2008; 24(8): 371-7. [http://dx.doi.org/10.1016/j.pt.2008.05.003] [PMID: 18603476]

[174] Fernández-Cotrina J, Iniesta V, Belinchón-Lorenzo S, *et al.* Experimental model for reproduction of canine visceral leishmaniosis by *Leishmania infantum.* Vet Parasitol 2013; 192(1-3): 118-28. [http://dx.doi.org/10.1016/j.vetpar.2012.10.002] [PMID: 23102507]

[175] Costa DJ, Carvalho RM, Abbehusen M, *et al.* Experimental infection of dogs with *Leishmania* and saliva as a model to study canine visceral Leishmaniasis. PLoS One 2013; 8(4): e60535. [http://dx.doi.org/10.1371/journal.pone.0060535] [PMID: 23577121]

[176] Foglia Manzillo V, Di Muccio T, Cappiello S, *et al.* Prospective study on the incidence and progression of clinical signs in naïve dogs naturally infected by *Leishmania infantum.* PLoS Negl Trop Dis 2013; 7(5): e2225.

[http://dx.doi.org/10.1371/journal.pntd.0002225] [PMID: 23675551]

[177] Oliveira F, Rowton E, Aslan H, *et al.* A sand fly salivary protein vaccine shows efficacy against vector-transmitted cutaneous leishmaniasis in nonhuman primates. Sci Transl Med 2015; 7(290): 290ra90.
[http://dx.doi.org/10.1126/scitranslmed.aaa3043] [PMID: 26041707]

[178] Silverman JM, Chan SK, Robinson DP, *et al.* Proteomic analysis of the secretome of *Leishmania donovani.* Genome Biol 2008; 9(2): R35.
[http://dx.doi.org/10.1186/gb-2008-9-2-r35] [PMID: 18282296]

[179] Silverman JM, Clos J, deOliveira CC, *et al.* An exosome-based secretion pathway is responsible for protein export from *Leishmania* and communication with macrophages. J Cell Sci 2010; 123(Pt 6): 842-52.
[http://dx.doi.org/10.1242/jcs.056465] [PMID: 20159964]

[180] Lambris JD, Ricklin D, Geisbrecht BV. Complement evasion by human pathogens. Nat Rev Microbiol 2008; 6(2): 132-42.
[http://dx.doi.org/10.1038/nrmicro1824] [PMID: 18197169]

[181] Puentes SM, Da Silva RP, Sacks DL, Hammer CH, Joiner KA. Serum resistance of metacyclic stage *Leishmania major* promastigotes is due to release of C5b-9. J Immunol 1990; 145(12): 4311-6.
[PMID: 2147941]

[182] Franke ED, McGreevy PB, Katz SP, Sacks DL. Growth cycle-dependent generation of complement-resistant *Leishmania promastigotes.* J Immunol 1985; 134(4): 2713-8.
[PMID: 3973390]

[183] Moreno I, Molina R, Toraño A, Laurin E, García E, Domínguez M. Comparative real-time kinetic analysis of human complement killing of *Leishmania infantum* promastigotes derived from axenic culture or from *Phlebotomus perniciosus*. Microbes Infect 2007; 9(14-15): 1574-80.
[http://dx.doi.org/10.1016/j.micinf.2007.09.009] [PMID: 18023393]

[184] Brittingham A, Morrison CJ, McMaster WR, McGwire BS, Chang KP, Mosser DM. Role of the *Leishmania* surface protease gp63 in complement fixation, cell adhesion, and resistance to complement-mediated lysis. J Immunol 1995; 155(6): 3102-11.
[PMID: 7673725]

[185] Joshi PB, Kelly BL, Kamhawi S, Sacks DL, McMaster WR. Targeted gene deletion in *Leishmania major* identifies leishmanolysin (GP63) as a virulence factor. Mol Biochem Parasitol 2002; 120(1): 33-40.
[http://dx.doi.org/10.1016/S0166-6851(01)00432-7] [PMID: 11849703]

[186] Späth GF, Garraway LA, Turco SJ, Beverley SM. The role(s) of lipophosphoglycan (LPG) in the establishment of *Leishmania major* infections in mammalian hosts. Proc Natl Acad Sci USA 2003; 100(16): 9536-41.
[http://dx.doi.org/10.1073/pnas.1530604100] [PMID: 12869694]

[187] Pearson RD, Steigbigel RT. Mechanism of lethal effect of human serum upon *Leishmania donovani.* J Immunol 1980; 125(5): 2195-201.
[PMID: 7430624]

[188] Domínguez M, Toraño A. Immune adherence-mediated opsonophagocytosis: the mechanism of *Leishmania* infection. J Exp Med 1999; 189(1): 25-35. [http://dx.doi.org/10.1084/jem.189.1.25] [PMID: 9874561]

[189] Domínguez M, Moreno I, López-Trascasa M, Toraño A. Complement interaction with trypanosomatid promastigotes in normal human serum. J Exp Med 2002; 195(4): 451-9. [http://dx.doi.org/10.1084/jem.20011319] [PMID: 11854358]

[190] Domínguez M, Moreno I, Aizpurua C, Toraño A. Early mechanisms of *Leishmania* infection in human blood. Microbes Infect 2003; 5(6): 507-13. [http://dx.doi.org/10.1016/S1286-4579(03)00071-6] [PMID: 12758280]

[191] Santos IK, Costa CH, Krieger H, *et al.* Mannan-binding lectin enhances susceptibility to visceral leishmaniasis. Infect Immun 2001; 69(8): 5212-5. [http://dx.doi.org/10.1128/IAI.69.8.5212-5215.2001] [PMID: 11447210]

[192] Cavalcante RR, Pereira MH, Gontijo NF. Anti-complement activity in the saliva of phlebotomine sand flies and other haematophagous insects. Parasitology 2003; 127(Pt 1): 87-93. [http://dx.doi.org/10.1017/S0031182003003329] [PMID: 12885192]

[193] Dunkelberger JR, Song WC. Complement and its role in innate and adaptive immune responses. Cell Res 2010; 20(1): 34-50. [http://dx.doi.org/10.1038/cr.2009.139] [PMID: 20010915]

[194] Shio MT, Hassani K, Isnard A, *et al.* Host cell signalling and Leishmania mechanisms of evasion. J Tropical Med 2012; 2012: 819512.

[195] Nandan D, Yi T, Lopez M, Lai C, Reiner NE. *Leishmania* EF-1alpha activates the Src homology 2 domain containing tyrosine phosphatase SHP-1 leading to macrophage deactivation. J Biol Chem 2002; 277(51): 50190-7. [http://dx.doi.org/10.1074/jbc.M209210200] [PMID: 12384497]

[196] Lambertz U, Silverman JM, Nandan D, *et al.* Secreted virulence factors and immune evasion in visceral leishmaniasis. J Leukoc Biol 2012; 91(6): 887-99. [http://dx.doi.org/10.1189/jlb.0611326] [PMID: 22442494]

[197] Nandan D, Tran T, Trinh E, Silverman JM, Lopez M. Identification of *leishmania* fructose-1, 6-bisphosphate aldolase as a novel activator of host macrophage Src homology 2 domain containing protein tyrosine phosphatase SHP-1. Biochem Biophys Res Commun 2007; 364(3): 601-7. [http://dx.doi.org/10.1016/j.bbrc.2007.10.065] [PMID: 18028878]

[198] Gomez MA, Contreras I, Hallé M, Tremblay ML, McMaster RW, Olivier M. *Leishmania* GP63 alters host signaling through cleavage-activated protein tyrosine phosphatases. Sci Signal 2009; 2(90): ra58. [http://dx.doi.org/10.1126/scisignal.2000213] [PMID: 19797268]

[199] Gómez MA, Olivier M. Proteases and phosphatases during *Leishmania*-macrophage interaction: paving the road for pathogenesis. Virulence 2010; 1(4): 314-8. [http://dx.doi.org/10.4161/viru.1.4.12194] [PMID: 21178462]

[200] Contreras I, Gómez MA, Nguyen O, Shio MT, McMaster RW, Olivier M. *Leishmania*-induced inactivation of the macrophage transcription factor AP-1 is mediated by the parasite metalloprotease GP63. PLoS Pathog 2010; 6(10): e1001148.

[http://dx.doi.org/10.1371/journal.ppat.1001148] [PMID: 20976196]

[201] Jaramillo M, Gomez MA, Larsson O, *et al. Leishmania* repression of host translation through mTOR cleavage is required for parasite survival and infection. Cell Host Microbe 2011; 9(4): 331-41. [http://dx.doi.org/10.1016/j.chom.2011.03.008] [PMID: 21501832]

[202] Medzhitov R. Recognition of microorganisms and activation of the immune response. Nature 2007; 449(7164): 819-26. [http://dx.doi.org/10.1038/nature06246] [PMID: 17943118]

[203] Broz P, Monack DM. Newly described pattern recognition receptors team up against intracellular pathogens. Nat Rev Immunol 2013; 13(8): 551-65. [http://dx.doi.org/10.1038/nri3479] [PMID: 23846113]

[204] Khadem F, Uzonna JE. Immunity to visceral leishmaniasis: implications for immunotherapy. Future Microbiol 2014; 9(7): 901-15. [http://dx.doi.org/10.2217/fmb.14.43] [PMID: 25156379]

[205] Faria MS, Reis FC, Azevedo-Pereira RL, Morrison LS, Mottram JC, Lima AP. *Leishmania* inhibitor of serine peptidase 2 prevents TLR4 activation by neutrophil elastase promoting parasite survival in murine macrophages. J Immunol 2011; 186(1): 411-22. [http://dx.doi.org/10.4049/jimmunol.1002175] [PMID: 21098233]

[206] Silverman JM, Clos J, Horakova E, *et al. Leishmania* exosomes modulate innate and adaptive immune responses through effects on monocytes and dendritic cells. J Immunol 2010; 185(9): 5011-22. [http://dx.doi.org/10.4049/jimmunol.1000541] [PMID: 20881185]

[207] Hassani K, Antoniak E, Jardim A, Olivier M. Temperature-induced protein secretion by *Leishmania mexicana* modulates macrophage signalling and function. PLoS One 2011; 6(5): e18724. [http://dx.doi.org/10.1371/journal.pone.0018724] [PMID: 21559274]

[208] Schorey JS, Cheng Y, Singh PP, Smith VL. Exosomes and other extracellular vesicles in host-pathogen interactions. EMBO Rep 2015; 16(1): 24-43. [http://dx.doi.org/10.15252/embr.201439363] [PMID: 25488940]

[209] Silverman JM, Reiner NE. Exosomes and other microvesicles in infection biology: organelles with unanticipated phenotypes. Cell Microbiol 2011; 13(1): 1-9. [http://dx.doi.org/10.1111/j.1462-5822.2010.01537.x] [PMID: 21040357]

[210] Hassani K, Shio MT, Martel C, Faubert D, Olivier M. Absence of metalloprotease GP63 alters the protein content of *Leishmania* exosomes. PLoS One 2014; 9(4): e95007. [http://dx.doi.org/10.1371/journal.ponc.0095007] [PMID: 24736445]

[211] Kaye PM. Costimulation and the regulation of antimicrobial immunity. Immunol Today 1995; 16(9): 423-7. [http://dx.doi.org/10.1016/0167-5699(95)80018-2] [PMID: 7546205]

[212] Overath P, Aebischer T. Antigen presentation by macrophages harboring intravesicular pathogens. Parasitol Today (Regul Ed) 1999; 15(8): 325-32. [http://dx.doi.org/10.1016/S0169-4758(99)01473-8] [PMID: 10407380]

[213] Antoine JC, Lang T, Prina E, Courret N, Hellio R. H-2M molecules, like MHC class II molecules, are targeted to parasitophorous vacuoles of *Leishmania*-infected macrophages and internalized by

amastigotes of *L. amazonensis* and *L. mexicana*. J Cell Sci 1999; 112(Pt 15): 2559-70. [PMID: 10393812]

[214] Jutras I, Desjardins M. Phagocytosis: at the crossroads of innate and adaptive immunity. Annu Rev Cell Dev Biol 2005; 21: 511-27. [http://dx.doi.org/10.1146/annurev.cellbio.20.010403.102755] [PMID: 16212505]

[215] Matheoud D, Moradin N, Bellemare-Pelletier A, *et al. Leishmania* evades host immunity by inhibiting antigen cross-presentation through direct cleavage of the SNARE VAMP8. Cell Host Microbe 2013; 14(1): 15-25. [http://dx.doi.org/10.1016/j.chom.2013.06.003] [PMID: 23870310]

[216] Gupta G, Oghumu S, Satoskar AR. Mechanisms of immune evasion in leishmaniasis. Adv Appl Microbiol 2013; 82: 155-84. [http://dx.doi.org/10.1016/B978-0-12-407679-2.00005-3] [PMID: 23415155]

[217] Louis J, Himmelrich H, Parra-Lopez C, Tacchini-Cottier F, Launois P. Regulation of protective immunity against *Leishmania major* in mice. Curr Opin Immunol 1998; 10(4): 459-64. [http://dx.doi.org/10.1016/S0952-7915(98)80121-0] [PMID: 9722923]

[218] Thiakaki M, Kolli B, Chang KP, Soteriadou K. Down-regulation of gp63 level in *Leishmania amazonensis* promastigotes reduces their infectivity in BALB/c mice. Microbes Infect 2006; 8(6): 1455-63. [http://dx.doi.org/10.1016/j.micinf.2006.01.006] [PMID: 16698300]

[219] Arango Duque G, Descoteaux A. *Leishmania* survival in the macrophage: where the ends justify the means. Curr Opin Microbiol 2015; 26: 32-40. [http://dx.doi.org/10.1016/j.mib.2015.04.007] [PMID: 25988701]

[220] Cordeiro-Da-Silva A, Borges MC, Guilvard E, Ouaissi A. Dual role of the *Leishmania major* ribosomal protein S3a homologue in regulation of T- and B-cell activation. Infect Immun 2001; 69(11): 6588-96. [http://dx.doi.org/10.1128/IAI.69.11.6588-6596.2001] [PMID: 11598026]

[221] Feng GJ, Goodridge HS, Harnett MM, *et al.* Extracellular signal-related kinase (ERK) and p38 mitogen-activated protein (MAP) kinases differentially regulate the lipopolysaccharide-mediated induction of inducible nitric oxide synthase and IL-12 in macrophages: *Leishmania* phosphoglycans subvert macrophage IL-12 production by targeting ERK MAP kinase. J Immunol 1999; 163(12): 6403-12. [PMID: 10586030]

[222] Cameron P, McGachy A, Anderson M, *et al.* Inhibition of lipopolysaccharide-induced macrophage IL-12 production by *Leishmania mexicana* amastigotes: the role of cysteine peptidases and the NF-kappaB signaling pathway. J Immunol 2004; 173(5): 3297-304. [http://dx.doi.org/10.4049/jimmunol.173.5.3297] [PMID: 15322192]

[223] Pollock KG, McNeil KS, Mottram JC, *et al.* The *Leishmania mexicana* cysteine protease, CPB2.8, induces potent Th2 responses. J Immunol 2003; 170(4): 1746-53. [http://dx.doi.org/10.4049/jimmunol.170.4.1746] [PMID: 12574338]

[224] Mottram JC, Coombs GH, Alexander J. Cysteine peptidases as virulence factors of *Leishmania.* Curr Opin Microbiol 2004; 7(4): 375-81.

[http://dx.doi.org/10.1016/j.mib.2004.06.010] [PMID: 15358255]

[225] Denise H, Poot J, Jiménez M, *et al.* Studies on the CPA cysteine peptidase in the *Leishmania infantum* genome strain JPCM5. BMC Mol Biol 2006; 7: 42.
[http://dx.doi.org/10.1186/1471-2199-7-42] [PMID: 17101050]

[226] Hatzigeorgiou DE, Geng J, Zhu B, *et al.* Lipophosphoglycan from *Leishmania* suppresses agonist-induced interleukin 1 beta gene expression in human monocytes *via* a unique promoter sequence. Proc Natl Acad Sci USA 1996; 93(25): 14708-13.
[http://dx.doi.org/10.1073/pnas.93.25.14708] [PMID: 8962119]

[227] Peacock CS, Seeger K, Harris D, *et al.* Comparative genomic analysis of three *Leishmania* species that cause diverse human disease. Nat Genet 2007; 39(7): 839-47.
[http://dx.doi.org/10.1038/ng2053] [PMID: 17572675]

[228] Zhang WW, Matlashewski G. Screening *Leishmania donovani*-specific genes required for visceral infection. Mol Microbiol 2010; 77(2): 505-17.
[http://dx.doi.org/10.1111/j.1365-2958.2010.07230.x] [PMID: 20545850]

CHAPTER 8

Virulence Factors and Immune Evasion in *Trypanosoma cruzi*

Jorge González[1,*], **Bessy Gutiérrez**[1], **José L. Vega**[2] and **Jorge E. Araya**[1]

[1] *Molecular Parasitology Unit, Medical Technology Department, Faculty of Health Sciences, University of Antofagasta, Antofagasta, Chile*

[2] *Experimental Physiology Laboratory, Instituto Antofagasta, University of Antofagasta, Antofagasta, Chile*

Abstract: Among the pathogens that have developed a variety of strategies to overcome the host immune system, is the causative agent of Chagas disease, *Trypanosoma cruzi*. During a long co-evolution process, the parasite has learned how to live in many different environments, including vertebrate and invertebrate hosts. The parasite has also evolved many invasive strategies, including several different ways to enter the host and also the capacity to target different host tissues. An acute systemic response arises in the host after the rapid parasite colonization, interfering with both innate and adaptive immunity. The capacity of *T. cruzi* to interfere with humoral and cellular immune responses is demonstrated by the expression of different sets of molecules called virulence factors. Among them, the role of antioxidant enzymes, cruzipain, the Tc85/transialidase superfamily, mucins, MASPs, GPI anchors, complement regulatory proteins and others are discussed in this chapter. The expression of parasite-specific virulence factors allows *T. cruzi* to overcome host immunity successfully and also to invade and disseminate in many different mammalian hosts. However, the picture that has emerged suggests that the basis and mechanisms of parasite virulence could be more complex than expected. Different aspects such as parasite genetic diversity, the effects of polyparasitism and the potential effects that vertebrate and invertebrate hosts have on parasite virulence and the outcome of natural or experimental infection by *T. cruzi* should be taken into account in futures studies to understand *T. cruzi* virulence.

* **Corresponding author Jorge González:** University of Antofagasta, Antofagasta, Chile; Tel: 56-55- 2 637376; Fax: 56- 55- 2 637802; E-mail: jagcparasitologiamolecular@gmail.com

Marcelo Santos da Silva & Maria Isabel N. Cano (Eds.)

Keywords: Complement regulatory proteins, CRP, GPI anchor, Immune evasion, MASPs, Mucins, Siglec, Small mucin-like gene, Tc85, *Trans*-sialidase.

INTRODUCTION

Trypanosoma cruzi is a kinetoplastid flagellate protozoan and the etiologic agent of Chagas disease. *T. cruzi* has a digenetic life cycle involving an invertebrate and a vertebrate host, including the human species. The parasite probably appeared in the region that originated the American and infected primitive mammals over 150 million years ago [1]. Thus, the wild cycle of *T. cruzi* has existed in nature for millions of years [2] and according to several authors [3, 4], humans made contact with the *T. cruzi* transmission cycle after acquiring sedentary habits that also included the domestication of animals and plants. In this scenario, several pieces of evidence suggest the hypothesis that *T. cruzi* infection is perhaps as antique as man's stay in the American continent. Consistent with this, PCR-positive samples from mummies from pre-Colombian Chile and Peru, were positive for *T. cruzi* infection indicating the occurrence of Chagas disease 9,000 years ago [5, 6]. This suggests that all along these years, *T. cruzi* has performed life cycles in triatomines, humans, domestic and wild mammals. During this long co-evolution process, the parasite has learned how to survive in many different hosts, developing an extensive repertoire of molecules that allows it to infect different triatomine species from the *Reduviidae* family and nearly all the tissues of more than one hundred species of mammals. Another significant issue of *T. cruzi* adaptive biology regards to the many ways that *T. cruzi* use to be transmitted to mammalian hosts. The infectious metacyclic trypomastigote (MT) stage can be transmitted principally by blood transfusion, by contamination of the skin and mucous membranes and by congenital and oral contact. The biological plasticity of *T. cruzi* facilitates the variability, complexity and distinct natural transmission cycles on a temporal and spatial scale. The complexity of its natural transmission cycles shows that over the time, humans may have contacted *T. cruzi* on several opportunities and through many different ways, according to principally how people interacted with the environment [7].

T. cruzi Developmental Cycle

The *T. cruzi* developmental cycle is divided into two different hosts. It can be initiate with a hematophagous triatomine, which during its blood meal could be infected with blood trypomastigotes from the mammalian host. After the triatomine infection, parasite migrates through the digestive tract of the triatomine and differentiate into the replicative, non-infectious form named epimastigote. In the insect gut, epimastigotes multiply by binary fission and migrate to the rectum to differentiate into the non-replicative and infective, MT form, which is excreted with the insect feces. During this transformation process known as metacyclogenesis, parasite develops adaptations in biochemical and morphological ways that allow it to invade nucleated cells and resist the immune attack of the vertebrate host.

The MT excreted by the triatomines penetrates the mucosal of the vertebrate host or the puncture mark at the wound generated by the insect. Inside the vertebrate host, MT can invade different cell types, such as fibroblasts, macrophages and muscle cells [8], and evades the highly oxidative phagocytic environment, starting the mammalian cycle. Upon surviving this hostile environment, MT escapes from the phagolysosomal vacuole and in the cytoplasm transform into amastigotes, which will multiply and converts into blood trypomastigotes [8]. Finally, trypomastigotes reach the blood vessels in order to disseminate the parasite infection in different cells and organs. Details about parasite life cycle and the morphology of different parasite life stage are described in Chapters 1 and 2.

THE MECHANISMS OF *T. CRUZI VIRULENCE*

For parasitologists, virulence is always described as the increase in host lethality resulted by pathogen invasion [9]. Even though, this is explicit and quantifiable, it does not consider several aspects of parasite biology that result in damage to the cells and organs without host death. For clinical microbiologists, virulence is comprehended as injury or sickness to the hostal though it does not explain how this damage is caused and which virulence factors could be involved in this harm [10].

Virulence factors are described as molecules synthesized and expressed by a pathogen. The lack or inhibition of these factors is detrimental for the pathogen virulence but not for its viability [10]. Virulence factors help the pathogen to invade, create an infection and develop in the host causing illness, and to evade host immune response [10].

It has been proposed that virulence can be controlled by both the host and parasite genetic variability, by parasite tropism for a specific host tissue and the capacity of the parasite to develop a long-term infection in the mammalian host. For the parasite, virulence is a complex feature that implies in its capacity (*e.g. T. cruzi*) to differentiate into an infective form, penetrate the epithelium, enter host cells, proliferate, and avoid host's immune system [8]. Several *T. cruzi* strains have been described and characterized, and the biological, biochemical and genetic variety between them have been reported. Many studies have shown that *T. cruzi* strains isolated from different geographic areas and originated from different vertebrate and invertebrate reservoirs show distinct biological features, such as tissue tropism, proliferation rate, infectivity capacity, complement sensitivity and drug susceptibility [11]. It has been proposed that the biochemical and structural heterogeneity of *T. cruzi* isolates may modulate *T. cruzi* pathogenicity [12].

Briefly, what we have learned about *T. cruzi* virulence is that it relies on a variety of genotypic and phenotypic parasite factors. Furthermore, the heterogeneity hypothesis and the theory that populations are composed of several subpopulations with different biological properties have been strengthened by studies based on cloned and non-cloned *T. cruzi* populations [13]. Therefore, the distinctive performance of one isolate would be the result of the interaction of many different strains or subpopulations, since the coexistence of multiple *T. cruzi* strains in mammalian and triatomine hosts has been reported. In addition, when the biological isolate is obtained and the laboratory isolation is done , the procedures could also act as selective factors. Several passages in mice or in axenic culture medium might also facilitate or inhibit one or more subpopulations present in the original isolate [13]. It has also been described that the procedures used to cultivate *T. cruzi in vitro*, affect MT genes and antigenic expression [14]. All of these mentioned factors could account either for the increase in *T. cruzi* populations virulence after *in vivo* passages or in the attenuation process after long-term *in vitro* culture.

However, the exact mechanisms used by *T. cruzi* to cause harm in the mammalian host are not well known. Even more, no experimental evidence allows us to have a general idea about the virulence displayed by *T. cruzi*. Although many different virulence factors have already been reported, their real input to parasite virulence and pathogenesis remains elusive.

The distinct *T. cruzi* life forms express several specific molecules that may act sequentially and in a coordinated manner. At the site of primary infection, MT invades mainly local fibroblasts, macrophages, and different cells in the mesenchymal tissues [15], where they resist oxidative stress by expressing several antioxidant enzymes. During and after intracellular replication as amastigotes and differentiation into bloodstream trypomastigotes, *T. cruzi* express different surface molecules that enable them to evade the innate and adaptive host humoral immune response. The life cycle continues when circulating trypomastigotes initiate new rounds of invasion, continually expressing surface antigens or secreting molecules that will allow them to attach to the host cell and induce signaling pathways that facilitate parasite cell penetration [16]. Many different molecules that participate in each of these processes have already been reported. However, their exact input to *T. cruzi* virulence remains elusive. In this chapter, we centered our review on the virulence factors that are mainly involved in evasion of the parasite from host immune response as well as in cell invasion, escape from the phagolysosome and parasite differentiation.

T. CRUZI EXPRESS VIRULENCE FACTORS INVOLVED IN RESISTANCE AND EVASION FROM THE HOST IMMUNE SYSTEM

T. cruzi has evolved sophisticated strategies to overcome host immunity.Trypanosome has an arsenal of evasion strategies linked to the alternation between intracellular proliferative forms (amastigotes and blood trypomastigotes) and the non-proliferative infective extracellular form (MT trypomastigotes). Immune evasion by *T. cruzi* relies primarily on subverting the complement system and the inhibitory effects of the mononuclear phagocyte system [17]. However, many different parasite molecules can interact with immune cells like T lymphocytes, macrophages and dendritic cells modulating the host immune response [17].

The *T. cruzi*-Phagocytic Cell Interaction, a First Barrier to Jump

The role of phagocytic cells and phagocytosis in *T. cruzi* infection has been widely described [18]. Downregulation of the phagocytic activity is a fundamental component of the invasive strategies of *T. cruzi*. In contrast to other protozoan parasites that inhibit the maturation of phagolysosomes, *T. cruzi* evades the microbicidal activity of macrophages by escaping from the phagolysosome to the host cell cytoplasm, where it replicates [19].

Virulence Factors Involved in T. cruzi Resistance to Oxidative Damage

The *T. cruzi* antioxidant defenses are very efficient and well-compartmentalized biochemical pathways that fulfill the fundamental function of detoxification of reactive oxygen and nitrogen species (ROS and NOS, respectively), generated by parasite–host cell interactions [20].

A macrophage membrane-associated NAD(P)H oxidase is activated when *T. cruzi* start the infective process, resulting in superoxide (O2·-) production. The superoxide (O2·-) can be dismutate to hydrogen peroxide (H_2O_2) or react with inducible nitric oxide synthase (iNOS)-derived nitric oxide (NO) to yield peroxynitrite (ONOO−), which is a powerful oxidant and has a strong cytotoxic effect against *T. cruzi* [21]. Parasite expresses antioxidative defenses at the onset of macrophage invasion, inclining the balance toward pathogen survival and promoting its exit from the vacuole and the establishment of infection. Overexpression of *T. cruzi* cytosolic tryparedoxin (TXN) peroxidase neutralizes macrophage-derived ONOO−dependent cytotoxicity and favors infection in an animal model. These observations directly support the effect of ONOO− as an intraphagosomal cytotoxin against pathogens, and the assumption that microbial peroxiredoxins promote infectivity *via* breakdown of macrophage-released ONOO− [22].

The antioxidant defenses of *T. cruzi* rely on a complex pathway in which reducing equivalents from NADPH are given to a variety of enzymatic detoxification systems through dithiol trypanothione (T(SH)2, N1,N8-bisglutathionyl-spermidine) and the thioredoxin homolog TXN. *T. cruzi* has five distinct peroxidases, which display different subcellular localization and substrate specificity. Glutathione peroxidase-I (GPXI) and Glutathione peroxidase-II (GPXII), localized respectively in the cytosol, GPXI in the glycosome and GPXII in the endoplasmic reticulum, provide resistance against exogenous hydroperoxides [23, 24]. The mitochondrial and cytosolic TXN peroxidases (mTXN and cTXN, respectively) detoxify H_2O_2 and small-chain organic hydroperoxides [23, 24]. *In vivo* detoxification of exogenous (macrophage-derived) and endogenous (parasite metabolism-derived) ONOO−, depend on the activity of both mTXN and cTXN [23]. In addition, in *T.cruzi*, resistance against H_2O_2 defiance, is provided by the plant-like ascorbate-dependent heme peroxidase (TcAPX) also located in the endoplasmic reticulum [25] (Fig. **1**).

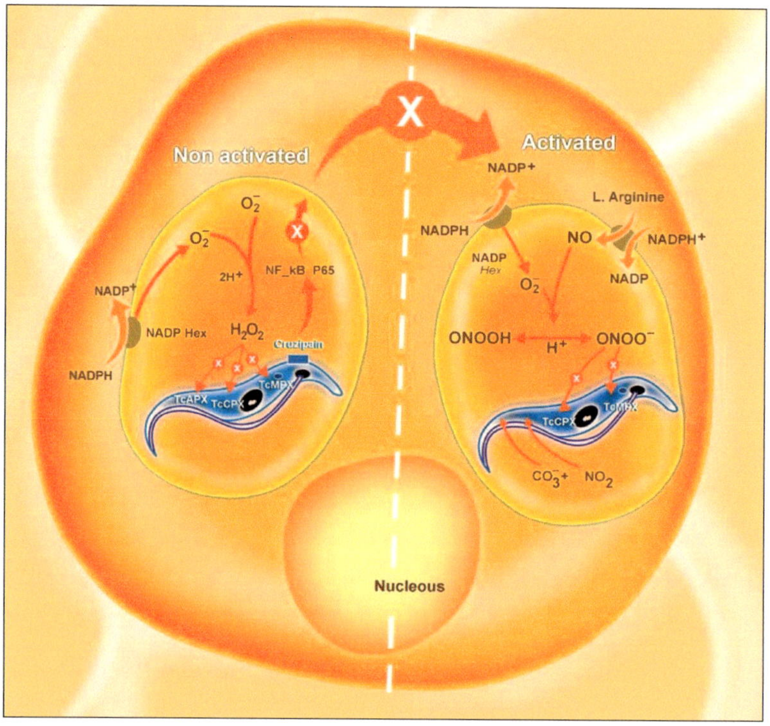

Fig. (1). Virulence factors expressed by *Trypanosoma cruzi* during macrophage invasion. At the first step of host cell invasion, *T. cruzi* must evade oxidative stress inside the macrophage phagolysosomal vacuole, where different ROS such as hydrogen peroxide, singlet oxygen and peroxynitrite are released. To escape this barrier, *T. cruzi* expresses cytoplasmatic tryparedoxin peroxidase (TcCPX), mitochondrial tryparedoxin peroxidases (TcMPX) and an endoplasmic reticulum ascorbate-dependent heme peroxidase (TcAPX). In both non-activated or activated macrophage, *T. cruzi* also express cruzipain responsible for inducing proteolysis of NF-κB P65, leading to unresponsiveness of the host macrophage. NO_2 (nitric oxide); CO_3^{-2} (carbonate).

Also, *T. cruzi* has four iron superoxide dismutases (Fe-SOD) that detoxify $O_2{\cdot}-$ produced in the cytoplasm, glycosomes, and mitochondria [26]. In the infective MT, proteomic studies have shown the positive regulation of *T. cruzi* antioxidant enzymes with respect to the non-infective epimastigote [27]. Thus, in all strains analyzed so far, the transformation of epimastigotes into the MT correlates with higher expression of mTXN, cTXN, and trypanothione synthetase (TryS). This biological condition makes MT less susceptible to ONOO− challenge [23,24].

According to Piacenza *et al.* [24], in the MT forms of virulent isolates, peroxiredoxins are expressed at high levels when compared to the attenuated parasite strains. In contrast, in infected mice, it was observed a strong correlation between the expression of TcCPX, TcMPX and TcTSy and the parasitemia showed by several isolates evaluated. These observations strongly suggest that *T. cruzi* antioxidant enzymes must be considered as new virulence factors with a critical role in the establishment of the parasite infection. Details about how *T. cruzi* maintains genome stability during oxidative stress are described in Chapter 5.

T. cruzi Complement Resistance and Evasion from Lytic Activity

The capacity of *T. cruzi* to survive and develop in different mammalian hosts is based on their ability to block or avoid the host immune response. At the beginning of the infection process, one of the initial barriers encountered by *T. cruzi* is the complement system. In fact, the activation of complement system cascade plays a central role in lysis and removal of pathogens, prior to the development of the adaptive host immune response. As was previously reported, MT and blood trypomastigotes are highly resistant to the complement lysis, whereas epimastigotes are highly susceptible and quickly lysed by the alternative complement pathway [28]. The strong complement resistance of *T. cruzi* trypomastigotes relays on the developmentally regulated expression of various parasite membrane glycoproteins in the trypomastigote forms, which inhibit complement activation [29].

Trypomastigote Decay-Accelerating Factor

The resistance of *T. cruzi* to complement-mediated lysis (CML) depends on the expression of a glycosylphosphatidylinositol (GPI) anchored surface glycoprotein with a molecular weight estimated in 87- to 93-kDa, known as the trypomastigote decay-accelerating factor (T-DAF). T-DAF is functionally similar to its human counterpart (mammalian complement regulatory component, decay-accelerating factor), and it accelerates the dissociation of C2a and Bb fragments from their respective ligation sites, by blocking the efficient assembly of the C3 convertases of both the classical and the alternative pathways [30].

Complement Regulatory Proteins

Complement regulatory proteins (CRPs) are also parasite glycoproteins that block the classical and alternative pathways of complement activation. CRPs are expressed in trypomastigotes cell surface but not in epimastigotes [29]. Norris [31], reported that stable CRP-transfected epimastigotes were protected from CML, suggesting that CRP expression is enough to give a virulence-associated property to epimastigotes. *T. cruzi* genome contains multiple copies of CRPs genes that share sequence similarities with members of the *T. cruzi trans*-sialidase (TS) superfamily. Among the TS superfamily, CRPs are one of the three functionally distinct subfamilies that do not present enzymatic activity but are functionally related to the mammalian and microbial proteins involved in the regulation of the complement activation [32] (Fig. **2**).

The Complement C2 Receptor Inhibitor Trispanning Protein

Cestari *et al.* [33], described by the first time that *T. cruzi* activates the lectin complement pathway, where the mannan-binding lectins, H-ficolin and L-ficolin bind fast to *T. cruzi* cell surface. According to these authors, trypomastigotes express a membrane protein called complement C2 receptor inhibitor trispanning protein (CRIT), which is present in *T. cruzi* and is evolutionarily conserved since it was also found in many other microorganisms. CRIT is a complement regulatory receptor, expressed in hematopoietic cells and various tissues, whose function is to block the lectin activation pathway and CML by indirectly inhibiting C2 cleavage, preventing the classical C3 convertase pathway formation. Therefore, parasites overexpressing CRIT were extremely resistant to CML [33].

Several pieces of evidence, suggest that the ability to resist the CML differs among different *T. cruzi* strains. For example, the Y strain appears to be more resistant to the classical and lectin complement activation pathways than the Colombian strain. Then, Cestari *et al.* [34] have reported that epimastigotes from Y strain overexpressing the CRIT gene were less susceptible to normal human serum (NHS). In addition, overexpression of the Y strain CRIT gene in Colombian strain epimastigotes increased *T. cruzi* ability to resist to CML by the classical and lectin pathways (Fig. **2**).

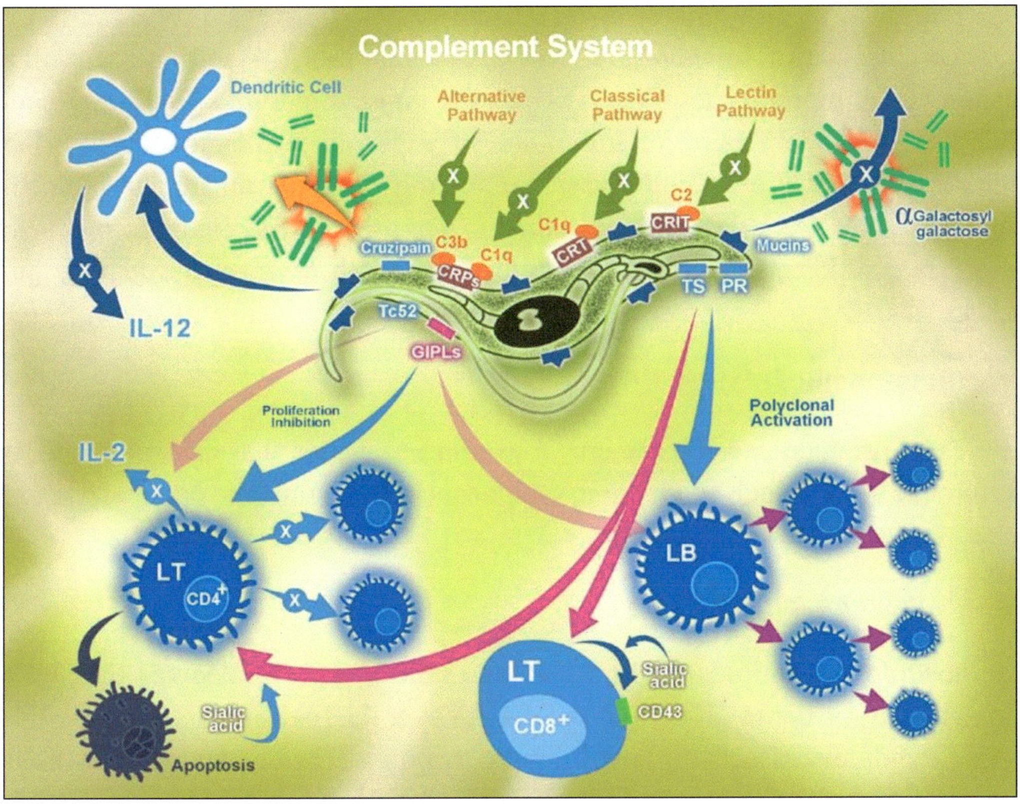

Fig. (2). Virulence factors involved in *Trypanosoma cruzi* immune evasion during bloodstream passage. After being released from the host cell, the parasite circulates in blood vessels coming into contact with the complement system. The parasite can resist complement attack, by expressing CRPs, calreticulin (CRT) and CRIT. *T. cruzi* releases molecules such as TS and proline racemases (PR), which induce polyclonal activation of B lymphocytes. *Trans*-silaidase (TS) also interfere with sialylation of CD4+ and CD8+ lymphocytes. Parasite glycoinositolphospholipids (GIPIs) induces polyclonal activation of B lymphocytes and Tc52 inhibits T cell proliferation. Cruzipain (CZP) is involved in immune evasion by proteolytic degradation of antibodies and sialylated mucins protect the parasite against lysis by anti-α-galactosyl antibodies impairing host dendritic cell function by inhibiting the production of IL-12.

Limited complement activation is a relevant immune evasion strategy provided by some proteins and utilized by many different pathogens, including parasites. [35] (Fig. **2**).

Calreticulin

Calreticulin (CRT) is a 45 kDa surface protein that binds the host C1q and blocks the activation of the classical complement pathway [36]. *T. cruzi* CRT also binds complement C1, enabling parasite infection. Most significant, preincubation of trypomastigotes with C1q enhanced *T. cruzi* invasion in RAW cell line, as well as parasitemia and mortality in mice, whereas the immunoglobulin Fab'2 fragments inhibit C1q-dependent parasite infection. In addition, unlike to other surface proteins engaged in *T. cruzi* cell invasion, CRT utilizes C1 as an adaptor molecule to bind host cells. As expected, experiments using Fab'2 fragments to block the C1/CRT interactions, reduce *T. cruzi* infectivity, both *in vitro* and *in vivo* [37] (Fig. **2**).

T. cruzi Immunomodulatory Effects

T. cruzi can produce and release enzymes and molecules that can modulate the host immune system such as glycoinositolphospholipids, mucins, cruzipain among others.

Glycoinositolphospholipids

Glycoinositolphospholipids (GIPLs) are free GPI anchors molecules that act as components of the dense glycolipid layer that cover the parasite cell surface. They present modulatory effects on host immune responses [38] by suppressing $CD4^+$ T-lymphocyte activation *in vitro* and *in vivo* through its ceramide domain [38]. Also, GIPLs suppress IL-2, but not IL-4 secretion, suggesting that GIPLs shift the immune response toward a Th2 profile [39]. GIPLs activate B-cell Ig secretion polyclonally, using the glycan moiety [40]. In addition, GIPLs down-regulate LPS-induced expression of co-stimulatory molecules and proinflammatory cytokine secretion in human macrophages and dendritic cells, functioning as a general suppressive factor in cell-mediated immunity [41]. GIPLs seems also to be involved in the alteration of the B cell compartment homeostasis and act. as virulence factors that function as TLR4 agonists with proinflammatory effect [38, 40, 42].

Mucin-Like Molecules and Gp35/50

A dense coat of mucin-type glycoconjugates covers *T. cruzi* cell surface. These parasite molecules are expressed to resist the potential mammalian and/or triatomine-derived defense mechanisms and to guarantee that the parasite target and invade specific cell types and organs. Even considering the complex structure and heterogeneity of mucins from different parasite isolates, *T. cruzi* mucins can be divided into two major groups: TcMUC I and TcMUC II, although other variants, such as the *T. cruzi* small mucin-like gene (TcSMUG) family, have been described [43]. The *T. cruzi* mucin gene repertoire is highly notable for its complexity and versatility. Taking into account the variation in the core mucin polypeptide expressed in the mammalian stages, Buscaglia *et al.* [43] suggested that this is one of the mechanisms by which *T. cruzi* may evade the host immune system. In fact, sialylation of mucins in *T. cruzi* trypomastigotes protects the parasite against the host humoral immune response and may give the parasite a negative charge [17]. Although the *O*-oligosaccharide composition of mucins has not yet been completely clarified, it is known that during the mammalian stage, mucins expose their terminal epitopes containing Gal(α1,3)Gal, considered the main targets of the host humoral immune system during parasite infection [44]. Nevertheless, the negative charge conferred by sialic acid residues, acquired by TS, neutralizes any complement-independent lysis induced by anti-α-galactosyl antibodies, allowing the parasite to avoid lysis [44]. Mucins can also impair host dendritic cell, as demonstrated by the inhibition of IL-12 production [43], which may occur at the transcriptional level since *T. cruzi*-derived mucins can inhibit transcription of the IL-2 gene in T cells [45]. This inhibition can also occur when the activation and proliferation of T cells are blocked in response to mitogens and antigens [46]. The parasite sialoglycoproteins also inhibit early events in T cell activation, in particular, the tyrosine phosphorylation of the adapter protein SLP-76 and the tyrosine kinase ZAP-70 [47].

The inhibitory effects of the *T. cruzi* mucins *in vivo* have been examined by Nunes *et al.* [46], after exposing a murine model of Chagas disease to these mucins during experimental infection. In this case, mice displayed increased susceptibility to infection, with enhanced parasitemia and heart damage. These effects were associated with a reduction in IFN-γ-producing CD4+ and CD8+ T cell responses, together with decreased levels of both splenic IFN-γ and TNF-α [46]. It has been shown that parasite-derived mucins bind to the mammalian acid-binding Tg-like lectin, sialic acid binding immunoglobulin-like lectin E (Siglec-E/CD33) [48, 49], suggesting that the binding of Siglec-E by *T. cruzi* mucins inhibits the mitogenic response in CD4+ T cells. The impairment of TCR/CD3-mediated activation of CD4+ T cells correlated with G1/S arrest and the interaction of the terminal sialyl residues of *T. cruzi* mucins with CD4+ T cells. In its turn, the interaction of mucins with CD4+ T cells led to the induction of p27/Kip1, a cell cycle regulator that blocks the transition from the G1 to the S phase of the cell cycle [45 - 47, 50].

Recent studies have shown that there is high variability in the expression and processing of TcSMUGs between parasite strains [51]. It seems that the high variable (HV) region of TcSMUGs has a mosaic of variants that are simultaneously expressed in the -parasite and are recognized by the serum of infected hosts, which can lead to anergy in the protective response of CD4+ cells [52].

Mucins can also induce the expression of some pro-inflammatory cytokines such as IL-12 and TNF-α, as well as the production of NO by macrophages. The GPI anchors motifs of these proteins are responsible for the immune stimulation of the proinflammatory response by binding specifically to Toll-like receptor 2 (TLR2) at the macrophage surface, although other TLRs may also be involved [53]. The HV region of mucins can provide binding ligands for the parasite on a variety of cell types. Treatment with monoclonal antibodies against mucins inhibits the entry of *T. cruzi* [54], although the actual role of these mucins in cell invasion may be to cooperate with other trypomastigote surface molecules such as the TSs because they are highly expressed during the trypomastigote and amastigote stages [55]. However, Yoshida *et al.* [56] reported that in MT, the mucin-like Gp35/50 surface antigen is implicated in parasite cell invasion. Similarly to Gp82 [57], Gp35/50 adheres to target cells in a receptor-mediated fashion and induce a bidirectional Ca^{2+} signaling cascade [14]. Gp35/50 mucins can also modulate host immune response, showing that it can destabilize TNF and cyclooxygenase-2 mRNA by blocking the activation of p38 mitogen-activated protein kinase (MAPK) and its substrate, MAPK-activated protein kinase-2 (MAPKAP-K2) [58].

Cruzipain

Cysteine proteases appear to develop a key role in the life cycles of several protozoan and helminth. The enzyme, catalyze many different virulence-related events besides to functions in catabolism and protein processing. Cysteine proteolytic enzymes have been involved in tissue, immune and cellular invasion, in excystment/encystment, and are considered relevant virulence factors in many different protozoan parasites, such as *Entamoeba histolytica*, *Trichomonas vaginalis*, *Giardia lamblia* (at present named *Giardia duodenalis*), *Leishmania, Toxoplasma gondii* and *Schistosoma* [59, 60].

In *T. cruzi*, cruzipain (CZP) is a main cysteine protease that plays clear roles in mammalian cell invasion, intracellular development and evasion of host immune system [61]. This protease is expressed in lysosome-related organelles in all parasite forms and some of its isoforms are expressed on the parasite cell membrane, probably through a GPI anchor [62]. The expression of CZP at the membrane surface is especially relevant in amastigotes and since it can be released from the cell, presumably it could fulfill some pathogenic role in the host cell cytoplasm. It has been reported that trypomastigotes can release some CZP isoforms into the medium, which is quite relevant to its role as a virulence factor during *in vivo T. cruzi* infection [62].

CZP is also involved in the proteolytic breakdown of host tissues and participates on evasion of the host immune system [63]. CZP may supply a strategy by which *T. cruzi* could prevent the humoral host immune response by digestion of all human IgG subclasses at the "hinge". At a neutral pH, the fragments produced from IgG1 and IgG3 were Fab and Fc, whereas IgG2 and IgG4 yielded Fab2 and Fc [63]. This pattern is compatible with the mechanism known as "fabulation", when the Fab2 fragment remains to coat the antigenic epitope on the external parasite membrane. In fact, these fragments are unable of opsonizing the pathogen for phagocytosis or of triggering the complement cascade system. Furthermore, alternative activation of macrophages associated with an increase in *T. cruzi* survival was observed when murine macrophages were incubated with CZP [64]. This alternative activation was characterized by an increase of the arginase-1 activity and release of TGF-β and IL-10 [64]. Studying the role of CZP in *T. cruzi* pathogenicity has been challenging and limited by genetic redundancy and the failure to attempt gene deletion. To evaluate the biological function of CZP protease-deficient parasites that displays less than 1% of the wild-type CZP enzyme activity were tested and showed that they were able to induce a fast activation of host macrophage *via* NF-κB P65, although they were not able to proliferate and develop inside of the macrophages. In contrast, infection with wild-type trypomastigotes results in the CZP-dependent proteolytic degradation of NF-κB P65, leading to an anergy of the host macrophage in the beginning of acute mice infection. This parasite strategy might help the replication and survival of *T. cruzi* by increasing the systemic spread of parasites [65]. It has also been reported that during *T. cruzi* infection, CZP may contact mannose receptors, increasing mannose receptor recycling, which leads to the up-regulation of arginase activity and intracellular parasite growth [66]. In the same way, sulfates from CZP, containing sialylated glycoproteins, were shown to interact with Siglec-E, an ortholog of human Siglec-9. This observation reinforces the idea that CZP might be engaged in the modulation of the mammalian immune response, inducing an increase of parasitemia and *T. cruzi* persistence [67].

The participation of CZP in tissue culture trypomastigotes (TCTs) cell invasion is related to its capacity to produce bradykinin [68]. Purified CZP increased *T. cruzi* invasion and induced Ca^{2+} mobilization in CHO cells overexpressing bradykinin B2 receptor (B_2R) (CHO–B_2R), which was abolished by the B2R antagonist HOE 140 or by the cysteine proteinase inhibitor E-64. This observation suggests that CZP performs a role in producing the kinin agonist from cell-bound kininogen, favoring an increase in the cytosolic calcium levels in the host cell, which is a key step in the parasite invasion process [68]. Recently, it has been reported that CZP was able to activate latent TGF-β from host cells and promote *T. cruzi* cell invasion [69]. Similarly, Maeda *et al.* [70], have shown that the *T. cruzi* cysteine proteinase expressed by MT forms is implicated in host cell invasion through its fibronectin-degrading activity.

It has been suggested that CZP may also fulfill a critical function in parasite metacyclogenesis, which was inhibited by treatment with cell-permeable inhibitors of cysteine proteinases [71]. This result was confirmed by Tomas and co-workers [72], who demonstrated that the over-expression of CZP increased the transformation of epimastigotes to MT.

Moreover, CZP seems to play a role in parasite metabolism and has been identified as both as a prominent candidate for vaccine development and as a target to develop trypanocidal drugs against *T. cruzi* [62]. In immunized mice, the CZP's sulfated oligosaccharides participate in host tissue injury and are the main epitopes recognized by host humoral immune response [62, 73].

Proline Racemases

The conversion of L- and D-proline enantiomers are catalyzed by enzymes called proline racemases (PRs) [74]. In *T. cruzi*, PRs are codified by two paralogous genes named TcPRACA and TcPRACB, which generate the secreted and cytoplasmic enzyme isoforms, respectively [75]. PRs are key molecules for *T. cruzi* immune evasion and parasite survival (Fig. **3**) [74]. TcPRACA is mitogenic for B lymphocytes and is able to induce non-specific polyclonal lymphocyte activation. The overexpression of the TcPRACA isoforms leads to increase in *T. cruzi* cell differentiation and invasiveness [76]. In the same way, the bioavailability of D-proline in the cytoplasm allows for the further addition of D-proline to parasite peptide chains. Thus, D-proline-having peptides, analogous to the mucopeptide layer of bacterial cell walls, may be useful to *T. cruzi* by conferring protection against host protease activity [77]. *In vitro* studies reported by Coutinho *et al.* [77] have been shown that both anti-*T. cruzi* PR antibodies and the PR inhibitor pyrrole-2-carboxylic acid, strongly reduced *T. cruzi* infectivity and parasite transformation into amastigotes.

Tc52

Tc52 is a 52-kDa secreted protein that can induce a suppression of murine T-cell response (Fig. **3**) [78]. Monoallelic disruption of Tc52 (Tc52+/−) results in a low virulent (*in vitro* and *in vivo*) *T. cruzi* clone that can not produce MT forms [79]. Tc52 also appears able to immunomodulate the macrophage expression of cytokines and iNOS, to modulate the NO synthesis and to inhibit human dendritic cells functions [80]. Consequently, Tc52 may be taken into account as a virulence factor that has a relevant function in the modulation of human and murine immune response [79 - 81].

Immune Evasion Mediated by Microvesicles

It has been recently reported a new immune evasion strategy of *T. cruzi*, which is mediated by vesicles devired from the host cell plasma membrane [82]. This mechanism is used by *T. cruzi* MT to be released in microvesicles from blood cells early in infection. Upon their release, microvesicles form a complex with the complement C3 convertase at the *T. cruzi* surface, leading to its stabilization and inhibition. The formation of this complex results in increased parasite survival. In addition, it was described that the release of microvesicles bearing TGF-β- from monocytes and lymphocytes promote rapid cell invasion by *T. cruzi*, also contributing to the escape of parasites from the complement attack. Moreover, *in vivo* infection with *T. cruzi* results in a rapid increase of microvesicle levels in mouse plasma, and infection with exogenous microvesicles lead to increased *T. cruzi* parasitemia [82, 83]. On the other hand, it has been shown that *T. cruzi* trypomastigotes continuously shed vesicles enriched in glycoproteins such as Gp85/TS superfamily, mucins, CZP, among others [83]. BALB/c mice treated with a vesicle fraction before *T. cruzi* infection show increased mortality, with 100% of the mice dying on the 20th-day post-infection, whereas 20% of untreated animals survived for more than 30 days. Treated animals also developed severe heart pathology, with an intense inflammatory reaction containing a high number of amastigote nests and high levels of IL-4 and IL-10 mRNAs detected in the hearts. Treatment of mice with neutralizing anti-IL-10 or anti-IL-4 antibodies precluded the vesicles pre-inoculation effects on infection. These results suggest that the components that *T. cruzi* shed from its membrane increase parasitism and inflammation *via* stimulation of IL-4 and IL-10 and thus, may play a central role in the pathogenesis of acute Chagas disease [84].

MT AND TCTS FACTORS INVOLVED WITH ADHESION AND CELL INVASION

The mechanisms involved in *T. cruzi* cell invasion have been largely studied using *in vitro* cell cultures infected with axenic MTs and tissue-culture trypomastigotes (TCTs) forms because they are considered biologically related models with the parasite forms that live respectively in the insect vector and in the blood of the mammalian host. Among the many different surface molecules expressed by both parasite forms, some of them appear to fulfill a central role in MT cell invasion [14]. According to that, we would like first to highlight the following MT factors.

Gp82

The 82 kDa surface glycoprotein (Gp82) is expressed in MT and works to facilitate parasite adhesion and cell penetration [57]. The penetration of cells induced by Gp82 is due to the activation of signaling pathways and increase of cytoplasmic Ca^{2+} levels in both host and parasite cells [14]. There is a strong correlation between the expression of Gp82 at parasite surface and the MT capacity to invade eukaryotic cells [14]. Studies with the poorly infectious *T. cruzi* CL-14 clone, reinforced the role played by Gp82 in MT infection, because when compared to the parental CL isolate, MTs derived from CL-14 clone are deficient in Gp82 expression [85].

Gp82 is also involved in the selective attachment to gastric mucin during *T. cruzi* oral infection [86].

Gp 30

T. cruzi strains deficient in Gp82, expresses a 30 kDa MT surface glycoprotein (gp30), that is also involved in the invasion of host cells [87]. Similarly to Gp82, attachment of gp30 to host cells trigger a Ca^{2+} response and regulates lysosome exocytosis, probably through the activation of signaling cascades with the participation of PKC, mTOR, and PI3K [87].

Gp 90

Malaga and Yoshida [88] reported that *T. cruzi* from different strains can express variable levels of distinct isoforms of a 90 kDa glycoprotein exclusively expressed at MT surface. In contrast to other parasite surface molecules, gp90 appears to be a negative regulator of *T. cruzi* cell infection, since high levels of gp90 expression are always correlated with a lower cell infectivity [89]. As opposed to Gp82 or gp30, and consistent with its role as a negative regulator of MT invasion, gp90 does not induce Ca^{2+} signaling after the adhesion of parasite to the host cell membrane [90].

Gp35/50

MT forms of poorly infective *T. cruzi* strains express a 35/50 kDa surface glycoprotein (Gp35/50), which is considered a highly glycosylated mucin-like glycoprotein, enriched in sialic acid and galactose residues. Gp35/50 may interact with the mammalian cells through its carbohydrate portion [56], triggering the elevation of intracellular Ca^{2+}, but to a lower degree compared to Gp82 [90]. Removal of its sialic acid stimulates Gp35/50 property to trigger a cellular Ca^{2+} response and enhance MT invasion [91].

Among the main TCTs proteins that participate in cell invasion we highlight the following.

Oligopeptidase B

Oligopeptidase B (OPB) is a stage-specific alkaline protease of 120-kDa, whose biological role is not entirely understood, although it has been suggested that OPB cleavage produces a Ca^{2+} agonist that is released by the infective stage, and is able to bind to a receptor on the surface of the target cells. This binding induces the activation of phospholipase C (PLC) and generates inositol triphosphate (IP_3), which in its turn, induces Ca^{2+} release from the endoplasmic reticulum [92].

The molecular characterization of OPB shows that the protease is cytosolic and closely related to the members of the prolyl oligopeptidase (POP) family of serine endopeptidases [93]. It has been reported that trypomastigotes OPB null mutants are less efficient to mobilizes Ca^{2+} from thapsigargin-sensitive stores to the host cells and are not able to set *in vitro* and *in vivo* infection [94]. OPB seems to be ubiquitous, and it is also present in Gram-negative bacteria and many different flagellates, like *Leishmania amazonensis*, *Leishmania major*, *Trypanosoma congolense* and *Trypanosoma brucei* [95]. In African trypanosomes, OPB is released during host infection by bloodstream circulating parasites . by. The role of OPB as a virulent factor appears to be wider than expected since this protease can degrade host serum hormones and regulatory peptides such as atrial natriuretic factor [95]. Even more, in *Leishmania*, OPB can mediate the activation of leishporins, which are cytosolic proteins that induce the formation of transmembrane pores in the macrophage cell membrane. In the presence of an oligopeptide inhibitor, OPB hydrolyzes the oligopeptide after the release of the active leishporin [48].

A member of the prolyl oligopeptidase (POP) family of serine endopeptidases was identified in cell-free extracts obtained from all stages of *T. cruzi* [59]. This 80-kDa protease, called POP Tc80, was characterized by its property of degrading fibronectin. In addition, the presence of POP selective inhibitors irreversibly inhibited cell invasion by TCT [49].

Gp85/TS Family

Gp85, an 85 kDa stage-specific surface glycoprotein is the best-characterized *T. cruzi* surface molecule. Gp85 is expressed by TCTs and participates in cell invasion [96] by using its conserved FLY domain (VTVXNVFLYNR), which has a high affinity for host cell laminin [97] and intermediate filaments such as cytokeratins and vimentin [98]. The gp85/TS family through its conserved FLY domain appears to be involved in the activation of extracellular signal-regulated kinases in host cell, facilitating *T. cruzi* infection [99]. The participation of the FLY domain in *T. cruzi* interaction with endothelial cells from different organs was investigated using bacteriophages expressing the FLY peptide, which was able to bind to these cells with high preference to the heart vasculature and to intermediate cell filaments, helping to explain why *T. cruzi* can infect different cell types. Also, it was observed that BALB/c mice inoculated with a single dose of FLY-synthetic peptide were most vulnerable to *T. cruzi* infection, showing high parasitemia especially in the heart (3-fold), small intestine (3·6-fold) and bladder (7·6-fold) and mortality. The elevated amount of parasites in these tissues suggest that the FLY peptide should be a key enhancer of tissue tropism [98]. FLY can also promotes *T. cruzi* infectivity *in vivo* working in association with other factors that enhance parasite survival, affecting the development of Chagas disease pathology [100]. Recently, it was reported that *T. cruzi* can bind to cytokeratin using conserved motifs found in the laminin-G-like domain of the gp85/TS glycoproteins [101].

TS Superfamily

The TS superfamily is composed of proteins anchored by GPI- at the surface of trypanosome parasites from the species *T. cruzi*, *T. brucei*, and *T. rangeli*. From the 1,430 genes belonging to the TS superfamily, half are apparently functional [102], including genes from different subfamilies that share a distinctive sialidase domain at the amino terminus, such as active TS, CRPs, and proteins of unknown function, including enzymatically inactive members of the TS family [103,104].

TS members are involved in the transfer of sialic acid from host glycoconjugates to the mucins at the parasite surface. Since, *T. cruzi* cannot synthesize sialic acid or the donor CMP-sialic acid, the role played by TS proteins is a very important step of the infectious process in Chagas disease [103].

TS seems to be involved in several aspects associated with parasite virulence, such as immunosuppression, complement resistance, thrombocytopenia and apoptosis of mouse thymocytes [105,106], because they are released into the bloodstream of the mammalian host by the action of a *T. cruzi* phosphatidylinositol–phospholipase C (PI-PLC). Circulating TS can target a broad range of cell types and systems and can interact directly with host lymphocytes through desialylation and resialylation of acceptor glycoproteins [107]. For example, TS co-stimulates $CD4^+$ T-cell proliferation and increases cytokine production by interacting with the lymphocyte mucin CD43 [108], but TS can induce lymphocyte apoptosis through targeting cell resialylation [107]. After T cell activation, the expression of sialic acid is down-regulated, which increases cellular contacts and enhance the reactivity of $CD8^+$ T cells [109]. One important target of *T. cruzi* TS appears to be $CD8^+$ T cells since treatment with TS resialylated CD43 and $CD8^+$ T cells *in vitro* and *in vivo* [110]. Increased resialylation compromises the ability of activated $CD8^+$ T cells to kill targets bearing *T. cruzi* epitopes [110].

It was also proposed that TS negatively regulates *T. cruzi* infection. In fact, the presence of monoclonal or polyclonal antibodies against TS increase TCT host cell invasion [111]. However, when experiments were carried out using TS+ or TS− parasite populations, it was observed that only TS+ trypomastigotes were highly infective, whereas TS− parasites displayed a poor infectivity to fibroblasts and epithelial cells. Moreover, the addition of small amounts of purified TS into suspensions of non-invasive TS− trypomastigotes transformed them into highly infective trypomastigotes [112]. The essential functions of TS in *T. cruzi* cell invasion have been well established by the observation that CHO (Lcc2) cells, which are deficient in sialic acid, are fairly resistant to *T. cruzi* invasion [113]. Specific TS inhibitors were also able to inhibit parasite cell invasion [114]. Incubation of cells with modified precursors of sialic acid reduces parasite invasiveness [115]. TS may also play a role as a counter-receptor for *T. cruzi* binding to alpha 2,3-sialyl receptors on host cells as a previous step of TCT invasion. Moreover, after the attachment of TCTs to target cells through another parasite molecule, secreted TS may transfer sialic acid from the surface of the eukaryotic cell to *T. cruzi* mucins [113] (Fig. **3**). Enzymatically active or inactive TS can attach to the surface of vertebrate cells, such as heart cells or T lymphocytes, in a lectin-like fashion [116,117]. Of particular relevance in the biology of Chagas disease is the binding of trypanosomes by their mucin-bound sialic acid coat to siglecs. The binding of sialoadhesin (siglec-1)-positive macrophages to sialylated ligands in *T. cruzi* surface might be necessary for the initial trypomastigote infection in mice [118].

The mechanisms of TS virulence are still not completely defined, but a recent study indicates that the TS-derived peptide FLY is a key player since it can increase the number of $CD4^{+}CD25^{+}FoxP3^{+}$ regulatory T cell factors, increase parasitemia in the hearts of infected animals and induces mortality [100].

According to the classification of *T. cruzi* strains used before (Details about the most recent classification of *T. cruzi* in discrete typing units, (DTUs are described in Chapter 9), two major phylogenetic groups (I and II) were defined [120]. At that time, it was demonstrated that infection with *T. cruzi* II isolates induced

100% mortality in mice, whereas infection with *T. cruzi* I isolates induced mice to enter the chronic phase, because trypomastigotes from *T. cruzi* II strains express and shed significantly higher amounts of TS than do those from the *T. cruzi* I lineage. In addition, after infection with *T. cruzi* II strains, it was possible to observe disorganization of the thymus histoarchitecture associated with circulating TS. In contrast, mice infected with *T. cruzi* I strain showed transient thymus lesions. Therefore, TS was considered the first *T. cruzi* virulence factor identified that is differentially expressed by the main parasite groups, partly explaining their different performance in the pathogenesis of Chagas disease [119].

Mucin-Associated Surface Proteins

The mucin-associated surface proteins (MASPs) multigene family, is composed of approximately 1,300 genes, and its discovery was an important finding of the *T. cruzi* genome project [102]. MASPs family members present high level of polymorphism which is associated with its localization at the surface of infective parasite forms. MASPs probably participate in host–parasite interactions and may substantially contribute to increasing the repertoire of parasite polypeptides that are exposed to the host cell immune system. MASP proteins are characterized by the presence of a highly variable and repetitive central region composed of peptides that are shared among all MASP members [121]. These peptides have a repertoire that may also contribute to cell-type-specific interactions since the polymorphism of *T. cruzi* surface proteins has been suggested to be a major factor in the parasite ability to infect multiple cell types [122]. The molecular characterization of a few members of MASP family showed that MASPs proteins despite being expressed at the parasite surface (circulating infective forms) they can be shed into the extracellular medium [121]. Differential expression of a few MASPs genes between trypomastigotes derived from epithelial and myoblast cell lines has also been reported [123]. In this context, the more pronounced MASPs expression changes were observed between bloodstream and TCT and between bloodstream forms obtained from sequential passages in acutely infected mice. In addition, different MASPs members could elicit different antibody responses, including variation in the antibody level against a specific peptide. The authors speculated that changes in the vast repertoire of MASPs peptides during an infection may contribute to the evasion of host immune system during the acute phase of Chagas disease [123]. Recently, a synthetic peptide originated from a *T. cruzi* MASP has been tested as a vaccine candidate against Chagas disease [124]. This vaccine candidate was able to induce 86% survival in the immunized group after challenging with a highly lethal dose of trypomastigotes. Immunized animals also showed the lowest parasite load in the heart, liver, and spleen compared to control animals as demonstrated by real-time PCR. Moreover, immunized animals produced trypanolytic, protective antibodies and showed a cytokine profile consistent with resistance against parasite infection [124]. According to De Pablos *et al.* [125], IgG antibodies specific to the catalytic region of the MASP52 protein significantly reduced the parasite capacity to infect host cells. This protein was also overexpressed in the infectious parasitic stages, demonstrating that the MASPs family plays a significant role in the mechanism of *T. cruzi* host cell invasion [125]. Furthermore, overexpression of MASPs in intracellular parasites starts 24 h postinfection and before amastigotes begin their replication, suggesting that proteins of the MASP family are key players in the survival and multiplication of intracellular amastigotes [125].

Calcineurin

T. cruzi contains different isoforms of calcineurin (CaN) which is composed of subunit A and subunit B. One of the *T. cruzi* CaN homolog is the CaN catalytic subunit A (TcCaNA2), which is expressed in the cytoplasm of all parasite developmental forms. The protein has phosphatase activity and participates in host cell invasion by the MT forms [126]. The other homolog is the CL strain CaNB gene (TcCaNB-CL), which encodes the subunit B that also plays a role in parasite infectivity. Treatment of MTs or TCTs with the CaN inhibitors cyclosporin A (CsA) or cypermethrin firmly blocked (62–64%) parasite invasion of HeLa cells. Also, incubation of trypomastigotes with antisense phosphorothioate oligomers raised against TcCaNB-CL decreased the expression of TcCaNB, reduced TcCaN enzymatic activity and resulted in a 50% inhibition of cell invasion [127]. In *Leishmania major*, it has been reported that deletion of the regulatory CanB subunit of CaN induced loss of thermotolerance, increased the susceptibility to compounds that generate unfolded protein response, caused alterations in membrane fluidity, as well as absolute loss of virulence to susceptible BALB/c mice [128].

Peptidyl-prolyl Cis-trans Isomerase

T. cruzi secrets a protein with Peptidyl-prolyl cis-trans isomerase (PPIase) activity, which is blocked by the immunosuppressant FK506 and related drugs. *T.cruzi* PPIase belongs to a protein family previously reported as macrophage infectivity potentiators (MIPs) and seems to increase parasite cell invasion [129].

Phospholipases

Phospholipases are a sophisticated and relevant group of enzymes, widely distributed in nature, that fulfills critical functions in many different cellular events. They are classified as phospholipases A_1(PLA_1), A2 (PLA_2), C (PLC), and D (PLD). PLA_1 and PLA_2 activities have been linked to pathogenesis in various microorganisms, and in pathogenic protozoa, they have been implicated in cell invasion [130].

Phospholipase A1

A PLA_1-enzymatic activity is expressed in all *T. cruzi* developmental forms. However, enzyme activity appears to be up to 20-fold higher in trypomastigotes and amastigotes than in epimastigotes. It was reported that either intact trypomastigotes or purified PLA_1 significantly alter the lipid pattern of the host cell, with the production of second lipid messengers and activation of protein kinase C, suggesting that *T. cruzi* PLA_1 could play a critical role in the early events of parasite–host cell interaction [130]. According to Belauzarán *et al.* [131] and contrary to the non-infective epimastigote forms, high levels of PLA_1 expression were detected in amastigotes and trypomastigotes forms of the highly virulent RA strain. The same expression pattern, but different PLA_1 enzymatic activity, was detected in bloodstream trypomastigotes from Cvd and RA lethal strains and from K98 non-virulent *T. cruzi* strain collected at their respective parasitemia peaks. This was probably the result of the modulation of the enzymatic activity, due to the presence of different levels of anti-*T. cruzi* PLA_1 antibodies in sera of infected mice. Furthermore, these antibodies significantly reduced trypomastigote infectivity *in vitro*, suggesting that *T. cruzi* PLA_1 must be considered a novel parasite virulence factor [131].

Gp 63

A 63 kDa surface glycoprotein (Gp63) gene family composed of several groups and isoforms with metalloprotease activity has been described in *T. cruzi*. Two of these groups, named TcGp63-I and II, are present in high-copy-number in parasite genome. The TcGp63-I group is characterized by the expression of a ~78 kDa surface protein, which is differentially expressed in different parasite stages [132]. The protein appears to be involved in parasite cell invasion, since pre-incubation of trypomastigotes with TcGp63 antiserum, or with purified TcGp63 C-terminal subfragments or with polyclonal antibodies raised against the TcGp63-I group, abolished parasite infection of myoblasts or of Vero cells [132].

Gp63 and its homologs have been reported in all trypanosomatids investigated so far [133]. Homologs of Gp63 also perform multiple relevant functions in the virulence of *Leishmania*, such as resistance to CML and promotion of promastigotes phagocytosis by macrophage [133]. In African trypanosomes, Gp63 participates, in association with PLC, in the release of the VSG coat when bloodstream forms trypanosomes transform into procyclic trypanosomes [133].

VIRULENCE FACTORS INVOLVED IN *T. CRUZI* PHAGOLYSOSOMAL ESCAPE

Invasion of mammalian cells by *T. cruzi* is characterized by the formation of parasitophorous vacuoles (PVs), from which the trypomastigote must be released to develop a cell differentiation program and then replicate free in the host cell cytoplasm.

The process initiates during parasite cell penetration when the lysosomes fuse with the emerging phagosome that forms around the parasite. The internal phagosome membrane expresses highly sialylated lysosomal glycoproteins (LGPs) that appear to fulfill a function in protecting the internal lysosomal membrane from the action of many lysosomal enzymes and are also potential substrates for *T. cruzi* TS [134]. TS could also participate in a parasite strategy focus on escaping from the phagosome. The phagolysosomal membrane ruptures after cell invasion and the parasite is released into the cytoplasm. In *T. cruzi,* a pore-forming protein, named Tc-tox, is responsible for the rupture of the vacuolar membrane [135], which is more active on desialylated membranes. Hall and Joiner [136] early proposed that when the enzyme releases sialic acid from the LGPs, it may help in the disruption of the phagosomal membrane by Tc-tox. *T. cruzi* quickly escapes from the phagosomes of sialic acid-deficient (Lec2) cells than from the phagosomes of wild-type cells. Moreover, in MT, the expression of TS by trypomastigotes also enhance their exit from the PVs. This reinforces the idea that TS overexpression plays a key role in the parasite early and fast escape from the PV to the cytoplasm, a necessary step in the subsequent trypomastigote transformation into amastigotes [137] (Fig. **3**).

VIRULENCE FACTORS INVOLVED IN *T. CRUZI* CELL DIFFERENTIATION OR PROLIFERATION

The pathways and enzymes that are involved in the differentiation of trypomastigotes to amastigotes in *T. cruzi*, are not well understood. Here, we describe the enzymes that have a central role in this process.

Proteasomes

The biological function of proteasomes in the differentiation of trypomastigotes to amastigotes was studied using the specific proteasome inhibitor Lactacystin β-lactone [138,139]. The presence of this proteasome inhibitor strongly abolished the differentiation process when parasites were cultivated in DMEM medium at pH 5.0. Using C^{14}-Leucine metabolic labeled trypomastigotes, it was possible to demonstrate that protein degradation takes place during *T. cruzi* remodeling and that proteolysis was abolished by proteasome inhibitors but not by classical inhibitors of serine and cysteine proteinases. These results suggest that the proteolysis process that takes place during *T. cruzi* cell differentiation depends on the 20 S proteasome or the 26S proteasome activities [138, 139], although the proteolysis due to the action of the 26S proteasome is ATP-dependent and occurs at acidic pH [139] (Fig. **3**).

Phosphatidylinositol Phospholipase C

The PI-PLCs expression is developmentally regulated and exclusive to amastigotes cell membrane. Its enzymatic activity is enhanced during parasite differentiation [140] (Fig. **3**). Okura *et al.* [141] showed that inhibition of PI-PLC expression by antisense oligonucleotides blocks the differentiation of trypomastigotes into amastigotes, whereas PI-PLC overexpression promotes it. PI-PLC is found at the extracellular side of the amastigotes cell membrane, and its N-terminal acyl modifications work as a molecular addressing system that directs PI-PLC to the outer cell surface [142]. PI-PLC is expressed at the same time as the amount of phosphatidylinositol 4,5-bisphosphate (PIP_2) diminishes in the host cell membrane, and the levels of its product, inositol 1,4,5-trisphosphate, increases. The expression of PI-PLC in the *T. cruzi* outer surface of the cytoplasmic membrane and its ability to be released and to modify phospholipids from the host are among the new virulence strategies of *T.cruzi* [143].

Protein Phosphatase 2A

Protein phosphatase 2A (PP2A) is implicated in *T. cruzi* cell differentiation from trypomastigotes to amastigotes [144] when parasites are cultivated in DMEM medium at pH 5.0 (Fig. **3**). This transformation is inhibited by okadaic acid, a classical PP2A inhibitor at concentrations lower than 0.1 μM. The purified PP2A enzyme shows biochemical activity against ^{32}P-labeled phosphorylase and was blocked by okadaic acid in a dose-dependent manner. *T. cruzi* PP2A seems to play the same biological role as *Giardia lamblia* PP2A [145]. In addition, in other trypanosomatids like *Leishmania*, several protein kinases, belonging to the MAP [146] and TOR families [147] and a tyrosine-specific phosphatase [148] appear to be critical for protozoan virulence. These observations support the idea that protein phosphorylation/dephosphorylation play a central function in the regulation of protozoan cell differentiation and stress responses [128].

Calpains

Calpain plays a role in parasite cell proliferation (Fig. **3**) as confirmed by the use of calpain inhibitor MDL28170, able to reduce the growth rate of *T. cruzi* epimastigotes [149]. MDL28170 was also effective against bloodstream trypomastigotes and was efficient during *in vitro* macrophage infection with trypomastigotes, showing a remarkable effect on parasite intracellular infectivity when they were treated before macrophage infection [150].

In addition, a calpain-like cysteine proteinase was identified by two-dimensional gel electrophoresis followed by liquid chromatography and mass spectrometry only in *T. cruzi* cells resistant to benznidazole. [151]. These observations strongly indicate that calpain can function as a virulence factor involved in *T. cruzi* resistance to benznidazole and cell proliferation.

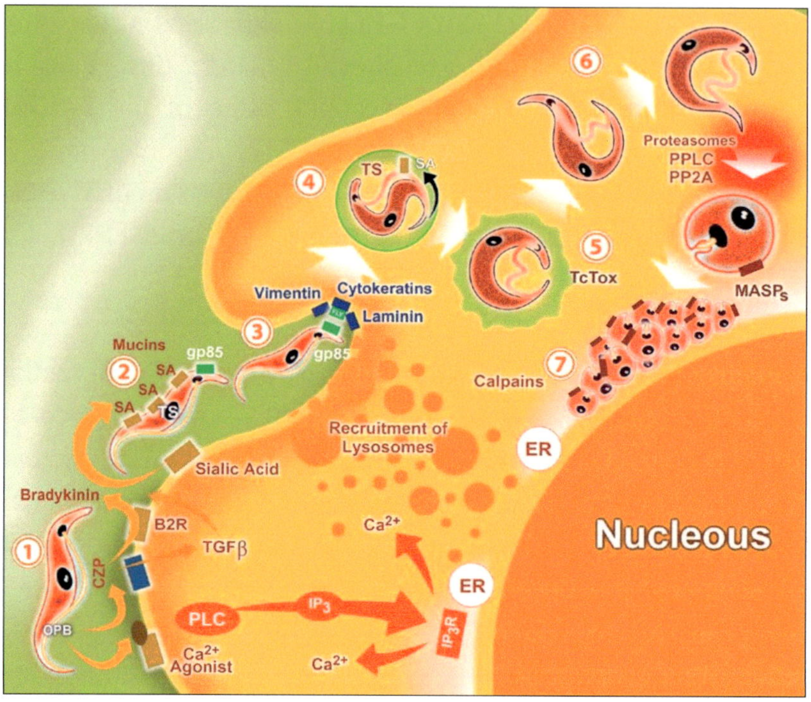

Fig. (3). Virulence factors expressed by *Trypanosoma cruzi* during host cell invasion and differentiation. After successful resistance to lytic complement activity, the parasite must invade nucleated host cells to continue its life cycle. *T. cruzi* displays an arsenal of surface molecules. (1) Cytoplasmic proteases like OPB are involved in calcium signaling and lysosomal regulated exocytosis. CZP is associated with its ability to generate bradykinin acting on the B type bradykinin receptor (B_2R) and is also able to activate latent TGF-β from host cells and to promote *T. cruzi* cell invasion; (2–3) The TS/gp85 glycoprotein family and the TS FLY motif participate in cell invasion by binding the host cell laminin, vimentin and cytokeratins; (4–5) Escape from PV is attributed to TS activity and to the expression of the hemolysin TcTox; (6) Differentiation of trypomastigotes to the replicative amastigote form is performed by phospholipase C (PPLC), protein phosphatase 2A (PP2A) and proteasomes; (7) overexpression of MASPs in amastigotes begin 24 h postinfection; (8) Calpains are involved in amastigote proliferation.

CONCLUDING REMARKS

During the last 30 years, we have gained tremendous knowledge about *T. cruzi* virulence factors. Indeed, seminal discoveries have allowed the understanding and characterization of different molecules that appear to play a central role in parasite immune evasion and cell invasion.

However, looking deeply at the *T. cruzi* virulence, several fundamental aspects remain to be investigated. For example, which molecules are essential to *T. cruzi* virulence? Which are the molecular divergences among a "virulent strain" that severely injure or kills mice and those "avirulent strains" that show poor infectivity with lower or undetectable parasitemia? Which are the biological differences among a strain that induce severe cardiac compromise, a strain that causes digestive mega syndromes and a strain that does not cause any systemic damage?

Moreover, the knowledge that we have about *T. cruzi* virulence is piecemeal, even more taking into account the roles in parasite pathogenesis that have been attributed to many different parasite proteins or glycoproteins reported so far. However, there still are no published research that incorporates all of this background in a comprehensive way that let us figure out the molecular basis that defines *T. cruzi* virulence and pathogenesis in a full dimension. Also, it is possible to think that besides all the virulence factors already reported in the literature, there may be new not yet described molecules that could be involved in *T. cruzi* virulence, which may be identified and characterized in deep. Among these, molecules are those engaged in metabolic pathways important to parasite survival in the vertebrate host and that could also be included in the arsenal of potential new virulence factors.

On the other hand, the picture that has emerged appears to suggest that the basis and mechanisms of parasite virulence could be more complex than expected. In fact, different studies have reached to a determined conclusion using a particular host-parasite interaction model or using a particular strain or DTU in a particular model of cell culture or mouse strain. These approaches have let us gained relevant knowledge about *T. cruzi* virulence mode but do not necessarily represent the overall reality. Furthermore, at least three additional aspects must be considered regarding *T. cruzi* virulence.One of them is related to the parasite's ability to display a multiclonal population structure with high intraspecific genetic diversity [152]. This variability is represented by at least six genetic lineages or DTUs, termed TcI through TcVI [153]. The current evidence strongly suggests that some virulence factors could be differentially expressed according to a particular DTU [154], reinforcing the idea that *T. cruzi* virulence should be studied using at least one representative strain or clone from each DTU. Nonetheless, for each DTU, one or a few virulence factors could be more relevant than others during the invasion or immune evasion.

A second aspect to be considered is related to the fact that vertebrate and invertebrate *T. cruzi* hosts could be naturally coinfected with different parasite strains [155]. This situation is of real concern because coinfections could affect the host susceptibility to acquiring infection and could cause variability in disease progression following infection [156]. Although concurrent infections appear to be the norm rather than an anomaly, the precise interactions between parasites causing a mixed infection, if they are synergistic or competitive, remain largely unknown [156]. Thus, a better understanding of parasite genetic diversity and the effects of polyparasitism is crucial to understand *T. cruzi* virulence.

Finally, a third aspect that must be considered is the potential effect that vertebrate and invertebrate hosts have on parasite virulence and the outcome of natural or experimental infection by *T. cruzi*. When triatomines bite and infect their hosts, they encounter a portion of blood in which more than one parasite population could be present. It is possible that all populations may adapt to the insect, but it is likely that one or only a few of them could adapt better to proliferate and dominate over the others. Similarly, when triatomines contact mammals and infect them with *T. cruzi*, the parasite subpopulation present in the infecting inoculum should adapt to mammals in many different ways, resulting in a host-parasite dependent adaptation. Finally, the characteristics that predominate after the first interaction of the parasite with the host immune system will lead to an infecting population that will present unique features as the result of that particular interaction. This should mean that *T. cruzi* virulence and pathogenicity would be in part dependent on the genetic background of the parasite. Nonetheless, the parasite has the ability to express select molecules that allow them to succeed despite the host immune system, and in doing so, some subpopulations would be eliminated favoring the proliferation of others. If the latter occurs, this would mean that *T. cruzi* virulence and disease progression could be the result of each unique host-parasite interaction.

T. cruzi is a complex protozoan that has coevolved with its human and animals hosts over millions of years. For this reason, to develop and to adapt to the challenge of confronting a host that is also changing and trying to find a manner to tolerate and survive the hostile vertebrate environment, *T. cruzi* should express an arsenal of virulence factors that allow it to succeed. Thus, *T. cruzi* virulence depends on the expression of a complex array of molecules that may guarantee the development of its life cycle. However, how gene expression is regulated to enable the parasite to resist, evade, escape or invade remains to be investigated.

Therefore, there is still a lot to learn about what virulence means for *T. cruzi* and which parasite molecules or arrays of molecules are required to make *T. cruzi* one of the most successful pathogens that invades, survives and persists in a mammalian host.

CONFLICT OF INTEREST

The authors confirm that this chapter content has no conflict of interest.

ACKNOWLEDGEMENTS

This work was supported by FONDECYT grants 1131007 (to JG) and 11130013 (to JLV). We would like to thanks, Mr. You Garmendia for preparing the figures.

ABBREVIATIONS

AMPs	Antimicrobial peptides
APX	Ascorbate-dependent heme peroxidase
BDKRB2	Bradykinin B2 receptor
CaN	Calcineurin
CML	Complement-mediated lysis
CRIT	Complement C2 receptor inhibitor trispanning

CRPs	Complement regulatory proteins
CRT	Calreticulin
CsA	Cyclosporin A
cTXN	Cytosolic tryparedoxin
CZP	Cruzipain
DTUs	Discrete typing units
Fe-SOD	Iron superoxide dismutase
GIPLs	Glycoinositolphospholipids
GPI	Glycosylphosphatidylinositol
Gp 63	63 kDa MT surface glycoprotein
Gp 30	30 kDa MT surface glycoprotein
Gp 35/50	35/50 kDa MT surface glycoprotein
Gp 82	82 kDa MT surface glycoprotein
Gp 85	85 kDa MT surface glycoprotein
Gp 90	90 kDa MT surface glycoprotein
GPXI	Glutathione peroxidase-I
GPXII	Glutathione peroxidase-II
H_2O_2	Hydrogen peroxide
HV	High variable
iNOS	Inducible nitric oxide synthase
IP_3	Inositol triphosphate
LGPs	Lysosomal glycoproteins
MAPK	p38 mitogen-activated protein kinase
MAPKAP-K2	MAPK activated protein kinase-2
MAPSs	Mucin-associated surface proteins
MIPs	Macrophage infectivity potentiators
PPIase	Peptidyl-prolyl cis-trans isomerase
MT	Metacyclic trypomastigotes
mTXN	Mitochondrial tryparedoxin
NHS	Normal human serum
NO	Nitric oxide
$O2^-$	Superoxide
$ONOO^-$	Peroxynitrite
OPB	Oligopeptidase B
PIP2	Phosphatidylinositol 4,5-bisphosphate

PI-PLC	Phosphatidylinositol phospholipase C
PLA$_1$	Phospholipase A1
POP	Prolyl oligopeptidase
PP2A	Protein phosphatase 2A
PLC	Phospholipase C
PRs	Proline racemases
PVs	Parasitophorous vacuoles
Siglec-E/CD33	Sialic acid-binding immunoglobulin-like lectin E
T-DAF	Trypomastigote decay-accelerating factor
Tc52	52 kDa protein of *Trypanosona.cruzi*
Tc85	85 kDa protein of *Trypanosoma .cruzi*
TcSMUG	*Trypanosoma cruzi* small mucin-like gene
TCTs	Tissue-culture trypomastigotes
TryS	Trypanothione synthetase
TLR2	Toll-like receptor 2
TS	Trans-sialidase
TXN	Tryparedoxin

REFERENCES

[1] Briones MR, Souto RP, Stolf BS, Zingales B. The evolution of two *Trypanosoma cruzi* subgroups inferred from rRNA genes can be correlated with the interchange of American mammalian faunas in the Cenozoic and has implications to pathogenicity and host specificity. Mol Biochem Parasitol 1999; 104(2): 219-32.
[http://dx.doi.org/10.1016/S0166-6851(99)00155-3] [PMID: 10593177]

[2] Coura JR, Viñas PA, Junqueira AC. Ecoepidemiology, short history and control of Chagas disease in the endemic countries and the new challenge for non-endemic countries. Mem Inst Oswaldo Cruz 2014; 109(7): 856-62.
[http://dx.doi.org/10.1590/0074-0276140236] [PMID: 25410988]

[3] Rothhammer F, Allison MJ, Núñez L, Standen V, Arriaza B. Chagas disease in pre-Columbian South America. Am J Phys Anthropol 1985; 68(4): 495-8.
[http://dx.doi.org/10.1002/ajpa.1330680405] [PMID: 3936363]

[4] Stevens JR, Gibson WC. The evolution of pathogenic trypanosomes. Cad Saude Publica 1999; 15(4): 673-84.
[http://dx.doi.org/10.1590/S0102-311X1999000400002] [PMID: 10633190]

[5] Guhl F, Jaramillo C, Vallejo GA, *et al.* Isolation of *Trypanosoma cruzi* DNA in 4,000-year-old mummified human tissue from northern Chile. Am J Phys Anthropol 1999; 108(4): 401-7.
[http://dx.doi.org/10.1002/(SICI)1096-8644(199904)108:4<401::AID-AJPA2>3.0.CO;2-P] [PMID: 10229385]

[6] Aufderheide AC, Salo W, Madden M, *et al.* A 9,000-year record of Chagas disease. Proc Natl Acad Sci USA 2004; 101(7): 2034-9.
[http://dx.doi.org/10.1073/pnas.0307312101] [PMID: 14766963]

[7] Coura JR, Borges-Pereira J. Chagas disease: 100 years after its discovery. A systemic review. Acta Trop 2010; 115(1-2): 5-13.
[http://dx.doi.org/10.1016/j.actatropica.2010.03.008] [PMID: 20382097]

[8] Andrade LO, Andrews NW. The *Trypanosoma cruzi*-host-cell interplay: location, invasion, retention. Nat Rev Microbiol 2005; 3(10): 819-23.
[http://dx.doi.org/10.1038/nrmicro1249] [PMID: 16175174]

[9] Anderson RM, May RM. Coevolution of hosts and parasites. Parasitology 1982; 85(Pt 2): 411-26.
[http://dx.doi.org/10.1017/S0031182000055360] [PMID: 6755367]

[10] Brown SP, Cornforth DM, Mideo N. Evolution of virulence in opportunistic pathogens: generalism, plasticity, and control. Trends Microbiol 2012; 20(7): 336-42.
[http://dx.doi.org/10.1016/j.tim.2012.04.005] [PMID: 22564248]

[11] Miles MA, Llewellyn MS, Lewis MD, *et al.* The molecular epidemiology and phylogeography of *Trypanosoma cruzi* and parallel research on *Leishmania*: looking back and to the future. Parasitology 2009; 136(12): 1509-28.
[http://dx.doi.org/10.1017/S0031182009990977] [PMID: 19691868]

[12] Brisse S, Barnabé C, Bañuls AL, Sidibé I, Noël S, Tibayrenc M. A phylogenetic analysis of the *Trypanosoma cruzi* genome project CL Brener reference strain by multilocus enzyme electrophoresis and multiprimer random amplified polymorphic DNA fingerprinting. Mol Biochem Parasitol 1998; 92(2): 253-63.
[http://dx.doi.org/10.1016/S0166-6851(98)00005-X] [PMID: 9657330]

[13] Pérez Brandán C, Padilla AM, Diosque P, Basombrío MA. *Trypanosoma cruzi*: infectivity modulation of a clone after passages through different hosts. Exp Parasitol 2006; 114(2): 89-93.
[http://dx.doi.org/10.1016/j.exppara.2006.02.014] [PMID: 16564526]

[14] Yoshida N. Molecular basis of mammalian cell invasion by *Trypanosoma cruzi*. An Acad Bras Cienc 2006; 78(1): 87-111.
[http://dx.doi.org/10.1590/S0001-37652006000100010] [PMID: 16532210]

[15] Epting CL, Coates BM, Engman DM. Molecular mechanisms of host cell invasion by *Trypanosoma cruzi*. Exp Parasitol 2010; 126(3): 283-91.
[http://dx.doi.org/10.1016/j.exppara.2010.06.023] [PMID: 20599990]

[16] Osorio L, Ríos I, Gutiérrez B, González J. Virulence factors of *Trypanosoma cruzi*: who is who? Microbes Infect 2012; 14(15): 1390-402.
[http://dx.doi.org/10.1016/j.micinf.2012.09.003] [PMID: 23006853]

[17] Flávia-Nardy A, Freire-de-Lima CG, Morrot A. Immune evasion strategies of *Trypanosoma cruzi*. J Immunol Res 2015; 2015: 178947.

[18] Brener Z. Immunity to *Trypanosoma cruzi*. Adv Parasitol 1980; 18: 247-92.
[http://dx.doi.org/10.1016/S0065-308X(08)60401-7] [PMID: 6159781]

[19] Bogdan C, Röllinghoff M. How do protozoan parasites survive inside macrophages? Parasitol Today (Regul Ed) 1999; 15(1): 22-8.
[http://dx.doi.org/10.1016/S0169-4758(98)01362-3] [PMID: 10234174]

[20] Irigoín F, Cibils L, Comini MA, Wilkinson SR, Flohé L, Radi R. Insights into the redox biology of *Trypanosoma cruzi*: Trypanothione metabolism and oxidant detoxification. Free Radic Biol Med 2008; 45(6): 733-42.
[http://dx.doi.org/10.1016/j.freeradbiomed.2008.05.028] [PMID: 18588970]

[21] Alvarez MN, Piacenza L, Irigoín F, Peluffo G, Radi R. Macrophage-derived peroxynitrite diffusion and toxicity to *Trypanosoma cruzi*. Arch Biochem Biophys 2004; 432(2): 222-32.
[http://dx.doi.org/10.1016/j.abb.2004.09.015] [PMID: 15542061]

[22] Alvarez MN, Peluffo G, Piacenza L, Radi R. Intraphagosomal peroxynitrite as a macrophage-derived cytotoxin against internalized *Trypanosoma cruzi*: consequences for oxidative killing and role of microbial peroxiredoxins in infectivity. J Biol Chem 2011; 286(8): 6627-40.
[http://dx.doi.org/10.1074/jbc.M110.167247] [PMID: 21098483]

[23] Piacenza L, Peluffo G, Alvarez MN, Kelly JM, Wilkinson SR, Radi R. Peroxiredoxins play a major role in protecting *Trypanosoma cruzi* against macrophage- and endogenously-derived peroxynitrite. Biochem J 2008; 410(2): 359-68.
[http://dx.doi.org/10.1042/BJ20071138] [PMID: 17973627]

[24] Piacenza L, Zago MP, Peluffo G, Alvarez MN, Basombrio MA, Radi R. Enzymes of the antioxidant network as novel determiners of *Trypanosoma cruzi* virulence. Int J Parasitol 2009; 39(13): 1455-64.
[http://dx.doi.org/10.1016/j.ijpara.2009.05.010] [PMID: 19505468]

[25] Wilkinson SR, Obado SO, Mauricio IL, Kelly JM. *Trypanosoma cruzi* expresses a plant-like ascorbate-dependent hemoperoxidase localized to the endoplasmic reticulum. Proc Natl Acad Sci USA 2002; 99(21): 13453-8.
[http://dx.doi.org/10.1073/pnas.202422899] [PMID: 12351682]

[26] Mateo H, Marín C, Pérez-Cordón G, Sánchez-Moreno M. Purification and biochemical characterization of four iron superoxide dismutases in *Trypanosoma cruzi*. Mem Inst Oswaldo Cruz 2008; 103(3): 271-6.
[http://dx.doi.org/10.1590/S0074-02762008000300008] [PMID: 18592100]

[27] Parodi-Talice A, Monteiro-Goes V, Arrambide N, *et al.* Proteomic analysis of metacyclic trypomastigotes undergoing *Trypanosoma cruzi* metacyclogenesis. J Mass Spectrom 2007; 42(11): 1422-32.
[http://dx.doi.org/10.1002/jms.1267] [PMID: 17960573]

[28] Kipnis TL, David JR, Alper CA, Sher A, da Silva WD. Enzymatic treatment transforms trypomastigotes of *Trypanosoma cruzi* into activators of alternative complement pathway and potentiates their uptake by macrophages. Proc Natl Acad Sci USA 1981; 78(1): 602-5.
[http://dx.doi.org/10.1073/pnas.78.1.602] [PMID: 6454138]

[29] Norris KA, Bradt B, Cooper NR, So M. Characterization of a *Trypanosoma cruzi* C3 binding protein with functional and genetic similarities to the human complement regulatory protein, decay-accelerating factor. J Immunol 1991; 147(7): 2240-7.
[PMID: 1717552]

[30] Tambourgi DV, Kipnis TL, da Silva WD, *et al.* A partial cDNA clone of trypomastigote decay-accelerating factor (T-DAF), a developmentally regulated complement inhibitor of *Trypanosoma cruzi*, has genetic and functional similarities to the human complement inhibitor DAF. Infect Immun 1993; 61(9): 3656-63.
[PMID: 7689538]

[31] Norris KA. Stable transfection of *Trypanosoma cruzi* epimastigotes with the trypomastigote-specific complement regulatory protein cDNA confers complement resistance. Infect Immun 1998; 66(6): 2460-5.
[PMID: 9596703]

[32] Beucher M, Norris KA. Sequence diversity of the *Trypanosoma cruzi* complement regulatory protein family. Infect Immun 2008; 76(2): 750-8.
[http://dx.doi.org/10.1128/IAI.01104-07] [PMID: 18070905]

[33] Cestari IdosS, Krarup A, Sim RB, Inal JM, Ramirez MI. Role of early lectin pathway activation in the complement-mediated killing of *Trypanosoma cruzi*. Mol Immunol 2009; 47(2-3): 426-37.
[http://dx.doi.org/10.1016/j.molimm.2009.08.030] [PMID: 19783051]

[34] Cestari IdosS, Evans-Osses I, Freitas JC, Inal JM, Ramirez MI. Complement C2 receptor inhibitor trispanning confers an increased ability to resist complement-mediated lysis in *Trypanosoma cruzi*. J Infect Dis 2008; 198(9): 1276-83.
[http://dx.doi.org/10.1086/592167] [PMID: 18781865]

[35] Blom AM, Hallström T, Riesbeck K. Complement evasion strategies of pathogens-acquisition of inhibitors and beyond. Mol Immunol 2009; 46(14): 2808-17.
[http://dx.doi.org/10.1016/j.molimm.2009.04.025] [PMID: 19477524]

[36] Valck C, Ramírez G, López N, *et al.* Molecular mechanisms involved in the inactivation of the first component of human complement by *Trypanosoma cruzi* calreticulin. Mol Immunol 2010; 47(7-8): 1516-21.
[http://dx.doi.org/10.1016/j.molimm.2010.01.019] [PMID: 20153898]

[37] Ramírez G, Valck C, Molina MC, *et al.* *Trypanosoma cruzi* calreticulin: a novel virulence factor that binds complement C1 on the parasite surface and promotes infectivity. Immunobiology 2011; 216(1-2): 265-73.
[http://dx.doi.org/10.1016/j.imbio.2010.04.001] [PMID: 20472323]

[38] Previato JO, Wait R, Jones C, *et al.* Glycoinositolphospholipid from *Trypanosoma cruzi*: structure, biosynthesis and immunobiology. Adv Parasitol 2004; 56: 1-41.
[http://dx.doi.org/10.1016/S0065-308X(03)56001-8] [PMID: 14710995]

[39] Gomes NA, Previato JO, Zingales B, Mendonça-Previato L, DosReis GA. Down-regulation of T lymphocyte activation *in vitro* and *in vivo* induced by glycoinositolphospholipids from *Trypanosoma cruzi*. Assignment of the T cell-suppressive determinant to the ceramide domain. J Immunol 1996; 156(2): 628-35.
[PMID: 8543814]

[40] Bento CA, Melo MB, Previato JO, Mendonça-Previato L, Peçanha LM. Glycoinositolphospholipids purified from *Trypanosoma cruzi* stimulate Ig production *in vitro*. J Immunol 1996; 157(11): 4996-5001.

[PMID: 8943406]

[41] Brodskyn C, Patricio J, Oliveira R, *et al.* Glycoinositolphospholipids from *Trypanosoma cruzi* interfere with macrophages and dendritic cell responses. Infect Immun 2002; 70(7): 3736-43. [http://dx.doi.org/10.1128/IAI.70.7.3736-3743.2002] [PMID: 12065516]

[42] Oliveira AC, Peixoto JR, de Arruda LB, *et al.* Expression of functional TLR4 confers proinflammatory responsiveness to *Trypanosoma cruzi* glycoinositolphospholipids and higher resistance to infection with *T. cruzi.* J Immunol 2004; 173(9): 5688-96. [http://dx.doi.org/10.4049/jimmunol.173.9.5688] [PMID: 15494520]

[43] Buscaglia CA, Campo VA, Frasch AC, Di Noia JM. *Trypanosoma cruzi* surface mucins: host-dependent coat diversity. Nat Rev Microbiol 2006; 4(3): 229-36. [http://dx.doi.org/10.1038/nrmicro1351] [PMID: 16489349]

[44] Pereira-Chioccola VL, Acosta-Serrano A, Correia de Almeida I, *et al.* Mucin-like molecules form a negatively charged coat that protects *Trypanosoma cruzi* trypomastigotes from killing by human anti-alpha-galactosyl antibodies. J Cell Sci 2000; 113(Pt 7): 1299-307. [PMID: 10704380]

[45] Kierszenbaum F, Fresno M, Sztein MB. The *Trypanosoma cruzi* membrane glycoprotein AGC10 inhibits human lymphocyte activation by a mechanism preceding translation of both, interleukin-2 and its high-affinity receptor subunits. Mol Biochem Parasitol 2002; 125(1-2): 91-101. [http://dx.doi.org/10.1016/S0166-6851(02)00217-7] [PMID: 12467977]

[46] Nunes MP, Fortes B, Silva-Filho JL, *et al.* Inhibitory effects of *Trypanosoma cruzi* sialoglycoproteins on CD4+ T cells are associated with increased susceptibility to infection. PLoS One 2013; 8(10): e77568. [http://dx.doi.org/10.1371/journal.pone.0077568] [PMID: 24204874]

[47] Alcaide P, Fresno M. The *Trypanosoma cruzi* membrane mucin AgC10 inhibits T cell activation and IL-2 transcription through L-selectin. Int Immunol 2004; 16(10): 1365-75. [http://dx.doi.org/10.1093/intimm/dxh138] [PMID: 15314038]

[48] Swenerton RK, Zhang S, Sajid M, *et al.* The oligopeptidase B of *Leishmania* regulates parasite enolase and immune evasion. J Biol Chem 2011; 286(1): 429-40. [http://dx.doi.org/10.1074/jbc.M110.138313] [PMID: 20961853]

[49] Bastos IM, Grellier P, Martins NF, *et al.* Molecular, functional and structural properties of the prolyl oligopeptidase of *Trypanosoma cruzi* (POP Tc80), which is required for parasite entry into mammalian cells. Biochem J 2005; 388(Pt 1): 29-38. [http://dx.doi.org/10.1042/BJ20041049] [PMID: 15581422]

[50] Kierszenbaum F, de Diego JL, Fresno M, Sztein MB. Inhibitory effects of the *Trypanosoma cruzi* membrane glycoprotein AGC10 on the expression of IL-2 receptor chains and secretion of cytokines by subpopulations of activated human T lymphocytes. Eur J Immunol 1999; 29(5): 1684-91. [http://dx.doi.org/10.1002/(SICI)1521-4141(199905)29:05<1684::AID-IMMU1684>3.0.CO;2-7] [PMID: 10359123]

[51] Urban I, Santurio LB, Chidichimo A, *et al.* Molecular diversity of the *Trypanosoma cruzi* TcSMUG family of mucin genes and proteins. Biochem J 2011; 438(2): 303-13. [http://dx.doi.org/10.1042/BJ20110683] [PMID: 21651499]

[52] Pollevick GD, Di Noia JM, Salto ML, *et al. Trypanosoma cruzi* surface mucins with exposed variant epitopes. J Biol Chem 2000; 275(36): 27671-80.
[PMID: 10843987]

[53] Almeida IC, Gazzinelli RT. Proinflammatory activity of glycosylphosphatidylinositol anchors derived from *Trypanosoma cruzi*: structural and functional analyses. J Leukoc Biol 2001; 70(4): 467-77.
[PMID: 11590183]

[54] Schenkman S, Jiang MS, Hart GW, Nussenzweig V. A novel cell surface trans-sialidase of *Trypanosoma cruzi* generates a stage-specific epitope required for invasion of mammalian cells. Cell 1991; 65(7): 1117-25.
[http://dx.doi.org/10.1016/0092-8674(91)90008-M] [PMID: 1712251]

[55] Acosta-Serrano A, Almeida IC, Freitas-Junior LH, Yoshida N, Schenkman S. The mucin-like glycoprotein super-family of *Trypanosoma cruzi*: structure and biological roles. Mol Biochem Parasitol 2001; 114(2): 143-50.
[http://dx.doi.org/10.1016/S0166-6851(01)00245-6] [PMID: 11378194]

[56] Yoshida N, Mortara RA, Araguth MF, Gonzalez JC, Russo M. Metacyclic neutralizing effect of monoclonal antibody 10D8 directed to the 35- and 50-kilodalton surface glycoconjugates of *Trypanosoma cruzi*. Infect Immun 1989; 57(6): 1663-7.
[PMID: 2656520]

[57] Ramirez MI, Ruiz RdeC, Araya JE, Da Silveira JF, Yoshida N. Involvement of the stage-specific 82-kilodalton adhesion molecule of *Trypanosoma cruzi* metacyclic trypomastigotes in host cell invasion. Infect Immun 1993; 61(9): 3636-41.
[PMID: 8359886]

[58] Alcaide P, Fresno M. AgC10, a mucin from *Trypanosoma cruzi*, destabilizes TNF and cyclooxygenase-2 mRNA by inhibiting mitogen-activated protein kinase p38. Eur J Immunol 2004; 34(6): 1695-704.
[http://dx.doi.org/10.1002/eji.200324660] [PMID: 15162440]

[59] McKerrow JH, Caffrey C, Kelly B, Loke P, Sajid M. Proteases in parasitic diseases. Annu Rev Pathol 2006; 1: 497-536.
[http://dx.doi.org/10.1146/annurev.pathol.1.110304.100151] [PMID: 18039124]

[60] Sajid M, McKerrow JH. Cysteine proteases of parasitic organisms. Mol Biochem Parasitol 2002; 120(1): 1-21.
[http://dx.doi.org/10.1016/S0166-6851(01)00438-8] [PMID: 11849701]

[61] Meirelles MN, Juliano L, Carmona E, *et al.* Inhibitors of the major cysteinyl proteinase (GP57/51) impair host cell invasion and arrest the intracellular development of *Trypanosoma cruzi in vitro*. Mol Biochem Parasitol 1992; 52(2): 175-84.
[http://dx.doi.org/10.1016/0166-6851(92)90050-T] [PMID: 1620157]

[62] Alvarez VE, Niemirowicz GT, Cazzulo JJ. The peptidases of *Trypanosoma cruzi*: digestive enzymes, virulence factors, and mediators of autophagy and programmed cell death. Biochim Biophys Acta 2012; 1824(1): 195-206.
[http://dx.doi.org/10.1016/j.bbapap.2011.05.011] [PMID: 21621652]

[63] Berasain P, Carmona C, Frangione B, Cazzulo JJ, Goñi F. Specific cleavage sites on human IgG subclasses by cruzipain, the major cysteine proteinase from *Trypanosoma cruzi*. Mol Biochem Parasitol 2003; 130(1): 23-9.
[http://dx.doi.org/10.1016/S0166-6851(03)00139-7] [PMID: 14550893]

[64] Stempin C, Giordanengo L, Gea S, Cerbán F. Alternative activation and increase of *Trypanosoma cruzi* survival in murine macrophages stimulated by cruzipain, a parasite antigen. J Leukoc Biol 2002; 72(4): 727-34.
[PMID: 12377942]

[65] Doyle PS, Zhou YM, Hsieh I, Greenbaum DC, McKerrow JH, Engel JC. The *Trypanosoma cruzi* protease cruzain mediates immune evasion. PLoS Pathog 2011; 7(9): e1002139.
[http://dx.doi.org/10.1371/journal.ppat.1002139] [PMID: 21909255]

[66] Garrido VV, Dulgerian LR, Stempin CC, Cerbán FM. The increase in mannose receptor recycling favors arginase induction and *Trypanosoma cruzi* survival in macrophages. Int J Biol Sci 2011; 7(9): 1257-72.
[http://dx.doi.org/10.7150/ijbs.7.1257] [PMID: 22110379]

[67] Ferrero MR, Heins AM, Soprano LL, Acosta DM, Esteva MI, Jacobs T, *et al.* Involvement of sulfates from cruzipain, a major antigen of *Trypanosoma cruzi*, in the interaction with immunomodulatory molecule. Immunobiology 2015. (Epub ahead, 2015)

[68] Scharfstein J, Schmitz V, Morandi V, *et al.* Host cell invasion by *Trypanosoma cruzi* is potentiated by activation of bradykinin B(2) receptors. J Exp Med 2000; 192(9): 1289-300.
[http://dx.doi.org/10.1084/jem.192.9.1289] [PMID: 11067878]

[69] Ferrão PM, dAvila-Levy CM, Araujo-Jorge TC, *et al.* Cruzipain activates latent TGF-β from host cells during *T. cruzi* invasion. PLoS One 2015; 10(5): e0124832.
[http://dx.doi.org/10.1371/journal.pone.0124832] [PMID: 25938232]

[70] Maeda FY, Cortez C, Izidoro MA, Juliano L, Yoshida N. Fibronectin-degrading activity of *Trypanosoma cruzi* cysteine proteinase plays a role in host cell invasion. Infect Immun 2014; 82(12): 5166-74.
[http://dx.doi.org/10.1128/IAI.02022-14] [PMID: 25267835]

[71] Franke de Cazzulo BM, Martínez J, North MJ, Coombs GH, Cazzulo JJ. Effects of proteinase inhibitors on the growth and differentiation of *Trypanosoma cruzi*. FEMS Microbiol Lett 1994; 124(1): 81-6.
[http://dx.doi.org/10.1111/j.1574-6968.1994.tb07265.x] [PMID: 8001773]

[72] Tomas AM, Miles MA, Kelly JM. Overexpression of cruzipain, the major cysteine proteinase of *Trypanosoma cruzi*, is associated with enhanced metacyclogenesis. Eur J Biochem 1997; 244(2): 596-603.
[http://dx.doi.org/10.1111/j.1432-1033.1997.t01-1-00596.x] [PMID: 9119029]

[73] Acosta DM, Arnaiz MR, Esteva MI, *et al.* Sulfates are main targets of immune responses to cruzipain and are involved in heart damage in BALB/c immunized mice. Int Immunol 2008; 20(4): 461-70.
[http://dx.doi.org/10.1093/intimm/dxm149] [PMID: 18195050]

[74] Reina-San-Martín B, Degrave W, Rougeot C, *et al.* A B-cell mitogen from a pathogenic trypanosome

is a eukaryotic proline racemase. Nat Med 2000; 6(8): 890-7.
[http://dx.doi.org/10.1038/78651] [PMID: 10932226]

[75] Chamond N, Grégoire C, Coatnoan N, *et al.* Biochemical characterization of proline racemases from the human protozoan parasite *Trypanosoma cruzi* and definition of putative protein signatures. J Biol Chem 2003; 278(18): 15484-94.
[http://dx.doi.org/10.1074/jbc.M210830200] [PMID: 12735293]

[76] Chamond N, Goytia M, Coatnoan N, *et al. Trypanosoma cruzi* proline racemases are involved in parasite differentiation and infectivity. Mol Microbiol 2005; 58(1): 46-60.
[http://dx.doi.org/10.1111/j.1365-2958.2005.04808.x] [PMID: 16164548]

[77] Coutinho L, Ferreira MA, Cosson A, *et al.* Inhibition of *Trypanosoma cruzi* proline racemase affects host-parasite interactions and the outcome of *in vitro* infection. Mem Inst Oswaldo Cruz 2009; 104(8): 1055-62.
[PMID: 20140365]

[78] Ouaissi MA, Dubremetz JF, Schöneck R, *et al. Trypanosoma cruzi*: a 52-kDa protein sharing sequence homology with glutathione S-transferase is localized in parasite organelles morphologically resembling reservosomes. Exp Parasitol 1995; 81(4): 453-61.
[http://dx.doi.org/10.1006/expr.1995.1138] [PMID: 8542986]

[79] Allaoui A, François C, Zemzoumi K, Guilvard E, Ouaissi A. Intracellular growth and metacyclogenesis defects in *Trypanosoma cruzi* carrying a targeted deletion of a Tc52 protein-encoding allele. Mol Microbiol 1999; 32(6): 1273-86.
[http://dx.doi.org/10.1046/j.1365-2958.1999.01440.x] [PMID: 10383767]

[80] Ouaissi A, Guilvard E, Delneste Y, *et al.* The *Trypanosoma cruzi* Tc52-released protein induces human dendritic cell maturation, signals *via* Toll-like receptor 2, and confers protection against lethal infection. J Immunol 2002; 168(12): 6366-74.
[http://dx.doi.org/10.4049/jimmunol.168.12.6366] [PMID: 12055254]

[81] Fernandez-Gomez R, Esteban S, Gomez-Corvera R, Zoulika K, Ouaissi A. *Trypanosoma cruzi*: Tc52 released protein-induced increased expression of nitric oxide synthase and nitric oxide production by macrophages. J Immunol 1998; 160(7): 3471-9.
[PMID: 9531308]

[82] Cestari I, Ansa-Addo E, Deolindo P, Inal JM, Ramirez MI. *Trypanosoma cruzi* immune evasion mediated by host cell-derived microvesicles. J Immunol 2012; 188(4): 1942-52.
[http://dx.doi.org/10.4049/jimmunol.1102053] [PMID: 22262654]

[83] Torrecilhas AC, Schumacher RI, Alves MJ, Colli W. Vesicles as carriers of virulence factors in parasitic protozoan diseases. Microbes Infect 2012; 14(15): 1465-74.
[http://dx.doi.org/10.1016/j.micinf.2012.07.008] [PMID: 22892602]

[84] Trocoli Torrecilhas AC, Tonelli RR, Pavanelli WR, *et al. Trypanosoma cruzi*: parasite shed vesicles increase heart parasitism and generate an intense inflammatory response. Microbes Infect 2009; 11(1): 29-39.
[http://dx.doi.org/10.1016/j.micinf.2008.10.003] [PMID: 19028594]

[85] Atayde VD, Neira I, Cortez M, Ferreira D, Freymüller E, Yoshida N. Molecular basis of non-virulence of *Trypanosoma cruzi* clone CL-14. Int J Parasitol 2004; 34(7): 851-60.

[http://dx.doi.org/10.1016/j.ijpara.2004.03.003] [PMID: 15157768]

[86] Staquicini DI, Martins RM, Macedo S, *et al.* Role of Gp82 in the selective binding to gastric mucin during oral infection with *Trypanosoma cruzi*. PLoS Negl Trop Dis 2010; 4(3): e613. [http://dx.doi.org/10.1371/journal.pntd.0000613] [PMID: 20209152]

[87] Cortez C, Yoshida N, Bahia D, Sobreira TJ. Structural basis of the interaction of a *Trypanosoma cruzi* surface molecule implicated in oral infection with host cells and gastric mucin. PLoS One 2012; 7(7): e42153. [http://dx.doi.org/10.1371/journal.pone.0042153] [PMID: 22860068]

[88] Málaga S, Yoshida N. Targeted reduction in expression of *Trypanosoma cruzi* surface glycoprotein gp90 increases parasite infectivity. Infect Immun 2001; 69(1): 353-9. [http://dx.doi.org/10.1128/IAI.69.1.353-359.2001] [PMID: 11119524]

[89] Yoshida N. Molecular basis of mammalian cell invasion by *Trypanosoma cruzi*. An Acad Bras Cienc 2006; 78(1): 87-111. [http://dx.doi.org/10.1590/S0001-37652006000100010] [PMID: 16532210]

[90] Ruiz RC, Favoreto S Jr, Dorta ML, *et al.* Infectivity of *Trypanosoma cruzi* strains is associated with differential expression of surface glycoproteins with differential Ca2+ signalling activity. Biochem J 1998; 330(Pt 1): 505-11. [http://dx.doi.org/10.1042/bj3300505] [PMID: 9461549]

[91] Yoshida N, Dorta ML, Ferreira AT, *et al.* Removal of sialic acid from mucin-like surface molecules of *Trypanosoma cruzi* metacyclic trypomastigotes enhances parasite-host cell interaction. Mol Biochem Parasitol 1997; 84(1): 57-67. [http://dx.doi.org/10.1016/S0166-6851(96)02783-1] [PMID: 9041521]

[92] Burleigh BA, Andrews NW. A 120-kDa alkaline peptidase from *Trypanosoma cruzi* is involved in the generation of a novel Ca(2+)-signaling factor for mammalian cells. J Biol Chem 1995; 270(10): 5172-80. [http://dx.doi.org/10.1074/jbc.270.10.5172] [PMID: 7890627]

[93] Burleigh BA, Caler EV, Webster P, Andrews NW. A cytosolic serine endopeptidase from *Trypanosoma cruzi* is required for the generation of Ca2+ signaling in mammalian cells. J Cell Biol 1997; 136(3): 609-20. [http://dx.doi.org/10.1083/jcb.136.3.609] [PMID: 9024691]

[94] Caler EV, Vaena de Avalos S, Haynes PA, Andrews NW, Burleigh BA. Oligopeptidase B-dependent signaling mediates host cell invasion by *Trypanosoma cruzi*. EMBO J 1998; 17(17): 4975-86. [http://dx.doi.org/10.1093/emboj/17.17.4975] [PMID: 9724634]

[95] Coetzer TH, Goldring JP, Huson LE. Oligopeptidase B: a processing peptidase involved in pathogenesis. Biochimie 2008; 90(2): 336-44. [http://dx.doi.org/10.1016/j.biochi.2007.10.011] [PMID: 18029266]

[96] Alves MJ, Abuin G, Kuwajima VY, Colli W. Partial inhibition of trypomastigote entry into cultured mammalian cells by monoclonal antibodies against a surface glycoprotein of *Trypanosoma cruzi*. Mol Biochem Parasitol 1986; 21(1): 75-82. [http://dx.doi.org/10.1016/0166-6851(86)90081-2] [PMID: 3534565]

[97] Giordano R, Fouts DL, Tewari D, Colli W, Manning JE, Alves MJ. Cloning of a surface membrane glycoprotein specific for the infective form of *Trypanosoma cruzi* having adhesive properties to laminin. J Biol Chem 1999; 274(6): 3461-8. [http://dx.doi.org/10.1074/jbc.274.6.3461] [PMID: 9920891]

[98] Tonelli RR, Giordano RJ, Barbu EM, *et al.* Role of the gp85/trans-sialidases in *Trypanosoma cruzi* tissue tropism: preferential binding of a conserved peptide motif to the vasculature *in vivo.* PLoS Negl Trop Dis 2010; 4(11): e864. [http://dx.doi.org/10.1371/journal.pntd.0000864] [PMID: 21072227]

[99] Magdesian MH, Tonelli RR, Fessel MR, *et al.* A conserved domain of the gp85/trans-sialidase family activates host cell extracellular signal-regulated kinase and facilitates *Trypanosoma cruzi* infection. Exp Cell Res 2007; 313(1): 210-8. [http://dx.doi.org/10.1016/j.yexcr.2006.10.008] [PMID: 17101128]

[100] Tonelli RR, Torrecilhas AC, Jacysyn JF, Juliano MA, Colli W, Alves MJ. *In vivo* infection by *Trypanosoma cruzi*: the conserved FLY domain of the gp85/trans-sialidase family potentiates host infection. Parasitology 2011; 138(4): 481-92. [http://dx.doi.org/10.1017/S0031182010001411] [PMID: 21040619]

[101] Teixeira AA, de Vasconcelos VdeC, Colli W, Alves MJ, Giordano RJ. *Trypanosoma cruzi* binds to cytokeratin through conserved peptide motifs found in the laminin-g-like domain of the gp85/trans-sialidase proteins. PLoS Negl Trop Dis 2015; 9(9): e0004099. [http://dx.doi.org/10.1371/journal.pntd.0004099] [PMID: 26398185]

[102] El-Sayed NM, Myler PJ, Bartholomeu DC, *et al.* The genome sequence of *Trypanosoma cruzi*, etiologic agent of Chagas disease. Science 2005; 309(5733): 409-15. [http://dx.doi.org/10.1126/science.1112631] [PMID: 16020725]

[103] Giorgi ME, de Lederkremer RM. Trans-sialidase and mucins of *Trypanosoma cruzi*: an important interplay for the parasite. Carbohydr Res 2011; 346(12): 1389-93. [http://dx.doi.org/10.1016/j.carres.2011.04.006] [PMID: 21645882]

[104] Schenkman S, Eichinger D, Pereira ME, Nussenzweig V. Structural and functional properties of *Trypanosoma* trans-sialidase. Annu Rev Microbiol 1994; 48: 499-523. [http://dx.doi.org/10.1146/annurev.mi.48.100194.002435] [PMID: 7826016]

[105] Dc-Rubin SS, Schenkman S. *T rypanosoma cruzi* trans-sialidase as a multifunctional enzyme in Chagas disease. Cell Microbiol 2012; 14(10): 1522-30. [http://dx.doi.org/10.1111/j.1462-5822.2012.01831.x] [PMID: 22747789]

[106] Tribulatti MV, Mucci J, Van Rooijen N, Leguizamón MS, Campetella O. The trans-sialidase from *Trypanosoma cruzi* induces thrombocytopenia during acute Chagas disease by reducing the platelet sialic acid contents. Infect Immun 2005; 73(1): 201-7. [http://dx.doi.org/10.1128/IAI.73.1.201-207.2005] [PMID: 15618155]

[107] Mucci J, Risso MG, Leguizamón MS, Frasch AC, Campetella O. The trans-sialidase from *Trypanosoma cruzi* triggers apoptosis by target cell sialylation. Cell Microbiol 2006; 8(7): 1086-95. [http://dx.doi.org/10.1111/j.1462-5822.2006.00689.x] [PMID: 16819962]

[108] Todeschini AR, Nunes MP, Pires RS, *et al.* Costimulation of host T lymphocytes by a trypanosomal

trans-sialidase: involvement of CD43 signaling. J Immunol 2002; 168(10): 5192-8. [http://dx.doi.org/10.4049/jimmunol.168.10.5192] [PMID: 11994475]

[109] Padilla A, Xu D, Martin D, Tarleton R. Limited role for CD4+ T-cell help in the initial priming of *Trypanosoma cruzi*-specific CD8+ T cells. Infect Immun 2007; 75(1): 231-5. [http://dx.doi.org/10.1128/IAI.01245-06] [PMID: 17043105]

[110] Freire-de-Lima L, Alisson-Silva F, Carvalho ST, *et al. Trypanosoma cruzi* subverts host cell sialylation and may compromise antigen-specific CD8+ T cell responses. J Biol Chem 2010; 285(18): 13388-96. [http://dx.doi.org/10.1074/jbc.M109.096305] [PMID: 20106975]

[111] Prioli RP, Mejia JS, Pereira ME. Monoclonal antibodies against *Trypanosoma cruzi* neuraminidase reveal enzyme polymorphism, recognize a subset of trypomastigotes, and enhance infection *in vitro*. J Immunol 1990; 144(11): 4384-91. [PMID: 2111345]

[112] Pereira ME, Zhang K, Gong Y, Herrera EM, Ming M. Invasive phenotype of *Trypanosoma cruzi* restricted to a population expressing trans-sialidase. Infect Immun 1996; 64(9): 3884-92. [PMID: 8751943]

[113] Schenkman RP, Vandekerckhove F, Schenkman S. Mammalian cell sialic acid enhances invasion by *Trypanosoma cruzi*. Infect Immun 1993; 61(3): 898-902. [PMID: 8381772]

[114] Carvalho ST, Sola-Penna M, Oliveira IA, *et al.* A new class of mechanism-based inhibitors for *Trypanosoma cruzi* trans-sialidase and their influence on parasite virulence. Glycobiology 2010; 20(8): 1034-45. [http://dx.doi.org/10.1093/glycob/cwq065] [PMID: 20466651]

[115] Lieke T, Gröbe D, Blanchard V, *et al.* Invasion of *Trypanosoma cruzi* into host cells is impaired by N-propionylmannosamine and other N-acylmannosamines. Glycoconj J 2011; 28(1): 31-7. [http://dx.doi.org/10.1007/s10719-010-9321-2] [PMID: 21240549]

[116] Alcântara-Neves NM, Ribeiro-dos-Santos R, Amor AL, *et al.* Parasite-derived trans-sialidase binds to heart tissue in *Trypanosoma cruzi*-infected animals. Microb Pathog 2004; 37(5): 273-8. [http://dx.doi.org/10.1016/j.micpath.2004.03.003] [PMID: 15519048]

[117] Todeschini AR, Girard MF, Wieruszeski JM, *et al.* trans-Sialidase from *Trypanosoma cruzi* binds host T-lymphocytes in a lectin manner. J Biol Chem 2002; 277(48): 45962-8. [http://dx.doi.org/10.1074/jbc.M203185200] [PMID: 12237289]

[118] Monteiro VG, Lobato CS, Silva AR, *et al.* Increased association of *Trypanosoma cruzi* with sialoadhesin positive mice macrophages. Parasitol Res 2005; 97(5): 380-5. [http://dx.doi.org/10.1007/s00436-005-1460-1] [PMID: 16151743]

[119] Risso MG, Garbarino GB, Mocetti E, *et al.* Differential expression of a virulence factor, the trans-sialidase, by the main *Trypanosoma cruzi* phylogenetic lineages. J Infect Dis 2004; 189(12): 2250-9. [http://dx.doi.org/10.1086/420831] [PMID: 15181573]

[120] Tibayrenc M. Genetic epidemiology of parasitic protozoa and other infectious agents: the need for an integrated approach. Int J Parasitol 1998; 28(1): 85-104.

[http://dx.doi.org/10.1016/S0020-7519(97)00180-X] [PMID: 9504337]

[121] Bartholomeu DC, Cerqueira GC, Leão AC, *et al.* Genomic organization and expression profile of the mucin-associated surface protein (masp) family of the human pathogen *Trypanosoma cruzi*. Nucleic Acids Res 2009; 37(10): 3407-17.
[http://dx.doi.org/10.1093/nar/gkp172] [PMID: 19336417]

[122] Macedo AM, Machado CR, Oliveira RP, Pena SD. *Trypanosoma cruzi*: genetic structure of populations and relevance of genetic variability to the pathogenesis of chagas disease. Mem Inst Oswaldo Cruz 2004; 99(1): 1-12.
[http://dx.doi.org/10.1590/S0074-02762004000100001] [PMID: 15057339]

[123] dos Santos SL, Freitas LM, Lobo FP, *et al.* The MASP family of *Trypanosoma cruzi*: changes in gene expression and antigenic profile during the acute phase of experimental infection. PLoS Negl Trop Dis 2012; 6(8): e1779.

[124] Serna C, Lara JA, Rodrigues SP, Marques AF, Almeida IC, Maldonado RA. A synthetic peptide from *Trypanosoma cruzi* mucin-like associated surface protein as candidate for a vaccine against Chagas disease. Vaccine 2014; 32(28): 3525-32.
[http://dx.doi.org/10.1016/j.vaccine.2014.04.026] [PMID: 24793944]

[125] De Pablos LM, González GG, Solano Parada J, *et al.* Differential expression and characterization of a member of the mucin-associated surface protein family secreted by *Trypanosoma cruzi*. Infect Immun 2011; 79(10): 3993-4001.
[http://dx.doi.org/10.1128/IAI.05329-11] [PMID: 21788387]

[126] Orrego PR, Olivares H, Cordero EM, *et al.* A cytoplasmic new catalytic subunit of calcineurin in *Trypanosoma cruzi* and its molecular and functional characterization. PLoS Negl Trop Dis 2014; 8(1): e2676.
[http://dx.doi.org/10.1371/journal.pntd.0002676] [PMID: 24498455]

[127] Araya JE, Cornejo A, Orrego PR, *et al.* Calcineurin B of the human protozoan parasite *Trypanosoma cruzi* is involved in cell invasion. Microbes Infect 2008; 10(8): 892-900.
[http://dx.doi.org/10.1016/j.micinf.2008.05.003] [PMID: 18657458]

[128] Naderer T, Dandash O, McConville MJ. Calcineurin is required for *Leishmania major* stress response pathways and for virulence in the mammalian host. Mol Microbiol 2011; 80(2): 471-80.
[http://dx.doi.org/10.1111/j.1365-2958.2011.07584.x] [PMID: 21320183]

[129] Moro A, Ruiz-Cabello F, Fernández-Cano A, Stock RP, González A. Secretion by *Trypanosoma cruzi* of a peptidyl-prolyl cis-trans isomerase involved in cell infection. EMBO J 1995; 14(11): 2483-90.
[PMID: 7540135]

[130] Belaunzaran ML, Lammel EM, de Isola EL. Phospholipases a in trypanosomatids. Enzym Res 2011; 2011: 392082.
[http://dx.doi.org/10.4061/2011/392082]

[131] Belaunzarán ML, Wilkowsky SE, Lammel EM, *et al.* Phospholipase A1: a novel virulence factor in *Trypanosoma cruzi*. Mol Biochem Parasitol 2013; 187(2): 77-86.
[http://dx.doi.org/10.1016/j.molbiopara.2012.12.004] [PMID: 23275096]

[132] Cuevas IC, Cazzulo JJ, Sánchez DO. Gp63 homologues in *Trypanosoma cruzi*: surface antigens with

metalloprotease activity and a possible role in host cell infection. Infect Immun 2003; 71(10): 5739-49. [http://dx.doi.org/10.1128/IAI.71.10.5739-5749.2003] [PMID: 14500495]

[133] Yao C. Major surface protease of trypanosomatids: one size fits all? Infect Immun 2010; 78(1): 22-31. [http://dx.doi.org/10.1128/IAI.00776-09] [PMID: 19858295]

[134] Hall BF, Webster P, Ma AK, Joiner KA, Andrews NW. Desialylation of lysosomal membrane glycoproteins by *Trypanosoma cruzi*: a role for the surface neuraminidase in facilitating parasite entry into the host cell cytoplasm. J Exp Med 1992; 176(2): 313-25. [http://dx.doi.org/10.1084/jem.176.2.313] [PMID: 1500849]

[135] Andrews NW, Abrams CK, Slatin SL, Griffiths G. A T. cruzi-secreted protein immunologically related to the complement component C9: evidence for membrane pore-forming activity at low pH. Cell 1990; 61(7): 1277-87. [http://dx.doi.org/10.1016/0092-8674(90)90692-8] [PMID: 2194668]

[136] Hall BF, Joiner KA. Developmentally-regulated virulence factors of *Trypanosoma cruzi* and their relationship to evasion of host defences. J Eukaryot Microbiol 1993; 40(2): 207-13. [http://dx.doi.org/10.1111/j.1550-7408.1993.tb04905.x] [PMID: 8461894]

[137] Rubin-de-Celis SS, Uemura H, Yoshida N, Schenkman S. Expression of trypomastigote trans-sialidase in metacyclic forms of *Trypanosoma cruzi* increases parasite escape from its parasitophorous vacuole. Cell Microbiol 2006; 8(12): 1888-98. [http://dx.doi.org/10.1111/j.1462-5822.2006.00755.x] [PMID: 16824037]

[138] González J, Ramalho-Pinto FJ, Frevert U, *et al.* Proteasome activity is required for the stage-specific transformation of a protozoan parasite. J Exp Med 1996; 184(5): 1909-18. [http://dx.doi.org/10.1084/jem.184.5.1909] [PMID: 8920878]

[139] de Diego JL, Katz JM, Marshall P, *et al.* The ubiquitin-proteasome pathway plays an essential role in proteolysis during *Trypanosoma cruzi* remodeling. Biochemistry 2001; 40(4): 1053-62. [http://dx.doi.org/10.1021/bi001659k] [PMID: 11170428]

[140] Furuya T, Kashuba C, Docampo R, Moreno SN. A novel phosphatidylinositol-phospholipase C of *Trypanosoma cruzi* that is lipid modified and activated during trypomastigote to amastigote differentiation. J Biol Chem 2000; 275(9): 6428-38. [http://dx.doi.org/10.1074/jbc.275.9.6428] [PMID: 10692446]

[141] Okura M, Fang J, Salto ML, Singer RS, Docampo R, Moreno SN. A lipid-modified phosphoinositide-specific phospholipase C (TcPI-PLC) is involved in differentiation of trypomastigotes to amastigotes of *Trypanosoma cruzi*. J Biol Chem 2005; 280(16): 16235-43. [http://dx.doi.org/10.1074/jbc.M414535200] [PMID: 15710612]

[142] de Paulo Martins V, Okura M, Maric D, *et al.* Acylation-dependent export of *Trypanosoma cruzi* phosphoinositide-specific phospholipase C to the outer surface of amastigotes. J Biol Chem 2010; 285(40): 30906-17. [http://dx.doi.org/10.1074/jbc.M110.142190] [PMID: 20647312]

[143] Martins VdeP, Galizzi M, Salto ML, Docampo R, Moreno SN. Developmental expression of a *Trypanosoma cruzi* phosphoinositide-specific phospholipase C in amastigotes and stimulation of host phosphoinositide hydrolysis. Infect Immun 2010; 78(10): 4206-12. [http://dx.doi.org/10.1128/IAI.00473-10] [PMID: 20643853]

[144] González J, Cornejo A, Santos MR, *et al.* A novel protein phosphatase 2A (PP2A) is involved in the transformation of human protozoan parasite *Trypanosoma cruzi*. Biochem J 2003; 374(Pt 3): 647-56. [http://dx.doi.org/10.1042/bj20030215] [PMID: 12737627]

[145] Lauwaet T, Davids BJ, Torres-Escobar A, *et al.* Protein phosphatase 2A plays a crucial role in *Giardia lamblia* differentiation. Mol Biochem Parasitol 2007; 152(1): 80-9. [http://dx.doi.org/10.1016/j.molbiopara.2006.12.001] [PMID: 17204341]

[146] Wiese M. *Leishmania* MAP kinasesfamiliar proteins in an unusual context. Int J Parasitol 2007; 37(10): 1053-62. [http://dx.doi.org/10.1016/j.ijpara.2007.04.008] [PMID: 17548090]

[147] Madeira da Silva L, Beverley SM. Expansion of the target of rapamycin (TOR) kinase family and function in *Leishmania* shows that TOR3 is required for acidocalcisome biogenesis and animal infectivity. Proc Natl Acad Sci USA 2010; 107(26): 11965-70. [http://dx.doi.org/10.1073/pnas.1004599107] [PMID: 20551225]

[148] Nascimento M, Zhang WW, Ghosh A, *et al.* Identification and characterization of a protein-tyrosine phosphatase in *Leishmania*: Involvement in virulence. J Biol Chem 2006; 281(47): 36257-68. [http://dx.doi.org/10.1074/jbc.M606256200] [PMID: 17005559]

[149] Sangenito LS, Ennes-Vidal V, Marinho FA, *et al.* Arrested growth of *Trypanosoma cruzi* by the calpain inhibitor MDL28170 and detection of calpain homologues in epimastigote forms. Parasitology 2009; 136(4): 433-41. [http://dx.doi.org/10.1017/S0031182009005629] [PMID: 19250597]

[150] Ennes-Vidal V, Menna-Barreto RF, Santos AL, Branquinha MH, dAvila-Levy CM. Effects of the calpain inhibitor MDL28170 on the clinically relevant forms of *Trypanosoma cruzi in vitro*. J Antimicrob Chemother 2010; 65(7): 1395-8. [http://dx.doi.org/10.1093/jac/dkq154] [PMID: 20457672]

[151] Andrade HM, Murta SM, Chapeaurouge A, Perales J, Nirdé P, Romanha AJ. Proteomic analysis of *Trypanosoma cruzi* resistance to Benznidazole. J Proteome Res 2008; 7(6): 2357-67. [http://dx.doi.org/10.1021/pr700659m] [PMID: 18435557]

[152] Macedo AM, Pena SD. Genetic Variability of *Trypanosoma cruzi*: implications for the pathogenesis of chagas disease. Parasitol Today (Regul Ed) 1998; 14(3): 119-24. [http://dx.doi.org/10.1016/S0169-4758(97)01179-4] [PMID: 17040719]

[153] Zingales B, Miles MA, Campbell DA, *et al.* The revised *Trypanosoma cruzi* subspecific nomenclature: rationale, epidemiological relevance and research applications. Infect Genet Evol 2012; 12(2): 240-53. [http://dx.doi.org/10.1016/j.meegid.2011.12.009] [PMID: 22226704]

[154] De Pablos LM, Osuna A. Multigene families in *Trypanosoma cruzi* and their role in infectivity. Infect Immun 2012; 80(7): 2258-64. [http://dx.doi.org/10.1128/IAI.06225-11] [PMID: 22431647]

[155] Perez CJ, Lymbery AJ, Thompson RC. Chagas disease: the challenge of polyparasitism? Trends Parasitol 2014; 30(4): 176-82. [http://dx.doi.org/10.1016/j.pt.2014.01.008] [PMID: 24581558]

[156] Araújo CA, Cabello PH, Jansen AM. Growth behaviour of two *Trypanosoma cruzi* strains in single

and mixed infections: *In vitro* and in the intestinal tract of the blood-sucking bug, *Triatoma brasiliensis*. Acta Trop 2007; 101(3): 225-31.
[http://dx.doi.org/10.1016/j.actatropica.2007.02.004] [PMID: 17374352]

CHAPTER 9

Molecular Tools and Strategies for Diagnosis of Chagas Disease and Leishmaniasis

Alejandro G. Schijman[*,1], **Juan M. Burgos**[2] and **Paula L. Marcet**[3]

[1] *Laboratorio de Biología Molecular de la Enfermedad de Chagas, Instituto de Investigaciones en Ingeniería Genética y Biología Molecular "Dr. Hector Torres" (INGEBI-CONICET), Ciudad de Buenos Aires, Argentina*

[2] *Universidad Nacional de San Martín, Provincia de Buenos Aires, Argentina*

[3] *Division of Parasitic Diseases and Malaria, Centers for Disease Control and Prevention (CDC), Atlanta, USA*

Abstract: Chagas disease and leishmaniasis, caused by the kinetoplastid protozoans *Trypanosoma cruzi* and *Leishmania* spp., respectively, affect millions of people worldwide, most of them belonging to neglected populations. Diagnostic tests for Chagas disease are employed during epidemiological surveys of vectorial and oral transmission, blood bank screening, analysis of pregnant women and their newborns, and in individual cases. However, the currently available assays need improvement. The different phases of the disease, the transmission mode and the high genetic variability of the parasite increase the difficulties of making diagnostic kits with different markers suitable for the diverse scenarios of *T. cruzi* infection. Different *Leishmania* species cause diverse clinical features and sequelae and require different clinical management. In contrast to Chagas disease diagnosis, molecular diagnosis for leishmaniasis requires not only confirmation of the infection but also the genotyping of complexes, species or subspecies. Precise diagnosis and rapid species identification can facilitate decision-making respect to treatment and follow-up of parasite spread. The aim of this chapter is to summarize the most commonly used molecular tools described to date to detect *T. cruzi* infection and to detect and genotype *Leishmania* spp.

[*] **Corresponding author Alejandro G. Schijman:** Laboratorio de Biología Molecular de la Enfermedad de Chagas, Instituto de Investigaciones en Ingeniería Genética y Biología Molecular "Dr Hector Torres" (INGEBI-CONICET), Buenos Aires, Argentina; Tel: 005411 47832871; Fax: 47868578; E-mail: schijman@dna.uba.ar

Marcelo Santos da Silva & Maria Isabel N. Cano (Eds.)

Keywords: Benznidazole, Chagas disease, HIV co-infection, Internal transcribed spacer, Isoenzymes, Leishmaniasis, Loop-mediated amplification, Molecular diagnosis, Recombinant antigens, Serodiagnosis.

CHAGAS DISEASE

Introduction

Chagas disease (CD), probably "the most neglected of the neglected diseases" is caused by the kinetoplastid parasite *Trypanosoma cruzi* [1]. Despite it has traditionally been treated as an endemic disease in the tropical and subtropicals areas of South and Central America, and Mexico, it is an emerging global concern due to the increasing migration of affected people to non-endemic areas, and the perpetuation of infection through congenital or transfusional transmission [2]. The outcome of the infection is a consequence of pathogen-host interactions modulated by their genetic composition, immunological host response and eco-epidemiological factors.

The acute phase of CD presents variable symptoms which generally decline spontaneously some weeks after infection. Prompt etiologic treatment usually eliminates the parasite during acute CD, although symptoms are only detected in 1-2% of infected individuals. The chronic phase is asymptomatic in around 70% of seropositive persons and 30% manifests digestive and/or cardiac pathologies years or decades later, that can evolve to heart failure and sudden death. *T. cruzi* infected immunosuppressed patients may develop necrotizing inflammatory injury in the central nervous system. Prognostic markers of disease progression remain to be identified [3].

Investment devoted to CD has been insufficient to provide efficient diagnosis and treatment tools for patient management and control programs [4].

Genetic Organization of T. cruzi

T. cruzi populations present a complex structure, consequence of clonal propagation [5, 6] as well as rare events of gene exchange [7]. Different biological, biochemical and molecular studies have demonstrated many genetic and protein polymorphisms [5, 8, 9]. By different markers, natural populations can be classified into six discrete typing units (DTUs Tc I to Tc VI), which are composed of sets of parasite stocks genetically closer to one another than to any other one. These DTUs are identifiable by specific molecular markers [10], they show different geographical distribution [11], DNA content and gene dosage [12 - 15]. All DTUs can cause CD as they are infective to humans. Different histotropism were reported in patients infected with DTU mixtures [16, 17]. *T. cruzi* genetic diversity should be taken into account when developing diagnostic tests for worldwide use and any new test should be validated with strains representing all DTUs. Indeed, some molecular targets used for recombinant antigen and nucleic acid amplification based diagnoses are polymorphic and present different gene dosage in strains belonging to different DTUs and in some instances in strains from the same DTU.

Current Situation of Chagas Disease Diagnosis

Diagnostic tests may be used during epidemiological surveys, blood bank screening, analysis of pregnant women and individual cases with suspicion of infection, but the available assays need improvement. In particular, no effective tools are available for population screening, the point of care settings or follow-up of anti-parasitic treatment.

The different transmission modes, the disease phases and the high genetic variability of the parasite increase the difficulties of making diagnostic kits with most appropriate markers for the diverse CD epidemiological settings.

The Polymerase Chain Reaction (PCR) has opened new opportunities to detect *T. cruzi* infection and to evaluate trypanocidal chemotherapy [18 - 20]. However, PCR procedures may have highly variable sensitivity, specificity and accuracy, depending on the type and volume of clinical sample, the storage conditions, the DNA purification methods, the parasite genes used as targets, primers and probes, the reagents as well as the thermocycling parameters. The intermittent blood-stream parasitic load is another factor responsible for the PCR clinical sensitivity. False negative results can also occur because of PCR inhibitors that can be co-purified during DNA extraction procedures whereas false positives may arise due to carry over DNA contamination [21]. The issues mentioned above have made it difficult to interpret and compare PCR results obtained in different laboratories settings. Due to the high variability in the performance of published PCR methods and the urgent need of a surrogate marker of parasitological response to anti-parasitic drugs, the Special Programme for Research and Training in Tropical Diseases (TDR-WHO) supported a comparative study of PCR procedures for the detection of *T. cruzi* DNA in peripheral blood samples stabilized in a chaotropic buffer [22]. In the context of an international collaborative study, forty-eight different procedures were compared for their ability to detect *T. cruzi* DNA from a blind panel containing DNA from different strains, including spiked blood samples with serial concentrations of parasite cells as well as blood samples from seropositive patients and non-infected controls. Four methods were selected for standardization and intra-laboratory validation; three were based on satellite repetitive DNA (SatDNA) sequences and one on DNA from kinetoplast minicircle. Two satellite DNA-based tests used real-time PCR, allowing potential quantitative applications, one used SYBR Green technology, based on the method reported by Duffy *et al.* [21], and the other one used TaqMan probes, based on the method published by Piron and coworkers [23]. Since then, a few real-time PCR strategies were optimized for *T. cruzi* DNA detection and quantification in CD patients [24 - 26]. In the framework of the "Small Grants Programme" (Joined initiative of "Communicable Diseases Research/Pan-American Health Organization (PAHO/WHO)"), and The "Special Programme for Research and Training in Tropical Diseases" (TDR/UNDP/UNICEF/World Bank/WHO) an international workshop was carried out to validate standardized quantitative Real-Time PCR strategies for detection and quantification of bloodstream parasite loads from CD patients. The workshop counted with the participation of twenty-six laboratories from 14 countrics with expertise in PCR [25]. The validated strategies included DNA extraction from peripheral blood previously treated with guanidine hydrochloride-EDTA buffer using a commercial kit based on glass-fiber columns. Two duplex qPCR methodologies using TaqMan probes were assayed: 1) SatDNA qPCR, and 2) kinetoplast DNA (kDNA) qPCR (Table **1**). Both methods included an internal amplification control (IAC): a normalized amount of a plasmid containing a heterologous sequence that is loaded in the blood sample, just before DNA extraction and used as internal standard for monitoring the whole procedure (Table **1**) [24, 27].

The following analytical parameters: reportable dynamic range, analytical sensitivity, detection limit, precision and trueness and limit of quantification were determined as recommended in international guidelines using seronegative human blood matrix spiked with known numbers of parasite cells. The parameters "inclusivity and selectivity" were determined using purified DNA from reference samples representing the different *T. cruzi* DTUs, *Trypanosoma rangeli* and *Leishmania* spp, respectively. Both methods were employed against 156 clinical samples from acute and chronic patients infected by different transmission routes in different countries. Analytical sensitivity of kDNA qPCR was higher than that of SatDNA qPCR, with limits of detection of 0.234 and 0.698 par. eq./mL, respectively. High concordance was observed between both methods regarding sensitivity and parasitic loads.

These efforts allowed production of standardized operative procedures for performing qualitative [22], as well as quantitative PCR [25] to assess parasitic response to treatment [28, 29] (see section Molecular-based monitoring of treatment response).

Table 1. Primers and TaqMan probes sequences for *T. cruzi* validated qPCRs.

Target Sequence	Primers / Probes	Sequences (5'-3')	Final Concentration (µM)
T. cruzi Satellite DNA	cruzi1 cruzi2 **cruzi3**	ASTCGGCTGATCGTTTTCGA AATTCCTCCAAGCAGCGGATA **FAM-CACACACTGGACACCAA-NFQ-MGB**	0.75 0.75 0.05
T. cruzi kDNA	32F 148R **71P**	TTTGGGAGGGGCGTTCA ATATTACACCAACCCCAATCGAA **FAM-CATCTCACCCGTACATT-BHQ1***	0.40 0.40 0.05
Internal Amplification Control (IAC)	IAC-Fw IAC-Rv **IAC-Tq**	ACCGTCATGGAACAGCACGTA CTCCCGCAACAAACCCTATAAAT **VIC-AGCATCTGTTCTTGAAGGT-NFQ-MGB**	0.10 0.10 0.05

Sequences of primers and TaqMan probes (in BOLD) that were analytically validated in the context of the international study and workshop launched by WHO/TDR/PAHO, aiming to harmonize reliable qPCR methods for monitoring parasitological response to treatment of Chagas disease. *C LNA nucleotides. Reference: Ramirez *et al.*, 2015 [25].

Diagnosis of *T. cruzi* Infection in Different Scenarios

Strategies of diagnosis differ depending if the infection is the result of vectorial, oral, congenital, blood transfusion or organ transplantation from an infected donor.

Acute Chagas Disease from Vectorial and Oral Transmission

The acute phase of disease derived from vectorial infection is non-frequently detected except when acute symptoms are severe. Thus, a large proportion of cases remains undiagnosed and evolves to the chronic phase.

Direct or indirect detection of circulating parasites is the diagnostic approach of choice. Microscopical observation of fresh blood can reveal alive parasites due to their motility. Stained blood smears permit detection of *T. cruzi* morphological characteristics and its differentiation from *T. rangeli*, which is especially important in regions where geographical distribution of these protozoans overlaps. Nevertheless, microscopic analysis of fresh or stained blood smears renders only 70% sensitivity. When parasitemia is low, a concentration method must be employed for reliable diagnosis [30].

Few serological methods have been developed for the diagnosis of acute infection. Currently, the IgM type humoral response against Shed Acute Phase Antigen (SAPA), derived from the *trans*-sialidase multigene family has been mostly searched. Studies determined that this antigen recognizes 90% of acute patients evaluated [31, 32].

Disease during the acute phase after oral transmission may be more virulent, which could lead to 12.5% mortality 30 days post-infection [33, 34]. Molecular methods have been used to detect acute infections in the first weeks after suspected oral transmission [35, 36].

A recent comparison of the analytical performance of two standardized qPCR methods [25] showed that qPCR targeted to the conserved region of minicircle molecules (kDNA) displayed the highest sensitivity for *T. cruzi* DNA detection. However, in some Central and South American countries where *T. rangeli* might be the cause of confounding diagnosis with *T. cruzi* [37], kDNA qPCR could lead to false positive results due to *T. rangeli* because the amplified minicircle region annealing with primers and the probe is highly conserved and repeated in both trypanosomatid species. Therefore, in these regions, alternative standardized qPCR assays, such as the one based on SatDNA sequences should be indicated, because *T. rangeli* presents very low copy number of SatDNA [27].

Transmission by Organ Transplantation

Organ transplantation is yet another way of transmission of CD, possibly facilitated by immunosuppressive therapy. The number of patients who can derive profit from organ transplants remarkably exceeds the number of available deceased donors, thus the main limitation of organ transplantation worldwide is the access to an allograft. Therefore, every effort is done to increase the number of available organs that can be safely transplanted. The use of organs from infected donors in endemic areas has always been a subject of controversy. Although it has been extended to non-endemic regions, which makes them more likely to become transplant candidates or organ donors, the number of infected people continues to grow. The primary-acquired *T. cruzi* infection in recipients of organs from an infected donor typically presents fever, hepatosplenomegaly, myocarditis, and/or congestive cardiac failure.

Organ transplant immunosuppressive drugs can induce abrogation of IgG and its isotypes, making the search for parasites the method of choice for diagnosis of infection during immunosuppression [41]. The availability of PCR monitoring to early detect *T. cruzi* transmission allows prompt treatment of donor-derived infection [38 - 40]. PCR was able to detect acute infection in transplanted recipients between 13 and 224 days earlier than conventional serological assays and between 28 and 47 days earlier than Strout [42]. The use of organs from seropositive donors in the Americas could potentially be expanded if PCR-based systematic monitoring of recipients is established in organ transplant units, which would have a significant impact in shortening the waiting period on the organ transplant lists.

Congenital Transmission

Between 5-6% of newborns from infected mothers can acquire the disease. Early diagnosis allows newborns to be treated successfully almost all the time [43]. Diagnosis of congenital CD close to delivery is optimal for prompt and efficient etiological treatment.

However, in a proportion of neonates parasitemia may increase during their first months of life and with it the sensitivity of parasitological tests done after that period [44]. If parasitological methods do not detect the infection soon after birth, screening for infant antibodies to the parasite is recommended at 9 months of age or greater to detect congenital infection. However, due to economic and social constraints, especially in rural areas, it is problematic for the mothers to carry their babies to the health centers for confirmation of diagnosis. Because treatment closer to delivery is almost 100% effective and with fewer toxic side effects [44], studies to improve early diagnosis are a priority for public health.

Current diagnostic techniques in newborns include direct examination of bloodstream parasites. The microhematocrit (MH) method [45] has been mainly recommended for early diagnosis of congenital Chagas disease [1]. Blood samples are collected in six to eight capillary tubes, and each tube is examined for at least 20 minutes. Its sensitivity is low unless the test is performed by experienced personnel [46]. An alternative method is the microStrout that uses polypropylene microtubes, which is safer for the operator.

Xenodiagnosis and hemoculture are traditional indirect parasitological methods; their sensitivity is very dependent on the parasitic load in the sample and can only be performed in specialized laboratories. Because of some patients present bug bite skin reactions, artificial xenodiagnosis is recommended, which has a higher positivity than the routine xenodiagnostic method [47, 48].

Conventional IgG serological assays are recommended after nine months of life when maternal antibodies fade [49]. Before that age, the use of SAPA antigen has been proposed [31, 32], especially in health care facilities and hospitals where DNA amplification techniques are not available [50].

Usefulness of PCR for Early Diagnosis of Congenital T. cruzi Infection

T. cruzi DNA amplification has been evaluated in different types of samples such as cord blood, peripheral blood, and serum samples, from neonates and infants suspected of congenital CD [44, 51 - 53]. Moreover, congenital *T. cruzi* infection was detected by PCR in amniotic fluid from an infected mother with a premature delivery [54]. Most qualitative PCR procedures have been targeted to the 330 bp four-high variable region of the kDNA [19, 55, 56] or the 195 bp satellite repeat [20, 44] and depicted the highest analytical sensitivities [22, 51, 52]. In a longitudinal study of 152 consecutively admitted infants born to CD mothers in Buenos Aires city, non-endemic for *T. cruzi* infection, PCR showed a sensitivity of 100% in babies younger than six months of life that decreased to 73.8% in older infants, considered in the early chronic phase, due to the decrease of parasitic load during natural evolution of the infection [51]. In an active transmission area of Bolivia, the PCR showed a sensitivity of 93.6% among infected children [57]. Russomando and coworkers [58] evaluated the PCR for early diagnosis of congenital CD in newborns from Paraguay; the rate of transmission was 3% using direct microscopic observation, but it increased to 10% using PCR, which corresponded to the incidence rate detected later by serologic monitoring in the chronic phase. Recent studies have employed real-time PCR allowing quantification of parasitic loads in newborns and neonates [24, 44]. However, it is still necessary to test standardized real-time procedures using low volumes of peripheral blood or using cord blood from newborns. Analysis of newborns and neonates requires the use of a much lower blood sample volumes than those already validated for adult chronic CD cohorts [22, 24, 25, 59]. Heel prick blood could be a sample of choice, especially if collected in solid supports such as filter paper. However, at the moment, laboratory conditions to achieve high PCR sensitivity in filter paper have been set up only in the screening of triatomine specimens [60]. Umbilical cord blood might be ideal because it does not present restrictions in sample volume; However, its use has been controversial because it might contain maternal parasite DNA. A kit prototype based on duplex TaqMan Real-Time PCR (qPCR) for molecular diagnosis of congenital CD has been constructed and is currently under field validation. The DNA is extracted from 1 mL of peripheral blood mixed with a DNA stabilizer solution. Analytical validation of the prototype kit was done following Clinical Laboratory Standard Institute guidelines [61 - 63]. Once standardized and validated, PCR can be a valuable tool for early and rapid diagnosis of infection at birth or during the first days of life, allowing prompt initiation of treatment before the mother leaves the hospital.

Chronic Phase of Chagas Disease

The primary diagnostic method for chronic CD is serological, which is used routinely both in clinical diagnosis and in blood banks and assesses the humoral response developed by patients against different defined antigens of the parasite. Three conventional test formats based on detection of parasite-specific antibodies are widely used because of their good balance between simplicity, cost, and performance regarding specificity and sensitivity: “enzyme-linked immunosorbent assay” (ELISA), “indirect haemagglutination” (IHA) and “indirect immuno-fluorescence” (IIF). Serodiagnostic kits usually employ crude antigenic parasite preparations, semi-purified fractions from the epimastigote form, or recombinant antigens. Most of them showed sensitivity of 94 and 99.5% and specificity of 94 and 96%. Currently, the World Health Organization (WHO) recommends running two immunological tests in parallel (one of them being an ELISA), with the use of a third one when discordant results are obtained, reaching an accurate serological status in more than 98% of sera [64].

An inconclusive serological result occurs when the results are not clearly positive or negative. The diagnosis systems currently sold are effective at diagnosis of patients with high antibody titers, but results are often questionable when antibody titers are low or undetectable, even if the patient tested positive in other tests or by PCR [65 - 68]. Concomitant infections are a potential source of inconclusive and false-positive results, particularly in patients infected by *Leishmania* spp. or *T. rangeli* [69 - 71]. This was observed in a blood donor screening test approved by the Food and Drug Administration of the United States (FDA) that use extracts of total parasite lysates with high sensitivity but slightly lower specificity because of the potential for cross-reactions [72]. The addition of recombinant antigens and/or synthetic peptides in diagnostic kits has meant an advance in specificity [67], but the use of them in single assays reduced sensitivity compared with conventional tests. This limitation was overcome with the application of cocktails of recombinant antigens, mixtures of synthetic peptides or multiepitope antigens. Several laboratories have identified, cloned and appointed *T. cruzi* antigens that indeed correspond to the same protein generating redundant information. A good summary of equivalence among antigens is depicted in da Silveira *et al.* [32]. Among the recombinant antigens found to have better performance are proteins from cytoplasm (CRA / JL8), trypomastigote surface (B13/Ag2), *trans*-sialidases family (TS/SAPA), cytoskeleton (FRA/JL7), and microtubule-associated proteins (MAP/JL9), and ribosomal (JL5), and flagellar proteins [73].

Several multi-antigen tests have been evaluated in several countries with good performance for diagnosing chronic CD [74 - 77]. The lower sensitivity and specificity of some tests could be caused by the antigenic mixes being absorbed to the support base of the reaction plates, which could disguise some of the epitopes of these antigens being recognized by the antibodies of the serum [32].

Numerous studies have also observed some marked differences among recombinant antigens. Antigens with the ability to distinguish between acute and chronic CD [74, 78], discrimination of clinical forms at heart disease [75], and digestive disease [79], as well as variation of sensitivity in different patient populations [80] have been observed.

Comparative tests between commercial kits have shown that some have good sensitivity and specificity, such as the radioimmunoassay test, RIPA [81]. The system consists of the co-precipitation of two glycoproteins, one of 72 and the other of 90 kDa, by the specific IgG in the sera of patients. However, the use of radioactive matters and the complexity of the procedures and the high cost may result in low applicability in practice.

Elevated concentrations of antibodies against an antigen can be caused by the existence of tandem repeats (TR), which are defined by possessing one or more copies of an amino acid pattern and are characteristic of multigene families [82]. These TR are found in highly conserved regions and their greater exposure to the immune system induces specific antibodies in the serum of infected patients with high affinity for the epitopes of the TR [32]. The immunodominance of these motifs is caused mainly by the molar ratio and/or their particular spatial structure [83]. This immunodominance is responsible for the response of B lymphocytes to *T. cruzi* [73, 82 - 84]. In trypanosomatids, there are TR in very large gene sequences (10,000 bp approximately) codifying proteins with known functions, such as the "zinc finger proteins" motifs that bind to DNA [82]. Besides of protein sequences containing TR in *T. cruzi*, 45% possess a signal peptide, and 53% transmembrane domains, constituting membrane proteins with domains on the parasite's external surface. These proteins belong to the three large gene families: *trans*-sialidase (TS), mucins and mucin-associate surface proteins (MASPs) [82]. One example is the SAPA in the C-terminus of TS [85]. The SAPA is made up of a TR of 12 amino acids (DSSAHSTPSTPA), which is linked to the induction of neutralizing antibodies as it remains active for some time in blood. These epitopes are thought to be an adaptation strategy to maintain virulence [86, 87]. The fact that these SAPA regions are considered a B epitope may be due to their light secondary and tertiary structures, being conformational epitopes targeted by antibodies [85]. As mentioned in the section on acute CD diagnosis, this SAPA epitope is defined as an acute phase antigen with a rapid induction of IgM and IgG, capable of inducing high titers in the initial stages of the disease. In contrast, SAPA is recognized in only 60% of chronic phase sera [85]. Other antigens with TR are FRA (flagellar repetitive antigen) and CRA (cytoplasmatic repetitive antigen) [88]. The gene encoding CRA contains 23 copies of a repeated segment of 42 bp and the one encoding FRA 14 copies of 204 bp, likely constituting recognizable epitopes in these TR regions.

The MASP (mucin-associated surface proteins) gene and protein family was described when sequencing the *T. cruzi* genome [89]. It is composed of 1,377 genes and 433 pseudogenes organized in arrays of up to 600 kb, although this number could be higher, due the collapse of annotations of the genome with sequencing families containing repetitive motifs [90]. The distribution of these proteins in other gene families corresponding to surface genes possibly contributes to maintaining genetic diversity, avoiding the homogenization of MASP gene sequences [91]. Up to 92 members of this family have been described as being expressed simultaneously in the trypomastigotes of some particular strains [92]. Many have been found in the membrane of trypomastigotes, but some may be part of the secretome [91, 93]. Expression studies using qPCR on *T.cruzi* strains belonging to different parasite DTUs demonstrated that the highest expression occurs in the trypomastigote form [93]. This parasite phase has the highest level of exposure to the immune system and therefore the expression of MASP protein variants in the parasite membrane may be a mechanism to evade the immune response [91].

The disaccharide gal-alpha-1,3-gal-beta is an immunodominant glycotope of a synthetic glycoarray containing nonreducing α-galactopyranosyl moieties associated to *O*-glycan mucin, analyzed using chemiluminescent enzyme-linked immunosorbent assay. It has shown its usefulness in the diagnosis of the infection in chronic stages [94].

A summary of recombinant antigens and peptides is shown in Table **2**.

PCR strategies have been recommended as an alternative tool to support CD diagnosis particularly in cases with indeterminate serological findings [95]. Because of the low number of circulating parasites in the peripheral blood during the indeterminate phase, PCR techniques are less sensitive than serological tests [22]. The control exerted by the host immune response as well as parasite factors such as the genetic background of the circulating strains, which in turn may determine tissue tropism and replication rates, could be responsible for the lower PCR sensitivity in detection of chronic infection.

Analysis of samples from chronic CD patients, including 70 asymptomatic individuals, and 75 patients with cardiac and/or digestive syndromes, using two standardized qPCR assays [25] showed that 117 (80.69%), and 122 (84.14%) out of 145 samples were positive using SatDNA- and kDNA- qPCR, respectively; 113 of these samples were positive by both qPCR methods, and 4 and 9 samples were positive only by SatDNA, or kDNA qPCRs, respectively; Kappa Index was 0.691 ($CI_{95\%}$: 0.535-0.847).

Analytical sensitivity was more uniform among the different *T. cruzi* DTUs for kDNA qPCR than for SatDNA qPCR, being this latter method less sensitive for some TcI and TcIV strains, indicative of a lower gene dosage in their genomes [13, 21].

Table 2. ***Trypanosoma cruzi*** **recombinant proteins or synthetic peptides with clinical/epidemiological use.**

Antigens	Remarks	Aplications	Refs.
CRA / Ag30 / JL8 /TCR27	Cytoplasmatic antigen	Chronic infections/ Digestive forms	[78, 83, 139 - 142]
FRA / Ag1 / JL7 / H49	Cytoskeleton-associated protein	Chronic infections/ Digestive forms	[79, 81, 83, 137, 139, 141 - 144]
B13 / Ag2 / TCR39 / PEP2 / TSSA	Trypomastigote surface protein	Chronic infections	[81, 137 - 140, 142 - 147]
Ag36 / JL9 / MAP	Microtubule-associated protein	Chronic / acute infections	[78, 139, 142]
SAPA / TCNA / TS	Trans-sialidases family	Acute / congenital infections	[137, 139, 142, 144, 148, 149]
FcaBP / 1F8 / Tc-24 / Tc-28	Flagellar Ca++ binding protein	Chronic infections/ monitoring of cure	[127, 139, 150]
TcE / JL5 / Tcp0	Ribosomal protein	Chronic infections/ cardiac forms	[78, 139, 140, 142, 151, 152]
cy-hsp70 / mt-hsp70 / grp-hsp78	Heat shock proteins	Chronic infections/ monitoring of cure	[82, 139, 153]
FL160 / CEA / CRP	Flagellum-associated surface protein	Chronic infections/ monitoring of cure	[154 - 157]
16 *T. cruzi* antigens, SAPA, Ag 1, 2, 13, 30, 36, KMP11, HSP70, PFR2, Tgp63	Antigens used in grouped panels	Chronic infections/ monitoring of treatment follow up	[34, 129, 132 - 136]

Several different antigen names given to identical or similar peptides are grouped. Refs. indicate original studies and others where antigens were evaluated in simple or mixed reactions. This table was modified from da Silveira *et al.*, 2001 [34]. **Abbreviations**: CRA, cytoplasmatic repetitive antigen; FRA, flagellar repetitive antigen; MAP, microtubule associated protein; SAPA, shed acute-phase antigen; CEA, chronic axoantigen; CRP, complement regulatory protein; cy-hsp, cytoplasmatic heat shock protein; grp-hsp, endoplasmatic reticulum heat shock protein; mt-hsp, mitochondrion heat shock protein, TCNA, *Trypanosoma cruzi* neuraminidase; TS, trans-sialidase; TSSA, trypomastigote small surface antigen. Usefulness and Limitations of PCR in chronic Chagas disease.

No significant differences were found between the parasitic loads from chronic CD patients infected with strains belonging to different DTUs using SatDNA qPCR. However, using kDNA qPCR, the patients infected with TcI strains presented higher parasitic loads (2.55 (2.08-2.87) Log_{10} par. eq./10 mL) than patients infected with strains belonging to the other DTUs. The persistence of *T. cruzi* in chronic chagasic patients was detected for the first time in biopsies taken from inflammatory gingival foci processed by PCR. Using three different PCR procedures, parasite DNA was detected in 7 out of 31 (22.5%) gum samples from chronically infected patients presenting different degrees of gingival inflammation. These findings suggest the establishment of small secondary foci for the maintenance of inapparent *T. cruzi* infection. Moreover, the non-invasive strategy to collect samples for PCR is easy and low-risk, which may add to the use of gingival biopsy as an alternative diagnostic tool to confirm *T. cruzi* infection in chronic CD cases [96].

Reactivation of Chronic Chagas Disease Due to Immunosuppression

Immunosuppression has gained increasing clinical relevance because of neoplastic and other systemic diseases, including HIV/AIDS and the use of immunosuppressive drugs [97]. Immunosuppression often alters the natural prognosis of other diseases, predisposing patients to opportunistic infections or reactivation of latent pathogens, with severe and often deadly, clinical manifestations [98]. CD in the immunosuppressed patient may present severe manifestations, atypical clinical forms and significant morbidity and mortality [99]. Reactivation of CD is characterized by the presence of trypomastigotes in blood and cerebrospinal fluid, that even in the absence of clinical symptoms, can be detected by direct parasitological and/or PCR techniques. In addition, subcutaneous nodules (chagomas), panniculitis, myocarditis with signs of heart failure, meningitis, and encephalitis can be seen [100, 101]. HIV-*T. cruzi* coinfected patients are more likely to have higher parasitemia than HIV-negative individuals, which may increase prior to reactivation and clinical manifestations [102]. Most reactivations occur in advanced immunosuppression and are considered an AIDS-defining condition [103], which may be the first clinical manifestation of Chagas disease. The risk of *T. cruzi* reactivation in patients with HIV ranges from 15 to 35% [101], where the majority of episodes were seen in patients with AIDS whose mean CD4 cell count was 98 (1-340 cells/mm^3). The involvement of the CNS as a target in reactivation has been described in most cases (74%), followed by myocardial involvement (17%) [104], but also skin, digestive tract, pericardium and cervix and/or multiple organs involvements [105, 106]. Reactivation episodes have a high mortality in the absence of antiparasitic treatment, but mortality may be reduced if drugs are started promptly [106 - 108]. Several guidelines for the treatment of immunosuppressed patients with CD have been published [39, 109, 110], but there is no consensus on the management of Chagas reactivation in these patients.

Molecular-based Monitoring of Treatment Response

Therapeutic success in CD seems to depend on the interplay among drug susceptibility of *T. cruzi* strains, drug access and accumulation in different environments, and the host immune response. Following years of little advance in research and development of new anti-*T. cruzi* compounds, significant progress, has recently been achieved [1, 111]. The harmonization and standardization of drug screening procedures and the development of high-throughput methodologies for *in vitro* testing have been improved. Two recent Phase IIa clinical trials that evaluated the antiparasitic activity of the ergosterol biosynthesis inhibitors posaconazole and ravuconazole (E1224, prodrug) in chronic CD patients have shown disappointing findings [59, 112]. A cysteine protease (cruzipain) inhibitor (vinyl sulfone derivative K777), the 5-nitroimidazole fexinidazole and a non-toxic fungicide fenarimol, among others, are in pre-clinical studies. Importantly, alternative treatment regimens are being considered, such as combination and intermittent therapies with available and new anti-*T. cruzi* drugs [113 - 115] aiming to achieve better safety, synergy, to avoid drug resistance and to allow the use of backup drugs if resistance arises. Chapter 10 contains information about new therapeutical tools against Chagas disease.

Chemotherapy efficacy is currently measured using anti-*T.cruzi* antibody titers and/or detection of parasitemia. A therapeutic failure is defined by the persistence of the parasite, treatment success is measured by the reduction of antibody titers and cure is defined by the disappearance of anti-*T. cruzi* antibodies. However, a decrease in *T. cruzi*-specific antibody titers often takes many years.

Use of Serologic Response to Parasite Antigens to Follow-up Treatment

<u>Studies in Children</u>: Therapeutic efficacy in children with early chronic CD has been studied. Sera from 130 children treated with benznidazole (BZN) or placebo were analyzed using a purified trypomastigote glycoconjugate in a chemiluminescent ELISA [116]. At 6 months post-treatment, 37 of 64 treated patients (56%)

compared with 3 of 65(5%) placebo cases were negative [117]. At 6 years post-treatment, the ELISA was negative in 47 out of 53 treated (88.7%) and in 12 of 46 placebo subjects (26.1%).

A short-term follow-up study of 44 chronically infected children treated with BZN used an ELISA test based on the recombinant *T. cruzi* flagellar calcium-binding protein (F29) [118], and detected 35.2 and 62.1% seroconversion by F29 after 6 and 48 months post-treatment, respectively.

Studies in Adults: Conventional serology has been compared with a complement-mediated lysis test (CoML) that employs live trypomastigotes at different periods after treatment [119]. Seroconversion of the CoML occurred in 38% of patients between 6 and 24 months following treatment, in 19% of patients between 24 and 36 months and one patient within 4 years post-treatment. However, the CoML assay requires living infective trypomastigotes, which makes its use problematical for the clinical practice. An alternative to this test is an ELISA technique based on the recombinant protein rTc24 [120]. All patients with positive CoML recognized rTc24 using ELISA and western blot, while 80% of seropositive patients with negative CoML were seronegative to rTc24. There was a decrease in anti-rTc24 reactivity measured by ELISA in 38% of patients between 6 and 24 months post-treatment and in 19% of patients at 36 months post-treatment.

Three groups of *T. cruzi*-infected patients, untreated cases, patients with treatment failure and successfully treated patients, were tested for antiparasite antibodies using an immunofluorescence assay of fixed trypomastigotes (referred as ISIFA) [121]. A successfully treated patient was defined as a case with undetectable parasitemia using xenodiagnosis at 6 years post-treatment. ISIFA was able to differentiate successfully treated cases from those untreated or with treatment failure [122].

Serological response using an ELISA based on the recombinant complement regulatory protein (rCRP) was evaluated in 31 patients before and after treatment [123], and revealed that seropositivity diminished from 100% at baseline to 70.3, 62.5, 71.4, and 33.4% at the first, second, third and fourth year post-treatment, respectively.

Antibody levels against *T. cruzi* ribosomal acidic protein P2b (Tc-P P2b) were analyzed in: 30 asymptomatic CD patients who received specific treatment with clinical follow-up for more than 20 years (group A); 37 asymptomatic patients who did not receive treatment (group B); and 11 untreated symptomatic chronic CD patients (group C) [124]. Antibody levels against TcP2b were significantly lower only in patients from group A.

Studies Using Detection of Serologic Response to Antigenic Panels: More recent studies were designed using groups of antigens for better accuracy. Antibody levels against SAPA, and *T. cruzi* antigens 1, 2, 13, 30, 36 were evaluated in sera from chronically infected patients before and after 3 years follow-up post-treatment. Antigen 13 was shown to be a good marker of treatment efficacy using ELISA, since seroconversion occurred in 67% (6 of 9) of patients ($p = 0.002$) [32, 125, 126].

A panel consisting of four recombinant proteins, named KMP11, HSP70, PFR2, and Tgp63 was used to follow up BZN-treated patients [127]. A significant decrease in reactivity against KMP11 occurred 6 months post-treatment in 26 out of 35 patients (74%), at 9 months post-treatment against PFR2 in 26 out of 35 (74%) and against HSP70 in 25 out of 35 (71%) patients. When the response against only two of these antigens was analyzed, the reduction in specific antibody titers happened in 80% of patients at 9 months post-treatment and continued during the 2-year post-treatment monitoring. The global reduction in titers against KMP11, HSP70, and PFR2 24 months post-treatment was 67, 50, and 34% of patients, respectively.

Sera from patients living in endemic areas depicted a distinctive and highly consistent reactivity pattern against 16 *T. cruzi* proteins arranged in a multiplex platform [128, 129]. A decreased response to the panel was observed in six patients monitored during 36 months after treatment. Two cases of treatment failure did not show a change in reactivity pattern over time. A strong correlation was observed between conventional serological tests (IA, IFI, and ELISA) and the above-mentioned multiplex system in 53 BZN-treated and 89 untreated chronic CD patients, with a median post-treatment follow-up of 36 months [128]. A decrease in conventional serology titers was observed in 64% *versus* 21% of treated and untreated patients, respectively, and seroconversion was achieved in 40% *versus* 7% of treated and untreated patients, respectively.

Post-treatment fluctuations in biochemical and non-conventional serological and immune parameters [130] may be employed as surrogate indicators of therapeutic efficacy. A reduction in anti-*T. cruzi* gamma interferon (IFN-)-producing cells [131], *T. cruzi* antigen-specific antibody titers detected using multiplex arrays [128] as well as in the serological reactivity against complement regulatory protein, recombinant *trans*-sialidase or kinetoplastid antigen was detected after treatment with BZN [123, 127].

Parasitological Tests

Parasitological tests have limited sensitivity and accordingly a negative result does not necessary mean the absence of parasites, but in contrast, a positive result indicates treatment failure. A highly sensitive and reliable method to detect parasitological cure is urgently needed.

Molecular Tests as Surrogate Markers of Treatment Response

Conventional and qPCR techniques have opened promising perspectives to estimate bloodstream parasite burden and to detect therapeutic failure or response [59, 112, 151, 152]. Using qPCR before and after etiological treatment with BZN, a long-term follow-up study conducted in a non-endemic country [153] demonstrated two key findings: before treatment 68% of chronic CD adult patients were PCR positive, and all became PCR negative immediately after treatment. Moreover, PCR-negative results were steady in 90% of patients 1 year after treatment.

Upon harmonization of qPCR techniques for monitoring patients under treatment, standardized PCR was applied in several clinical trials, namely:

The BENEFIT Trial: At baseline, conventional PCR was performed on blood samples from 1896 CD patients and 60.5% were PCR positive. At the end of treatment, 66.2% of patients treated with BZN converted to negative PCR, whereas in the placebo group only 33.5% converted to negative PCR . The PCR conversion to negative was 46.7% for BZN and 33.1% for placebo after 5 years of follow-up ($p < 0.001$). Interestingly, the rate of PCR conversion was different in cohorts from different countries: in Argentina and Bolivia, the odds ratio for PCR conversion was 2.79 (95% CI, 1.99 to 3.92) for 5 or more years of follow-up; in Brazil, it was 1.87 (95% CI, 1.33 to 2.63) while in Colombia and El Salvador it was 0.96 (95% CI, 0.63 to 1.45). However, the rates of PCR conversion to negative were not associated with the clinical outcome ($p = 0.16$ for interaction).

TRAENA (Treatment in adult patients) was a randomized clinical trial conducted in Argentina during 1999-2012 in adult patients, including subjects in the indeterminate phase of chronic Chagas disease. Among the endpoints, this trial aimed to evaluate the impact of BZN treated patients on the parasitological and patient serological profiles. The serological evaluation was performed through ELISA targeting the recombinant antigen F29 [154]. The quantification of parasitemia was achieved using standardized qPCR [24]. The BZN arm showed serological negativization in 46% of patients. The qPCR showed an overall negativization index (no detectable qPCR/total qPCR) of 90% and 42% in the BZN and placebo arms ($p < 0.001$) respectively, at the end of follow-up (Riarte Adelina and Prado Nilda, personal communication).

The Chagasazol Trial: This prospective randomized trial assessed the efficacy and safety of posaconazole in comparison with BZN in chronic CD adults. Antiparasitic activity was studied using qPCR-based on SatDNA during the treatment period and 10 months after the end of treatment. During treatment, all patients were PCR negative after day 14, except for 2 cases in the low-dose posaconazole branch, who were PCR positive on day 60. During post-treatment monitoring, in the intention-to-treat analysis, PCR was positive in 92% of patients who received low-dose posaconazole, and in 81% treated with high-dose posaconazole, as compared with 38% receiving BZN. In the per protocol analysis, post-treatment PCR was positive in 90% of patients who received low-dose posaconazole, in 80% receiving high-dose posaconazole and in only 6% treated with BZN. In sum, treatment failure was higher in the posaconazole groups than in the BZN arm [59].

***The E1224 Trial*:** This trial assessed the efficacy and safety of E1224, a prodrug of ravuconazole, in comparison with BZN in adults with chronic CD. The parasitological response was determined by serial negative qualitative PCR: 3 negative PCR results from 3 samples collected over 7 days at the end of treatment (EOT). The serial examination was used to increase sensitivity, reaching more than 90% of *T. cruzi* DNA bloodstream detection in chronic CD cohorts. Secondary efficacy endpoints included sustained parasitological eradication, determined by consistently negative serial qualitative PCR results at 4, 6 and 12 months of follow-up. Qualitative and qPCR measured parasite eradication and changes in parasite load on days 8, 15, 36, EOT, and at 4, 6, and 12 months' follow-up. The TaqMan-based qPCR assay had a detection limit of 0.69 par. equivalents eq./mL and a limit of quantification of 1.531 par. eq./mL [24]. Serological response was assessed at EOT and 4, 6, and 12 months after treatment. Regarding clearance of parasitic load at EOT on Day 65 (the primary efficacy endpoint), a significant difference was seen in all active treatment arms *vs.* placebo ($p < 0.001$). The highest clearance was observed in the BZN arm (91%). The proportion of patients achieving sustained response (SR, defined as D65 clearance with no positive samples available after that until the end of follow-up) in the E1224 Short dose (SD) and Low Dose (LD) arms, respectively was 10.9% and 8.3%, similar to placebo values (8.5%). SR values for BZN were 81.0% compared with 28.9% in the E1224 High Dose (HD) arm. A significant difference (one-sided $p < 0.025$) was measured for the comparison of E1224 HD and BZN arms *vs.* placebo for SR at 12 months. Thus, unlike BZN, E1224 achieved parasite DNA clearance during the treatment phase, but response declined after EOT. Kaplan-Meier analysis estimated that the time to reach 75% parasite DNA clearance was 11 days in the SD E1224 (95% CI, 10.0–11.0), HD E1224 (95% CI, 10.0–17.0), and BZN (95% CI, 10.0–11.0) treatment arms, whereas in the LD E1224 arm, the time to reach 75% parasite DNA clearance was 16 days (95% CI, 11.0–18.0). After one week of treatment, mean qPCR repeated measures analyses showed a significant difference in parasite load *vs.* placebo in all treatment arms. All BZN-treated patients had cleared parasitic loads after 2 weeks of treatment. After EOT, parasitic load remained undetectable in the BZN arm. Parasite DNA levels in the LD and SD E1224 arms gradually returned to placebo levels. HD E1224 remained significantly different from placebo, with no statistical difference from BZN, though a trend was observed. A lower risk of relapse was observed with BZN (HR=0.06 (95% CI 0.02, 0.21)), and HD E1224

(HR=0.60 (95% CI 0.26, 1.37)), compared to placebo. Interestingly, a higher baseline parasite load was independently associated with a significantly increased hazard of parasi-tological relapse.

Target Product Profiles for Chagas Disease Diagnosis

The PAHO, in collaboration with the "Drugs for Neglected Diseases initiative" (DNDi), "Médecins sans Frontières" (MSF) and "The Special Programme for Research and Training in Tropical Diseases" (TDR), convened a multidisciplinary group of experts in Rio de Janeiro, Brazil (April 2010) aiming to reach consensus on the desirable product profiles for diagnosis of Chagas disease. The discussion group proposed TPPs for point of care diagnosis associated with acute transmission and to the chronic phase and assessment of response to anti-parasitic treatment of chronic CD [155].

Important features of the diagnostic methods to be considered included current needs related to individual health care, different population groups and scenarios of use, medical criteria to apply a certain diagnostic tool, sampling of biological specimens, volume, conservation conditions, infrastructure needed, shipping procedures, number and fractionation of samples, technical skills of operators, testing time to reporting results, need for qualitative or quantitative outcomes, capacity to discriminate parasite strains as well as sensitivity and specificity.

Regardless of its specific characteristics and potential applications, an optimal diagnostic test for public health use should include low cost, simple manu-facturing and distribution methods, sustainable production, and supply requirements. It should also contain stable materials for sample collection and perform the assays in prevailing climatic conditions. Disposal should be in agreement with biosecurity standards and appropriateness by the health care and target population.

Ideal and acceptable TPPs proposed for the diagnosis of *T. cruzi* infection acquired *via* different transmission routes, in different disease phases and patients under treatment can be found in recent publications [151, 155].

LEISHMANIASIS

Introduction

Leishmaniasis, caused by trypanosomatid species of the *Leishmania* and *Viannia* subgenera, is endemic in focal areas in the tropics, subtropics, and southern Europe [156]. The disease encompasses three main clinical syndromes: visceral leishmaniasis, which can be life-threatening, cutaneous and mucocutaneous leishmaniasis [157, 158]. Due to a high diversity of *Leishmania* species, phlebotomine vectors and reservoir hosts, the disease displays diverse epidemiology and significant clinical pleomorphism [159].

At least 20 species are known to infect humans. *Leishmania* distribution splits into the Old world (East-Africa, the Mediterranean region, and the Indian subcontinent) and the New world (Central and South-America) groups. *L. major, L. aethiopica, L. tropica, L. donovani, and L. infantum* are endemic in the Old World; the latter two species are members of the so-called *L. donovani* complex. In the New World, two subgenera, *L.(Viannia)* and *L. (Leishmania)* are causative agents of leishmaniasis. The subgenus *Viannia* is composed by *L. (V.) braziliensis*, *L. (V.) panamensis, L. (V.) guyanensis* and other less frequently identified species. The *Leishmania* subgenus includes *L.(L.) mexicana*, and *L.(L.) amazonensis*, that cause localized or diffuse cutaneous disease [160]; and *L. (L.) chagasi* (considered a synonym of *L. (L.)infantum)* [161], which is the etiological agent of American visceral leishmaniasis [162].

Precise and rapid diagnosis and species identification favor prompt decisions regarding optimal treatment and parasite spread monitoring [163 - 166]. Traditionally, identification relied on the geographic area where the infection occurred, but this strategy became unreliable due to the dynamics of *Leishmania* species distribution being influenced by the world climate changes and demographic movements [167].

Classical diagnosis is based on the clinical presentation in a specific epidemiological context, confirmed by conventional laboratory diagnoses, such as serology and microscopic examination of Giemsa-stained smears from the sample and *in vitro* culture of the parasite. Microscopic observation of amastigote forms in tissue specimens is the gold standard for diagnosis; it is carried out in skin lesions (CL and other cutaneous syndromes, mucosa (ML), bone marrow, spleen and lymph nodes (for VL). However, it involves invasive sampling, sensitivity is variable and detection requires technical expertise. Indeed, the sensitivity of VL diagnosis using a direct microscopic examination of lymph node aspirates ranges from 52 to 58% and from 60 to 85% in bone marrow aspirates. The highest sensitivity is achieved using splenic aspirates (93%-98%), but the procedure can be associated with hemorrhage and thus is hazardous to be performed in remote field settings [168].

Although microscopic examination is inexpensive, fast and relatively easy to perform, it requires some expertise and lacks sensitivity due to the low parasite load in tissue samples [169, 170].

Subsequent strain typing is usually achieved by culturing *Leishmania* parasites from biopsies or aspirates from skin lesions, bone marrow, lymph node or spleen [1]. However, culture is labor-intensive and time-consuming, requires a sophisticated laboratory setup, it harbors the risk of contamination, and the final result might not be obtained for weeks [169, 171]. Culturing of promastigotes in Novy-MacNeal-Nicolle (NNN) agar or other suitable media has been considered a standard for leishmanial diagnosis, but sensitivity depends on the variable growth parameters of species, and there is a constant risk of contamination.

As stated before, the precise taxonomic identification of *Leishmania* species allows for easier clinical management and determines whether/which treatment is indicated. The gold standard for species identification has been isoenzyme analysis and multilocus enzyme electrophoresis (MLEE, also known as zymodeme typing) [172]. It has been the most widely used approach for the last 30 years for the identificationof *Leishmania* at the species and subspecies levels. However, MLEE analysis involves the use of specialized equipment, large parasite cultures and it has not enough resolution to detect variability within a zymodeme [159]. On the other hand, because different enzyme complexes are found in different epidemiological scenarios (*i.e.* Europe *vs* the Americas), results are hard to compare among research groups [159]. These constraints make the technique not generally applicable [159, 173].

Serological Diagnosis

Visceral Leishmaniasis (VL): Several serological assays (*e.g.* immunofluorescence-based ELISA) have been assessed in different settings with heterogeneous results. All methods have limitations as indirect serodiagnosis cannot distinguish between past and current infections as seroreactivity remains positive many months after cure. Also, depending on the antigens, cross-reactivity with other pathogens could further complicate the results interpretation. Another disadvantage is their inadequate sensitivity because antibody titers may vary with the parasite species, its histotropism and the host immunocompetence [169, 174].

Direct agglutination test (DAT) in plasma or serum samples was the first VL serological test employed in epidemiological surveys and control programs. It uses whole, trypsinized, Coomassie-stained promastigotes, either as a suspension or in a freeze-dried form suitable for field applications [175]. It is easy to perform, reproducible and has high sensitivity (70%-100%) and specificity (90%-100%) [176]. However, titers measured by DAT may remain positive for up to 5 years after recovery in >50% VL patients and is not helpful to detect relapse.

Currently, the most used method for VL diagnostics is an ELISA, which is generally based on the use of crude soluble antigen (CSA) and a recombinant antigen (rk39) (a conserved portion of a kinesin-related protein). The CSA-ELISA sensitivity is 80-100%, but specificity is poor due to cross-reactivity with serum samples from patients with trypanosomiasis, toxoplasmosis and tuber-culosis [175].

Rapid diagnostic kit prototypes (RDTs) have also been assayed for the diagnosis of VL; they can provide a point-of-care strategy in remote endemic villages. Their sensitivities and specificities vary according to the supplier and the geographic origin of the antigens/sera. A recent multicenter study carried out with samples from endemic regions of India, East Africa and Brazil, under the coordination of WHO/TDR, evaluated an RDT based on immunochromatography using the recombinant antigen rK39 (ICT/RDTs). The test performed very well in the Indian territory with 92.8 to 100% of sensitivity and 96 to 100% of specificity, but in sera samples from East Africa and Brazil these parameters were significantly lower (61%-91%) [1].

The "antigen detection in urine" (ADU) [177], a latex agglutination based method, was the first assay developed to detect VL markers in urine specimens. It is specific for the diagnosis of patients with active disease; results quickly turn negative after effective treatment, which is especially useful to establish post-treatment remission [178]. This test is particularly advantageous in cases of coinfection with other diseases, *i.e.* HIV-AIDS (that makes the patients more susceptible to relapse after treatment) or other health situations of deficient humoral immune response [174, 178]. However, sensitivity is highly variable from 47 to 100% and has further limitations because it is operator-dependent, requires boiled urine and the length of time that the antigen remains in the urine is unknown [178, 179]. New formats of the test are currently being developed [168].

Recently, a promising recombinant protein-based assay to detect *L. infantum* antigens has been developed [180], which has the potential to detect VL infection markers in urine and other human body fluids. Research is also underway to develop monoclonal antibodies and to evaluate the assay in different epidemiological contexts.

Cutaneous and Mucocutaneous Leishmaniasis (MCL): The leishmanin skin test (LST) based on detection of cell-mediated immunity is easy to perform and shows adequate sensitivity and specificity for both CL and MCL diagnosis. However, due to delayed expression of specific cellular immune response in VL, it can not be applied in this clinical form [181].

Molecular Tools for *Leishmania* Diagnosis and Genotyping

Nucleic acid amplification strategies are usually reliable, reproducible and user-friendly tools with high sensitivity, which allow direct detection of parasites in diverse human, reservoir and vector tissues, allows species determination specificity and the possibility of having results available within days. In contrast to CD diagnosis, molecular diagnosis for leishmaniasis requires not only confirmation of the infection but also the genotyping of complexes, species or subspecies. In this sense, the characterization of *Leishmania* species through molecular methods is even more advantageous in areas where different pathogens overlap or where the disease has been recently established, has a complex travel history or influx of people and serological results are negative or equivocal [169, 171, 182 - 186]. Moreover, bloodstream based PCR is not only beneficial but it decreases the need to use invasive sample collection such as bone marrow, lymph node, or spleen aspirates. It has also been used for diagnosis and post-treatment evaluation of VL, with a range of sensitivities of 55 to 100% [187 - 190].

However, numerous challenges remain to be transposed in order to standardize PCR assays for worldwide applicability, including the conditions of specimen collection, preservation, and transportation. Also, culture-positive specimens may give negative PCR results due to low and/or non-uniform distribution of parasites in the starting specimen used as a template. Therefore, the clinical benefit of available molecular tests remains to be demonstrated [1]. Multicenter studies are yet necessary to establish standardized operative procedures that could be applied to the different settings of leishmaniasis syndromes.

Several PCR-based assays were developed for diagnostic of VL and CL from samples collected from tissues, blood and body secretions. While these assays had significantly contributed to improve parasite species identification and differentiation, they have not yet been validated for routine use in diagnostic laboratories [169]. Although some approaches allow for the identification of *Leishmania* at the genus level [191,192] they might not provide enough specificity to differentiate parasite species. Others were designed for a particular species complex characterization [193,194], and thus, can only be applied in restricted geographical areas. On the other hand, some assays involve the combination of PCR amplification followed by DNA probe hybridization to enhance sensitivity [195 - 197], which are cumbersome and time-consuming procedures. Species identification techniques are based on qPCR amplification and the comparison of melting curves with reference values [198, 199]. However, because different species have the same melting point, this method is not always reliable to unequivocally identify all species and thus, to make the best therapeutic decision [199]. An alternative strategy is to perform a multiplex PCR, including independent amplification reactions with specific primer/probe combinations [200].

Molecular Markers Proposed for Diagnosis and Genotyping

Several nuclear and kinetoplast DNA sequences have been evaluated as targets for amplification, such as the small subunit rRNA (SSU or 18S) gene [201 - 203], miniexon genes [204, 205], repetitive sequences [206], internal transcribed spacer (ITS) regions [207, 208], the tubulin gene [209], the hsp 70 gene [210, 211], the gp63 gene locus [212], microsatellite DNA [213], and the repetitive kinetoplast DNA (kDNA) minicircles [193, 214]. Examples of some of these targets are given below and in Table **3**.

Minicircle DNA : One of the first PCR techniques developed for the diagnosis of leishmaniasis was based on the conserved region of 120-150 bp of highly repetitive minicircle sequences (800 bp) found in the kinetoplast. This PCR approach is sensitive to detect the presence of *Leishmania* spp, but lacks specificity to directly distinguish between subgenera and species. Species typing can be achieved *via* subsequent hybridization of the PCR products using specific molecular probes [160, 195]. Later on, several kDNA-based PCR were developed to target different species and were optimized according to the sample type (biopsy specimens, cultures, dog tissue, vectors). Among widely used examples, the marker was developed for detection of the *Leishmania donovani* complex [215] and *L. infantum chagasi* in sandflies [216] and was later optimized for multiple species detection [217, 218].

Ribosomal Intergenic Spacers : Within the *Leishmania* spp. genome, there are an estimated 20 to 200 identical copies of the ribosomal intergenic spacer 1 (ITS1) making it a good target for analyzing low parasite quantities [158]. A PCR assay based on rDNA-ITS1 was described for species identification and the discrimination of *L. aethiopica, L. tropica, L. major*, and the *L. donovani* complex [158]. That study considered the ITS1 sequences of 24 isolates, used to validate the four species. Remarkably, PCR amplification of this target allows discrimination of the major complexes under the same cycling conditions. Moreover, the addition of an internal amplification control allows the distinction of false negative results, and was capable of detecting 0.1 pg of leishmanial DNA and was selective for species identification [158].

ITS1 has also has been employed for species discrimination using restriction fragment length polymorphisms [207, 219 - 221], reverse hybridization assays [222], direct sequencing, nucleotide divergence comparison [223] and melting curve analysis [224, 225].

De Almeida and coworkers designed an assay that allows the detection and identification of multiple *Leishmania* species based on the amplification and sequencing analysis of a fragment of the rRNA internal transcribed spacer 2 (ITS2) [157]. The method was proposed as a diagnostic tool for reference laboratories with adequate infrastructure to perform molecular characterization of pathogens. Authors validated the technique with clinical specimens and the comparison with *in vitro* culture, including typing with isoenzyme analysis. Eight *Leishmania* species were identified among 159 patients that resulted positive by both approaches.

18s Ribosomal RNA Gene: A PCR-enzyme-linked immunosorbent assay (ELISA) was evaluated in bone marrow aspirates (BMA) and blood samples (BS) to VL in human patients coinfected with HIV. The assay was set-up using oligonucleotides targeting the 18S rRNA gene as primers and its performance was compared with that of direct observation and *in vitro* culturing [226]. All 79 specimens (33 BMA and 46 BS) collected from 77 human subjects without VL had negative results. Fifty-three specimens (9 BMA and 44 BS) were collected from 13 patients with VL: 6 samples obtained during treatment tested negative with all used techniques, and 47 (9 BMA and 38 BS) were positive with at least one methodology. Sensitivities of direct examination, culture and PCR were 51%, 81% and 98%, respectively. In sum, PCR-ELISA was more sensitive than the other assayed techniques for diagnosing VL in HIV-coinfected patients, and blood sampling was better for follow-up studies [226].

Cytochrome B Gene: Among *Leishmania* species the cytochrome b (cyt b) gene exhibits extensive polymorphism, (including 245 variable nucleotide positions and 190 parsimony informative sites), and thus sequence analysis of this gene has the potential to distinguish among the most common 13 human-pathogenic species causing leishmaniasis in both the New and Old Worlds [227]. Cyt *b* sequencing was successfully used to identify species in Argentina [228] and

Pakistan [229]. A diagnostic algorithm was set up, including an initial diagnostic step based on real-time amplification of the 18S rRNA gene region, and the species identification step was achieved based on cyt b gene sequencing of the 18S rDNA positive samples. This method allows rapid and direct species identification in the clinical sample avoiding the need for parasite isolation. A nested PCR of cyt b was later developed to increase sensitivity and allow the processing of biological samples without the need for culture [218].

Miniexon Gene: This nuclear gene is represented in 100 to 200 tandemly arrayed repeats, which makes it useful as genotyping marker [204, 205]. Each repeat consists of a transcribed region comprising a highly conserved 39-nucleotide exon; a moderately conserved intron, approximately 55 to 101 bp; and a non-transcribed, intergenic spacer of 51 to 1,350 bp depending on the genus and species [230]. A single-step multiplex PCR assay targeted to the mini-exon gene repeats was developed to differentiate the three complexes of New World *Leishmania* (*L. (V.) braziliensis, L. mexicana,* and *L. donovani*) [231]. The assay correctly identifies broadly distributed species in Latin America. It is unique to detect *Leishmania* genus without amplifying related kinetoplastid protozoa, such as *T. cruzi, T. brucei,* and *Crithidia fasciculata.* Depending on the complex analyzed, the sensitivity of the reaction ranged from 1 fg to 10 pg of DNA (0.01 to 100 parasites) [231].

The optimization of the miniexon PCR assay for parasite detection in naturally infected clinical samples allows for routine diagnosis of leishmaniasis [169, 232]. The genotyping capability of the miniexon RFLP system was validated in parallel reactions with different typing schemes, based on an uncharacterized repetitive genomic sequence of Old World *Leishmania* [206] and the ITS [169].

Gp63 Gene: The *Leishmania* glycoprotein GP63, the major surface protease of the promastigote, is a metalloprotease with a role in parasite infectivity, survival and virulence [233]. It is encoded by tandemly repeated genes, from 6 described in *L.* (*L.*) *major* [234] up to 71 in *L.* (*V.*) *braziliensis* [235]. The use of the gp63 locus has been explored as a target for detection and genetic characterization of *Leishmania* because it is repetitive and polymorphic at both specific and intraspecific levels [235 - 238]. PCR-RFLP of gp63 loci was optimized to discriminate the first four *Viannia* species *L. braziliensis, L. peruviana, L. guyanensis*, *and L. lainsoni* [212] and for intraspecific distinction within *L. peruviana*. This approach revealed more polymorphism than isoenzyme analysis (of 13 different enzymes), as every stock analyzed represented a distinct genotype, whereas several of these stocks showed the same isoenzyme profile. It also represented a substantial operational improvement, combining sensitivity and discrimination of species associated with different clinical patterns of the disease and therefore may have prognostic value [186]. More recently, a highly accurate and sensitive rt-PCR assay targeting gp63 was developed to detect and quantify *Leishmania* parasites [239]. In a single assay, the authors were able to detect and quantify several species of *Leishmania* with diverse disease symptomatology with a detection level down to 0.1 parasites (*i.e.*, for *L. donovani* and *L. major*). Remarkably, the assay allowed for parasite load quantification without overt lesions and revealed a systemic distribution of *Leishmania*. Additionally experiments in CBA/CaJ mice, which are resistant to *L. major* the assay allowed to follow the course of the cutaneous disease. Moreover, sequencing analysis of the gp63 gene was later applied to compare evolutive and phylogenetic traits of different species and strains, genotype strains of *L. donovani* complex and characterize polymorphisms for particular strains [240].

Markers combination strategy: A remarkable example of pipeline work, in which several techniques are combined to achieve reliable diagnosis and *Leishmania* species typing is the approach developed by Dr. Kato and collaborators [217, 218, 241 - 243]. The method was designed to easily process a large number of samples, and thus provides a powerful tool for rapidly identifying prevalent *Leishmania* species as well as for monitoring the infection rate in hosts and sand fly populations in leishmaniasis-endemic areas. For parasite detection, PCR primers (Table **3**) were designed to amplify a 700 bp DNA fragment of the *Leishmania* minicircle kinetoplast conserved region among species. The assay was determined to detect 0.1−1 parasites from all the *Leishmania* species evaluated [217], a sensitivity determined to be sufficiently high to detect *Leishmania* promastigotes within individual sand flies. Species identification is achieved by sequence analysis of the cyt b gene [227].

Additionally, the authors evaluated the performance of the FTA card (Whatman) for its potential as a molecular epidemiology device to analyze samples in Peru and Venezuela [243, 244]. This approach allows fast DNA extraction from clinical samples, avoiding the use of organic solvents and specialized infrastructure. The samples are directly aspirated or scraped from patients' lesions and dotted onto the cards. After cell lysis, the nucleic acids are fixed and thus protected from nuclease, oxidative, and UV damage and growth of contaminating microorganisms is prevented [244]. The cards are ideal for long-term storage and shipment at room temperature. DNA fixed is amplified by nested PCR targeting the cyt b gene, and the amplification product is sequenced [218].

The card has various merits: 1) permits the collection of samples directly from patients' lesions; 2) is less invasive for patients, substantially reducing pain and suffering associated with obtaining skin-biopsy specimens; 3) is easy to handle for both physicians and other medical personnel to handle; 4) allows for easy intra- and/or intercountry transportation for analysis; 5) allows the study of different forms of leishmanial syndromes and 6) reduces the possibility of contamination of *Leishmania* isolates during short- and/or long-term passages of culture *in vitro*. Further research is needed to implement such technology in different epidemiological settings.

Several studies are underway attempting to standardize the criteria for *Leishmania* spp. detection and typing, and thereby to compare the performance of various previously described markers in a variety of epidemiological scenarios and using different sample types [245 - 248]. A comprehensive literature review (n=40 pre-selected publications according to previously defined criteria) was performed to compare available diagnostic methods and determine their relative efficiency (sensitivity and specificity) for diagnosing VL, considering different infecting *Leishmania* spp, the geographic regions where the studies were performed, and the types of study designs. The authors found that sensitivities ranged from 29 to 100%, and specificities ranged from 25 to 100% [248], but their analysis did not detect statistically significant differences among sensitivity and specificity according to the type of samples (blood *versus* bone marrow) or the molecular method used. However, this assessment was done comparing the molecular method *versus* parasitologic or serologic detection. Therefore the performance of molecular methods per se was not evaluated. Nevertheless, the authors were able to detect substantial heterogeneity among different epidemiologic contexts and protocols used in various research facilities, concluding that follow-up confirmatory molecular testing should be conducted in most cases [248].

Although variable results are obtained according to the strain/geographical area/sample type, it seems that the combination of kDNA for detection and DNA sequencing analysis of the Cyt b, mini-exon or heat shock protein 70 genes for strain typing, are among the most sensitive and specific methods for diagnosis and strain typing directly from biological samples [218, 228, 246, 249].

Table 3. Molecular markers currently used for detection and identification of *Leishmania* spp.

Marker	Reference	Marker type	Primer Sequences (5'- 3')	Applications	
Minicircle	[250]	PCR	D1: CCAGTTTCCCGCCCCG D2: GGGGTTGGTGGTGTAAAATAG	detection *Leishmania* spp	Culture
	[217]	PCR	L.MC-1S: CTRGGGGTTGGTGTAAAATAG L.MC-1R: TWTGAACGGGRTTTCTG	detection *Leishmania* spp	Human/sand flies
	[160, 182, 195]	PCR/ hibridization	150 F: GGGKAGGGGCGTTCT SCGAA 152 R: SSSWCTATWTTACACCAACCCC	detection *Leishmania* spp/typing	Human samples, dogs, vector
	[251]	PCR/ hibridization	13AF: GTGGGGGAGGGGCGTTCT 13BR: ATTTTA CACCAACCCCCAGTT	*Leishmania* spp detection + sps ID	Human/vector/dogs
Miniexon	[204]	PCR/seq	S1629: GGAATTCAATAWAGTACAGAAACTG S1630: GGGAAGCTTCTGTACTWTATTGGTA	*Leishmania* sub genus typing	Culture
	[169, 232]	PCR/RFLP	Fme: TATTGGTATGCGAAACTTCCG Rme: ACAGAAACTGATACTTATATAGCG ez: EaeI ; Eco72 ; HaeIII ; Mbo I ; MluI ; NcoI ; Not I; RsaI; SmaI	*Leishmania* sp ID	Culture/ clinical samples
18S	[201]		R221: GGTTCCTTTCCTTGATTTAGC R332: GGCCGGTAAAGGCCGAATAG	*Leishmania* spp detection	Culture/samples (vectors)
	[171]	rtPCR	LEI-18S1: ACTCACGGCCTCTAGGAATGA LEI-18S2: TCGATCTCCACACTTTGGTTCT LEI-18S3: AGAATTTCACCTCTGACGCCCCAGT (5'LCRed640-3 Ph) LEI-18S4: GCTGTAGTTCGTCTTGGTGCGGTCT-(3'FITC)	*Leishmania* sp ID	Clinical specimens (blood, bone marrow, skin biopsy)

(Table 3) contd.....

Marker	Reference	Marker type	Primer Sequences (5'- 3')	Applications	
cytB	[171]	PCR/seq	LEI-CYTB9: TTATGGTGTAGGTTTTAGTYTAGGTT LEI-CYTB10: CCATCCGAACTCATAAAATAATGT LEI-CYTB11: TTTGTTATTGAATWTGAGGWAGTGA LEI-CYTB12: TGCTAAAAAACCACTCATAAATATACT	*Leishmania* sp ID	Clinical specimens (blood, bone marrow, skin biopsy)
	[218, 244]	PCR nested sequence (FTA card)	L.cyt-AS: GCGGAGAGRARGAAAAGGC L.cyt-AR: CCACTCATAAATATACTATA inner primers (L.cyt-S: GGTGTAGGTTTTAGTYTAGG L.cyt-R: CTACAATAAACAAATCATAATATRCAATT	CL *Leishmania* sp ID	Skin samples
	[227]	PCR/seq	L.cyt-S: GGTGTAGGTTTTAGTYTAGG L.cyt-R: CTACAATAAACAAATCATAATATRCAATT	*Leishmania* sp ID / phylogenetics	Culture/clinical samples
GP63	[212, 235]	PCR/RFLP	TDM1: GTCTCCACCGCAGACCTCACGGA TDM2: TGATGTAGCTGCCATTCACGAAG(TDM2),	*Leishmania* sp ID	Culture
rRNA ITS1	[252]	PCR/seq PCR/RFLP	LITS1: CTGGATCATTTTCCGATG 5.8s: TGATACCACTTATCGCACTT	*Leishmania* sp ID	Clinical samples
rRNA ITS2	[157]	PCR/seq	LGITSF2: GCATGCCATATTCTCAGTGTC LGITSR2: GGCCAACGCGAAGTTGAATTC LGITS2R1: GAATTCTCGTTTTGGTTTTTTG	*Leishmania* sub genus typing	Culture/ clinical samples

A summary of widely used molecular diagnostic and *Leishmania* spp typing markers. References: **Minicircle:** kinetoplast minicircle DNA (kDNA); **miniexon:** non-transcribed spacer of the mini-exon gene; **18S:** small subunit ribosomal RNA (SSU-rRNA) gene; **Cyt b:** Cytochrome B gene; **GP63: Rodgers** glycoprotein GP63; **rRNA ITS1 & ITS2:** Ribosomal intergenic transcribed Spacers 1 and 2; **PCR/seq**: amplification through PCR and posterior sequencing of the amplicon. Species are typified through sequence comparison or blast. **PCR/RFLP**: amplification through PCR and posterior enzyme digestion of the amplicon. Species are typified through fragment pattern polymorphism.

Novel Technologies with Potential Application in Diagnosis of CD and Leishmaniasis

An alternative to PCR for molecular diagnosis of *T. cruzi* and *Leishmania* spp infections is the Loop-mediated isothermal amplification (LAMP). Through an intricate design of primers, auto-strand displacement DNA synthesis and Bst DNA polymerase, this technique allows the amplification of large amounts of DNA within 30-60 minutes. Among the advantages of this promising technology, the reactions occur between 60 and 65°C, so they can be completed without the use of a thermocycler [253]. The amplification product can be visualized directly through simple detection methods such as photometry [254], which allows easy visualization by the naked eye. Moreover, the assays are highly specific as the design includes the use of six primers. The reaction can also be followed in real-time either by measuring the turbidity or by fluorescence using intercalating dyes. In-tube detection of DNA amplification is possible using manganese loaded calcein which starts fluorescing upon complexation of manganese by pyrophosphate during *in vitro* DNA synthesis. An additional advantage of this methodology relies on the fact that LAMP reagents remain stable at room temperatures up to 37 °C, without the constraint of keeping the cold chain. Thus far, LAMP has been implemented for protozoan parasitic diseases such as *Plasmodium* and *Trypanosoma* with high efficacy and has become a powerful tool for point-of-care diagnostics.

As a global health initiative, the Foundation for Innovative New Diagnostics (FIND) has embarked on discovery and development program for new assays dedicated to the diagnosis of neglected tropical diseases, making use of LAMP methodology in an attempt to translate the technologies into user-friendly formats for endemic areas [255, 256]. Recently, several projects have been initiated, aiming to develop and validating LAMP procedures for diagnosis of CD envisioned for point-of-care settings. A prototype kit has been developed by Eiken Company and is currently under validation for its application in early detection of congenital CD (http://www.finddiagnostics.org/programs/hat-ond/chagas/ lamp-for-chagas.html; Schijman AG, unpublished results).

Final Remarks

Assessment of the biological value of PCR results regarding the viability of parasites has been poorly explored [257]. Studies are underway to assess the negative predictive value of PCR to evaluate cure in CD (Isabela Ribeiro, unpublished results). While the treatment goal for infectious diseases is or should be pathogen elimination, there are other equally important therapeutic outcomes to be considered [258].

In the case of American cutaneous leishmaniasis, the clinical cure is rarely associated with sterile cure. DNA specific for *Leishmania* (*Viannia*) was detected in scars of 30 (93.7%) of 32 treated and clinically cured patients. In specimens from 3 of the scars, *Leishmania* parasites could be isolated by culture [259]. The implications of the persistence of parasites for the clinical evolution, relapse, and transmission of leishmaniasis deserve further study. Control and reduction of the pathogen burden are well-recognized strategies for some infections, such as AIDS, to convert a fatal disease into a chronically controlled disease with the administration of appropriate treatment.

Although molecular based techniques are proven useful in research laboratories, reference health care centers and clinical trial settings [59, 152, 153], it was difficult to envision their application, in their present forms, in the monitoring of large populations in the field.

Prospective studies must still be done to determine the optimal algorithm for molecular diagnosis of these infections in different epidemiological and/or clinical scenarios. A necessary next step is the development of commercial kits based on standardized methods. Developing new diagnostic tools adapted to the needs of affected populations and the particular situations of the health systems in different countries will significantly improve the ability to control CD and leishmaniasis [1]. The elimination of these endemic parasitic diseases will require long-term commitments and multidimensional strategies.

CONFLICT OF INTEREST

The authors confirm that they have no conflict of interest to declare for this publication.

ACKNOWLEDGEMENTS

Schijman A.G. and Burgos J.M. are members of the "Carrera del Investigador Científico", from the Consejo Nacional de Investigaciones Científicas y Técnicas de Argentina (CONICET). The conclusions given in this chapter are those of the authors and do not necessarily represent the views of the Centers for Disease Control and Prevention (CDC).

ABBREVIATIONS

BZN	Benznidazole
CD	Chagas disease
CL	Cutaneous Leishmaniasis
Cyt b	Cytochrome b
DNA	Deoxyribonucleic acid
DNDi	Drugs for Neglected Diseases initiative
DTU	Discrete typing unit
EDTA	Ethylenediaminetetraacetic acid
ELISA	Enzyme-linked immunosorbent assay
EOT	End of treatment
FDA	Food and Drug Administration
HD	High Dose
HIV	Human immunodeficiency virus
IAC	Internal amplification control
IHA	Haemagglutination
IIF	Indirect immunofluorescence
ITS1/ITS2	Ribosomal intergenic spacer 1 and 2
kDNA	Kinetoplast DNA
LAMP	Loop-mediated isothermal amplification
LD	Low dose
MH	Microhaematocrit concentration method
MLEE	Multilocus enzyme electrophoresis
OD	Optical density (serology)
PAHO	Panamerican Health Organization
PCR	Polymerase chain reaction
qPCR	Quantitative PCR (Real-Time)
RDTs	Rapid diagnostic tests
RFLP	Restriction Fragment Length Polymorphism
SatDNA	Satellite DNA
SAPA	Antigen: shed acute phase antigen
SD	Short dose
TDR	Research and Training in Tropical Diseases (WHO)
TPP	Target product profiles

VL	Visceral Leishmaniasis
WHO	World Health Organization

REFERENCES

[1] WHO, Research priorities for Chagas disease, human African trypanosomiasis and Leishmaniasis. World Health Organization technical report series/Geneva, Switzerland: World Health Organization/ Special Programme for Research and Training in Tropical Diseases (TDR), 2012(975):v-xii,1-100.

[2] Schmunis GA, Yadon ZE. Chagas disease: a Latin American health problem becoming a world health problem. Acta Trop 2010; 115(1-2): 14-21. [http://dx.doi.org/10.1016/j.actatropica.2009.11.003] [PMID: 19932071]

[3] Prata A. Clinical and epidemiological aspects of Chagas disease. Lancet Infect Dis 2001; 1(2): 92-100. [http://dx.doi.org/10.1016/S1473-3099(01)00065-2] [PMID: 11871482]

[4] Reithinger R, Tarleton RL, Urbina JA, Kitron U, Gürtler RE. Eliminating Chagas disease: challenges and a roadmap. BMJ 2009; 338: b1283. [http://dx.doi.org/10.1136/bmj.b1283] [PMID: 19366739]

[5] Buscaglia CA, Di Noia JM. *Trypanosoma cruzi* clonal diversity and the epidemiology of Chagas disease. Microbes Infect 2003; 5(5): 419-27. [http://dx.doi.org/10.1016/S1286-4579(03)00050-9] [PMID: 12737998]

[6] Tibayrenc M, Ward P, Moya A, Ayala FJ. Natural populations of *Trypanosoma cruzi*, the agent of Chagas disease, have a complex multiclonal structure. Proc Natl Acad Sci USA 1986; 83(1): 115-9. [http://dx.doi.org/10.1073/pnas.83.1.115] [PMID: 3510428]

[7] Sturm NR, Campbell DA. Alternative lifestyles: the population structure of *Trypanosoma cruzi*. Acta Trop 2010; 115(1-2): 35-43. [http://dx.doi.org/10.1016/j.actatropica.2009.08.018] [PMID: 19695212]

[8] Macedo AM, Machado CR, Oliveira RP, Pena SD. *Trypanosoma cruzi*: genetic structure of populations and relevance of genetic variability to the pathogenesis of chagas disease. Mem Inst Oswaldo Cruz 2004; 99(1): 1-12. [http://dx.doi.org/10.1590/S0074-02762004000100001] [PMID: 15057339]

[9] Pena SD, Machado CR, Macedo AM. *Trypanosoma cruzi*: ancestral genomes and population structure. Mem Inst Oswaldo Cruz 2009; 104 (Suppl. 1): 108-14. [http://dx.doi.org/10.1590/S0074-02762009000900016] [PMID: 19753465]

[10] Tibayrenc M. Integrated genetic epidemiology of infectious diseases: the Chagas model. Mem Inst Oswaldo Cruz 1998; 93(5): 577-80. [http://dx.doi.org/10.1590/S0074-02761998000500003] [PMID: 9830520]

[11] Miles MA, Llewellyn MS, Lewis MD, *et al.* The molecular epidemiology and phylogeography of *Trypanosoma cruzi* and parallel research on *Leishmania*: looking back and to the future. Parasitology 2009; 136(12): 1509-28. [http://dx.doi.org/10.1017/S0031182009990977] [PMID: 19691868]

[12] Ienne S, Pedroso A, Carmona E Ferreira R, Briones MR, Zingales B. Network genealogy of 195-bp

satellite DNA supports the superimposed hybridization hypothesis of *Trypanosoma cruzi* evolutionary pattern. Infect Genet Evol 2010; 10(5): 601-6.
[http://dx.doi.org/10.1016/j.meegid.2010.04.007] [PMID: 20433949]

[13] Lewis MD, Llewellyn MS, Gaunt MW, Yeo M, Carrasco HJ, Miles MA. Flow cytometric analysis and microsatellite genotyping reveal extensive DNA content variation in *Trypanosoma cruzi* populations and expose contrasts between natural and experimental hybrids. Int J Parasitol 2009; 39(12): 1305-17.
[http://dx.doi.org/10.1016/j.ijpara.2009.04.001] [PMID: 19393242]

[14] Telleria J, Lafay B, Virreira M, Barnabé C, Tibayrenc M, Svoboda M. *Trypanosoma cruzi*: sequence analysis of the variable region of kinetoplast minicircles. Exp Parasitol 2006; 114(4): 279-88.
[http://dx.doi.org/10.1016/j.exppara.2006.04.005] [PMID: 16730709]

[15] Vargas N, Pedroso A, Zingales B. Chromosomal polymorphism, gene synteny and genome size in *T. cruzi* I and *T. cruzi* II groups. Mol Biochem Parasitol 2004; 138(1): 131-41.
[http://dx.doi.org/10.1016/j.molbiopara.2004.08.005] [PMID: 15500924]

[16] Burgos JM, Begher S, Silva HM, *et al.* Molecular identification of *Trypanosoma cruzi* I tropism for central nervous system in Chagas reactivation due to AIDS. Am J Trop Med Hyg 2008; 78(2): 294-7.
[PMID: 18256432]

[17] Burgos JM, Diez M, Vigliano C, *et al.* Molecular identification of *Trypanosoma cruzi* discrete typing units in end-stage chronic Chagas heart disease and reactivation after heart transplantation. Clin Infect Dis 2010; 51(5): 485-95.
[http://dx.doi.org/10.1086/655680] [PMID: 20645859]

[18] Avila HA, Pereira JB, Thiemann O, *et al.* Detection of *Trypanosoma cruzi* in blood specimens of chronic chagasic patients by polymerase chain reaction amplification of kinetoplast minicircle DNA: comparison with serology and xenodiagnosis. J Clin Microbiol 1993; 31(9): 2421-6.
[PMID: 8408566]

[19] Britto C, Cardoso MA, Vanni CM, *et al.* Polymerase chain reaction detection of *Trypanosoma cruzi* in human blood samples as a tool for diagnosis and treatment evaluation. Parasitology 1995; 110(Pt 3): 241-7.
[http://dx.doi.org/10.1017/S0031182000080823] [PMID: 7724232]

[20] Moser DR, Kirchhoff LV, Donelson JE. Detection of *Trypanosoma cruzi* by DNA amplification using the polymerase chain reaction. J Clin Microbiol 1989; 27(7): 1477-82.
[PMID: 2504769]

[21] Duffy T, Bisio M, Altcheh J, *et al.* Accurate real-time PCR strategy for monitoring bloodstream parasitic loads in chagas disease patients. PLoS Negl Trop Dis 2009; 3(4): e419.
[http://dx.doi.org/10.1371/journal.pntd.0000419] [PMID: 19381287]

[22] Schijman AG, Bisio M, Orellana L, *et al.* International study to evaluate PCR methods for detection of *Trypanosoma cruzi* DNA in blood samples from Chagas disease patients. PLoS Negl Trop Dis 2011; 5(1): e931.
[http://dx.doi.org/10.1371/journal.pntd.0000931] [PMID: 21264349]

[23] Piron M, Fisa R, Casamitjana N, *et al.* Development of a real-time PCR assay for *Trypanosoma cruzi* detection in blood samples. Acta Trop 2007; 103(3): 195-200.
[http://dx.doi.org/10.1016/j.actatropica.2007.05.019] [PMID: 17662227]

[24] Duffy T, Cura CI, Ramirez JC, *et al.* Analytical performance of a multiplex Real-Time PCR assay using TaqMan probes for quantification of *Trypanosoma cruzi* satellite DNA in blood samples. PLoS Negl Trop Dis 2013; 7(1): e2000.

[25] Ramírez JC, Cura CI, da Cruz Moreira O, *et al.* Analytical validation of quantitative real-time pcr methods for quantification of *Trypanosoma cruzi* DNA in blood samples from chagas disease patients. J Mol Diagn 2015; 17(5): 605-15.
[http://dx.doi.org/10.1016/j.jmoldx.2015.04.010] [PMID: 26320872]

[26] Qvarnstrom Y, Schijman AG, Veron V, *et al.* Sensitive and specific detection of *Trypanosoma cruzi* DNA in clinical specimens using a multi-target real-time PCR approach. PLoS Negl Trop Dis 2012; 6(7): e1689.

[27] Burns MJ, Nixon GJ, Foy CA, Harris N. Standardisation of data from real-time quantitative PCR methods - evaluation of outliers and comparison of calibration curves. BMC Biotechnol 2005; 5: 31.
[http://dx.doi.org/10.1186/1472-6750-5-31] [PMID: 16336641]

[28] Riarte A, Velazquez E, Prado N, *et al.* Traena" tratamiento en pacientes Adultos. Una evaluación preliminar de un ensayo clínico aleatorizado con Benznidazol en la enfermedad de Chagas crónica. VIII Workshop on Imported Chagas Disease. 5 March 2012. Available at: https://www.isglobal.org/documents/10179/25254/VIII+taller+sobre+la+enfermedad+de+Chagas+importada.pdf/c3ab08f9-d161-412e-b6b0-f7e65c7e0f55

[29] Marin-Neto JA, Rassi A Jr, Morillo CA, *et al.* Rationale and design of a randomized placebo-controlled trial assessing the effects of etiologic treatment in Chagas cardiomyopathy: the BENznidazole Evaluation For Interrupting Trypanosomiasis (BENEFIT). Am Heart J 2008; 156(1): 37-43.
[http://dx.doi.org/10.1016/j.ahj.2008.04.001] [PMID: 18585495]

[30] De Rissio AM, Riarte AR, García MM, Esteva MI, Quaglino M, Ruiz AM. Congenital *Trypanosoma cruzi* infection. Efficacy of its monitoring in an urban reference health center in a non-endemic area of Argentina. Am J Trop Med Hyg 2010; 82(5): 838-45.
[http://dx.doi.org/10.4269/ajtmh.2010.08-0383] [PMID: 20439964]

[31] Reyes MB, Lorca M, Muñoz P, Frasch AC. Fetal IgG specificities against *Trypanosoma cruzi* antigens in infected newborns. Proc Natl Acad Sci USA 1990; 87(7): 2846-50.
[http://dx.doi.org/10.1073/pnas.87.7.2846] [PMID: 2108442]

[32] da Silveira JF, Umezawa ES, Luquetti AO. Chagas disease: recombinant *Trypanosoma cruzi* antigens for serological diagnosis. Trends Parasitol 2001; 17(6): 286-91.
[http://dx.doi.org/10.1016/S1471-4922(01)01897-9] [PMID: 11378036]

[33] Yoshida N. *Trypanosoma cruzi* infection by oral route: how the interplay between parasite and host components modulates infectivity. Parasitol Int 2008; 57(2): 105-9.
[http://dx.doi.org/10.1016/j.parint.2007.12.008] [PMID: 18234547]

[34] Benchimol-Barbosa PR. Predictors of mortality in Chagas disease: the impact of atrial fibrillation and oral transmission on infected population. Int J Cardiol 2009; 133(2): 275-7.
[http://dx.doi.org/10.1016/j.ijcard.2007.11.053] [PMID: 18199497]

[35] Alarcón de Noya B, Díaz-Bello Z, Colmenares C, *et al.* Large urban outbreak of orally acquired acute

Chagas disease at a school in Caracas, Venezuela. J Infect Dis 2010; 201(9): 1308-15. [http://dx.doi.org/10.1086/651608] [PMID: 20307205]

[36] Blanchet D, Brenière SF, Schijman AG, *et al.* First report of a family outbreak of Chagas disease in French Guiana and posttreatment follow-up. Infect Genet Evol 2014; 28: 245-50. [http://dx.doi.org/10.1016/j.meegid.2014.10.004] [PMID: 25448161]

[37] Guhl F, Vallejo GA. *Trypanosoma* (Herpetosoma) *rangeli* Tejera, 1920: an updated review. Mem Inst Oswaldo Cruz 2003; 98(4): 435-42. [http://dx.doi.org/10.1590/S0074-02762003000400001] [PMID: 12937750]

[38] Casadei D. Chagas disease and solid organ transplantation. Transplant Proc 2010; 42(9): 3354-9. [http://dx.doi.org/10.1016/j.transproceed.2010.09.019] [PMID: 21094779]

[39] Pinazo MJ, Miranda B, Rodríguez-Villar C, *et al.* Recommendations for management of Chagas disease in organ and hematopoietic tissue transplantation programs in nonendemic areas. Transplant Rev (Orlando) 2011; 25(3): 91-101. [http://dx.doi.org/10.1016/j.trre.2010.12.002] [PMID: 21530219]

[40] McCormack L, Quiñónez E, Goldaracena N, *et al.* Liver transplantation using Chagas-infected donors in uninfected recipients: a single-center experience without prophylactic therapy. Am J Transplant 2012; 12(10): 2832-7. [http://dx.doi.org/10.1111/j.1600-6143.2012.04160.x] [PMID: 22813351]

[41] Cura CI, Lattes R, Nagel C, *et al.* Early molecular diagnosis of acute Chagas disease after transplantation with organs from *Trypanosoma cruzi*-infected donors. Am J Transplant 2013; 13(12): 3253-61. [http://dx.doi.org/10.1111/ajt.12487] [PMID: 24266974]

[42] Riarte A, Luna C, Sabatiello R, *et al.* Chagas disease in patients with kidney transplants: 7 years of experience 19891996. Clin Infect Dis 1999; 29(3): 561-7. [http://dx.doi.org/10.1086/598634] [PMID: 10530448]

[43] Pérez-López FR, Chedraui P. Chagas disease in pregnancy: a non-endemic problem in a globalized world. Arch Gynecol Obstet 2010; 282(6): 595-9. [http://dx.doi.org/10.1007/s00404-010-1553-7] [PMID: 20559648]

[44] Bua J, Volta BJ, Perrone AE, *et al.* How to improve the early diagnosis of *Trypanosoma cruzi* infection: relationship between validated conventional diagnosis and quantitative DNA amplification in congenitally infected children. PLoS Negl Trop Dis 2013; 7(10): e2476. [http://dx.doi.org/10.1371/journal.pntd.0002476] [PMID: 24147166]

[45] Freilij H, Altcheh J. Congenital Chagas disease: diagnostic and clinical aspects. Clin Infect Dis 1995; 21(3): 551-5. [http://dx.doi.org/10.1093/clinids/21.3.551] [PMID: 8527542]

[46] Schijman AG, Altcheh J, Burgos JM, *et al.* Aetiological treatment of congenital Chagas disease diagnosed and monitored by the polymerase chain reaction. J Antimicrob Chemother 2003; 52(3): 441-9. [http://dx.doi.org/10.1093/jac/dkg338] [PMID: 12917253]

[47] Mora MC, Sanchez Negrette O, Marco D, *et al.* Early diagnosis of congenital *Trypanosoma cruzi*

infection using PCR, hemoculture, and capillary concentration, as compared with delayed serology. J Parasitol 2005; 91(6): 1468-73.
[http://dx.doi.org/10.1645/GE-549R.1] [PMID: 16539033]

[48] Schenone H, Xenodiagnosis SH. Mem Inst Oswaldo Cruz 1999; 94: 289-94.

[49] dos Santos AH, da Silva IG, Rassi A. A comparative study between natural and artificial xenodiagnosis in chronic Chagas disease patients. Rev Soc Bras Med Trop 1995; 28(4): 367-73.
[PMID: 8668837]

[50] Blanco SB, Segura EL, Cura EN, *et al.* Congenital transmission of *Trypanosoma cruzi*: an operational outline for detecting and treating infected infants in north-western Argentina. Trop Med Int Health 2000; 5(4): 293-301.
[http://dx.doi.org/10.1046/j.1365-3156.2000.00548.x] [PMID: 10810029]

[51] Volta BJ, Russomando G, Bustos PL, *et al.* Diagnosis of congenital *Trypanosoma cruzi* infection: A serologic test using Shed Acute Phase Antigen (SAPA) in mother-child binomial samples. Acta Trop 2015; 147: 31-7.
[http://dx.doi.org/10.1016/j.actatropica.2015.03.026] [PMID: 25847262]

[52] Virreira M, Torrico F, Truyens C, *et al.* Comparison of polymerase chain reaction methods for reliable and easy detection of congenital *Trypanosoma cruzi* infection. Am J Trop Med Hyg 2003; 68(5): 574-82.
[PMID: 12812349]

[53] Russomando G, Figueredo A, Almirón M, Sakamoto M, Morita K. Polymerase chain reaction-based detection of *Trypanosoma cruzi* DNA in serum. J Clin Microbiol 1992; 30(11): 2864-8.
[PMID: 1452655]

[54] Nilo ME, Alvarado J, Ramirez M, *et al.* Hallazgo de tripomastigoto en estudio citoquimico de liquido amniotico. Parasitología al día 2000; 24(1/2): 49-51.
[http://dx.doi.org/10.4067/S0716-07202000000100008]

[55] Sturm NR, Degrave W, Morel C, Simpson L. Sensitive detection and schizodeme classification of *Trypanosoma cruzi* cells by amplification of kinetoplast minicircle DNA sequences: use in diagnosis of Chagas disease. Mol Biochem Parasitol 1989; 33(3): 205-14.
[http://dx.doi.org/10.1016/0166-6851(89)90082-0] [PMID: 2565018]

[56] Avila HA, Sigman DS, Cohen LM, Millikan RC, Simpson L. Polymerase chain reaction amplification of *Trypanosoma cruzi* kinetoplast minicircle DNA isolated from whole blood lysates: diagnosis of chronic Chagas disease. Mol Biochem Parasitol 1991; 48(2): 211-21.
[http://dx.doi.org/10.1016/0166-6851(91)90116-N] [PMID: 1662334]

[57] Wincker P, Telleria J, Bosseno MF, *et al.* PCR-based diagnosis for Chagas disease in Bolivian children living in an active transmission area: comparison with conventional serological and parasitological diagnosis. Parasitology 1997; 114(Pt 4): 367-73.
[http://dx.doi.org/10.1017/S0031182096008554] [PMID: 9107023]

[58] Russomando G, de Tomassone MM, de Guillen I, *et al.* Treatment of congenital Chagas disease diagnosed and followed up by the polymerase chain reaction. Am J Trop Med Hyg 1998; 59(3): 487-91.
[PMID: 9749649]

[59] Molina I, Gómez i Prat J, Salvador F, *et al.* Randomized trial of posaconazole and benznidazole for chronic Chagas disease. N Engl J Med 2014; 370(20): 1899-908. [http://dx.doi.org/10.1056/NEJMoa1313122] [PMID: 24827034]

[60] Braz LM, Raiz-Júnior R, Alárcon RS, Gakiya E, Amato-Neto V, Okay TS. Suitability of a rapid DNA isolation and amplification for detection of *Trypanosoma cruzi* in Triatoma infestans dry fecal spots collected on filter paper. Parasite 2008; 15(4): 595-8. [http://dx.doi.org/10.1051/parasite/2008154595] [PMID: 19202767]

[61] NCCLS, protocols for determination of limits of detection and limits of quantification: approved guideline. Wayne, Pennsylvania. 2004.

[62] NCCLS, evaluation of precision performance of quantitative measurement methods: approved guideline-second edition. Wayne, Pennsylvania. 2004.

[63] Burd EM. Validation of laboratory-developed molecular assays for infectious diseases. Clin Microbiol Rev 2010; 23(3): 550-76. [http://dx.doi.org/10.1128/CMR.00074-09] [PMID: 20610823]

[64] Control of Chagas Disease: Second report of the WHO Expert Comitee. WHO Technical Report Series. Geneva: World Health Organization 2002; p. 109.

[65] Moncayo A, Ortiz Yanine MI. An update on Chagas disease (human American trypanosomiasis). Ann Trop Med Parasitol 2006; 100(8): 663-77. [http://dx.doi.org/10.1179/136485906X112248] [PMID: 17227647]

[66] Silveira-Lacerda EP, Silva AG, Junior SF, *et al.* Chagas disease: application of TESA-blot in inconclusive sera from a Brazilian blood bank. Vox Sang 2004; 87(3): 204-7. [http://dx.doi.org/10.1111/j.1423-0410.2004.00571.x] [PMID: 15569074]

[67] Caballero ZC, Sousa OE, Marques WP, Saez-Alquezar A, Umezawa ES. Evaluation of serological tests to identify *Trypanosoma cruzi* infection in humans and determine cross-reactivity with *Trypanosoma rangeli* and *Leishmania* spp. Clin Vaccine Immunol 2007; 14(8): 1045-9. [http://dx.doi.org/10.1128/CVI.00127-07] [PMID: 17522327]

[68] Salomone OA, Basquiera AL, Sembaj A, *et al. Trypanosoma cruzi* in persons without serologic evidence of disease, Argentina. Emerg Infect Dis 2003; 9(12): 1558-62. [http://dx.doi.org/10.3201/eid0912.030008] [PMID: 14720396]

[69] Umezawa ES, Shikanai-Yasuda MA, Stolf AM. Changes in isotype composition and antigen recognition of anti-*Trypanosoma cruzi* antibodies from acute to chronic Chagas disease. J Clin Lab Anal 1996; 10(6): 407-13. [http://dx.doi.org/10.1002/(SICI)1098-2825(1996)10:6<407::AID-JCLA16>3.0.CO;2-0] [PMID: 8951611]

[70] Frank FM, Fernández MM, Taranto NJ, *et al.* Characterization of human infection by *Leishmania* spp. in the Northwest of Argentina: immune response, double infection with *Trypanosoma cruzi* and species of *Leishmania* involved. Parasitology 2003; 126(Pt 1): 31-9. [http://dx.doi.org/10.1017/S0031182002002585] [PMID: 12613761]

[71] Saldaña A, Samudio F, Miranda A, *et al.* Predominance of *Trypanosoma rangeli* infection in children from a Chagas disease endemic area in the west-shore of the Panama canal. Mem Inst Oswaldo Cruz

2005; 100(7): 729-31.
[http://dx.doi.org/10.1590/S0074-02762005000700008] [PMID: 16410959]

[72] Tobler LH, Contestable P, Pitina L, *et al.* Evaluation of a new enzyme-linked immunosorbent assay for detection of Chagas antibody in US blood donors. Transfusion 2007; 47(1): 90-6.
[http://dx.doi.org/10.1111/j.1537-2995.2007.01068.x] [PMID: 17207235]

[73] Frasch AC. Trans-sialidase, SAPA amino acid repeats and the relationship between *Trypanosoma cruzi* and the mammalian host. Parasitology 1994; 108(108) (Suppl.): S37-44.
[http://dx.doi.org/10.1017/S0031182000075703] [PMID: 8084653]

[74] Umezawa ES, Luquetti AO, Levitus G, *et al.* Serodiagnosis of chronic and acute Chagas disease with *Trypanosoma cruzi* recombinant proteins: results of a collaborative study in six Latin American countries. J Clin Microbiol 2004; 42(1): 449-52.
[http://dx.doi.org/10.1128/JCM.42.1.449-452.2004] [PMID: 14715803]

[75] Longhi SA, Brandariz SB, Lafon SO, *et al.* Evaluation of in-house ELISA using *Trypanosoma cruzi* lysate and recombinant antigens for diagnosis of Chagas disease and discrimination of its clinical forms. Am J Trop Med Hyg 2012; 87(2): 267-71.
[http://dx.doi.org/10.4269/ajtmh.2012.11-0533] [PMID: 22855757]

[76] Villagrán-Herrera ME, Sánchez-Moreno M, Rodríguez-Méndez AJ, *et al.* Comparative serology techniques for the diagnosis of *Trypanosoma cruzi* infection in a rural population from the state of Querétaro, Mexico. Mem Inst Oswaldo Cruz 2014; 109(7): 967-72.
[http://dx.doi.org/10.1590/0074-0276130413] [PMID: 25411005]

[77] Vega Benedetti AF, Cimino RO, Cajal PS, *et al.* Performance of different *Trypanosoma cruzi* antigens in the diagnosis of Chagas disease in patients with American cutaneous leishmaniasis from a co-endemic region in Argentina. Trop Med Int Health 2013; 18(9): 1103-9.
[http://dx.doi.org/10.1111/tmi.12144] [PMID: 23837449]

[78] Flechas ID, Cuellar A, Cucunubá ZM, *et al.* Characterising the KMP-11 and HSP-70 recombinant antigens humoral immune response profile in chagasic patients. BMC Infect Dis 2009; 9: 186.
[http://dx.doi.org/10.1186/1471-2334-9-186] [PMID: 19939275]

[79] Vasconcelos RH, Amaral FN, Cavalcanti MG, *et al.* Increased levels of IgA antibodies against CRA and FRA recombinant antigens of *Trypanosoma cruzi* differentiate digestive forms of Chagas disease. Hum Immunol 2010; 71(10): 964-7.
[http://dx.doi.org/10.1016/j.humimm.2010.07.004] [PMID: 20659514]

[80] Verani JR, Seitz A, Gilman RH, *et al.* Geographic variation in the sensitivity of recombinant antigen-based rapid tests for chronic *Trypanosoma cruzi* infection. Am J Trop Med Hyg 2009; 80(3): 410-5.
[PMID: 19270291]

[81] Otani MM, Vinelli E, Kirchhoff LV, *et al.* WHO comparative evaluation of serologic assays for Chagas disease. Transfusion 2009; 49(6): 1076-82.
[http://dx.doi.org/10.1111/j.1537-2995.2009.02107.x] [PMID: 19290995]

[82] Goto Y, Carter D, Reed SG. Immunological dominance of *Trypanosoma cruzi* tandem repeat proteins. Infect Immun 2008; 76(9): 3967-74.
[http://dx.doi.org/10.1128/IAI.00604-08] [PMID: 18625739]

[83] Schofield L. On the function of repetitive domains in protein antigens of *Plasmodium* and other eukaryotic parasites. Parasitol Today (Regul Ed) 1991; 7(5): 99-105. [http://dx.doi.org/10.1016/0169-4758(91)90166-L] [PMID: 15463458]

[84] Alvarez P, Leguizamón MS, Buscaglia CA, Pitcovsky TA, Campetella O. Multiple overlapping epitopes in the repetitive unit of the shed acute-phase antigen from *Trypanosoma cruzi* enhance its immunogenic properties. Infect Immun 2001; 69(12): 7946-9. [http://dx.doi.org/10.1128/IAI.69.12.7946-7949.2001] [PMID: 11705983]

[85] Cazzulo JJ, Frasch AC. SAPA/*trans*-sialidase and cruzipain: two antigens from *Trypanosoma cruzi* contain immunodominant but enzymatically inactive domains. FASEB J 1992; 6(14): 3259-64. [PMID: 1426764]

[86] Buscaglia CA, Alfonso J, Campetella O, Frasch AC. Tandem amino acid repeats from *Trypanosoma cruzi* shed antigens increase the half-life of proteins in blood. Blood 1999; 93(6): 2025-32. [PMID: 10068675]

[87] Pitcovsky TA, Buscaglia CA, Mucci J, Campetella O. A functional network of intramolecular cross-reacting epitopes delays the elicitation of neutralizing antibodies to *Trypanosoma cruzi* trans-sialidase. J Infect Dis 2002; 186(3): 397-404. [http://dx.doi.org/10.1086/341463] [PMID: 12134236]

[88] Lafaille JJ, Linss J, Krieger MA, Souto-Padrón T, de Souza W, Goldenberg S. Structure and expression of two *Trypanosoma cruzi* genes encoding antigenic proteins bearing repetitive epitopes. Mol Biochem Parasitol 1989; 35(2): 127-36. [http://dx.doi.org/10.1016/0166-6851(89)90115-1] [PMID: 2475776]

[89] El-Sayed NM, Myler PJ, Bartholomeu DC, *et al.* The genome sequence of *Trypanosoma cruzi*, etiologic agent of Chagas disease. Science 2005; 309(5733): 409-15. [http://dx.doi.org/10.1126/science.1112631] [PMID: 16020725]

[90] Arner E, Kindlund E, Nilsson D, *et al.* Database of *Trypanosoma cruzi* repeated genes: 20,000 additional gene variants. BMC Genomics 2007; 8: 391. [http://dx.doi.org/10.1186/1471-2164-8-391] [PMID: 17963481]

[91] Bartholomeu DC, Cerqueira GC, Leão AC, *et al.* Genomic organization and expression profile of the mucin-associated surface protein (masp) family of the human pathogen *Trypanosoma cruzi*. Nucleic Acids Res 2009; 37(10): 3407-17. [http://dx.doi.org/10.1093/nar/gkp172] [PMID: 19336417]

[92] Osuna CE, Gonzalez AM, Chang HH, *et al.* TCR affinity associated with functional differences between dominant and subdominant SIV epitope-specific CD8+ T cells in Mamu-A*01+ rhesus monkeys. PLoS Pathog 2014; 10(4): e1004069. [http://dx.doi.org/10.1371/journal.ppat.1004069] [PMID: 24743648]

[93] De Pablos LM, González GG, Solano Parada J, *et al.* Differential expression and characterization of a member of the mucin-associated surface protein family secreted by *Trypanosoma cruzi*. Infect Immun 2011; 79(10): 3993-4001. [http://dx.doi.org/10.1128/IAI.05329-11] [PMID: 21788387]

[94] Ashmus RA, Schocker NS, Cordero-Mendoza Y, *et al.* Potential use of synthetic α-galactosy-

-containing glycotopes of the parasite *Trypanosoma cruzi* as diagnostic antigens for Chagas disease. Org Biomol Chem 2013; 11(34): 5579-83. [http://dx.doi.org/10.1039/c3ob40887f] [PMID: 23863943]

[95] Gascón J. Diagnosis and treatment of imported Chagas disease. Med Clin (Barc) 2005; 125(6): 230-5. [PMID: 16022837]

[96] Añez N, Crisante G, Caraballo F, Delgado W, Parada H. *Trypanosoma cruzi* persistence at oral inflammatory foci in chronic chagasic patients. Acta Trop 2011; 117(3): 207-11. [http://dx.doi.org/10.1016/j.actatropica.2010.12.010] [PMID: 21215250]

[97] UNAIDS WHO, AIDS and epidemic update. UNAIDS/09.36E/JC1700E 2009. Available at: http://data.unaids.org/pub/Report/2009/JC1700_Epi_Update_2009_en.pdf

[98] da Costa SC. Immunocompromised host: from the early events until the impact of acquired immunodeficiency syndrome. Mem Inst Oswaldo Cruz 2000; 95(95) (Suppl. 1): 141-4. [http://dx.doi.org/10.1590/S0074-02762000000700023] [PMID: 11142703]

[99] Bern C. Chagas disease in the immunosuppressed host. Curr Opin Infect Dis 2012; 25(4): 450-7. [http://dx.doi.org/10.1097/QCO.0b013e328354f179] [PMID: 22614520]

[100] Pinazo MJ. Immunosuppression and Chagas disease: a management challenge. PLoS Negl Trop Dis 2013; 7(1): e1965.

[101] Martinez-Perez A, Norman FF, Monge-Maillo B, Perez-Molina JA, Lopez-Velez R. An approach to the management of *Trypanosoma cruzi* infection (Chagas disease) in immunocompromised patients. Expert Rev Anti Infect Ther 2014; 12(3): 357-73. [http://dx.doi.org/10.1586/14787210.2014.880652] [PMID: 24484076]

[102] de Freitas VL, da Silva SC, Sartori AM, *et al.* Real-time PCR in HIV/*Trypanosoma cruzi* coinfection with and without Chagas disease reactivation: association with HIV viral load and CD4 level. PLoS Negl Trop Dis 2011; 5(8): e1277. [http://dx.doi.org/10.1371/journal.pntd.0001277] [PMID: 21912712]

[103] PAHO, Panamerican Health Organization. OPS/FCH/HI/04-09.E. 2009, 2009.

[104] Almeida EA, Ramos Júnior AN, Correia D, Shikanai-Yasuda MA. Co-infection *Trypanosoma cruzi*/HIV: systematic review (19802010). Rev Soc Bras Med Trop 2011; 44(6): 762-70. [http://dx.doi.org/10.1590/S0037-86822011000600021] [PMID: 22231251]

[105] Almeida EA, Lima JN, Lages-Silva E, *et al.* Chagas disease and HIV co-infection in patients without effective antiretroviral therapy: prevalence, clinical presentation and natural history. Trans R Soc Trop Med Hyg 2010; 104(7): 447-52. [http://dx.doi.org/10.1016/j.trstmh.2010.02.004] [PMID: 20303560]

[106] Sartori AM, Ibrahim KY, Nunes Westphalen EV, *et al.* Manifestations of Chagas disease (American trypanosomiasis) in patients with HIV/AIDS. Ann Trop Med Parasitol 2007; 101(1): 31-50. [http://dx.doi.org/10.1179/136485907X154629] [PMID: 17244408]

[107] Cordova E, Boschi A, Ambrosioni J, Cudos C, Corti M. Reactivation of Chagas disease with central nervous system involvement in HIV-infected patients in Argentina, 19922007. Int J Infect Dis 2008; 12(6): 587-92. [http://dx.doi.org/10.1016/j.ijid.2007.12.007] [PMID: 18337139]

[108] Altclas JD, Barcan L, Nagel C, Lattes R, Riarte A. Organ transplantation and Chagas disease. JAMA 2008; 299(10): 1134.
[http://dx.doi.org/10.1001/jama.299.10.1134-a] [PMID: 18334687]

[109] Le Loup G, Pialoux G, Lescure FX. Update in treatment of Chagas disease. Curr Opin Infect Dis 2011; 24(5): 428-34.
[http://dx.doi.org/10.1097/QCO.0b013e32834a667f] [PMID: 21857512]

[110] Pérez-Molina JA, Rodríguez-Guardado A, Soriano A, *et al.* Guidelines on the treatment of chronic coinfection by *Trypanosoma cruzi* and HIV outside endemic areas. HIV Clin Trials 2011; 12(6): 287-98.
[http://dx.doi.org/10.1310/hct1206-287] [PMID: 22189148]

[111] Buckner FS. Experimental chemotherapy and approaches to drug discovery for *Trypanosoma cruzi* infection. Adv Parasitol 2011; 75: 89-119.
[http://dx.doi.org/10.1016/B978-0-12-385863-4.00005-8] [PMID: 21820553]

[112] Molina I, Salvador F, Sánchez-Montalvá A. Posaconazole *versus* benznidazole for chronic Chagas disease. N Engl J Med 2014; 371(10): 966.
[PMID: 25184871]

[113] Benaim G, Sanders JM, Garcia-Marchán Y, *et al.* Amiodarone has intrinsic anti-*Trypanosoma cruzi* activity and acts synergistically with posaconazole. J Med Chem 2006; 49(3): 892-9.
[http://dx.doi.org/10.1021/jm050691f] [PMID: 16451055]

[114] Bustamante JM, Craft JM, Crowe BD, Ketchie SA, Tarleton RL. New, combined, and reduced dosing treatment protocols cure *Trypanosoma cruzi* infection in mice. J Infect Dis 2014; 209(1): 150-62.
[http://dx.doi.org/10.1093/infdis/jit420] [PMID: 23945371]

[115] Diniz LdeF, Urbina JA, de Andrade IM, *et al.* Benznidazole and posaconazole in experimental Chagas disease: positive interaction in concomitant and sequential treatments. PLoS Negl Trop Dis 2013; 7(8): e2367.
[http://dx.doi.org/10.1371/journal.pntd.0002367] [PMID: 23967360]

[116] de Andrade AL, Zicker F, de Oliveira RM, *et al.* Randomised trial of efficacy of benznidazole in treatment of early *Trypanosoma cruzi* infection. Lancet 1996; 348(9039): 1407-13.
[http://dx.doi.org/10.1016/S0140-6736(96)04128-1] [PMID: 8937280]

[117] Andrade AL, Martelli CM, Oliveira RM, *et al.* Short report: benznidazole efficacy among *Trypanosoma cruzi*-infected adolescents after a six-year follow-up. Am J Trop Med Hyg 2004; 71(5): 594-7.
[PMID: 15569790]

[118] Sosa Estani S, Segura EL, Ruiz AM, Velazquez E, Porcel BM, Yampotis C. Efficacy of chemotherapy with benznidazole in children in the indeterminate phase of Chagas disease. Am J Trop Med Hyg 1998; 59(4): 526-9.
[PMID: 9790423]

[119] Galvao LM, Nunes RM, Cançado JR, Brener Z, Krettli AU. Lytic antibody titre as a means of assessing cure after treatment of Chagas disease: a 10 years follow-up study. Trans R Soc Trop Med Hyg 1993; 87(2): 220-3.

[http://dx.doi.org/10.1016/0035-9203(93)90501-G] [PMID: 8337734]

[120] Krautz GM, Galvão LM, Cançado JR, Guevara-Espinoza A, Ouaissi A, Krettli AU. Use of a 24-kilodalton *Trypanosoma cruzi* recombinant protein to monitor cure of human Chagas disease. J Clin Microbiol 1995; 33(8): 2086-90.
[PMID: 7559953]

[121] Levy AM, Boainain E, Kloetzel JK. *In situ* indirect fluorescent antibody: a new specific test to detect ongoing chagasic infections. J Clin Lab Anal 1996; 10(2): 98-103.
[http://dx.doi.org/10.1002/(SICI)1098-2825(1996)10:2<98::AID-JCLA7>3.0.CO;2-H] [PMID: 8852362]

[122] Cancado JR. Long term evaluation of etiological treatment of chagas disease with benznidazole. Rev Inst Med Trop Sao Paulo 2002; 44(1): 29-37.
[http://dx.doi.org/10.1590/S0036-46652002000100006] [PMID: 11896410]

[123] Meira WS, Galvão LM, Gontijo ED, Machado-Coelho GL, Norris KA, Chiari E. Use of the *Trypanosoma cruzi* recombinant complement regulatory protein to evaluate therapeutic efficacy following treatment of chronic chagasic patients. J Clin Microbiol 2004; 42(2): 707-12.
[http://dx.doi.org/10.1128/JCM.42.2.707-712.2004] [PMID: 14766840]

[124] Fabbro DL, Olivera V, Bizai ML, *et al.* Humoral immune response against P2β from *Trypanosoma cruzi* in persons with chronic Chagas disease: its relationship with treatment against parasites and myocardial damage. Am J Trop Med Hyg 2011; 84(4): 575-80.
[http://dx.doi.org/10.4269/ajtmh.2011.10-0261] [PMID: 21460013]

[125] Ibañez CF, Affranchino JL, Macina RA, *et al.* Multiple *Trypanosoma cruzi* antigens containing tandemly repeated amino acid sequence motifs. Mol Biochem Parasitol 1988; 30(1): 27-33.
[http://dx.doi.org/10.1016/0166-6851(88)90129-6] [PMID: 3135494]

[126] Sánchez Negrette O, Sánchez Valdéz FJ, Lacunza CD, *et al.* Serological evaluation of specific-antibody levels in patients treated for chronic Chagas disease. Clin Vaccine Immunol 2008; 15(2): 297-302.
[http://dx.doi.org/10.1128/CVI.00106-07] [PMID: 18057184]

[127] Fernández-Villegas A, Pinazo MJ, Marañón C, *et al.* Short-term follow-up of chagasic patients after benzonidazole treatment using multiple serological markers. BMC Infect Dis 2011; 11: 206.
[http://dx.doi.org/10.1186/1471-2334-11-206] [PMID: 21801456]

[128] Viotti R, Vigliano C, Alvarez MG, *et al.* Impact of aetiological treatment on conventional and multiplex serology in chronic Chagas disease. PLoS Negl Trop Dis 2011; 5(9): e1314.
[http://dx.doi.org/10.1371/journal.pntd.0001314] [PMID: 21909451]

[129] Cooley G, Etheridge RD, Boehlke C, *et al.* High throughput selection of effective serodiagnostics for *Trypanosoma cruzi* infection. PLoS Negl Trop Dis 2008; 2(10): e316.
[http://dx.doi.org/10.1371/journal.pntd.0000316] [PMID: 18841200]

[130] Pinazo MJ, Tàssies D, Muñoz J, *et al.* Hypercoagulability biomarkers in *Trypanosoma cruzi* -infected patients. Thromb Haemost 2011; 106(4): 617-23.
[http://dx.doi.org/10.1160/TH11-04-0251] [PMID: 21866301]

[131] Laucella SA, Mazliah DP, Bertocchi G, *et al.* Changes in *Trypanosoma cruzi*-specific immune

responses after treatment: surrogate markers of treatment efficacy. Clin Infect Dis 2009; 49(11): 1675-84.
[http://dx.doi.org/10.1086/648072] [PMID: 19877967]

[132] Frasch AC, Cazzulo JJ, Aslund L, Pettersson U. Comparison of genes encoding *Trypanosoma cruzi* antigens. Parasitol Today (Regul Ed) 1991; 7(6): 148-51.
[http://dx.doi.org/10.1016/0169-4758(91)90284-U] [PMID: 15463477]

[133] Hernández P, Heimann M, Riera C, *et al.* Highly effective serodiagnosis for Chagas disease. Clin Vaccine Immunol 2010; 17(10): 1598-604.
[http://dx.doi.org/10.1128/CVI.00489-08] [PMID: 20668136]

[134] Bottino CG, Gomes LP, Pereira JB, Coura JR, Provance DW Jr, De-Simone SG. Chagas disease-specific antigens: characterization of epitopes in CRA/FRA by synthetic peptide mapping and evaluation by ELISA-peptide assay. BMC Infect Dis 2013; 13: 568.
[http://dx.doi.org/10.1186/1471-2334-13-568] [PMID: 24299278]

[135] Levin MJ, Franco da Silveira J, Frasch AC, *et al.* Recombinant *Trypanosoma cruzi* antigens and Chagas disease diagnosis: analysis of a workshop. FEMS Microbiol Immunol 1991; 4(1): 11-9.
[http://dx.doi.org/10.1111/j.1574-6968.1991.tb04965.x] [PMID: 1815706]

[136] Umezawa ES, Bastos SF, Coura JR, *et al.* An improved serodiagnostic test for Chagas disease employing a mixture of *Trypanosoma cruzi* recombinant antigens. Transfusion 2003; 43(1): 91-7.
[http://dx.doi.org/10.1046/j.1537-2995.2003.00279.x] [PMID: 12519436]

[137] Floridia-Yapur N, Vega-Benedetti AF, Rumi MM, *et al.* Evaluation of recombinant antigens of *Trypanosoma cruzi* to diagnose infection in naturally infected dogs from Chaco region, Argentina. Parasite Immunol 2014; 36(12): 694-9.
[http://dx.doi.org/10.1111/pim.12144] [PMID: 25201522]

[138] Gruber A, Zingales B. *Trypanosoma cruzi*: characterization of two recombinant antigens with potential application in the diagnosis of Chagas disease. Exp Parasitol 1993; 76(1): 1-12.
[http://dx.doi.org/10.1006/expr.1993.1001] [PMID: 8467895]

[139] Peralta JM, Teixeira MG, Shreffler WG, *et al.* Serodiagnosis of Chagas disease by enzyme-linked immunosorbent assay using two synthetic peptides as antigens. J Clin Microbiol 1994; 32(4): 971-4.
[PMID: 8027352]

[140] Balouz V, Cámara MdeL, Cánepa GE, *et al.* Mapping antigenic motifs in the trypomastigote small surface antigen from *Trypanosoma cruzi*. Clin Vaccine Immunol 2015; 22(3): 304-12.
[http://dx.doi.org/10.1128/CVI.00684-14] [PMID: 25589551]

[141] Brenière SF, Bosseno MF, Noireau F, *et al.* Integrate study of a Bolivian population infected by *Trypanosoma cruzi*, the agent of Chagas disease. Mem Inst Oswaldo Cruz 2002; 97(3): 289-95.
[http://dx.doi.org/10.1590/S0074-02762002000300002] [PMID: 12048553]

[142] Lorca M, Gonzalez A, Veloso C, Reyes V, Vergara U. Immunodetection of antibodies in sera from symptomatic and asymptomatic Chilean Chagas disease patients with *Trypanosoma cruzi* recombinant antigens. Am J Trop Med Hyg 1992; 46(1): 44-9.
[PMID: 1536383]

[143] Abate T, Rosenberg M, Aldao RR, Slezynger TC. Cloning and partial characterization of a 28 kDa

antigenic protein of *Trypanosoma cruzi*. Biol Res 1993; 26(1-2): 121-30. [PMID: 7670524]

[144] Houghton RL, Benson DR, Reynolds LD, *et al.* A multi-epitope synthetic peptide and recombinant protein for the detection of antibodies to *Trypanosoma cruzi* in radioimmunoprecipitation-confirmed and consensus-positive sera. J Infect Dis 1999; 179(5): 1226-34. [http://dx.doi.org/10.1086/314723] [PMID: 10191227]

[145] Levin MJ, Mesri E, Benarous R, *et al.* Identification of major *Trypanosoma cruzi* antigenic determinants in chronic Chagas heart disease. Am J Trop Med Hyg 1989; 41(5): 530-8. [PMID: 2479275]

[146] Krautz GM, Peterson JD, Godsel LM, Krettli AU, Engman DM. Human antibody responses to *Trypanosoma cruzi* 70-kD heat-shock proteins. Am J Trop Med Hyg 1998; 58(2): 137-43. [PMID: 9502594]

[147] Cetron MS, Hoff R, Kahn S, Eisen H, Van Voorhis WC. Evaluation of recombinant trypomastigote surface antigens of *Trypanosoma cruzi* in screening sera from a population in rural northeastern Brazil endemic for Chagas disease. Acta Trop 1992; 50(3): 259-66. [http://dx.doi.org/10.1016/0001-706X(92)90082-9] [PMID: 1348602]

[148] Jazin EE, Bontempi EJ, Sanchez DO, *et al. Trypanosoma cruzi* exoantigen is a member of a 160 kDa gene family. Parasitology 1995; 110(Pt 1): 61-9. [http://dx.doi.org/10.1017/S0031182000081051] [PMID: 7845713]

[149] Norris KA, Schrimpf JE, Szabo MJ. Identification of the gene family encoding the 160-kilodalton *Trypanosoma cruzi* complement regulatory protein. Infect Immun 1997; 65(2): 349-57. [PMID: 9009282]

[150] Meira WS, Galvão LM, Gontijo ED, Machado-Coelho GL, Norris KA, Chiari E. *Trypanosoma cruzi* recombinant complement regulatory protein: a novel antigen for use in an enzyme-linked immunosorbent assay for diagnosis of Chagas disease. J Clin Microbiol 2002; 40(10): 3735-40. [http://dx.doi.org/10.1128/JCM.40.10.3735-3740.2002] [PMID: 12354873]

[151] Pinazo MJ, Thomas MC, Bua J, *et al.* Biological markers for evaluating therapeutic efficacy in Chagas disease, a systematic review. Expert Rev Anti Infect Ther 2014; 12(4): 479-96. [http://dx.doi.org/10.1586/14787210.2014.899150] [PMID: 24621252]

[152] Morillo CA, Marin-Neto JA, Avezum A, *et al.* Randomized trial of benznidazole for chronic chagas cardiomyopathy. N Engl J Med 2015; 373(14): 1295-306. [http://dx.doi.org/10.1056/NEJMoa1507574] [PMID: 26323937]

[153] Murcia L, Carrilero B, Muñoz MJ, Iborra MA, Segovia M. Usefulness of PCR for monitoring benznidazole response in patients with chronic Chagas disease: a prospective study in a non-diseas--endemic country. J Antimicrob Chemother 2010; 65(8): 1759-64. [http://dx.doi.org/10.1093/jac/dkq201] [PMID: 20542903]

[154] Porcel BM, Bontempi EJ, Henriksson J, *et al. Trypanosoma rangeli* and *Trypanosoma cruzi*: molecular characterization of genes encoding putative calcium-binding proteins, highly conserved in trypanosomatids. Exp Parasitol 1996; 84(3): 387-99. [http://dx.doi.org/10.1006/expr.1996.0127] [PMID: 8948328]

[155] Porrás AI, Yadon ZE, Altcheh J, *et al.* Target product profile (TPP) for chagas disease point-of-care diagnosis and assessment of response to treatment. PLoS Negl Trop Dis 2015; 9(6): e0003697. [http://dx.doi.org/10.1371/journal.pntd.0003697] [PMID: 26042730]

[156] Herwaldt BL. Leishmaniasis. Lancet 1999; 354(9185): 1191-9. [http://dx.doi.org/10.1016/S0140-6736(98)10178-2] [PMID: 10513726]

[157] de Almeida ME, Steurer FJ, Koru O, Herwaldt BL, Pieniazek NJ, da Silva AJ. Identification of *Leishmania* spp. by molecular amplification and DNA sequencing analysis of a fragment of rRNA internal transcribed spacer 2. J Clin Microbiol 2011; 49(9): 3143-9. [http://dx.doi.org/10.1128/JCM.01177-11] [PMID: 21752983]

[158] Odiwuor SO, Saad AA, De Doncker S, *et al.* Universal PCR assays for the differential detection of all Old World *Leishmania* species. Eur J Clin Microbiol Infect Dis 2011; 30(2): 209-18. [http://dx.doi.org/10.1007/s10096-010-1071-3] [PMID: 20936316]

[159] Schönian G, Kuhls K, Mauricio IL. Molecular approaches for a better understanding of the epidemiology and population genetics of *Leishmania.* Parasitology 2011; 138(4): 405-25. [http://dx.doi.org/10.1017/S0031182010001538] [PMID: 21078222]

[160] Ashford RW. The leishmaniases as emerging and reemerging zoonoses. Int J Parasitol 2000; 30(12-13): 1269-81. [http://dx.doi.org/10.1016/S0020-7519(00)00136-3] [PMID: 11113254]

[161] Kuhls K, Alam MZ, Cupolillo E, *et al.* Comparative microsatellite typing of new world *Leishmania* infantum reveals low heterogeneity among populations and its recent old world origin. PLoS Negl Trop Dis 2011; 5(6): e1155. [http://dx.doi.org/10.1371/journal.pntd.0001155] [PMID: 21666787]

[162] Lainson R, Shaw JJ. New world Leishmaniasis. The neotropical *Leishmania* species, A.B. L. Collier, M. Sussman (Eds.), Editor. Topley and Wilson's Microbiology and Microbial Infectious Diseases. Arnold, London, 1998, 241–266.

[163] Marzochi MC, Marzochi KB. Tegumentary and visceral leishmaniases in Brazil: emerging anthropozoonosis and possibilities for their control. Cad Saude Publica 1994; 10 (Suppl. 2): 359-75. [http://dx.doi.org/10.1590/S0102-311X1994000800014] [PMID: 15042226]

[164] Passos VM, Fernandes O, Lacerda PA, *et al. Leishmania* (*Viannia*) *braziliensis* is the predominant species infecting patients with American cutaneous leishmaniasis in the State of Minas Gerais, Southeast Brazil. Acta Trop 1999; 72(3): 251-8. [http://dx.doi.org/10.1016/S0001-706X(98)00100-4] [PMID: 10232781]

[165] Reithinger R, Dujardin JC. Molecular diagnosis of leishmaniasis: current status and future applications. J Clin Microbiol 2007; 45(1): 21-5. [http://dx.doi.org/10.1128/JCM.02029-06] [PMID: 17093038]

[166] Kevric I, Cappel MA, Keeling JH. New world and old world *Leishmania* infections: a practical review. Dermatol Clin 2015; 33(3): 579-93. [http://dx.doi.org/10.1016/j.det.2015.03.018] [PMID: 26143433]

[167] Palumbo E. Current treatment for cutaneous leishmaniasis: a review. Am J Ther 2009; 16(2): 178-82. [http://dx.doi.org/10.1097/MJT.0b013e3181822e90] [PMID: 19300044]

[168] Showler AJ, Boggild AK. Cutaneous leishmaniasis in travellers: a focus on epidemiology and treatment in 2015. Curr Infect Dis Rep 2015; 17(7): 489.
[http://dx.doi.org/10.1007/s11908-015-0489-2] [PMID: 26031962]

[169] Marfurt J, Nasereddin A, Niederwieser I, Jaffe CL, Beck HP, Felger I. Identification and differentiation of *Leishmania* species in clinical samples by PCR amplification of the miniexon sequence and subsequent restriction fragment length polymorphism analysis. J Clin Microbiol 2003; 41(7): 3147-53.
[http://dx.doi.org/10.1128/JCM.41.7.3147-3153.2003] [PMID: 12843055]

[170] Deniau M, Cañavate C, Faraut-Gambarelli F, Marty P. The biological diagnosis of leishmaniasis in HIV-infected patients. Ann Trop Med Parasitol 2003; 97(97) (Suppl. 1): 115-33.
[http://dx.doi.org/10.1179/000349803225002598] [PMID: 14678639]

[171] Foulet F, Botterel F, Buffet P, *et al.* Detection and identification of *Leishmania* species from clinical specimens by using a real-time PCR assay and sequencing of the cytochrome B gene. J Clin Microbiol 2007; 45(7): 2110-5.
[http://dx.doi.org/10.1128/JCM.02555-06] [PMID: 17475750]

[172] Rioux JA, Lanotte G, Serres E, Pratlong F, Bastien P, Perieres J. Taxonomy of *Leishmania*. Use of isoenzymes. Suggestions for a new classification. Ann Parasitol Hum Comp 1990; 65(3): 111-25.
[http://dx.doi.org/10.1051/parasite/1990653111] [PMID: 2080829]

[173] Laurent T, Van der Auwera G, Hide M, *et al.* Identification of Old World *Leishmania* spp. by specific polymerase chain reaction amplification of cysteine proteinase B genes and rapid dipstick detection. Diagn Microbiol Infect Dis 2009; 63(2): 173-81.
[http://dx.doi.org/10.1016/j.diagmicrobio.2008.10.015] [PMID: 19097841]

[174] Alvar J, Cañavate C, Gutiérrez-Solar B, *et al. Leishmania* and human immunodeficiency virus coinfection: the first 10 years. Clin Microbiol Rev 1997; 10(2): 298-319.
[PMID: 9105756]

[175] Sundar S, Rai M. Laboratory diagnosis of visceral leishmaniasis. Clin Diagn Lab Immunol 2002; 9(5): 951-8.
[PMID: 12204943]

[176] Singh R, Subba Raju BV, Jain RK, Salotra P. Potential of direct agglutination test based on promastigote and amastigote antigens for serodiagnosis of post-kala-azar dermal leishmaniasis. Clin Diagn Lab Immunol 2005; 12(10): 1191-4.
[PMID: 16210482]

[177] Attar ZJ, Chance ML, el-Safi S, *et al.* Latex agglutination test for the detection of urinary antigens in visceral leishmaniasis. Acta Trop 2001; 78(1): 11-6.
[http://dx.doi.org/10.1016/S0001-706X(00)00155-8] [PMID: 11164745]

[178] Riera C, Fisa R, Lopez P, *et al.* Evaluation of a latex agglutination test (KAtex) for detection of *Leishmania* antigen in urine of patients with HIV-*Leishmania* coinfection: value in diagnosis and post-treatment follow-up. Eur J Clin Microbiol Infect Dis 2004; 23(12): 899-904.
[PMID: 15599651]

[179] El-Safi SH, Abdel-Haleem A, Hammad A, *et al.* Field evaluation of latex agglutination test for

detecting urinary antigens in visceral leishmaniasis in Sudan. East Mediterr Health J 2003; 9(4): 844-55.
[PMID: 15748081]

[180] Abeijon C, Kashino SS, Silva FO, *et al.* Identification and diagnostic utility of *Leishmania infantum* proteins found in urine samples from patients with visceral leishmaniasis. Clin Vaccine Immunol 2012; 19(6): 935-43.
[http://dx.doi.org/10.1128/CVI.00125-12] [PMID: 22518013]

[181] Sadeghian G, Momeni A, Siadat AH, Yousefi P. Evaluation of leishmanin skin test and its relationship with the clinical form and duration of cutaneous leishmaniasis. Dermatol Online J 2006; 12(7): 3.
[PMID: 17459289]

[182] Gomes AH, Ferreira IM, Lima ML, *et al.* PCR identification of *Leishmania* in diagnosis and control of canine Leishmaniasis. Vet Parasitol 2007; 144(3-4): 234-41.
[http://dx.doi.org/10.1016/j.vetpar.2006.10.008] [PMID: 17196339]

[183] Manual de vigilância e controle da leishmaniose visceral (visceral leishmaniasis: surveillance and control. Technical manual). 2010. Available from: http://bvsms.saude.gov.br/bvs/publicacoes/manual_vigilancia_controle_leishmaniose_visceral.pdf in Ministerio da Saude do Brasil (Health Ministry of Brazil)

[184] Almeida AB, Sousa VR, Gasparetto ND, *et al.* Canine visceral leishmaniasis: diagnostic approaches based on polymerase chain reaction employing different biological samples. Diagn Microbiol Infect Dis 2013; 76(3): 321-4.
[http://dx.doi.org/10.1016/j.diagmicrobio.2013.03.017] [PMID: 23619344]

[185] Weiss JB. DNA probes and PCR for diagnosis of parasitic infections. Clin Microbiol Rev 1995; 8(1): 113-30.
[PMID: 7704890]

[186] Dujardin JC, Victoir K, De Doncker S, Guerbouj S, Arévalo J, Le Ray D. Molecular epidemiology and diagnosis of *Leishmania*: what have we learnt from genome structure, dynamics and function? Trans R Soc Trop Med Hyg 2002; 96 (Suppl. 1): S81-6.
[http://dx.doi.org/10.1016/S0035-9203(02)90056-8] [PMID: 12055856]

[187] Fisa R, Riera C, López-Chejade P, *et al. Leishmania* infantum DNA detection in urine from patients with visceral leishmaniasis and after treatment control. Am J Trop Med Hyg 2008; 78(5): 741-4.
[PMID: 18458307]

[188] Fisa R, Riera C, Ribera E, Gállego M, Portús M. A nested polymerase chain reaction for diagnosis and follow-up of human visceral leishmaniasis patients using blood samples. Trans R Soc Trop Med Hyg 2002; 96 (Suppl. 1): S191-4.
[http://dx.doi.org/10.1016/S0035-9203(02)90075-1] [PMID: 12055837]

[189] Lachaud L, Dereure J, Chabbert E, *et al.* Optimized PCR using patient blood samples for diagnosis and follow-up of visceral Leishmaniasis, with special reference to AIDS patients. J Clin Microbiol 2000; 38(1): 236-40.
[PMID: 10618093]

[190] Bossolasco S, Gaiera G, Chini D, *et al.* Real-time PCR assay for clinical management of human immunodeficiency virus-infected patients with visceral Leishmaniasis. J Clin Microbiol 2004; 42(4):

1858-8.
[http://dx.doi.org/10.1128/JCM.42.4.1858.2004] [PMID: 14605142]

[191] Bhattacharyya R, Das K, Sen S, Roy S, Majumder HK. Development of a genus specific primer set for detection of *Leishmania* parasites by polymerase chain reaction. FEMS Microbiol Lett 1996; 135(2-3): 195-200.
[http://dx.doi.org/10.1111/j.1574-6968.1996.tb07989.x] [PMID: 8595858]

[192] Mathis A, Deplazes P. PCR and *in vitro* cultivation for detection of *Leishmania* spp. in diagnostic samples from humans and dogs. J Clin Microbiol 1995; 33(5): 1145-9.
[PMID: 7615719]

[193] de Bruijn MH, Barker DC. Diagnosis of New World leishmaniasis: specific detection of species of the *Leishmania braziliensis* complex by amplification of kinetoplast DNA. Acta Trop 1992; 52(1): 45-58.
[http://dx.doi.org/10.1016/0001-706X(92)90006-J] [PMID: 1359760]

[194] Eresh S, McCallum SM, Barker DC. Identification and diagnosis of *Leishmania mexicana* complex isolates by polymerase chain reaction. Parasitology 1994; 109(Pt 4): 423-33.
[http://dx.doi.org/10.1017/S0031182000080677] [PMID: 7800410]

[195] Degrave W, Fernandes O, Campbell D, Bozza M, Lopes U. Use of molecular probes and PCR for detection and typing of *Leishmania*a mini-review. Mem Inst Oswaldo Cruz 1994; 89(3): 463-9.
[http://dx.doi.org/10.1590/S0074-02761994000300032] [PMID: 7476234]

[196] Ramos A, Maslov DA, Fernandes O, Campbell DA, Simpson L. Detection and identification of human pathogenic *Leishmania* and *Trypanosoma* species by hybridization of PCR-amplified mini-exon repeats. Exp Parasitol 1996; 82(3): 242-50.
[http://dx.doi.org/10.1006/expr.1996.0031] [PMID: 8631375]

[197] Rodríguez N, Guzman B, Rodas A, Takiff H, Bloom BR, Convit J. Diagnosis of cutaneous leishmaniasis and species discrimination of parasites by PCR and hybridization. J Clin Microbiol 1994; 32(9): 2246-52.
[PMID: 7814554]

[198] Nicolas L, Milon G, Prina E. Rapid differentiation of Old World *Leishmania* species by LightCycler polymerase chain reaction and melting curve analysis. J Microbiol Methods 2002; 51(3): 295-9.
[http://dx.doi.org/10.1016/S0167-7012(02)00099-4] [PMID: 12223289]

[199] Roussel M, Nacher M, Frémont G, *et al.* Comparison between one and two injections of pentamidine isethionate, at 7 mg/kg in each injection, in the treatment of cutaneous leishmaniasis in French Guiana. Ann Trop Med Parasitol 2006; 100(4): 307-14.
[http://dx.doi.org/10.1179/136485906X105561] [PMID: 16762111]

[200] Wortmann G, Hochberg L, Houng HH, *et al.* Rapid identification of *Leishmania* complexes by a real-time PCR assay. Am J Trop Med Hyg 2005; 73(6): 999-1004.
[PMID: 16354801]

[201] van Eys GJ, Schoone GJ, Kroon NC, Ebeling SB. Sequence analysis of small subunit ribosomal RNA genes and its use for detection and identification of *Leishmania* parasites. Mol Biochem Parasitol 1992; 51(1): 133-42.
[http://dx.doi.org/10.1016/0166-6851(92)90208-2] [PMID: 1565128]

[202] Bossolasco S, Gaiera G, Olchini D, *et al.* Real-time PCR assay for clinical management of human immunodeficiency virus-infected patients with visceral leishmaniasis. J Clin Microbiol 2003; 41(11): 5080-4.
[http://dx.doi.org/10.1128/JCM.41.11.5080-5084.2003] [PMID: 14605142]

[203] Wortmann G, Sweeney C, Houng HS, *et al.* Rapid diagnosis of leishmaniasis by fluorogenic polymerase chain reaction. Am J Trop Med Hyg 2001; 65(5): 583-7.
[PMID: 11716118]

[204] Fernandes O, Murthy VK, Kurath U, Degrave WM, Campbell DA. Mini-exon gene variation in human pathogenic *Leishmania* species. Mol Biochem Parasitol 1994; 66(2): 261-71.
[http://dx.doi.org/10.1016/0166-6851(94)90153-8] [PMID: 7808476]

[205] Hassan MQ, Ghosh A, Ghosh SS, *et al.* Enzymatic amplification of mini-exon-derived RNA gene spacers of *Leishmania donovani*: primers and probes for DNA diagnosis. Parasitology 1993; 107(Pt 5): 509-17.
[http://dx.doi.org/10.1017/S0031182000068086] [PMID: 8295790]

[206] Piarroux R, Fontes M, Perasso R, *et al.* Phylogenetic relationships between Old World *Leishmania* strains revealed by analysis of a repetitive DNA sequence. Mol Biochem Parasitol 1995; 73(1-2): 249-52.
[http://dx.doi.org/10.1016/0166-6851(95)00097-K] [PMID: 8577334]

[207] Cupolillo E, Grimaldi Júnior G, Momen H, Beverley SM. Intergenic region typing (IRT): a rapid molecular approach to the characterization and evolution of *Leishmania.* Mol Biochem Parasitol 1995; 73(1-2): 145-55.
[http://dx.doi.org/10.1016/0166-6851(95)00108-D] [PMID: 8577322]

[208] Eisenberger CL, Jaffe CL. *Leishmania*: identification of Old World species using a permissively primed intergenic polymorphic-polymerase chain reaction. Exp Parasitol 1999; 91(1): 70-7.
[http://dx.doi.org/10.1006/expr.1999.4355] [PMID: 9920044]

[209] Luis L, Ramírez A, Aguilar CM, Eresh S, Barker DC, Mendoza-León A. The genomic fingerprinting of the coding region of the beta-tubulin gene in *Leishmania* identification. Acta Trop 1998; 69(3): 193-204.
[http://dx.doi.org/10.1016/S0001-706X(97)00128-9] [PMID: 9638272]

[210] Fraga J, Montalvo AM, De Doncker S, Dujardin JC, Van der Auwera G. Phylogeny of *Leishmania* species based on the heat-shock protein 70 gene. Infect Genet Evol 2010; 10(2): 238-45.
[http://dx.doi.org/10.1016/j.meegid.2009.11.007] [PMID: 19913110]

[211] Van der Auwera G, Maes I, De Doncker S, *et al.* Heat-shock protein 70 gene sequencing for *Leishmania* species typing in European tropical infectious disease clinics. Euro Surveill 2013; 18(30): 20543.
[http://dx.doi.org/10.2807/1560-7917.ES2013.18.30.20543] [PMID: 23929181]

[212] Victoir K, Bañuls AL, Arevalo J, *et al.* The gp63 gene locus, a target for genetic characterization of *Leishmania* belonging to subgenus *Viannia.* Parasitology 1998; 117(Pt 1): 1-13.
[PMID: 9695095]

[213] Russell R, Iribar MP, Lambson B, *et al.* Intra and inter-specific microsatellite variation in the

Leishmania subgenus *Viannia.* Mol Biochem Parasitol 1999; 103(1): 71-7.
[http://dx.doi.org/10.1016/S0166-6851(99)00117-6] [PMID: 10514082]

[214] Belli A, Rodriguez B, Aviles H, Harris E. Simplified polymerase chain reaction detection of new world *Leishmania* in clinical specimens of cutaneous leishmaniasis. Am J Trop Med Hyg 1998; 58(1): 102-9.
[PMID: 9452300]

[215] Colombo FA, Odorizzi RM, Laurenti MD, Galati EA, Canavez F, Pereira-Chioccola VL. Detection of *Leishmania (Leishmania) infantum* RNA in fleas and ticks collected from naturally infected dogs. Parasitol Res 2011; 109(2): 267-74.
[http://dx.doi.org/10.1007/s00436-010-2247-6] [PMID: 21221638]

[216] Bezerra-Vasconcelos DR, Melo LM, Albuquerque ES, Luciano MC, Bevilaqua CM. Real-time PCR to assess the *Leishmania* load in *Lutzomyia longipalpis* sand flies: screening of target genes and assessment of quantitative methods. Exp Parasitol 2011; 129(3): 234-9.
[http://dx.doi.org/10.1016/j.exppara.2011.08.010] [PMID: 21864530]

[217] Kato H, Uezato H, Katakura K, *et al.* Detection and identification of *Leishmania* species within naturally infected sand flies in the andean areas of ecuador by a polymerase chain reaction. Am J Trop Med Hyg 2005; 72(1): 87-93.
[PMID: 15728872]

[218] Kato H, Uezato H, Gomez EA, *et al.* Establishment of a mass screening method of sand fly vectors for *Leishmania* infection by molecular biological methods. Am J Trop Med Hyg 2007; 77(2): 324-9.
[PMID: 17690406]

[219] Schönian G, Nasereddin A, Dinse N, *et al.* PCR diagnosis and characterization of *Leishmania* in local and imported clinical samples. Diagn Microbiol Infect Dis 2003; 47(1): 349-58.
[http://dx.doi.org/10.1016/S0732-8893(03)00093-2] [PMID: 12967749]

[220] Bensoussan E, Nasereddin A, Jonas F, Schnur LF, Jaffe CL. Comparison of PCR assays for diagnosis of cutaneous leishmaniasis. J Clin Microbiol 2006; 44(4): 1435-9.
[http://dx.doi.org/10.1128/JCM.44.4.1435-1439.2006] [PMID: 16597873]

[221] Ajaoud M, Es-Sette N, Charrel RN, *et al. Phlebotomus sergenti* in a cutaneous leishmaniasis focus in Azilal province (High Atlas, Morocco): molecular detection and genotyping of *Leishmania tropica*, and feeding behavior. PLoS Negl Trop Dis 2015; 9(3): e0003687.
[http://dx.doi.org/10.1371/journal.pntd.0003687] [PMID: 25826399]

[222] Nasereddin A, Bensoussan-Hermano E, Schönian G, Baneth G, Jaffe CL. Molecular diagnosis of Old World cutaneous leishmaniasis and species identification by use of a reverse line blot hybridization assay. J Clin Microbiol 2008; 46(9): 2848-55.
[http://dx.doi.org/10.1128/JCM.00951-08] [PMID: 18614659]

[223] Abbasi I, Aramin S, Hailu A, *et al.* Evaluation of PCR procedures for detecting and quantifying *Leishmania donovani* DNA in large numbers of dried human blood samples from a visceral leishmaniasis focus in Northern Ethiopia. BMC Infect Dis 2013; 13: 153.
[http://dx.doi.org/10.1186/1471-2334-13-153] [PMID: 23530965]

[224] Talmi-Frank D, Nasereddin A, Schnur LF, *et al.* Detection and identification of old world *Leishmania* by high resolution melt analysis. PLoS Negl Trop Dis 2010; 4(1): e581.

[http://dx.doi.org/10.1371/journal.pntd.0000581] [PMID: 20069036]

[225] Aghaei AA, Rassi Y, Sharifi I, *et al.* First report on natural *Leishmania* infection of Phlebotomus sergenti due *Leishmania tropica* by high resolution melting curve method in South-eastern Iran. Asian Pac J Trop Med 2014; 7(2): 93-6.
[http://dx.doi.org/10.1016/S1995-7645(14)60002-X] [PMID: 24461520]

[226] Costa JM, Durand R, Deniau M, *et al.* PCR enzyme-linked immunosorbent assay for diagnosis of leishmaniasis in human immunodeficiency virus-infected patients. J Clin Microbiol 1996; 34(7): 1831-3.
[PMID: 8784604]

[227] Luyo-Acero GE, Uezato H, Oshiro M, *et al.* Sequence variation of the cytochrome b gene of various human infecting members of the genus *Leishmania* and their phylogeny. Parasitology 2004; 128(Pt 5): 483-91.
[http://dx.doi.org/10.1017/S0031182004004792] [PMID: 15180316]

[228] Marco JD, Uezato H, Mimori T, *et al.* Are cytochrome B gene sequencing and polymorphism-specific polymerase chain reaction as reliable as multilocus enzyme electrophoresis for identifying *Leishmania* spp. from Argentina? Am J Trop Med Hyg 2006; 75(2): 256-60.
[PMID: 16896128]

[229] Marco JD, Bhutto AM, Soomro FR, *et al.* Multilocus enzyme electrophoresis and cytochrome B gene sequencing-based identification of *Leishmania* isolates from different foci of cutaneous leishmaniasis in Pakistan. Am J Trop Med Hyg 2006; 75(2): 261-6.
[PMID: 16896129]

[230] Ullu E, Tschudi C, Günzl A. Trans-splicing in trypanosomatid protozoa, in In: Molecular biology of parasitic protozoa. D.F. Smith and M. Parsons (ed). IRL Press at Oxford University Press, Oxford, England 1996.

[231] Harris E, Kropp G, Belli A, Rodriguez B, Agabian N. Single-step multiplex PCR assay for characterization of New World *Leishmania* complexes. J Clin Microbiol 1998; 36(7): 1989-95.
[PMID: 9650950]

[232] Marfurt J, Niederwieser I, Makia ND, Beck HP, Felger I. Diagnostic genotyping of old and new world *Leishmania* species by PCR-RFLP. Diagn Microbiol Infect Dis 2003; 46(2): 115-24.
[http://dx.doi.org/10.1016/S0732-8893(03)00040-3] [PMID: 12812715]

[233] Chakrabarty R, Mukherjee S, Lu HG, McGwire BS, Chang KP, Basu MK. Kinetics of entry of virulent and avirulent strains of *Leishmania donovani* into macrophages: a possible role of virulence molecules (gp63 and LPG). J Parasitol 1996; 82(4): 632-5.
[http://dx.doi.org/10.2307/3283790] [PMID: 8691373]

[234] Button LL, McMaster WR. Molecular cloning of the major surface antigen of *leishmania.* J Exp Med 1988; 167(2): 724-9.
[http://dx.doi.org/10.1084/jem.167.2.724] [PMID: 3346625]

[235] Victoir K, Dujardin JC, de Doncker S, *et al.* Plasticity of gp63 gene organization in *Leishmania (Viannia) braziliensis* and *Leishmania (Viannia) peruviana.* Parasitology 1995; 111(Pt 3): 265-73.
[http://dx.doi.org/10.1017/S0031182000081828] [PMID: 7567095]

[236] Steinkraus HB, Greer JM, Stephenson DC, Langer PJ. Sequence heterogeneity and polymorphic gene arrangements of the *Leishmania guyanensis* gp63 genes. Mol Biochem Parasitol 1993; 62(2): 173-85. [http://dx.doi.org/10.1016/0166-6851(93)90107-9] [PMID: 7908120]

[237] Dujardin JC, De Doncker S, Victoir K, Le Ray D, Hamers R, Arevalo J. Size polymorphism of chromosomes bearing gp63 genes in *Leishmania* of the *Braziliensis* complex: indication of a rearrangement of the gp63 genes in *L. braziliensis* and *L. peruviana.* Ann Trop Med Parasitol 1994; 88(4): 445-8.
[PMID: 7545964]

[238] Espinoza JR, Skinner AC, Davies CR, *et al.* Extensive polymorphism at the Gp63 locus in field isolates of *Leishmania peruviana.* Mol Biochem Parasitol 1995; 72(1-2): 203-13. [http://dx.doi.org/10.1016/0166-6851(95)00079-G] [PMID: 8538690]

[239] Tupperwar N, Vineeth V, Rath S, Vaidya T. Development of a real-time polymerase chain reaction assay for the quantification of *Leishmania* species and the monitoring of systemic distribution of the pathogen. Diagn Microbiol Infect Dis 2008; 61(1): 23-30. [http://dx.doi.org/10.1016/j.diagmicrobio.2007.12.013] [PMID: 18255247]

[240] Mauricio IL, Gaunt MW, Stothard JR, Miles MA. Glycoprotein 63 (gp63) genes show gene conversion and reveal the evolution of Old World *Leishmania.* Int J Parasitol 2007; 37(5): 565-76. [http://dx.doi.org/10.1016/j.ijpara.2006.11.020] [PMID: 17280675]

[241] Kato H, Gomez EA, Cáceres AG, *et al.* Natural infections of man-biting sand flies by *Leishmania* and *Trypanosoma* species in the northern Peruvian Andes. Vector Borne Zoonotic Dis 2011; 11(5): 515-21.
[http://dx.doi.org/10.1089/vbz.2010.0138] [PMID: 20954867]

[242] Kato H, Uezato H, Sato H, *et al.* Natural infection of the sand fly *Phlebotomus kazeruni* by *Trypanosoma* species in Pakistan. Parasit Vectors 2010; 3: 10. [http://dx.doi.org/10.1186/1756-3305-3-10] [PMID: 20184773]

[243] Kato H, Watanabe J, Mendoza Nieto I, Korenaga M, Hashiguchi Y. *Leishmania* species identification using FTA card sampling directly from patients cutaneous lesions in the state of Lara, Venezuela. Trans R Soc Trop Med Hyg 2011; 105(10): 561-7. [http://dx.doi.org/10.1016/j.trstmh.2011.05.009] [PMID: 21907375]

[244] Kato H, Cáceres AG, Mimori T, *et al.* Use of FTA cards for direct sampling of patients lesions in the ecological study of cutaneous leishmaniasis. J Clin Microbiol 2010; 48(10): 3661-5. [http://dx.doi.org/10.1128/JCM.00498-10] [PMID: 20720027]

[245] Freitas-Lidani KC, Messias-Reason IJ, Ishikawa EA. A comparison of molecular markers to detect *Lutzomyia longipalpis* naturally infected with *Leishmania* (*Leishmania*) *infantum.* Mem Inst Oswaldo Cruz 2014; 109(4): 442-7.
[http://dx.doi.org/10.1590/0074-0276130285] [PMID: 25004147]

[246] Van der Auwera G, Ravel C, Verweij JJ, Bart A, Schönian G, Felger I. Evaluation of four single-locus markers for *Leishmania* species discrimination by sequencing. J Clin Microbiol 2014; 52(4): 1098-104.
[http://dx.doi.org/10.1128/JCM.02936-13] [PMID: 24452158]

[247] Colombo FA, Pereira-Chioccola VL, Meira CdaS, *et al.* Performance of a real time PCR for leishmaniasis diagnosis using a *L. (L.) infantum* hypothetical protein as target in canine samples. Exp Parasitol 2015; 157: 156-62.
[http://dx.doi.org/10.1016/j.exppara.2015.08.014] [PMID: 26297683]

[248] de Ruiter CM, van der Veer C, Leeflang MM, Deborggraeve S, Lucas C, Adams ER. Molecular tools for diagnosis of visceral leishmaniasis: systematic review and meta-analysis of diagnostic test accuracy. J Clin Microbiol 2014; 52(9): 3147-55.
[http://dx.doi.org/10.1128/JCM.00372-14] [PMID: 24829226]

[249] Asato Y, Oshiro M, Myint CK, *et al.* Phylogenic analysis of the genus *Leishmania* by cytochrome b gene sequencing. Exp Parasitol 2009; 121(4): 352-61.
[http://dx.doi.org/10.1016/j.exppara.2008.12.013] [PMID: 19159626]

[250] Smyth AJ, Ghosh A, Hassan MQ, *et al.* Rapid and sensitive detection of *Leishmania* kinetoplast DNA from spleen and blood samples of kala-azar patients. Parasitology 1992; 105(Pt 2): 183-92.
[http://dx.doi.org/10.1017/S0031182000074096] [PMID: 1333582]

[251] Rodgers MR, Popper SJ, Wirth DF. Amplification of kinetoplast DNA as a tool in the detection and diagnosis of *Leishmania.* Exp Parasitol 1990; 71(3): 267-75.
[http://dx.doi.org/10.1016/0014-4894(90)90031-7] [PMID: 2170165]

[252] el Tai NO, Osman OF, el Fari M, Presber W, Schönian G. Genetic heterogeneity of ribosomal internal transcribed spacer in clinical samples of *Leishmania donovani* spotted on filter paper as revealed by single-strand conformation polymorphisms and sequencing. Trans R Soc Trop Med Hyg 2000; 94(5): 575-9.
[http://dx.doi.org/10.1016/S0035-9203(00)90093-2] [PMID: 11132393]

[253] Adams ER, Schoone GJ, Ageed AF, Safi SE, Schallig HD. Development of a reverse transcriptase loop-mediated isothermal amplification (LAMP) assay for the sensitive detection of *Leishmania* parasites in clinical samples. Am J Trop Med Hyg 2010; 82(4): 591-6.
[http://dx.doi.org/10.4269/ajtmh.2010.09-0369] [PMID: 20348505]

[254] Mori Y, Nagamine K, Tomita N, Notomi T. Detection of loop-mediated isothermal amplification reaction by turbidity derived from magnesium pyrophosphate formation. Biochem Biophys Res Commun 2001; 289(1): 150-4.
[http://dx.doi.org/10.1006/bbrc.2001.5921] [PMID: 11708792]

[255] Poon LL, Wong BW, Ma EH, *et al.* Sensitive and inexpensive molecular test for falciparum malaria: detecting *Plasmodium falciparum* DNA directly from heat-treated blood by loop-mediated isothermal amplification. Clin Chem 2006; 52(2): 303-6.
[http://dx.doi.org/10.1373/clinchem.2005.057901] [PMID: 16339303]

[256] Njiru ZK, Mikosza AS, Matovu E, *et al.* African trypanosomiasis: sensitive and rapid detection of the sub-genus Trypanozoon by loop-mediated isothermal amplification (LAMP) of parasite DNA. Int J Parasitol 2008; 38(5): 589-99.
[http://dx.doi.org/10.1016/j.ijpara.2007.09.006] [PMID: 17991469]

[257] Zhang L, Tarleton RL. Parasite persistence correlates with disease severity and localization in chronic Chagas disease. J Infect Dis 1999; 180(2): 480-6.
[http://dx.doi.org/10.1086/314889] [PMID: 10395865]

[258] Viotti R, Alarcón de Noya B, Araujo-Jorge T, *et al.* Towards a paradigm shift in the treatment of chronic Chagas disease. Antimicrob Agents Chemother 2014; 58(2): 635-9. [http://dx.doi.org/10.1128/AAC.01662-13] [PMID: 24247135]

[259] Mendonça MG, de Brito ME, Rodrigues EH, Bandeira V, Jardim ML, Abath FG. Persistence of *leishmania* parasites in scars after clinical cure of American cutaneous leishmaniasis: is there a sterile cure? J Infect Dis 2004; 189(6): 1018-23.

CHAPTER 10

New Chemotherapy Against Trypanosomiasis and Leishmaniasis

Jair L. de Siqueira-Neto[*, 1,2]

[1] *Skaggs School of Pharmacy and Pharmaceutical Sciences, University of California San Diego, La Jolla, USA*

[2] *Center for Discovery and Innovation in Parasitic Diseases, University of California San Diego, La Jolla, USA*

Abstract: Chagas disease is caused by the protozoan parasite *Trypanosoma cruzi* affecting mostly the American continent eventually leading to chronic cardiomyopathy or digestive syndromes. Human African Trypanosomiasis is caused by the related parasite *T. brucei*, endemic in the African continent and being characterized by invasion and damage in the central nervous system. Leishmaniasis is caused by a number of species from the genera *Leishmania* and can manifest with different clinical outcomes including skin ulcers and visceral organ damage, being endemic in 88 countries in tropical areas of the globe. Despite having different geographical distribution and unique clinical symptoms, these diseases are all caused by related protozoan parasites from the order Kinetoplastida. Another aspect shared by these diseases is related to the treatment options currently available, unfortunately, all inadequate. Serious problems are toxicity and inefficacy due parasite acquired resistance or lack of natural susceptibility. The population affected by these three diseases does not represent an attractive economic market, reflecting on little pharmaceutical industry interest in developing better chemotherapies. However, in the last decade, the situation has dramatically improved, with the active engagement of philanthropic financial support, national government organizations, research centers, the pharmaceutical industry and the Drugs for Neglected Diseases Initiative (DND*i*). After the development of screening assays, millions of molecules have been tested, and some have reached Clinical Trials stage. A milestone has been set for the year 2020, by the London Declaration: control of Chagas disease and leishmaniasis, and elimination

[*] **Corresponding author Jair L. de Siqueira-Neto:** Skaggs School of Pharmacy and Pharmaceutical Sciences / Center for Discovery and Innovation in Parasitic Diseases, University of California San Diego, La Jolla, CA 92093, USA; Tel/Fax: +1 (858) 822-5595; E-mail: jairlage@ucsd.edu

Marcelo Santos da Silva & Maria Isabel N. Cano (Eds.)

of Human African Trypanosomiasis. Advances and progress to achieve these goals are presented in this chapter.

Keywords: African Trypanosomiasis, Chagas Disease, Chemotherapy, Drug development, High-throughput hit selection, Lead optimization, Leishmaniasis, Medicinal chemistry, Neglected Diseases, Pharmacokinetics.

INTRODUCTION

Drugs used today to treat trypanosomiasis (Chagas disease – CD and Human African Trypanosomiasis – HAT, or sleeping sickness) and leishmaniasis (both Visceral Leishmaniasis – VL and Cutaneous Leishmaniasis – CL) are old and not adapted to the field. All have drawbacks, especially toxicity, resistance and inefficacy. The fact that no effective vaccines exist against these diseases (see Chapter 11 for details), supports the urgent need for new and better treatments. Since the affected population does not represent an attractive economic market, the pharmaceutical industry has neglected these diseases for decades [1]. The World Health Organization (WHO) named Neglected Tropical Diseases (NTDs) a group of 17 diseases in a similar situation. In addition to the three caused by kinetoplastid parasites focused in the chapter (Chagas disease, HAT and leishmaniasis), the other diseases are dengue, rabies, trachoma, Buruli ulcer, endemic treponematoses, leprosy, taeniasis/cysticercosis, dracunculiasis (guinea worm disease), echinococcosis/hydatidosis, foodborne trematodes, lymphatic filariasis, onchocerciasis (river blindness), schistosomiasis, and soil-transmitted helminthiases [2]. Despite the significant burden caused by these diseases, for decades not much has been done regarding funding, research, and public health policy. As a group, they significantly increase morbidity and mortality in tropical areas of the globe, representing a substantial economic burden on the affected endemic countries [3, 4]. For three decades starting from the 70's, only 21 drugs out of 1,556 (1.3%) were approved specifically to treat NTDs, despite the fact that NTDs affect 11.4% of the disease burden in the world.

Branding these diseases with a collective name (NTDs) was a wise marketing strategy to raise funds and awareness according to Dr. Hotez [5, 6]. The World Health Organization has recently set ambitious goals to control or eliminate 10 of the NTDs by the year 2020 and significantly reduce the burden of the other 7 NTDs. A meeting held in London in 2012 brought together representatives from the pharmaceutical industry, donors, and politicians who all agreed on a series of commitments to provide more drugs, research, and funds to achieve the 2020 goals. The document with the mission statements is known as the "London Declaration" [7]. Regarding the diseases caused by kinetoplastid parasites, the goal is to control CD and leishmaniasis and to eliminate HAT by 2020. A roadmap was launched in this meeting to guide this global plan [8].

The drug discovery process, in general, long and expensive (Fig. **1**), is a highly multidisciplinary endeavor associated with high risk, the high attrition rate of candidate compounds and requires significant amounts of resources [9]. These challenges combined with low perspectives of market profitability make these tropical diseases unattractive for pharmaceutical companies [10, 11] unless special incentives are given [12].

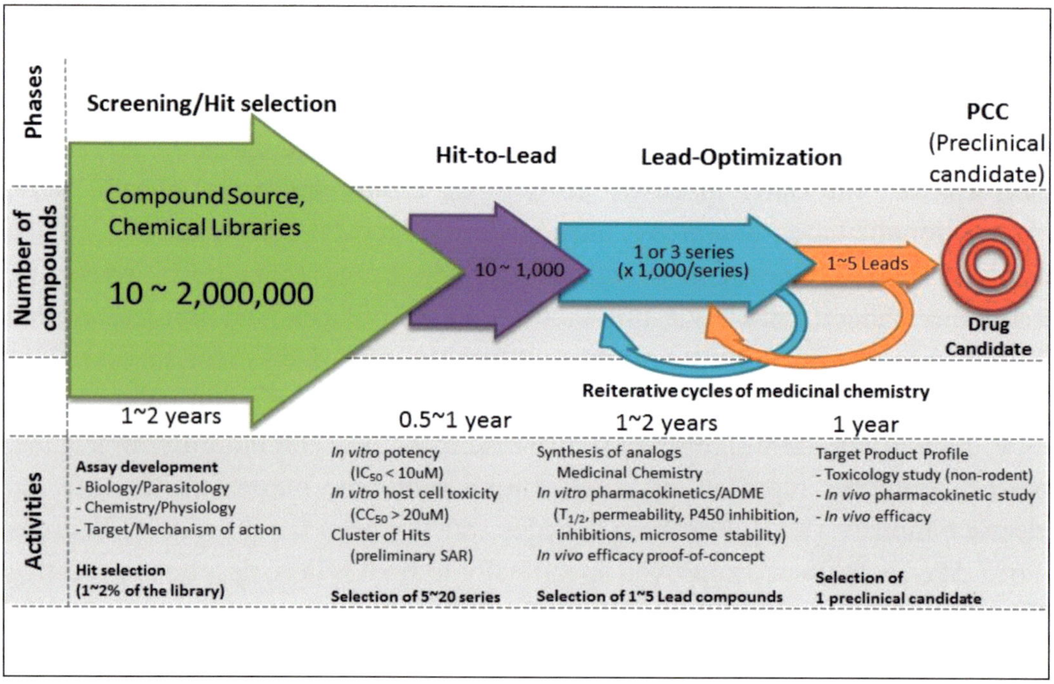

Fig. (1). Drug Discovery Pipeline process with decision-making tools and key decision points.

Before the early 2000's there were no significant initiatives for drug discovery for the diseases caused by kinetoplastid parasites. Pharmaceutical industry demonstrated no interest in CD, leishmaniasis or HAT, and academic laboratories were responsible for practically all the research done to find new active molecules against these parasites. Since the drug discovery and development process is highly complex, multidisciplinary and can cost millions to hundred millions of dollars, it was extremely improbable that an academic research center would be able to discover and develop a new treatment for one of these diseases. After identifying this demand for translational research, the philanthropic association Doctors Without Borders (Medicins Sans Frontiers - MSF) brought together a team of international experts to propose a solution for the crisis in drug research and development for neglected tropical diseases. The outcome of the experts meeting was the concept of the Drugs for Neglected Diseases Initiative (DND*i*), inspired in the Medicines for Malaria Venture (MMV) model [13 - 15]. DND*i* was founded and launched in 2003 after 7 international organizations joined forces to support the initiative: the Oswaldo Cruz Foundation (Fiocruz) from Brazil, the Indian Council for Medical Research, the Kenya Medical Research Institute, the Ministry of Health of Malaysia, the Institut Pasteur, the UNDP/World Bank/WHO's Special Program for Research and Training in Tropical Diseases (TDR) and, of course, the Doctors Without Borders (MSF). DND*i* does not have its own laboratories, only offices instead, working like a virtual organization [9]. They hold a team of experts in different stages of the drug discovery and development. The organization was able to secure funds from national governments, international agencies and philanthropic organizations. With the intellectual critical mass and financial resources, DND*i* was able to establish partnerships and collaborations with academic laboratories, research organizations with expertise in parasite biology, and also with thc pharmaceutical industry sector containing the chemical asset to be tested as well as drug development expertise. By bringing all these players together and promoting interaction between experts in the biological assays and pharmaceutical industry, the process was catalyzed in a very efficient way, and significant progress has been made in the past decade (from 2005 to 2015). Thanks to DND*i*, the method from drug discovery switched from testing few molecules in the academic environment to screening libraries of hundreds of thousands to millions of compounds in a systematic fashion through partnerships between research institutes and the major pharmaceutical companies. The first automated high-throughput phenotypic screening assays against the relevant intracellular *Leishmania* sp. and intracellular *T. cruzi* were developed at the Institut Pasteur Korea [16 - 18] with the support of DND*i*. Some pharmaceutical companies including Pfizer, Merck, Sanofi, GSK, AstraZeneca, Anacor and Synexis signed agreements to screen part of their compound libraries against neglected diseases through DND*i*.

WIPO (World Intellectual Property Organization) Re: Search is another entity with a significant role in the development of drugs for neglected tropical diseases. WIPO promotes partnerships with different players in the drug discovery process providing access to intellectual property that includes chemical compounds, technologies, know-how and data available for research and development [19].

GlaxoSmithKline (GSK), inspired by the potential to reduce the burden of tropical diseases that affected poor populations of the globe, created the Open Labs initiative and the Drugs for Developing World (DDW). The Open Labs accept proposals to develop and implement relevant projects in drug discovery at the GSK site in Tres Cantos – Spain with the help, expertise, funds and infrastructure offered by GSK. The DDW is a division at GSK focused on tropical diseases, being divided into three sub-groups: one focused on Malaria, one focused on Tuberculosis and the third, a bit behind, focused on diseases caused by Kinetoplastid parasites: CD, VL and HAT.

With high-throughput screening assays functionally running in academic centers and available external financial support, the major pharmaceutical companies gained interest in testing their library of compounds against NTDs. The new approach did not require any upfront investment from their end and could easily add value to their chemical assets also bringing a positive image in case a drug were to be found and developed for a neglected disease. The DND*i* model indeed represented a game changer in the field.

The chemical space made available by the pharmaceutical industry relies mostly on small molecules synthetic compounds. These have drug-like properties and represent the start building blocks in a drug discovery initiative. The libraries can be focused on specific chemotype, or focused on a specific molecular target, or could be simply based on diversity. Natural products represent a different chemical space compared to the small molecules and have been investigated for a number of different diseases. More than 50% of all drugs developed between 1981 and 2010 are natural products or synthetic small molecules derived from a natural product [20]. They have the advantage of being selected by natural evolution to have a biological effect. However, natural products often have resupply challenges and complex chemical conformation making it very hard to be synthesized in the laboratory. The interested in marine natural products started in the 1960s and had been continuously growing [21]. The first marine product approved by the FDA was Cytarabine (1β-arabinofuranosylcytosine), developed in 1969 to treat cancer. Six other drugs were approved shortly after Cytarabine: vidarabine to treat cancer and viral diseases, ziconotide as an analgesic, eribulin mesylate also to treat cancer, brentuximab vedotin for the treatment of Hodgkin's lymphoma and large cell lymphoma, and the omega-3-ethyl ester preparations, lovaza and vascepa developed as triglyceride lowering agents. In addition, one marine-derived drug was approved by the European Medicines Agency (EMA): trabectedin, also drawn up to treat cancer [22 - 28]. No marine natural products or derivatives have entered preclinical development for trypanosomatid diseases, but several have been reported to exhibit anti-trypanosomal activity representing a promising chemical space with good potential [29 - 31].

Finding a Drug Candidate

There are different approaches to search for active chemical compounds to treat diseases. Three common methods applied for neglected tropical diseases drug discovery are: "*de novo* synthesis", "piggy-back" and "repurposing" [32]. "*De novo* synthesis" focuses on the identification of new chemical entities through extensive compound screening. "Piggy-back" discovery takes advantage of chemical scaffolds developed for another disease that may share some mechanistic identity or a common molecular target. It facilitates chemical start points to be followed by some analog molecules ranging from few hundreds to few thousands available to perform first structural activity relationship (SAR) studies, reducing the time and cost for the lead optimization phase. An example is the use of kinase inhibitor research data from cancer treatment to provide shortcuts for the development of a kinase inhibitor more selective towards an essential parasitic kinase. "Repurposing" or "label extension" is the third and certainly, the most cost-effective approach, as the name says is based on repurposing an existing drug to be used in the treatment of neglected diseases. This approach saves considerable time and resources for the approval processes after efficacy confirmation in preclinical studies for the repurposing indication, since toxicological data and clinical tests are readily available for the initial indication. When the therapeutic dose, regimen and formulation are the same or safer than the initially approved use of the drug, the Clinical Phase I trial can be skipped and the drug can be tested directly in Clinical Phase II-A, in a small group of patients to demonstrate efficacy against the neglected disease indication. Examples of successful application of this strategy are miltefosine for leishmaniasis [33, 34], praziquantel for schistosomiasis [35] and ivermectin for filariasis/onchocerciasis [36]. More recently, Auranofin, an approved drug for rheumatoid arthritis, has been identified as an amebicide [37] and a clinical trial is being launched in Bangladesh.

Two different approaches that can be taken in the "*de novo* synthesis" to search for new active chemical entities, usually based on a diversity library of compounds. One uses a pre-defined molecular target, and the other relies on a cell-based phenotypic approach. The first (target-based) depends on previously discovered and characterized molecular target, chemically and/or genetically validated. The molecular target is usually a protein, often an enzyme that can be inhibited, blocked, activated or enhanced by a chemical entity that will interfere in the biology of the parasite disrupting an essential mechanism, and compromising its viability. The exposure effect of the active compound will either lead to deficient parasite growth or jeopardize its ability to establish a successful infection. The genetic validation of a target can be assessed by artificially modifying its expression *in vitro*. Lethal gene knockout or compromised viability of gene knockdown are both ways to validate a parasite protein as a drug target. The traditional method to achieve a gene-knockout relies on homologous recombination of the targeted gene with a selection marker - a coding sequence of a gene that confers resistance to a given antibiotic used to select parasites. It is important to verify in the genome sequence of the parasite of interest how many copies are there for the gene of interest. Multiple copies are too complicated and often impossible to be a knockout by recombination method. In the case of duplicate copies of the gene of interest, it is necessary two subsequent homologous recombination steps with two distinct selection markers. To perform the homologous recombination, it is necessary to clone the gene coding the selection marker between two homologous regions of the locus to be eliminated (containing the gene of interest), in a way that when the DNA is transfected into the parasite, homologous recombination occurs with the substitution of the endogenous gene by the exogenous selection marker. Unfortunately, parasites have usually low potential for recombination and slow growth kinetics, making transgenic techniques cumbersome [38]. Unlike *T. brucei*, *T. cruzi* does not have functional RNAi machinery due to the apparent loss of its components during evolution [39, 40]. Most recently, the use of CRISPR-Cas9 technology has been reported in *T. cruzi* with reduction of gene expression of an enzyme gene family consisting of 65 members [41].

Chemical validation of a target requires specific and selective inhibitor or probe for a given target. Both genetic and chemical approaches, therefore, have limitations. After performing the experimental target validation, inhibitors can be screened using biochemical assays. These assays usually make use of recombinant proteins. In the past, many of the labs working with kinetoplastid parasites have screened only a small number of compounds, from synthetic or natural origin. The scenario changed considerably with the development of high-throughput technology based on automated protocols, assay miniaturization and streamlined, saving time, reducing costs and leading to quality and reproducible data. These scaffolds are starting points for further optimization and highlight the advantages of combining biological and bioinformatics analysis for prioritization of molecules after performing a massive screening campaign. In another example of target-based drug discovery, CYP51 from *Mycobacterium tuberculosis* was screened against a library of 20,000 organic compounds and resulted in two active compounds [42]. One of these compounds (ChemDiv C155-0123) later showed selective inhibitory activity against the *T. cruzi* orthologous enzyme [43]. The major limitations of molecular target approaches are the possibility of poor disease linkage, low or impossible druggability of the target, the risk of off-target effects that may translate into significant toxicity, and the chance of redundant research by different groups since there are so few targets characterized [44].

To overcome target-based assay limitations, the second "*de novo* synthesis" approach is based on phenotypic readouts. Instead of a single molecular target, whole cells are tested directly in the presence of compounds and selection is made based on the observance of the required phenotype. This allows the selection of only those compounds that are active against the parasite as a whole cell, despite their mechanism of action. Also, it readily selects those compounds with the minimal pharmacodynamic and pharmacokinetic properties needed: proper intracellular distribution and accumulation, physiological binding and inhibition to target, *etc.* Their properties are difficult to predict with target-based strategies.

The most recent trend in whole-cell assays has employed the use of high-content screening software analysis – automated microscopes that can image many conditions (wells) in clear bottom culture microplates. When the technology appeared one assay was developed that made use of mammalian cells expressing GFP and parasites stained with DAPI, and both manual and automated data analysis were performed [45]. Because of limitations of genetically modified parasites/host cells, another screening technology was developed using wild-type cells and parasites, without the need of any genetic transformation [16, 17]. The development of analysis software further automated the campaigns and allowed an early assessment of host cell toxicity as well.

Balancing the benefits and disadvantages of different strategies in hit identification for parasitic diseases, phenotypic approaches emerged as the most promising methodology [32]. Particularly for intracellular parasites, as the case of *Leishmania* sp. and *T. cruzi*, a phenotypic assay allows for the selection compounds showing antiparasitic activity without affecting the host cell. Some recent first-in-class new drugs were discovered by phenotypic screening [46].

Usually, in large screening campaigns, compounds are tested at a single concentration, as single points or duplicates. After comparing the antiparasitic activity with positive and negative controls, a threshold is defined, and the selected active compounds are then cherry-picked and re-tested in dose response to determine potency measured by the IC_{50}, the concentration to achieve 50% of its maximal effect (parasite inhibition). General criteria for hit selection for infectious diseases of the developing world have been recently published [47]. An IC_{50} lower than 10 μM (pIC_{50} >5) is considered a good starting value. It is good practice to confirm activity using a different batch of the compound and to quality-control the structure using mass spectrometry, for example. When available, confirmed hits should also be subject to alternative (secondary) assays to assess antiparasitic activity using different readouts.

Once the hits are identified, it is important to make few chemical modifications in representative molecules to have a sense of chemical tractability of the scaffold. The question: "Can we optimize this scaffold for activity and reduced toxicity?" has to be answered to define which series should progress in the development pipeline. The medicinal chemistry necessary to optimize a lead optimization is very costly and constitutes a real obstacle to many drug discovery efforts. Few in academia have the resources or access to the synthetic chemistry capacity necessary to produce the tens to hundreds of compounds usually required for lead optimization. A serious lead optimization campaign for one chemical scaffold should have at least 5 chemists (ideally 8 to 10 chemists) and would have few thousand of molecules produced over the course of 1 to 2 years. All these molecules would be tested in the biological assays, and the results would represent a quick feedback for the chemists to decide on the next chemical modifications need to be explored. At this point, in addition to the biological activity, it is also important that the select molecules present adequate pharmacokinetic profile, including good solubility, aLogP, stability, affordable production cost, safety (hERG) and being non-mutagenic. A set of criteria for both hit classification and lead selection have been recently published [47].

In the case of *T. cruzi* and *T. brucei*, it is especially important to test the activity of the compound against different strains of parasites, since they may have very different susceptibility [48]. The same holds true for different species of *Leishmania* sp. always taking into consideration if the drug will be developed to treat visceral leishmaniasis, cutaneous (or mucocutaneous) or both.

Another important aspect of the drug discovery and development is written in a document named Target Product Profile (TPP) [49]. This document lists a set of ideal and acceptable parameters the future drug must have. It sets the goals and helps driving the development of the molecules for a particular disease. DND*i* makes available the TPP for CD [50], Leishmaniasis [51] and HAT (www.dndi.org). An optimized lead compound that fits the TPP can be defined a preclinical candidate and enter the preclinical phase.

CHAGAS DISEASE (CD)

Two distinct phases characterize CD: the acute phase takes place in the first few weeks post-infection, followed by the chronic phase. During the acute stage, the infection can be asymptomatic, or it can be associated with flu-like symptoms, including fever, abdominal pain, *etc.* Blood parasitemia defined by a number of trypomastigote parasites in the peripheral blood circulation is characteristic of the acute phase. After the first weeks, the host immune system controls the blood parasitemia, reducing it to undetectable levels. While asymptomatic, this phase is named indeterminate chronic phase. Parasites have been described in different tissues at that phase, infecting cells from the heart muscle, smooth muscle from the digestive system, adipocytes, liver and eventually the central nervous system. About 30% of the infected individuals develop cardiomyopathy, characterized by the formation of fibrotic tissue in the heart as a consequence of an inflammatory response against the parasite present in that tissue. The fibrotic heart leads to heart failure, slowly progressing and eventually culminating in death. Another 10% of the Chagas positive individuals in the indeterminate chronic phase will develop a digestive syndrome characterized by the formation of fibrosis in the colon or esophagus as a consequence of an inflammatory process against parasites located in those tissues. The malfunction of these muscles jeopardizes the proper flow in the digestive system, progressing to the characteristic megacolon or megaeso-phagus syndromes. For yet unknown reasons, 60% of the people once infected with *T. cruzi* will never develop any pathology associated with the parasite infection. Unfortunately, there are no good biomarkers to predict which infected individuals will develop a particular pathology or not, make it complex to define an efficient treatment plan for the patients. Details about *T. cruzi* virulent factors and immune evasion are described in Chapter 8.

Determining parasite burden in the peripheral blood through real-time PCR based on parasitic DNA traces in the circulation is the method currently used to assess the efficacy of antiparasitic drugs in clinical trials. A positive result for the presence of parasitic DNA in the circulation is an undisputable evidence of parasite presence and is used to define failure for a parasitological sterile cure in clinical trials. A negative result, on the other hand, cannot be used to define the parasitological sterile cure because the levels of parasitic DNA can simply be below the detectable levels at the time of material collection. Therefore, this method based on real-time PCR cannot be used to assess treatment success in clinical trials. The test is usually performed at the end of the treatment and repeated after 6 months, 1 year and 2 years. Repeated negative results provide stronger indications of parasitological cure, but there is always the chance of late rebound. Details about the use of molecular tools to detect parasite in blood samples are available in Chapter 9.

The biology of *T. cruzi* and interaction with the host should be taken into account when developing a new drug. An effective drug should be able to eliminate parasite both in the acute and chronic phases of the disease. Moreover, if a biomarker that predicts the manifestation of pathology in the chronic phase is developed, the approach to managing Chagas patients will likely no longer be a parasitological sterile cure, but merely preventing the expression of the cardiomyopathy or the digestive syndromes assessed by this biomarker.

Current Chemotherapy for Chagas Disease

The only two available drugs to treat CD are benznidazole and nifurtimox. These are both nitro-heterocyclic compounds developed decades ago and associated with some adverse effects. Even though the disease is prevalent in the United States with more than 300,000 estimated cases [52], in Europe with 120,000 cases [53], in Japan with estimated 4,000 cases [54] and even in Australia [55], neither of these drugs are approved by the respective regulatory agencies in these locations due to the toxicity of treatment. The efficacy of these medications in the chronic phase was not clear until recent results of a clinical trial using benznidazole which demonstrated that it is not able to prevent the progress of cardiomyopathy in chronic patients [56].

Benznidazole

Benznidazole (*N*-benzyl-2-(2-nitro-1*H*-imidazol-1-yl)acetamide) was developed in the seventies by Roche and marketed as Rochagan™ or Rodani™. The drug is metabolized in eukaryotic cells, being reduced by nitroreductases and generating very reactive free radical intermediates and electrophilic metabolites that are responsible for the trypanocidal activity. The reactive elements of reduced benznidazole nonspecifically target parasitic proteins, lipids and nucleic acid material, disrupting the normal cell function and metabolism. There are also scientific reports suggesting that *T. cruzi* NADH-fumarate reductase inhibition, phagocytosis improvement and death by INF- γ are also implicated in the antiparasitic effect of benznidazole [57]. The host cells also contain nitro-reductase, activating the drug in non-specific tissues and leading to side effects [58]. The human liver NADPH, xanthine oxidase, aldehyde oxidase and cyto-chrome P-450 reductase are some of the enzymes associated with the activation of benznidazole in the human host, leading to the observable adverse effects of the treatment [59].

The most common side effects are peripheral neuropathy and dermatitis with the appearance of skin rashes from hypersensitivity to the drug in 20 to 25% of the patients, usually after 10 days of the treatment. Weekly monitoring is recom-mended for this reason. Other associated but less frequent adverse effects include digestive intolerance, depression of bone marrow, toxic hepatitis and lymphomas. Treatment length is up to 60 days, and half of the patients cannot handle the entire course of therapy due to adverse effects. When used in the acute phase of the disease, it reduces symptom severity and clears parasitemia [60]. Upon treatment of congenitally infected infants in the first year of life, the clinical cure rates are very high in the 90% range [61]. In non-infant cases of acute CD, the clinical cure is reported for more than 60% of the cases [62].

Patients in the chronic phase of CD have been treated with benznidazole in different parts of the world, even though the efficacy of the treatment had not been clinically assessed. Some reports indicated 20% cure rates at this stage [63]. Recently, a large clinical trial called BENEFIT (Benznidazole Evaluation For Interrupting Trypanosomiasis, ClinicaTrial.gov, ID: NCT00123916) with 2,854 patients from different endemic countries had its results published. This was a multicenter, double-blind, randomized, placebo-controlled trial with the goal to evaluate the ability of benznidazole treatment with daily doses during 40 to 80 days of treatment to reduce mortality and morbidity in patients with chronic Chagas cardiomyopathy. The results indicated that benznidazole was indeed efficient to reduce the parasite activity in these patients, but unfortunately, that did not correlate with reduction or stagnation of heart complications over a 5-year period [56]. The lack of benefit in reducing the progression of cardiomyopathy may have been related to strain susceptibility or insufficient long-term activity against the parasite. The lesson learned in this clinical trial is that reduction of parasitic activity in the chronic cardiac phase of the disease is not sufficient to prevent the progress of the cardiomyopathy. Strains resistant to benznidazole are another concern and have been reported *in vitro*. *T. cruzi* Colombian strain, for example, was reported to cure only 16% of mice infected with different clones [64]. *In vitro* results using real-time PCR suggest nitroreductases (NTRs) as the principal mechanism of resistance *in vitro*, probably due to loss of an NTR gene copy [65, 66]. A recent study warns of the relative ease in which benznidazole can develop resistance *in vitro* by a couple of different mechanisms such as chromosome loss and different point mutations in the NTR gene, all arising from a single population [67].

Nifurtimox

Nifurtimox is the other drug in use to treat CD. It is a 5-nitrofuran derivative (N-(3-methyl-1,1-dioxo-1,4-thiazinan-4-yl)-1-(5-nitrofuran-2-yl)methenamine developed by Bayer in the sixties and named Bayer 2502 during its development, now marketed as Lampit™. It was the first treatment available to treat Chagas patients. The mechanism of action of nifurtimox is very similar to benznidazole since it is also a nitroheterocyclic compound [68]. The drug is reduced by nitroreductases generating reactive free radicals followed by redox cycling. Differently from benznidazole, however, in the case of nifurtimox, it seems like reactive oxygen species are more involved in the anti-parasitic activity, including superoxide ion and hydrogen peroxide. The absence of catalase or peroxidase activity and reduced superoxide dismutase activity, turns *T. cruzi* sensitive to oxidative stress [69].

The efficacy of treatment is similar to benznidazole, however, it is associated with more severe and more frequent adverse effects [59]. Reports indicate that 98% of patients suffer some form of side effect [70]. Less than half of the Chagas patients treat with nifurtimox can complete the 60-day course treatment. About a third of the patients will not even reach the initial 30 days. Most common side effects are digestive symptoms. Nifurtimox is blood-brain permeable, so neurological alterations are also present and perhaps the most serious adverse effects due to nifurtimox administration, including psychic alterations, is excitability and sleepiness. Other frequent adverse effects observed are anorexia, loss of weight, digestive manifestations such as nausea or vomiting, and occasionally intestinal colic and diarrhea. Case studies indicate that 7% of patients had other severe adverse effects including angioedema, myocarditis, and grade-3 anaphylactic reactions [71].

Compared to benznidazole, it is relatively easy to evolve *T. cruzi* resistance to nifurtimox *in vitro*. Parasite nitroreductases seem to be associated with the resistance mechanism, suggesting that cross-resistance with benznidazole is possible [65]. It is known that some strains of *T. cruzi* are less susceptible to benznidazole or nifurtimox (ex.: Colombian and Y strains), but that seems to be an intrinsic characteristic of the strains and not an acquired resistance after drug exposure.

Discovery Initiatives for Chagas Disease

Similar to all the other NTDs, the scenario for drug discovery and development for CD was not very positive ten years ago, with initiatives depending mostly on academic laboratories working on few molecule candidates [72, 73]. Being a very complex and multidisciplinary process, it was highly improbable that one compound would make its way to the patients without industry partnership. The quality of the available compounds was often poor and associated with druggability issues [74, 75]. Putative targets and pathways of potential essentiality for *T. cruzi* such as cysteine protease, trypanothione reductase, and ergosterol biosynthesis were addressed for a new chemotherapy. Phenotypic assays were also applied, but with limited throughput [76, 77]. The scenario has dramatically changed and recently there are many research groups involved in the drug discovery and development for CD, including both academic groups and the pharmaceutical industry sector. Investment from national governments and philanthropic entities provided the necessary resources to boost the field, gaining attraction from pharmaceutical companies that could offer chemical matter and expertise in the process of drug development. Technological advancements have been made enabling high-throughput screening of large compound libraries in a short time. One example is the Chagas Drug Discovery Consortium created in 2010 funded by the U.S. National Institutes of Health (NIH) including laboratories from the United States to discover candidate drugs for the disease [78]. All these advances allowed for the first clinical trials of new chemical entities for CD in the past 40 years [79, 80].

The Global Health Innovative Technology (GHIT) Fund recently gathered experts from the Medicines for Malaria Venture, the DND*i*, the TB Alliance, representatives from the Bill & Melinda Gates Foundation, aiming the definition of disease-specific criteria for hits and leads for tropical infectious diseases including CD. For CD the criteria for hits included cellular potency characterized by a half-maximal inhibitory concentration (IC_{50}) <10 μM against intracellular *T. cruzi*. Key lead selection criteria: in an acute mouse model of CD, a hit should demonstrate at least 80% reduction of a number of parasites in organs as well as an increase in lifespan compared to untreated controls. The outcomes have to be achieved with up to 50 mg per kg delivered orally for a maximum of 10 days [47]. The Target Product Profile for CD is presented in Table **1**.

Table 1. The Target Product Profile (TPP) for Chagas disease as currently proposed by DND*i* [50].

	Acceptable	**Ideal**
Target population	Chronic	Chronic and Acute (Reactivations)
Efficacy	Non inferior to benznidazole standard dose* in all regions (parasitological)	Superiority to benznidazole standard dose to different phases of disease (acute and chronic) (parasitological)
Clinical efficacy	Non inferior to benznidazole in all endemic regions (parasitological)	Superiority to benznidazole to different phases of disease (acute and chronic) (parasitological)

(Table 1) contd.....

	Acceptable	Ideal
Safety	Superiority to benznidazole* in the frequency of definitive treatment discontinuations for medical indication (clinical and laboratory)**	Superiority to benznidazole* in the frequency of definitive treatment discontinuations for medical indication (clinical and laboratory)**
Contraindications	Pregnancy/lactation	None
Precautions	No genotoxicity; No significant pro-arrythmic potential	No genotoxicity; No teratogenicity; No negative inotropic effect; significant pro-arrythmic potential
Interactions	No clinically significant interaction with antihypertensive, antiarrhythmic and anticoagulants drugs	None
Presentation	Oral/Parenteral (short POC)*** Age-adapted	Oral Age-adapted
Stability	3 years, climatic zone IV	5 years, climatic zone IV
Dosing regime	Oral - any duration. Parenteral <7 days	<30 days

*As per WHO recommendation; ** No genotoxicity is a condition only for NCEs; *** Need for parenteral treatment for the severe disease.

Some targets have been investigated for drug discovery purposes. One of the most promising is shared with fungi pathogenic organisms: the sterol 14 α-demethylase (CYP51), involved in the biosynthesis of sterols in *T. cruzi* [81]. The target was identified by Roberto Docampo in the early 1980's [82]. The enzyme catalyzes the removal of the C14 methyl group from the intermediate sterol eburicol originating 14α-demethyl-14dihydroeburicol [83]. Sterols are the end products of this pathway and have essential functions including a role in the membrane fluidity and permeability, regulation of enzyme activity bound to the membrane and signal. Plants, fungi and protozoan organisms produce ergosterol, the equivalent of the cholesterol produced by the other eukaryotes. The differences are a second double bond in the B ring and a fully saturated side chain with a methyl group at C24 in cholesterol. After the validation of this pathway in *T. cruzi*, compounds that were originally developed as antifungals were naturally tested against the protozoan parasite. The early generation of imidazoles (*e.g.* miconazole, ketoconazole) and triazole (*e.g.* itraconazole, fluconazole) sterol biosynthesis inhibitors confirmed antiparasitic activity, however, they failed to achieve parasitological sterile cure [84]. The newer generation of azoles to treat fungi infections was also tested for anti-*T. cruzi* activity. Posaconazole was one of them and represented one of the most promising molecules of the past decade to become a CD drug. This triazole originally marketed as Noxafil by Schering-Plough pharmaceutical is indicated against *Candida* spp. and *Aspergillus* spp., and shows potent and specific *in vitro* activity against *T. cruzi*, in particular against the amastigote stage. Effect on acute and chronic mice models was curative, as some earlier tested antifungal compounds demonstrated [85]. Posaconazole also proved to have trypanocidal activity against benznidazole and nifurtimox resistant strains developed in the lab, even in immunosuppressed mouse models [86]. A comparative study between posaconazole and benznidazole in a mouse model of CD showed both drugs led to 100% survival rates, suppression of parasitemia and negative *T. cruzi* antibodies. However, when assessed 54 days after infection only, posaconazole-treated mice had completely negative hemoculture compared to benznidazole-treated that had 50% parasite reappearance in hemoculture. Also, enzymatic plasma assessment of cardiac lesion was indistinguishable from uninfected control for posaconazole but significantly higher for benznidazole [87]. In addition to the antiparasitic activity, posaconazole has a great pharmacokinetic and safety profiles with a long half-life and a large volume of distribution, coupled with the fact that this lipophilic drug accumulates in cell membranes. This is expected to give high local concentrations of the drug to interact with the membrane-bound CYP51 target [88]. A drawback is the difficult synthesis of posaconazole and the associated costs of about €8,000 per treatment, a value that clearly is incompatible with the economically impoverished majority of the population affected [89]. These promising results led to two clinical trials: one Phase 2 Clinical Trial sponsored by Hospital Universitairi Vall d'Hebron Research Institute (ClinicalTrials.gov, ID: NCT00349271, CHAGASOL) that evaluated posaconazole and benznidazole for the treatment of CD chronic infection [90] and a Phase 2 Clinical Trial by Schering-Plough (now Merck & Co.) for the treatment of asymptomatic CD. In a reported case study, a woman with chronic CD and systemic lupus erythematosus (requiring immunosuppression) was treated with posaconazole eliminating *T. cruzi* completely, whereas benznidazole treatment failed in this patient [89]. The first clinical trial of posaconazole in humans did not replicate the results reported for this first patient with lupus erythematosus. The trial enrolled patients in the indeterminate chronic phase with positive real-time

PCR results and absence of cardiomyopathy or digestive syndrome. Posaconazole treatment was conducted for 60 days with 100 mg and 400 mg twice a day, compared to benznidazole 150 mg twice a day. Its efficacy was assessed through real-time PCR done on peripheral blood collected in the course of the treatment, and another follow up collected 10 months after the end of treatment. At the end of the treatment, all the patients that were treated for at least 14 days with any of the drugs or doses were negative for real-time PCR reaction. However, 10 months after treatment, traces of *T. cruzi* DNA were detected by real-time PCR in 92% of patients from the low-dose posaconazole group and 81% from the high-dose posaconazole group, compared to the surprisingly low failure of 38% for the patients treated with benznidazole [79]. One Clinical Trial currently in course aims to assess the efficacy of posaconazole in combination with benznidazole (ClinicalTrials.gov, ID: NCT01377480, STOP CHAGAS) [91].

Another ergosterol biosynthesis inhibitor that reached patients in the clinical study is Ravuconazole prodrug: E1224, from Eisai Co. E1224 is the mono lysine salt of ravuconazole with a short half-life with a simpler and cheaper synthesis compared to posaconazole. In this trial sponsored by DND*i* and Eisai, adults with chronic indeterminate CD were given a placebo, E1124 or benznidazole (ClinicalTrials.gov, ID: NCT01489228). A series of examinations were then carried out in the following months to evaluate parasitological sterile cures [92]. Same to what happened with posaconazole, E1224 failed to maintain sustained efficacy 1 year after the end of treatment.

Fenarimol is another fungicide inhibitor of CYP51 with anti-*T. cruzi* activity *in vitro* and *in vivo* in a mouse model of CD, reducing parasitemia after 20 days treatment to negligible levels. However, reactivation was observed after three cycles of immunosuppression [93]. Novel compounds designed to target TcCYP51 are still being pursued by some academic groups [94, 95]. In addition, other enzymes of the ergosterol biosynthesis pathway may be targeted as potential drug targets for CD, including squalene synthase, the enzyme responsible for the first step of ergosterol biosynthesis [96, 97]. The effective and potent squalene synthase inhibitor 4-phenoxyphenoxyethyl thiocyanate effective against epimastigote proliferation produces an accumulation of mevalonate pathway intermediates, being an example of a compound targeting this enzyme [98, 99]. SQ109 is another example of drug lead for CD targeting squalene synthase. E5700 is a drug in development by Eisai as human cholesterol lowering agent, also targeting squalene synthase and is showing efficacy against *T. cruzi* [100]. Amiodarone also inhibits ergosterol biosynthesis and is currently in clinical trials (as well as dronedarone) against the chronic phase of the disease [101]. This antiarrhythmic drug is used in the treatment of cardiac failure in chronic chagasic patients and has been found to act in a synergistic manner with azoles in disrupting *T. cruzi* metabolism. Amiodarone interferes with the calcium homeostasis but also inhibits ergosterol biosynthesis, while posaconazole or itraconazole also affects calcium homeostasis, suggesting a viable and advantageous drug combination [102 - 104]. Even though so many initiatives are still investing on ergosterol biosynthesis inhibitors to treat Chagas patients, this target has been deprioritized by DND*i*, some research institutes, and pharmaceutical companies. An assay for rapid triage was recently developed by scientists from Dundee University and GSK to assess ergosterol biosynthesis inhibition quickly, as a way to filter out the active ones [105].

Fexinidazole, which is currently in late-stage clinical trial in Africa to treat sleeping sickness, is a nongenotoxic 5-nitroimidazole with demonstrated curative properties in a mouse model of CD [106 - 108]. The renewed interest in nitro-heterocycles has spurred research into finding alternative nitro-heterocycles, these include heteroallyl-containing 5-nitrofuranes, 5-nitrofuryl containing thiosemi-carbazones and 2- or 3-nitro-1H-imidazole-based amides and sulfonamides. Nitro-based compounds are well known to be associated with adverse effects including digestive toxicity and neuropathy (central or peripherical). To reduce the off-target effect and side effects, one should design compounds that will be reduced selectively by parasitic reductase enzyme to activate the nitro- group. Another issue is that parasites eventually resistant to benznidazole or nifurtimox would have a good chance to be also resistant to other nitro-based drugs. Some recently synthesized novel nitrofuran amides show excellent selectivity toward the parasites and have a trypanocidal activity that seems to be independent of nitroreductase activity. A recent review highlights the potential of re-discovering nitro drugs to treat trypanosomatid diseases [109].

Trypanothione reductase is a drug target on its own involved in the thiol-dependent redox metabolism [110]. This is essential for detoxification of free radicals and maintenance of the intracellular reducing environment. Recently, a study validated the trypanothione pathway as drug target by a metabolic modeling approach. All of the constituent enzymes and transporters present are essential for proper pathway function, but not all of them have therapeutic potential [111]. Many reductase inhibitors have been identified, including polyamine derivatives, crystal violet, acridine-based tricyclics, phenothiazine, benzodiazepine, isoalloxazine and pyridoquinoline [112 - 115]. *In vivo* testing of a trypanothione reductase inhibitor, thioridazine, promoted heart protection but failed to eradicate the parasite [116] completely.

Protease inhibitors have also been successfully tested against *T. cruzi*. Cruzain, also known as cruzipain), is a cathepsin L-like cysteine protease identified in the early 1990's at UC San Francisco by James McKerrow and Charles Craik [117, 118]. The enzyme is responsible for the majority of proteolytic activity in all the stages of *T. cruzi* and is involved in a number of metabolic pathways, metacyclogenesis, immune evasion, invasion of host cells and has been validated as a target for drug discovery [118-120], see Chapter 8 for more details about cruzipain and other *T. cruzi* virulent factors). One of the most advanced inhibitors for this target is the K777, a vinyl sulfone cysteine protease inhibitor originally synthesized at Khepri Pharmaceuticals for anti-inflammatory indication. During early tests against *T. cruzi*, K777 demonstrated to alter the parasite Golgi complex ultrastructure and function, leading to a lethal phenotype [121]. Later, another work demonstrated the proof-of-concept for *in vivo* efficacy to treat CD, first in a mouse model being able to rescue mice from lethal infection displaying repetitive hemoculture, and indicative of parasitological cure [122]. It was also effective to treat mice with the non-functional immune system, indicating that its activity was independent of the host's immune system [123]. K777 also abrogated myocardial damage in beagle dogs treated orally for seven days [124]. After confirming safety in a non-human primate toxicology study, K777 will be filed for first-in-human Clinical Trial Phase I, contingent on the US-FDA authorization. Other classes of cruzain inhibitors have been reported as potential drug leads, including selenosemicarbazones [125], amidines bearing benzofuroxan or benzimidazole [126], and other scaffolds [127, 128]. Effective nitrile inhibitors of cruzain have also been identified and serve to validate this target [129] chemically. A high-throughput screen for 197,861 compounds against cruzain yielded 921 active compounds. In parallel, a computational docking analysis excluded false positives, finally selecting 146 specific competitive ligands, representing 5 different chemical scaffolds [130]. Thiosemicarbazones are another class of cruzain covalent inhibitors originated from a screening against cruzain. Thiosemicarbazones inhibit Cysteine proteases by forming a reversible tetrahedral adduct of the Cys thiolate with the carbon of the thiocarbonyl group [131]. Some thiosemicarbazones were inactive against *T. cruzi* enzyme, but even so were lethal to *T. cruzi* parasites in a cell-based assay, indicating an alternative mechanism of action in addition to cruzain inhibition - polypharmacology [132 - 134]. Later this class of compounds revealed to have binding promiscuity, being deprioritized because of off-target toxicity.

Pyrophosphate metabolism in *T. cruzi* is another pathway that has gained attention as a target. The metabolic activity happens in the parasitic specific organelle called acidocalcisomes involved in calcium homeostasis, response to cell stress, osmoregulation, and energy transduction [135]. Some bisphosphonates currently used to treat osteoporosis accumulate in the acidocalcisomes and can inhibit a key enzyme of pyrophosphate metabolism – farnesyl pyrophosphate synthase [136]. The first report of this activity in animal models demonstrated that risedronate demonstrated antiparasitic effect and reductions in mortality, even though complete parasitological cures were not achieved [137]. Cardiopathy could not be avoided or prevented in infected mice with the treatment, and Tulahuen strain infected C57BL/6 mice had no improvement in mortality [138].

Based on all the recent initiatives, it is possible to note a trend moving from target-based screening to target-free phenotypic assays, with most of the identified targets being not well validated [139].

Using the target-free phenotypic screening approach combined with a focused library of compounds based on one or few scaffolds can accelerate the hit to lead phase with a preliminary SAR done from the primary screening, guiding the future promising synthesis. A typical example of this exercise is the oxaborole series produced by Anacor [140, 141]. Combining a target-free screening with focused library sets represent a strategy to advance rapidly to a lead candidate since some pharmacokinetic properties of the compounds are frequently available for these target libraries including solubility, metabolism data, and plasma exposure. One preclinical oxaborole candidate is still being developed for CD.

The first large target-free phenotypic screening campaign to test small synthetic chemical compounds against *T. cruzi* in a cell-based assay and made publicly available was performed by the BROAD Institute of MIT and Harvard and MIT. They tested 303,224 molecules and selected 35 compounds based on antiparasitic activity, selectivity, potency and chemical tractability. After reducing the active ones to three scaffolds of interest, the team selected the probe ML157 (CID-16190867) was selected the lead candidate with 100-fold selective inhibition, a drug-like structure and no chemical liabilities [142]. In 2013, the Broad Institute partnered with Eisai to further develop compounds of interest. Funds from the Global Health Innovative Technology Fund (GHIT) will support the optimization of diversity-oriented synthesis–derived small molecules [143, 144].

Another relevant academic/industry partnership was established between the Drug Discovery Unit at Dundee University and the GlaxoSmithKline (GSK) from Tres Cantos, Spain. They were able to secure funds from the Welcome Trust to screen compounds against *T. cruzi* as well as for *T. brucei* and *Leishmania* [145]. The results of the interrogation of the GSK library of 1.8 million compounds were recently published [146]. A set of 222 hits against *T. cruzi* were made available for the scientific community to help prioritize the most interesting ones. In addition, the GSK Open Lab Foundation at Tres Cantos – Spain provides opportunities to researchers to identify and optimize compounds with the support of GSK assets and expertise (http://www.openlabfoundation.org/).

The European scientific community is certainly engaged in fighting the three tropical diseases caused by kinetoplastid. The European Framework Program 7 (FP7) provided generous funds to support multidisciplinary and multi-institutional consortia targeting CD, leishmaniasis and sleeping sickness. Four consortia are currently active: KINDRED, NMTrypI, PDE4NPD and A-PARADISE (CORDIS; Seventh Framework Programme (FP7) http://cordis.europa.eu/projects/home_en.html). Promising results will soon be shared with the community, and new drug candidates are optimistically expected.

In the American continent, the US National Institute of Health supported a partnership between the University of California and the Genomics Institute of the Novartis Research Foundation targeting *T. cruzi*. The initiative lead to the discovery of antiparasitic activity of xanthine scaffold [147]. The collaboration also resulted in the identification of a novel drug target in the parasite: cytochrome b [148].

Further advanced in the discovery pipeline, DND*i* Latin America recently started the Lead Optimization Latin America Initiative (LOLA). Academic groups from Brazilian universities UNICAMP (University of Campinas) and USP (University of Sao Paulo, Sao Carlos campus) will be responsible for providing medicinal chemistry synthesis combined with pharmacokinetic studies to develop preclinical candidates for CD and leishmaniasis. In the long-term, goals and expectations grow to meet the need for innovative discoveries for the treatment of neglected diseases. This partnership is consolidated based on the chemistry experience of UNICAMP and USP, and DNDi's access to pharmaceutical and biotechnology partners' libraries of compounds and expertise in evaluation and safety. The TDR / WHO is also supporting the initiative. The LOLA project aims to build upon and enhance the research and development potential in the region. With an international collaborative approach, this 'virtual laboratory' sets a precedent for all emerging neglected disease-endemic countries. For the first time, early stage R&D activities are rooted in Latin America, an endemic area for both CD and leishmaniasis (http://www.dndi.org/media-centre/news/1657-first- early-stag--research-latin-america.html).

Another significant advance in the field of drug discovery for CD is the development of more sophisticated animal models to assess therapy efficacy *in vivo*. After selecting initial hits and lead candidates, it is important to demonstrate a proof of concept for efficacy *in vivo*. The translation of results observed in mouse models of CD and the clinical outcomes is limited to only 4 drugs: benznidazole and nifurtimox (reference drugs used in patient's treatment), posaconazole and E1224 Ravuconazole prodrug (both CYP51 inhibitors tested in Phase II Clinical Trials). Existing animal models address both disease pathology and treatment efficacy. Posaconazole was shown to be not inferior to benznidazole in murine models [149 - 151]. This observation was later proved a different situation in human clinical trials, demonstrating that posaconazole performed significantly worse than benznidazole, with more than 80% of patients had parasite reappearance 10 months after treatment compared to only 35% of the patients treated with benznidazole testing positive for the presence of parasites [79]. Models currently available in the pre-clinical stage cannot predict efficacy of drug candidates in clinical trials with confidence [149]. Recently, the group lead by John Kelly at the London School of Hygiene and Tropical Medicine developed a mouse model infected *T. cruzi* parasites expressing red-shifted luciferase. This was the only method able to indicate the better performance of benznidazole compared to posaconazole to treat CD [152]. The method was first published in January/2015 [153], after the posaconazole clinical trial study was concluded. New drug candidates should be tested in this mouse model before being tested in clinical trial studies.

Some alternative animal models have been tested and could represent different approaches studying the disease biology, but probably not to screen or predict drug efficacy due to significant cost elevation compared to mice models. Rats, for example, have been tried developing a mild and slow pathology, suggesting low susceptibility to *T. cruzi* infection [154]. Rabbits develop characteristics of the chronic alterations such as focal myocarditis with a fibrous component, but do not show more severe forms of chronic myocarditis or the histological lesions in the digestive track and skeletal muscles found in common infections [155]. The Syrian hamster has also been proposed as an animal model for chronic Chagas with cardiomyopathy but was not able to display all the characteristics findings of human cases [156]. Dogs, on the other hand, develop most of the clinical aspects of the human CD, in particular, the indeterminate form characterized by a latent infection, without disease symptoms, and with normal electrocardiograms; just a fraction of the animals develop chronic phase symptoms [157, 158].

Future Perspectives for Chagas Disease Chemotherapy

The lack of reliable biomarkers of a cure for CD is a challenge for drug development. Recent clinical trials assessed failure through real-time PCR performed on peripheral blood from patients [79]. Immunoassays using recombinant proteins or peptides could be alternatives, but results have not been satisfactory. Recent studies indicate fragments of the human apolipoprotein A1 (APOA1), a component of the high-density cholesterol, explicitly associated with CD [159]. Further assessment is currently under investigation and could become the new standard to diagnose and evaluate successful treatment/cure of CD.

Results of the recent clinical trials with inhibitors of the biosynthesis of ergosterol (posaconazole and E1224) have once again raised the discussion about the existence of latent parasites in CD. Based on the mechanism of action of these inhibitors targeting the production of a membrane component, one can imagine its efficacy is dependent on the parasite demand for replication. In other words, if some parasites were able to stay in a non-dividing (latent) state, they would not be affected by inhibitors of ergosterol biosynthesis. That would explain why the patients became PCR negative right after the treatment with posaconazole or E1224 but became again positive after months of follow-up post-treatment. At the end of the treatment, the drugs should have eliminated all actively dividing parasites, but parasites in a hypothetical non-dividing state would survive the treatment, eventually being re-activated in the future. There is no scientific evidence to date supporting this hypothetical non-dividing state, but the research community has addressed this question. Independently from the inexistence of a scientific explanation for the disappointing results of posaconazole and E1224 trials, some organizations including DND*i*, universities, and pharmaceutical companies are deprioritizing compounds targeting the biosynthesis of ergosterol to treat CD [146].

The BENEFIT trial with results recently published taught us another valuable lesson. The study confirmed that benznidazole has a significant trypanocidal activity even in the chronic phase of CD. Unexpectedly, though, the reduction of parasites did not prevent or reduced the cardiomyopathy of the patients. This finding has profound implications in the decision of care for the chronic patients with some signs of cardiac manifestation. The cardiomyopathy will have to be addressed independently of, or at least in addition to directly fighting the parasites. So, the Target Product Profile for CD would need to be updated, stating for the "ideal" treatment a combination therapy addressing both the parasite and the prevention or recovery of tissue damage/fibrosis.

HUMAN AFRICAN TRYPANOSOMIASIS (HAT)

Human African trypanosomiasis (HAT) or sleeping sickness is caused by the protozoan parasite *Trypanosoma brucei* and transmitted by the bite of infected tsetse flies (*Glossina* spp.). Differently from *T. cruzi*, the parasites are extracellular in both stages in the insect vector and in blood and tissue fluids of their human host. Cases of the disease are limited to the geographical area where tsetse flies are present, mainly the sub-Saharan Africa [160], including 36 countries. Most of the transmissions occur in rural areas [161]. Two strains with different geographical distributions are responsible for the disease in humans: *T. brucei rhodesiense* present in East and Southern Africa, and *T. brucei gambiense* in West and Central Africa. Infection with *T. b. rhodesiense* progresses rapidly from an acute form of the disease characterized by parasites in the blood circulation and lymphatic system (stage 1), to the stage 2 characterized by parasites in the central nervous system. Infection with *T. b. gambiense* stays chronic in stage 1 for a longer period. Symptoms of stage 1 with parasites restricted to the blood and lymph system include fever, headache, joint pain and itching [162]. The clinical manifestations of stage 2 of the disease characterized by trypanosomes crossing the blood-brain barrier are neurological and endocrinal disorders [163]. When left untreated sleeping sickness is lethal within months if caused by *T. b. rhodesiense* or years if caused by *T. b. gambiense*. The disease causes a huge humanitarian and economic burden in Africa, being hypothesized as one of the factors jeopardizing the development of the continent [164, 165]. The most devastating epidemics of sleeping sickness happened from 1896 and 1906 and resulted in up to 500,000 deaths. It affected the Congo Basin and the Busoga focus in Uganda and Kenya [166]. More details about HAT clinical manifestations and epidemiology and parasite life cycle and taxonomy are described in Chapter 2. As a consequence, efforts on drug discovery demanded by colonial administrators led to the development of the first anti-sleeping sickness drugs.

Current Chemotherapy for HAT

The efforts to optimize a chemical compound for a specific disease applying reasonable chemical modifications in the molecules was used for the first time by Paul Ehrlich to find an effective treatment for trypanosomiasis [167]. He took advantage of the well-developed dyestuff industry by the end of XIX century. After screening more than 100 compounds, he found only one with trypanocidal activity, a benzopurpurine compound he named Nagana Red. He made some chemical modifications optimizing the chemical series. Trypan Blue was another compound derived from the benzopurpurine scaffold with trypanocidal activity [168, 169]. Chemists Oskar Dressel and Richard Kothe working in Wilhelm Roehl's team (a former assistant of Paul Ehrlich) found in 1917 a colorless compound with exceptional efficacy to treat trypanosomiasis in both experimental animals and humans after the synthesis and screening of more than 1000 naphthalene ureas [170]. The secret compound was named Bayer 205, and later Germanin. In 1924, the French pharmacist Ernest Fourneau published the structure of Bayer 205 [171], and four years later, the Bayer Company confirmed it was identical with Germanin's structure. The drug was subsequently renamed suramin and is still the only drug in use in the therapy of early-stage *T. b. rhodesiense* sleeping sickness. Unfortunately, there are severe adverse effects associated with suramin including urticarial rash, adrenal cortical gland damage, kidney damage and exfoliative dermatitis associated with skin urticarial rashes [172]. A study with HIV/AIDS patients had to be suspended due to high fatal occurrences during the trial with suramin [173].

In 1937, the English chemist Arthur James Ewins synthesized several aromatic diamidines, being Stilbaminidine and pentamidine the most promising ones to treat HAT at least regarding the antiparasitic efficacy [174, 175]. However, the first caused serious neurological effects in some patients and was abandoned [176, 177]. Pentamidine, on the other hand, successfully made it through the development process and is still used to treat stage 1 caused by *T. b. gambiense*. It is also used nowadays with other indications including leishmaniasis, *Pneumocystis jirovecii* pneumonia, mostly in AIDS patients [178]. Pentamidine cannot cross the blood-brain barrier, therefore has no efficacy against stage 2 of HAT. Compared to suramin, pentamidine shows much less adverse effects and is considered a safer treatment. Unfortunately, it is not effective against stage 1 HAT caused by *T. b. rhodesiense*, probably due to the absence of efficient membrane transporters to internalize the drug in this more virulent strain [179].

The treatment of the second stage of sleeping sickness caused by *T. b. rhodesiense* is probably the worst regarding toxicity compared to any other treatment in use for a disease caused by a kinetoplastid protozoan parasite. The only available drug is Merlasoprol, an arsenic based chemical responsible for the death of up to 5% of the patients that undergo treatment. It was first suggested for the treatment of sleeping sickness by the Scottish missionary and explorer David Livingston in 1858 [180], even though the causative agent of the disease was not known by that time. Paul Ehrlich also worked with arsenic-based compounds. Together with his colleague Sahachiro Hata, a Japanese bacteriologist, Ehrlich accidently discovered the first effective treatment for syphilis, arsphenamine, named Salvarsan, which means 'the arsenic that saves' [181]. Another arsenic compound - named tryparsamide - was synthesized in 1919 by the American chemist Walter Jacobs and the American immunologist Michael Heidelberger [182]. It had a key property to treat sleeping sickness, the ability to enter the central nervous system and became the first drug for treating the second stage of sleeping sickness being the drug of choice for that indication until 1960, despite the adverse effects including damage to the optical nerve [182]. In 1938, Ernst Friedheim synthesized melarsen with a melamine moiety in the molecules. Melarsen was less toxic but more expensive than other existing drugs. So Friedheim kept working on the molecules trying to reduce the toxicity profile having synthesized a trivalent analog of melarsen, melarsen oxide. Trivalent arsenicals were much more potent against the parasites than pentavalent ones, however, the toxicity issues remained [184]. In a serendipity moment, Friedheim observed that eating Swiss cheese reduced the side effects of melarsen oxide in some patients [183]. He then came up with the idea to use it in combination with dimercaprol, an arsenic antidote developed to prevent arsenic poisoning during the World War II. That idea leads to the discovery of Merlasoprol, which was 100 times less cytotoxic but only 2.5 times less trypanocidal than melarsen oxide. Merlasoprol was introduced in 1949 for the treatment of stage 2 of sleeping sickness, substituting tryparsamide [185]. Melarsoprol was not toxic to the optic nerve, but also had serious adverse effects, causing encephalopathy in 5-10% of patients, with 1-5% mortality [162]. More than 65 years after its introduction, Melarsoprol is still the only effective drug to treat the second stage of sleeping sickness caused by the more aggressive parasites: *T. b. rhodesiense*. It was also the only option to treat the second stage caused by *T. b. gambiense*, but in 1990 another drug was approved for that purpose: eflornithine [186].

Eflornithine, a fluorinated analog of the amino acid is also a repurposed drug, originally developed in the 1970s as an anti-cancer drug [187]. It is an irreversible inhibitor of the enzyme ornithine decarboxylase involved in the synthesis of polyamines putrescine and spermidine [188]. Its use in cancer was not successful because of the rapid turnover rate of the human ornithine decarboxylase with a short half-life of 10-30 minutes [188, 189]. Cyrus Bacchi, an American biologist, was the one who found the antiparasitic activity of eflornithine in 1980 in a trypanosomiasis mouse model, without evident signs of toxicity *in vivo* [190]. Following his findings, human clinical trials demonstrated therapeutic efficacy against stage 2 sleeping sickness caused by *T. b. gambiense* refractory to merlasoprol treatment [191 - 194]. However, the drug did not show the same activity against *T. b. rhodesiense*. Later it was discovered that the reason for this discrepancy is related to the turnover of the targeted enzyme in the parasites. Ornithine decarboxylase in *T. b. gambiense* has a half-life of 18-19 hours, while for *T. b. rhodesiense* it is only 4.3 hours [195]. It does also affect trypanothione, an antioxidant involved in the antiparasitic effect of nitro-heterocyclic compounds like nifurtimox [196], supporting the idea of combination therapy with synergistic effects. Nifurtimox is a nitrofuran derivative developed in the 1960s by the German Bayer Company, in use since 1967 to treat CD caused by *T. cruzi*. The compound is reduced by nitroreductases in the parasite generating reactive oxygen species, which damage cellular components such as DNA, membrane lipids, and proteins [197]. The reduction mechanism is not unique, so the compound does also get activated in human host cells leading to a number of adverse effects [59]. The first attempts to repurpose nifurtimox to treat melarsoprol-resistant *T. b. gambiense* sleeping sickness patients showed that high dosage was necessary to achieve a cure, jeopardizing the approach due to the severe adverse effects associated with that required effective dose [198, 199]. The idea of combining nifurtimox and eflornithine (NECT or Nifurtimox-Eflornithine Combination Therapy) is currently being tested in clinical trials against stage 2 caused by *T. b. gambiense* and represents one of the greatest advances in the treatment of sleeping sickness [200].

As an interesting side note, Ethidium Bromide currently used as DNA staining in molecular biology laboratories was discovered and developed as an antitrypanosomal compound. It is still used these days to treat cattle trypanosomiasis [201]. A complete review of the history of sleeping sickness drug development was published by Dr. Dietmar Steverding [202].

In summary, the current therapy recommended for sleeping sickness is pentamidine for stage 1 caused by *T. b. gambiense* and NECT (as an alternative to merlasoprol) in stage 2. To treat sleeping sickness caused by *T. b. rhodesiense* the only available options are suramin for stage 1 and merlasoprol for stage 2. Hopefully, this situation is close of being changed thanks to an ongoing clinical trial using another nitro-heteroaromatic compound, fexinidazole.

Discovery Initiatives for HAT

Compared to leishmaniasis and CD – both intracellular parasites – sleeping sickness has the easiest model to be assayed *in vitro* for drug screening. Since the introduction of multi-well colorimeter or fluorimeter reader instruments, it was possible to develop simple viability assays amenable to high-throughput scale to test anti-sleeping sickness compounds in large scale [203]. Common methods to screen compounds against *T. brucei* use well established whole cell viability measurements including reduction of resazurin by mitochondrial metabolism, luminescence (luciferase) readings of ATP and total DNA quantification [204 - 208]. All these methods have been vastly applied in the past decade leading to the discovery of more than 1,000 molecules, with only three achieving Clinical Phase: pafuramidine (DB289), fexinidazole and SCYX-7158 [209]. Parafuramidine (DB289) is a diamidine prodrug after an effort to increase the efficacy and tolerability of pentamidine. The drug is orally bioavailable being converted in the liver to the active compound furamidine (DB75) [210]. DB289 produced 100% cures in murine models of HAT and also in a *T. b. rhodesiense* primate model following oral administration [211, 212]. No acute toxicity was observed, and the compound entered clinical development and in a Phase IIa trial, 95% of patients with early stage *T. b. gambiense* infections were cured with the remaining 5% of patients lost at follow-up [210]. Despite these promising initial results, the identification of acute renal toxicity in the patients led to the clinical development of DB289 being suspended. Fexinidazole was identified through the screening of nitroheterocyclic molecules sponsored by DND*i*. Fexinidazole was an orally available nitroimidazole prodrug with *in vivo* activity in both acute *T. b. rhodesiense* and *T. b. brucei* central nervous system (CNS) murine models and is the second compound to enter clinical development for HAT [108, 213]. In Phase I, first-in-man trial to determine the optimum dosing schedule, fexinidazole was well tolerated at doses up to 3,600 mg with headaches and vomiting, which resolved spontaneously, being the only adverse effects reported [214]. The optimal treatment regime required to obtain effective drug concentrations, while maintaining a good therapeutic index, was determined to be 1,800 mg daily for 4 days followed by 1,200 mg daily for 6 days [214]. A Phase II/III evaluating this treatment regime in patients with CNS infections commenced in 2012 but to date, preliminary outcomes from this trial have not been reported. In addition to fexinidazole, SCYX-7158 is also currently in the DND*i* clinical development pipeline for HAT. The compound was identified through the high-throughput screening (HTS) of a library of benzoxaboroles and a subsequent lead optimization program. The compound is orally efficacious in both acute and CNS murine models of HAT and pharmacokinetic studies suggest that once a day dosing would be feasible due to the maintenance of drug concentrations above the minimum trypanocidal concentration for 14 to 20 h [215]. A Phase I, the first-in-human study commenced in 2012 but had to be temporarily suspended in 2013 for additional studies to be completed in canines due to the half-life in human plasma being longer than expected (http://www.dndi.org/diseases-projects/portfolio/scyx-7158/). The trial recommenced later in 2013 and safety profiling in additional subjects is ongoing.

Recently, the Dundee focused protein kinase library containing 3,885 compounds [216] was screened at the Swiss Tropical and Public Health Institute (STPH) against *T. b rhodesiense* and identified initially 121 active compounds. The most potent inhibitor showed antiparasitic activity at 80 nM and is not cytotoxic to the mammalian cell MRC5 up to 50 μM, used as counter screening. After selecting some series to be explored with medicinal chemistry, the hits were triaged in a Static-Cidal assay to aid hit prioritization [217]. The molecular target for the confirmed active compounds is not yet known despite the fact that the compounds are targeting kinases, and showed selectivity towards the parasitic enzymes [218].

Some other targets in *T. brucei* are also well validated as essential for the parasite virulence. One of the most studied is the Cathepsin L enzyme rhodesain [219]. A recent screening focused on inhibitors of this major protease from the parasite led to the discovery of active compounds from the Malaria Box [220]. As the authors mentioned, the compounds identified with adequate efficacy represent an interesting chemical space for optimization. Another group, led by Dr. Gonzalez, found that some dipeptidyl enoates are selective Rhodesain reversible inhibitors at very low concentrations and irreversible inhibitors at higher concentration [221]. This was the first time a dual mechanism of action has been reported for this parasitic protease. Differently from the peptidyl enoates, thiosemicarbazones target cathepsin B from various pathogenic organisms, including rhodesain (*T.* brucei), cruzain (*T. cruzi*) and SmCB1 (*Schistosoma mansoni*) [222].

The first large-scale screening initiative against sleeping sickness led by a pharmaceutical company was recently published [146]. The GSK site at Tres Cantos, Spain, screened their 1.8 million compounds library and found 192 active and selective compounds against *T. brucei*. The compounds were made available to the scientific community willing to further investigate and develop the hits.

Natural products represent a different chemical space yet to be explored in a more systematic way as it has been done for synthetic small molecules in the past few years. Marine natural products were recently tested and confirmed promising antiparasitic activity [30]. A novel approach used genome-directed lead discovery to identify *in vitro* activity of lobosamide against *T. brucei*, a polyene macrolactam family member [223]. All these works with natural products are very recent and early in the discovery pipeline but hold a good potential for further development as alternative chemotherapy for sleeping sickness. One of the greatest challenges is finding active compounds able to cross the blood-brain barrier to address parasites in stage 2 of the disease.

Mice models are available to assess compound activity in stage 2. A recent non-invasive model developed by Jeremy Mottram and collaborators uses *T. brucei brucei* strain GVR35 expressing luciferase. With the utilization of an *In vivo* Imaging System (IVIS) that contains a very sensitive camera, the parasites can be indirectly detected in the mice based on photons readout upon injection of luciferin. This method allowed the detection of trypanosomes in the brain meninges 5 days post infection, but these could be killed by diminazene drugs. After day 21, the parasites enter the parenchyma and become no longer susceptible to diminazene. All bioluminescent parasites were permanently eliminated after treatment with melarsoprol and DB829 [224], compounds known to cure stage 2 of the disease. The optimized bioluminescent *T. brucei* parasite enhanced the application of *in vivo* imaging to study stage 2 African trypanosomiasis in animal models represents a valuable tool to study host-parasite relationships, specially regarding the central nervous system infection, ideally suiting drug evaluation programs [225]. Other sleeping sickness animal models were reviewed by Dr. Kennedy in 2007 [226].

The criteria for hit classification and lead selection may vary from one laboratory to the other, but there are not many divergences. For a hit, the compound should have antiparasitic $IC_{50} < 10$ μM, preferentially < 1 μM, achieving 100% activity. A counter-screening against a panel of human cells should demonstrate selectivity towards the parasite target of at least 10x, ideally >100x. A lead compound has to demonstrate adequate pharmacokinetic profile, being the most important characteristics oral bioavailability. It is also important that the final drug candidate keeps stable at room temperature in tropical areas and presents reduced production cost [227]. Table **2A** presents the target product profile for a future drug to treat both stages 1 and 2, and Table **2B** lists the ideal and acceptable features of a drug to treat just stage 1 if failed for stage 2.

Table 2A. The Target Product Profile (TPP) for Human African Trypanosomiasis (sleeping sickness) Stages 1 and 2 as currently proposed by the DND*i* (http://www.dndi.org/diseases- projects/ hat/target-product-profile/).

Ideal	Acceptable: improvement to current stage 2 treatment	NECT
Effective against stage 1 and 2	Effective against stage 1+2 (used stage 2 only)	stage 1+2 (used stage 2 only)
Broad Spectrum (*gambiense* and *rhodesiense*)	Efficacy against *gambiense* only	*gambiense*
Clinical efficacy > 95% at 18 months follow up		clinical efficacy: 96.5% (ITT NECT Study)
Effective in melarsoprol refractory patients		effective
<0.1% drug related mortality	<1% drug related mortality	1.2% possibly related mortality (NECT Field)
Safe during pregnancy and for lactating women		no event in babies born or being breastfed after treatment
Adult and pediatric formulations		DFMO pediatric dosing available + Nifurtimox 5 mg tablets to be cut
No monitoring for AEs	Weekly simple lab testing (field testing)	hospitalization required
< 7 days oral once daily (DOT)	10 days oral (up to tid)	7 days IV infusion (bid) + 10 days po (tid)
< 7 days intramuscular once daily	10 days intramuscular once daily	
Stability in Zone 4 for > 3 years	Stability in Zone 4 for > 12 months	Stability in Zone 4 for > 24 months
Cidal		DFMO static + Nifurtimox cidal
Multi-target	Unique target (but not uptake *via* P2-transporter only)	
< 30 € / course* (only drug cost)	< 100 €* / course	222.5 € / course (in 4 treatments kits; WHO)
	< 200 €* / course ok if very good on other criteria	

Table 2B. The Target Product Profile (TPP) for Human African Trypanosomiasis (sleeping sickness) if a stage 2 treatment fails to show efficacy (http://www.dndi.org/diseases-projects/ hat/target- product-profile/).

Ideal	Acceptable: Improvement to current stage 2 treatment
Effective against stage 1	Effective against stage 1
Broad spectrum (*gambiense* and *rhodesiense*)	Efficacy against *gambiense* only
Clinical efficacy > 95% at 18 months follow-up	Clinical efficacy no worse than pentamidine
0% drug related mortality	0% drug related mortality
≤ 3 days treatment	≤ 7 days treatment
Safe during pregnancy, for breastfeeding women and children	Safe during pregnancy, for breastfeeding women and children
Adult and pediatric formulations	Adult and pediatric formulations
No monitoring for AEs	No monitoring for AEs
Single dose oral or intramuscular	2-3 daily doses oral or intramuscular
(single dose in animal models, long half-life)	
Stability in Zone 4 for > 4 years	Stability in Zone 4 for > 2 years
Cidal	Cidal
Multi-target	1 target but resistance not readily inducible
< 10 € / course*	< 30 €* / course

Future Perspectives for HAT Chemotherapy

Recent attempts to reduce the burden caused by sleeping sickness have been made. A panel of WHO experts gathered in 2013 aiming to establish guiding principles targeting the elimination of the disease. Among the recommendations, new diagnostic and chemotherapy options were highlighted as high priority. A study defined the risk of a population exposure to HAT to be hard to predict, as there are many important factors to be considered. The best estimate suggests 70 million people from 24 endemic African countries to be in the risk area. Thanks to the newly implemented surveillance system integrated with health systems, in 2012 all the diagnosed cases of HAT were treated with drugs donated by Sanofi and Bayer [228]. This integration will be key to achieve the elimination goal. One of the biggest challenges is the poor health systems in rural areas where HAT is endemic, potentially jeopardizing the approach. Moreover, one last aspect to be considered for the elimination relies on a better understanding of the animal reservoirs as well as the healthy carriers role in the transmission of HAT. The goal of eliminating HAT is feasible, but not probable in short-term [228].

LEISHMANIASIS

Leishmaniasis is caused by the *Leishmania* sp. protozoan parasite and is transmitted by the sandfly insect vector. Depending on the parasite species and on the host immune system the disease can present different symptoms, ranging from cutaneous manifestation (skin ulcers or damage in mucocutaneous tissues) to life-threatening visceral organs damage (especially liver, spleen and bone marrow). Cutaneous leishmaniasis (CL) usually heals without leaving the scars. It can evolve into diffuse cutaneous leishmaniasis, relapsing leishmaniasis, or mucocutaneous leishmaniasis (MCL). Visceral leishmaniasis (VL), also called kala-azar, is the most severe form of the disease, being lethal if left untreated. Some treated cases of VL can progress to a cutaneous form known as post-kala-azar dermal leishmaniasis (PKDL), a more resistant form of the disease being hard to treat. More than 88 countries in tropical areas of the globe are endemic to leishmaniasis, exposing more than 350 million people to the risk of infection. The majority of MCL cases occur in South America, more specifically in Bolivia, Brazil, and Peru. CL concentrates 90% of the cases in only six countries: Afghanistan, Brazil, Iran, Peru, Saudi Arabia and Syria. More details about leishmaniasis epidemiology are described in Chapter 2.

Since the host immune system plays an important role in the control of leishmaniasis, immunosuppressed individuals like HIV carriers may develop more aggressive manifestations of leishmaniasis. For example, an immunosuppressed patient can develop VL upon infection caused by a species that commonly lead to the cutaneous form of the disease. Since the immune system plays a major role in the clinical outcomes of leishmaniasis, immunocompromised HIV patients represent a severe risk group, being the co-infection with *Leishmania* parasites not uncommon in areas where both diseases are present [229].

Current Chemotherapy for Leishmaniasis

Antimonial-based compounds have been used as the first line treatment for all forms of leishmaniasis since the 1950s. Actually, antimony was introduced by Paracelsus in the 16th century as a general panacea (as published in Leipzig in 1604). Plimmer and Thompson showed the activities of sodium and potassium tartrate against trypanosomes in 1905, and that represented the beginning of the use of antimonials to treat leishmaniasis in modern ages [230 - 232]. In India and Brazil, increasing numbers of resistance to antimonials have been reported [233, 234]. In some areas of India for example, 65% of previously untreated patients fail to respond to antimonials [235]. In addition to resistance cases, instability in tropical areas and high toxicity characterized by a cough, chest pain, and significant depression are the major concerns and support the use of alternative treatments [236, 237]. Surprisingly, the molecular target or mechanisms of action are not completely elucidated. Due to their amorphous state, structures of two complexes of pentavalent antimony with N-methyl-D-glucamine (meglumine antimoniate or Glucantime) or sodium gluconate (sodium stibogluconate or Pentostam) remained unknown for decades. NMR and mass spectrometric approaches have recently allowed progress in this field [238]. More recently, mass spectroscopic approaches reveal the accumulation of pentavalent and trivalent forms of antimony in both stages of the parasite. Accumulation of pentavalent form was found to be higher in axenic amastigotes than in promastigotes in some species [239, 240]. It is not clear, though, which form of antimony is the active one against the parasites: the pentavalent or the trivalent. One hypothesis is that pentavalent antimony behaves as a prodrug, being reduced to much more active/toxic trivalent form of antimony that exhibits antileishmanial activity. Antimony transport to the parasite has been associated with its efficacy and was first studied in *L. mexicana* and *L. donovani* parasites using (^{125}Sb) Pentostam [241]. Because gluconate competitively inhibits uptake of pentavalent antimony in axenic amastigotes, it is speculated that the antimony enters into the parasites *via* a protein that recognizes a sugar moiety-like structure shared with gluconate [242]. After being internalized the compound is reduced in the cytoplasm. If the mechanism of reduction is enzymatic or nonenzymatic, remain controversial. Amastigotes can reduce pentavalent to trivalent antimony in a much more efficient way than promastigotes, supporting why the first are more susceptible to pentavalent form than the latter [243 - 247]. Another hypothesis suggests that macrophage enzymes drive the reduction, entering the parasites already as reduced trivalent antimony. Important to note that macrophages are highly sensitive to trivalent antimony which can kill 50% of THP-1 macrophage cell line *in vitro* at 25 μg/mL [248, 249]. Reduction of pentavalent to trivalent antimony likely occurs at both sites, and the parasite plays a major role in the generation of higher, lethal concentrations of trivalent antimony. Thiol compounds are involved in the conversion of pentavalent to trivalent forms both in mammalian cells and inside the parasite [250 - 252]. Parasite-specific thiol compound is a complex of glutathione and spermidine (trypanothione (T(SH)2)) [253, 254]. One example is the enzyme TDR1, a tetramer protein with glutathione S-transferases (GSTs) domains that use glutathione (GSH) as the reductant. Despite the high abundance of in the amastigote stage of the parasite, no correlation between enzyme activity and antimony sensitivity was found. It is likely that distinct mechanisms reduce antimony in the cells. The parasitic trypanothione reductase (TR) and the glutathione reductase (GR) have structural and mechanistic similarities, and the selective inhibition is due to differences in the disulfide-binding site between the enzymes. Trivalent antimonials inhibit TR and induce a rapid efflux of intracellular T(SH)2 and GSH into *Leishmania* cells [255, 256]. Another putative mechanism of action may also be associated with a zinc-finger protein. Trivalent antimony has also been suggested to bind a CCHC zinc finger peptide, leading to the ejection of Zn(II) [257]. This may implicate in impairment of protein-nucleic acid and protein-protein interactions in the parasite [257]. Furthermore, studies also suggested that sodium stibogluconate inhibits glycolysis, type I DNA topoisomerase and fatty acid beta-oxidation, affecting the energy metabolism, replication and inhibiting macromolecular biosynthesis in amastigotes [258, 259]. Treatment of *Leishmania* amastigotes with trivalent antimony has been found to induce apoptosis *via* induction of the oxidative stress and increase in intracellular (Ca^{2+}) [260, 261]. This is probably an unspecific consequence of the poly-pharmacological nature of antimony in leishmaniasis. Sodium Antimony Gluconate (SAG) induces reactive oxygen species (ROS) and nitric oxide (NO) production consequently killing intracellular *Leishmania* parasites within infected macrophage, which also requires phosphorylation of ERK *via* the PI3K-PK--Ras/Raf pathway. In addition, the activation of the PI3K/Akt pathway and p38MAPK are essential for induction of NO production. These events lead *Leishmania* deathin infected macrophage following SAG treatment. It was further shown that p38MAPK mediated generation of NO by SAG treatment is an indirect mechanism. It has been shown that p38MAPK can induce the production of TNFα, inducing iNOS2 expression and consequent production of NO. SAG-mediated NO generation and parasite killing could be abolished by treatment with anti-TNFα neutralizing antibody [262].

Second-line drugs to treat leishmaniasis include pentamidine and amphotericin B [263]. Amphotericin B deoxycholate (Fungizone) is a systemic antifungal also toxic to the host, requiring careful and slow IV administration. Amphotericin B complexes with parasitic membrane components like ergosterol, inducing the formation of pores which alter the ion balance leading to cell death [264]. Resistance has been only observed *in vitro* [265]. Lipid formulations of amphotericin B have been developed to improve its bioavailability and pharmacokinetic properties, considerably reducing side effects [266, 267]. The liposomal formulation (AmBisome) is an approved treatment for VL, has a better half-life and a high level of efficacy, with 90% cure rate, besides reducing the host toxicity. In VL models, AmBisome had hepatic accumulation, reaches therapeutic levels faster than antimonials and has a longer half-life [267 - 269]. Limitations include high cost, administration route (infusion required) and lack of stability at high temperature (cold chain is needed). Pentamidine is the other alternative to antimony therapy to treat leishmaniasis patients. In India, the use of pentamidine as monotherapy for visceral leishmaniasis was abandoned due to high toxicity and risk of resistance development [235], but the drug is still being recommended for combination therapies [270]. The cellular target of pentamidine is unknown, but it seems to play a role in the mitochondria, as it accumulates in this organelle [271].

Miltefosine (hexadecylphosphocholine) is the first oral drug for leishmaniasis and was originally developed as an anticancer agent [272]. Another pharmacokinetic advantage is its long half-life (120 hours) and can be used in case of antimony resistance [273]. The main drawbacks for miltefosine are the high cost incompatible with the patient population and teratogenicity, making it not recommended for a woman in childbearing age, or administered in combination with contraceptives during the treatment course and for two months after that [274]. It is the most recent antileishmanial drug on the market and the first effective oral treatment against VL, being recommended as first line drug for childhood VL [275]. The mechanism of action of miltefosine can be a direct action against the parasite by impairing the lipid metabolism [276] and causing parasite apoptosis [277]. The drug action can also be a host cell-dependent mechanism, modulating the immune system by activating the macrophages to stimulate inducible nitric oxide synthetase 2 (iNOS2) that catalyzes the generation of nitric oxide (NO) to kill the parasite [278].

Paromomycin is an aminoglycoside antibiotic with antileishmanial activity used in topical treatment for CL and as an intravenously for VL [279]. The mechanism of action is suggested to be related to the mitochondrial membrane potential impairment, protein synthesis inhibition, and respiratory dysfunction. It also alters membrane fluidity and lipid metabolism [280]. As it is not used for routine treatment of patients, resistance cases have not yet been reported. Nevertheless, mechanisms of resistance have already been described, such as modified uptake, mutated ribosome, or production of aminoglycoside-modifying enzymes [280].

Discovery Initiatives for Leishmaniasis

Analyses of the genome of *Leishmania* parasites have been used to identify putative drug targets [281, 282]. Some of these targets were identified and proved to be essential for parasite survival [270], including the parasite dihydrofolate reductase (DHFR) [283, 284] and cyclin-dependent kinases [285], but no significant progress has been made. The TDR targets database (https://tdrtargets.org) was created based on genetic, biochemical and pharmacological data related to tropical disease pathogens in addition to computationally predicted drugability [286] and contains a list of speculated molecular targets. *In silico* methods applied for the prioritization of these targets ranked cysteine peptidase C, TR and DHFR as top-three targets for *Leishmania* [287]. Cyclin-dependent kinase CRK3 complexed with CYC6 cyclin is another listed target. Inhibitors of CRK3-CYC6 were discovered on a target-based high-throughput screening (HTS) campaign, confirming *in vitro* activity against promastigotes and intracellular amastigotes infecting macrophage [288]. Overall, the target-based screening approach for antileishmanial drug discovery has not shown promising results in the past years. The high failure rate with target-based biochemical screenings against *Leishmania* parasites may be supported by the difficult access to the parasite in a physiological condition. In the human host the parasites are located in the parasitophorous vacuole inside the macrophage where the pH is lower than in the surrounding environment. Several membranes have to cross through a pH gradient to make a drug accessible to the parasite. Moreover, leishmaniasis would not be the first infectious disease model with a low success rate for a target-based strategy for drug discovery [139]. Targeting a pathway instead of a molecular target was already suggested as an alternative strategy [289]. The robustness of critical cellular metabolism relies on a complex and highly linked adjustable network with redundant pathways to maintain a normal flow of molecules in the cell [290]. If the target pathway could be translated to a stable readable phenotype, a screening assay could be developed using a cell-based approach taking into account the compound accessibility to the intracellular parasite to select the active molecules. As a first filter strategy, if the targeted pathway is conserved in both life cycle stages of the parasite (promastigote and amastigote), the insect stage, which is extracellular, and grows axenically could be used in primary screening. In other words, if a pathway itself can be defined as the target, it would be a way to select chemical compounds that are able to interfere with the pathway, no matter the individual targets. The target product profile (TPP) with the ideal and acceptable characteristics for a novel drug to treat CL and VL are listed in Tables **3.1** and **3.2**, respectively.

Table 3.1. The Target Product Profile (TPP) for Cutaneous Leishmaniasis (CL) as currently proposed by the DND*i*; (http://www.dndi.org/diseases-projects/leishmaniasis/tpp-cl/).

Attribute	Ideal	Acceptable
Target Species		
Species dependent treatment response	One treatment for All species of *Leishmania*	*L. tropica or L. braziliensis*
Safety		
Safety monitoring requirement	None	Primary Health Care (minimal contact).
		No major safety concerns.
Tolerability	Well tolerated	Well tolerated in >95% of patients treated.
Common Terminology Criteria for AE[1]	All AR's < grade 1	Systemic AR > grade 2 in <5%.
		Local AR ≤ grade 2 in <30%.

(Table 3.1) contd.....

Attribute	Ideal	Acceptable
		No treatment-induced mortality.
Contraindications	None	Can be assessed at PHC level.
Pregnancy	Safe in pregnancy and lactation	Category B[2]
Efficacy		
Complete clinical cure at 3 months from treatment onset	>95% patients	60% for *L. tropica*
100% epithelialization / flattening (primary outcome)		70% for *L. braziliensis*
Improved scar formation	Minimal scar	No worse than natural healing
Prevention of relapse and recidivans	No relapse or recidivans/ML	<5% rate of relapse or recidivans/ML at 1 year
Parasitological endpoint requirement	None	None
Drug / Treatment schedule		
Route of administration	Topical / oral	Non-parenteral or few doses if parenteral
Topical	14 days	28 days
Oral	< 7 days	Oral: bid for 28 days
Parenteral	No	3 injections
Target Population		
Age	No restrictions	> 9 months of age
Gender	No restrictions	No restrictions
Use in Pregnancy	Yes	No
Efficacy in immunocompromised patients	Yes	No
Product Characteristics		
Stability	No cold chain, at least 3 years at 37ºC	2 years at 4-8ºC
Cost		
Cost of products / procedures per treatment	US$5	US$50

A target-free screening would not define any specific molecule or even pathways as a target, but merely select compounds that can eliminate the parasite. It would be the least biased approach possible, where the parasite will reveal its most appropriate pathways to be targeted. From 1999 to 2008, 75 first-in-class drugs with new molecular mechanisms of action were approved by the Food and Drug Administration (FDA), being 28 drugs discovered from phenotypic screening, 17 drugs from target-based approaches and the other 30 based on other methodologies [46]. A phenotypic screening is typically a high-content screening (HCS), a versatile technology collecting information on several parameters during the screening, enabling thus the identification and categorization of complex cellular phenomena (phenotypes). This information is commonly interpreted by automated software that classifies and quantifies the cellular components and characteristics [291 - 293]. For intracellular parasites as the case on *Leishmania* and *T. cruzi*, this is considered the most suitable strategy for drug screening.

Table 3.2. The Target Product Profile (TPP) for Visceral Leishmaniasis (VL) as currently proposed by the DND*i*; (http://www.dndi.org/diseases-projects/leishmaniasis/tpp-vl/).

	Optimal Target Profile	Minimal Target Profile
Target Label	VL and PKDL	VL
Species	All species	*L. donovani*
Distribution	All areas	Either India or Africa
Target Population	Immunocompetent and immunosuppressed	Immunocompetent
Clinical Efficacy	> 95%	> 90%
Resistance	Active against resistant strains	
Safety and Tolerability	No AEs requiring monitoring	1 monitoring visit in mid/end - point
Contraindications	None	Pregnancy/lactation
Interactions	None - Compatible for combination therapy	None for malaria, TB, and HIV therapies
Formulation	Oral / intramuscular depot	Oral / intramuscular depot
Treatment Regimen	1/day for 10 days oral/ 3 shots over 10 days	bid for <10 days oral; or <3 shots over 10 days
Stability	3 years in zone 4	Stable under conditions that can be reasonably achieved in the target region (> 2 years)
Cost	< $10 / course	< $80 / course

The first high-throughput screening assays for *Leishmania* were developed in the early 2000's using different stages of the parasite, transgenic parasites, different readouts, different formats and different host cells [18, 294 - 300]. *In vitro* susceptibilities of *L. donovani* promastigote and amastigote stages to anti-leishmanial reference drugs were also addressed by the scientific community, indicating the practical relevance of stage-specific differences [301, 302].

The Genomic Institute of the Novartis Research Foundation (GNF) screened 1.8 million compounds from their diversity library using axenic amastigotes in a 1,536 well plates format [78]. Although it is recognized that this form of the parasite is not entirely representative of the intracellular amastigote [303], some starting points for drug discovery were identified. A recent campaign has also revealed the lack of activity translation between axenic amastigote and intracellular amastigotes screens on a variety of compounds other than reference drugs. GSK has also screened 1.8 million compounds from their library, also using axenic amastigotes for the primary screening, but confirming the hits in intracellular parasites infecting macrophages [146]. A list of 192 active compounds has been made available for the scientific community to help further development and target identification.

The development of a robust HCS/HTS method using what is considered the most relevant form of the parasite – intracellular amastigotes infecting human macrophage – was firstly reported in 2010 as a secondary assay to confirm activity of compounds selected as actives against the promastigote stage [18]. In 2011, it was adapted to a 96 well plate format to screen a few hundred compounds [301]. The Institut Pasteur Korea (IPK), together with the DND*i* developed an HTS/HCS visual screen in a 384 well plate format to find promising new chemical compound classes active against *Leishmania*. The human macrophage cell line THP-1 (human acute monocytic leukemia) was used as the host cell and infected with *L. donovani* metacyclic promastigotes. The readout of the assay was image-based, and customized software was developed for image analysis, allowing access to information such as infection ratio, the number of cells per well, the number of parasites per infected cell, and the number of parasites per cell cytoplasm area among others. This assay was used for screening over 300,000 compounds and more than 350 hits were identified with IC_{50} lower than 10 μM against intracellular parasites, and no cytotoxicity at 30 μM based on host cell counting.

After hits are identified, usually its pharmacokinetic profile is determined followed by *in vivo* efficacy proof-of-concept. A number of animal models exist for leishmaniasis, including mouse, rat, hamster, dog and non-human primates. None of these models, however, completely reproduces the pathology and immunology of the human disease. The *in vivo* efficacy models for VL have been shortly reviewed by Gupta and Nishi [304]. Balb/c mouse is commonly used for acute model and the endpoint of the assay is the parasite load in the liver of the animals (amastigotes per liver cell nuclei) following 5 to 14 days of treatment [305]. A more stringent and relevant chronic hamster model for VL uses the outbred Syrian golden hamster (*Mesocricetus auratus*), described to have more characteristics of the human pathology with synchronous infection in the liver and spleen [306 - 309]. For cutaneous leishmaniasis, there are also different options regarding the targeted tissue to be infected. The parasites develop lesions in the footpad, tail, back or ear tissue in mice species. *Leishmania major* expressing luciferase was recently used in a mouse model of CL using IVIS for drug efficacy assessment [310]. A recent review describes different animal model options for leishmaniasis drug in details [311].

Future Perspectives for Leishmaniasis Chemotherapy

Similarly to the current situation for CD and sleeping sickness, there is no adequate therapy to treat the different forms of leishmaniasis. Substantial progress has been made in the recent years in terms of advances in technology (development of relevant high-throughput screening assays), the involvement of pharmaceutical industry and number of promising candidates discovered in the last 5 years. However, there is still a long way to go. These recently identified active chemotypes have to be optimized and lead compounds, as well as preclinical candidates, should be announced in the next 5 to 10 years. The real measure of success will remain the access of these future drugs to the patients in need, and this depends on government plans of action.

One initiative to control and attempt to eliminate VL involved Bangladesh, India and Nepal, all the three countries showing a high incidence of VL. The goal of the initiative was to eliminate the disease by 2015. Even though the goal was not achieved, a significant reduction in incidence and actual elimination in a number of sub-districts can be considered a positive outcome from the initiative [228]. Bangladesh adopted a single dose AmBisome as a policy guideline to treat VL. Vector control also played an important role in the reduction in a number of cases In East Africa, another initiative similar to the one in Central Asia attempted to monitor the number of cases of VL outbreak in South Sudan. Improved access to diagnostic and treatment is thought to be directly linked to the incidence declined in all three major endemic countries in 2013 [228].

Some of the remaining challenges include early and specific diagnose in the point of care, as well as having available an efficient and safe chemotherapy option today represented by liposomal Amphotericin B [228].

CONCLUDING REMARKS

Improvements in the past decade including more financial opportunities, the involvement of the pharmaceutical industry, governments and philanthropists (Gates Foundation, Carlos Slim), and the key participation and leadership of DND*i* as an independent, not-for-profit initiative has dramatically changed the perspectives for the future of CD, HAT and leishmaniasis. The concept of "open science" seems to be an alternative to accelerate the development of new and better chemotherapies. Even major pharmaceutical sciences are adhering to the trend, as the case of GlaxoSmithKline GSK, making available to the scientific community their hit compounds after screening 1.8 million compounds library [146].

Simply identifying an active compound *in vitro* will not move a drug candidate to the patients. It is important to have a cascade of events and tests with clear pass/fail criteria. This goes from the selection of active compounds, to hit identification, lead scaffold and optimization to achieve the best possible pharmacokinetic, pharmacodynamics and safety (toxicology) profiles, finally identifying a preclinical candidate. Resources are limited, so one has to be wise in the selection of candidate scaffolds to move forward in this long process of drug development. In this context, it is necessary for example, to know the concept and avoid the PAINS: Pan Assay Interference Compounds [312, 313]. These are classes of promiscuous molecules, especially covalent modifiers and redox cyclers that comprise 5-12% of academic libraries and are identified as hits in several assays. These false positive compounds end up polluting the literature and wasting precious resources in the drug development field. Good medicinal chemists should be able to spot the PAINS to be excluded from the follow-up list.

Coming back to the London Declaration, how realistic it is to expect success and meet the goals set for 2020: control of CD and leishmaniasis and elimination of HAT? From 2012 to 2015, which accounts for the first third of the time to reach the deadline, significant progress has been made. There is lots of hope in the current clinical trials with fexinidazole to achieve the goal of elimination for sleeping sickness, with expectations for efficacy against both *T. b. gambiense* and *T. b. rhodesiense* and both stages, and most importantly, being safe for the patients without any apparent acute signs of toxicity. A failure of this drug would tremendously jeopardize the plans for 2020 elimination since there are no other good advanced candidates in the pipeline. Another limitation factor is setting up the clinical trials in Africa, due to logistics challenges and the current small number of patients available for clinical studies enrollment.

The situation for CD is unpredictable. In the past two years (2014 and 2015), three clinical trial results were published changing some paradigms of the disease. First, the disappointing results from the inhibitors of ergosterol biosynthesis posaconazole and E1224 (Ravuconazole prodrug) being unable to eliminate the parasites sustainably. Now, more recently the release of the BENEFIT trial results demonstrating that benznidazole is indeed effective to reduce the parasite numbers even in the chronic phase, but unexpectedly this fact was not enough to recover or prevent the progress of the cardiomyopathy. This was a clear indication that CD has to be addressed both from the parasite perspective but also from the cardiomyopathy (and probably also from the digestive syndrome) perspective for patients management. It is yet to be proven if reducing the parasite levels before cardiac manifestation would be effective in preventing cardiomyopathy. New biomarkers as early offsets to predict digestive manifestation or cardiomyopathy will be an important tool to be used in clinical studies in addition to the parasite levels assessed by real-time PCR.

Finally, leishmaniasis is probably the hardest one to meet the goals of control by

2020. Political and economic unstable situation in the Middle East where the disease is endemic has significantly increased the number of cases [314 - 317]. Combination therapies successes in India represent some hope, but unfortunately, there are no good advanced new chemical entity candidates in the pipeline to enter clinical studies shortly.

CONFLICT OF INTEREST

The authors confirm that they have no conflict of interest to declare for this publication.

ACKNOWLEDGEMENTS

The author would like to thank the US National Institute of Health, Bill & Melinda Gates Foundation and the European Union Seventh Framework Programme (602773-KINDRED) for the research support. The author would also like to kindly thank Dr. Miriam Giardini for proof-reading and comments, and the editors Dr. Maria Isabel N. Cano and Dr. Marcelo S. da Silva for the invitation to write this chapter.

ABBREVIATIONS

AEs	Adverse Events
AIDS	Acquired Immune Deficiency Syndrome
APOA1	Apolipoprotein A1
AR	Adverse Reaction
ATP	Adenosine Triphosphate
BENEFIT	Benznidazole Evaluation For Interrupting Trypanosomiasis
Bid	*bis in die* (twice a day)
CD	Chagas disease
CE	Clinical Evaluation
CL	Cutaneous Leishmaniasis
CNS	Central Nervous System
CRISPR	Clustered Regularly Interspaced Short Palindromic Repeats
CYP51	Sterol 14-Alpha-Demethylase
DAPI	4',6-diamidino-2-phenylindole
DDW	Drugs for Developing World

DFMO	2-Difluoromethylornithine
DHFR	Dihydrofolate Reductase
DNA	Deoxyribonucleic Acid
DND*i*	Drugs for Neglected Diseases Initiative
DOT	Directly Observed Therapy
E1224	Ravuconazole prodrug
EMA	European Medicines Agency
ERK	Extracellular Signal-Regulated Kinase
FDA	U.S. Food and Drug Administration
Fiocruz	Oswaldo Cruz Foundation
FP7	European Framework Program 7
GFP	Green Fluorescent Protein
GHIT	Global Health Innovative Technology
GNF	Genomic Institute of the Novartis Research Foundation
GR	Glutathione Reductase
GSH	Glutathione
GSK	GlaxoSmithKline
GST	Glutathione S Transferase
HAT	Human African Trypanosomiasis
HCS	High-Content Screening
HIV	Human Immunodeficiency Virus
HTS	High-Throughput Screening
INF-γ	Interferon gamma
iNOS2	Inducible Nitric Oxide Synthase
IPK	Institut Pasteur Korea
ITT	Intention-to-treat
IV	Intravenous
IVIS	*In vivo* Imaging System
LE	Laboratory Evaluation
LOLA	Lead Optimization Latin America initiative
MAPK	Mitogen-Activated Protein Kinase
MCL	Mucocutaneous Leishmaniasis
MIT	Massachusetts Institute of Technology
ML	Mucosal Lesion
MMV	Medicines for Malaria Venture

MSF	Medicins Sans Frontiers or Doctors Without Borders
NADH	Nicotinamide Adenine Dinucleotide (reduced)
NADPH	Nicotinamide Adenine Dinucleotide Phosphate (reduced)
NECT	Nifurtimox-Eflornithine Combination Therapy
NIH	U.S. National Institutes of Health
NMR	Nuclear Magnetic Resonance
NO	Nitric Oxide
NTDs	Neglected Tropical Diseases
NTRs	Nitroreductases
PAINS	Pan Assay Interference Compounds
PCC	Preclinical candidate (Fig. **1**)
PCR	Polymerase Chain Reaction
PHC	Primary Health Care
PKDL	Post-Kala-azar Dermal Leishmaniasis
R&D	Research and Development
RNAi	Ribonucleic Acid Interference, or RNA Interference
ROS	Reactive Oxygen Species
SAG	Sodium Antimony Gluconate
SAR	Structural Activity Relationship
STPH	Swiss Tropical and Public Health Institute
T(SH)2	Trypanothione
TB	Tuberculosis
TcI	*T. cruzi* Group I
TcII	*T. cruzi* Group II
TcV	*T. cruzi* Group V
TcVI	*T. cruzi* Group VI
TDR	Research and Training in Tropical Diseases
TDR1	Thiol-Dependent Reductase I
THP1	Human Acute Monocytic Leukemia cell line
Tid	*ter in die* (three times a day)
TNFα	Tumor Necrosis factor alpha
TPP	Target Product Profile
TR	Trypanothione Reductase
UC	University of California
UNICAMP	State University of Campinas

USP	University of Sao Paulo
VL	Visceral Leishmaniasis
WHO	World Health Organization
WIPO	World Intellectual Property Organization

REFERENCES

[1] Hotez PJ, Fenwick A, Savioli L, Molyneux DH. Rescuing the bottom billion through control of neglected tropical diseases. Lancet 2009; 373(9674): 1570-5. [http://dx.doi.org/10.1016/S0140-6736(09)60233-6] [PMID: 19410718]

[2] Johnston KL, Ford L, Taylor MJ. Overcoming the challenges of drug discovery for neglected tropical diseases: the A·WOL experience. J Biomol Screen 2014; 19(3): 335-43. [http://dx.doi.org/10.1177/1087057113511270] [PMID: 24241712]

[3] Conteh L, Engels T, Molyneux DH. Socioeconomic aspects of neglected tropical diseases. Lancet 2010; 375(9710): 239-47. [http://dx.doi.org/10.1016/S0140-6736(09)61422-7] [PMID: 20109925]

[4] Mathers CD, Ezzati M, Lopez AD. Measuring the burden of neglected tropical diseases: the global burden of disease framework. PLoS Negl Trop Dis 2007; 1(2): e114. [http://dx.doi.org/10.1371/journal.pntd.0000114] [PMID: 18060077]

[5] Hotez PJ. NTDs V.2.0: blue marble healthneglected tropical disease control and elimination in a shifting health policy landscape. PLoS Negl Trop Dis 2013; 7(11): e2570. [http://dx.doi.org/10.1371/journal.pntd.0002570] [PMID: 24278496]

[6] Molyneux DH. The Neglected Tropical Diseases: now a brand identity; responsibilities, context and promise. Parasit Vectors 2012; 5: 23. [http://dx.doi.org/10.1186/1756-3305-5-23] [PMID: 22289579]

[7] World Health Organization. Global Plan to Combat Neglected Tropical Diseases 2008-2015. 2007. http://appswhoint/iris/bitstream/10665/69708/1/WHO_CDS_NTD_20073_engpdf

[8] World Health Organization. Accelerating Work to Obercome the Global Impact of Neglected Tropical Diseases: A Roadmap for Implementation. 2012. http://whqlibdocwhoint/hq/2012/WHO_HTM_NTD_20122011_engpdf

[9] Chatelain E, Ioset JR. Drug discovery and development for neglected diseases: the DNDi model. Drug Des Devel Ther 2011; 5: 175-81. [PMID: 21552487]

[10] Nwaka S, Ridley RG. Virtual drug discovery and development for neglected diseases through public-private partnerships. Nat Rev Drug Discov 2003; 2(11): 919-28. [http://dx.doi.org/10.1038/nrd1230] [PMID: 14668812]

[11] Dumonteil E, Bottazzi ME, Zhan B, *et al.* Accelerating the development of a therapeutic vaccine for human Chagas disease: rationale and prospects. Expert Rev Vaccines 2012; 11(9): 1043-55. [http://dx.doi.org/10.1586/erv.12.85] [PMID: 23151163]

[12] Hotez PJ, Dumonteil E, Heffernan MJ, Bottazzi ME. Innovation for the bottom 100 million: eliminating neglected tropical diseases in the Americas. Adv Exp Med Biol 2013; 764: 1-12. [http://dx.doi.org/10.1007/978-1-4614-4726-9_1] [PMID: 23654053]

[13] Nwaka S, Riopel L, Ubben D, Craft JC. Medicines for malaria venture new developments in antimalarials. Travel Med Infect Dis 2004; 2(3-4): 161-70. [http://dx.doi.org/10.1016/j.tmaid.2004.03.002] [PMID: 17291977]

[14] Bathurst I, Hentschel C. Medicines for malaria venture: sustaining antimalarial drug development. Trends Parasitol 2006; 22(7): 301-7. [http://dx.doi.org/10.1016/j.pt.2006.05.011] [PMID: 16757213]

[15] MMV: New Medicines for Malaria Venture. TDR News. 1999 (58): 2, 4.

[16] Moon S, Siqueira-Neto JL, Moraes CB, *et al.* An image-based algorithm for precise and accurate high throughput assessment of drug activity against the human parasite *Trypanosoma cruzi*. PLoS One 2014; 9(2): e87188. [http://dx.doi.org/10.1371/journal.pone.0087188] [PMID: 24503652]

[17] Siqueira-Neto JL, Moon S, Jang J, Yang G, *et al.* An image-based high-content screening assay for compounds targeting intracellular *Leishmania donovani* amastigotes in human macrophages. PLoS Negl Trop Dis 2012; 6(6): e1671.

[18] Siqueira-Neto JL, Song OR, Oh H, *et al.* Antileishmanial high-throughput drug screening reveals drug candidates with new scaffolds. PLoS Negl Trop Dis 2010; 4(5): e675. [http://dx.doi.org/10.1371/journal.pntd.0000675] [PMID: 20454559]

[19] Dent J, Ramamoorthi R, Graef K, Nelson LM, Wichard JC. WIPO Re:Search: a consortium catalyzing research and product development for neglected tropical diseases. Pharm Pat Anal 2013; 2(5): 591-6. [http://dx.doi.org/10.4155/ppa.13.49] [PMID: 24237166]

[20] Newman DJ, Cragg GM. Natural products as sources of new drugs over the 30 years from 1981 to 2010. J Nat Prod 2012; 75(3): 311-35. [http://dx.doi.org/10.1021/np200906s] [PMID: 22316239]

[21] Blunt JW, Copp BR, Keyzers RA, Munro MH, Prinsep MR. Marine natural products. Nat Prod Rep 2013; 30(2): 237-323. [http://dx.doi.org/10.1039/C2NP20112G] [PMID: 23263727]

[22] Werner Bergmann RJ. The isolation of a new thymine pentoside from sponges. J Am Chem Soc 1950; 72(6): 2809-10. [http://dx.doi.org/10.1021/ja01162a543]

[23] Werner Bergmann AN. Contributions to the study of marine products. component acids of lipids sponges. J Org Chem 1951; 16(8): 1206-21. [http://dx.doi.org/10.1021/jo50002a005]

[24] ODay DM, Poirier RH, Jones DB, Elliott JH. Vidarabine therapy of complicated herpes simplex keratitis. Am J Ophthalmol 1976; 81(5): 642-9. [http://dx.doi.org/10.1016/0002-9394(76)90131-8] [PMID: 1275044]

[25] Pavan-Langston D, Hess F. Ocular and systemic antiviral activity of vidarabine. Compr Ther 1977; 3(7): 42-8.

[PMID: 195767]

[26] Mori J, Tsubokura M, Kami M. Cytarabine dose for acute myeloid leukemia. N Engl J Med 2011; 364(22): 2166-7.
[http://dx.doi.org/10.1056/NEJMc1104343] [PMID: 21631342]

[27] Fox BW. Pharmacology and chemistry of some inhibitors of herpes replication. J Antimicrob Chemother 1977; 3(Suppl A): 23-32.

[28] Gedik CM, Collins AR. The mode of action of 1-beta-D-arabinofuranosylcytosine in inhibiting DNA repair; new evidence using a sensitive assay for repair DNA synthesis and ligation in permeable cells. Mutat Res 1991; 254(3): 231-7.
[http://dx.doi.org/10.1016/0921-8777(91)90061-S] [PMID: 1904985]

[29] Singh N, Mishra BB, Bajpai S, Singh RK, Tiwari VK. Natural product based leads to fight against leishmaniasis. Bioorg Med Chem 2014; 22(1): 18-45.
[http://dx.doi.org/10.1016/j.bmc.2013.11.048] [PMID: 24355247]

[30] Jones AJ, Grkovic T, Sykes ML, Avery VM. Trypanocidal activity of marine natural products. Mar Drugs 2013; 11(10): 4058-82.
[http://dx.doi.org/10.3390/md11104058] [PMID: 24152565]

[31] Wink M. Medicinal plants: a source of anti-parasitic secondary metabolites. Molecules 2012; 17(11): 12771-91.
[http://dx.doi.org/10.3390/molecules171112771] [PMID: 23114614]

[32] Nwaka S, Hudson A. Innovative lead discovery strategies for tropical diseases. Nat Rev Drug Discov 2006; 5(11): 941-55.
[http://dx.doi.org/10.1038/nrd2144] [PMID: 17080030]

[33] Croft SL, Snowdon D, Yardley V. The activities of four anticancer alkyllysophospholipids against *Leishmania donovani*, *Trypanosoma cruzi* and *Trypanosoma brucei*. J Antimicrob Chemother 1996; 38(6): 1041-7.
[http://dx.doi.org/10.1093/jac/38.6.1041] [PMID: 9023651]

[34] Kuhlencord A, Maniera T, Eibl H, Unger C. Hexadecylphosphocholine: oral treatment of visceral leishmaniasis in mice. Antimicrob Agents Chemother 1992; 36(8): 1630-4.
[http://dx.doi.org/10.1128/AAC.36.8.1630] [PMID: 1329624]

[35] Gönnert R, Andrews P. Praziquantel, a new board-spectrum antischistosomal agent. Z Parasitenkd 1977; 52(2): 129-50.
[http://dx.doi.org/10.1007/BF00389899] [PMID: 410178]

[36] Campbell WC. Efficacy of the avermectins against filarial parasites: a short review. Vet Res Commun 1982; 5(3): 251-62.
[PMID: 7051527]

[37] Debnath A, Parsonage D, Andrade RM, *et al.* A high-throughput drug screen for *Entamoeba histolytica* identifies a new lead and target. Nat Med 2012; 18(6): 956-60.
[http://dx.doi.org/10.1038/nm.2758] [PMID: 22610278]

[38] Taylor MC, Huang H, Kelly JM. Genetic techniques in *Trypanosoma cruzi*. Adv Parasitol 2011; 75: 231-50.

[http://dx.doi.org/10.1016/B978-0-12-385863-4.00011-3] [PMID: 21820559]

[39] Kolev NG, Tschudi C, Ullu E. RNA interference in protozoan parasites: achievements and challenges. Eukaryot Cell 2011; 10(9): 1156-63.
[http://dx.doi.org/10.1128/EC.05114-11] [PMID: 21764910]

[40] DaRocha WD, Otsu K, Teixeira SM, Donelson JE. Tests of cytoplasmic RNA interference (RNAi) and construction of a tetracycline-inducible T7 promoter system in *Trypanosoma cruzi*. Mol Biochem Parasitol 2004; 133(2): 175-86.
[http://dx.doi.org/10.1016/j.molbiopara.2003.10.005] [PMID: 14698430]

[41] Peng D, Kurup SP, Yao PY, Minning TA, Tarleton RL. CRISPR-Cas9-mediated single-gene and gene family disruption in *Trypanosoma cruzi*. MBio 2014; 6(1): e02097-14.
[http://dx.doi.org/10.1128/mBio.02097-14] [PMID: 25550322]

[42] Podust LM, von Kries JP, Eddine AN, *et al.* Small-molecule scaffolds for CYP51 inhibitors identified by high-throughput screening and defined by X-ray crystallography. Antimicrob Agents Chemother 2007; 51(11): 3915-23.
[http://dx.doi.org/10.1128/AAC.00311-07] [PMID: 17846131]

[43] Chen CK, Doyle PS, Yermalitskaya LV, *et al. Trypanosoma cruzi* CYP51 inhibitor derived from a *Mycobacterium tuberculosis* screen hit. PLoS Negl Trop Dis 2009; 3(2): e372.
[http://dx.doi.org/10.1371/journal.pntd.0000372] [PMID: 19190730]

[44] Brown D. Unfinished business: target-based drug discovery. Drug Discov Today 2007; 12(23-24): 1007-12.
[http://dx.doi.org/10.1016/j.drudis.2007.10.017] [PMID: 18061878]

[45] Nohara LL, Lema C, Bader JO, Aguilera RJ, Almeida IC. High-content imaging for automated determination of host-cell infection rate by the intracellular parasite *Trypanosoma cruzi*. Parasitol Int 2010; 59(4): 565-70.
[http://dx.doi.org/10.1016/j.parint.2010.07.007] [PMID: 20688189]

[46] Swinney DC, Anthony J. How were new medicines discovered? Nat Rev Drug Discov 2011; 10(7): 507-19.
[http://dx.doi.org/10.1038/nrd3480] [PMID: 21701501]

[47] Katsuno K, Burrows JN, Duncan K, *et al.* Hit and lead criteria in drug discovery for infectious diseases of the developing world. Nat Rev Drug Discov 2015; 14(11): 751-8.
[http://dx.doi.org/10.1038/nrd4683] [PMID: 26435527]

[48] Moraes CB, Giardini MA, Kim H, *et al.* Nitroheterocyclic compounds are more efficacious than CYP51 inhibitors against *Trypanosoma cruzi*: implications for Chagas disease drug discovery and development. Sci Rep 2014; 4: 4703.
[http://dx.doi.org/10.1038/srep04703] [PMID: 24736467]

[49] US Food and Drug Administration. Target Product Profile - A Strategic Development Process Toll. http://wwwfdagov/downloads/drugs/guidancecomplianceregulatoryinformation/guidances/ucm080593 pdf. 2007.

[50] Porrás AI, Yadon ZE, Altcheh J, *et al.* Target product profile (TPP) for chagas disease point-of-care diagnosis and assessment of response to treatment. PLoS Negl Trop Dis 2015; 9(6): e0003697.

[http://dx.doi.org/10.1371/journal.pntd.0003697] [PMID: 26042730]

[51] Freitas-Junior LH, Chatelain E, Kim HA, Siqueira-Neto JL. Visceral leishmaniasis treatment: What do we have, what do we need and how to deliver it? Int J Parasitol Drugs Drug Resist 2012; 2: 11-9. [http://dx.doi.org/10.1016/j.ijpddr.2012.01.003] [PMID: 24533267]

[52] Hotez PJ, Dumonteil E, Betancourt Cravioto M, *et al.* An unfolding tragedy of Chagas disease in North America. PLoS Negl Trop Dis 2013; 7(10): e2300. [http://dx.doi.org/10.1371/journal.pntd.0002300] [PMID: 24205411]

[53] Jackson Y, Varcher Herrera M, Gascon J. Economic crisis and increased immigrant mobility: new challenges in managing Chagas disease in Europe. Bull World Health Organ 2014; 92(10): 771-2. [http://dx.doi.org/10.2471/BLT.13.134072] [PMID: 25378732]

[54] Imai K, Maeda T, Sayama Y, *et al.* Chronic Chagas disease with advanced cardiac complications in Japan: Case report and literature review. Parasitol Int 2015; 64(5): 240-2. [http://dx.doi.org/10.1016/j.parint.2015.02.005] [PMID: 25744336]

[55] Pinto A, Pett S, Jackson Y. Identifying Chagas disease in Australia: an emerging challenge for general practitioners. Aust Fam Physician 2014; 43(7): 440-2. [PMID: 25006603]

[56] Morillo CA, Marin-Neto JA, Avezum A, *et al.* Randomized trial of benznidazole for chronic chagas cardiomyopathy. N Engl J Med 2015; 373(14): 1295-306. [http://dx.doi.org/10.1056/NEJMoa1507574] [PMID: 26323937]

[57] Turrens JF, Watts BP Jr, Zhong L, Docampo R. Inhibition of *Trypanosoma cruzi* and *T. brucei* NADH fumarate reductase by benznidazole and anthelmintic imidazole derivatives. Mol Biochem Parasitol 1996; 82(1): 125-9. [http://dx.doi.org/10.1016/0166-6851(96)02722-3] [PMID: 8943158]

[58] Carrilero B, Murcia L, Martinez-Lage L, Segovia M. Side effects of benznidazole treatment in a cohort of patients with Chagas disease in non-endemic country. Revista espanola de quimioterapia : publicacion oficial de la Sociedad Espanola de Quimioterapia 2011; 24(3): 123-6.

[59] Castro JA, Diaz de Toranzo EG. Toxic effects of nifurtimox and benznidazole, two drugs used against American trypanosomiasis (Chagas disease). Biomed Environ Sci 1988; 1(1): 19-33. [PMID: 3151755]

[60] de Andrade AL, Zicker F, de Oliveira RM, *et al.* Randomised trial of efficacy of benznidazole in treatment of early *Trypanosoma cruzi* infection. Lancet 1996; 348(9039): 1407-13. [http://dx.doi.org/10.1016/S0140-6736(96)04128-1] [PMID: 8937280]

[61] Schijman AG, Altcheh J, Burgos JM, *et al.* Aetiological treatment of congenital Chagas disease diagnosed and monitored by the polymerase chain reaction. J Antimicrob Chemother 2003; 52(3): 441-9. [http://dx.doi.org/10.1093/jac/dkg338] [PMID: 12917253]

[62] Sosa Estani S, Segura EL, Ruiz AM, Velazquez E, Porcel BM, Yampotis C. Efficacy of chemotherapy with benznidazole in children in the indeterminate phase of Chagas disease. Am J Trop Med Hyg 1998; 59(4): 526-9. [PMID: 9790423]

[63] Viotti R, Vigliano C, Armenti H, Segura E. Treatment of chronic Chagas disease with benznidazole: clinical and serologic evolution of patients with long-term follow-up. Am Heart J 1994; 127(1): 151-62.
[http://dx.doi.org/10.1016/0002-8703(94)90521-5] [PMID: 8273735]

[64] Mejía-Jaramillo AM, Fernández GJ, Montilla M, Nicholls RS, Triana-Chávez O. *Trypanosoma cruzi* strains resistant to benznidazole occurring in Colombia. Biomedica 2012; 32(2): 196-205.
[PMID: 23242293]

[65] Wilkinson SR, Taylor MC, Horn D, Kelly JM, Cheeseman I. A mechanism for cross-resistance to nifurtimox and benznidazole in trypanosomes. Proc Natl Acad Sci USA 2008; 105(13): 5022-7.
[http://dx.doi.org/10.1073/pnas.0711014105] [PMID: 18367671]

[66] Campos MC, Leon LL, Taylor MC, Kelly JM. Benznidazole-resistance in *Trypanosoma cruzi*: evidence that distinct mechanisms can act in concert. Mol Biochem Parasitol 2014; 193(1): 17-9.
[http://dx.doi.org/10.1016/j.molbiopara.2014.01.002] [PMID: 24462750]

[67] Mejia AM, Hall BS, Taylor MC, *et al.* Benznidazole-resistance in *Trypanosoma cruzi* is a readily acquired trait that can arise independently in a single population. J Infect Dis 2012; 206(2): 220-8.
[http://dx.doi.org/10.1093/infdis/jis331] [PMID: 22551809]

[68] Boiani M, Piacenza L, Hernández P, *et al.* Mode of action of nifurtimox and N-oxide-containing heterocycles against *Trypanosoma cruzi*: is oxidative stress involved? Biochem Pharmacol 2010; 79(12): 1736-45.
[http://dx.doi.org/10.1016/j.bcp.2010.02.009] [PMID: 20178775]

[69] Taylor MC, Lewis MD, Fortes Francisco A, Wilkinson SR, Kelly JM. The *Trypanosoma cruzi* vitamin C dependent peroxidase confers protection against oxidative stress but is not a determinant of virulence. PLoS Negl Trop Dis 2015; 9(4): e0003707.
[http://dx.doi.org/10.1371/journal.pntd.0003707] [PMID: 25875298]

[70] Pinazo MJ, Guerrero L, Posada E, Rodríguez E, Soy D, Gascon J. Benznidazole-related adverse drug reactions and their relationship to serum drug concentrations in patients with chronic chagas disease. Antimicrob Agents Chemother 2013; 57(1): 390-5.
[http://dx.doi.org/10.1128/AAC.01401-12] [PMID: 23114763]

[71] Jackson Y, Alirol E, Getaz L, Wolff H, Combescure C, Chappuis F. Tolerance and safety of nifurtimox in patients with chronic chagas disease. Clin Infect Dis 2010; 51(10): e69-75.
[http://dx.doi.org/10.1086/656917] [PMID: 20932171]

[72] Chatelain E. Chagas disease drug discovery: toward a new era. J Biomol Screen 2015; 20(1): 22-35.
[http://dx.doi.org/10.1177/1087057114550585] [PMID: 25245987]

[73] Clayton J. Chagas disease: pushing through the pipeline. Nature 2010; 465(7301): S12-5.
[http://dx.doi.org/10.1038/nature09224] [PMID: 20571548]

[74] Soeiro MdeN, de Castro SL. Screening of Potential anti-*Trypanosoma cruzi* Candidates: *In vitro* and *In vivo* Studies. Open Med Chem J 2011; 5: 21-30.
[http://dx.doi.org/10.2174/1874104501105010021] [PMID: 21629508]

[75] Buckner FS, Navabi N. Advances in Chagas disease drug development: 2009 2010. Curr Opin Infect Dis 2010; 23(6): 609-16.

[http://dx.doi.org/10.1097/QCO.0b013e3283402956] [PMID: 20885320]

[76] Moreira DR, Leite AC, dos Santos RR, Soares MB. Approaches for the development of new anti-*Trypanosoma cruzi* agents. Curr Drug Targets 2009; 10(3): 212-31. [http://dx.doi.org/10.2174/138945009787581140] [PMID: 19275558]

[77] Duschak VG. A decade of targets and patented drugs for chemotherapy of Chagas disease. Recent Pat Antiinfect Drug Discov 2011; 6(3): 216-59. [http://dx.doi.org/10.2174/157489111796887864] [PMID: 21824073]

[78] Bustamante JM, Park HJ, Tarleton RL. Report of the 2nd Chagas Drug Discovery Consortium meeting, held on 3 November 2010; Atlanta GA, USA. Expert Opin Drug Discov 2011; 6(9): 965-73. [http://dx.doi.org/10.1517/17460441.2011.602063] [PMID: 22646217]

[79] Molina I, Gómez i Prat J, Salvador F, *et al.* Randomized trial of posaconazole and benznidazole for chronic Chagas disease. N Engl J Med 2014; 370(20): 1899-908. [http://dx.doi.org/10.1056/NEJMoa1313122] [PMID: 24827034]

[80] Drugs for Neglected Diseases Initiative. Drug trial for leading parasitic killer of the Americas shows mixed results but provides new evidence for improved therapy. http://www.dndi.org/2013/ media-centre/press-releases/e1224/. 2013.

[81] de Souza W, Rodrigues JC. Sterol biosynthesis pathway as target for anti-trypanosomatid drugs. Interdiscip Perspect Infect Dis 2009; 2009: 642502. [http://dx.doi.org/10.1155/2009/642502]

[82] Docampo R, Moreno SN, Turrens JF, Katzin AM, Gonzalez-Cappa SM, Stoppani AO. Biochemical and ultrastructural alterations produced by miconazole and econazole in *Trypanosoma cruzi.* Mol Biochem Parasitol 1981; 3(3): 169-80. [http://dx.doi.org/10.1016/0166-6851(81)90047-5] [PMID: 6265775]

[83] Lepesheva GI, Villalta F, Waterman MR. Targeting *Trypanosoma cruzi* Sterol 14α-Demethylase (CYP51). In: Parasitol A, Ed. Louis M Weiss HBT, Louis VK. Academic Press 2011; Vol. 75: pp. 65-87. [http://dx.doi.org/10.1016/B978-0-12-385863-4.00004-6]

[84] Campos R, Amato Neto V, Moreira AA, *et al.* Evaluation of the therapeutic activity of fluconazole in acute experimental infection caused by *Trypanosoma cruzi.* Rev Hosp Clin Fac Med Sao Paulo 1992; 47(4): 174-5. [PMID: 1340597]

[85] Urbina JA, Payares G, Contreras LM, *et al.* Antiproliferative effects and mechanism of action of SCH 56592 against *Trypanosoma (Schizotrypanum) cruzi*: *in vitro* and *in vivo* studies. Antimicrob Agents Chemother 1998; 42(7): 1771-7. [PMID: 9661019]

[86] Molina J, Martins-Filho O, Brener Z, Romanha AJ, Loebenberg D, Urbina JA. Activities of the triazole derivative SCH 56592 (posaconazole) against drug-resistant strains of the protozoan parasite Trypanosoma (Schizotrypanum) cruzi in immunocompetent and immunosuppressed murine hosts. Antimicrob Agents Chemother 2000; 44(1): 150-5. [http://dx.doi.org/10.1128/AAC.44.1.150-155.2000] [PMID: 10602737]

[87] Olivieri BP, Molina JT, de Castro SL, *et al.* A comparative study of posaconazole and benznidazole in the prevention of heart damage and promotion of trypanocidal immune response in a murine model of Chagas disease. Int J Antimicrob Agents 2010; 36(1): 79-83. [http://dx.doi.org/10.1016/j.ijantimicag.2010.03.006] [PMID: 20452188]

[88] Campoli P, Al Abdallah Q, Robitaille R, *et al.* Concentration of antifungal agents within host cell membranes: a new paradigm governing the efficacy of prophylaxis. Antimicrob Agents Chemother 2011; 55(12): 5732-9. [http://dx.doi.org/10.1128/AAC.00637-11] [PMID: 21930891]

[89] Pinazo MJ, Espinosa G, Gállego M, López-Chejade PL, Urbina JA, Gascón J. Successful treatment with posaconazole of a patient with chronic Chagas disease and systemic lupus erythematosus. Am J Trop Med Hyg 2010; 82(4): 583-7. [http://dx.doi.org/10.4269/ajtmh.2010.09-0620] [PMID: 20348503]

[90] Chollet C, Crousse B, Bories C, Bonnet-Delpon D, Loiseau PM. *In vitro* antileishmanial activity of fluoro-artemisinin derivatives against *Leishmania donovani*. Biomed Pharmacother 2008; 62(7): 462-5. [http://dx.doi.org/10.1016/j.biopha.2008.04.003] [PMID: 18538529]

[91] Clinicaltrials.gov registry number NCT01377480 [online]. Available from URL: http://wwwclinicaltrialsgov [Accessed 2012 May 17].

[92] Torrico F. Rationale and design of a proof-of-concept phase II clinical study of E1224, a new drug candidate for chronic chagas disease. Trop Med Int Health 2011; 16: 21-2.

[93] Keenan M, Abbott MJ, Alexander PW, *et al.* Analogues of fenarimol are potent inhibitors of *Trypanosoma cruzi* and are efficacious in a murine model of Chagas disease. J Med Chem 2012; 55(9): 4189-204. [http://dx.doi.org/10.1021/jm2015809] [PMID: 22536986]

[94] Hoekstra WJ, Hargrove TY, Wawrzak Z, *et al.* Antiparasitic effect *in vitro*, activity in a murine model of Chagas disease, and structural characterization in complex with the target enzyme CYP51 from *Trypanosoma cruzi* of the potent clinical candidate VT-1161. Antimicrob Agents Chemother 2015.

[95] Choi JY, Podust LM, Roush WR. Drug strategies targeting CYP51 in neglected tropical diseases. Chem Rev 2014; 114(22): 11242-71. [http://dx.doi.org/10.1021/cr5003134] [PMID: 25337991]

[96] Shang N, Li Q, Ko TP, *et al.* Squalene synthase as a target for Chagas disease therapeutics. PLoS Pathog 2014; 10(5): e1004114. [http://dx.doi.org/10.1371/journal.ppat.1004114] [PMID: 24789335]

[97] Urbina JA, Concepcion JL, Rangel S, Visbal G, Lira R. Squalene synthase as a chemotherapeutic target in *Trypanosoma cruzi* and *Leishmania mexicana*. Mol Biochem Parasitol 2002; 125(1-2): 35-45. [http://dx.doi.org/10.1016/S0166-6851(02)00206-2] [PMID: 12467972]

[98] Szajnman SH, Yan W, Bailey BN, Docampo R, Elhalem E, Rodriguez JB. Design and synthesis of aryloxyethyl thiocyanate derivatives as potent inhibitors of *Trypanosoma cruzi* proliferation. J Med Chem 2000; 43(9): 1826-40. [http://dx.doi.org/10.1021/jm9905007] [PMID: 10794699]

[99] Urbina JA, Concepcion JL, Montalvetti A, Rodriguez JB, Docampo R. Mechanism of action of 4-phenoxyphenoxyethyl thiocyanate (WC-9) against *Trypanosoma cruzi*, the causative agent of Chagas disease. Antimicrob Agents Chemother 2003; 47(6): 2047-50. [http://dx.doi.org/10.1128/AAC.47.6.2047-2050.2003] [PMID: 12760897]

[100] Urbina JA, Concepcion JL, Caldera A, *et al. In vitro* and *in vivo* activities of E5700 and ER-119884, two novel orally active squalene synthase inhibitors, against *Trypanosoma cruzi*. Antimicrob Agents Chemother 2004; 48(7): 2379-87. [http://dx.doi.org/10.1128/AAC.48.7.2379-2387.2004] [PMID: 15215084]

[101] Clayton J. Chagas disease: pushing through the pipeline. Nature. 2010; 465(n7301_supp): S12-S5. [http://dx.doi.org/10.1038/nature09224]

[102] Benaim G, Sanders JM, Garcia-Marchán Y, *et al.* Amiodarone has intrinsic anti-*Trypanosoma cruzi* activity and acts synergistically with posaconazole. J Med Chem 2006; 49(3): 892-9. [http://dx.doi.org/10.1021/jm050691f] [PMID: 16451055]

[103] Paniz-Mondolfi AE, Pérez-Alvarez AM, Lanza G, Márquez E, Concepción JL. Amiodarone and itraconazole: a rational therapeutic approach for the treatment of chronic Chagas disease. Chemotherapy 2009; 55(4): 228-33. [http://dx.doi.org/10.1159/000219436] [PMID: 19451712]

[104] Adesse D, Azzam EM, Meirelles MdeN, Urbina JA, Garzoni LR. Amiodarone inhibits *Trypanosoma cruzi* infection and promotes cardiac cell recovery with gap junction and cytoskeleton reassembly *in vitro*. Antimicrob Agents Chemother 2011; 55(1): 203-10. [http://dx.doi.org/10.1128/AAC.01129-10] [PMID: 21078932]

[105] Riley J, Brand S, Voice M, Caballero I, Calvo D, Read KD. Development of a fluorescence-based *Trypanosoma cruzi* CYP51 inhibition assay for effective compound triaging in drug discovery programmes for chagas disease. PLoS Negl Trop Dis 2015; 9(9): e0004014. [http://dx.doi.org/10.1371/journal.pntd.0004014] [PMID: 26394211]

[106] Bahia MT, Nascimento AF, Mazzeti AL, *et al.* Antitrypanosomal activity of fexinidazole metabolites, potential new drug candidates for Chagas disease. Antimicrob Agents Chemother 2014; 58(8): 4362-70. [http://dx.doi.org/10.1128/AAC.02754-13] [PMID: 24841257]

[107] Bahia MT, de Andrade IM, Martins TA, *et al.* Fexinidazole: a potential new drug candidate for Chagas disease. PLoS Negl Trop Dis 2012; 6(11): e1870.

[108] Torreele E, Bourdin Trunz B, Tweats D, *et al.* Fexinidazolea new oral nitroimidazole drug candidate entering clinical development for the treatment of sleeping sickness. PLoS Negl Trop Dis 2010; 4(12): e923. [http://dx.doi.org/10.1371/journal.pntd.0000923] [PMID: 21200426]

[109] Patterson S, Wyllie S. Nitro drugs for the treatment of trypanosomatid diseases: past, present, and future prospects. Trends Parasitol 2014; 30(6): 289-98. [http://dx.doi.org/10.1016/j.pt.2014.04.003] [PMID: 24776300]

[110] Krauth-Siegel RL, Coombs GH. Enzymes of parasite thiol metabolism as drug targets. Parasitol Today (Regul Ed) 1999; 15(10): 404-9.

[http://dx.doi.org/10.1016/S0169-4758(99)01516-1] [PMID: 10481152]

[111] Olin-Sandoval V, González-Chávez Z, Berzunza-Cruz M, *et al.* Drug target validation of the trypanothione pathway enzymes through metabolic modelling. FEBS J 2012; 279(10): 1811-33. [http://dx.doi.org/10.1111/j.1742-4658.2012.08557.x] [PMID: 22394478]

[112] de Paula da Silva CH, Bernardes LS, da Silva VB, Zani CL, Carvalho I. Novel aryl β-aminocarbonyl derivatives as inhibitors of *Trypanosoma cruzi* trypanothione reductase: binding mode revised by docking and GRIND2-based 3D-QSAR procedures. J Biomol Struct Dyn 2012; 29(6): 702-16. [http://dx.doi.org/10.1080/07391102.2011.672633] [PMID: 22546000]

[113] de Oliveira RB, Vaz AB, Alves RO, *et al.* Arylfurans as potential *Trypanosoma cruzi* trypanothione reductase inhibitors. Mem Inst Oswaldo Cruz 2006; 101(2): 169-73. [http://dx.doi.org/10.1590/S0074-02762006000200009] [PMID: 16830710]

[114] da Rocha Pita SS, Albuquerque MG, Rodrigues CR, Castro HC, Hopfinger AJ. Receptor-dependent 4D-QSAR analysis of peptidemimetic inhibitors of *Trypanosoma cruzi* trypanothione reductase with receptor-based alignment. Chem Biol Drug Des 2012; 79(5): 740-8. [http://dx.doi.org/10.1111/j.1747-0285.2012.01338.x] [PMID: 22269140]

[115] Edd T, Saric M, St Phillips M, Karney N, O'Sullivan M. Synthesis of spermidine and spermine derivatives as potential inhibitors of *Trypanosoma cruzi* trypanothione reductase. Abstr Pap Am Chem Soc 2009; 237.

[116] Lo Presti MS, Rivarola HW, Bustamante JM, *et al.* Thioridazine treatment prevents cardiopathy in *Trypanosoma cruzi* infected mice. Int J Antimicrob Agents 2004; 23(6): 634-6. [http://dx.doi.org/10.1016/j.ijantimicag.2003.10.006] [PMID: 15194137]

[117] Eakin AE, McGrath ME, McKerrow JH, Fletterick RJ, Craik CS. Production of crystallizable cruzain, the major cysteine protease from *Trypanosoma cruzi*. J Biol Chem 1993; 268(9): 6115-8. [PMID: 8454586]

[118] Eakin AE, Mills AA, Harth G, McKerrow JH, Craik CS. The sequence, organization, and expression of the major cysteine protease (cruzain) from *Trypanosoma cruzi*. J Biol Chem 1992; 267(11): 7411-20. [PMID: 1559982]

[119] Engel JC, Doyle PS, Hsieh I, McKerrow JH. Cysteine protease inhibitors cure an experimental *Trypanosoma cruzi* infection. J Exp Med 1998; 188(4): 725-34. [http://dx.doi.org/10.1084/jem.188.4.725] [PMID: 9705954]

[120] Li R, Chen X, Gong B, *et al.* Structure-based design of parasitic protease inhibitors. Bioorg Med Chem 1996; 4(9): 1421-7. [http://dx.doi.org/10.1016/0968-0896(96)00136-8] [PMID: 8894100]

[121] Engel JC, Doyle PS, Palmer J, Hsieh I, Bainton DF, McKerrow JH. Cysteine protease inhibitors alter Golgi complex ultrastructure and function in *Trypanosoma cruzi*. J Cell Sci 1998; 111(Pt 5): 597-606. [PMID: 9454733]

[122] Engel JC, Doyle PS, Hsieh I, McKerrow JH. Cysteine protease inhibitors cure an experimental *Trypanosoma cruzi* infection. J Exp Med 1998; 188(4): 725-34. [http://dx.doi.org/10.1084/jem.188.4.725] [PMID: 9705954]

[123] Doyle PS, Zhou YM, Engel JC, McKerrow JH. A cysteine protease inhibitor cures Chagas disease in an immunodeficient-mouse model of infection. Antimicrob Agents Chemother 2007; 51(11): 3932-9. [http://dx.doi.org/10.1128/AAC.00436-07] [PMID: 17698625]

[124] Barr SC, Warner KL, Kornreic BG, *et al.* A cysteine protease inhibitor protects dogs from cardiac damage during infection by *Trypanosoma cruzi*. Antimicrob Agents Chemother 2005; 49(12): 5160-1. [http://dx.doi.org/10.1128/AAC.49.12.5160-5161.2005] [PMID: 16304193]

[125] Pizzo C, Faral-Tello P, Salinas G, *et al.* Selenosemicarbazones as potent cruzipain inhibitors and their antiparasitic properties against *Trypanosoma cruzi*. MedChemComm 2012; 3(3): 362-8. [http://dx.doi.org/10.1039/c2md00283c]

[126] Merlino A, Benitez D, Campillo NE, *et al.* Amidines bearing benzofuroxan or benzimidazole 1,3-dioxide core scaffolds as *Trypanosoma cruzi*-inhibitors: structural basis for their interactions with cruzipain. MedChemComm 2012; 3(1): 90-101. [http://dx.doi.org/10.1039/C1MD00223F]

[127] Beaulieu C, Isabel E, Fortier A, *et al.* Identification of potent and reversible cruzipain inhibitors for the treatment of Chagas disease. Bioorganic &amp. Medicinal Chemistry Letters 2010; 20(24): 7444-9. [http://dx.doi.org/10.1016/j.bmcl.2010.10.015]

[128] de Oliveira C, Santana LA, Carmona AK, *et al.* Structure of cruzipain/cruzain inhibitors isolated from Bauhinia bauhinioides seeds. Biol Chem 2001; 382(5): 847-52. [http://dx.doi.org/10.1515/bchm.2001.382.5.847] [PMID: 11517940]

[129] Ndao M, Beaulieu C, Black WC, Isabel E. Reversible cysteine protease inhibitors show promise for a chagas disease cure. Antimicrob Agents Chemother 2014; 58(2): 1167-78. [http://dx.doi.org/10.1128/AAC.01855-13]

[130] Ferreira RS, Simeonov A, Jadhav A, *et al.* Complementarity between a docking and a high-throughput screen in discovering new cruzain inhibitors. J Med Chem 2010; 53(13): 4891-905. [http://dx.doi.org/10.1021/jm100488w] [PMID: 20540517]

[131] Du X, Guo C, Hansell E, *et al.* Synthesis and structure-activity relationship study of potent trypanocidal thio semicarbazone inhibitors of the trypanosomal cysteine protease cruzain. J Med Chem 2002; 45(13): 2695-707. [http://dx.doi.org/10.1021/jm010459j] [PMID: 12061873]

[132] Moreira DRM, de Oliveira ADT. Conformational restriction of aryl thiosemicarbazones produces potent and selective anti-*Trypanosoma cruzi* compounds which induce apoptotic parasite death. Eur J Med Chem 2014; 75: 467-78.

[133] Greenbaum DC, Mackey Z, Hansell E, *et al.* Synthesis and structure-activity relationships of parasiticidal thiosemicarbazone cysteine protease inhibitors against *Plasmodium falciparum*, *Trypanosoma brucei*, and *Trypanosoma cruzi*. J Med Chem 2004; 47(12): 3212-9. [http://dx.doi.org/10.1021/jm030549j] [PMID: 15163200]

[134] Blau L, Menegon RF, Trossini GHG, Molino JVD. Design, synthesis and biological evaluation of new aryl thiosemicarbazone as antichagasic candidates. Eur J Med Chem 2013; 67: 142-51. [http://dx.doi.org/10.1016/j.ejmech.2013.04.022]

[135] Docampo R, de Souza W, Miranda K, Rohloff P, Moreno SN. Acidocalcisomes - conserved from

bacteria to man. Nat Rev Microbiol 2005; 3(3): 251-61.
[http://dx.doi.org/10.1038/nrmicro1097] [PMID: 15738951]

[136] Montalvetti A, Bailey BN, Martin MB, Severin GW, Oldfield E, Docampo R. Bisphosphonates are potent inhibitors of *Trypanosoma cruzi* farnesyl pyrophosphate synthase. J Biol Chem 2001; 276(36): 33930-7.
[http://dx.doi.org/10.1074/jbc.M103950200] [PMID: 11435429]

[137] Garzoni LR, Waghabi MC, Baptista MM, *et al.* Antiparasitic activity of risedronate in a murine model of acute Chagas disease. Int J Antimicrob Agents 2004; 23(3): 286-90.
[http://dx.doi.org/10.1016/j.ijantimicag.2003.07.019] [PMID: 15164970]

[138] Bouzahzah B, Jelicks LA, Morris SA, Weiss LM, Tanowitz HB. Risedronate in the treatment of Murine Chagas disease. Parasitol Res 2005; 96(3): 184-7.
[http://dx.doi.org/10.1007/s00436-005-1331-9] [PMID: 15844009]

[139] Payne DJ, Gwynn MN, Holmes DJ, Pompliano DL. Drugs for bad bugs: confronting the challenges of antibacterial discovery. Nat Rev Drug Discov 2007; 6(1): 29-40.
[http://dx.doi.org/10.1038/nrd2201] [PMID: 17159923]

[140] Bustamante JM, Craft JM, Crowe BD, Ketchie SA, Tarleton RL. New, combined, and reduced dosing treatment protocols cure *Trypanosoma cruzi* infection in mice. J Infect Dis 2014; 209(1): 150-62.
[http://dx.doi.org/10.1093/infdis/jit420] [PMID: 23945371]

[141] Jacobs RT, Plattner JJ, Keenan M. Boron-based drugs as antiprotozoals. Curr Opin Infect Dis 2011; 24(6): 586-92.
[http://dx.doi.org/10.1097/QCO.0b013e32834c630e] [PMID: 22001943]

[142] Carmody LC, Germain A, Barker D, *et al.* Identification of small-molecule inhibitors of Trypansoma cruzi infection. Bethesda, MD: Probe Reports from the NIH Molecular Libraries Program 2010.

[143] Eisai Press Release. Eisai enters global agreement with broad institute to develop new drugs for neglected tropical diseases and tuberculosis. Drug discovery project for chagas disease awarded ghit fund grant. http://www.eisai.com/news/news201364.html. 2014.

[144] Keenan M, Alexander PW, Chaplin JH, *et al.* Selection and optimization of hits from a high-throughput phenotypic screen against *Trypanosoma cruzi*. Future Med Chem 2013; 5(15): 1733-52.
[http://dx.doi.org/10.4155/fmc.13.139] [PMID: 24144410]

[145] University of Dundee - Press Releaese. £10 million boost for neglected tropical disease research, Dundee University, GSK and Wellcome Trust enter five-year collaboration for neglected diseases. 2013.

[146] Peña I, Pilar Manzano M, Cantizani J, *et al.* New compound sets identified from high throughput phenotypic screening against three kinetoplastid parasites: an open resource. Sci Rep 2015; 5: 8771.
[http://dx.doi.org/10.1038/srep08771] [PMID: 25740547]

[147] Neitz RJ, Chen S, Supek F, *et al.* Lead identification to clinical candidate selection: drugs for Chagas disease. J Biomol Screen 2015; 20(1): 101-11.
[http://dx.doi.org/10.1177/1087057114553103] [PMID: 25281737]

[148] Khare S, Roach SL, Barnes SW, *et al.* Utilizing chemical genomics to identify cytochrome b as a novel drug target for chagas disease. PLoS Pathog 2015; 11(7): e1005058.

[http://dx.doi.org/10.1371/journal.ppat.1005058] [PMID: 26186534]

[149] Chatelain E, Konar N. Translational challenges of animal models in Chagas disease drug development: a review. Drug Des Devel Ther 2015; 9: 4807-23.
[http://dx.doi.org/10.2147/DDDT.S90208] [PMID: 26316715]

[150] Olivieri BP, Molina JT, de Castro SL, *et al.* A comparative study of posaconazole and benznidazole in the prevention of heart damage and promotion of trypanocidal immune response in a murine model of Chagas disease. Int J Antimicrob Agents 2010; 36(1): 79-83.
[http://dx.doi.org/10.1016/j.ijantimicag.2010.03.006] [PMID: 20452188]

[151] Molina J, Martins-Filho O, Brener Z, Romanha AJ, Loebenberg D, Urbina JA. Activities of the triazole derivative SCH 56592 (posaconazole) against drug-resistant strains of the protozoan parasite *Trypanosoma (Schizotrypanum) cruzi* in immunocompetent and immunosuppressed murine hosts. Antimicrob Agents Chemother 2000; 44(1): 150-5.
[http://dx.doi.org/10.1128/AAC.44.1.150-155.2000] [PMID: 10602737]

[152] Francisco AF, Lewis MD, Jayawardhana S, Taylor MC, Chatelain E, Kelly JM. Limited ability of posaconazole to cure both acute and chronic *Trypanosoma cruzi* infections revealed by highly sensitive *in vivo* imaging. Antimicrob Agents Chemother 2015; 59(8): 4653-61.
[http://dx.doi.org/10.1128/AAC.00520-15] [PMID: 26014936]

[153] Lewis MD, Francisco AF, Taylor MC, Kelly JM. A new experimental model for assessing drug efficacy against *Trypanosoma cruzi* infection based on highly sensitive *in vivo* imaging. J Biomol Screen 2015; 20(1): 36-43.
[http://dx.doi.org/10.1177/1087057114552623] [PMID: 25296657]

[154] Rivera-Vanderpas MT, Rodriguez AM, Afchain D, Bazin H, Capron A. *Trypanosoma cruzi*: variation in susceptibility of inbred strains of rats. Acta Trop 1983; 40(1): 5-10.
[PMID: 6134453]

[155] da Silva AM, Eduardo Ramirez L, Vargas M, Chapadeiro E, Brener Z. Evaluation of the rabbit as a model for Chagas disease-II. Histopathologic studies of the heart, digestive tract and skeletal muscle. Mem Inst Oswaldo Cruz 1996; 91(2): 199-206.
[http://dx.doi.org/10.1590/S0074-02761996000200015] [PMID: 8736091]

[156] Bilate AM, Salemi VM, Ramires FJ, *et al.* The Syrian hamster as a model for the dilated cardiomyopathy of Chagas disease: a quantitative echocardiographical and histopathological analysis. Microbes Infect 2003; 5(12): 1116-24.
[http://dx.doi.org/10.1016/j.micinf.2003.07.001] [PMID: 14554253]

[157] Andrade ZA. The canine model of Chagas' disease. Mem Inst Oswaldo Cruz 1984.
[http://dx.doi.org/10.1590/S0074-02761984000500016]

[158] Guedes PM, Veloso VM, Tafuri WL, *et al.* The dog as model for chemotherapy of the Chagas disease. Acta Trop 2002; 84(1): 9-17.
[http://dx.doi.org/10.1016/S0001-706X(02)00139-0] [PMID: 12387906]

[159] Miao Q, Santamaria C, Bailey D, Genest J, Ward BJ, Ndao M. Apolipoprotein A-I truncations in Chagas disease are caused by cruzipain, the major cysteine protease of *Trypanosoma cruzi*. Am J Pathol 2014; 184(4): 976-84.
[http://dx.doi.org/10.1016/j.ajpath.2013.12.018] [PMID: 24495738]

[160] Ratmanov P, Mediannikov O, Raoult D. Vectorborne diseases in West Africa: geographic distribution and geospatial characteristics. Trans R Soc Trop Med Hyg 2013; 107(5): 273-84. [http://dx.doi.org/10.1093/trstmh/trt020] [PMID: 23479360]

[161] World Health Organization. Control of human African trypanosomiasis: a strategy for the African region. AFRO, AFR/RC55/11. 2005.

[162] Kuzoe FA. Current situation of African trypanosomiasis. Acta Trop 1993; 54(3-4): 153-62. [http://dx.doi.org/10.1016/0001-706X(93)90089-T] [PMID: 7902654]

[163] World Health Organization. African trypanosomiasis (sleeping sickness) http://www.who.int/mediacentre/factsheets/fs259/en/. World Health Organ Fact Sheet. 2006; 259.

[164] Sutherland CS, Yukich J, Goeree R, Tediosi F. A literature review of economic evaluations for a neglected tropical disease: human African trypanosomiasis (sleeping sickness). PLoS Negl Trop Dis 2015; 9(2): e0003397. [http://dx.doi.org/10.1371/journal.pntd.0003397] [PMID: 25654605]

[165] Reid H, Kibona S, Rodney A, *et al.* Assessment of the burden of human African trypanosomiasis by rapid participatory appraisal in three high-risk villages in Urambo District, Northwest Tanzania. Afr Health Sci 2012; 12(2): 104-13. [PMID: 23056014]

[166] Steverding D. The history of African trypanosomiasis. Parasit Vectors 2008; 1(1): 3. [http://dx.doi.org/10.1186/1756-3305-1-3] [PMID: 18275594]

[167] Sneader W. Drug discovery: a history. Wiley 2005. [http://dx.doi.org/10.1002/0470015535]

[168] Nicolle MM. Traitement des trypanosomiases par les couleurs de benzidine. Premiére partie - etude chemique. Ann Inst Pasteur (Paris) 1906; 20: 417-48.

[169] Mesnil FN. Traitement des trypanosomiases par les couleurs de benzidine. Second partie - etude expérimentale. Ann Inst Pasteur (Paris) 1906; 20: 513-38.

[170] Dressel JO. The discovery of Germanin by Oskar Dressel and Richard Kothe. J Chem Educ 1961; 38: 620-1. [http://dx.doi.org/10.1021/ed038p620]

[171] Fourneau ET, Vallée J. Recherches de chimiothérapie dans la série du 205 Bayer. Urées des acides aminobenzoylaminonaphtaléniques. Ann Inst Pasteur (Paris) 1924; 38: 81-114.

[172] McGeary RP, Bennett AJ, Tran QB, Cosgrove KL, Ross BP. Suramin: clinical uses and structure-activity relationships. Mini Rev Med Chem 2008; 8(13): 1384-94. [http://dx.doi.org/10.2174/138955708786369573] [PMID: 18991754]

[173] Kaplan LD, Wolfe PR, Volberding PA, *et al.* Lack of response to suramin in patients with AIDS and AIDS-related complex. Am J Med 1987; 82(3 Spec No): 615-20.

[174] Lourie EM. Studies in chemotherapy. XXI. The trypanocidal action of certain aromatic diamidines. Ann Trop Med Parasitol 1939; 33: 289-304. [http://dx.doi.org/10.1080/00034983.1939.11685073]

[175] Ashley JN, Ewins AJ, Newbery G, Self AD. A chemotherapeutic comparison of the trypanocidal

action of some aromatic diamidines. J Chem Soc 1942; 103-16.
[http://dx.doi.org/10.1039/jr9420000103]

[176] Napier LE. SGP. A peculiar neurological sequel to administration of 4:4'-diamidino-diphenyl-ethylene (M&B 744). Ind Med Gaz 1942; 77: 71-4.

[177] Collard PJ, Hargreaves WH. Neuropathy after stilbamidine; treatment of kala-azar. Lancet 1947; 2(6480): 686-8.
[http://dx.doi.org/10.1016/S0140-6736(47)90716-2] [PMID: 20271209]

[178] Soeiro MN, De Souza EM, Stephens CE, Boykin DW. Aromatic diamidines as antiparasitic agents. Expert Opin Investig Drugs 2005; 14(8): 957-72.
[http://dx.doi.org/10.1517/13543784.14.8.957] [PMID: 16050790]

[179] Damper D, Patton CL. Pentamidine transport and sensitivity in brucei-group trypanosomes. J Protozool 1976; 23(2): 349-56.
[http://dx.doi.org/10.1111/j.1550-7408.1976.tb03787.x] [PMID: 6797]

[180] D. L. Arsenic as a remedy for tsetse bite. BMJ 1858; 70: 360-1.

[181] Bosch F, Rosich L. The contributions of Paul Ehrlich to pharmacology: a tribute on the occasion of the centenary of his Nobel Prize. Pharmacology 2008; 82(3): 171-9.
[http://dx.doi.org/10.1159/000149583] [PMID: 18679046]

[182] Jacobs WA. Aromatic arsenic compounds v. N-substituted glycylarsanilic acids. J Am Chem Soc 1919; 41: 1809-21.
[http://dx.doi.org/10.1021/ja02232a012]

[183] Friedheim EA. Some approaches to the development of chemotherapeutic compounds. Ann Trop Med Parasitol 1959; 53(1): 1-9.
[http://dx.doi.org/10.1080/00034983.1959.11685892] [PMID: 13650482]

[184] Friedheim EA. Melarsen oxide in the treatment of human trypanosomiasis. Ann Trop Med Parasitol 1948; 42(3-4): 357-63.
[http://dx.doi.org/10.1080/00034983.1948.11685383] [PMID: 18110349]

[185] Friedheim EA. Mel B in the treatment of human trypanosomiasis. Am J Trop Med Hyg 1949; 29(2): 173-80.
[PMID: 18116843]

[186] Nightingale SL. From the food and drug administration. JAMA 1991; 265(10): 1229.
[http://dx.doi.org/10.1001/jama.1991.03460100029008] [PMID: 1995961]

[187] Meyskens FL Jr, Gerner EW. Development of difluoromethylornithine (DFMO) as a chemoprevention agent. Clin Cancer Res 1999; 5(5): 945-51.
[PMID: 10353725]

[188] Oredsson S, Anehus S, Heby O. Inhibition of cell proliferation by DL-alpha-difluoromethylornithine, a catalytic irreversible inhibitor of ornithine decarboxylase. Acta Chem Scand, B, Org Chem Biochem 1980; 34(6): 457-8.
[http://dx.doi.org/10.3891/acta.chem.scand.34b-0457] [PMID: 6781187]

[189] Tabor CW, Tabor H. Polyamines. Annu Rev Biochem 1984; 53: 749-90.
[http://dx.doi.org/10.1146/annurev.bi.53.070184.003533] [PMID: 6206782]

[190] Bacchi CJ, Nathan HC, Hutner SH, McCann PP, Sjoerdsma A. Polyamine metabolism: a potential therapeutic target in trypanosomes. Science 1980; 210(4467): 332-4.
[http://dx.doi.org/10.1126/science.6775372] [PMID: 6775372]

[191] Eozenou P, Jannin J, Ngampo S, Carme B, Tell GP, Schechter PJ. A trial treatment with eflornithine of trypanosomiasis caused by *Trypanosoma brucei gambiense* in the Peoples Republic of the Congo. Med Trop (Mars) 1989; 49(2): 149-54.
[PMID: 2507863]

[192] Pepin J, Milord F, Guern C, Schechter PJ. Difluoromethylornithine for arseno-resistant *Trypanosoma brucei gambiense* sleeping sickness. Lancet 1987; 2(8573): 1431-3.
[http://dx.doi.org/10.1016/S0140-6736(87)91131-7] [PMID: 2891995]

[193] Doua F, Boa FY, Schechter PJ, *et al.* Treatment of human late stage gambiense trypanosomiasis with alpha-difluoromethylornithine (eflornithine): efficacy and tolerance in 14 cases in Côte dIvoire. Am J Trop Med Hyg 1987; 37(3): 525-33.
[PMID: 3120607]

[194] Van Nieuwenhove S, Schechter PJ, Declercq J, Boné G, Burke J, Sjoerdsma A. Treatment of gambiense sleeping sickness in the Sudan with oral DFMO (DL-alpha-difluoromethylornithine), an inhibitor of ornithine decarboxylase; first field trial. Trans R Soc Trop Med Hyg 1985; 79(5): 692-8.
[http://dx.doi.org/10.1016/0035-9203(85)90195-6] [PMID: 3938090]

[195] Iten M, Mett H, Evans A, Enyaru JC, Brun R, Kaminsky R. Alterations in ornithine decarboxylase characteristics account for tolerance of *Trypanosoma brucei* rhodesiense to D,L-alph--difluoromethylornithine. Antimicrob Agents Chemother 1997; 41(9): 1922-5.
[PMID: 9303385]

[196] Fairlamb AH. Chemotherapy of human African trypanosomiasis: current and future prospects. Trends Parasitol 2003; 19(11): 488-94.
[http://dx.doi.org/10.1016/j.pt.2003.09.002] [PMID: 14580959]

[197] Docampo R, Moreno SN, Stoppani AO, *et al.* Mechanism of nifurtimox toxicity in different forms of *Trypanosoma cruzi*. Biochem Pharmacol 1981; 30(14): 1947-51.
[http://dx.doi.org/10.1016/0006-2952(81)90204-5] [PMID: 7023488]

[198] Pépin J, Milord F, Meurice F, Ethier L, Loko L, Mpia B. High-dose nifurtimox for arseno-resistant *Trypanosoma brucei gambiense* sleeping sickness: an open trial in central Zaire. Trans R Soc Trop Med Hyg 1992; 86(3): 254-6.
[http://dx.doi.org/10.1016/0035-9203(92)90298-Q] [PMID: 1412646]

[199] Pepin J, Milord F, Mpia B, *et al.* An open clinical trial of nifurtimox for arseno-resistant *Trypanosoma brucei gambiense* sleeping sickness in central Zaire. Trans R Soc Trop Med Hyg 1989; 83(4): 514-7.
[http://dx.doi.org/10.1016/0035-9203(89)90270-8] [PMID: 2694491]

[200] Eperon G, Balasegaram M, Potet J, Mowbray C, Valverde O, Chappuis F. Treatment options for second-stage gambiense human African trypanosomiasis. Expert Rev Anti Infect Ther 2014; 12(11): 1407-17.
[http://dx.doi.org/10.1586/14787210.2014.959496] [PMID: 25204360]

[201] Roy Chowdhury A, Bakshi R, Wang J, *et al.* The killing of African trypanosomes by ethidium

bromide. PLoS Pathog 2010; 6(12): e1001226.
[http://dx.doi.org/10.1371/journal.ppat.1001226] [PMID: 21187912]

[202] Steverding D. The development of drugs for treatment of sleeping sickness: a historical review. Parasit Vectors 2010; 3(1): 15.
[http://dx.doi.org/10.1186/1756-3305-3-15] [PMID: 20219092]

[203] Fairlamb AH, Opperdoes FR, Borst P. New approach to screening drugs for activity against African trypanosomes. Nature 1977; 265(5591): 270-1.
[http://dx.doi.org/10.1038/265270a0] [PMID: 834274]

[204] Faria J, Moraes CB, Song R, *et al.* Drug discovery for human African trypanosomiasis: identification of novel scaffolds by the newly developed HTS SYBR Green assay for *Trypanosoma brucei*. J Biomol Screen 2015; 20(1): 70-81.
[http://dx.doi.org/10.1177/1087057114556236] [PMID: 25342146]

[205] Sykes ML, Avery VM. A luciferase based viability assay for ATP detection in 384-well format for high throughput whole cell screening of *Trypanosoma brucei* bloodstream form strain 427. Parasit Vectors 2009; 2(1): 54.
[http://dx.doi.org/10.1186/1756-3305-2-54] [PMID: 19909542]

[206] Sykes ML, Avery VM. Development of an Alamar Blue viability assay in 384-well format for high throughput whole cell screening of *Trypanosoma brucei* bloodstream form strain 427. Am J Trop Med Hyg 2009; 81(4): 665-74.
[http://dx.doi.org/10.4269/ajtmh.2009.09-0015] [PMID: 19815884]

[207] Mackey ZB, Baca AM, Mallari JP, *et al.* Discovery of trypanocidal compounds by whole cell HTS of *Trypanosoma brucei*. Chem Biol Drug Des 2006; 67(5): 355-63.
[http://dx.doi.org/10.1111/j.1747-0285.2006.00389.x] [PMID: 16784460]

[208] Docampo R, Moreno SN. Current chemotherapy of human African trypanosomiasis. Parasitol Res 2003; 90 (Suppl. 1): S10-3.
[PMID: 12811544]

[209] Jones AJ, Avery VM. Future treatment options for human African trypanosomiasis. Expert Rev Anti Infect Ther 2015; 13(12): 1429-32.
[http://dx.doi.org/10.1586/14787210.2015.1094374] [PMID: 26414688]

[210] Yeates C. DB-289 Immtech International. IDrugs 2003; 6(11): 1086-93.
[PMID: 14600842]

[211] Mdachi RE, Thuita JK, Kagira JM, *et al.* Efficacy of the novel diamidine compound 2,5-Bis(--amidinophenyl)- furan-bis-O-Methlylamidoxime (Pafuramidine, DB289) against *Trypanosoma brucei* rhodesiense infection in vervet monkeys after oral administration. Antimicrob Agents Chemother 2009; 53(3): 953-7.
[http://dx.doi.org/10.1128/AAC.00831-08] [PMID: 19064893]

[212] Thuita JK, Karanja SM, Wenzler T, *et al.* Efficacy of the diamidine DB75 and its prodrug DB289, against murine models of human African trypanosomiasis. Acta Trop 2008; 108(1): 6-10.
[http://dx.doi.org/10.1016/j.actatropica.2008.07.006] [PMID: 18722336]

[213] Kaiser M, Bray MA, Cal M, Bourdin Trunz B, Torreele E, Brun R. Antitrypanosomal activity of

fexinidazole, a new oral nitroimidazole drug candidate for treatment of sleeping sickness. Antimicrob Agents Chemother 2011; 55(12): 5602-8.
[http://dx.doi.org/10.1128/AAC.00246-11] [PMID: 21911566]

[214] Tarral A, Blesson S, Mordt OV, *et al.* Determination of an optimal dosing regimen for fexinidazole, a novel oral drug for the treatment of human African trypanosomiasis: first-in-human studies. Clin Pharmacokinet 2014; 53(6): 565-80.
[http://dx.doi.org/10.1007/s40262-014-0136-3] [PMID: 24535888]

[215] Jacobs RT, Nare B, Wring SA, *et al.* SCYX-7158, an orally-active benzoxaborole for the treatment of stage 2 human African trypanosomiasis. PLoS Negl Trop Dis 2011; 5(6): e1151.
[http://dx.doi.org/10.1371/journal.pntd.0001151] [PMID: 21738803]

[216] Johnson SL, Chen LH, Pellecchia M. A high-throughput screening approach to anthrax lethal factor inhibition. Bioorg Chem 2007; 35(4): 306-12.
[http://dx.doi.org/10.1016/j.bioorg.2006.12.005] [PMID: 17320146]

[217] De Rycker M, O'Neill S, Joshi D, Campbell L, Gray DW, Fairlamb AH. A static-cidal assay for *Trypanosoma brucei* to aid hit prioritisation for progression into drug discovery programmes. PLoS Negl Trop Dis 2012; 6(11): e1932.

[218] Woodland A, Thompson S, Cleghorn LA, *et al.* Discovery of inhibitors of *Trypanosoma brucei* by phenotypic screening of a focused protein kinase library. ChemMedChem 2015; 10(11): 1809-20.
[http://dx.doi.org/10.1002/cmdc.201500300] [PMID: 26381210]

[219] Steverding D, Sexton DW, Wang X, Gehrke SS, Wagner GK, Caffrey CR. *Trypanosoma brucei*: chemical evidence that cathepsin L is essential for survival and a relevant drug target. Int J Parasitol 2012; 42(5): 481-8.
[http://dx.doi.org/10.1016/j.ijpara.2012.03.009] [PMID: 22549023]

[220] Jefferson T, McShan D, Warfield J, Ogungbe IV. Screening and identification of inhibitors of *Trypanosoma brucei* cathepsin L with antitrypanosomal activity. Chem Biol Drug Des 2015.
[PMID: 26242248]

[221] Royo S, Rodríguez S, Schirmeister T, Kesselring J, Kaiser M, González FV. Dipeptidyl enoates As potent rhodesain inhibitors that display a dual mode of action. ChemMedChem 2015; 10(9): 1484-7.
[http://dx.doi.org/10.1002/cmdc.201500204] [PMID: 26179752]

[222] Fonseca NC, da Cruz LF, da Silva Villela F, *et al.* Synthesis of a sugar-based thiosemicarbazone series and structure-activity relationship *versus* the parasite cysteine proteases rhodesain, cruzain, and *Schistosoma mansoni* cathepsin B1. Antimicrob Agents Chemother 2015; 59(5): 2666-77.
[http://dx.doi.org/10.1128/AAC.04601-14] [PMID: 25712353]

[223] Schulze CJ, Donia MS, Siqueira-Neto JL, *et al.* Genome-directed lead discovery: biosynthesis, structure elucidation, and biological evaluation of two families of polyene macrolactams against *Trypanosoma brucei*. ACS Chem Biol 2015; 10(10): 2373-81.
[http://dx.doi.org/10.1021/acschembio.5b00308] [PMID: 26270237]

[224] Wenzler T, Yang S, Braissant O, Boykin DW, Brun R, Wang MZ. Pharmacokinetics, *Trypanosoma brucei gambiense* efficacy, and time of drug action of DB829, a preclinical candidate for treatment of second-stage human African trypanosomiasis. Antimicrob Agents Chemother 2013; 57(11): 5330-43.
[http://dx.doi.org/10.1128/AAC.00398-13] [PMID: 23959303]

[225] McLatchie AP, Burrell-Saward H, Myburgh E, *et al.* Highly sensitive *in vivo* imaging of *Trypanosoma brucei* expressing red-shifted luciferase. PLoS Negl Trop Dis 2013; 7(11): e2571. [http://dx.doi.org/10.1371/journal.pntd.0002571] [PMID: 24278497]

[226] Kennedy PG. Animal models of human African trypanosomiasisvery useful or too far removed? Trans R Soc Trop Med Hyg 2007; 101(11): 1061-2. [http://dx.doi.org/10.1016/j.trstmh.2007.05.001] [PMID: 17548098]

[227] Nwaka S, Ramirez B, Brun R, Maes L, Douglas F, Ridley R. Advancing drug innovation for neglected diseases-criteria for lead progression. PLoS Negl Trop Dis 2009; 3(8): e440. [http://dx.doi.org/10.1371/journal.pntd.0000440] [PMID: 19707561]

[228] World Health Organization. Delivering on Promises and Driving Progress. http://unitingtocombatntds.org/sites/default/files/document/NTD_report_04102014_v4_singles.pdf. Uniting to Combat NTDs. 2014; 2014 Report.

[229] Badaró R, Gonçalves LO, Gois LL, Maia ZP, Benson C, Grassi MF. Leishmaniasis as a manifestation of immune reconstitution inflammatory syndrome (IRIS) in HIV-infected patients: a literature review. J Int Assoc Provid AIDS Care 2015; 14(5): 402-7. [http://dx.doi.org/10.1177/2325957414555225] [PMID: 25331225]

[230] G. V. Tratamento da leishmaniose tegumentar por injeções intravenosas de tártaro emético. 7 Congresso Brasileiro de Medicina Tropical de São Paulo. 1912; 4: 426-8.

[231] Cristina GD, Caronia G. Sulla terapia della leishmaniosi interna. Pathologica 1915; 7: 82-3.

[232] Cook GC. Leonard rogers KCSI FRCP FRS (1868-1962) and the founding of the calcutta school of tropical medicine notes and records of the Royal Society 2006; 60(2): 171-81.

[233] Lira R, Sundar S, Makharia A, *et al.* Evidence that the high incidence of treatment failures in Indian kala-azar is due to the emergence of antimony-resistant strains of *Leishmania donovani*. J Infect Dis 1999; 180(2): 564-7. [http://dx.doi.org/10.1086/314896] [PMID: 10395884]

[234] Faraut-Gambarelli F, Piarroux R, Deniau M, *et al. In vitro* and *in vivo* resistance of *Leishmania* infantum to meglumine antimoniate: a study of 37 strains collected from patients with visceral leishmaniasis. Antimicrob Agents Chemother 1997; 41(4): 827-30. [PMID: 9087498]

[235] Sundar S. Drug resistance in Indian visceral leishmaniasis. Trop Med Int Health 2001; 6(11): 849-54. [http://dx.doi.org/10.1046/j.1365-3156.2001.00778.x] [PMID: 11703838]

[236] Brahmachari UN. Chemotherapy of antimonial compounds in kala-azar infection. Part IV. Further observations on the therapeutic values of urea stibamine. By U.N. Brahmachari, 1922. Indian J Med Res 1989; 89: 393-404. [PMID: 2644171]

[237] Shortt HE. Recent research on kala-azar in India. Trans R Soc Trop Med Hyg 1945; 39(1): 13-41. [http://dx.doi.org/10.1016/0035-9203(45)90030-7] [PMID: 20293977]

[238] Frézard F, Demicheli C, Ribeiro RR. Pentavalent antimonials: new perspectives for old drugs. Molecules 2009; 14(7): 2317-36. [http://dx.doi.org/10.3390/molecules14072317] [PMID: 19633606]

[239] Brochu C, Haimeur A, Ouellette M. The heat shock protein HSP70 and heat shock cognate protein HSC70 contribute to antimony tolerance in the protozoan parasite *Leishmania*. Cell Stress Chaperones 2004; 9(3): 294-303.
[http://dx.doi.org/10.1379/CSC-15R1.1] [PMID: 15544167]

[240] Croft SL, Neame KD, Homewood CA. Accumulation of [125Sb]sodium stibogluconate by *Leishmania mexicana amazonensis* and *Leishmania donovaniin vitro*. Comp Biochem Physiol C Comp Pharmacol 1981; 68C(1): 95-8.
[http://dx.doi.org/10.1016/0306-4492(81)90043-5] [PMID: 6108831]

[241] Rosen BP. Transport and detoxification systems for transition metals, heavy metals and metalloids in eukaryotic and prokaryotic microbes. Comp Biochem Physiol A Mol Integr Physiol 2002; 133(3): 689-93.
[http://dx.doi.org/10.1016/S1095-6433(02)00201-5] [PMID: 12443926]

[242] Brochu C, Wang J, Roy G, *et al.* Antimony uptake systems in the protozoan parasite *Leishmania* and accumulation differences in antimony-resistant parasites. Antimicrob Agents Chemother 2003; 47(10): 3073-9.
[http://dx.doi.org/10.1128/AAC.47.10.3073-3079.2003] [PMID: 14506011]

[243] Goyard S, Segawa H, Gordon J, *et al.* An *in vitro* system for developmental and genetic studies of *Leishmania donovani* phosphoglycans. Mol Biochem Parasitol 2003; 130(1): 31-42.
[http://dx.doi.org/10.1016/S0166-6851(03)00142-7] [PMID: 14550894]

[244] Shaked-Mishan P, Ulrich N, Ephros M, Zilberstein D. Novel Intracellular SbV reducing activity correlates with antimony susceptibility in *Leishmania donovani*. J Biol Chem 2001; 276(6): 3971-6.
[http://dx.doi.org/10.1074/jbc.M005423200] [PMID: 11110784]

[245] Ephros M, Bitnun A, Shaked P, Waldman E, Zilberstein D. Stage-specific activity of pentavalent antimony against *Leishmania donovani* axenic amastigotes. Antimicrob Agents Chemother 1999; 43(2): 278-82.
[PMID: 9925518]

[246] Ephros M, Waldman E, Zilberstein D. Pentostam induces resistance to antimony and the preservative chlorocresol in *Leishmania donovani* promastigotes and axenically grown amastigotes. Antimicrob Agents Chemother 1997; 41(5): 1064-8.
[PMID: 9145870]

[247] Callahan HL, Roberts WL, Rainey PM, Beverley SM. The PGPA gene of *Leishmania major* mediates antimony (SbIII) resistance by decreasing influx and not by increasing efflux. Mol Biochem Parasitol 1994; 68(1): 145-9.
[http://dx.doi.org/10.1016/0166-6851(94)00154-5] [PMID: 7891738]

[248] Wyllie S, Fairlamb AH. Differential toxicity of antimonial compounds and their effects on glutathione homeostasis in a human leukaemia monocyte cell line. Biochem Pharmacol 2006; 71(3): 257-67.
[http://dx.doi.org/10.1016/j.bcp.2005.10.043] [PMID: 16318845]

[249] Sereno D, Cavaleyra M, Zemzoumi K, Maquaire S, Ouaissi A, Lemesre JL. Axenically grown amastigotes of *Leishmania infantum* used as an *in vitro* model to investigate the pentavalent antimony mode of action. Antimicrob Agents Chemother 1998; 42(12): 3097-102.
[PMID: 9835497]

[250] Yan S, Li F, Ding K, Sun H. Reduction of pentavalent antimony by trypanothione and formation of a binary and ternary complex of antimony(III) and trypanothione. J Biol Inorg Chem 2003; 8(6): 689-97. [http://dx.doi.org/10.1007/s00775-003-0468-1] [PMID: 12827457]

[251] Ferreira CdosS, Martins PS, Demicheli C, Brochu C, Ouellette M, Frézard F. Thiol-induced reduction of antimony(V) into antimony(III): a comparative study with trypanothione, cysteinyl-glycine, cysteine and glutathione. Biometals 2003; 16(3): 441-6. [http://dx.doi.org/10.1023/A:1022823605068] [PMID: 12680707]

[252] Frézard F, Demicheli C, Ferreira CS, Costa MA. Glutathione-induced conversion of pentavalent antimony to trivalent antimony in meglumine antimoniate. Antimicrob Agents Chemother 2001; 45(3): 913-6. [http://dx.doi.org/10.1128/AAC.45.3.913-916.2001] [PMID: 11181379]

[253] Denton H, McGregor JC, Coombs GH. Reduction of anti-leishmanial pentavalent antimonial drugs by a parasite-specific thiol-dependent reductase, TDR1. Biochem J 2004; 381(Pt 2): 405-12. [http://dx.doi.org/10.1042/BJ20040283] [PMID: 15056070]

[254] Fairlamb AH, Cerami A. Metabolism and functions of trypanothione in the Kinetoplastida. Annu Rev Microbiol 1992; 46: 695-729. [http://dx.doi.org/10.1146/annurev.mi.46.100192.003403] [PMID: 1444271]

[255] Wyllie S, Cunningham ML, Fairlamb AH. Dual action of antimonial drugs on thiol redox metabolism in the human pathogen *Leishmania donovani*. J Biol Chem 2004; 279(38): 39925-32. [http://dx.doi.org/10.1074/jbc.M405635200] [PMID: 15252045]

[256] Cunningham ML, Fairlamb AH. Trypanothione reductase from *Leishmania donovani*. Purification, characterisation and inhibition by trivalent antimonials. Eur J Biochem 1995; 230(2): 460-8. [http://dx.doi.org/10.1111/j.1432-1033.1995.tb20583.x] [PMID: 7607216]

[257] Demicheli C, Frézard F, Mangrum JB, Farrell NP. Interaction of trivalent antimony with a CCHC zinc finger domain: potential relevance to the mechanism of action of antimonial drugs. Chem Commun (Camb) 2008; (39): 4828-30. [http://dx.doi.org/10.1039/b809186b] [PMID: 18830507]

[258] Berman JD, Gallalee JV, Best JM. Sodium stibogluconate (Pentostam) inhibition of glucose catabolism *via* the glycolytic pathway, and fatty acid beta-oxidation in *Leishmania mexicana amastigotes*. Biochem Pharmacol 1987; 36(2): 197-201. [http://dx.doi.org/10.1016/0006-2952(87)90689-7] [PMID: 3028425]

[259] Berman JD, Waddell D, Hanson BD. Biochemical mechanisms of the antileishmanial activity of sodium stibogluconate. Antimicrob Agents Chemother 1985; 27(6): 916-20. [http://dx.doi.org/10.1128/AAC.27.6.916] [PMID: 2411217]

[260] Sudhandiran G, Shaha C. Antimonial-induced increase in intracellular Ca2+ through non-selective cation channels in the host and the parasite is responsible for apoptosis of intracellular *Leishmania donovani amastigotes*. J Biol Chem 2003; 278(27): 25120-32. [http://dx.doi.org/10.1074/jbc.M301975200] [PMID: 12707265]

[261] Mukherjee SB, Das M, Sudhandiran G, Shaha C. Increase in cytosolic Ca2+ levels through the activation of non-selective cation channels induced by oxidative stress causes mitochondrial

depolarization leading to apoptosis-like death in *Leishmania donovani promastigotes*. J Biol Chem 2002; 277(27): 24717-27.
[http://dx.doi.org/10.1074/jbc.M201961200] [PMID: 11983701]

[262] Mookerjee Basu J, Mookerjee A, Sen P, *et al.* Sodium antimony gluconate induces generation of reactive oxygen species and nitric oxide *via* phosphoinositide 3-kinase and mitogen-activated protein kinase activation in *Leishmania donovani*-infected macrophages. Antimicrob Agents Chemother 2006; 50(5): 1788-97.
[http://dx.doi.org/10.1128/AAC.50.5.1788-1797.2006] [PMID: 16641451]

[263] Mishra M, Biswas UK, Jha DN, Khan AB. Amphotericin *versus* pentamidine in antimony-unresponsive kala-azar. Lancet 1992; 340(8830): 1256-7.
[http://dx.doi.org/10.1016/0140-6736(92)92952-C] [PMID: 1359322]

[264] Roberts CW, McLeod R, Rice DW, Ginger M, Chance ML, Goad LJ. Fatty acid and sterol metabolism: potential antimicrobial targets in apicomplexan and trypanosomatid parasitic protozoa. Mol Biochem Parasitol 2003; 126(2): 129-42.
[http://dx.doi.org/10.1016/S0166-6851(02)00280-3] [PMID: 12615312]

[265] Singh AK, Papadopoulou B, Ouellette M. Gene amplification in amphotericin B-resistant *Leishmania tarentolae*. Exp Parasitol 2001; 99(3): 141-7.
[http://dx.doi.org/10.1006/expr.2001.4663] [PMID: 11846524]

[266] Gangneux JP, Sulahian A, Garin YJ, Derouin F. Lipid formulations of amphotericin b in the treatment of experimental visceral leishmaniasis due to *Leishmania infantum*. Trans R Soc Trop Med Hyg 1996; 90(5): 574-7.
[http://dx.doi.org/10.1016/S0035-9203(96)90330-2] [PMID: 8944278]

[267] Yardley V, Croft SL. Activity of liposomal amphotericin B against experimental cutaneous leishmaniasis. Antimicrob Agents Chemother 1997; 41(4): 752-6.
[PMID: 9087483]

[268] Berman JD, Badaro R, Thakur CP, *et al.* Efficacy and safety of liposomal amphotericin B (AmBisome) for visceral leishmaniasis in endemic developing countries. Bull World Health Organ 1998; 76(1): 25-32.
[PMID: 9615494]

[269] Sundar S, Mehta H, Suresh AV, Singh SP, Rai M, Murray HW. Amphotericin B treatment for Indian visceral leishmaniasis: conventional *versus* lipid formulations. Clin Infect Dis 2004; 38(3): 377-83.
[http://dx.doi.org/10.1086/380971] [PMID: 14727208]

[270] Croft SL, Coombs GH. *Leishmaniasis* current chemotherapy and recent advances in the search for novel drugs. Trends Parasitol 2003; 19(11): 502-8.
[http://dx.doi.org/10.1016/j.pt.2003.09.008] [PMID: 14580961]

[271] Basselin M, Denise H, Coombs GH, Barrett MP. Resistance to pentamidine in *Leishmania mexicana* involves exclusion of the drug from the mitochondrion. Antimicrob Agents Chemother 2002; 46(12): 3731-8.
[http://dx.doi.org/10.1128/AAC.46.12.3731-3738.2002] [PMID: 12435669]

[272] Unger C, Damenz W, Fleer EA, *et al.* Hexadecylphosphocholine, a new ether lipid analogue. Studies on the antineoplastic activity *in vitro* and *in vivo.* Acta Oncol 1989; 28(2): 213-7.

[http://dx.doi.org/10.3109/02841868909111249] [PMID: 2736110]

[273] Sundar S, Gupta LB, Makharia MK, *et al.* Oral treatment of visceral leishmaniasis with miltefosine. Ann Trop Med Parasitol 1999; 93(6): 589-97. [http://dx.doi.org/10.1080/00034989958096] [PMID: 10707104]

[274] Sundar S, Olliaro PL. Miltefosine in the treatment of leishmaniasis: Clinical evidence for informed clinical risk management. Ther Clin Risk Manag 2007; 3(5): 733-40. [PMID: 18472998]

[275] Bhattacharya SK, Jha TK, Sundar S, *et al.* Efficacy and tolerability of miltefosine for childhood visceral leishmaniasis in India. Clin Infect Dis 2004; 38(2): 217-21. [http://dx.doi.org/10.1086/380638] [PMID: 14699453]

[276] Croft SL, Neal RA, Pendergast W, Chan JH. The activity of alkyl phosphorylcholines and related derivatives against *Leishmania donovani*. Biochem Pharmacol 1987; 36(16): 2633-6. [http://dx.doi.org/10.1016/0006-2952(87)90543-0] [PMID: 3606662]

[277] Paris C, Loiseau PM, Bories C, Bréard J. Miltefosine induces apoptosis-like death in *Leishmania donovani promastigotes*. Antimicrob Agents Chemother 2004; 48(3): 852-9. [http://dx.doi.org/10.1128/AAC.48.3.852-859.2004] [PMID: 14982775]

[278] Wadhone P, Maiti M, Agarwal R, Kamat V, Martin S, Saha B. Miltefosine promotes IFN-gamm--dominated anti-leishmanial immune response. J Immunol 2009; 182(11): 7146-54. [http://dx.doi.org/10.4049/jimmunol.0803859] [PMID: 19454711]

[279] Scott JA, Davidson RN, Moody AH, *et al.* Aminosidine (paromomycin) in the treatment of leishmaniasis imported into the United Kingdom. Trans R Soc Trop Med Hyg 1992; 86(6): 617-9. [http://dx.doi.org/10.1016/0035-9203(92)90151-2] [PMID: 1287915]

[280] Jhingran A, Chawla B, Saxena S, Barrett MP, Madhubala R. Paromomycin: uptake and resistance in *Leishmania donovani*. Mol Biochem Parasitol 2009; 164(2): 111-7. [http://dx.doi.org/10.1016/j.molbiopara.2008.12.007] [PMID: 19146886]

[281] Ivens AC, Peacock CS, Worthey EA, *et al.* The genome of the kinetoplastid parasite, *Leishmania major*. Science 2005; 309(5733): 436-42. [http://dx.doi.org/10.1126/science.1112680] [PMID: 16020728]

[282] Peacock CS, Seeger K, Harris D, *et al.* Comparative genomic analysis of three *Leishmania* species that cause diverse human disease. Nat Genet 2007; 39(7): 839-47. [http://dx.doi.org/10.1038/ng2053] [PMID: 17572675]

[283] Cunningham ML, Beverley SM. Pteridine salvage throughout the *Leishmania* infectious cycle: implications for antifolate chemotherapy. Mol Biochem Parasitol 2001; 113(2): 199-213. [http://dx.doi.org/10.1016/S0166-6851(01)00213-4] [PMID: 11295174]

[284] Gilbert IH. Inhibitors of dihydrofolate reductase in *Leishmania* and trypanosomes. Biochim Biophys Acta 2002; 1587(2-3): 249-57. [http://dx.doi.org/10.1016/S0925-4439(02)00088-1] [PMID: 12084467]

[285] Doerig C, Meijer L, Mottram JC. Protein kinases as drug targets in parasitic protozoa. Trends Parasitol 2002; 18(8): 366-71. [http://dx.doi.org/10.1016/S1471-4922(02)02321-8] [PMID: 12377287]

[286] Agüero F, Al-Lazikani B, Aslett M, *et al.* Genomic-scale prioritization of drug targets: the TDR Targets database. Nat Rev Drug Discov 2008; 7(11): 900-7.
[http://dx.doi.org/10.1038/nrd2684] [PMID: 18927591]

[287] Crowther GJ, Shanmugam D, Carmona SJ, *et al.* Identification of attractive drug targets in neglected-disease pathogens using an *in silico* approach. PLoS Negl Trop Dis 2010; 4(8): e804.
[http://dx.doi.org/10.1371/journal.pntd.0000804] [PMID: 20808766]

[288] Walker RG, Thomson G, Malone K, *et al.* High throughput screens yield small molecule inhibitors of *Leishmania* CRK3:CYC6 cyclin-dependent kinase. PLoS Negl Trop Dis 2011; 5(4): e1033.
[http://dx.doi.org/10.1371/journal.pntd.0001033] [PMID: 21483720]

[289] Hellerstein MK. A critique of the molecular target-based drug discovery paradigm based on principles of metabolic control: advantages of pathway-based discovery. Metab Eng 2008; 10(1): 1-9.
[http://dx.doi.org/10.1016/j.ymben.2007.09.003] [PMID: 17962055]

[290] Fischer E, Sauer U. Large-scale *in vivo* flux analysis shows rigidity and suboptimal performance of *Bacillus subtilis* metabolism. Nat Genet 2005; 37(6): 636-40.
[http://dx.doi.org/10.1038/ng1555] [PMID: 15880104]

[291] Bickle M. The beautiful cell: high-content screening in drug discovery. Anal Bioanal Chem 2010; 398(1): 219-26.
[http://dx.doi.org/10.1007/s00216-010-3788-3] [PMID: 20577725]

[292] Haney SA, LaPan P, Pan J, Zhang J. High-content screening moves to the front of the line. Drug Discov Today 2006; 11(19-20): 889-94.
[http://dx.doi.org/10.1016/j.drudis.2006.08.015] [PMID: 16997138]

[293] Korn K, Krausz E. Cell-based high-content screening of small-molecule libraries. Curr Opin Chem Biol 2007; 11(5): 503-10.
[http://dx.doi.org/10.1016/j.cbpa.2007.08.030] [PMID: 17931958]

[294] Ashutosh GS, Gupta S, Ramesh , Sundar S, Goyal N. Use of *Leishmania donovani* field isolates expressing the luciferase reporter gene in *in vitro* drug screening. Antimicrob Agents Chemother 2005; 49(9): 3776-83.
[http://dx.doi.org/10.1128/AAC.49.9.3776-3783.2005] [PMID: 16131481]

[295] Buckner FS, Wilson AJ. Colorimetric assay for screening compounds against *Leishmania amastigotes* grown in macrophages. Am J Trop Med Hyg 2005; 72(5): 600-5.
[PMID: 15891135]

[296] Lang T, Goyard S, Lebastard M, Milon G. Bioluminescent *Leishmania* expressing luciferase for rapid and high throughput screening of drugs acting on amastigote-harbouring macrophages and for quantitative real-time monitoring of parasitism features in living mice. Cell Microbiol 2005; 7(3): 383-92.
[http://dx.doi.org/10.1111/j.1462-5822.2004.00468.x] [PMID: 15679841]

[297] Mandal S, Maharjan M, Ganguly S, *et al.* High-throughput screening of amastigotes of *Leishmania donovani* clinical isolates against drugs using a colorimetric beta-lactamase assay. Indian J Exp Biol 2009; 47(6): 475-9.
[PMID: 19634714]

[298] Osorio Y, Travi BL, Renslo AR, Peniche AG, Melby PC. Identification of small molecule lead compounds for visceral leishmaniasis using a novel *ex vivo* splenic explant model system. PLoS Negl Trop Dis 2011; 5(2): e962.
[http://dx.doi.org/10.1371/journal.pntd.0000962] [PMID: 21358812]

[299] Sharlow ER, Close D, Shun T, *et al.* Identification of potent chemotypes targeting *Leishmania major* using a high-throughput, low-stringency, computationally enhanced, small molecule screen. PLoS Negl Trop Dis 2009; 3(11): e540.
[http://dx.doi.org/10.1371/journal.pntd.0000540] [PMID: 19888337]

[300] Singh N, Dube A. Short report: fluorescent Leishmania: application to anti-leishmanial drug testing. Am J Trop Med Hyg 2004; 71(4): 400-2.
[PMID: 15516633]

[301] De Muylder G, Ang KK, Chen S, Arkin MR, Engel JC, McKerrow JH. A screen against *Leishmania* intracellular amastigotes: comparison to a promastigote screen and identification of a host cell-specific hit. PLoS Negl Trop Dis 2011; 5(7): e1253.
[http://dx.doi.org/10.1371/journal.pntd.0001253] [PMID: 21811648]

[302] Vermeersch M, da Luz RI, Toté K, Timmermans JP, Cos P, Maes L. *In vitro* susceptibilities of *Leishmania donovani promastigote* and amastigote stages to antileishmanial reference drugs: practical relevance of stage-specific differences. Antimicrob Agents Chemother 2009; 53(9): 3855-9.
[http://dx.doi.org/10.1128/AAC.00548-09] [PMID: 19546361]

[303] Rochette A, Raymond F, Corbeil J, Ouellette M, Papadopoulou B. Whole-genome comparative RNA expression profiling of axenic and intracellular amastigote forms of *Leishmania infantum*. Mol Biochem Parasitol 2009; 165(1): 32-47.
[http://dx.doi.org/10.1016/j.molbiopara.2008.12.012] [PMID: 19393160]

[304] Gupta S. Visceral leishmaniasis: experimental models for drug discovery. Indian J Med Res 2011; 133: 27-39.
[PMID: 21321417]

[305] Croft SL, Seifert K, Yardley V. Current scenario of drug development for leishmaniasis. Indian J Med Res 2006; 123(3): 399-410.
[PMID: 16778319]

[306] Farrell JP. *Leishmania donovani*: acquired resistance to visceral leishmaniasis in the golden hamster. Exp Parasitol 1976; 40(1): 89-94.
[http://dx.doi.org/10.1016/0014-4894(76)90069-2] [PMID: 950004]

[307] Gifawesen C, Farrell JP. Comparison of T-cell responses in self-limiting *versus* progressive visceral *Leishmania donovani* infections in golden hamsters. Infect Immun 1989; 57(10): 3091-6.
[PMID: 2528507]

[308] Melby PC, Chandrasekar B, Zhao W, Coe JE. The hamster as a model of human visceral leishmaniasis: progressive disease and impaired generation of nitric oxide in the face of a prominent Th1-like cytokine response. J Immunol 2001; 166(3): 1912-20.
[http://dx.doi.org/10.4049/jimmunol.166.3.1912] [PMID: 11160239]

[309] Nieto A, Domínguez-Bernal G, Orden JA, De La Fuente R, Madrid-Elena N, Carrión J. Mechanisms

of resistance and susceptibility to experimental visceral leishmaniosis: BALB/c mouse *versus* Syrian hamster model. Vet Res 2011; 42(1): 39.
[http://dx.doi.org/10.1186/1297-9716-42-39] [PMID: 21345200]

[310] Fortin A, Caridha DP, Leed S, *et al.* Direct comparison of the efficacy and safety of oral treatments with oleylphosphocholine (OlPC) and miltefosine in a mouse model of *L. major* cutaneous leishmaniasis. PLoS Negl Trop Dis 2014; 8(9): e3144.
[http://dx.doi.org/10.1371/journal.pntd.0003144] [PMID: 25210745]

[311] Fischer MR, John D, Kautz-Neu K, Schermann AI, Schwonberg K, von Stebut E. Animal model for cutaneous leishmaniasis. Methods Mol Biol 2013; 961: 389-402.
[http://dx.doi.org/10.1007/978-1-62703-227-8_26] [PMID: 23325659]

[312] Baell J, Walters MA. Chemistry: Chemical con artists foil drug discovery. Nature 2014; 513(7519): 481-3.
[http://dx.doi.org/10.1038/513481a] [PMID: 25254460]

[313] Baell JB, Holloway GA. New substructure filters for removal of pan assay interference compounds (PAINS) from screening libraries and for their exclusion in bioassays. J Med Chem 2010; 53(7): 2719-40.
[http://dx.doi.org/10.1021/jm901137j] [PMID: 20131845]

[314] Inci R, Ozturk P, Mulayim MK, Ozyurt K, Alatas ET, Inci MF. Effect of the syrian civil war on prevalence of cutaneous leishmaniasis in southeastern anatolia, Turkey. Med Sci Monit 2015; 21: 2100-4.
[http://dx.doi.org/10.12659/MSM.893977] [PMID: 26190279]

[315] Alawieh A, Musharrafieh U, Jaber A, Berry A, Ghosn N, Bizri AR. Revisiting leishmaniasis in the time of war: the Syrian conflict and the Lebanese outbreak. Int J Infect Dis 2014; 29: 115-9.
[http://dx.doi.org/10.1016/j.ijid.2014.04.023] [PMID: 25449245]

[316] Fawcett JM, Hay RJ. Cutaneous leishmaniasis and human conflict. Acta Derm Venereol 2015; 95(1): 3-4.
[http://dx.doi.org/10.2340/00015555-1994] [PMID: 25363408]

[317] Hayani K, Dandashli A, Weisshaar E. Cutaneous leishmaniasis in Syria: clinical features, current status and the effects of war. Acta Derm Venereol 2015; 95(1): 62-6.
[http://dx.doi.org/10.2340/00015555-1988] [PMID: 25342106]

CHAPTER 11

Recombinant Vaccines Against Pathogenic Trypanosomatids

Priscila Martins Andrade Denapoli[1,2], **Alba Marina Gimenez**[1,3] and **Maurício Martins Rodrigues**[1,3,*]

[1] *Centro de Terapia Celular e Molecular (CTCMol), Escola Paulista de Medicina, UNIFESP, São Paulo, SP, Brazil*

[2] *Departamento de Imunologia, Instituto de Ciências Biomédicas, Universidade de São Paulo, USP, São Paulo, SP, Brazil*

[3] *Departamento de Microbiologia, Imunologia e Parasitologia, Escola Paulista de Medicina, Universidade Federal de São Paulo, São Paulo, SP, Brazil*

Abstract: The *Leishmania* spp, *Trypanosoma cruzi* and *Trypanosoma brucei* spp are the causative agents of tropical infections, and over 20 million people worldwide suffer from these neglected diseases. During the last century, vaccine development has had an undeniable impact on public health and may offer some alternatives for the control of parasitic diseases. Immune protection against experimental infection with these parasites has been studied and many types of immunogens have been used. Use of new technologies has allowed the development of recombinant proteins and DNA-based vaccines against those protozoans, aiming to generate both humoral and cellular protective responses. A large amount of data have been obtained from preclinical model systems which gave us promising results. The main challenge at the present is to translate what has been succeeded in these models into efficient human vaccines. The objective of this review is to summarize the efforts of the science community about the development of recombinant vaccines against trypanosomatids.

Keywords: African trypanosomiasis, Cellular response, Chagas disease, DNA vaccine, Humoral response, Immunotherapy, *Leishmania* spp, Leishmaniasis, *Trypanosoma brucei*, *Trypanosoma cruzi*.

* **Corresponding author Maurício Martins Rodrigues:** Escola Paulista de Medicina, Universidade Federal de São Paulo, São Paulo, Brazil; Tel/Fax: 55 11 50899218; E-mail: mrodrigues@unifesp.br

Marcelo Santos da Silva & Maria Isabel N. Cano (Eds.)

INTRODUCTION

Trypanosomatids are unicellular protozoans and etiologic agents of infectious diseases in humans as well as livestock. *Leishmania* spp, *Trypanosoma brucei* and *Trypanosoma cruzi* are parasites of medical and economic importance since they are responsible for diseases that cause numerous infections in mammals. In human health, over 20 million people are estimated to be affected, primarily in developing and under-developed countries. The complete genome sequences of *T. brucei* [1], *T. cruzi* [2], and *L. major* [3], known as the TriTryp genome, constitutes an important advance in the study of these parasites. However, human treatment of the diseases is limited to drugs that frequently have adverse side effects, and to date, no suitable vaccines are available [4].

First Generation Prophylactic Vaccines

In general, prophylactic vaccines can be grouped into 3 categories: containing live attenuated microbes, containing killed microbes and containing microbial subunits. In the first group, attenuated agents are obtained by selection of mutants under unfavorable conditions; in the second, inactivated vaccines are composed of parts of the microorganism (carbohydrates and proteins), generally obtained from the surface of the killed pathogens. In both cases, the objective is the induction of antibodies and to evoke cellular responses. The latter group is composed of vaccine formulations containing one or a few components purified from a microorganism, frequently delivered with an adjuvant and administered employing a prime/boost strategy [5].

The First generation group of vaccines, including live attenuated or killed microbes in their formulation, was widely used and allowed a huge advance in the combat of parasitic diseases. Currently, however, there are several inconveniences associated with these vaccines. They frequently have a limited shelf-life, require chilling or freezing, and their production is cumbersome and time-consuming. For live vaccines, moreover, there is the risk of pathogenicity reversion of the attenuated vaccine strains, and/or changes in the parasite population that can lead to reduced protection. For killed or inactivated vaccines, the cellular immune responses are frequently poor.

Second Generation Prophylactic Vaccines

Subunit vaccines, when formulated with purified (native or recombinant) antigens, are considered a Second generation group of vaccines. Use of new technologies has allowed different cells types (bacteria, yeast, animal and insect cells) to be used as tools for the production of recombinant antigenic proteins. These formulations have several distinct advantages over the native proteins, which include easier production, possibility of manipulation using genetic engineering techniques, superior stability, much safer and less expensive. In this scenario, new information about innate immunity processes, of the different antibody functions, and of both $CD4^+$ and $CD8^+$ T cells functions in response to vaccination, is essential for antigen designs [5]. However, the efficient production, purification and implementation of recombinant proteins for vaccine development can be harder than it appears to be.

Different systems have been employed to produce recombinant proteins for experimental use or biotechnological applications. They include chemical peptide synthesis, microorganisms, animal cells, plant cells, transgenic plants and transgenic animals (reviewed in [6]). Microorganisms, such as bacteria, were the first to be used to produce recombinant proteins. This approach was highly efficient for some proteins, mainly produced in *E. coli* and was used in a broad spectrum of applications, including vaccine development. However, bacteria cannot fold a number of proteins properly (as they are unable of performing adequate post-translational modifications) or can form inclusion bodies precluding an easy purification. In addition, some proteins are toxic for bacteria and cannot be prepared in this way. An alternative is the use of yeast as *Pichia pastoris*, which may be easily transformed and is capable of producing folded proteins and exporting them, facilitating the subsequent purification. The negative point is that yeast often produces a limited amount of recombinant proteins, which are not glycosylated or unduly glycosylated [6]. Still, yeast is the currently preferred system for the production of many recombinant proteins.

Concerning animal and plant as bioreactors, most of the recombinant proteins used as pharmaceuticals have been prepared from animal cells. However, this system is costly and poorly flexible compared to plant cells, in which the yield of recombinant proteins that can be obtained is almost immeasurable with a low cost. In addition, plant cells are able to produce properly folded proteins and associated subunits as those forming antibodies, as efficiently as animal cells. Moreover, proteins prepared from plants contain few or none pathogens for humans or animals, although this system presents the concern of uncontrolled dissemination of the proteins when plants are cultured in open fields. Another approach is the generation of transgenic animals as a source of recombinant proteins. Even though this process is relatively laborious and costly, proteins can be purified from blood, milk, egg white, seminal plasma, urine and silk gland [6].

Adjuvants Used in Vaccination

Even with the production and purification problems solved, isolated recombinant proteins are commonly less potent to induce an immune response than the whole pathogens, mainly if they are alive. The reason is the lack of important features present in pathogens like lipopolysaccharides (LPS) or unmethylated CpG-containing-DNA that are able to stimulate the innate immune system and shape the adaptive immune response [6]. Adjuvants are needed to activate innate immune responses that could involve B and T cells, generating increased adaptive immunity.

There is a vast list of adjuvants used in vaccination approaches. Here we will mention just a few commonly used in recombinant vaccine formulations. Some compounds show potent adjuvant efficacy, but lack appropriate safety properties; therefore they can be used only in experimental trials. Among them, the most widely used are Freund's adjuvant; water-in-oil emulsions containing or not killed *M. tuberculosis* (Complete Freund's adjuvant - CFA and Incomplete Freund's adjuvant - IFA, respectively). These formulations are strong adjuvants having the capacity to stimulate both Th1 (CFA) and Th2 (IFA) responses. Bacterial toxins such as cholera toxin (CT) or *E. coli* heat labile enterotoxin (LT) has been successfully employed as mucosal adjuvants, whereas killed bacteria (*P. acnes*) exerts an adjuvant effect on the differentiation of B-1 mice lymphocytes [7]. Aluminum-based mineral salt (alum) is a classical adjuvant, well tolerated and safe; however, it is a relatively weak adjuvant for antibody induction against recombinant vaccines, inducing mainly a Th2 immune response [8]. For use both in pre-clinical and clinical trials, several formulations have been developed based on these adjuvants. For example, emulsion adjuvants such as MF59, AS03, AF03 and Montanide™ (Seppic), have been incorporated into vaccines to increase antibody and Th responses [9]. The saponin Quil A, derived from the bark of *Quillaja saponaria*, or purified extracts (QS-21) have been evaluated as alternatives to alum for cell-mediated responses activation [10]. Several agonists of Toll-like receptors (TLRs) such as CpG dinucleotides-containing oligo-deoxynucleotides (CpG ODN), flagellin and double-stranded RNA (such as Poly (I:C)) were shown to be efficient in different vaccinc formulations. CpG ODN are very potent at orienting the immune system toward a Th1-biased response and can, therefore, be of primary interest for vaccines where a Th1-biased reaction is needed to achieve protective immunity [6]. Another TLR agonist widely used is monophosphoryl lipid A (MPL), an LPS mimetic compound with low toxicity which promotes the release of pro-inflammatory cytokines like TNF-α, IL-6, IL-10 and IFN-γ, which will ultimately enhance the adaptive immune response [11].

Immunostimulatory complexes (ISCOMS) are adjuvants composed of hydrophobically associated cholesterol, phospholipids and quillaja saponins that are able to induce strong antibody and cellular immune response [12]. Some adjuvants are composed of a mix of others, such as AS01 (MPL, QS21, liposomes), AS02 (MPL, QS21, emulsion), AS04 (MPL, aluminum salt), AS15 (MPL, QS21, CpG, liposomes) and IC31 (CpG, cationic peptide) [9]. In addition, cytokines can also be used to increase immune responses or to regulate them, enabling Th1 or Th2 pathways; recombinant IL-12 and granulocyte macrophage colony-stimulating factor (GM-CSF) are some examples.

Virus-like Particles (VLP)

A relatively new alternative to adjuvants for the Second generation of vaccine formulations is the use of Virus-like particles (VLP) (recently reviewed in [13]). VLPs are multimeric nanostructures that are assembled from viral structural proteins, lacking any genetic material, and that emulate the conformation of authentic viruses. Thus, these structural proteins help to invade host cells and provide a delivery system highly immunogenic and they are very safe. Generally, VLPs combine both B cell and T cell epitopes and are capable of dendritic cell

(DC) activation and maturation, inducing innate and cognate immune responses. These properties promoted VLPs as good candidates for foreign epitopes presentation. This is achieved through modification of the VLP gene sequence so that fusions of VLP components with a foreign epitope are assembled into VLPs during expression [13]. Currently, several vaccines are using VLP in their formulation. Prophylactic vaccines against Human papillomavirus (HPV), Hepatitis B and E have already been approved and are commercially available, whereas VLP-based vaccines against Influenza, Norwalk virus, Ebola, Hepatitis C, Human Immunodeficiency Virus (HIV) and Malaria are under development (reviewed in [14]). Nonetheless, there are several unsolved questions about VLP vaccines; for example the requirement (or not) of TLR agonist as an adjuvant, the number of effective doses, the possible (or not) use of enveloped viruses and the actual capacity of generating memory B cells [15]. Still, the success of this type of vaccines indicates that this approach will see increased use and perhaps lead to new efficacious formulations shortly [14].

Monoclonal Antibodies

Another method developed using recombinant technology and sometimes employed to obtain effective vaccines, is the generation of recombinant or monoclonal antibodies (Abs). They represent an emerging class of effective therapeutics, however, most recombinant Abs currently available do not target infectious agents. Some of them have been developed against potential bioweapons (bacteria, viruses and toxins) at risk of intentional dissemination [16]. An alternative for vaccine development is the use of Nanobodies® (Nbs; Vrije Universiteit Brussel, Belgium); they are naturally evolved heavy chain antibodies (hcAbs) derived from camelids, with unique antigen recognizing and biophysical properties, as the antigen-binding unit is reduced to only one single variable domain [17]. As recombinant antibodies, Nbs have been expressed in bacteria, fungi, mammalian cell lines, plants, and insects. They are mostly used in fundamental research (as GFP-binding protein, protein purification or inactivation), as diagnostics tools (sensitive detection probes, preclinical noninvasive diagnostic imaging) and as "magic bullets" for therapy [18].

Third Generation Prophylactic Vaccines

Subunit vaccines containing DNA coding the target antigen instead of the protein itself compose the Third generation group of vaccines. DNA vaccines, when compared to a peptide or recombinant protein-based subunit vaccines, are even more cost-effective, stable and easily transportable in the lyophilized form. They comprise circular bacterial plasmids with the desired antigen gene under the transcriptional control of viral or eukaryotic promoter along with a poly-adenylation (polyA) signal sequence and bacterial origin of replication. Promoter sequences most commonly used are from viral origin especially of Rous Sarcoma Virus (RSV), Simian Virus 40 (SV40) and Cytomegalovirus (CMV) which is the preferred promoter ([19], and references therein). Chimeric plasmids are capable of *in vivo* expression and presentation of desired viral, bacterial or parasitic antigens, and to induce specific immune responses.

DNA vaccines are capable of inducing a complete immune response against an encoded antigen. Professional antigen presenting cells (APCs) such as DC play a pivotal role in the generation of this immune response. They are primed either by direct transfection with plasmid DNA or indirectly by the transfected myocytes *via* cross-presentation, leading to antigen presentation to $CD8^+$ and $CD4^+$ T cells by class I or II major histocompatibility complex (MHC), respectively.

DNA vaccines have many advantages that could be extremely beneficial for the development of vaccines against tropical parasitic diseases. In the first place, they have an excellent safety profile as they lack pathogenic microbes in their composition, which could eventually revert their virulence. As others DNA vaccines, they have significant advantages in manufacturing, storage and distribution over other types of vaccines and biological therapeutics. One of them is the easy manipulation, storage, and distribution in tropical endemic regions, mainly when compared to recombinant or live attenuated vaccines. Finally, the administration is also simpler and it is possible the elaboration of multivalent vaccines by a combination of multiple plasmids.

Immune responses to DNA vaccines have been found to be relatively weak compared to that elicited by conventional live vaccines. Therefore, several studies have assessed strategies to intensify vaccine efficacy, including the use of cytokine/chemokine (such as IL-12 and GM-CSF) encoding plasmids to increase the immune response evoked by the vaccine (reviewed in [20]). Also, DNA vaccine priming plus adjuvant or with recombinant proteins or adenoviruses boosting in heterologous prime-boost regime improves immune response against several antigens from viruses, protozoans, and bacteria. In addition, there are vector design innovations that can improve DNA vaccine efficacy, safety, and manufacturing, through increasing antigen-specific adaptive immunity and improve plasmid expression, between other approaches [21].

The next sections present our current knowledge about the control of diseases caused by trypanosomatids, particularly in the field of immunology and vaccine development, and will comment on remarkable perspectives that can be exploited for the rational design of new and effective vaccines.

VACCINES CANDIDATES FOR TRYPANOSOMATIDS

Leishmania spp. and Leishmaniasis

Characteristics of Disease and Treatment

As already detailed in previous chapters, leishmaniasis are a group of diseases caused by the protozoa parasite *Leishmania*. Several regions in the world are affected by different species, including sub-regions of Africa, Latin America, Asia and Mediterranean; affecting over 10 million people worldwide (information from World Health Organization: www.who.int/leishmaniasis/en/). Most of the affected population lives in poor regions and often suffers from other disorders, such as malnutrition and weak immune system because of the poor sanitary condition in these areas. Over 300 million people are at risk of acquiring a *Leishmania* infection [22]. Annually, 1.5 million people develop the symptomatic disease, and World Health Organization has estimated that 50,000 people die each year, mostly children.

The transmission occurs by the bite of infected phlebotomine sandflies. Two different forms of transmission are described in the literature: anthroponotic (human to vector to human) or zoonotic (animal to vector to human). Zoonotic hosts include a variety of animals, although infections are more common in rodents and dogs. In some countries of America and Europe, dogs are considered the main reservoirs causing veterinary concern [23].

Different aspects are involved in the interaction parasite-host, such as species; these various interactions are responsible for various forms of three clinical manifestations: cutaneous leishmaniasis (CL), mucosal leishmaniasis (ML) and visceral leishmaniasis (VL). In the spite of CL being the most common leishmaniasis form, characterized by lesions skins and ulcers, VL is the most severe form, characterized by fever, weakness, lost weight, cachexia, hepatosplenomegaly and hypergammaglobulinemia [24].

The current treatment of leishmaniasis is based on chemotherapy. Nowadays, the main drugs used are pentavalent antimonials, amphotericin B, oral miltefosine, liposomal amphotericin B and paramomycin. The primary treatment consists on the two pentavalent antimonials: sodium stibogluconate (Pentostam) and meglumine antimoniate (Glucantime). These drugs have been used since 1950 with several levels of efficacy described, from 10% to 90%; however continuous use is responsible for parasite resistance in endemic regions [25]. As the effectiveness of pentavalent antimonials is very low, they have been replaced by other more efficient drugs, like the liposomal form of amphotericin B, miltefosine and paramomycin. These drugs showed promising results in early stages of the disease, even so, there are reports of decreased effectiveness of these drugs in certain patients [26 - 28]. Combination drug therapy has been an alternative for drug resistance in endemic areas [29, 30] but recently a study has shown that *L. donovani* can develop resistance even when drugs are used in combination [30]. Others problems associated with these drugs are cost, the need for multiple injcctions, length and duration of treatment and the route of injection (systemic or intralesional) which is painful and toxic [25]. Therefore, the lack of suitable chemotherapeutic options has stressed the need for the development of new therapeutic strategies. An important strategy to control canine visceral leishmaniasis would be the development of preventive or prophylactic vaccines. Moreover, this tool also would help to prevent health concern of *L. chagasi* in humans [31]. In addition, a vaccine against leishmaniasis might be beneficial to increase the rate of patient cure and decrease the duration of drugs treatment.

Vaccine Candidates

The knowledge about immune response against *Leishmania* spp. parasites is essential to generate an effective vaccine. Immune response plays an important role in the establishment of infection or protection. About 80% of infected people remains asymptomatic or develop a subclinical infection, and this is associated with robust cell-mediated immunity [32 - 34]. Initially, Th1/Th2 dichotomy was described. The resistance of host against *Leishmania* spp. infection is associated with activation of Th1 response (IL-12, IFN- γ, IL-2, and TNF-α) whereas the Th2 cell response [IL-4, IL-5, IL-6, IL-10, and TGF-β) favors the disease [35].

Inflammatory cytokines such as IFN- γ and TNF-α are elicited by $CD4^+$and $CD8^+$ T cells. They are necessary for the production of reactive oxygen species, thus infected macrophages are capable of eliminating parasites [36]. For details about virulence factors and immune evasion in *Leishmania* see Chapter 7.

In this context, many scientists have explored traits to develop a possible vaccine against visceral and cutaneous leishmaniasis [37 - 39]. To generate a protective vaccine, it is necessary to elicit immunological memory. Experiments with *L. major* showed that T cell memory (TCM) cells are maintained after elimination of parasite, while effector T cells are eliminated at this point. These results demonstrated the importance of expansion of TCM population as a target to develop effective vaccine candidates [40]. Numerous attempts to the development of an efficient vaccine have been done, and there are several categories of vaccine candidates. These formulations include first generation (live and killed *Leishmania* parasites); second generation (purified *Leishmania* antigens or live recombinant bacteria expressing *Leishmania* antigens); and DNA vaccines (plasmids coding *Leishmania* antigens) [41,42] (Table **1**).

The First generation vaccine was demonstrated to elicit some alterations in the immune response when a combination with chemotherapy promoted an intolerable toxicity [102]. Other problems were associated with a limitation of a variable immune response in humans [103] and difficulty in producing those vaccines consistently on a large scale. Most of the vaccines under investigation today are based on recombinant and DNA vaccines. These vaccines have been managed to provide prophylactic protection against visceral leishmaniasis, both *in vitro* and *in vivo* studies [71, 104, 105]. In the last years, several efforts to identify the major polypeptides naturally secreted by promastigotes and amastigotes of *L. amazonensis* and *L. infantum* have been performed [106 - 108]. These studies are very useful to identify new candidates for a next generation vaccine design.

Table 1. Summary of the main antigens and their corresponding antigen delivery system used for vaccination trials against *Leishmania* spp infection.

Antigen	Adjuvant	Delivery System	Model/Trial	Reference
A2	E6 gene Saponin, alum, and IL-12 Saponin Alum and rIL-12	Plasmid (pcDNA3) Recombinant protein Recombinant protein Prime-boost (DNA/protein; Ad5-A2/protein)	Mice/Pre-clinical Dogs/Pre-clinical Dogs/Pre-clinical Monkey/Pre-clinical	[43] [44] [45] [46]
Cocktail (H2A, H2B, H3, H4)	-	Plasmid (pcDNA3)	Mice/Pre-clinical	[47]
CPa/CPb	-	Prime-boost (DNA/protein)	Mice/Pre-clinical	[48]
CPIII	-	Prime-boost (DNA/DNA)	Mice/Pre-clinical	[49]
F14	MPL	Recombinant protein	Hamster/Pre-clinical	[50]
g-GCS	-	Plasmid (pVax)	Mice/Pre-clinical	[51]
Gp63	BCG CpG-ODN - -	Recombinant protein Recombinant protein Plasmid (pcDNA3) Prime-boost (DNA/protein; DNA/DNA; DNA/protein)	Mice/Pre-clinical Mice/Pre-clinical Mice/Pre-clinical Mice/Pre-clinical	[52] [53] [54] [55]
HASPB1	IL-12	Recombinant protein	Mice/Pre-clinical	[56]
HbR	-	Plasmid (pcDNA3.1)	Hamster/Pre-clinical	[57]
KMP-11	- CpG-ODN - - - HSP70	Prime-boost (DNA/rVV) Recombinant protein Plasmid (pCMV-LIC) Plasmid (pMOK) Plasmid (pCMV-LIC) Recombinant protein	Mice and Hamster/ Pre-clinical Mice/Pre-clinical Mice/Pre-clinical Dogs/Pre-clinical Hamster/Pre-clinical Dogs/Pre-clinical	[58] [59] [60] [61] [62] [63]

(Table 1) contd.....

Antigen	Adjuvant	Delivery System	Model/Trial	Reference
LACK	- - IL-12/IL-18 - - rIL-12	Plasmid (pCMV3ISS) Prime-boost (DNA/VV) Prime-boost (DNA/VV) Prime-boost (DNA/WR; DNA/MVA) Prime-boost (DNA/MVA; DNA/rVV) Recombinant protein	Mice/Pre-clinical Dogs/Pre-clinical Mice/Pre-clinical Mice/Pre-clinical Dogs/Pre-clinical Mice/Pre-clinical	[64] [65] [66] [67] [68] [69]
LCR1	Ribi	Recombinant protein	Mice/Pre-clinical	[70]
Ldccys1	BCG/*P. acnes*	Recombinant protein	Mice/Pre-clinical	[71]
LdP1	-	Recombinant protein or Plasmid (pVAX)	Hamster/Pre-clinical	[72]
Ldp45	-	Recombinant protein	Mice/Pre-clinical	[73]
LdPDI	-	Recombinant protein	Hamster/Pre-clinical	[74]
LdTPR	BCG	Recombinant protein	Hamster/Pre-clinical	[75]
Leish-111f (Polyprotein vaccine)	MPL-SE and/or Glucantime	Recombinant protein	Dogs/Pre-clinical Human/Clinical (I-II)	[76] [77 - 80]
Leishdnavax (KMP-11, TSA, CPA, CPB and P74)	-	Plasmid (MIDGE-Th1 vectors)	Mice/Pre-clinical Mice and Rats/ Pre-clinical	[81] [82]
LelF-2	-	Recombinant protein	Hamster/Pre-clinical	[83]
LiHyp1	Saponin	Recombinant protein	Mice/Pre-clinical	[84]
LiP0	-	Prime-boost (DNA/DNA; DNA/Recombinant protein)	Hamster/Pre-clinical	[85]
LmSTI1	-	Plasmid (pcDNA3)	Mice/Pre-clinical	[86]
LPG3	-	Prime-boost (DNA/DNA; DNA/Recombinant protein)	Mice/Pre-clinical	[87, 88]
Meta1	CFA	Recombinant protein	Mice/Pre-clinical	[89]
Multicomponent DNA (10 Antigens)	IL-12 or GM-CSF(DNA)	Plasmid (pcDNA3)	Dogs/Pre-clinical	[90]

(Table 1) contd.....

Antigen	Adjuvant	Delivery System	Model/Trial	Reference
NH36	-	Plasmid (VR1012)	Mice/Pre-clinical	[91]
ORFF	- - IL-12 (DNA)	Plasmid (pcDNA3) Prime-boost (DNA/Protein) Recombinant protein	Mice/Pre-clinical Mice/Pre-clinical Mice/Pre-clinical	[92] [93] [94]
P4	HSP70 or IL-12 (DNA)	Plasmid (pcDNA3)	Mice/Pre-clinical	[95]
papLe22	-	Plasmid (pcDNA3)	Hamster/Pre-clinical	[96]
PPG	-	Plasmid (pcDNA3)	Hamster/Pre-clinical	[97]
PFR2	- HSP70	Plasmid (pcDNA3) Recombinant protein	Hamster/Pre-clinical Dogs/Pre-clinical	[98] [63]
PSA-2	-	Plasmid (pCMV3ISS)	Mice/Pre-clinical	[64]
SMT	MPL-SE	Recombinant protein	Mice/Pre-clinical	[37]
TRYP	- - -	Plasmid (pMOK) Prime-boost (DNA/MVA) Prime-boost (DNA/MVA)	Dogs/Pre-clinical Dogs/Pre-clinical Mice/Pre-clinical	[61] [99] [100]
UBQ-ORFF	-	Plasmid (pUMVC3)	Mice/Pre-clinical	[101]

Antigens: A2, amastigote specific protein; ccys1, cysteine proteinase s1; CP; cysteine proteinase; elF-2, elongation factor-2; γ-GCS, gamma-glutamyl cysteine synthetase; Gp63, surface glycoprotein; H(2A, 2B, 3, 4), nucleosomal histones; HASPB1,. hydrophilicacylated surface protein B1; HbR, hemoglobina receptor; Hyp1, hypericin biosynthesis; KMP-11, kinetoplastid membrane protein-11; LACK, *Leishmania* homolog of receptors for activated C-kinase; LCR1, L. chagasic immunogenic gene; LeIF, *Leishmania* elongation initiation factor; LPG3, Lipophosphoglycan 3; MDP, muramyl di peptide; NH, nucleoside hydrolase; ORFF, open Reading frame F; P0, ribosomal acidic protein P0; P1, ribosomal P1 gene; p45, partial coding region of methionine aminopeptidase; papLe22, *L. infantum* potentially aggravating protein; PDI, protein disulfide isomerase; PFR, paraflegellar rod; PPG, proteophosphoglycan; PSA, promastigote surface antigens; SMT, sterol 24-c-methyltransferase; STI1, stress-inducible protein 1; TPI, triose phosphate isomerase; TPR, trypanothione reductase; TRYP, Tryparedoxin peroxidase; TSA, trans-sialidase; UBQ, ubiquitin. **Adjuvants**: BCG, *Mycobacterium bovis* Bacillus Calmette-Guerin; CFA: complete Freund's adjuvant; CpG, CpG dinucleotides; GM-CSF, granulocyte macrophage colony-stimulating factor; HSP70, Heat Shock Protein 70; MPL, monophosphoryl lipid; ODN, oligodeoxynucleotides; rIL-12, recombinant IL-12.

Several recombinant antigens have been studied in mice and dog models, including amastigote specific protein A2 [59, 109], surface expressed glycoprotein gp63 [52], nucleoside hydrolase [110], *Leishmania*-activated C kinase [39], cysteine proteinase B [111], *L. braziliensis* elongation and initiation factor [112], sterol 24-c-methyltransferase [37] and protein disulfide isomerase [113]. Gupta *et al.* [73] tested recombinant *L. donovani* protein p45, and this protein was able to generate a Th1-type immune response confirmed by increased level of expression of mRNA for IFN-γ, IL-12 and TNF-α associated with a decrease in IL-4 and TGF-β. The immunization with rLdp45 exerted considerable prophylactic efficacy, protecting approximately 85% of infected mice. The control of *Leishmania* parasite burden is associated with Th1 immune response, and several authors have also related protective response corresponding to this profile. Carrillo *et al.* [63] evaluated the immunogenicity of the *L. infantum* heat shock protein (HSP)-70, paraflagellar rod protein (PFR)-2 and kinetoplastid membrane protein (KMP)-11 recombinant proteins in dogs experimentally infected. The vaccination with theses recombinant proteins was able to produce an increase of IFN-γ and TNF-α with simultaneous decrease of IL-4.

Long-term protection against leishmaniasis is related to the development of $CD4^+$ and $CD8^+$ T TCM response [114]. A serological screening of patient serum infected with VL identified an enzyme involved in the synthesis of ergosterol, sterol 24-c-methyltransferase (SMT), as a possible candidate antigen to *Leishmania* vaccine. Mice immunized with recombinant SMT plus MPL®-SE (commercial adjuvant) developed antigen-specific $CD4^+$ and $CD8^+$ T response, and parasite burden was lower in different organs when compared to non-immunized mice [37]. An antigen specific for amastigote form of *Leishmania* parasite A2, when formulated in recombinant form with saponin, alum, and IL-12, showed to be effective as a vaccine in animals since protection against leishmaniasis was associated with specific $CD4^+$and $CD8^+$ T cells [44].

The use of adjuvants such as rIL-12, CpG, and MPL, could enhance immuno-genicity generating a strong immune response, resulting in protection against experimental infection. This effect was observed in a vaccination model with rORFF antigens co-administered with a plasmid encoding both p35 and p40 subunits of IL-12. In this case, the response profile in mice´s spleen and liver was also correlated to Th1 profile and treatment was able to confer protection against experimental infection [94]. On the other hand, rHASPB1 was tested without adjuvant, achieving a partial protection at the liver [56]. Vaccination with LACK (*Leishmania* activated C kinase) antigen administrated with recombinant IL-12 (rIL-12) elicited $CD4^+$ T cells and induced protection in mice against *L. major* challenge; the protection was correlated to the increment of IFN-γ and reduction of IL-4 [69].

Two Second generation vaccines against leishmaniasis have been developed. A vaccine preparation mixing saponin and the recombinant protein A2, named Leish-Tec® [45], has been tested in dogs living in endemic areas for CVL and demonstrated a protection in 92.9% of vaccinated dogs [44]. This strategy has been helpful in the control of the domestic reservoir of parasites in Brazil. Concerning to clinical trials in humans, a promising formulation is Leish F1 (or Leish-111f), a fusion of three preserved *L. major* proteins: thiol-specific antioxidant, *L. major* stress-inducible protein 1 (LmSTI1) and *Leishmania* elongation initiation factor (LeIF) which was first tested in dogs [115]. Leish-111F/MPL-SE formulation was approved to be tested in phase I and phase II clinical trials in developing countries, as Brazil, Peru and India [77 - 80]. In these studies, healthy volunteers were recruited and two features had been analyzed: immunogenicity and security. Patients showed the absence of side effects and demonstrated activation of cellular and humoral responses, with an increase in the frequency of specific memory $CD4^+$ T cells after vaccination, which was associated with clinical cure [78, 79]. Similar results were described in the Indian study [80]. In these clinical trials Leish-F1 plus MPL-SE formulation was safe and immunogenic in no-infected and asymptomatic infected individuals. This formulation is very promising to be tested in phase III clinical trials.

Third-generation vaccines have been developed to increase vaccination specificity and potency using several strategies including DNA encoding *Leishmania* proteins and prime-boost strategies. The ribosomal P1 gene of *L. donovani* (LdP1) has been used as DNA vaccine. Masih *et al.* [72] compared its use as peptide DNA vaccine (single or double dose) against leishmaniasis challenge. The recombinant protein rLdP1 or DNA vaccine (plasmid) encoding LdP1 antigen, in a single dose were not protective; however a double dose system elicited a Th1 response and secretion of cytokines in the vaccinated group. This response was associated with a parasite burden reduction in spleen and with an increase of survival rate. Recently, Pereira *et al.* [85] investigated the protective capacity of *L. infantum* nucleosomal histones (HIS) and ribosomal acidic protein P0 (LiP0) against experimental leishmaniasis, using two types of immunization:

homologous (DNA) or heterologous (DNA + protein + adjuvant). The authors demonstrated that only one of these antigens (LiP0) was able to protect hamsters against lethal challenge. The immunized group showed polarization of Th1/Th2 response: increased secretion of IFN-γ and decreased levels of IL-10 and TGF-β. The parasitic burden was evaluated and a reduction in different organs was observed, mainly in the lymph node. These findings suggest that LiP0 DNA vaccine is a good candidate against visceral leishmaniasis.

The prime-boost heterologous system has been used because the difficult of DNA immunizations to induce a robust immune response against antigens in humans [116, 117]. In this context, an heterologous system using two different vectors, plasmid DNA for prime and vaccinia virus (rVV) for boost, both expressing KMP-11 antigen has been tested in mice and hamsters [58]. This system conferred protection, eliciting $CD4^+$ and $CD8^+$ T cells response. Cytokine production such as IFN-γ, IL-2 and TNF-α has been detected. Other antigens also have been tested in this system, such as the lipophosphoglycan 3 (LPG3), an endoplasmic reticulum chaperone involved in the synthesis of essential parasite molecules. The prime-boost heterologous DNA/protein regime was found to be more efficient in stimulating an immunogenic response in mice compared to DNA/DNA regime [87]. Similar results appeared when three doses of plasmid encoding LPG3 were administered; the Th1/Th2 response was elicited, but there was no protection against *Leishmania* challenge, even when a recombinant HSP70 was added to the vaccine formulation [88]. The administration of recombinant HSP70 as an adjuvant has demonstrated to increase protection against canine leishmaniasis, through the production of IFN-γ [63], but it was not effective in this case.

Some of these vaccination protocols have been tested in non-human primate models, for both visceral and cutaneous leishmaniasis, to check: immunogenicity, efficiency, and safety. Grimaldi *et al.* [46] tested A2 antigen in the prime-boost heterologous system in a macaque model. In this protocol, DNA or Ad5 was used for priming and recombinant protein A2 conjugated with alum plus recombinant IL-12 was used for boost. The results showed encouraging protective effects and reduction of clinical symptoms.

Several platform optimization strategies that improve antigen expression, potency and immunogenicity have been developed. One of them is the Minimalistic Immunogenically Defined Gene Expression (MIDGE) vectors. It represents one of the best concepts for improvement, as MIDGE-TH1 vectors have attached a dodecapeptide with the simian virus 40 large T antigen nuclear localization sequence PKKKRKVEDPYC that is meant to enhance immune responses to express antigens. Das *et al.* [81] developed an MIDGE-TH1-based T cell-directed DNA vaccine against leishmaniasis. In this study, they selected five MIDGE-Th1 vectors with different encoding *Leishmania* antigens: KMP-11, trans-sialidase (TSA), cysteine proteases A (CPA) and B (CPB) and the elongation factor P74 (P74), forming the Leishdnavax vaccine. Several studies were performed using this promising formulation. Leishdnavax showed to be very effective in a test against visceral leishmaniasis. The protection level was associated with the development of $CD4^+$ and $CD8^+$ T cell responses. In naïve and infected mice and rats, this formulation demonstrated to be safe and well tolerated [82]. The antigens clearance was evaluated and occurred very fast from most of the organs, persisting only in dermal and lymph nodes even after recurrent intradermal immunizations. These results have shown that Leishdnavax is safe and therefore is an excellent candidate to be tested in humans.

Although some results from the human vaccine development are encouraging, some aspects need to be discussed. DNA vaccines have confirmed efficacy in mice, but the results of DNA vaccination in human have so far been disappointing, probably due to the low level of antigen expression, even when 1000-fold higher doses are used [118]. The heterologous prime/boost regime, effective in eliciting stronger and prolonged immunity responses could be employed as a good alternative to these promising DNA vaccines.

Trypanosoma cruzi and Chagas Disease

Characteristics of Disease and Treatment

Chagas disease is a parasitic infection caused by intracellular protozoan *Trypanosoma cruzi*. Currently, the World Health Organization estimates that 8 million people worldwide are infected with this parasite (www.who.int). Most of these people live in Latin America countries; however the number of infected people in developed countries of America as Canada and the United States, or in European countries, is increasing every year. One of the main reasons for this phenomenon is the increased migration of the South American people to the other countries of the world [119,120]. In recent years, United States reported 300,000 people infected residing in this country [121]. Nowadays, the most common form of transmission occurs through the vector-borne (triatomine), although others forms can also occur: infected blood transfusion, congenital transmission, organ transplantation and oral infection (through food contaminated with feces of triatomine) [122].

The acute phase of Chagas disease, which can last weeks or months, is characterized by an elevated parasitemia but often is symptoms free. When the symptoms occur, they usually include: a headache, fever, nausea, fatigue, loss of appetite, diarrhea or vomiting and rarely is lethal. After this stage, the infected patients may remain asymptomatic for many years in chronic phase. A 20-40% of the patients present a symptomatic chronic phase, which signs and symptoms may occur 10 to 20 years after initial infection, and they may include: heart complications, as chagasic cardiomyopathy, arrhythmias, and heart failure; and/or gastro-digestives complications, as megacolon and megaesophagus [123].

The pharmacological treatment with benznidazole or nifurtimox, used since the 1960s, demonstrates efficacy only in acute phase. Furthermore, the continuous use can be toxic, and treatment cause severe side effects in up to 30-50% of treated individuals and 15% of them interrupt the treatment for this reason. The drug reactions can be mild, as urticarial and petechial rashes; to moderate and severe as peripheral neuritis, immunosuppression and hepatitis [124]. Others problems with these drugs are that interfering during pregnancy [125] and have high cost during the acute phase ($591.80 annual) [126]. The expenses of the treatment are even more elevated in chronic phase; each patient can spend about $3,000-11,000 depending on the stage disease [127].

Vaccine Candidates

In Latin America, vector control interventions have significantly reduced Chagas disease frequency and prevalence [128]. However, it seems likely that a low rate of vectorial transmission will be maintained in many regions as some triatomine species has been difficult to be controlled [129]. In this scenario, the development of a prophylactic and therapeutic vaccine may be useful to monitor any form of *T. cruzi* infection. A safe and effective vaccine will represent a decrease in costs of drugs and vector combat. This tool will affect the economy in endemic areas significantly. Initial studies have tried to develop a vaccine by immunizing with live attenuated [130] or killed *T. cruzi* parasites [131]. Other strategies, as cross-reaction induced by mutual antigens of related trypanosomatids (tomato parasite), have been unsuccessfully tested [132].

One method used to identify and characterize potential vaccine targets is to assay human immune or infected serum against parasite recognized antigens. The molecular characterization of different antigens allowed the scientific community to design recombinant proteins and to clone specific genes as potential vaccine candidates. These studies have identified several groups of proteins. One of them contains members of TSA family, a group of proteins expressed mainly in infective forms. A second protein group belongs to cysteine protease family present in all *T. cruzi* forms. Thus, experimental studies examined the vaccine potential of these purified *T. cruzi* proteins. Several proteins were tested, including GP90, GP82, GP56, paraflagellar rod (PFR), cruzipain (Cz), complement-regulatory (CRP), TC52, amastigote surface proteins (ASPs), among others (reviewed in [133]). These studies showed that most of these proteins were able to induce antibody responses which generated protection against lethal challenge; however, the clearance is dependent on cellular responses because antibodies are ineffective to eliminate intracellular parasites. $CD4^+$ and $CD8^+$ T cells play a major role in this control. Activation of Th1 responses triggers IFN-γ production and nitric oxide (NO) release, enhancing the expression of MHC class I molecules in infected cells. Consequently, these cells are eliminated by cytotoxic activity of $CD8^+$ T cells. Thus, a good vaccine candidate should be able to induce such responses, control *T. cruzi* parasitemia and lessening the diseases symptoms. In this context, recombinant proteins, plasmid DNA, and recombinant viruses and bacteria have become attractive strategies because they are able to elicit a complete immune response. Several antigens have been tested in these delivery systems. The main antigens used for experimental vaccination formulations are listed in Table **2**. Details about the mechanism of action of these proteins during parasite infection are available in Chapter 8.

Table 2. Summary of the main antigens and the corresponding antigen delivery system used for experimental vaccination trials against *T. cruzi* infection.

Antigen	Adjuvant	Delivery System	Model/Trial	Reference
ASP-1	-	Plasmid (pCMVI.UBF3/2)	Mice/Pre-clinical	[134]
ASP-2	- - - Alum/CpG - - - - -	Prime-boost (Flu/Adenovírus) rSEV Yellow Fever 17D virus Recombinant protein Recombinant protein Plasmid (pCMVI.UBF3/2) Plasmid (pcDNA3) Adenovirus Prime-boost (DNA/Adenovirus)	Mice/Pre-clinical Mice/Pre-clinical Mice/Pre-clinical Mice/Pre-clinical Mice/Pre-clinical Mice/Pre-clinical Mice/Pre-clinical Mice/Pre-clinical Mice/Pre-clinical	[135] [136] [137] [138] [139] [134] [140] [141] [142]
ASP-3	-	Plasmid (pcDNA3)	Mice/Pre-clinical	[143]
ASP-4	-	Plasmid (pcDNA3)	Mice/Pre-clinical	[143]
CRP	-	Plasmid (pBC12BI)	Mice/Pre-clinical	[144]
Cruzipain	Anti-IL4/IL-12 MPL-2 -	Recombinant protein Recombinant protein Recombinant *Salmonella*	Mice/Pre-clinical Mice/Pre-clinical Mice/Pre-clinical	[145] [146] [147]
FcaBP/Tc24	IL12/GM-CSF	Plasmid (pCMV)	Mice/Pre-clinical	[148]
KMP-11	HSP70	Plasmid (gene fusion)	Mice/Pre-clinical	[149]
LyT1	IL-12/GM-CSF	Plasmid (pCMV)	Mice/Pre-clinical	[148]
PFR mix	rIL-12 or Alum	Recombinant protein	Mice/Pre-clinical	[150]
SSP4	-	Recombinant protein or DNA	Mice/Pre-clinical	[151]
Tc13	-	Plasmid (pcDNA3)	Mice/Pre-clinical	[152]
TcG1	IL-12/GM-CSF	Plasmid (pCDNA 3.1)	Mice/Pre-clinical	[153]
TcG2	IL-12/GM-CSF	Plasmid (pCDNA 3.1)	Mice/Pre-clinical	[153]
TcG4	IL-12/GM-CSF	Plasmid (pCDNA 3.1)	Mice/Pre-clinical	[153]
TcVac1 (TcG1, TcG2, TcG4)	IL-12/GM-CSF	Plasmid (pcDNA3)	Mice and Dogs/ Pre-clinical	[153, 154]
TcVac2 (TcG1, TcG2, TcG4)	IL-12/GM-CSF	Prime-boost (DNA/Recombinant protein)	Mice/Pre-clinical	[155]
TcVac3 (TcG2, TcG4)	IL-12/GM-CFS	Prime-boost (DNA/MVA)	Mice/Pre-clinical	[156]
Tcβ3	IL-12/GM-CSF	Plasmid (pCMV)	Mice/Pre-clinical	[148]

Antigen	Adjuvant	Delivery System	Model/Trial	Reference
TSA	CFA/IFA	Recombinant protein	Mice/Pre-clinical	[157]
	CpG-ODN	Recombinant protein	Mice/Pre-clinical	[158]
	-	Adenovirus	Mice/Pre-clinical	[141]
	IL-15	Plasmid (pcDNA)	Mice/Pre-clinical	[159]
	-	Plasmid (pCDNA 3.1)	Mice/Pre-clinical	[160]
	IL-12	Plasmid (pCM)	Mice/Pre-clinical	[161]
	RANK-L	Prime-boost	Mice/Pre-clinical	[162]
	-	(Adenovirus/rMVA)	Mice/Pre-clinical	[163]
	-	Plasmid (VR1012)	Mice/Pre-clinical	[164]

Antigens: ASP, amastigote surface protein; CRP, complement regulatory protein; FcaBP, flagellar calcium-binding protein; KMP-11, kinetoplastid membrane protein 11; LyT1, lytic protein; PFR, paraflegellar rod; SSP4, amastigote-specific surface protein; Tc13, *trans*-sialidase family protein; Tc24, trypomastigote excretory-secretory protein; Tcβ3, human b3 subunit of AP-3 complex homolog; TcG, membrane-associated GPI proteins; TSA, trans-sialidase. **Adjuvants**: BCG, *Mycobacterium bovis* Bacillus Calmette-Guerin; CFA: complete Freund's adjuvant; CpG, CpG dinucleotides; GM-CSF, granulocyte macrophage colony-stimulating factor; HSP70, Heat Shock Protein 70; IFA, incomplete Freund's adjuvant; MPL monophosphoryl lipid; ODN,oligodeoxynucleotides; RANK-L, receptor activator of nuclear factor kappa-B ligand; rIL-12, recombinant IL-12.

Bioinformatic approaches have been employed to identify antigenic sequences in order to develop a vaccine against *T. cruzi* [165]. This study selected 71 potential sequences targeting proteins exclusively expressed in this parasite. Eight of them had conserved domains in clinically relevant parasite strains. Only three antigens (TcG1, TcG2 and TcG4; encoding membrane surface proteins) were able to elicit lytic antibodies against *T. cruzi* [153]. The protective efficacy of TcG1, TcG2 and TcG4 was tested in a prime-boost heterologous regime (DNA/recombinant protein) associated or not with adjuvants IL-12 and GM-CSF. This system, named TcVac2, demonstrated to be efficient as parasite burden was decreased during the acute phase and absent in the chronic phase. This protective response was associated with specific antibody response, with an increase of IgG2b and IgG1 after challenge [155]. As previously discussed, several aspects are involved in the control of acute parasitemia, as Th1 helper cells activities and the presence of macrophages and cytotoxic T lymphocytes. Vaccination with TcVac2 can induce a strong TCD8^{+} response in the acute phase. The efficacy of heterologous DNA/protein boost TcVac2 against *T. cruzi* proved to elicit better protective immunity than did DNA-prime/DNA boost vaccine (TcVac1) [153], which provided a degree of control but it was not enough to inhibit chronic myocarditis in mice and canines [154]. Approaches have been employed in this technology to decrease the complexity of the TcVac2 system. For example, TcVac3 is composed of TcG2 and TcG4 proteins. In this new system, a prime-boost heterologous vaccine (DNA- prime and MVA (Modified Vaccinia Ankara) boost) was used. IL-12 and GM-CSF were used as adjuvants, but this strategy was unsuccessful as B and T cell responses were not increased [156]. On the other hand, TcVac3 was able to prevent inflammatory infiltration in the heart that normally occurs as a result of parasite persistence; therefore this vaccine should be further tested as therapeutic for the chronic disease phase.

Nowadays, oral transmission has become a route of emerging relevance [166, 167]. After oral infection, trypomastigotes invade the gastrointestinal tract and rapidly become systemic. Two antigens, cruzipain [147] and Tc52 [168], have been used with attenuated *Salmonella* bacteria. Vaccines against several parasitic diseases have been used with attenuated bacteria as a delivery system for DNA antigens [169 - 171]. This approach has been showed efficient because its ability to elicit a mucosal humoral immune response. In these systems the oral vaccination was able to induce enhanced mucosal antigen-specific immunity (IgA) and $CD8^+$ T cells specific response, improving the mucosal and systemic immunity. Intranasal immunization using recombinant cruzipain co-administered with MALP-2 triggered a strong systemic and mucosal Ig response and IFN-γ producing cellular immunity but failed to protect against a lethal challenge [146].

The majority of candidate genes cited in Table **2** elicited varying levels of Th1 cytokines, $CD8^+$ CTL activity, and antibody responses; also, the vaccinated animals showed a reduction in the acute phase parasitemia and mortality rate. In most studies, protective immunity against *T. cruzi* has been achieved only after several doses of DNA or recombinant viral vectors. The use of recombinant Sendai virus as a delivery system of ASP-2 gene was able to develop protection with a single intranasal vaccination [136]. The protection was associated with antigen-specific $CD8^+$ T cells response. This feature of Sendai virus, generating protection with a single dose, is a powerful tool for clinical vaccination.

Many preclinical and clinical immunization protocols have used human adenovirus type 5 codifying parasitic antigens [172 - 174]. One advantage of the use of the adenoviral vector for a vaccine approach is the vector ability to infect a broad variety of cells types. In addition, this vector can induce strong Th1-type immune responses. Myahira *et al.* [163] evoked a protective immune response using a heterologous prime-boost approach with recombinant adenoviral vector and vaccinia virus coding TSA antigen. Another prime-boost heterologous system that has been used is DNA immunization followed by recombinant adenovirus 5 boosts. The success of this tool was associated with induction of $CD8^+$ and $CD4^+$ cell-specific responses against intracellular pathogens that include *T. cruzi* ([142, 175 - 177]). This protocol, using different vectors coding ASP-2 gene, was able to elicit $CD4^+$ and $CD8^+$ T cells associated with burden parasite decrease and survival increase. The $CD8^+$ T cells, induced by a prime-boost protocol, the expression of IFN-γ/TNF-α or IFN-γ/$CD107^+$, which are highly cytotoxic *in vitro* [142]; moreover, these cells maintained the T effector memory (TEM)-like phenotype, for up to 6 months after the experimental infection with *T. cruzi* [176].

The concept that TEM-like cells can persist for a long time and thus protect against reinfection is an important aspect for improvement of new vaccines. In fact, these results support the hypothesis that TEM cells have an important role during the protective response against *T. cruzi* and may contribute to the control of systemic infection [142]. *T. cruzi* infection develops a suboptimal response that is associated with a CD95 expression on $CD8^+$ T cells whereas vaccination with adenoviral ASP-2 antigen not only fails to induce the expression of CD95 but also blocks the expression of this generating $CD8^+$ T cell response that can control the lethal infection [178]. These data strongly suggest that the pro- and anti -apoptotic pathways may be involved in lymphocyte TEM maintaining long life mechanism induced by vaccination of heterologous prime-boost type.

Other genes, such as those involved in cellular cycle regulation or transcription factors may be crucial in maintaining of these cells, and a future transcriptome analysis may provide important insights into immunity and vaccine development against *T. cruzi* and others intracellular pathogens.

It is known that an efficient immune response to *T. cruzi* infection requires T cell activation. Currently, however, several crucial gaps about the implications of this knowledge remain open. For example, the mechanisms of specific $CD8^+$ T cells induction *in vivo*; the main effectors pathways which act against the different parasite stages and the possibility of different parasite strain use different mechanisms of immune evasion. In humans, there are also several technical restrictions; therefore, vaccination trials directed to inducing TCM in non-human primates have to be performed in order to achieve new knowledge applicable to develop the first vaccines against *T. cruzi* in the future.

Hopefully, new information about the immune response evoked by vaccination procedures using DNA or recombinant proteins will help the improvement of existing strategies or the design of new tools to combat *T. cruzi* infections.

Trypanosoma brucei and African Trypanosomiasis

Characteristics of Disease and Treatment

As already detailed in previous chapters, African Trypanosomiasis (AT) is a parasitic disease in mammals, characterized by anemia, loss of appetite and immunosuppression, which can be fatal if untreated. It is caused by an infection of the protozoa *Trypanosoma congolense, Trypanosoma vivax and Trypanosoma brucei*. Human African trypanosomiasis (HAT), also known as sleeping sickness, can be caused by two main subspecies of *T. brucei*, which are infective for humans and are transmitted by the hematophagous tsetse fly (*Glossina* spp.): *Trypanosoma brucei gambiense*, the causative agent of endemic disease in West and Central Africa, and *Trypanosoma brucei rhodesiense*, responsible for acute disease cases in East and Southern Africa. In contrast, *T. brucei brucei, T. congolense and T. evansi* usually do not infect humans, although there have been sporadic reports of diseases in humans caused by nonhuman-pathogenic trypanosome species [179].

HAT is a disease of public health importance mainly in Africa, where rural populations are the most affected with 10,000 cases per year (World Health Organization reports, http://www.who.int/trypanosomiasis_african/en/). In sub-Saharan Africa, there are 70 million people estimated to be at risk of infection, causing a severe social and economic problem.

This disease has two main stages. In the first one, haemolymphatic phase, trypanosomes multiply locally, establishing infections in both local lymph nodes and the bloodstream. This phase is categorized by chronic and sporadic fever, headache, pruritus, lymphadenopathy, and sometimes, hepatosplenomegaly. In the late or meningoencephalitic phase, trypanosomes can cross the blood-brain barrier and invade the central nervous system. This generates sleep disturbances and neuropsychiatric disorders, leading to unconsciousness and death if not treated [179, 180].

Few treatments are currently available for HAT. For the first stage of the disease, pentamidine and suramin are the drugs of choice for infection by *T. b. gambiense* and *T. b. rhodesiense*, respectively [179]. The late-stage is treated with highly toxic melarsoprol [181], which remains the only choice for the meningo-encephalitic phase caused by *T. b. rhodesiense* infection. This is also the most extensively used drug for treatment of the chronic phase produced by *T. b. gambiense* in underdeveloped countries. The most significant adverse reaction is an encephalopathic syndrome, which can be lethal.

In the last five decades, only one new drug has been developed for the treatment of HAT. Eflornithine is the recommended drug and the first option of treatment for second-stage of *T. b. gambiense* disease [179], as this treatment has shown a significative reduction in the mortality rate when compared to melarsoprol. However, it is more expensive and difficult to administer [182], and adverse reactions are comparable to those of other cytostatic drugs, including anemia, leucopenia, and thrombocytopenia [183].

Vaccine Candidates

Over the last 30 years, some different approaches have been proposed and tested with the objective of developing a protective vaccine against AT. However, up to the present time, all implemented strategies have been unsuccessful. Trypanosomes have achieved the ability to protect themselves through evasion mechanisms against host antibodies and to eliminate the host B cell memory reservoir. Therefore, the existing vaccine development technologies are ineffective, and new knowledge about the development of immune memory will be necessary to solve these problems.

The main feature preventing the host from develops effective sterile anti- African trypanosomes immunity is parasite antigenic variation [184, 185]. Trypanosomes surface is coated with a dense layer of 10^7 identical copies of the variant-specific glycoproteins (VSGs) that protect parasites from lytic factors in human plasma. This protein, highly immunogenic, is recognized by the host's immune system, evoking IgM and IgG humoral and type 2 cellular responses. These immune mechanisms, with the production of interleukin 10 and interleukin 4 or interleukin-1, have an anti-inflammatory effect and antibodies neutralize the majority of the parasites, leading to decreased parasitemia [186,187]. For this reason, host's immune system is capable of controlling parasitemia in a high degree mainly by neutralization of VSGs with antibodies [188]. However, some of the trypanosomes will be able to escape from these antibodies as they have modifications in their surface coats, by expressing another variant glycoprotein that is not recognized and therefore will be not neutralized. These trypanosomes can proliferate until other neutralizing antibodies being produced. As in every cycle, few parasites will escape from immune control because the host is unable to eliminate all the trypanosomes [179]. Thus, it is believed that vaccination against glycoprotein surface of trypanosomes will never result in sterile immunity and protection against AT [185]. Recently, a computational analysis was performed to design DNA vaccine candidates, taking VSG sequence variants of different strains of *T. brucei* [189]. However, clinical trials are required to validate the efficacy of these formulations.

Infections produced by intradermic injections of few African trypanosomes are controlled by innate immune response involving iNOS and TNF-α, that require neither antibodies nor T cells for protection [190]. Thus, parasitemia control seems to be mediated by T-cell-independent IgM responses, since the contribution of T cells is believed to be associated with immunosuppression. After several studies, it has been proposed that the presence of viable trypanosomes results in the depletion of the host B cell memory reservoir. In fact, experimental infection of mice with *T. brucei* produce reduced bone marrow B cell expansion, eliminates splenic B cell maturation and leads to loss of humoral protection, including memory development induced by vaccination [191]. Therefore, in order to an anti-trypanosome vaccine be effective, it will be necessary to address the problem of generation of immunosuppression by the trypanosomes injected into the skin by infected tsetse flies [192]. Moreover, an effective vaccine should have the capacity to neutralize all circulating parasites prior to B cell memory suppression or depletion by living and dividing parasites [184]. In Table **3**, the vaccine formulations developed to date against AT are described.

A liposome-based VSG-glycosylphosphatidyl inositol (GPI) anchor vaccination strategy has been developed [203, 204] to avoid excessive immune response upon infection, as GPI was identified to be the main inflammatory TNF-inducing molecule [205]. This strategy resulted in a decrease of infection-elicited inflammatory type I response, due to the increase of the anti-inflammatory cytokine IL-10 and diminished levels of inflammatory cytokines TNF and IL-6. Moreover, this GPI-based approach was successful in preventing several host clinical indicators of *T. brucei* infections like anemia, acidosis, elevated serum AST levels and weight loss. However, this response was found to be short lived and lacking B cell contribution and memory development, as a similar behavior was observed both in wild-type C57BL/6 and B-cell deficient mice, and the influence of GPI-based treatment on anemia was independent of the humoral response. Probably, the effectiveness of the GPI-based approach is related to the lessening of type I inflammation evoked by a parasite, particularly at the level of macrophage activation [203], but would not be applicable for preventive vaccine development.

Table 3. Summary of the different antigens and the corresponding antigen delivery system used for experimental vaccination trials against *T. brucei* infection.

Antigen	Adjuvant	Delivery System	Model/Trial	Reference
Actin	CFA/IFA	Recombinant protein	Mice/Pre-clinical	[193]
Beta-tubulin	CFA/IFA	Recombinant protein	Mice/Pre-clinical	[194]
FP membranes	CFA 10% alum CFA/IFA	Native proteins Native proteins Native proteins	Mice/Pre-clinical Cattle/Pre-clinical Mice/Pre-clinical	[195] [196] [197]
ISG75	- -	Recombinant protein Plasmid (pVAX1)	Mice/Pre-clinical Mice/Pre-clinical	[184, 185] [198]
MAP p15	CFA/IFA -	Native protein Adenovirus	Mice/Pre-clinical Mice/Pre-clinical	[199] [199]
MAP p52, aldolase and GAPDH	CFA/IFA	Native proteins	Mice/Pre-clinical	[200]

(Table 3) contd.....

Antigen	Adjuvant	Delivery System	Model/Trial	Reference
TSA	-	Plasmid (pVAX1)	Mice/Pre-clinical	[201]
Tubulin	CFA/IFA	Native proteins	Mice/Pre-clinical	[202]
VSG-GPI preparation	8-MG	Native proteins + Liposomes	Mice/Pre-clinical	[203]

Antigens: FP, flagellar pocket; GPI, glycosylphosphatidyl inostitol; ISG, invariant surface glycoproteins; MAP, microtubule-associated protein; TSA, *trans*-sialidase;VSG, variant-specific glycoproteins. **Adjuvants**: CFA, completeFreund's adjuvant; IFA, incomplete Freund's adjuvant; 8-MG, 8-mercaptoguanosine.

Another vaccination strategy was developed based on trypanosome flagellar pocket (FP), an organelle specialized in the binding and endocytosis of host molecules. The FP contains surface-exposed proteins conserved among different trypanosome species, being a potential vaccine target. In fact, early studies showed that mice immunization with FP membrane preparations could confer at least partial protection against infection [195]. Later, an antigen derived from the FP of *T. b. rhodesiense* was used in a vaccine formulation containing BSA as the carrier and 10% alum as the adjuvant. Immunized cattle were challenged by natural exposure to trypanosome infected tsetse flies; unfortunately, FP vaccination only resulted in partial protection [196]. In another approach, an FP preparation was used for mice vaccination in a scheme of 3 doses applied with complete/incomplete Freund adjuvants (CFA/IFA). Despite this formulation being capable of eliciting high IgG2a and IgG1 antibody responses against FP, just a partial (60%) protection was observed after a challenge with a low dose of 500 bloodstream form of *T. brucei* EATRO1125 AnTaR 1 parasites. In addition, screening for cytokine production (IL-2, IFN-γ, IL-4 and IL-10) did not show vaccination-induced polarized Th1 immunity, suggesting that partial protection could result from a lack of parasitemia initiation, rather than from the vaccination of the mice [197].

Non-variable antigens such as parasite cytoskeleton proteins (tubulin heterodimers, microtubule-associated proteins, and actin) are considered possible targets for vaccination protocols. Vaccination with purified and renatured *T. brucei* tubulin was proved to induce protection in mice challenged both with the homologous *T. brucei* UTRO 020191B strain and heterologous strain of *T. b. brucei, T. congolense* and *T. b. rhodesiense*, suggesting that VSG proteins (probably co-purified with tubulin) were not responsible for the immunoprotection observed [202]. Concerning to microtubule-associated proteins (MAP), Balaban *et al.* [200] showed that mice vaccinated with 3 doses of a formulation containing co-purified MAP p52, aldolase and GAPDH of *T. b. brucei*, were fully protected against the challenge with a homologous infection. Moreover, Rasooly and Balaban [199] showed that mice immunized with both native and recombinant MAP p15 were protected against a challenge with a heterologous strain of *T. b. brucei*, likely helped by $CD8^+$ T cell proliferation induced by an adenovirus. Despite these hopeful perspectives, there has not been a positive follow-up on these results, probably because vaccination/challenge studies were performed in very short periods or with few parasites, leading to unrealistic positive outcomes [184]. Also, trials using renatured recombinant beta-tubulin and recombinant actin, derived from *T. evansi* and produced in *E. coli*, showed both partial and similar degrees of protection against *T. evansi, T. equiperdum* and *T. brucei* infections in mice [193, 194]. However, no further trials have been performed.

Besides VSG, parasite surface exposes several proteins to the environment, as transferrin receptor [206] and invariant surface glycoproteins ISG64 [207], ISG65 [208], ISG75 [209] and ISG100 [210]. A recombinant trypanosome invariant surface protein based on the extracellular domain of ISG75 was expressed in *E. coli*, recovered highly purified and with appropriate conformation, as demonstrated by HPLC and circular dichroism respectively [211]. Some efforts have focused on mice immunization with this recombinant protein. However, despite generating high titers of specific antibodies, no protection was obtained upon challenge with *T. brucei* parasites and titers decreased in short time to a level similar to that found in non-vaccinated infected mice [184, 185].

More recently, by using a DNA vaccine strategy with the objective of eliciting an immune response capable of generating protection against the parasite, it has been demonstrated that ISG75 could be used in DNA vaccine formulations against the experimental model of AT. In these experiments, it was shown that immunizations with ISG75 produce mainly Th1-like IgG2a antibodies responses which generate partial immune protection [198]. Since further studies are necessary to improve the level of protection achieved, the heterologous prime-boost strategy seems to be attractive for this antigen.

TSA enzymes are suggested as factors of virulence in AT, and their involvement in parasite pathogenicity has been widely studied in *T. cruzi* (discussed above). In *T. brucei*, the enzyme was reported and characterized by Pontes de Carvalho *et al.* [212] and the gene coding TSA was cloned by Montagna *et al.* [213]. Their role has been well established in the insect stages, as it is expressed only on the surface of the procyclic forms [214]. However, Ammar *et al.* [215] demonstrated enzymatic TSA activity in *T. b. gambiense* BSF, which has a role in the activation of endothelial cells. Accordingly, recombinant TbTSA expressed in bacteria and a total protein extract purified from *T. b. brucei* BSF were specifically recognized by serum IgG antibodies elicited from mice immunized intramuscularly with a plasmid encoding N-terminal TbTSA (nTSApVAX1). This vaccination protocol was able to partially (60%) protect BALB-c mice against infection upon challenge with *T. b. brucei* infective parasites [201]. The authors state that this protection could result from a Th1-biased response evoked by immunization with plasmid DNA. Even though additional trials to improve the protection effect are required, TSA is a promising antigen target for vaccine development strategies in AT.

Another approach developed using recombinant technology and employed to obtain an efficient vaccine against AT, is the generation of Nanobodies® (Nbs). Nbs against a conserved VSG carbohydrate epitope were generated and used to target trypanotoxic or trypanolytic molecules towards the parasite membrane [186, 216]. Recently, a nanobody which recognizes the PFR protein of *T. evansi, T. brucei, T. congolense* and *T. vivax* was produced. As PFR proteins were proposed as attractive targets for elicit protective immunity against *T. cruzi*, this strategy could be potentially useful both for the development of vaccine formulations and diagnostic tools for trypanosomiasis [217]. For details about PFR localization and composition see Chapter 1.

Finally, an alternative approach for anti-trypanosome vaccination is the development of transmission-blocking vaccines, which use antigens from the insect vector stage. Trypanosoma forms in the tsetse fly have GPI-anchored glycoproteins other than VSG [218, 219] that could be used as targets for vaccines. In this regard, studies performed using *Glossina* spp. extracts for immunization protocols are somehow promising [220 - 222], even though antigens potentially effective remain to be identified.

CONCLUDING REMARKS AND FUTURE PERSPECTIVES

Development of preventive or therapeutic vaccines against leishmaniasis, Chagas disease, and African trypanosomiasis, may provide substantial health and economic benefits mainly in epidemic areas.

The best scenario is for leishmaniasis's vaccine. Currently, there is a veterinary vaccine against canine leishmaniasis (Leish-Tec®) and a formulation against leishmaniasis has been tested in clinical trials in human (Leish F1). Therefore, in the short-term at least one human vaccine against this disease could become commercially available. Furthermore, there are several other formulations, of Second and Third generation, with great possibilities of being successfully tested both in non-human primate models and in human trials. Since most of the published studies up to date have focused on CD8 T cell -mediated responses, especially in the cytotoxic response and memory generated by these cells, it should be interesting to study other immune responses as well.

As mentioned previously, in the last few years a number of vaccine candidates were able to elicit protective immunity against *T. cruzi* in mice and dogs, suggesting that a veterinary vaccine could be developed in the short-term. However, because of the complexity of Chagas disease, we believe that it is hard to predict the efficacy of these strategies in human. Vaccination trials using non-human primates have yet to be performed and should contribute to the knowledge about the usefulness of these approaches to advance in the development of the first vaccines against *T. cruzi.*

Additional studies are necessary to identify epitopes from different *T. cruzi* strains which show geographical diversity since they are differentially distributed in several countries of America, as have been conducted in leishmaniasis. These analyses could lead to the identification of epitopes of broad geographical coverage, with a wide spectrum of protective efficacy and different levels of drug resistance. Also, genes and proteins of *T. cruzi* involved in parasite survival, adaptation and host cell invasion have been described some examples in ([223-226], see Chapter 8 for details). Further analysis of these genes in different parasite stages may provide valuable insights into vaccine development against *T. cruzi*.

Different from *T. cruzi* and *Leishmania* spp, all developmental stages of *T. brucei* are extracellular. In the mammal host, *T. brucei* multiplies in the bloodstream; consequently, unlike other pathogenic trypanosomatids which are capable of hiding intracellularly, *T. brucei* needs to survive to the antibody response elicited by the parasite in the host. The sophisticated immune evasion protocol based on VSG switching developed by *T. brucei*, avoid that the host's immune system recognize other invariant surface proteins and thus the parasite can proliferate indeterminately in the bloodstream [4]. Therefore, besides the fact of VSG switching being a problem *per se* for the development of vaccines, the lack of genes involved in host cell invasion and intracellular survival in *T. brucei* challenges, even more, the prospect of a new vaccine. In addition to to the lack of knowledge about how to overcome this problem, a great delay when compared to leishmaniasis or Chagas disease research in vaccine development, is the lack of formulations that could be used in clinical trials. Despite some discouraging results obtained so far in all efforts to develop a prophylactic vaccine against African trypanosomiasis, it is time to move forward and develop new formulations suitable for clinical trials using, perhaps, some of the promising antigens here summarized.

New knowledge about parasites biology would help us to understand how variations in the expression levels of some genes could be associated with protective immune responses. Also, structural biology and reverse vaccinology could be useful tools to the design of more efficient antigens. Thus, enhancing our knowledge of how to elicit improved immune responses may be relevant for the development of new vaccines against these diseases.

CONFLICT OF INTEREST

The authors confirm that they have no conflict of interest to declare for this publication.

ACKNOWLEDGEMENTS

This chapter aims to contribute to the development of effective vaccines against neglected tropical diseases, and is dedicated in memory of Professor Mauricio Martins Rodrigues, a brilliant scientist who devoted his life to the study of parasite vaccines. We would like to thank Dr. Karina Ramalho Bortoluci for helpful comments and critical revision and Dr. Sheila McNeill Ingham for proofreading and language corrections.

This work was supported by the Fundação de Amparo à Pesquisa do Estado de São Paulo (FAPESP), Coordenação de Aperfeiçoamento Pessoal de Nível Superior (CAPES) e Conselho Nacional de Desenvolvimento Científico e Tecnológico (CNPq).

ABBREVIATIONS

8-MG	8-mercaptoguanosine
A2	Amastigote specific protein
Abs	Antibodies
Ad5	Adenovirus type 5
Alum	Aluminum-based mineral salt
APC	Antigen presenting cells
ASP	Amastigote surface proteins
AST	Aspartate aminotransferase
AT	African Trypanosomiasis
BCG	*Mycobacterium bovis* bacillus Calmette Guerin
BSF	Blood stream forms
CCYS1	Cysteine proteinase s1
CFA	Complete Freund's adjuvant
CL	Cutaneous Leishmaniasis
CMV	Cytomegalovirus
CP	Cysteine proteinase
CPA	Cysteine proteases A
CPB	Cysteine proteases B
CpG	CpG dinucleotides

CRP	Complement regulatory
CT	Cholera toxin
CZ	Cruzipain
ElF-2	Elongation factor-2
FP	Flagellar pocket
GM-CSF	Granulocyte macrophage colony-stimulating factor
GP	Glycosylphosphatidyl inostitol
Gp63	Surface glycoprotein 63
HASPB1	Hydrophilicacylated surface protein B1
HAT	Human African trypanosomiasis
HbR	Hemoglobina receptor
HPLC	High performance liquid chromatografy
HPV	Human papilloma vírus
HSP70	Heat Shock Protein 70
Hyp1	Hypericin biosynthesis
IFA	Incomplete Freund's adjuvante
IL	Interleukin
IFN	Interferon
ISCOMS	Immunostimulatory complexes
ISG	Invariant surface glycoproteins
KMP-11	Kinetoplastid membrane protein-11
LACK	*Leishmania* homolog of receptors for activated C-kinase
LCR1	*L. chagasic* immunogenic gene
LeIF	*Leishmania* elongation initiation factor
LmSTI1	*L. major* stress-inducible protein 1
LPG3	Lipophosphoglycan 3
LPS	Lipopolysaccharides
MAP	Microtubule-associated protein
MDP	Muramyl di peptide
ML	Mucosal Leishmaniasis
MPL	Monophosphoryl lipid A
NH	Nucleoside hydrolase
NO	Nitric oxide
ODN	Oligodeoxynucleotides
ORFF	Open Reading frame F

P0	Ribosomal acidic protein P0
P1	Ribosomal P1 gene
p45	Partial coding region of methionine aminopeptidase
papLe22	*L. infantum* potentially aggravating protein
PDI	Protein disulfide isomerase
PFR	Paraflegellar rod
polyA	Poly-adenylation
PPG	Proteophosphoglycan
PSA	Promastigote surface antigens
RANK-L	Receptor activator of nuclear factor kappa-B ligand
RSV	Rous Sarcoma Virus
SMT	Sterol 24-c-methyltransferase
STI1	Stress-inducible protein 1
SV40	Simian Virus 40
TCM	T cell memory
TEM	T effector memory
TGF	Tumor growth factor
TLR	Toll-like receptor
TNF	Tumor necrosis factor
TPI	Triose phosphate isomerase
TPR	Trypanothione reductase
TRYP	Tryparedoxin peroxidase
TSA	Trans-sialidase
UBQ	Ubiquitin.
VL	Visceral Leishmaniasis
VLP	Virus-like particles
VSG	Variant-specific glycoproteins
VV	Vaccínia vírus
γ-GCS	Gamma-glutamyl cysteine synthetase

REFERENCES

[1] Berriman M, Ghedin E, Hertz-Fowler C, *et al.* The genome of the African trypanosome *Trypanosoma brucei*. Science 2005; 309(5733): 416-22. [http://dx.doi.org/10.1126/science.1112642] [PMID: 16020726]

[2] El-Sayed NM, Myler PJ, Blandin G, *et al.* Comparative genomics of trypanosomatid parasitic

protozoa. Science 2005; 309(5733): 404-9.
[http://dx.doi.org/10.1126/science.1112181] [PMID: 16020724]

[3] Ivens AC, Peacock CS, Worthey EA, *et al.* The genome of the kinetoplastid parasite, *Leishmania major*. Science 2005; 309(5733): 436-42.
[http://dx.doi.org/10.1126/science.1112680] [PMID: 16020728]

[4] Bartholomeu DC, de Paiva RM, Mendes TA, DaRocha WD, Teixeira SM. Unveiling the intracellular survival gene kit of trypanosomatid parasites. PLoS Pathog 2014; 10(12): e1004399.
[http://dx.doi.org/10.1371/journal.ppat.1004399] [PMID: 25474314]

[5] Plotkin SA, Plotkin SL. The development of vaccines: how the past led to the future. Nat Rev Microbiol 2011; 9(12): 889-93.
[http://dx.doi.org/10.1038/nrmicro2668] [PMID: 21963800]

[6] Soler E, Houdebine LM. Preparation of recombinant vaccines. Biotechnol Annu Rev 2007; 13: 65-94.
[http://dx.doi.org/10.1016/S1387-2656(07)13004-0] [PMID: 17875474]

[7] Mussalem JS, Squaiella-Baptistão CC, Teixeira D, *et al.* Adjuvant effect of killed *Propionibacterium acnes* on mouse peritoneal B-1 lymphocytes and their early phagocyte differentiation. PLoS One 2012; 7(3): e33955.
[http://dx.doi.org/10.1371/journal.pone.0033955] [PMID: 22448280]

[8] Gupta RK. Aluminum compounds as vaccine adjuvants. Adv Drug Deliv Rev 1998; 32(3): 155-72.
[http://dx.doi.org/10.1016/S0169-409X(98)00008-8] [PMID: 10837642]

[9] Reed SG, Orr MT, Fox CB. Key roles of adjuvants in modern vaccines. Nat Med 2013; 19(12): 1597-608.
[http://dx.doi.org/10.1038/nm.3409] [PMID: 24309663]

[10] Petrovsky N. Novel human polysaccharide adjuvants with dual Th1 and Th2 potentiating activity. Vaccine 2006; 24(Suppl 2): S2-26-9.
[http://dx.doi.org/10.1016/j.vaccine.2005.01.107]

[11] Okemoto K, Kawasaki K, Hanada K, Miura M, Nishijima M. A potent adjuvant monophosphoryl lipid A triggers various immune responses, but not secretion of IL-1beta or activation of caspase-1. J Immunol 2006; 176(2): 1203-8.
[http://dx.doi.org/10.4049/jimmunol.176.2.1203] [PMID: 16394010]

[12] Sanders MT, Brown LE, Deliyannis G, Pearse MJ. ISCOM-based vaccines: the second decade. Immunol Cell Biol 2005; 83(2): 119-28.
[http://dx.doi.org/10.1111/j.1440-1711.2005.01319.x] [PMID: 15748208]

[13] Chroboczek J, Szurgot I, Szolajska E. Virus-like particles as vaccine. Acta Biochim Pol 2014; 61(3): 531-9.
[PMID: 25273564]

[14] Jain NK, Sahni N, Kumru OS, Joshi SB, Volkin DB, Russell Middaugh C. Formulation and stabilization of recombinant protein based virus-like particle vaccines. Adv Drug Deliv Rev 2015; 93: 42-55.
[http://dx.doi.org/10.1016/j.addr.2014.10.023] [PMID: 25451136]

[15] Schiller JT, Lowy DR. Raising expectations for subunit vaccine. J Infect Dis 2015; 211(9): 1373-5.

[http://dx.doi.org/10.1093/infdis/jiu648] [PMID: 25420478]

[16] Froude JW, Stiles B, Pelat T, Thullier P. Antibodies for biodefense. MAbs 2011; 3(6): 517-27. [http://dx.doi.org/10.4161/mabs.3.6.17621] [PMID: 22123065]

[17] Muyldermans S. Nanobodies: natural single-domain antibodies. Annu Rev Biochem 2013; 82: 775-97. [http://dx.doi.org/10.1146/annurev-biochem-063011-092449] [PMID: 23495938]

[18] De Meyer T, Muyldermans S, Depicker A. Nanobody-based products as research and diagnostic tools. Trends Biotechnol 2014; 32(5): 263-70. [http://dx.doi.org/10.1016/j.tibtech.2014.03.001] [PMID: 24698358]

[19] Dhama K, Mahendran M, Gupta PK, Rai A. DNA vaccines and their applications in veterinary practice: current perspectives. Vet Res Commun 2008; 32(5): 341-56. [http://dx.doi.org/10.1007/s11259-008-9040-3] [PMID: 18425596]

[20] Dumonteil E. DNA Vaccines against Protozoan Parasites: Advances and Challenges. J Biomed Biotechnol 2007; 2007(6): 90520. [PMID: 17710244]

[21] Williams JA. Improving DNA vaccine performance through vector design. Curr Gene Ther 2014; 14(3): 170-89. [http://dx.doi.org/10.2174/156652321403140819122538] [PMID: 25142448]

[22] Gazzinelli RT, Ropert C, Campos MA. Role of the Toll/interleukin-1 receptor signaling pathway in host resistance and pathogenesis during infection with protozoan parasites. Immunol Rev 2004; 201: 9-25. [http://dx.doi.org/10.1111/j.0105-2896.2004.00174.x] [PMID: 15361229]

[23] Cecílio P, Pérez-Cabezas B, Santarém N, Maciel J, Rodrigues V, Cordeiro da Silva A. Deception and manipulation: the arms of *Leishmania*, a successful parasite. Front Immunol 2014; 5: 480. [PMID: 25368612]

[24] Ready PD. Epidemiology of visceral leishmaniasis. Clin Epidemiol 2014; 6: 147-54. [http://dx.doi.org/10.2147/CLEP.S44267] [PMID: 24833919]

[25] Chakravarty J, Sundar S. Drug resistance in leishmaniasis. J Glob Infect Dis 2010; 2(2): 167-76. [http://dx.doi.org/10.4103/0974-777X.62887] [PMID: 20606973]

[26] Bhandari V, Kulshrestha A, Deep DK, Stark O, Prajapati VK, Ramesh V, *et al.* Drug susceptibility in *Leishmania* isolates following miltefosine treatment in cases of visceral leishmaniasis and post kala-azar dermal leishmaniasis. PLoS Negl Trop Dis 2012; 6(5): e1657.

[27] Rijal S, Ostyn B, Uranw S, *et al.* Increasing failure of miltefosine in the treatment of Kala-azar in Nepal and the potential role of parasite drug resistance, reinfection, or noncompliance. Clin Infect Dis 2013; 56(11): 1530-8. [http://dx.doi.org/10.1093/cid/cit102] [PMID: 23425958]

[28] Sundar S, Chakravarty J, Agarwal D, Rai M, Murray HW. Single-dose liposomal amphotericin B for visceral leishmaniasis in India. N Engl J Med 2010; 362(6): 504-12. [http://dx.doi.org/10.1056/NEJMoa0903627] [PMID: 20147716]

[29] Sundar S, Chakravarty J. Leishmaniasis: an update of current pharmacotherapy. Expert Opin Pharmacother 2013; 14(1): 53-63.

[http://dx.doi.org/10.1517/14656566.2013.755515] [PMID: 23256501]

[30] Garcia-Hernandez R, Manzano JI, Castanys S, Gamarro F. *Leishmania donovani* develops resistance to drug combinations. PLoS Negl Trop Dis 2013; 6(12): e1974.

[31] Hommel M, Jaffe CL, Travi B, Milon G. Experimental models for leishmaniasis and for testing anti-leishmanial vaccines. Ann Trop Med Parasitol 1995; 89 (Suppl. 1): 55-73. [PMID: 8745928]

[32] Jeronimo SM, Oliveira RM, Mackay S, *et al.* An urban outbreak of visceral leishmaniasis in Natal, Brazil. Trans R Soc Trop Med Hyg 1994; 88(4): 386-8. [http://dx.doi.org/10.1016/0035-9203(94)90393-X] [PMID: 7570812]

[33] Jeronimo SM, Teixeira MJ, Sousa Ad, Thielking P, Pearson RD, Evans TG. Natural history of *Leishmania* (Leishmania) chagasi infection in Northeastern Brazil: long-term follow-up. Clin Infect Dis 2000; 30(3): 608-9. [http://dx.doi.org/10.1086/313697] [PMID: 10722458]

[34] Gidwani K, Jones S, Kumar R, Boelaert M, Sundar S. Interferon-gamma release assay (modified QuantiFERON) as a potential marker of infection for *Leishmania donovani*, a proof of concept study. PLoS Negl Trop Dis 2011; 5(4): e1042. [http://dx.doi.org/10.1371/journal.pntd.0001042] [PMID: 21526219]

[35] Sacks D, Noben-Trauth N. The immunology of susceptibility and resistance to *Leishmania major* in mice. Nat Rev Immunol 2002; 2(11): 845-58. [http://dx.doi.org/10.1038/nri933] [PMID: 12415308]

[36] Kumar R, Engwerda C. Vaccines to prevent leishmaniasis. Clin Transl Immunology 2014; 3(3): e13. [http://dx.doi.org/10.1038/cti.2014.4] [PMID: 25505961]

[37] Goto Y, Bogatzki LY, Bertholet S, Coler RN, Reed SG. Protective immunization against visceral leishmaniasis using *Leishmania* sterol 24-c-methyltransferase formulated in adjuvant. Vaccine 2007; 25(42): 7450-8. [http://dx.doi.org/10.1016/j.vaccine.2007.08.001] [PMID: 17804125]

[38] Araújo MS, de Andrade RA, Sathler-Avelar R, *et al.* Immunological changes in canine peripheral blood leukocytes triggered by immunization with first or second generation vaccines against canine visceral leishmaniasis. Vet Immunol Immunopathol 2011; 141(1-2): 64-75. [http://dx.doi.org/10.1016/j.vetimm.2011.02.007] [PMID: 21439654]

[39] Benhnini F, Chenik M, Laouini D, Louzir H, Cazenave PA, Dellagi K. Comparative evaluation of two vaccine candidates against experimental leishmaniasis due to *Leishmania major* infection in four inbred mouse strains. Clin Vaccine Immunol 2009; 16(11): 1529-37. [http://dx.doi.org/10.1128/CVI.00153-09] [PMID: 19726616]

[40] Zaph C, Uzonna J, Beverley SM, Scott P. Central memory T cells mediate long-term immunity to *Leishmania major* in the absence of persistent parasites. Nat Med 2004; 10(10): 1104-10. [http://dx.doi.org/10.1038/nm1108] [PMID: 15448686]

[41] Palatnik-de-Sousa CB. Vaccines for leishmaniasis in the fore coming 25 years. Vaccine 2008; 26(14): 1709-24. [http://dx.doi.org/10.1016/j.vaccine.2008.01.023] [PMID: 18295939]

[42] Gradoni L. An update on antileishmanial vaccine candidates and prospects for a canine *Leishmania* vaccine. Vet Parasitol 2001; 100(1-2): 87-103.
[http://dx.doi.org/10.1016/S0304-4017(01)00486-1] [PMID: 11522409]

[43] Ghosh A, Zhang WW, Matlashewski G. Immunization with A2 protein results in a mixed Th1/Th2 and a humoral response which protects mice against *Leishmania donovani* infections. Vaccine 2001; 20(1-2): 59-66.
[http://dx.doi.org/10.1016/S0264-410X(01)00322-X] [PMID: 11567746]

[44] Fernandes CB, Junior JT, de Jesus C, *et al.* Comparison of two commercial vaccines against visceral leishmaniasis in dogs from endemic areas: IgG, and subclasses, parasitism, and parasite transmission by xenodiagnosis. Vaccine 2014; 32(11): 1287-95.
[http://dx.doi.org/10.1016/j.vaccine.2013.12.046] [PMID: 24406392]

[45] Testasicca MC, dos Santos MS, Machado LM, *et al.* Antibody responses induced by Leish-Tec®, an A2-based vaccine for visceral leishmaniasis, in a heterogeneous canine population. Vet Parasitol 2014; 204(3-4): 169-76.
[http://dx.doi.org/10.1016/j.vetpar.2014.04.025] [PMID: 24863572]

[46] Grimaldi G Jr, Teva A, Porrozzi R, *et al.* Clinical and parasitological protection in a *Leishmania infantum*-macaque model vaccinated with adenovirus and the recombinant A2 antigen. PLoS Negl Trop Dis 2014; 8(6): e2853.
[http://dx.doi.org/10.1371/journal.pntd.0002853] [PMID: 24945284]

[47] Iborra S, Soto M, Carrión J, Alonso C, Requena JM. Vaccination with a plasmid DNA cocktail encoding the nucleosomal histones of *Leishmania* confers protection against murine cutaneous leishmaniosis. Vaccine 2004; 22(29-30): 3865-76.
[http://dx.doi.org/10.1016/j.vaccine.2004.04.015] [PMID: 15364433]

[48] Rafati S, Salmanian AH, Taheri T, Vafa M, Fasel N. A protective cocktail vaccine against murine cutaneous leishmaniasis with DNA encoding cysteine proteinases of *Leishmania major*. Vaccine 2001; 19(25-26): 3369-75.
[http://dx.doi.org/10.1016/S0264-410X(01)00081-0] [PMID: 11348700]

[49] Khoshgoo N, Zahedifard F, Azizi H, Taslimi Y, Alonso MJ, Rafati S. Cysteine proteinase type III is protective against *Leishmania infantum* infection in BALB/c mice and highly antigenic in visceral leishmaniasis individuals. Vaccine 2008; 26(46): 5822-9.
[http://dx.doi.org/10.1016/j.vaccine.2008.08.065] [PMID: 18804512]

[50] Bhardwaj S, Vasishta RK, Arora SK. Vaccination with a novel recombinant *Leishmania* antigen plus MPL provides partial protection against *L. donovani* challenge in experimental model of visceral leishmaniasis. Exp Parasitol 2009; 121(1): 29-37.
[http://dx.doi.org/10.1016/j.exppara.2008.09.019] [PMID: 18983842]

[51] Carter KC, Henriquez FL, Campbell SA, *et al.* DNA vaccination against the parasite enzyme gamma-glutamylcysteine synthetase confers protection against *Leishmania donovani* infection. Vaccine 2007; 25(22): 4502-9.
[http://dx.doi.org/10.1016/j.vaccine.2007.03.014] [PMID: 17418459]

[52] Connell ND, Medina-Acosta E, McMaster WR, Bloom BR, Russell DG. Effective immunization against cutaneous leishmaniasis with recombinant bacille Calmette-Guérin expressing the *Leishmania*

surface proteinase gp63. Proc Natl Acad Sci USA 1993; 90(24): 11473-7. [http://dx.doi.org/10.1073/pnas.90.24.11473] [PMID: 8265576]

[53] Jaafari MR, Badiee A, Khamesipour A, *et al.* The role of CpG ODN in enhancement of immune response and protection in BALB/c mice immunized with recombinant major surface glycoprotein of *Leishmania* (rgp63) encapsulated in cationic liposome. Vaccine 2007; 25(32): 6107-17. [http://dx.doi.org/10.1016/j.vaccine.2007.05.009] [PMID: 17629372]

[54] Walker PS, Scharton-Kersten T, Rowton ED, *et al.* Genetic immunization with glycoprotein 63 cDNA results in a helper T cell type 1 immune response and protection in a murine model of leishmaniasis. Hum Gene Ther 1998; 9(13): 1899-907. [http://dx.doi.org/10.1089/hum.1998.9.13-1899] [PMID: 9741428]

[55] Mazumder S, Maji M, Das A, Ali N. Potency, efficacy and durability of DNA/DNA, DNA/protein and protein/protein based vaccination using gp63 against *Leishmania donovani* in BALB/c mice. PLoS One 2011; 6(2): e14644. [http://dx.doi.org/10.1371/journal.pone.0014644] [PMID: 21311597]

[56] Stäger S, Smith DF, Kaye PM. Immunization with a recombinant stage-regulated surface protein from *Leishmania donovani* induces protection against visceral leishmaniasis. J Immunol 2000; 165(12): 7064-71. [http://dx.doi.org/10.4049/jimmunol.165.12.7064] [PMID: 11120835]

[57] Guha R, Gupta D, Rastogi R, *et al.* Vaccination with *Leishmania* hemoglobin receptor-encoding DNA protects against visceral leishmaniasis. Sci Transl Med 2013; 5(202): 202ra121. [http://dx.doi.org/10.1126/scitranslmed.3006406] [PMID: 24027025]

[58] Guha R, Das S, Ghosh J, *et al.* Heterologous priming-boosting with DNA and vaccinia virus expressing kinetoplastid membrane protein-11 induces potent cellular immune response and confers protection against infection with antimony resistant and sensitive strains of *Leishmania (Leishmania) donovani*. Vaccine 2013; 31(15): 1905-15. [http://dx.doi.org/10.1016/j.vaccine.2013.02.025] [PMID: 23499564]

[59] Agallou M, Margaroni M, Karagouni E. Cellular vaccination with bone marrow-derived dendritic cells pulsed with a peptide of *Leishmania infantum* KMP-11 and CpG oligonucleotides induces protection in a murine model of visceral leishmaniasis. Vaccine 2011; 29(31): 5053-64. [http://dx.doi.org/10.1016/j.vaccine.2011.04.089] [PMID: 21569815]

[60] Santos DM, Carneiro MW, de Moura TR, *et al.* Towards development of novel immunization strategies against leishmaniasis using PLGA nanoparticles loaded with kinetoplastid membrane protein-11. Int J Nanomedicine 2012; 7: 2115-27. [PMID: 22619548]

[61] Rodríguez-Cortés A, Ojeda A, López-Fuertes L, *et al.* Vaccination with plasmid DNA encoding KMPII, TRYP, LACK and GP63 does not protect dogs against *Leishmania infantum* experimental challenge. Vaccine 2007; 25(46): 7962-71. [http://dx.doi.org/10.1016/j.vaccine.2007.08.023] [PMID: 17942199]

[62] Basu R, Bhaumik S, Basu JM, Naskar K, De T, Roy S. Kinetoplastid membrane protein-11 DNA vaccination induces complete protection against both pentavalent antimonial-sensitive and -resistant strains of *Leishmania donovani* that correlates with inducible nitric oxide synthase activity and IL-4

generation: evidence for mixed Th1- and Th2-like responses in visceral leishmaniasis. J Immunol 2005; 174(11): 7160-71.
[http://dx.doi.org/10.4049/jimmunol.174.11.7160] [PMID: 15905560]

[63] Carrillo E, Crusat M, Nieto J, *et al.* Immunogenicity of HSP-70, KMP-11 and PFR-2 leishmanial antigens in the experimental model of canine visceral leishmaniasis. Vaccine 2008; 26(15): 1902-11.
[http://dx.doi.org/10.1016/j.vaccine.2008.01.042] [PMID: 18321614]

[64] Ahmed SB, Bahloul C, Robbana C, Askri S, Dellagi K. A comparative evaluation of different DNA vaccine candidates against experimental murine leishmaniasis due to L. major. Vaccine 2004; 22(13-14): 1631-9.
[http://dx.doi.org/10.1016/j.vaccine.2003.10.046] [PMID: 15068845]

[65] Ramiro MJ, Zárate JJ, Hanke T, *et al.* Protection in dogs against visceral leishmaniasis caused by *Leishmania infantum* is achieved by immunization with a heterologous prime-boost regime using DNA and vaccinia recombinant vectors expressing LACK. Vaccine 2003; 21(19-20): 2474-84.
[http://dx.doi.org/10.1016/S0264-410X(03)00032-X] [PMID: 12744881]

[66] Tapia E, Pérez-Jiménez E, López-Fuertes L, Gonzalo R, Gherardi MM, Esteban M. The combination of DNA vectors expressing IL-12 + IL-18 elicits high protective immune response against cutaneous leishmaniasis after priming with DNA-p36/LACK and the cytokines, followed by a booster with a vaccinia virus recombinant expressing p36/LACK. Microbes Infect 2003; 5(2): 73-84.
[http://dx.doi.org/10.1016/S1286-4579(02)00077-1] [PMID: 12650765]

[67] Dondji B, Pérez-Jimenez E, Goldsmith-Pestana K, Esteban M, McMahon-Pratt D. Heterologous prime-boost vaccination with the LACK antigen protects against murine visceral leishmaniasis. Infect Immun 2005; 73(8): 5286-9.
[http://dx.doi.org/10.1128/IAI.73.8.5286-5289.2005] [PMID: 16041057]

[68] Ramos I, Alonso A, Marcen JM, *et al.* Heterologous prime-boost vaccination with a non-replicative vaccinia recombinant vector expressing LACK confers protection against canine visceral leishmaniasis with a predominant Th1-specific immune response. Vaccine 2008; 26(3): 333-44.
[http://dx.doi.org/10.1016/j.vaccine.2007.11.021] [PMID: 18093705]

[69] Rezvan H, Moafi M. An overview on *Leishmania* vaccines: A narrative review article. Vet Res Forum 2015; 6(1): 1-7.
[PMID: 25992245]

[70] Wilson ME, Young BM, Andersen KP, *et al.* A recombinant *Leishmania* chagasi antigen that stimulates cellular immune responses in infected mice. Infect Immun 1995; 63(5): 2062-9.
[PMID: 7729921]

[71] Ferreira JH, Gentil LG, Dias SS, Fedeli CE, Katz S, Barbiéri CL. Immunization with the cysteine proteinase Ldccys1 gene from *Leishmania (Leishmania)* chagasi and the recombinant Ldccys1 protein elicits protective immune responses in a murine model of visceral leishmaniasis. Vaccine 2008; 26(5): 677-85.
[http://dx.doi.org/10.1016/j.vaccine.2007.11.044] [PMID: 18160187]

[72] Masih S, Arora SK, Vasishta RK. Efficacy of *Leishmania donovani* ribosomal P1 gene as DNA vaccine in experimental visceral leishmaniasis. Exp Parasitol 2011; 129(1): 55-64.
[http://dx.doi.org/10.1016/j.exppara.2011.05.014] [PMID: 21640106]

[73] Gupta R, Kushawaha PK, Tripathi CD, Sundar S, Dube A. A novel recombinant *Leishmania donovani* p45, a partial coding region of methionine aminopeptidase, generates protective immunity by inducing a Th1 stimulatory response against experimental visceral leishmaniasis. Int J Parasitol 2012; 42(5): 429-35.
[http://dx.doi.org/10.1016/j.ijpara.2012.02.013] [PMID: 22502587]

[74] Kushawaha PK, Gupta R, Tripathi CD, Sundar S, Dube A. Evaluation of *Leishmania donovani* protein disulfide isomerase as a potential immunogenic protein/vaccine candidate against visceral Leishmaniasis. PLoS One 2012; 7(4): e35670.
[http://dx.doi.org/10.1371/journal.pone.0035670] [PMID: 22539989]

[75] Khare P, Jaiswal AK, Tripathi CD, Joshi S, Sundar S, Dube A. Efficacy of *Leishmania donovani* trypanothione reductase, identified as a potent Th1 stimulatory protein, for its immunogenicity and prophylactic potential against experimental visceral leishmaniasis. Parasitol Res 2014; 113(3): 851-62.
[http://dx.doi.org/10.1007/s00436-013-3716-5] [PMID: 24370734]

[76] Trigo J, Abbehusen M, Netto EM, *et al.* Treatment of canine visceral leishmaniasis by the vaccine Leish-111f+MPL-SE. Vaccine 2010; 28(19): 3333-40.
[http://dx.doi.org/10.1016/j.vaccine.2010.02.089] [PMID: 20206667]

[77] Vélez ID, Gilchrist K, Martínez S, *et al.* Safety and immunogenicity of a defined vaccine for the prevention of cutaneous leishmaniasis. Vaccine 2009; 28(2): 329-37.
[http://dx.doi.org/10.1016/j.vaccine.2009.10.045] [PMID: 19879995]

[78] Nascimento E, Fernandes DF, Vieira EP, *et al.* A clinical trial to evaluate the safety and immunogenicity of the LEISH-F1+MPL-SE vaccine when used in combination with meglumine antimoniate for the treatment of cutaneous leishmaniasis. Vaccine 2010; 28(40): 6581-7.
[http://dx.doi.org/10.1016/j.vaccine.2010.07.063] [PMID: 20688040]

[79] Llanos-Cuentas A, Calderón W, Cruz M, *et al.* A clinical trial to evaluate the safety and immunogenicity of the LEISH-F1+MPL-SE vaccine when used in combination with sodium stibogluconate for the treatment of mucosal leishmaniasis. Vaccine 2010; 28(46): 7427-35.
[http://dx.doi.org/10.1016/j.vaccine.2010.08.092] [PMID: 20851080]

[80] Chakravarty J, Kumar S, Trivedi S, *et al.* A clinical trial to evaluate the safety and immunogenicity of the LEISH-F1+MPL-SE vaccine for use in the prevention of visceral leishmaniasis. Vaccine 2011; 29(19): 3531-7.
[http://dx.doi.org/10.1016/j.vaccine.2011.02.096] [PMID: 21414377]

[81] Das S, Freier A, Boussoffara T, *et al.* Modular multiantigen T cell epitope-enriched DNA vaccine against human leishmaniasis. Sci Transl Med 2014; 6(234): 234ra56.
[http://dx.doi.org/10.1126/scitranslmed.3008222] [PMID: 24786324]

[82] Riede O, Seifert K, Oswald D, *et al.* Preclinical safety and tolerability of a repeatedly administered human leishmaniasis DNA vaccine. Gene Ther 2015; 22(8): 628-35.
[http://dx.doi.org/10.1038/gt.2015.35] [PMID: 25871827]

[83] Kushawaha PK, Gupta R, Sundar S, Sahasrabuddhe AA, Dube A. Elongation factor-2, a Th1 stimulatory protein of *Leishmania donovani*, generates strong IFN-γ and IL-12 response in cured Leishmania-infected patients/hamsters and protects hamsters against *Leishmania* challenge. J Immunol 2011; 187(12): 6417-27.

[http://dx.doi.org/10.4049/jimmunol.1102081] [PMID: 22079980]

[84] Martins VT, Chávez-Fumagalli MA, Costa LE, *et al.* Antigenicity and protective efficacy of a *Leishmania* amastigote-specific protein, member of the super-oxygenase family, against visceral leishmaniasis. PLoS Negl Trop Dis 2013; 7(3): e2148.
[http://dx.doi.org/10.1371/journal.pntd.0002148] [PMID: 23573301]

[85] Pereira L, Abbehusen M, Teixeira C, *et al.* Vaccination with *Leishmania infantum* acidic ribosomal P0 but not with nucleosomal histones proteins controls *Leishmania infantum* infection in hamsters. PLoS Negl Trop Dis 2015; 9(2): e0003490.
[http://dx.doi.org/10.1371/journal.pntd.0003490] [PMID: 25642946]

[86] Méndez S, Gurunathan S, Kamhawi S, *et al.* The potency and durability of DNA- and protein-based vaccines against *Leishmania major* evaluated using low-dose, intradermal challenge. J Immunol 2001; 166(8): 5122-8.
[http://dx.doi.org/10.4049/jimmunol.166.8.5122] [PMID: 11290794]

[87] Abdian N, Gholami E, Zahedifard F, Safaee N, Rafati S. Evaluation of DNA/DNA and prime-boost vaccination using LPG3 against *Leishmania major* infection in susceptible BALB/c mice and its antigenic properties in human leishmaniasis. Exp Parasitol 2011; 127(3): 627-36.
[http://dx.doi.org/10.1016/j.exppara.2010.12.007] [PMID: 21187087]

[88] Pirdel L, Hosseini AZ, Rasouli M. Immune response in susceptible BALB/c mice immunized with DNA encoding Lipophosphoglycan 3 of *Leishmania infantum*. Parasite Immunol 2014; 36(12): 700-7.
[http://dx.doi.org/10.1111/pim.12147] [PMID: 25244583]

[89] Serezani CH, Franco AR, Wajc M, *et al.* Evaluation of the murine immune response to *Leishmania* meta 1 antigen delivered as recombinant protein or DNA vaccine. Vaccine 2002; 20(31-32): 3755-63.
[http://dx.doi.org/10.1016/S0264-410X(02)00354-7] [PMID: 12399206]

[90] Saldarriaga OA, Travi BL, Park W, Perez LE, Melby PC. Immunogenicity of a multicomponent DNA vaccine against visceral leishmaniasis in dogs. Vaccine 2006; 24(11): 1928-40.
[http://dx.doi.org/10.1016/j.vaccine.2005.10.052] [PMID: 16310897]

[91] Gamboa-León R, Paraguai de Souza E, Borja-Cabrera GP, *et al.* Immunotherapy against visceral leishmaniasis with the nucleoside hydrolase-DNA vaccine of *Leishmania donovani*. Vaccine 2006; 24(22): 4863-73.
[http://dx.doi.org/10.1016/j.vaccine.2006.03.005] [PMID: 16635538]

[92] Sukumaran B, Tewary P, Saxena S, Madhubala R. Vaccination with DNA encoding ORFF antigen confers protective immunity in mice infected with *Leishmania donovani*. Vaccine 2003; 21(11-12): 1292-9.
[http://dx.doi.org/10.1016/S0264-410X(02)00352-3] [PMID: 12559811]

[93] Tewary P, Jain M, Sahani MH, Saxena S, Madhubala R. A heterologous prime-boost vaccination regimen using ORFF DNA and recombinant ORFF protein confers protective immunity against experimental visceral leishmaniasis. J Infect Dis 2005; 191(12): 2130-7.
[http://dx.doi.org/10.1086/430348] [PMID: 15898000]

[94] Tewary P, Saxena S, Madhubala R. Co-administration of IL-12 DNA with rORFF antigen confers long-term protective immunity against experimental visceral leishmaniaisis. Vaccine 2006; 24(13): 2409-16.

[http://dx.doi.org/10.1016/j.vaccine.2005.11.058] [PMID: 16413950]

[95] Campbell K, Diao H, Ji J, Soong L. DNA immunization with the gene encoding P4 nuclease of *Leishmania amazonensis* protects mice against cutaneous Leishmaniasis. Infect Immun 2003; 71(11): 6270-8.
[http://dx.doi.org/10.1128/IAI.71.11.6270-6278.2003] [PMID: 14573646]

[96] Fragaki K, Suffia I, Ferrua B, Rousseau D, Le Fichoux Y, Kubar J. Immunisation with DNA encoding *Leishmania infantum* protein papLe22 decreases the frequency of parasitemic episodes in infected hamsters. Vaccine 2001; 19(13-14): 1701-9.
[http://dx.doi.org/10.1016/S0264-410X(00)00398-4] [PMID: 11166894]

[97] Samant M, Gupta R, Kumari S, *et al.* Immunization with the DNA-encoding N-terminal domain of proteophosphoglycan of *Leishmania donovani* generates Th1-type immunoprotective response against experimental visceral leishmaniasis. J Immunol 2009; 183(1): 470-9.
[http://dx.doi.org/10.4049/jimmunol.0900265] [PMID: 19542458]

[98] Saravia NG, Hazbón MH, Osorio Y, *et al.* Protective immunogenicity of the paraflagellar rod protein 2 of *Leishmania mexicana*. Vaccine 2005; 23(8): 984-95.
[http://dx.doi.org/10.1016/j.vaccine.2004.07.044] [PMID: 15620471]

[99] Carson C, Antoniou M, Ruiz-Argüello MB, *et al.* A prime/boost DNA/Modified vaccinia virus Ankara vaccine expressing recombinant *Leishmania* DNA encoding TRYP is safe and immunogenic in outbred dogs, the reservoir of zoonotic visceral leishmaniasis. Vaccine 2009; 27(7): 1080-6.
[http://dx.doi.org/10.1016/j.vaccine.2008.11.094] [PMID: 19095029]

[100] Jayakumar A, Castilho TM, Park E, Goldsmith-Pestana K, Blackwell JM, McMahon-Pratt D. TLR1/2 activation during heterologous prime-boost vaccination (DNA-MVA) enhances CD8+ T Cell responses providing protection against *Leishmania (Viannia)*. PLoS Negl Trop Dis 2011; 5(6): e1204.
[http://dx.doi.org/10.1371/journal.pntd.0001204] [PMID: 21695103]

[101] Sharma A, Madhubala R. Ubiquitin conjugation of open reading frame F DNA vaccine leads to enhanced cell-mediated immune response and induces protection against both antimony-susceptible and -resistant strains of *Leishmania donovani*. J Immunol 2009; 183(12): 7719-31.
[http://dx.doi.org/10.4049/jimmunol.0900132] [PMID: 19933862]

[102] Jamshidi Sh, Avizeh R, Mohebali M, Bokaie S. Immunotherapy using autoclaved *L. major* antigens and M. vaccae with meglumine antimoniate, for the treatment of experimental canine visceral leishmaniasis. Iran J Parasitol 2011; 6(3): 26-34.
[PMID: 22347294]

[103] Choudhury R, Das P, De T, Chakraborti T. 115 kDa serine protease confers sustained protection to visceral leishmaniasis caused by *Leishmania donovani via* IFN-γ induced down-regulation of TNF-α mediated MMP-9 activity. Immunobiology 2013; 218(1): 114-26.
[http://dx.doi.org/10.1016/j.imbio.2012.02.008] [PMID: 22440312]

[104] Araújo MS, de Andrade RA, Sathler-Avelar R, *et al.* Immunological changes in canine peripheral blood leukocytes triggered by immunization with first or second generation vaccines against canine visceral leishmaniasis. Vet Immunol Immunopathol 2011; 141(1-2): 64-75.
[http://dx.doi.org/10.1016/j.vetimm.2011.02.007] [PMID: 21439654]

[105] Singh B, Sundar S. Leishmaniasis: vaccine candidates and perspectives. Vaccine 2012; 30(26): 3834-

42.
[http://dx.doi.org/10.1016/j.vaccine.2012.03.068] [PMID: 22475861]

[106] Lemesre JL, Holzmuller P, Cavaleyra M, Gonçalves RB, Hottin G, Papierok G. Protection against experimental visceral leishmaniasis infection in dogs immunized with purified excreted secreted antigens of *Leishmania infantum* promastigotes. Vaccine 2005; 23(22): 2825-40.
[http://dx.doi.org/10.1016/j.vaccine.2004.11.061] [PMID: 15780731]

[107] Lemesre JL, Holzmuller P, Gonçalves RB, *et al.* Long-lasting protection against canine visceral leishmaniasis using the LiESAp-MDP vaccine in endemic areas of France: double-blind randomised efficacy field trial. Vaccine 2007; 25(21): 4223-34.
[http://dx.doi.org/10.1016/j.vaccine.2007.02.083] [PMID: 17395339]

[108] Bras-Gonçalves R, Petitdidier E, Pagniez J, *et al.* Identification and characterization of new *Leishmania* promastigote surface antigens, LaPSA-38S and LiPSA-50S, as major immunodominant excreted/secreted components of *L. amazonensis* and *L. infantum*. Infect Genet Evol 2014; 24: 1-14.
[http://dx.doi.org/10.1016/j.meegid.2014.02.017] [PMID: 24614507]

[109] Fernandes AP, Coelho EA, Machado-Coelho GL, Grimaldi G Jr, Gazzinelli RT. Making an anti-amastigote vaccine for visceral leishmaniasis: rational, update and perspectives. Curr Opin Microbiol 2012; 15(4): 476-85.
[http://dx.doi.org/10.1016/j.mib.2012.05.002] [PMID: 22698479]

[110] Al-Wabel MA, Tonui WK, Cui L, Martin SK, Titus RG. Protection of susceptible BALB/c mice from challenge with *Leishmania major* by nucleoside hydrolase, a soluble exo-antigen of Leishmania. Am J Trop Med Hyg 2007; 77(6): 1060-5.
[PMID: 18165522]

[111] Rafati S, Zahedifard F, Nazgouee F. Prime-boost vaccination using cysteine proteinases type I and II of *Leishmania infantum* confers protective immunity in murine visceral leishmaniasis. Vaccine 2006; 24(12): 2169-75.
[http://dx.doi.org/10.1016/j.vaccine.2005.11.011] [PMID: 16325969]

[112] Skeiky YA, Kennedy M, Kaufman D, *et al.* LeIF: a recombinant *Leishmania* protein that induces an IL-12-mediated Th1 cytokine profile. J Immunol 1998; 161(11): 6171-9.
[PMID: 9834103]

[113] Kushawaha PK, Gupta R, Tripathi CD, *et al. Leishmania donovani* triose phosphate isomerase: a potential vaccine target against visceral leishmaniasis. PLoS One 2012; 7(9): e45766.
[http://dx.doi.org/10.1371/journal.pone.0045766] [PMID: 23049855]

[114] Kaech SM, Wherry EJ, Ahmed R. Effector and memory T-cell differentiation: implications for vaccine development. Nat Rev Immunol 2002; 2(4): 251-62.
[http://dx.doi.org/10.1038/nri778] [PMID: 12001996]

[115] Trigo J, Abbehusen M, Netto EM, *et al.* Treatment of canine visceral leishmaniasis by the vaccine Leish-111f+MPL-SE. Vaccine 2010; 28(19): 3333-40.
[http://dx.doi.org/10.1016/j.vaccine.2010.02.089] [PMID: 20206667]

[116] Stober CB, Lange UG, Roberts MT, Alcami A, Blackwell JM. Heterologous priming-boosting with DNA and modified vaccinia virus Ankara expressing tryparedoxin peroxidase promotes long-term memory against *Leishmania major* in susceptible BALB/c Mice. Infect Immun 2007; 75(2): 852-60.

[http://dx.doi.org/10.1128/IAI.01490-06] [PMID: 17101647]

[117] Wang R, Epstein J, Baraceros FM, *et al.* Induction of CD4(+) T cell-dependent CD8(+) type 1 responses in humans by a malaria DNA vaccine. Proc Natl Acad Sci USA 2001; 98(19): 10817-22. [http://dx.doi.org/10.1073/pnas.181123498] [PMID: 11526203]

[118] Endmann A, Baden M, Weisermann E, *et al.* Immune response induced by a linear DNA vector: influence of dose, formulation and route of injection. Vaccine 2010; 28(21): 3642-9. [http://dx.doi.org/10.1016/j.vaccine.2010.03.034] [PMID: 20362204]

[119] Schmunis GA, Yadon ZE. Chagas disease: a Latin American health problem becoming a world health problem. Acta Trop 2010; 115(1-2): 14-21. [http://dx.doi.org/10.1016/j.actatropica.2009.11.003] [PMID: 19932071]

[120] Gascon J, Bern C, Pinazo MJ. Chagas disease in Spain, the United States and other non-endemic countries. Acta Trop 2010; 115(1-2): 22-7. [http://dx.doi.org/10.1016/j.actatropica.2009.07.019] [PMID: 19646412]

[121] Bern C, Kjos S, Yabsley MJ, Montgomery SP. *Trypanosoma cruzi* and Chagas Disease in the United States. Clin Microbiol Rev 2011; 24(4): 655-81. [http://dx.doi.org/10.1128/CMR.00005-11] [PMID: 21976603]

[122] Rassi A Jr, Rassi A, Marin-Neto JA. Chagas disease. Lancet 2010; 375(9723): 1388-402. [http://dx.doi.org/10.1016/S0140-6736(10)60061-X] [PMID: 20399979]

[123] Quijano-Hernandez I, Dumonteil E. Advances and challenges towards a vaccine against Chagas disease. Hum Vaccin 2011; 7(11): 1184-91. [http://dx.doi.org/10.4161/hv.7.11.17016] [PMID: 22048121]

[124] Viotti R, Vigliano C, Lococo B, *et al.* Side effects of benznidazole as treatment in chronic Chagas disease: fears and realities. Expert Rev Anti Infect Ther 2009; 7(2): 157-63. [http://dx.doi.org/10.1586/14787210.7.2.157] [PMID: 19254164]

[125] Hotez PJ, Dumonteil E, Heffernan MJ, Bottazzi ME. Innovation for the bottom 100 million: eliminating neglected tropical diseases in the Americas. Adv Exp Med Biol 2013; 764: 1-12. [http://dx.doi.org/10.1007/978-1-4614-4726-9_1] [PMID: 23654053]

[126] Basombrío MA, Schofield CJ, Rojas CL, del Rey EC. A cost-benefit analysis of Chagas disease control in north-western Argentina. Trans R Soc Trop Med Hyg 1998; 92(2): 137-43. [http://dx.doi.org/10.1016/S0035-9203(98)90720-9] [PMID: 9764315]

[127] Vallejo M, Montenegro P, Reyes PA. How much does the medical treatment of chronic Chagas cardiopathy cost? Direct costs in a cardiology hospital. Arch Cardiol Mex 2002; 72(2): 129-37. [PMID: 12148332]

[128] Dias JC, Silveira AC, Schofield CJ. The impact of Chagas disease control in Latin America: a review. Mem Inst Oswaldo Cruz 2002; 97(5): 603-12. [http://dx.doi.org/10.1590/S0074-02762002000500002] [PMID: 12219120]

[129] Schofield CJ, Jannin J, Salvatella R. The future of Chagas disease control. Trends Parasitol 2006; 22(12): 583-8. [http://dx.doi.org/10.1016/j.pt.2006.09.011] [PMID: 17049308]

[130] Menezes H. Protective effect of an avirulent (cultivated) strain of *Trypanosoma cruzi* against experimental infection in mice. Rev Inst Med Trop Sao Paulo 1968; 10(1): 1-4. [PMID: 4971608]

[131] Basombrio MA. *Trypanosoma cruzi*: partial prevention of the natural infection of guinea pigs with a killed parasite vaccine. Exp Parasitol 1990; 71(1): 1-8. [http://dx.doi.org/10.1016/0014-4894(90)90002-T] [PMID: 2113004]

[132] Pinge-Filho P, Peron JP, de Moura TR, *et al.* Protective immunity against *Trypanosoma cruzi* provided by oral immunization with *Phytomonas serpens*: role of nitric oxide. Immunol Lett 2005; 96(2): 283-90. [http://dx.doi.org/10.1016/j.imlet.2004.09.010] [PMID: 15585334]

[133] Vázquez-Chagoyán JC, Gupta S, Garg NJ. Vaccine development against *Trypanosoma cruzi* and Chagas disease. Adv Parasitol 2011; 75: 121-46. [http://dx.doi.org/10.1016/B978-0-12-385863-4.00006-X] [PMID: 21820554]

[134] Garg N, Tarleton RL. Genetic immunization elicits antigen-specific protective immune responses and decreases disease severity in *Trypanosoma cruzi* infection. Infect Immun 2002; 70(10): 5547-55. [http://dx.doi.org/10.1128/IAI.70.10.5547-5555.2002] [PMID: 12228281]

[135] Barbosa RP, Filho BG, Dos Santos LI, *et al.* Vaccination using recombinants influenza and adenoviruses encoding amastigote surface protein-2 are highly effective on protection against *Trypanosoma cruzi* infection. PLoS One 2013; 8(4): e61795. [http://dx.doi.org/10.1371/journal.pone.0061795] [PMID: 23637908]

[136] Duan X, Yonemitsu Y, Chou B, *et al.* Efficient protective immunity against *Trypanosoma cruzi* infection after nasal vaccination with recombinant Sendai virus vector expressing amastigote surface protein-2. Vaccine 2009; 27(44): 6154-9. [http://dx.doi.org/10.1016/j.vaccine.2009.08.026] [PMID: 19712768]

[137] Nogueira RT, Nogueira AR, Pereira MC, *et al.* Recombinant yellow fever viruses elicit CD8+ T cell responses and protective immunity against *Trypanosoma cruzi*. PLoS One 2013; 8(3): e59347. [http://dx.doi.org/10.1371/journal.pone.0059347] [PMID: 23527169]

[138] Araújo AF, de Alencar BC, Vasconcelos JR, *et al.* CD8+-T-cell-dependent control of *Trypanosoma cruzi* infection in a highly susceptible mouse strain after immunization with recombinant proteins based on amastigote surface protein 2. Infect Immun 2005; 73(9): 6017-25. [http://dx.doi.org/10.1128/IAI.73.9.6017-6025.2005] [PMID: 16113322]

[139] de Alencar BC, Araújo AF, Penido ML, Gazzinelli RT, Rodrigues MM. Cross-priming of long lived protective CD8+ T cells against *Trypanosoma cruzi* infection: importance of a TLR9 agonist and CD4+ T cells. Vaccine 2007; 25(32): 6018-27. [http://dx.doi.org/10.1016/j.vaccine.2007.05.022] [PMID: 17629597]

[140] Claser C, Espíndola NM, Sasso G, Vaz AJ, Boscardin SB, Rodrigues MM. Immunologically relevant strain polymorphism in the Amastigote Surface Protein 2 of *Trypanosoma cruzi*. Microbes Infect 2007; 9(8): 1011-9. [http://dx.doi.org/10.1016/j.micinf.2007.04.006] [PMID: 17548222]

[141] Machado AV, Cardoso JE, Claser C, Rodrigues MM, Gazzinelli RT, Bruna-Romero O. Long-term

protective immunity induced against *Trypanosoma cruzi* infection after vaccination with recombinant adenoviruses encoding amastigote surface protein-2 and trans-sialidase. Hum Gene Ther 2006; 17(9): 898-908.
[http://dx.doi.org/10.1089/hum.2006.17.898] [PMID: 16972758]

[142] de Alencar BC, Persechini PM, Haolla FA, *et al.* Perforin and gamma interferon expression are required for CD4+ and CD8+ T-cell-dependent protective immunity against a human parasite, *Trypanosoma cruzi*, elicited by heterologous plasmid DNA prime-recombinant adenovirus 5 boost vaccination. Infect Immun 2009; 77(10): 4383-95.
[http://dx.doi.org/10.1128/IAI.01459-08] [PMID: 19651871]

[143] Silveira EL, Claser C, Haolla FA, Zanella LG, Rodrigues MM. Novel protective antigens expressed by *Trypanosoma cruzi* amastigotes provide immunity to mice highly susceptible to Chagas disease. Clin Vaccine Immunol 2008; 15(8): 1292-300.
[http://dx.doi.org/10.1128/CVI.00142-08] [PMID: 18579696]

[144] Sepulveda P, Hontebeyrie M, Liegeard P, Mascilli A, Norris KA. DNA-Based immunization with *Trypanosoma cruzi* complement regulatory protein elicits complement lytic antibodies and confers protection against *Trypanosoma cruzi* infection. Infect Immun 2000; 68(9): 4986-91.
[http://dx.doi.org/10.1128/IAI.68.9.4986-4991.2000] [PMID: 10948115]

[145] Schnapp AR, Eickhoff CS, Sizemore D, Curtiss R III, Hoft DF. Cruzipain induces both mucosal and systemic protection against *Trypanosoma cruzi* in mice. Infect Immun 2002; 70(9): 5065-74.
[http://dx.doi.org/10.1128/IAI.70.9.5065-5074.2002] [PMID: 12183554]

[146] Cazorla SI, Frank FM, Becker PD, Corral RS, Guzmán CA, Malchiodi EL. Prime-boost immunization with cruzipain co-administered with MALP-2 triggers a protective immune response able to decrease parasite burden and tissue injury in an experimental *Trypanosoma cruzi* infection model. Vaccine 2008; 26(16): 1999-2009.
[http://dx.doi.org/10.1016/j.vaccine.2008.02.011] [PMID: 18342408]

[147] Cazorla SI, Becker PD, Frank FM, *et al.* Oral vaccination with Salmonella enterica as a cruzipain-DNA delivery system confers protective immunity against *Trypanosoma cruzi*. Infect Immun 2008; 76(1): 324-33. [PubMed].
[http://dx.doi.org/10.1128/IAI.01163-07] [PMID: 17967857]

[148] Fralish BH, Tarleton RL. Genetic immunization with LYT1 or a pool of trans-sialidase genes protects mice from lethal *Trypanosoma cruzi* infection. Vaccine 2003; 21(21-22): 3070-80.
[http://dx.doi.org/10.1016/S0264-410X(03)00121-X] [PMID: 12798651]

[149] Planelles L, Thomas MC, Alonso C, López MC. DNA immunization with *Trypanosoma cruzi* HSP70 fused to the KMP11 protein elicits a cytotoxic and humoral immune response against the antigen and leads to protection. Infect Immun 2001; 69(10): 6558-63.
[http://dx.doi.org/10.1128/IAI.69.10.6558-6563.2001] [PMID: 11553607]

[150] Luhrs KA, Fouts DL, Manning JE. Immunization with recombinant paraflagellar rod protein induces protective immunity against *Trypanosoma cruzi* infection. Vaccine 2003; 21(21-22): 3058-69.
[http://dx.doi.org/10.1016/S0264-410X(03)00108-7] [PMID: 12798650]

[151] Arce-Fonseca M, Ramos-Ligonio A, López-Monteón A, Salgado-Jiménez B, Talamás-Rohana P, Rosales-Encina JL. A DNA vaccine encoding for TcSSP4 induces protection against acute and chronic

infection in experimental Chagas disease. Int J Biol Sci 2011; 7(9): 1230-8. [http://dx.doi.org/10.7150/ijbs.7.1230] [PMID: 22110377]

[152] García GA, Arnaiz MR, Laucella SA, *et al.* Immunological and pathological responses in BALB/c mice induced by genetic administration of Tc 13 Tul antigen of *Trypanosoma cruzi*. Parasitology 2006; 132(Pt 6): 855-66. [http://dx.doi.org/10.1017/S0031182005009753] [PMID: 16478565]

[153] Bhatia V, Garg NJ. Previously unrecognized vaccine candidates control *Trypanosoma cruzi* infection and immunopathology in mice. Clin Vaccine Immunol 2008; 15(8): 1158-64. [http://dx.doi.org/10.1128/CVI.00144-08] [PMID: 18550728]

[154] Aparicio-Burgos JE, Ochoa-García L, Zepeda-Escobar JA, *et al.* Testing the efficacy of a multi-component DNA-prime/DNA-boost vaccine against *Trypanosoma cruzi* infection in dogs. PLoS Negl Trop Dis 2011; 5(5): e1050. [http://dx.doi.org/10.1371/journal.pntd.0001050] [PMID: 21625470]

[155] Gupta S, Garg NJ. Prophylactic efficacy of TcVac2 against *Trypanosoma cruzi* in mice. PLoS Negl Trop Dis 2010; 4(8): e797. [http://dx.doi.org/10.1371/journal.pntd.0000797] [PMID: 20706586]

[156] Gupta S, Garg NJ. TcVac3 induced control of *Trypanosoma cruzi* infection and chronic myocarditis in mice. PLoS One 2013; 8(3): e59434. [http://dx.doi.org/10.1371/journal.pone.0059434] [PMID: 23555672]

[157] Fontanella GH, De Vusser K, Laroy W, *et al.* Immunization with an engineered mutant trans-sialidase highly protects mice from experimental *Trypanosoma cruzi* infection: a vaccine candidate. Vaccine 2008; 26(19): 2322-34. [http://dx.doi.org/10.1016/j.vaccine.2008.02.060] [PMID: 18403070]

[158] Hoft DF, Eickhoff CS, Giddings OK, Vasconcelos JR, Rodrigues MM. Trans-sialidase recombinant protein mixed with CpG motif-containing oligodeoxynucleotide induces protective mucosal and systemic *Trypanosoma cruzi* immunity involving CD8+ CTL and B cell-mediated cross-priming. J Immunol 2007; 179(10): 6889-900. [http://dx.doi.org/10.4049/jimmunol.179.10.6889] [PMID: 17982080]

[159] Eickhoff CS, Vasconcelos JR, Sullivan NL, *et al.* Co-administration of a plasmid DNA encoding IL-15 improves long-term protection of a genetic vaccine against *Trypanosoma cruzi*. PLoS Negl Trop Dis 2011; 5(3): e983. [http://dx.doi.org/10.1371/journal.pntd.0000983] [PMID: 21408124]

[160] Rodrigues MM, Ribeirão M, Pereira-Chioccola V, Renia L, Costa F. Predominance of CD4 Th1 and CD8 Tc1 cells revealed by characterization of the cellular immune response generated by immunization with a DNA vaccine containing a *Trypanosoma cruzi* gene. Infect Immun 1999; 67(8): 3855-63. [PMID: 10417149]

[161] Katae M, Miyahira Y, Takeda K, *et al.* Coadministration of an interleukin-12 gene and a *Trypanosoma cruzi* gene improves vaccine efficacy. Infect Immun 2002; 70(9): 4833-40. [http://dx.doi.org/10.1128/IAI.70.9.4833-4840.2002] [PMID: 12183527]

[162] Miyahira Y, Akiba H, Katae M, *et al.* Cutting edge: a potent adjuvant effect of ligand to receptor

activator of NF-kappa B gene for inducing antigen-specific CD8+ T cell response by DNA and viral vector vaccination. J Immunol 2003; 171(12): 6344-8.
[http://dx.doi.org/10.4049/jimmunol.171.12.6344] [PMID: 14662831]

[163] Miyahira Y, Takashima Y, Kobayashi S, *et al.* Immune responses against a single CD8+-T-cell epitope induced by virus vector vaccination can successfully control *Trypanosoma cruzi* infection. Infect Immun 2005; 73(11): 7356-65.
[http://dx.doi.org/10.1128/IAI.73.11.7356-7365.2005] [PMID: 16239534]

[164] Wizel B, Garg N, Tarleton RL. Vaccination with trypomastigote surface antigen 1-encoding plasmid DNA confers protection against lethal *Trypanosoma cruzi* infection. Infect Immun 1998; 66(11): 5073-81.
[PMID: 9784506]

[165] Bhatia V, Sinha M, Luxon B, Garg N. Utility of the *Trypanosoma cruzi* sequence database for identification of potential vaccine candidates by *in silico* and *in vitro* screening. Infect Immun 2004; 72(11): 6245-54.
[http://dx.doi.org/10.1128/IAI.72.11.6245-6254.2004] [PMID: 15501750]

[166] Coura JR, Junqueira AC, Fernandes O, Valente SA, Miles MA. Emerging Chagas disease in Amazonian Brazil. Trends Parasitol 2002; 18(4): 171-6.
[http://dx.doi.org/10.1016/S1471-4922(01)02200-0] [PMID: 11998705]

[167] Sánchez LV, Ramírez JD. Congenital and oral transmission of American trypanosomiasis: an overview of physiopathogenic aspects. Parasitology 2013; 140(2): 147-59.
[http://dx.doi.org/10.1017/S0031182012001394] [PMID: 23010131]

[168] Matos MN, Cazorla SI, Bivona AE, Morales C, Guzmán CA, Malchiodi EL. Tc52 amino-termina--domain DNA carried by attenuated *Salmonella enterica* serovar Typhimurium induces protection against a *Trypanosoma cruzi* lethal challenge. Infect Immun 2014; 82(10): 4265-75.
[http://dx.doi.org/10.1128/IAI.02190-14] [PMID: 25069980]

[169] Chen G, Dai Y, Chen J, *et al.* Oral delivery of the Sj23LHD-GST antigen by *Salmonella typhimurium* type III secretion system protects against *Schistosoma japonicum* infection in mice. PLoS Negl Trop Dis 2011; 5(9): e1313.
[http://dx.doi.org/10.1371/journal.pntd.0001313] [PMID: 21909450]

[170] Pompa-Mera EN, Yépez-Mulia L, Ocaña-Mondragón A, García-Zepeda EA, Ortega-Pierres G, González-Bonilla CR. *Trichinella spiralis*: intranasal immunization with attenuated *Salmonella enterica* carrying a gp43 antigen-derived 30mer epitope elicits protection in BALB/c mice. Exp Parasitol 2011; 129(4): 393-401.
[http://dx.doi.org/10.1016/j.exppara.2011.08.013] [PMID: 21907709]

[171] Jazayeri SD, Ideris A, Zakaria Z, Yeap SK, Omar AR. Improved immune responses against avian influenza virus following oral vaccination of chickens with HA DNA vaccine using attenuated *Salmonella typhimurium* as carrier. Comp Immunol Microbiol Infect Dis 2012; 35(5): 417-27.
[http://dx.doi.org/10.1016/j.cimid.2012.03.007] [PMID: 22512819]

[172] Ono T, Yamaguchi Y, Oguma T, *et al.* Actively induced antigen-specific CD8+ T cells by epitope-bearing parasite pre-infection but not prime/boost virus vector vaccination could ameliorate the course of *Plasmodium yoelii* blood-stage infection. Vaccine 2012; 30(44): 6270-8.

[http://dx.doi.org/10.1016/j.vaccine.2012.08.009] [PMID: 22902783]

[173] Casimiro DR, Chen L, Fu TM, *et al.* Comparative immunogenicity in rhesus monkeys of DNA plasmid, recombinant vaccinia virus, and replication-defective adenovirus vectors expressing a human immunodeficiency virus type 1 gag gene. J Virol 2003; 77(11): 6305-13.
[http://dx.doi.org/10.1128/JVI.77.11.6305-6313.2003] [PMID: 12743287]

[174] Shiver JW, Emini EA. Recent advances in the development of HIV-1 vaccines using replication-incompetent adenovirus vectors. Annu Rev Med 2004; 55: 355-72.
[http://dx.doi.org/10.1146/annurev.med.55.091902.104344] [PMID: 14746526]

[175] Haolla FA, Claser C, de Alencar BC, *et al.* Strain-specific protective immunity following vaccination against experimental *Trypanosoma cruzi* infection. Vaccine 2009; 27(41): 5644-53.
[http://dx.doi.org/10.1016/j.vaccine.2009.07.013] [PMID: 19635607]

[176] Rigato PO, de Alencar BC, de Vasconcelos JR, *et al.* Heterologous plasmid DNA prime-recombinant human adenovirus 5 boost vaccination generates a stable pool of protective long-lived CD8(+) T effector memory cells specific for a human parasite, *Trypanosoma cruzi*. Infect Immun 2011; 79(5): 2120-30.
[http://dx.doi.org/10.1128/IAI.01190-10] [PMID: 21357719]

[177] Dominguez MR, Silveira EL, de Vasconcelos JR, *et al.* Subdominant/cryptic CD8 T cell epitopes contribute to resistance against experimental infection with a human protozoan parasite. PLoS One 2011; 6(7): e22011.
[http://dx.doi.org/10.1371/journal.pone.0022011] [PMID: 21779365]

[178] Vasconcelos JR, Bruña-Romero O, Araújo AF, *et al.* Pathogen-induced proapoptotic phenotype and high CD95 (Fas) expression accompany a suboptimal CD8+ T-cell response: reversal by adenoviral vaccine. PLoS Pathog 2012; 8(5): e1002699.
[http://dx.doi.org/10.1371/journal.ppat.1002699] [PMID: 22615561]

[179] Brun R, Blum J, Chappuis F, Burri C. Human African trypanosomiasis. Lancet 2010; 375(9709): 148-59.
[http://dx.doi.org/10.1016/S0140-6736(09)60829-1] [PMID: 19833383]

[180] MacLean L, Myburgh E, Rodgers J, Price HP. Imaging African trypanosomes. Parasite Immunol 2013; 35(9-10): 283-94.
[http://dx.doi.org/10.1111/pim.12046] [PMID: 23790101]

[181] Kuepfer I, Schmid C, Allan M, *et al.* Safety and efficacy of the 10-day melarsoprol schedule for the treatment of second stage Rhodesiense sleeping sickness. PLoS Negl Trop Dis 2012; 6(8): e1695.

[182] Simarro PP, Cecchi G, Franco JR, *et al.* Estimating and mapping the population at risk of sleeping sickness. PLoS Negl Trop Dis 2012; 6(10): e1859.

[183] Burri C, Brun R. Eflornithine for the treatment of human African trypanosomiasis. Parasitol Res 2003; 90 (Suppl. 1): S49-52.
[PMID: 12811548]

[184] Magez S, Caljon G, Tran T, Stijlemans B, Radwanska M. Current status of vaccination against African trypanosomiasis. Parasitology 2010; 137(14): 2017-27.
[http://dx.doi.org/10.1017/S0031182010000223] [PMID: 20441680]

[185] Magez S, Radwanska M. African trypanosomiasis and antibodies: implications for vaccination, therapy and diagnosis. Future Microbiol 2009; 4(8): 1075-87.
[http://dx.doi.org/10.2217/fmb.09.65] [PMID: 19824795]

[186] Stijlemans B, Conrath K, Cortez-Retamozo V, *et al.* Efficient targeting of conserved cryptic epitopes of infectious agents by single domain antibodies. African trypanosomes as paradigm. J Biol Chem 2004; 279(2): 1256-61.
[http://dx.doi.org/10.1074/jbc.M307341200] [PMID: 14527957]

[187] Stijlemans B, Guilliams M, Raes G, Beschin A, Magez S, De Baetselier P. African trypanosomosis: from immune escape and immunopathology to immune intervention. Vet Parasitol 2007; 148(1): 3-13.
[http://dx.doi.org/10.1016/j.vetpar.2007.05.005] [PMID: 17560035]

[188] Taylor JE, Rudenko G. Switching trypanosome coats: whats in the wardrobe? Trends Genet 2006; 22(11): 614-20.
[http://dx.doi.org/10.1016/j.tig.2006.08.003] [PMID: 16908087]

[189] Akhoon BA, Slathia PS, Sharma P, Gupta SK, Verma V. *In silico* identification of novel protective VSG antigens expressed by *Trypanosoma brucei* and an effort for designing a highly immunogenic DNA vaccine using IL-12 as adjuvant. Microb Pathog 2011; 51(1-2): 77-87.
[http://dx.doi.org/10.1016/j.micpath.2011.01.011] [PMID: 21349321]

[190] Wei G, Bull H, Zhou X, Tabel H. Intradermal infections of mice by low numbers of african trypanosomes are controlled by innate resistance but enhance susceptibility to reinfection. J Infect Dis 2011; 203(3): 418-29.
[http://dx.doi.org/10.1093/infdis/jiq051] [PMID: 21208931]

[191] Lejon V, Mumba Ngoyi D, Kestens L, *et al.* Gambiense human african trypanosomiasis and immunological memory: effect on phenotypic lymphocyte profiles and humoral immunity. PLoS Pathog 2014; 10(3): e1003947.
[http://dx.doi.org/10.1371/journal.ppat.1003947] [PMID: 24603894]

[192] Tabel H, Wei G, Bull HJ. Immunosuppression: cause for failures of vaccines against African Trypanosomiases. PLoS Negl Trop Dis 2013; 7(3): e2090.
[http://dx.doi.org/10.1371/journal.pntd.0002090] [PMID: 23516643]

[193] Li SQ, Yang WB, Ma LJ, *et al.* Immunization with recombinant actin from *Trypanosoma evansi* induces protective immunity against *T. evansi*, *T. equiperdum* and *T. b. brucei* infection. Parasitol Res 2009; 104(2): 429-35.
[http://dx.doi.org/10.1007/s00436-008-1216-9] [PMID: 18923843]

[194] Li SQ, Fung MC, Reid SA, Inoue N, Lun ZR. Immunization with recombinant beta-tubulin from *Trypanosoma evansi* induced protection against *T. evansi*, *T. equiperdum* and *T. b. brucei* infection in mice. Parasite Immunol 2007; 29(4): 191-9.
[http://dx.doi.org/10.1111/j.1365-3024.2006.00933.x] [PMID: 17371456]

[195] Olenick JG, Wolff R, Nauman RK, McLaughlin J. A flagellar pocket membrane fraction from *Trypanosoma brucei rhodesiense*: immunogold localization and nonvariant immunoprotection. Infect Immun 1988; 56(1): 92-8.
[PMID: 3335412]

[196] Mkunza F, Olaho WM, Powell CN. Partial protection against natural trypanosomiasis after vaccination with a flagellar pocket antigen from *Trypanosoma brucei rhodesiense*. Vaccine 1995; 13(2): 151-4. [http://dx.doi.org/10.1016/0264-410X(95)93128-V] [PMID: 7625108]

[197] Radwanska M, Magez S, Dumont N, Pays A, Nolan D, Pays E. Antibodies raised against the flagellar pocket fraction of *Trypanosoma brucei* preferentially recognize HSP60 in cDNA expression library. Parasite Immunol 2000; 22(12): 639-50. [http://dx.doi.org/10.1046/j.1365-3024.2000.00348.x] [PMID: 11123756]

[198] Lança AS, de Sousa KP, Atouguia J, Prazeres DM, Monteiro GA, Silva MS. *Trypanosoma brucei*: immunisation with plasmid DNA encoding invariant surface glycoprotein gene is able to induce partial protection in experimental African trypanosomiasis. Exp Parasitol 2011; 127(1): 18-24. [http://dx.doi.org/10.1016/j.exppara.2010.06.017] [PMID: 20599996]

[199] Rasooly R, Balaban N. Trypanosome microtubule-associated protein p15 as a vaccine for the prevention of African sleeping sickness. Vaccine 2004; 22(8): 1007-15. [http://dx.doi.org/10.1016/j.vaccine.2003.08.041] [PMID: 15161078]

[200] Balaban N, Waithaka HK, Njogu AR, Goldman R. Intracellular antigens (microtubule-associated protein copurified with glycosomal enzymes)possible vaccines against trypanosomiasis. J Infect Dis 1995; 172(3): 845-50. [http://dx.doi.org/10.1093/infdis/172.3.845] [PMID: 7658080]

[201] Silva MS, Prazeres DM, Lança A, Atouguia J, Monteiro GA. Trans-sialidase from *Trypanosoma brucei* as a potential target for DNA vaccine development against African trypanosomiasis. Parasitol Res 2009; 105(5): 1223-9. [http://dx.doi.org/10.1007/s00436-009-1542-6] [PMID: 19582478]

[202] Lubega GW, Byarugaba DK, Prichard RK. Immunization with a tubulin-rich preparation from *Trypanosoma brucei* confers broad protection against African trypanosomosis. Exp Parasitol 2002; 102(1): 9-22. [http://dx.doi.org/10.1016/S0014-4894(02)00140-6] [PMID: 12615162]

[203] Stijlemans B, Baral TN, Guilliams M, *et al.* A glycosylphosphatidylinositol-based treatment alleviates trypanosomiasis-associated immunopathology. J Immunol 2007; 179(6): 4003-14. [http://dx.doi.org/10.4049/jimmunol.179.6.4003] [PMID: 17785839]

[204] Magez S, Stijlemans B, Baral T, De Baetselier P. VSG-GPI anchors of African trypanosomes: their role in macrophage activation and induction of infection-associated immunopathology. Microbes and infection / Institut Pasteur 2002; 4(9): 999-1006. [http://dx.doi.org/10.1016/S1286-4579(02)01617-9]

[205] Magez S, Stijlemans B, Radwanska M, Pays E, Ferguson MA, De Baetselier P. The glycosyl-inosito--phosphate and dimyristoylglycerol moieties of the glycosylphosphatidylinositol anchor of the trypanosome variant-specific surface glycoprotein are distinct macrophage-activating factors. J Immunol 1998; 160(4): 1949-56. [PMID: 9469458]

[206] Steverding D, Stierhof YD, Chaudhri M, *et al.* ESAG 6 and 7 products of *Trypanosoma brucei* form a transferrin binding protein complex. Eur J Cell Biol 1994; 64(1): 78-87. [PMID: 7957316]

[207] Jackson DG, Windle HJ, Voorheis HP. The identification, purification, and characterization of two invariant surface glycoproteins located beneath the surface coat barrier of bloodstream forms of *Trypanosoma brucei*. J Biol Chem 1993; 268(11): 8085-95.
[PMID: 8463323]

[208] Ziegelbauer K, Multhaup G, Overath P. Molecular characterization of two invariant surface glycoproteins specific for the bloodstream stage of *Trypanosoma brucei*. J Biol Chem 1992; 267(15): 10797-803.
[PMID: 1587856]

[209] Ziegelbauer K, Overath P. Identification of invariant surface glycoproteins in the bloodstream stage of *Trypanosoma brucei*. J Biol Chem 1992; 267(15): 10791-6.
[PMID: 1587855]

[210] Nolan DP, Jackson DG, Windle HJ, *et al.* Characterization of a novel, stage-specific, invariant surface protein in *Trypanosoma brucei* containing an internal, serine-rich, repetitive motif. J Biol Chem 1997; 272(46): 29212-21.
[http://dx.doi.org/10.1074/jbc.272.46.29212] [PMID: 9361000]

[211] Tran T, Büscher P, Vandenbussche G, Wyns L, Messens J, De Greve H. Heterologous expression, purification and characterisation of the extracellular domain of trypanosome invariant surface glycoprotein ISG75. J Biotechnol 2008; 135(3): 247-54.
[http://dx.doi.org/10.1016/j.jbiotec.2008.04.012] [PMID: 18538880]

[212] Pontes-de-Carvalho LC, Tomlinson S, Nussenzweig V. *Trypanosoma rangeli* sialidase lacks trans-sialidase activity. Mol Biochem Parasitol 1993; 62(1): 19-25.
[http://dx.doi.org/10.1016/0166-6851(93)90173-U] [PMID: 8114822]

[213] Montagna G, Cremona ML, Paris G, *et al.* The trans-sialidase from the african trypanosome *Trypanosoma brucei*. European Journal of Biochemistry/FEBS 2002; 269(12): 2941-50.

[214] Montagna GN, Donelson JE, Frasch AC. Procyclic *Trypanosoma brucei* expresses separate sialidase and trans-sialidase enzymes on its surface membrane. J Biol Chem 2006; 281(45): 33949-58.
[http://dx.doi.org/10.1074/jbc.M604951200] [PMID: 16956887]

[215] Ammar Z, Plazolles N, Baltz T, Coustou V. Identification of trans-sialidases as a common mediator of endothelial cell activation by African trypanosomes. PLoS Pathog 2013; 9(10): e1003710.
[http://dx.doi.org/10.1371/journal.ppat.1003710] [PMID: 24130501]

[216] Baral TN, Magez S, Stijlemans B, *et al.* Experimental therapy of African trypanosomiasis with a nanobody-conjugated human trypanolytic factor. Nat Med 2006; 12(5): 580-4.
[http://dx.doi.org/10.1038/nm1395] [PMID: 16604085]

[217] Obishakin E, Stijlemans B, Santi-Rocca J, *et al.* Generation of a nanobody targeting the paraflagellar rod protein of trypanosomes. PLoS One 2014; 9(12): e115893.
[http://dx.doi.org/10.1371/journal.pone.0115893] [PMID: 25551637]

[218] Roditi I, Lehane MJ. Interactions between trypanosomes and tsetse flies. Curr Opin Microbiol 2008; 11(4): 345-51.
[http://dx.doi.org/10.1016/j.mib.2008.06.006] [PMID: 18621142]

[219] Urwyler S, Studer E, Renggli CK, Roditi I. A family of stage-specific alanine-rich proteins on the

surface of epimastigote forms of *Trypanosoma brucei*. Mol Microbiol 2007; 63(1): 218-28. [http://dx.doi.org/10.1111/j.1365-2958.2006.05492.x] [PMID: 17229212]

[220] Caljon G, Van Den Abbeele J, Sternberg JM, Coosemans M, De Baetselier P, Magez S. Tsetse fly saliva biases the immune response to Th2 and induces anti-vector antibodies that are a useful tool for exposure assessment. Int J Parasitol 2006; 36(9): 1025-35. [http://dx.doi.org/10.1016/j.ijpara.2006.05.002] [PMID: 16777113]

[221] Caljon G, Van Den Abbeele J, Stijlemans B, Coosemans M, De Baetselier P, Magez S. Tsetse fly saliva accelerates the onset of *Trypanosoma brucei* infection in a mouse model associated with a reduced host inflammatory response. Infect Immun 2006; 74(11): 6324-30. [http://dx.doi.org/10.1128/IAI.01046-06] [PMID: 16954393]

[222] Kinyua JK, Nguu EK, Mulaa F, Ndungu JM. Immunization of rabbits with Glossina pallidipes tsetse fly midgut proteins: effects on the fly and trypanosome transmission. Vaccine 2005; 23(29): 3824-8. [http://dx.doi.org/10.1016/j.vaccine.2005.02.032] [PMID: 15893620]

[223] Gimenez AM, Santander VS, Villasuso AL, Pasquaré SJ, Giusto NM, Machado EE. Regulation of phosphatidic acid levels in *Trypanosoma cruzi*. Lipids 2011; 46(10): 969-79. [http://dx.doi.org/10.1007/s11745-011-3577-6] [PMID: 21667213]

[224] Hashimoto M, Nara T, Hirawake H, Morales J, Enomoto M, Mikoshiba K. Antisense oligonucleotides targeting parasite inositol 1,4,5-trisphosphate receptor inhibits mammalian host cell invasion by *Trypanosoma cruzi*. Sci Rep 2014; 4: 4231. [http://dx.doi.org/10.1038/srep04231] [PMID: 24577136]

[225] Gimenez AM, Gesumaría MC, Schoijet AC, *et al.* Phosphatidylinositol kinase activities in *Trypanosoma cruzi* epimastigotes. Mol Biochem Parasitol 2015; 203(1-2): 14-24. [http://dx.doi.org/10.1016/j.molbiopara.2015.10.002] [PMID: 26493613]

[226] Ritagliati C, Alonso VL, Manarin R, Cribb P, Serra EC. Overexpression of cytoplasmic TcSIR2RP1 and mitochondrial TcSIR2RP3 impacts on *Trypanosoma cruzi* growth and cell invasion. PLoS Negl Trop Dis 2015; 9(4): e0003725. [http://dx.doi.org/10.1371/journal.pntd.0003725] [PMID: 25875650]

SUBJECT INDEX

D

E

F

G

H

I

K

L

M

R

S

T

U

V

X

Made in the USA
Las Vegas, NV
03 November 2026

57424492R00366